STUDENT'S SOLUTIONS MANUAL

JUDITH A. PENNA
Indiana University Purdue University Indianapolis

ELEMENTARY AND INTERMEDIATE ALGEBRA: CONCEPTS AND APPLICATIONS

FOURTH EDITION

Marvin L. Bittinger
Indiana University Purdue University Indianapolis

David J. Ellenbogen
Community College of Vermont

Barbara L. Johnson
Indiana University Purdue University Indianapolis

Boston San Francisco New York
London Toronto Sydney Tokyo Singapore Madrid
Mexico City Munich Paris Cape Town Hong Kong Montreal

Reproduced by Pearson Addison-Wesley from electronic files supplied by the author.

Publishing as Pearson Addison-Wesley, 75 Arlington Street, Boston, MA 02116.

ISBN 0-321-28678-2

10 11 12 13 14 15 16 17 BB 15 14 13 12 11 10 09

Contents

Chapter 1

Introduction to Algebraic Expressions

Exercise Set 1.1

1. $4x+7$ does not contain an equals sign, so it is an expression.

3. $2x-5=9$ contains an equals sign, so it is an equation.

5. $38=2t$ contains an equals sign, so it is an equation.

7. $4a-5b$ does not contain an equals sign, so it is an expression.

9. $2x-3y=8$ contains an equals sign, so it is an equation.

11. $7-4rt$ does not contain an equals sign, so it is an expression.

13. Substitute 9 for a and multiply.

$$4a = 4\cdot 9 = 36$$

15. Substitute 7 for x and multiply.

$$8\cdot 7 = 56$$

17. $\dfrac{a}{b}=\dfrac{45}{9}=5$

19. $\dfrac{x+y}{4}=\dfrac{2+14}{4}=\dfrac{16}{4}=4$

21. $\dfrac{15+20}{7}=\dfrac{35}{7}=5$

23. $\dfrac{m-n}{2}=\dfrac{20-8}{2}=\dfrac{12}{2}=6$

25.
$$\begin{aligned} bh &= (6\text{ ft})(4\text{ ft})\\ &= (6)(4)(\text{ft})(\text{ft})\\ &= 24\text{ ft}^2,\ \text{or 24 square feet}\end{aligned}$$

27.
$$\begin{aligned} A &= \frac{1}{2}bh\\ &= \frac{1}{2}(5\text{ cm})(6\text{ cm})\\ &= \frac{1}{2}(5)(6)(\text{cm})(\text{cm})\\ &= \frac{5}{2}\cdot 6\text{ cm}^2\\ &= 15\text{ cm}^2,\ \text{or 15 square centimeters}\end{aligned}$$

29. $\dfrac{h}{a}=\dfrac{7}{19}\approx 0.368$

31. Let r represent Ron's age. Then we have $r+5$, or $5+r$.

33. $b+6$, or $6+b$

35. $c-9$

37. $6+q$, or $q+6$

39. Let s represent Phil's speed. Then we have $9s$, or $s\cdot 9$.

41. $y-x$

43. $x\div w$, or $\dfrac{x}{w}$

45. $n-m$

47. Let l and h represent the box's length and height, respectively. Then we have $l+h$, or $h+l$.

49. $9\cdot 2m$, or $2m\cdot 9$

51. Let y represent "some number." Then we have $\dfrac{1}{4}y$, or $\dfrac{y}{4}$, or $y/4$, or $y\div 4$.

53. Let x represent the number of women attending. Then we have 64% of x, or $0.64x$.

55.
$$\begin{array}{c|c} x+17 = 32 & \\ \hline 15+17 & 32 \\ 32 \ ? \ 32 & \end{array}$$

Writing the equation; Substituting 15 for x; $32=32$ is TRUE.

Since the left-hand and right-hand sides are the same, 15 is a solution.

57.
$$\begin{array}{c|c} a-28 = 75 & \\ \hline 93-28 & 75 \\ 65 \ ? \ 75 & \end{array}$$

Writing the equation; Substituting 93 for a; $65=75$ is FALSE.

Since the left-hand and right-hand sides are not the same, 93 is not a solution.

59.
$$\begin{array}{c|c} \frac{t}{7} = 9 & \\ \hline \frac{63}{7} & 9 \\ 9 \ ? \ 9 & \end{array}$$

$9=9$ is TRUE.

Since the left-hand and right-hand sides are the same, 63 is a solution.

61.
$$\begin{array}{c|c} \frac{108}{x} = 36 & \\ \hline \frac{108}{3} & 36 \\ 36 \ ? \ 36 & \end{array}$$

$36=36$ is TRUE.

Since the left-hand and right-hand sides are the same, 3 is a solution.

63. Let x represent the number.

What number added to 73 is 201?

Translating: $x \quad + \quad 73 = 201$

$x+73=201$

65. Let y represent the number.

Rewording: 42 times $\underbrace{\text{what number}}$ is 2352?

Translating: $42 \cdot y = 2352$

$42y = 2352$

67. Let s represent the number of unoccupied squares.

Rewording: $\underbrace{\text{The number of unoccupied squares}}$ $\underbrace{\text{added to}}$ 19 is 64.

Translating: $s + 19 = 64$

$s + 19 = 64$

69. Let x represent the total amount of waste generated, in millions of tons.

Rewording: $\underbrace{27\%}$ of $\underbrace{\text{the total amount of waste}}$ is $\underbrace{\text{56 million tons.}}$

Translating: $27\% \cdot x = 56$

$27\% \cdot x = 56$, or $0.27x = 56$

71. The sum of two numbers m and n is $m + n$, and twice the sum is $2(m + n)$. Choice (f) is the correct answer.

73. Two more than a number t is $t + 2$. If this expression is equal to 5, we have the equation $t + 2 = 5$. Choice (d) is the correct answer.

75. The sum of a number t and 5 is $t+5$, and 3 times the sum is $3(t + 5)$. Choice (g) is the correct answer.

77. The product of two numbers a and b is ab, and 1 less than this product is $ab - 1$. If this expression is equal to 49, we have the equation $ab - 1 = 49$. Choice (e) is the correct answer.

79. *Writing Exercise*

81. *Writing Exercise*

83. Area of sign: $A = \frac{1}{2}(3\text{ ft})(2.5\text{ ft}) = 3.75\text{ ft}^2$

Cost of sign: $\$90(3.75) = \$337.50$

85. When x is twice y, then y is one-half x, so $y = \frac{12}{2} = 6$.

$\frac{x-y}{3} = \frac{12-6}{3} = \frac{6}{3} = 2$

87. When a is twice b, then b is one-half a, so $b = \frac{16}{2} = 8$.

$\frac{a+b}{4} = \frac{16+8}{4} = \frac{24}{4} = 6$

89. The next whole number is one more than $w + 3$:

$w + 3 + 1 = w + 4$

91. $l + w + l + w$, or $2l + 2w$

93. If t is Molly's race time, then Joe's race time is $t + 3$ and Ellie's race time is

$t + 3 + 5 = t + 8$.

Exercise Set 1.2

1. Commutative

3. Associative

5. Distributive

7. Associative

9. Commutative

11. $x + 7$ Changing the order

13. $c + ab$

15. $3y + 9x$

17. $5(1 + a)$

19. $a \cdot 2$ Changing the order

21. ts

23. $5 + ba$

25. $(a + 1)5$

27. $a + (5 + b)$

29. $(r + t) + 7$

31. $ab + (c + d)$

33. $8(xy)$

35. $(2a)b$

37. $(3 \cdot 2)(a + b)$

39. a) $r + (t + 6) = (t + 6) + r$ Using the commutative law

$= (6 + t) + r$ Using the commutative law again

b) $r + (t + 6) = (t + 6) + r$ Using the commutative law

$= t + (6 + r)$ Using the associative law

Answers may vary.

41. a) $(17a)b = b(17a)$ Using the commutative law

$= b(a17)$ Using the commutative law again

b) $(17a)b = (a17)b$ Using the commutative law

$= a(17b)$ Using the associative law

Answers may vary.

43. $(5+x)+2$

$= (x+5)+2$ Commutative law

$= x+(5+2)$ Associative law

$= x+7$ Simplifying

45. $(m \cdot 3)7 = m(3 \cdot 7)$ Associative law

$= m \cdot 21$ Simplifying

$= 21m$ Commutative law

47. $4(a+3) = 4 \cdot a + 4 \cdot 3 = 4a + 12$

49. $6(1+x) = 6 \cdot 1 + 6 \cdot x = 6 + 6x$

51. $3(x+1) = 3 \cdot x + 3 \cdot 1 = 3x + 3$

53. $8(3+y) = 8 \cdot 3 + 8 \cdot y = 24 + 8y$

55. $9(2x+6) = 9 \cdot 2x + 9 \cdot 6 = 18x + 54$

57. $5(r+2+3t) = 5 \cdot r + 5 \cdot 2 + 5 \cdot 3t = 5r + 10 + 15t$

59. $(a+b)2 = a(2) + b(2) = 2a + 2b$

61. $(x+y+2)5 = x(5) + y(5) + 2(5) = 5x + 5y + 10$

63. $x + xyz + 19$

The terms are separated by plus signs. They are x, xyz, and 19.

65. $2a + \frac{a}{b} + 5b$

The terms are separated by plus signs. They are $2a$, $\frac{a}{b}$, and $5b$.

67. $2a + 2b = 2(a+b)$ The common factor is 2.

Check: $2(a+b) = 2 \cdot a + 2 \cdot b = 2a + 2b$

69. $7 + 7y = 7 \cdot 1 + 7 \cdot y$ The common factor is 7.

$= 7(1+y)$ Using the distributive law

Check: $7(1+y) = 7 \cdot 1 + 7 \cdot y = 7 + 7y$

71. $18x + 3 = 3 \cdot 6x + 3 \cdot 1 = 3(6x+1)$

Check: $3(6x+1) = 3 \cdot 6x + 3 \cdot 1 = 18x + 3$

73. $5x + 10 + 15y = 5 \cdot x + 5 \cdot 2 + 5 \cdot 3y = 5(x+2+3y)$

Check: $5(x+2+3y) = 5 \cdot x + 5 \cdot 2 + 5 \cdot 3y = 5x + 10 + 15y$

75. $12x + 9 = 3 \cdot 4x + 3 \cdot 3 = 3(4x+3)$

Check: $3(4x+3) = 3 \cdot 4x + 3 \cdot 3 = 12x + 9$

77. $3a + 9b = 3 \cdot a + 3 \cdot 3b = 3(a+3b)$

Check: $3(a+3b) = 3 \cdot a + 3 \cdot 3b = 3a + 9b$

79. $44x + 11y + 22z = 11 \cdot 4x + 11 \cdot y + 11 \cdot 2z = 11(4x + y + 2z)$

Check: $11(4x + y + 2z) = 11 \cdot 4x + 11 \cdot y + 11 \cdot 2z = 44x + 11y + 22z$

81. $st = s \cdot t$

The factors are s and t.

83. $3(x+y) = 3 \cdot (x+y)$

The factors are 3 and $(x+y)$.

85. The factors of $7 \cdot a$ are 7 and a.

87. $(a-b)(x-y) = (a-b) \cdot (x-y)$

The factors are $(a-b)$ and $(x-y)$.

89. *Writing Exercise*

91. Let k represent Kara's salary. Then we have $2k$.

93. *Writing Exercise*

95. The expressions are equivalent by the distributive law.

$8 + 4(a+b) = 8 + 4a + 4b = 4(2 + a + b)$

97. The expressions are equivalent by the commutative law of multiplication and the distributive law.

$(rt + st)5 = 5(rt + st) = 5 \cdot t(r+s) = 5t(r+s)$

99. The expressions are not equivalent.

Let $x = 1$ and $y = 0$. Then we have:

$30 \cdot 0 + 1 \cdot 15 = 0 + 15 = 15$, but

$5[2(1 + 3 \cdot 0)] = 5[2(1)] = 5 \cdot 2 = 10$.

101. *Writing Exercise*

Exercise Set 1.3

1. 9 is composite because it has more than two different factors. They are 1, 3, and 9.

3. 31 is prime because it has only two different factors, 31 and 1.

5. 25 is composite because it has more than two different factors. They are 1, 5, and 25.

7. 2 is prime because it has only two different factors, 2 and 1.

9. The terms "prime" and "composite" apply only to natural numbers. Since 0 is not a natural number, it is neither prime nor composite.

11. Since $35 = 5 \cdot 7$, choice (b) is correct.

13. Since 65 is an odd number and has more than two different factors, choice (d) is correct.

15. We write two factorizations of 50. There are other factorizations as well.

$2 \cdot 25,\ 5 \cdot 10$

List all of the factors of 50:

1, 2, 5, 10, 25, 50

17. We write two factorizations of 42. There are other factorizations as well.

$2 \cdot 21,\ 6 \cdot 7$

List all of the factors of 42:

1, 2, 3, 6, 7, 14, 21, 42

19. $26 = 2 \cdot 13$

21. We begin factoring 30 in any way that we can and continue factoring until each factor is prime.

$30 = 2 \cdot 15 = 2 \cdot 3 \cdot 5$

23. We begin by factoring 27 in any way that we can and continue factoring until each factor is prime.

$27 = 3 \cdot 9 = 3 \cdot 3 \cdot 3$

25. We begin by factoring 18 in any way that we can and continue factoring until each factor is prime.

$18 = 2 \cdot 9 = 2 \cdot 3 \cdot 3$

27. We begin by factoring 40 in any way that we can and continue factoring until each factor is prime.

$40 = 4 \cdot 10 = 2 \cdot 2 \cdot 2 \cdot 5$

29. 43 has exactly two different factors, 43 and 1. Thus, 43 is prime.

31. $210 = 2 \cdot 105 = 2 \cdot 3 \cdot 35 = 2 \cdot 3 \cdot 5 \cdot 7$

33. $115 = 5 \cdot 23$

35. $\frac{14}{21} = \frac{7 \cdot 2}{7 \cdot 3}$ Factoring numerator and denominator

$= \frac{7}{7} \cdot \frac{2}{3}$ Rewriting as a product of two fractions

$= 1 \cdot \frac{2}{3}$ $\frac{7}{7} = 1$

$= \frac{2}{3}$ Using the identity property of 1

37. $\frac{16}{56} = \frac{2 \cdot 8}{7 \cdot 8} = \frac{2}{7} \cdot \frac{8}{8} = \frac{2}{7} \cdot 1 = \frac{2}{7}$

39. $\frac{6}{48} = \frac{1 \cdot 6}{8 \cdot 6}$ Factoring and using the identity property of 1 to write 6 as $1 \cdot 6$

$= \frac{1}{8} \cdot \frac{6}{6}$

$= \frac{1}{8} \cdot 1 = \frac{1}{8}$

41. $\frac{49}{7} = \frac{7 \cdot 7}{1 \cdot 7} = \frac{7}{1} \cdot \frac{7}{7} = \frac{7}{1} \cdot 1 = 7$

43. $\frac{19}{76} = \frac{1 \cdot 19}{4 \cdot 19}$ Factoring and using the identity property of 1 to write 19 as $1 \cdot 19$

$= \frac{1 \cdot \cancel{19}}{4 \cdot \cancel{19}}$ Removing a factor equal to 1: $\frac{19}{19} = 1$

$= \frac{1}{4}$

45. $\frac{150}{25} = \frac{6 \cdot 25}{1 \cdot 25}$ Factoring and using the identity property of 1 to write 25 as $1 \cdot 25$

$= \frac{6 \cdot \cancel{25}}{1 \cdot \cancel{25}}$ Removing a factor equal to 1: $\frac{25}{25} = 1$

$= \frac{6}{1}$

$= 6$ Simplifying

47. $\frac{42}{50} = \frac{2 \cdot 21}{2 \cdot 25}$ Factoring the numerator and the denominator

$= \frac{\cancel{2} \cdot 21}{\cancel{2} \cdot 25}$ Removing a factor equal to 1: $\frac{2}{2} = 1$

$= \frac{21}{25}$

49. $\frac{120}{82} = \frac{2 \cdot 60}{2 \cdot 41}$ Factoring

$= \frac{\cancel{2} \cdot 60}{\cancel{2} \cdot 41}$ Removing a factor equal to 1: $\frac{2}{2} = 1$

$= \frac{60}{41}$

51. $\frac{210}{98} = \frac{2 \cdot 7 \cdot 15}{2 \cdot 7 \cdot 7}$ Factoring

$= \frac{\cancel{2} \cdot \cancel{7} \cdot 15}{\cancel{2} \cdot \cancel{7} \cdot 7}$ Removing a factor equal to 1: $\frac{2 \cdot 7}{2 \cdot 7} = 1$

$= \frac{15}{7}$

53. $\frac{1}{2} \cdot \frac{3}{7} = \frac{1 \cdot 3}{2 \cdot 7}$ Multiplying numerators and denominators

$= \frac{3}{14}$

55. $\frac{9}{2} \cdot \frac{3}{4} = \frac{9 \cdot 3}{2 \cdot 4} = \frac{27}{8}$

57. $\frac{1}{8} + \frac{3}{8} = \frac{1+3}{8}$ Adding numerators; keeping the common denominator

$= \frac{4}{8}$

$= \frac{1 \cdot \cancel{4}}{2 \cdot \cancel{4}} = \frac{1}{2}$ Simplifying

59. $\frac{4}{9}+\frac{13}{18}=\frac{4}{9}\cdot\frac{2}{2}+\frac{13}{18}$ Using 18 as the common denominator

$=\frac{8}{18}+\frac{13}{18}$

$=\frac{21}{18}$

$=\frac{7\cdot\cancel{3}}{6\cdot\cancel{3}}=\frac{7}{6}$ Simplifying

61. $\frac{3}{a}\cdot\frac{b}{7}=\frac{3b}{7a}$ Multiplying numerators and denominators

63. $\frac{4}{a}+\frac{3}{a}=\frac{7}{a}$ Adding numerators; keeping the common denominator

65. $\frac{3}{10}+\frac{8}{15}=\frac{3}{10}\cdot\frac{3}{3}+\frac{8}{15}\cdot\frac{2}{2}$ Using 30 as the common denominator

$=\frac{9}{30}+\frac{16}{30}$

$=\frac{25}{30}$

$=\frac{5\cdot\cancel{5}}{6\cdot\cancel{5}}=\frac{5}{6}$ Simplifying

67. $\frac{9}{7}-\frac{2}{7}=\frac{7}{7}=1$

69. $\frac{13}{18}-\frac{4}{9}=\frac{13}{18}-\frac{4}{9}\cdot\frac{2}{2}$ Using 18 as the common denominator

$=\frac{13}{18}-\frac{8}{18}$

$=\frac{5}{18}$

71. Note that $\frac{20}{30}=\frac{2}{3}$. Thus, $\frac{20}{30}-\frac{2}{3}=0$.
We can also do this exercise by finding a common denominator:
$\frac{20}{30}-\frac{2}{3}=\frac{20}{30}-\frac{20}{30}=0$

73. $\frac{7}{6}\div\frac{3}{5}=\frac{7}{6}\cdot\frac{5}{3}$ Multiplying by the reciprocal of the divisor

$=\frac{35}{18}$

75. $\frac{8}{9}\div\frac{4}{15}=\frac{8}{9}\cdot\frac{15}{4}=\frac{2\cdot\cancel{4}\cdot\cancel{3}\cdot 5}{\cancel{3}\cdot 3\cdot\cancel{4}}=\frac{10}{3}$

77. $12\div\frac{3}{7}=\frac{12}{1}\cdot\frac{7}{3}=\frac{4\cdot\cancel{3}\cdot 7}{1\cdot\cancel{3}}=28$

79. Note that we have a number divided by itself. Thus, the result is 1. We can also do this exercise as follows:
$\frac{7}{13}\div\frac{7}{13}=\frac{7}{13}\cdot\frac{13}{7}=\frac{7\cdot 13}{7\cdot 13}=1$

81. $\frac{\frac{2}{7}}{\frac{5}{3}}=\frac{2}{7}\div\frac{5}{3}=\frac{2}{7}\cdot\frac{3}{5}=\frac{2\cdot 3}{7\cdot 5}=\frac{6}{35}$

83. $\frac{9}{\frac{1}{2}}=9\div\frac{1}{2}=\frac{9}{1}\cdot\frac{2}{1}=\frac{9\cdot 2}{1\cdot 1}=18$

85. *Writing Exercise*

87. $5(x+3)=5(3+x)$ Commutative law of addition
Answers may vary.

89. *Writing Exercise*

91.

Product	56	63	36	72	140	96	168
Factor	7	7	2	36	14	8	8
Factor	8	9	18	2	10	12	21
Sum	15	16	20	38	24	20	29

93. $\frac{16\cdot 9\cdot 4}{15\cdot 8\cdot 12}=\frac{\cancel{4}\cdot\cancel{4}\cdot\cancel{3}\cdot\cancel{3}\cdot\cancel{2}\cdot 2}{\cancel{3}\cdot 5\cdot\cancel{2}\cdot\cancel{4}\cdot\cancel{3}\cdot\cancel{4}}=\frac{2}{5}$

95. $\frac{27pqrs}{9prst}=\frac{3\cdot\cancel{9}\cdot\cancel{p}\cdot q\cdot\cancel{r}\cdot\cancel{s}}{\cancel{9}\cdot\cancel{p}\cdot\cancel{r}\cdot\cancel{s}\cdot t}=\frac{3q}{t}$

97. $\frac{15\cdot 4xy\cdot 9}{6\cdot 25x\cdot 15y}=\frac{\cancel{15}\cdot\cancel{2}\cdot 2\cdot\cancel{x}\cdot\cancel{y}\cdot\cancel{3}\cdot 3}{\cancel{2}\cdot\cancel{3}\cdot 25\cdot\cancel{x}\cdot\cancel{15}\cdot\cancel{y}}=\frac{6}{25}$

99. $\frac{\frac{27ab}{15mn}}{\frac{18bc}{25np}}=\frac{27ab}{15mn}\div\frac{18bc}{25np}=\frac{27ab}{15mn}\cdot\frac{25np}{18bc}=$

$\frac{27ab\cdot 25np}{15mn\cdot 18bc}=\frac{\cancel{3}\cdot\cancel{9}\cdot a\cdot\cancel{b}\cdot\cancel{5}\cdot 5\cdot\cancel{n}\cdot p}{\cancel{3}\cdot\cancel{5}\cdot m\cdot\cancel{n}\cdot 2\cdot\cancel{9}\cdot\cancel{b}\cdot c}=\frac{5ap}{2mc}$

101. $\frac{5\frac{3}{4}rs}{4\frac{1}{2}st}=\frac{\frac{23}{4}rs}{\frac{9}{2}st}=\frac{\frac{23rs}{4}}{\frac{9st}{2}}=\frac{23rs}{4}\div\frac{9st}{2}=$

$\frac{23rs}{4}\cdot\frac{2}{9st}=\frac{23rs\cdot 2}{4\cdot 9st}=\frac{23\cdot r\cdot\cancel{s}\cdot\cancel{2}}{\cancel{2}\cdot 2\cdot 9\cdot\cancel{s}\cdot t}=\frac{23r}{18t}$

103. $A=lw=\left(\frac{4}{5}\text{ m}\right)\left(\frac{7}{9}\text{ m}\right)$

$=\left(\frac{4}{5}\right)\left(\frac{7}{9}\right)(\text{m})(\text{m})$

$=\frac{28}{45}\text{ m}^2$, or $\frac{28}{45}$ square meters

105. $P=4s=4\left(3\frac{5}{9}\text{ m}\right)=4\cdot\frac{32}{9}\text{ m}=\frac{128}{9}\text{ m}$, or $14\frac{2}{9}\text{ m}$

107. There are 12 edges, each with length $2\frac{3}{10}$ cm. We multiply to find the total length of the edges.

$12\cdot 2\frac{3}{10}\text{ cm}=12\cdot\frac{23}{10}\text{ cm}$

$=\frac{12\cdot 23}{10}\text{ cm}$

$=\frac{\cancel{2}\cdot 6\cdot 23}{\cancel{2}\cdot 5}\text{ cm}$

$=\frac{138}{5}\text{ cm}$, or $27\frac{3}{5}$ cm

Exercise Set 1.4

1. Since $\frac{4}{7} = 0.\overline{571428}$, the correct choice is "repeating."

3. The set of integers consists of all whole numbers along with their opposites, so the correct choice is "integer."

5. A "rational number" has the form described.

7. A "natural number" can be thought of as a counting number.

9. The real number -1349 corresponds to 1349 ft below sea level. The real number 29,035 corresponds to 29,035 ft above sea level

11. The real number 950,000,000 corresponds to 950 million°F. The real number -460 corresponds to 460°F below zero.

13. The real number 2 corresponds to two over par. The real number -6 corresponds to six under par.

15. The real number 750 corresponds to a \$750 deposit, and the real number -125 corresponds to a \$125 withdrawal.

17. The Jets are 34 pins behind, so the real number -34 corresponds to the situation from the Jets' viewpoint. The Strikers are 34 pins ahead, so the real number 34 corresponds to the situation from the Strikers' point of view.

19. Since $\frac{10}{3} = 3\frac{1}{3}$, its graph is $\frac{1}{3}$ of a unit to the right of 3.

$\frac{10}{3}$
-5 -4 -3 -2 -1 0 1 2 3 4 5

21. The graph of -4.3 is $\frac{3}{10}$ of a unit to the left of -4.

-4.3
-5 -4 -3 -2 -1 0 1 2 3 4 5

23.
-5 -4 -3 -2 -1 0 1 2 3 4 5

25. $\frac{7}{8}$ means $7 \div 8$, so we divide.

```
   0.8 7 5
8 ) 7.0 0 0
    6 4
      6 0
      5 6
        4 0
        4 0
          0
```

We have $\frac{7}{8} = 0.875$.

27. We first find decimal notation for $\frac{3}{4}$. Since $\frac{3}{4}$ means $3 \div 4$, we divide.

```
   0.7 5
4 ) 3.0 0
    2 8
      2 0
      2 0
        0
```

Thus, $\frac{3}{4} = 0.75$, so $-\frac{3}{4} = -0.75$.

29. $\frac{7}{6}$ means $7 \div 6$, so we divide.

```
   1.1 6 6
6 ) 7.0 0 0
    6
    1 0
      6
      4 0
      3 6
        4 0
        3 6
          4
```

We have $\frac{7}{6} = 1.1\overline{6}$.

31. $\frac{2}{3}$ means $2 \div 3$, so we divide.

```
   0.6 6 6 ...
3 ) 2.0 0 0
    1 8
      2 0
      1 8
        2 0
        1 8
          2
```

We have $\frac{2}{3} = 0.\overline{6}$.

33. We first find decimal notation for $\frac{1}{2}$. Since $\frac{1}{2}$ means $1 \div 2$, we divide.

```
   0.5
2 ) 1.0
    1 0
      0
```

Thus, $\frac{1}{2} = 0.5$, so $-\frac{1}{2} = -0.5$.

35. Since the denominator is 100, we know that $\frac{13}{100} = 0.13$. We could also divide 13 by 100 to find this result.

37. Since -9 is to the left of 4, we have $-9 < 4$.

39. Since 7 is to the right of 0, we have $7 > 0$.

41. Since -6 is to the left of 6, we have $-6 < 6$.

43. Since -8 is to the left of -5, we have $-8 < -5$.

45. Since -5 is to the right of -11, we have $-5 > -11$.

47. Since -12.5 is to the left of -9.4, we have $-12.5 < -9.4$.

49. We convert to decimal notation.
$\frac{5}{12} = 0.41\overline{6}$ and $\frac{11}{25} = 0.44$. Thus, $\frac{5}{12} < \frac{11}{25}$.

51. $-7 > x$ has the same meaning as $x < -7$.

53. $-10 \leq y$ has the same meaning as $y \geq -10$.

55. $-3 \geq -11$ is true, since $-3 > -11$ is true.

57. $0 \geq 8$ is false, since neither $0 > 8$ nor $0 = 8$ is true.

59. $-8 \leq -8$ is true because $-8 = -8$ is true.

61. $|-58| = 58$ since -58 is 58 units from 0.

63. $|17| = 17$ since 17 is 17 units from 0.

65. $|5.6| = 5.6$ since 5.6 is 5.6 units from 0.

67. $|329| = 329$ since 329 is 329 units from 0.

69. $\left|-\frac{9}{7}\right| = \frac{9}{7}$ since $-\frac{9}{7}$ is $\frac{9}{7}$ units from 0.

71. $|0| = 0$ since 0 is 0 units from itself.

73. $|x| = |-8| = 8$

75. $-83, -4.7, 0, \frac{5}{9}, 8.31, 62$

77. $-83, 0, 62$

79. All are real numbers.

81. *Writing Exercise*

83. $3xy = 3 \cdot 2 \cdot 7 = 42$

85. *Writing Exercise*

87. *Writing Exercise*

89. List the numbers as they occur on the number line, from left to right: $-23, -17, 0, 4$

91. Converting to decimal notation, we can write
$\frac{4}{5}, \frac{4}{3}, \frac{4}{8}, \frac{4}{6}, \frac{4}{9}, \frac{4}{2}, -\frac{4}{3}$ as
$0.8, 1.3\overline{3}, 0.5, 0.6\overline{6}, 0.4\overline{4}, 2, -1.3\overline{3}$, respectively. List the numbers (in fractional form) as they occur on the number line, from left to right:
$-\frac{4}{3}, \frac{4}{9}, \frac{4}{8}, \frac{4}{6}, \frac{4}{5}, \frac{4}{3}, \frac{4}{2}$

93. $|4| = 4$ and $|-7| = 7$, so $|4| < |-7|$.

95. $|23| = 23$ and $|-23| = 23$, so $|23| = |-23|$.

97. $|x| = 7$

x represents a number whose distance from 0 is 7. Thus, $x = 7$ or $x = -7$.

99. $2 < |x| < 5$

x represents an integer whose distance from 0 is greater than 2 and also less than 5. Thus, $x = -4, -3, 3, 4$

101. $0.9\overline{9} = 3(0.3\overline{3}) = 3 \cdot \frac{1}{3} = \frac{3}{3}$

103. $7.7\overline{7} = 70(0.1\overline{1}) = 70 \cdot \frac{1}{9} = \frac{70}{9}$

(See Exercise 100.)

105. *Writing Exercise*

Exercise Set 1.5

1. Choice (f), $-3n$, has the same variable factor as $8n$.

3. Choice (e), 9, is a constant as is 43.

5. Choice (b), $5x$, has the same variable factor as $-2x$.

7. Start at 5. Move 8 units to the left.

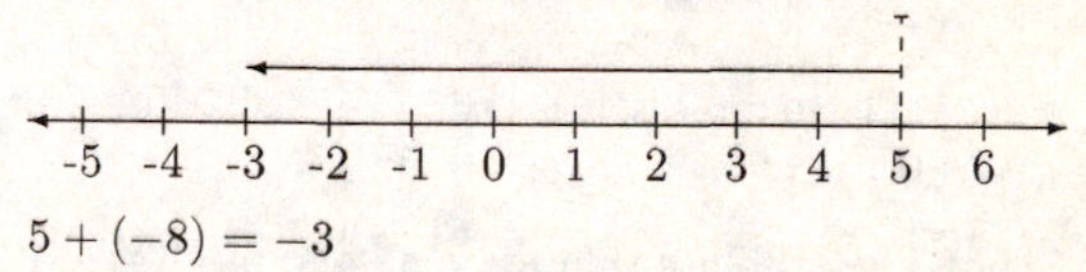

$5 + (-8) = -3$

9. Start at -5. Move 9 units to the right.

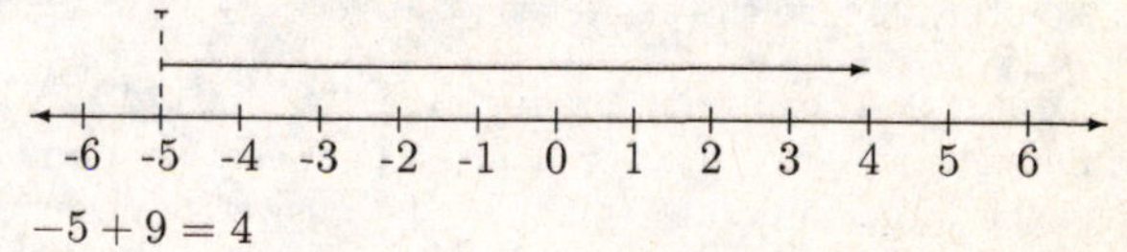

$-5 + 9 = 4$

11. Start at 8. Move 8 units to the left.

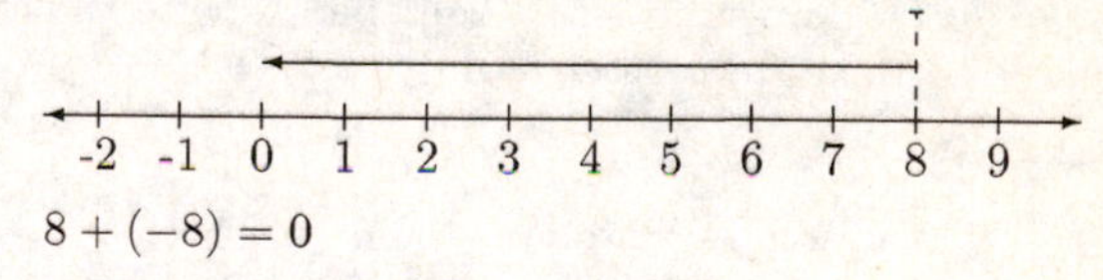

$8 + (-8) = 0$

13. Start at -3. Move 5 units to the left.

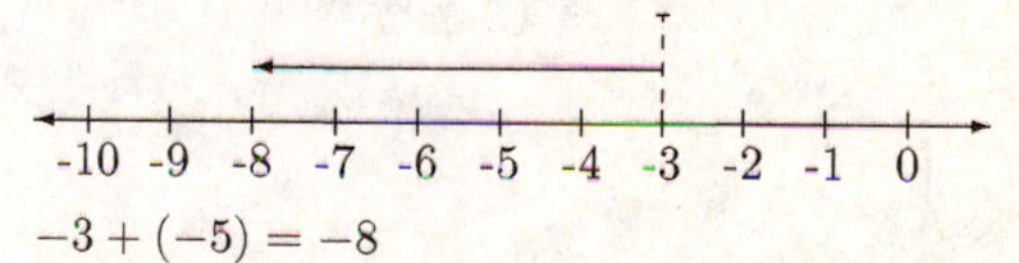

$-3 + (-5) = -8$

15. $-27+0$ One number is 0. The answer is the other number. $-27 + 0 = -27$

17. $0+(-8)$ One number is 0. The answer is the other number. $0 + (-8) = -8$

19. $12 + (-12)$ The numbers have the same absolute value. The sum is 0. $12 + (-12) = 0$

21. $-24 + (-17)$ Two negatives. Add the absolute values, getting 41. Make the answer negative.
$-24 + (-17) = -41$

23. $-13 + 13$ The numbers have the same absolute value. The sum is 0. $-13 + 13 = 0$

25. $18+(-11)$ The absolute values are 18 and 11. The difference is $18 - 11$, or 7. The positive number has the larger absolute value, so the answer is positive. $18 + (-11) = 7$

27. $10+(-12)$ The absolute values are 10 and 12. The difference is $12 - 10$, or 2. The negative number has the larger absolute value, so the answer is negative. $10+(-12) = -2$

29. $-3+14$ The absolute values are 3 and 14. The difference is $14-3$, or 11. The positive number has the larger absolute value, so the answer is positive. $-3+14=11$

31. $-14+(-19)$ Two negatives. Add the absolute values, getting 33. Make the answer negative.
$-14+(-19)=-33$

33. $19+(-19)$ The numbers have the same absolute value. The sum is 0. $19+(-19)=0$

35. $23+(-5)$ The absolute values are 23 and 5. The difference is $23-5$ or 18. The positive number has the larger absolute value, so the answer is positive. $23+(-5)=18$

37. $-31+(-14)$ Two negatives. Add the absolute values, getting 45. Make the answer negative.
$-31+(-14)=-45$

39. $40+(-40)$ The numbers have the same absolute value. The sum is 0. $40+(-40)=0$

41. $85+(-65)$ The absolute values are 85 and 65. The difference is $85-65$, or 20. The positive number has the larger absolute value, so the answer is positive. $85+(-65)=20$

43. $-3.6+1.9$ The absolute values are 3.6 and 1.9. The difference is $3.6-1.9$, or 1.7. The negative number has the larger absolute value, so the answer is negative. $-3.6+1.9=-1.7$

45. $-5.4+(-3.7)$ Two negatives. Add the absolute values, getting 9.1. Make the answer negative. $-5.4+(-3.7)=-9.1$

47. $\frac{-3}{5}+\frac{4}{5}$ The absolute values are $\frac{3}{5}$ and $\frac{4}{5}$. The difference is $\frac{4}{5}-\frac{3}{5}$, or $\frac{1}{5}$. The positive number has the larger absolute value, so the answer is positive.
$\frac{-3}{5}+\frac{4}{5}=\frac{1}{5}$

49. $\frac{-4}{7}+\frac{-2}{7}$ Two negatives. Add the absolute values, getting $\frac{6}{7}$. Make the answer negative.
$\frac{-4}{7}+\frac{-2}{7}=\frac{-6}{7}$

51. $-\frac{2}{5}+\frac{1}{3}$ The absolute values are $\frac{2}{5}$ and $\frac{1}{3}$. The difference is $\frac{6}{15}-\frac{5}{15}$, or $\frac{1}{15}$. The negative number has the larger absolute value, so the answer is negative.
$-\frac{2}{5}+\frac{1}{3}=-\frac{1}{15}$

53. $\frac{-4}{9}+\frac{2}{3}$ The absolute values are $\frac{4}{9}$ and $\frac{2}{3}$. The difference is $\frac{6}{9}-\frac{4}{9}$, or $\frac{2}{9}$. The positive number has the larger absolute value, so the answer is positive.
$\frac{-4}{9}+\frac{2}{3}=\frac{2}{9}$

55. $35+(-14)+(-19)+(-5)$
$=35+[(-14)+(-19)+(-5)]$ Using the associative law of addition
$=35+(-38)$ Adding the negatives
$=-3$ Adding a positive and a negative

57. $-4.9+8.5+4.9+(-8.5)$
Note that we have two pairs of numbers with different signs and the same absolute value: -4.9 and 4.9, 8.5 and -8.5. The sum of each pair is 0, so the result is $0+0$, or 0.

59. Rewording: First increase plus decrease plus second increase is change in price.
Translating: $5¢ + (-3¢) + 7¢ =$ change in price

Since $5+(-3)+7$
$=2+7$
$=9$,
the price rose 9¢ during the given period.

61. Rewording: July bill plus payment plus August charges is new balance.
Translating: $-82 + 50 + (-63) =$ new balance

Since $-82+50+(-63)=-32+(-63)$
$=-95$,
Maya's new balance was \$95.

63. Rewording: First try yardage plus second try yardage plus third try yardage is total gain or loss.
Translating: $13 + (-12) + 21 =$ total gain or loss

Since $13+(-12)+21$
$=1+21$
$=22$,
the total gain was 22 yd.

65. Rewording: Original balance plus first payment plus

Translating: $-470 + 45 +$

new charges plus second payment is new balance.

$-160 + 500 =$ new balance

Since $-470+45+(-160)+500$
$=[-470+(-160)]+(45+500)$
$=-630+545$
$=-85,$

Lyle owes \$85 on his credit card.

67. Rewording: Elevation of base plus total height is elevation of peak.

Translating: $-19,684 + 33,480 =$ elevation of peak.

Since $-19,684+33,480=13,796$, the elevation of the peak is 13,796 ft above sea level.

69. $8a+6a=(8+6)a$ Using the distributive law
$=14a$

71. $-3x+12x=(-3+12)x$ Using the distributive law
$=9x$

73. $5t+8t=(5+8)t=13t$

75. $7m+(-9m)=[7+(-9)]m=-2m$

77. $-5a+(-2a)=[-5+(-2)]a=-7a$

79. $-3+8x+4+(-10x)$
$=-3+4+8x+(-10x)$ Using the commutative law of addition
$=(-3+4)+[8+(-10)]x$ Using the distributive law
$=1-2x$ Adding

81. Perimeter $=8+5x+9+7x$
$=8+9+5x+7x$
$=(8+9)+(5+7)x$
$=17+12x$

83. Perimeter $=3t+3r+7+5t+9+4r$
$=3t+5t+3r+4r+7+9$
$=(3+5)t+(3+4)r+(7+9)$
$=8t+7r+16$

85. Perimeter $=9+6n+7+8n+4n$
$=9+7+6n+8n+4n$
$=(9+7)+(6+8+4)n$
$=16+18n$

87. *Writing Exercise*

89. $7(3z+y+2)=7\cdot 3z+7\cdot y+7\cdot 2=21z+7y+14$

91. *Writing Exercise*

93. Starting with the final value, we "undo" the rise and drop in value by adding their opposites. The result is the original value.

Rewording: Final value plus opposite of rise plus

Translating: $64.38 + (-2.38) +$

opposite of drop is original value.

$3.25 =$ original value.

Since $64.38+(-2.38)+3.25=62.00+3.25$
$=65.25,$

the stock's original value was \$65.25.

95. $4x+\_\_\_+(-9x)+(-2y)$
$=4x+(-9x)+\_\_\_+(-2y)$
$=[4+(-9)]x+\_\_\_+(-2y)$
$=-5x+\_\_\_+(-2y)$

This expression is equivalent to $-5x-7y$, so the missing term is the term which yields $-7y$ when added to $-2y$. Since $-5y+(-2y)=-7y$, the missing term is $-5y$.

97. $3m+2n+\_\_\_+(-2m)$
$=2n+\_\_\_+(-2m)+3m$
$=2n+\_\_\_+(-2+3)m$
$=2n+\_\_\_+m$

This expression is equivalent to $2n+(-6m)$, so the missing term is the term which yields $-6m$ when added to m. Since $-7m+m=-6m$, the missing term is $-7m$.

99. Note that, in order for the sum to be 0, the two missing terms must be the opposites of the given terms. Thus, the missing terms are $-7t$ and -23.

101. $-3+(-3)+2+(-2)+1=-5$

Since the total is 5 under par after the five rounds and $-5=-1+(-1)+(-1)+(-1)+(-1)$, the golfer was 1 under par on average.

Exercise Set 1.6

1. $-x$ is read "the opposite of x," so choice (d) is correct.

3. $12-(-x)$ is read "twelve minus the opposite of x," so choice (f) is correct.

5. $x-(-12)$ is read "x minus negative twelve," so choice (a) is correct.

7. $-x-x$ is read "the opposite of x minus x," so choice (b) is correct.

9. $4-10$ is read "four minus ten."

11. $2-(-9)$ is read "two minus negative nine."

13. $9-(-t)$ is read "nine minus the opposite of t."

15. $-x-y$ is read "the opposite of x minus y."

17. $-3-(-n)$ is read "negative three minus the opposite of n."

19. The opposite of 39 is -39 because $39+(-39)=0$.

21. The opposite of -9 is 9 because $-9+9=0$.

23. The opposite of -3.14 is 3.14 because $-3.14+3.14=0$.

25. If $x=23$, then $-x=-(23)=-23$. (The opposite of 23 is -23.)

27. If $x=-\frac{14}{3}$, then $-x=-\left(-\frac{14}{3}\right)=\frac{14}{3}$.

$\left(\text{The opposite of } -\frac{14}{3} \text{ is } \frac{14}{3}.\right)$

29. If $x=0.101$, then $-x=-(0.101)=-0.101$.

(The opposite of 0.101 is -0.101.)

31. If $x=-72$, then $-(-x)=-(-72)=72$

(The opposite of the opposite of 72 is 72.)

33. If $x=-\frac{2}{5}$, then $-(-x)=-\left[-\left(-\frac{2}{5}\right)\right]=-\frac{2}{5}$.

$\left(\text{The opposite of the opposite of } -\frac{2}{5} \text{ is } -\frac{2}{5}.\right)$

35. When we change the sign of -1 we obtain 1.

37. When we change the sign of 7 we obtain -7.

39. $6-8=6+(-8)=-2$

41. $0-5=0+(-5)=-5$

43. $3-9=3+(-9)=-6$

45. $0-10=0+(-10)=-10$

47. $-9-(-3)=-9+3=-6$

49. Note that we are subtracting a number from itself. The result is 0. We could also do this exercise as follows:

$$-8-(-8)=-8+8=0$$

51. $14-19=14+(-19)=-5$

53. $30-40=30+(-40)=-10$

55. $-7-(-9)=-7+9=2$

57. $-9-(-9)=-9+9=0$

(See Exercise 49.)

59. $5-5=5+(-5)=0$

(See Exercise 49.)

61. $4-(-4)=4+4=8$

63. $-7-4=-7+(-4)=-11$

65. $6-(-10)=6+10=16$

67. $-4-15=-4+(-15)=-19$

69. $-6-(-5)=-6+5=-1$

71. $5-(-12)=5+12=17$

73. $0-5=0+(-5)=-5$

75. $-5-(-2)=-5+2=-3$

77. $-7-14=-7+(-14)=-21$

79. $0-(-5)=0+5=5$

81. $-8-0=-8+0=-8$

83. $3-(-7)=3+7=10$

85. $2-25=2+(-25)=-23$

87. $-42-26=-42+(-26)=-68$

89. $-51-7=-51+(-7)=-58$

91. $3.2-8.7=3.2+(-8.7)=-5.5$

93. $0.072-1=0.072+(-1)=-0.928$

95. $\frac{2}{11}-\frac{9}{11}=\frac{2}{11}+\left(-\frac{9}{11}\right)=-\frac{7}{11}$

97. $\frac{-1}{5}-\frac{3}{5}=\frac{-1}{5}+\left(\frac{-3}{5}\right)=\frac{-4}{5}$, or $-\frac{4}{5}$

99. $-\frac{4}{17}-\left(-\frac{9}{17}\right)=-\frac{4}{17}+\frac{9}{17}=\frac{5}{17}$

101. We subtract the smaller number from the larger.

Translate: $3.8-(-5.2)$

Simplify: $3.8-(-5.2)=3.8+5.2=9$

103. We subtract the smaller number from the larger.

Translate: $114-(-79)$

Simplify: $114-(-79)=114+79=193$

105. $-21-37=-21+(-37)=-58$

107. $9-(-25)=9+25=34$

109. $25-(-12)-7-(-2)+9=25+12+(-7)+2+9=41$

111. $-31+(-28)-(-14)-17=(-31)+(-28)+14+(-17)=-62$

113. $-34-28+(-33)-44=(-34)+(-28)+(-33)+(-44)=-139$

115. $-93+(-84)-(-93)-(-84)$

Note that we are subtracting -93 from -93 and -84 from -84. Thus, the result will be 0. We could also do this exercise as follows:

$-93+(-84)-(-93)-(-84)=-93+(-84)+93+84=0$

117. $-7x-4y=-7x+(-4y)$, so the terms are $-7x$ and $-4y$.

119. $9-5t-3st=9+(-5t)+(-3st)$, so the terms are 9, $-5t$, and $-3st$.

121. $4x-7x$

$=4x+(-7x)$ Adding the opposite

$=(4+(-7))x$ Using the distributive law

$=-3x$

123. $7a-12a+4$

$=7a+(-12a)+4$ Adding the opposite

$=(7+(-12))a+4$ Using the distributive law

$=-5a+4$

125. $-8n-9+n$

$=-8n+(-9)+n$ Adding the opposite

$=-8n+n+(-9)$ Using the commutative law of addition

$=-7n-9$ Adding like terms

127. $3x+5-9x$

$=3x+5+(-9x)$

$=3x+(-9x)+5$

$=-6x+5$

129. $2-6t-9-2t$

$=2+(-6t)+(-9)+(-2t)$

$=2+(-9)+(-6t)+(-2t)$

$=-7-8t$

131. $5y+(-3x)-9x+1-2y+8$

$=5y+(-3x)+(-9x)+1+(-2y)+8$

$=5y+(-2y)+(-3x)+(-9x)+1+8$

$=3y-12x+9$

133. $13x-(-2x)+45-(-21)-7x$

$=13x+2x+45+21+(-7x)$

$=13x+2x+(-7x)+45+21$

$=8x+66$

135. We subtract the lower temperature from the higher temperature:

$25-(-125)=25+125=150$

The temperature range is 150°C.

137. We subtract the lower elevation from the higher elevation:

$$29,035-(-1349)=29,035+1349=30,384$$

The difference in elevation is 30,384 ft.

139. We subtract the lower elevation from the higher elevation:

$$-40-(-156)=-40+156=116$$

Lake Assal is 116 m lower than the Valdes Peninsula.

141. *Writing Exercise*

143. Area $= lw = (36\text{ ft})(12\text{ ft}) = 432\text{ ft}^2$

145. *Writing Exercise*

147. If the clock reads 8:00 A.M. on the day following the blackout when the actual time is 3:00 P.M., then the clock is 7 hr behind the actual time. This indicates that the power outage lasted 7 hr, so power was restored 7 hr after 4:00 P.M., or at 11:00 P.M. on August 14.

149. False. For example, let $m=-3$ and $n=-5$. Then $-3>-5$, but $-3+(-5)=-8\not>0$.

151. True. For example, for $m=4$ and $n=-4$, $4=-(-4)$ and $4+(-4)=0$; for $m=-3$ and $n=3$, $-3=-3$ and $-3+3=0$.

153. [(−)] [9] [−] [(−)] [7] [ENTER]

Exercise Set 1.7

1. The product of two reciprocals is 1.

3. The sum of a pair of additive inverses is 0.

5. The number 0 has no reciprocal.

7. The number 1 is the multiplicative identity.

9. A nonzero number divided by itself is 1.

11. $-3\cdot 8=-24$ Think: $3\cdot 8=24$, make the answer negative.

13. $-8\cdot 7=-56$ Think: $8\cdot 7=56$, make the answer negative.

15. $8\cdot(-3)=-24$

17. $-9\cdot 8=-72$

19. $-6\cdot(-7)=42$ Multiplying absolute values; the answer is positive.

21. $-5\cdot(-9)=45$ Multiplying absolute values; the answer is positive.

23. $-19\cdot(-10)=190$

25. $-12\cdot 12=-144$

27. $-25\cdot(-48)=1200$

29. $-3.5\cdot(-28)=98$

31. $6\cdot(-13)=-78$

33. $-7\cdot(-3.1)=21.7$

35. $\frac{2}{3}\cdot\left(-\frac{3}{5}\right)=-\left(\frac{2\cdot 3}{3\cdot 5}\right)=-\left(\frac{2}{5}\cdot\frac{3}{3}\right)=-\frac{2}{5}$

37. $-\frac{3}{8}\cdot\left(-\frac{2}{9}\right)=\frac{\cancel{3}\cdot\cancel{2}\cdot 1}{4\cdot\cancel{2}\cdot\cancel{3}\cdot 3}=\frac{1}{12}$

39. $(-5.3)(2.1)=-11.13$

41. $-\frac{5}{9}\cdot\frac{3}{4}=-\frac{5\cdot\cancel{3}}{\cancel{3}\cdot 3\cdot 4}=-\frac{5}{12}$

43. $3\cdot(-7)\cdot(-2)\cdot 6$

$=-21\cdot(-12)$ Multiplying the first two numbers and the last two numbers

$=252$

45. 0, The product of 0 and any real number is 0.

47. $-\frac{1}{3}\cdot\frac{1}{4}\cdot\left(-\frac{3}{7}\right)=-\frac{1}{12}\cdot\left(-\frac{3}{7}\right)=\frac{3}{12\cdot 7}=$

$\frac{\cancel{3}\cdot 1}{\cancel{3}\cdot 4\cdot 7}=\frac{1}{28}$

49. $-2\cdot(-5)\cdot(-3)\cdot(-5)=10\cdot 15=150$

51. 0, The product of 0 and any real number is 0.

53. $(-8)(-9)(-10)=72(-10)=-720$

55. $(-6)(-7)(-8)(-9)(-10)=42\cdot 72\cdot(-10)=$ $3024\cdot(-10)=-30,240$

57. $14\div(-2)=-7$ Check: $-7\cdot(-2)=14$

59. $\frac{36}{-9}=-4$ $-4\cdot(-9)=36$

61. $\frac{-56}{8}=-7$ Check: $-7\cdot 8=-56$

63. $\frac{-48}{-12}=4$ Check: $4(-12)=-48$

65. $\frac{-72}{9}=-8$ Check: $-8\cdot 9=-72$

67. $-100\div(-50)=2$ Check: $2(-50)=-100$

69. $-108\div 9=-12$ Check: $-12\cdot 9=-108$

71. $\frac{400}{-50}=-8$ Check: $-8\cdot(-50)=400$

73. Undefined

75. $-4.8\div 1.2=-4$ Check: $-4(1.2)=-4.8$

77. $\frac{0}{-9}=0$

79. $\frac{9.7(-2.8)0}{4.3}$

Since the numerator has a factor of 0, the product in the numerator is 0. The denominator is nonzero, so the quotient is 0.

81. $\frac{-8}{3}=\frac{8}{-3}$ and $\frac{-8}{3}=-\frac{8}{3}$

83. $\frac{29}{-35}=\frac{-29}{35}$ and $\frac{29}{-35}=-\frac{29}{35}$

85. $-\frac{7}{3}=\frac{-7}{3}$ and $-\frac{7}{3}=\frac{7}{-3}$

87. $\frac{-x}{2}=\frac{x}{-2}$ and $\frac{-x}{2}=-\frac{x}{2}$

89. The reciprocal of $\frac{4}{-5}$ is $\frac{-5}{4}$ $\left(\text{or equivalently, } -\frac{5}{4}\right)$ because $\frac{4}{-5}\cdot\frac{-5}{4}=1$.

91. The reciprocal of $-\frac{47}{13}$ is $-\frac{13}{47}$ because $-\frac{47}{13}\cdot\left(-\frac{13}{47}\right)=1$.

93. The reciprocal of -10 is $\frac{1}{-10}$ $\left(\text{or equivalently, } -\frac{1}{10}\right)$ because $-10\left(\frac{1}{-10}\right)=1$.

95. The reciprocal of 4.3 is $\frac{1}{4.3}$ because $4.3\left(\frac{1}{4.3}\right)=1$.

Since $\frac{1}{4.3}=\frac{1}{4.3}\cdot\frac{10}{10}=\frac{10}{43}$, the reciprocal can also be expressed as $\frac{10}{43}$.

97. The reciprocal of $\frac{-9}{4}$ is $\frac{4}{-9}$ $\left(\text{or equivalently, } -\frac{4}{9}\right)$ because $\frac{-9}{4}\cdot\frac{4}{-9}=1$.

99. The reciprocal of 0 does not exist. (There is no number n for which $0\cdot n=1$.)

101. $\left(\frac{-7}{4}\right)\left(-\frac{3}{5}\right)$

$=\left(-\frac{7}{4}\right)\left(-\frac{3}{5}\right)$ Rewriting $\frac{-7}{4}$ as $-\frac{7}{4}$

$=\frac{21}{20}$

103. $\left(\frac{-6}{5}\right)\left(\frac{2}{-11}\right)$

$=\left(\frac{-6}{5}\right)\left(\frac{-2}{11}\right)$ Rewriting $\frac{2}{-11}$ as $\frac{-2}{11}$

$=\frac{12}{55}$

105. $\frac{-3}{8}+\frac{-5}{8}=\frac{-8}{8}=-1$

107. $\left(\frac{-9}{5}\right)\left(\frac{5}{-9}\right)$

Note that this is the product of reciprocals. Thus, the result is 1.

109. $\left(-\frac{3}{11}\right)+\left(-\frac{6}{11}\right)=-\frac{9}{11}$

111. $\frac{7}{8}\div\left(-\frac{1}{2}\right)=\frac{7}{8}\cdot\left(-\frac{2}{1}\right)=-\frac{14}{8}=-\frac{7\cdot\cancel{2}}{\cancel{2}\cdot 4\cdot 1}=-\frac{7}{4}$

113. $\frac{9}{5}\cdot\frac{-20}{3}=\frac{9}{5}\left(-\frac{20}{3}\right)=-\frac{180}{15}=-\frac{\cancel{3}\cdot 3\cdot 4\cdot\cancel{5}}{\cancel{5}\cdot\cancel{3}\cdot 1}=-12$

115. $\left(-\frac{18}{7}\right)+\left(-\frac{3}{7}\right)=-\frac{21}{7}=-3$

117. $-\frac{5}{9} \div \left(-\frac{5}{9}\right)$

Note that we have a number divided by itself. Thus, the result is 1.

119. $-44.1 \div (-6.3) = 7$ Do the long division. The answer is positive.

121. $\frac{5}{9} - \frac{7}{9} = -\frac{2}{9}$

123. $\frac{-3}{10} + \frac{2}{5} = \frac{-3}{10} + \frac{2}{5} \cdot \frac{2}{2} = \frac{-3}{10} + \frac{4}{10} = \frac{1}{10}$

125. $\frac{7}{10} \div \left(\frac{-3}{5}\right) = \frac{7}{10} \div \left(-\frac{3}{5}\right) = \frac{7}{10} \cdot \left(-\frac{5}{3}\right) = -\frac{35}{30} = -\frac{7 \cdot \not{5}}{2 \cdot \not{5} \cdot 3} = -\frac{7}{6}$

127. $\frac{5}{7} - \frac{1}{-7} = \frac{5}{7} - \left(-\frac{1}{7}\right) = \frac{5}{7} + \frac{1}{7} = \frac{6}{7}$

129. $\frac{-4}{15} + \frac{2}{-3} = \frac{-4}{15} + \frac{-2}{3} = \frac{-4}{15} + \frac{-2}{3} \cdot \frac{5}{5} = \frac{-4}{15} + \frac{-10}{15} = \frac{-14}{15}$, or $-\frac{14}{15}$

131. *Writing Exercise*

133. $\frac{264}{468} = \frac{\not{2} \cdot \not{2} \cdot 2 \cdot \not{3} \cdot 11}{\not{2} \cdot \not{2} \cdot \not{3} \cdot 3 \cdot 13} = \frac{22}{39}$

135. *Writing Exercise*

137. Consider the sum $2 + 3$. Its reciprocal is $\frac{1}{2+3}$, or $\frac{1}{5}$, but $\frac{1}{2} + \frac{1}{3} = \frac{5}{6}$.

139. When n is negative, $-n$ is positive, so $\frac{m}{-n}$ is the quotient of a negative and a positive number and, thus, is negative.

141. When n is negative, $-n$ is positive, so $\frac{-n}{m}$ is the quotient of a positive and a negative number and, thus, is negative. When m is negative, $-m$ is positive, so $-m \cdot \left(\frac{-n}{m}\right)$ is the product of a positive and a negative number and, thus, is negative.

143. $m + n$ is the sum of two negative numbers, so it is negative; $\frac{m}{n}$ is the quotient of two negative numbers, so it is positive. Then $(m+n) \cdot \frac{m}{n}$ is the product of a negative and a positive number and, thus, is negative.

145. a) m and n have different signs;

b) either m or n is zero;

c) m and n have the same sign

147. *Writing Exercise*

Exercise Set 1.8

1. a) $4 + 8 \div 2 \cdot 2$

There are no grouping symbols or exponential expressions, so we multiply and divide from left to right. This means that we divide first.

b) $7 - 9 + 15$

There are no grouping symbols, exponential expressions, multiplications, or divisions, so we add and subtract from left to right. This means that we subtract first.

c) $5 - 2(3 + 4)$

We perform the operation in the parentheses first. This means that we add first.

d) $6 + 7 \cdot 3$

There are no grouping symbols or exponential expressions, so we multiply and divide from left to right. This means that we multiply first.

e) $18 - 2[4 + (3 - 2)]$

We perform the operation in the innermost grouping symbols first. This means that we perform the subtraction in the parentheses first.

f) $\frac{5 - 6 \cdot 7}{2}$

Since the denominator does not need to be simplified, we consider the numerator. There are no grouping symbols or exponential expressions, so we multiply and divide from left to right. This means that we multiply first.

3. $\underbrace{2 \cdot 2 \cdot 2}_{3 \text{ factors}} = 2^3$

5. $\underbrace{x \cdot x \cdot x \cdot x \cdot x \cdot x \cdot x}_{7 \text{ factors}} = x^7$

7. $3t \cdot 3t \cdot 3t \cdot 3t \cdot 3t = (3t)^5$

9. $3^2 = 3 \cdot 3 = 9$

11. $(-4)^2 = (-4)(-4) = 16$

13. $-4^2 = -(4 \cdot 4) = -16$

15. $4^3 = 4 \cdot 4 \cdot 4 = 16 \cdot 4 = 64$

17. $(-5)^4 = (-5)(-5)(-5)(-5) = 25 \cdot 25 = 625$

19. $7^1 = 7$ (1 factor)

21. $(3t)^4 = (3t)(3t)(3t)(3t) = 3 \cdot 3 \cdot 3 \cdot 3 \cdot t \cdot t \cdot t \cdot t = 81t^4$

23. $(-7x)^3 = (-7x)(-7x)(-7x) = (-7)(-7)(-7)(x)(x)(x) = -343x^3$

25. $5 + 3 \cdot 7 = 5 + 21$ Multiplying

$= 26$ Adding

27. $8 \cdot 7 + 6 \cdot 5 = 56 + 30$ Multiplying
$= 86$ Adding

29. $19 - 5 \cdot 3 + 3 = 19 - 15 + 3$ Multiplying
$= 4 + 3$ Subtracting and add-
$= 7$ ing from left to right

31. $9 \div 3 + 16 \div 8 = 3 + 2$ Dividing
$= 5$ Adding

33. $14 \cdot 19 \div (19 \cdot 14)$

Since $14 \cdot 19$ and $19 \cdot 14$ are equivalent, we are dividing the product $14 \cdot 19$ by itself. Thus the result is 1.

35. $3(-10)^2 - 8 \div 2^2$
$= 3(100) - 8 \div 4$ Simplifying the exponential expressions
$= 300 - 8 \div 4$ Multiplying and
$= 300 - 2$ dividing from left to right
$= 298$ Subtracting

37. $8 - (2 \cdot 3 - 9)$
$= 8 - (6 - 9)$ Multiplying inside the parentheses
$= 8 - (-3)$ Subtracting inside the parentheses
$= 8 + 3$ Removing parentheses
$= 11$ Adding

39. $(8 - 2)(3 - 9)$
$= 6(-6)$ Subtracting inside the parentheses
$= -36$ Multiplying

41. $13(-10)^2 + 45 \div (-5)$
$= 13(100) + 45 \div (-5)$ Simplifying the exponential expression
$= 1300 + 45 \div (-5)$ Multiplying and
$= 1300 - 9$ dividing from left to right
$= 1291$ Subtracting

43. $2^4 + 2^3 - 10 \div (-1)^4 = 16 + 8 - 10 \div 1 =$
$16 + 8 - 10 = 24 - 10 = 14$

45. $5 + 3(2 - 9)^2 = 5 + 3(-7)^2 = 5 + 3 \cdot 49 = 5 + 147 = 152$

47. $[2 \cdot (5 - 8)]^2 = [2 \cdot (-3)]^2 = (-6)^2 = 36$

49. $\dfrac{7 + 2}{5^2 - 4^2} = \dfrac{9}{25 - 16} = \dfrac{9}{9} = 1$

51. $8(-7) + |6(-5)| = -56 + |-30| = -56 + 30 = -26$

53. $\dfrac{(-2)^3 + 4^2}{3 - 5^2 + 3 \cdot 6} = \dfrac{-8 + 16}{3 - 25 + 3 \cdot 6} = \dfrac{8}{3 - 25 + 18} =$
$\dfrac{8}{-22 + 18} = \dfrac{8}{-4} = -2$

55. $\dfrac{27 - 2 \cdot 3^2}{8 \div 2^2 - (-2)^2} = \dfrac{27 - 2 \cdot 9}{8 \div 4 - 4} = \dfrac{27 - 18}{2 - 4} = \dfrac{9}{-2} = -\dfrac{9}{2}$

57. $9 - 4x = 9 - 4 \cdot 5$ Substituting 5 for x
$= 9 - 20$ Multiplying
$= -11$ Subtracting

59. $24 \div t^3$
$= 24 \div (-2)^3$ Substituting -2 for t
$= 24 \div (-8)$ Simplifying the exponential expression
$= -3$ Dividing

61. $45 \div 3 \cdot a = 45 \div 3 \cdot (-1)$ Substituting -1 for a
$= 15 \cdot (-1)$ Dividing
$= -15$ Multiplying

63. $5x \div 15x^2$
$= 5 \cdot 3 \div 15(3)^2$ Substituting 3 for x
$= 5 \cdot 3 \div 15 \cdot 9$ Simplifying the exponential expression
$= 15 \div 15 \cdot 9$ Multiplying and dividing
$= 1 \cdot 9$ in order from
$= 9$ left to right

65. $45 \div 3^2 x(x - 1)$
$= 45 \div 3^2 \cdot 3(3 - 1)$ Substituting 3 for x
$= 45 \div 3^2 \cdot 3(2)$ Subtracting inside the parentheses
$= 45 \div 9 \cdot 3(2)$ Evaluating the exponential expression
$= 5 \cdot 3(2)$ Dividing and
$= 15(2)$ multiplying
$= 30$ from left to right

67. $-x^2 - 5x = -(-3)^2 - 5(-3) = -9 - 5(-3) =$
$-9 + 15 = 6$

69. $\dfrac{3a - 4a^2}{a^2 - 20} = \dfrac{3 \cdot 5 - 4(5)^2}{(5)^2 - 20} = \dfrac{3 \cdot 5 - 4 \cdot 25}{25 - 20} =$
$\dfrac{15 - 100}{5} = \dfrac{-85}{5} = -17$

71. $-(9x + 1) = -9x - 1$ Removing parentheses and changing the sign of each term

73. $-[5 - 6x] = -5 + 6x$ Removing grouping symbols and changing the sign of each term

75. $-(4a - 3b + 7c) = -4a + 3b - 7c$

77. $-(3x^2 + 5x - 1) = -3x^2 - 5x + 1$

79. $8x - (6x + 7)$
$= 8x - 6x - 7$ Removing parentheses and changing the sign of each term
$= 2x - 7$ Collecting like terms

81. $2a - (5a - 9) = 2a - 5a + 9 = -3a + 9$

83. $2x + 7x - (4x + 6) = 2x + 7x - 4x - 6 = 5x - 6$

85. $9t - 5r - 2(3r + 6t) = 9t - 5r - 6r - 12t = -3t - 11r$

87. $15x - y - 5(3x - 2y + 5z)$

$= 15x - y - 15x + 10y - 25z$ Multiplying each term in parentheses by -5

$= 9y - 25z$

89. $3x^2 + 7 - (2x^2 + 5) = 3x^2 + 7 - 2x^2 - 5$

$= x^2 + 2$

91. $5t^3 + t - 3(t + 2t^3) = 5t^3 + t - 3t - 6t^3$

$= -t^3 - 2t$

93. $12a^2 - 3ab + 5b^2 - 5(-5a^2 + 4ab - 6b^2)$

$= 12a^2 - 3ab + 5b^2 + 25a^2 - 20ab + 30b^2$

$= 37a^2 - 23ab + 35b^2$

95. $-7t^3 - t^2 - 3(5t^3 - 3t)$

$= -7t^3 - t^2 - 15t^3 + 9t$

$= -22t^3 - t^2 + 9t$

97. $5(2x - 7) - [4(2x - 3) + 2]$

$= 5(2x - 7) - [8x - 12 + 2]$

$= 5(2x - 7) - [8x - 10]$

$= 10x - 35 - 8x + 10$

$= 2x - 25$

99. *Writing Exercise*

101. Let x represent "a number." Then we have $2x+9$, or $9+2x$

103. *Writing Exercise*

105. $5t - \{7t - [4r - 3(t - 7)] + 6r\} - 4r$

$= 5t - \{7t - [4r - 3t + 21] + 6r\} - 4r$

$= 5t - \{7t - 4r + 3t - 21 + 6r\} - 4r$

$= 5t - \{10t + 2r - 21\} - 4r$

$= 5t - 10t - 2r + 21 - 4r$

$= -5t - 6r + 21$

107. $\{x - [f - (f - x)] + [x - f]\} - 3x$

$= \{x - [f - f + x] + [x - f]\} - 3x$

$= \{x - [x] + [x - f] - 3x$

$= \{x - x + x - f\} - 3x$

$= x - f - 3x$

$= -2x - f$

109. *Writing Exercise*

111. True; $m - n = -n + m = -(n - m)$

113. False; let $m = 2$ and $n = 1$. Then $-2(1-2) = -2(-1) = 2$, but $-(2 \cdot 1 + 2^2) = -(2 + 4) = -6$.

115. $[x + 3(2 - 5x) \div 7 + x](x - 3)$

When $x = 3$, the factor $x - 3$ is 0, so the product is 0.

117. $4 \cdot 20^3 + 17 \cdot 20^2 + 10 \cdot 20 + 0 \cdot 2$

$= 4 \cdot 8000 + 17 \cdot 400 + 10 \cdot 20 + 0 \cdot 2$

$= 32,000 + 6800 + 200 + 0$

$= 39,000$

119. The tower is composed of cubes with sides of length x. The volume of each cube is $x \cdot x \cdot x$, or x^3. Now we count the number of cubes in the tower. The two lowest levels each contain 3×3, or 9 cubes. The next level contains one cube less than the two lowest levels, so it has $9 - 1$, or 8 cubes. The fourth level from the bottom contains one cube less than the level below it, so it has $8-1$, or 7 cubes. The fifth level from the bottom contains one cube less than the level below it, so it has $7 - 1$, or 6 cubes. Finally, the top level contains one cube less than the level below it, so it has $6-1$, or 5 cubes. All together there are $9 + 9 + 8 + 7 + 6 + 5$, or 44 cubes, each with volume x^3, so the volume of the tower is $44x^3$.

Chapter 2

Equations, Inequalities, and Problem Solving

Exercise Set 2.1

1. A solution is a replacement that makes an equation true.

3. The 9 in $9ab$ is a coefficient.

5. The multiplication principle is used to solve $\frac{2}{3} \cdot x = -4$.

7.
$$\begin{aligned} x + 6 &= 23 \\ x + 6 - 6 &= 23 - 6 \quad \text{Subtracting 6 from both sides} \\ x &= 17 \quad \text{Simplifying} \end{aligned}$$

Check: $x + 6 = 23$
$$17 + 6 \;|\; 23$$
$$23 \stackrel{?}{=} 23 \quad \text{TRUE}$$

The solution is 17.

9.
$$\begin{aligned} y + 7 &= -4 \\ y + 7 - 7 &= -4 - 7 \quad \text{Subtracting 7 from both sides} \\ t &= -11 \end{aligned}$$

Check: $y + 7 = -4$
$$-11 + 7 \;|\; -4$$
$$-4 \stackrel{?}{=} -4 \quad \text{TRUE}$$

The solution is -11.

11.
$$\begin{aligned} t + 9 &= -12 \\ t + 9 - 9 &= -12 - 9 \\ t &= -21 \end{aligned}$$

Check: $t + 9 = -12$
$$-21 + 9 \;|\; -12$$
$$-12 \stackrel{?}{=} -12 \quad \text{TRUE}$$

The solution is -21.

13.
$$\begin{aligned} -6 &= y + 25 \\ -6 - 25 &= y + 25 - 25 \\ -31 &= y \end{aligned}$$

Check: $-6 = y + 25$
$$-6 \;|\; -31 + 25$$
$$-6 \stackrel{?}{=} -6 \quad \text{TRUE}$$

The solution is -31.

15.
$$\begin{aligned} x - 8 &= 5 \\ x - 8 + 8 &= 5 + 8 \\ x &= 13 \end{aligned}$$

Check: $x - 8 = 5$
$$13 - 8 \;|\; 5$$
$$5 \stackrel{?}{=} 5 \quad \text{TRUE}$$

The solution is 13.

17.
$$\begin{aligned} 12 &= -7 + y \\ 7 + 12 &= 7 + (-7) + y \\ 19 &= y \end{aligned}$$

Check: $12 = -7 + y$
$$12 \;|\; -7 + 19$$
$$12 \stackrel{?}{=} 12 \quad \text{TRUE}$$

The solution is 19.

19.
$$\begin{aligned} -5 + t &= -9 \\ 5 + (-5) + t &= 5 + (-9) \\ t &= -4 \end{aligned}$$

Check: $-5 + t = -9$
$$-5 + (-4) \;|\; -9$$
$$-9 \stackrel{?}{=} -9 \quad \text{TRUE}$$

The solution is -4.

21.
$$\begin{aligned} r + \frac{1}{3} &= \frac{8}{3} \\ r + \frac{1}{3} - \frac{1}{3} &= \frac{8}{3} - \frac{1}{3} \\ r &= \frac{7}{3} \end{aligned}$$

Check: $r + \frac{1}{3} = \frac{8}{3}$
$$\frac{7}{3} + \frac{1}{3} \;\Big|\; \frac{8}{3}$$
$$\frac{8}{3} \stackrel{?}{=} \frac{8}{3} \quad \text{TRUE}$$

The solution is $\frac{7}{3}$.

23.
$$\begin{aligned} x + \frac{3}{5} &= -\frac{7}{10} \\ x + \frac{3}{5} - \frac{3}{5} &= -\frac{7}{10} - \frac{3}{5} \\ x &= -\frac{7}{10} - \frac{3}{5} \cdot \frac{2}{2} \\ x &= -\frac{7}{10} - \frac{6}{10} \\ x &= -\frac{13}{10} \end{aligned}$$

Check: $x + \frac{3}{5} = -\frac{7}{10}$

$x + \frac{3}{5}$	$-\frac{7}{10}$
$-\frac{13}{10} + \frac{3}{5}$	$-\frac{7}{10}$
$-\frac{13}{10} + \frac{6}{10}$	

$-\frac{7}{10} \stackrel{?}{=} -\frac{7}{10}$ TRUE

The solution is $-\frac{13}{10}$.

25.
$$x - \frac{5}{6} = \frac{7}{8}$$
$$x - \frac{5}{6} + \frac{5}{6} = \frac{7}{8} + \frac{5}{6}$$
$$x = \frac{7}{8} \cdot \frac{3}{3} + \frac{5}{6} \cdot \frac{4}{4}$$
$$x = \frac{21}{24} + \frac{20}{24}$$
$$x = \frac{41}{24}$$

Check: $x - \frac{5}{6} = \frac{7}{8}$

$x - \frac{5}{6}$	$\frac{7}{8}$
$\frac{41}{24} - \frac{5}{6}$	$\frac{7}{8}$
$\frac{41}{24} - \frac{20}{24}$	$\frac{21}{24}$

$\frac{21}{24} \stackrel{?}{=} \frac{21}{24}$ TRUE

The solution is $\frac{41}{24}$.

27.
$$-\frac{1}{5} + z = -\frac{1}{4}$$
$$\frac{1}{5} - \frac{1}{5} + z = \frac{1}{5} - \frac{1}{4}$$
$$z = \frac{1}{5} \cdot \frac{4}{4} - \frac{1}{4} \cdot \frac{5}{5}$$
$$z = \frac{4}{20} - \frac{5}{20}$$
$$z = -\frac{1}{20}$$

Check: $-\frac{1}{5} + z = -\frac{1}{4}$

$-\frac{1}{5} + z$	$-\frac{1}{4}$
$-\frac{1}{5} + \left(-\frac{1}{20}\right)$	$-\frac{1}{4}$
$-\frac{4}{20} + \left(-\frac{1}{20}\right)$	$-\frac{5}{20}$

$-\frac{5}{20} \stackrel{?}{=} -\frac{5}{20}$ TRUE

The solution is $-\frac{1}{20}$.

29.
$$m + 3.9 = 5.4$$
$$m + 3.9 - 3.9 = 5.4 - 3.9$$
$$m = 1.5$$

Check: $m + 3.9 = 5.4$

$m + 3.9$	5.4
$1.5 + 3.9$	5.4

$5.4 \stackrel{?}{=} 5.4$ TRUE

The solution is 1.5.

31.
$$-9.7 = -4.7 + y$$
$$4.7 + (-9.7) = 4.7 + (-4.7) + y$$
$$-5 = y$$

Check: $-9.7 = -4.7 + y$

-9.7	$-4.7 + y$
-9.7	$-4.7 + (-5)$

$-9.7 \stackrel{?}{=} -9.7$ TRUE

The solution is -5.

33. $5x = 70$

$\frac{5x}{5} = \frac{70}{5}$ Dividing both sides by 5

$1 \cdot x = 14$ Simplifying

$x = 14$ Identity property of 1

Check: $5x = 70$

$5x$	70
$5 \cdot 14$	70

$70 \stackrel{?}{=} 70$ TRUE

The solution is 14.

35. $9t = 36$

$\frac{9t}{9} = \frac{36}{9}$ Dividing both sides by 9

$1 \cdot t = 4$ Simplifying

$t = 4$ Identity property of 1

Check: $9t = 36$

$9t$	36
$9 \cdot 4$	36

$36 \stackrel{?}{=} 36$ TRUE

The solution is 4.

37. $84 = 7x$

$\frac{84}{7} = \frac{7x}{7}$ Dividing both sides by 7

$12 = 1 \cdot x$

$12 = x$

Check: $84 = 7x$

84	$7x$
84	$7 \cdot 12$

$84 \stackrel{?}{=} 84$ TRUE

The solution is 12.

39.
$$-x = 23$$
$$-1 \cdot x = 23$$
$$-1 \cdot (-1 \cdot x) = -1 \cdot 23$$
$$1 \cdot x = -23$$
$$x = -23$$

Check:

$-x = 23$	
$-(-23)$	23

$23 \stackrel{?}{=} 23$ TRUE

The solution is -23.

41. $-t = -8$

The equation states that the opposite of t is the opposite of 8. Thus, $t = 8$. We could also do this exercise as follows.

$-t = -8$

$-1(-t) = -1(-8)$ Multiplying both sides by -1

$t = 8$

Check:

$-t = -8$	
$-(8)$	-8

$-8 \stackrel{?}{=} -8$ TRUE

The solution is 8.

43.
$$7x = -49$$
$$\frac{7x}{7} = \frac{-49}{7}$$
$$1 \cdot x = -7$$
$$x = -7$$

Check:

$7x = -49$	
$7(-7)$	-49

$-49 \stackrel{?}{=} -49$ TRUE

The solution is -7.

45.
$$-1.3a = -10.4$$
$$\frac{-1.3a}{-1.3} = \frac{-10.4}{-1.3}$$
$$a = 8$$

Check:

$-1.3a = -10.4$	
$-1.3(8)$	-10.4

$-10.4 \stackrel{?}{=} -10.4$ TRUE

The solution is 8.

47.
$$\frac{y}{-8} = 11$$
$$-\frac{1}{8} \cdot y = 11$$
$$-8\left(-\frac{1}{8}\right) \cdot y = -8 \cdot 11$$
$$y = -88$$

Check:

$\frac{y}{-8} = 11$	
$-\frac{88}{-8}$	11

$11 \stackrel{?}{=} 11$ TRUE

The solution is -88.

49.
$$\frac{4}{4} = 16$$
$$\frac{5}{4} \cdot \frac{4}{5}x = \frac{5}{4} \cdot 16$$
$$x = \frac{5 \cdot \not{4} \cdot 4}{\not{4} \cdot 1}$$
$$y = 20$$

Check:

$\frac{4}{5}x = 16$	
$\frac{4}{5} \cdot 20$	16

$16 \stackrel{?}{=} 16$ TRUE

The solution is 20.

51.
$$\frac{-x}{6} = 9$$
$$-\frac{1}{6} \cdot x = 9$$
$$-6\left(-\frac{1}{6}\right) \cdot x = -6 \cdot 9$$
$$x = -54$$

Check:

$\frac{-x}{6} = 9$	
$\frac{-(-54)}{6}$	9
$\frac{54}{6}$	

$9 \stackrel{?}{=} 9$ TRUE

The solution is -54.

53.
$$\frac{1}{9} = \frac{z}{5}$$
$$\frac{1}{9} = \frac{1}{5} \cdot z$$
$$5 \cdot \frac{1}{9} = 5 \cdot \frac{1}{5} \cdot z$$
$$\frac{5}{9} = z$$

Check:

$\frac{1}{9} = \frac{z}{5}$	
$\frac{1}{9}$	$\frac{5/9}{5}$
	$\frac{5}{9} \cdot \frac{1}{5}$

$\frac{1}{9} \stackrel{?}{=} \frac{1}{9}$ TRUE

The solution is $\frac{5}{9}$.

55. $-\frac{3}{5}r = -\frac{3}{5}$

The solution of the equation is the number that is multiplied by $-\frac{3}{5}$ to get $-\frac{3}{5}$. That number is 1. We could also do this exercise as follows:

$$-\frac{3}{5}r = -\frac{3}{5}$$
$$-\frac{5}{3}\cdot\left(-\frac{3}{5}r\right) = -\frac{5}{3}\left(-\frac{3}{5}\right)$$
$$r = 1$$

Check: $-\frac{3}{5}r = -\frac{3}{5}$

$-\frac{3}{5}\cdot 1$	$-\frac{3}{5}$

$-\frac{3}{5} \stackrel{?}{=} -\frac{3}{5}$ TRUE

The solution is 1.

57.
$$\frac{-3r}{2} = -\frac{27}{4}$$
$$-\frac{3}{2}r = -\frac{27}{4}$$
$$-\frac{2}{3}\cdot\left(-\frac{3}{2}r\right) = -\frac{2}{3}\cdot\left(-\frac{27}{4}\right)$$
$$r = \frac{\not{2}\cdot\not{3}\cdot 3\cdot 3}{\not{3}\cdot\not{2}\cdot 2}$$
$$r = \frac{9}{2}$$

Check: $\frac{-3r}{2} = -\frac{27}{4}$

$-\frac{3}{2}\cdot\frac{9}{2}$	$-\frac{27}{4}$

$-\frac{27}{4} \stackrel{?}{=} -\frac{27}{4}$ TRUE

The solution is $\frac{9}{2}$.

59.
$$4.5 + t = -3.1$$
$$4.5 + t - 4.5 = -3.1 - 4.5$$
$$t = -7.6$$

The solution is -7.6.

61.
$$-8.2x = 20.5$$
$$\frac{-8.2x}{-8.2} = \frac{20.5}{-8.2}$$
$$x = -2.5$$

The solution is -2.5.

63.
$$x - 4 = -19$$
$$x - 4 + 4 = -19 + 4$$
$$x = -15$$

The solution is -15.

65.
$$3 + t = 21$$
$$-3 + 3 + t = -3 + 21$$
$$t = 18$$

The solution is 18.

67.
$$-12x = 72$$
$$\frac{-12x}{-12} = \frac{72}{-12}$$
$$1\cdot x = -6$$
$$x = -6$$

The solution is -6.

69.
$$48 = -\frac{3}{8}y$$
$$-\frac{8}{3}\cdot 48 = -\frac{8}{3}\left(-\frac{3}{8}y\right)$$
$$-\frac{8\cdot\not{3}\cdot 16}{\not{3}} = y$$
$$-128 = y$$

The solution is -128.

71.
$$a - \frac{1}{6} = -\frac{2}{3}$$
$$a - \frac{1}{6} + \frac{1}{6} = -\frac{2}{3} + \frac{1}{6}$$
$$a = -\frac{4}{6} + \frac{1}{6}$$
$$a = -\frac{3}{6}$$
$$a = -\frac{1}{2}$$

The solution is $-\frac{1}{2}$.

73.
$$-24 = \frac{8x}{5}$$
$$-24 = \frac{8}{5}x$$
$$\frac{5}{8}(-24) = \frac{5}{8}\cdot\frac{8}{5}x$$
$$-\frac{5\cdot\not{8}\cdot 3}{\not{8}\cdot 1} = x$$
$$-15 = x$$

The solution is -15.

75.
$$-\frac{4}{3}t = -16$$
$$-\frac{3}{4}\left(-\frac{4}{3}t\right) = -\frac{3}{4}(-16)$$
$$t = \frac{3\cdot\not{4}\cdot 4}{\not{4}}$$
$$t = 12$$

The solution is 12.

77.
$$-483.297 = -794.053 + t$$
$$-483.297 + 794.053 = -794.053 + t + 794.053$$
$$310.756 = t \quad \text{Using a calculator}$$

The solution is 310.756.

79. *Writing Exercise*

81. $9 - 2 \cdot 5^2 + 7$

$= 9 - 2 \cdot 25 + 7$ Simplifying the exponential expression

$= 9 - 50 + 7$ Multiplying

$= -41 + 7$ Subtracting and

$= -34$ Adding from left to right

83. $16 \div (2 - 3 \cdot 2) + 5$

$= 16 \div (2 - 6) + 5$ Simplifying inside

$= 16 \div (-4) + 5$ the parentheses

$= -4 + 5$ Dividing

$= 1$ Adding

85. *Writing Exercise*

87. $mx = 9.4m$

$$\frac{mx}{m} = \frac{9.4m}{m}$$

$x = 9.4$

The solution is 9.4.

89. $cx + 5c = 7c$

$cx + 5c - 5c = 7c - 5c$

$cx = 2c$

$$\frac{cx}{c} = \frac{2c}{c}$$

$x = 2$

The solution is 2.

91. $7 + |x| = 20$

$-7 + 7 + |x| = -7 + 20$

$|x| = 13$

x represents a number whose distance from 0 is 13. Thus $x = -13$ or $x = 13$.

93. $t - 3590 = 1820$

$t - 3590 + 3590 = 1820 + 3590$

$t = 5410$

$t + 3590 = 5410 + 3590$

$t + 3590 = 9000$

95. To "undo" the last step, divide 22.5 by 0.3.

$22.5 \div 0.3 = 75$

Now divide 75 by 0.3.

$75 \div 0.3 = 250$

The answer should be 250 not 22.5.

Exercise Set 2.2

1. $3x - 1 = 7$

$3x - 1 + 1 = 7 + 1$ Adding 1 to both sides

$3x = 7 + 1$

Choice (c) is correct.

3. $6(x - 1) = 2$

$6x - 6 = 2$ Using the distributive law

Choice (a) is correct.

5. $4x = 3 - 2x$

$4x + 2x = 3 - 2x + 2x$ Adding $2x$ to both sides

$4x + 2x = 3$

Choice (b) is correct.

7. $2x + 9 = 25$

$2x + 9 - 9 = 25 - 9$ Subtracting 9 from both sides

$2x = 16$ Simplifying

$$\frac{2x}{2} = \frac{16}{2}$$ Dividing both sides by 2

$x = 8$ Simplifying

Check: $2x + 9 = 25$

$2 \cdot 8 + 9$	25
$16 + 9$	

$25 \stackrel{?}{=} 25$ TRUE

The solution is 8.

9. $6z + 4 = 46$

$6z + 4 - 4 = 46 - 4$ Subtracting 4 from both sides

$6z = 42$ Simplifying

$$\frac{6z}{6} = \frac{42}{6}$$ Dividing both sides by 6

$z = 7$ Simplifying

Check: $6z + 4 = 46$

$6 \cdot 7 + 4$	46
$42 + 4$	

$46 \stackrel{?}{=} 46$ TRUE

The solution is 7.

11. $7t - 8 = 27$

$7t - 8 + 8 = 27 + 8$ Adding 8 to both sides

$7t = 35$

$$\frac{7t}{7} = \frac{35}{7}$$ Dividing both sides by 7

$t = 5$

Check: $7t - 8 = 27$

$7 \cdot 5 - 8$	27
$35 - 8$	

$27 \stackrel{?}{=} 27$ TRUE

The solution is 5.

13. $3x - 9 = 33$

$3x - 9 + 9 = 33 + 9$

$3x = 42$

$$\frac{3x}{3} = \frac{42}{3}$$

$x = 14$

Check: $3x - 9 = 33$

$$\begin{array}{r|l} 3 \cdot 14 - 9 & 33 \\ 42 - 9 & \end{array}$$

$33 \stackrel{?}{=} 33$ TRUE

The solution is 14.

15.
$$\begin{aligned} 8z + 2 &= -54 \\ 8z + 2 - 2 &= -54 - 2 \\ 8z &= -56 \\ \frac{8z}{8} &= \frac{-56}{8} \\ z &= -7 \end{aligned}$$

Check: $8z + 2 = -54$

$$\begin{array}{r|l} 8(-7) + 2 & -54 \\ -56 + 2 & \end{array}$$

$-54 \stackrel{?}{=} -54$ TRUE

The solution is -7.

17.
$$\begin{aligned} -91 &= 9t + 8 \\ -91 - 8 &= 9t + 8 - 8 \\ -99 &= 9t \\ \frac{-99}{9} &= \frac{9t}{9} \\ -11 &= t \end{aligned}$$

Check: $-91 = 9t + 8$

$$\begin{array}{r|l} -91 & 9(-11) + 8 \\ & -99 + 8 \end{array}$$

$-91 \stackrel{?}{=} -91$ TRUE

The solution is -11.

19.
$$\begin{aligned} 12 - 4x &= 108 \\ -12 + 12 - 4x &= -12 + 108 \\ -4x &= 96 \\ \frac{-4x}{-4} &= \frac{96}{-4} \\ x &= -24 \end{aligned}$$

Check: $12 - 4x = 108$

$$\begin{array}{r|l} 12 - 4(-24) & 108 \\ 12 + 96 & \end{array}$$

$108 \stackrel{?}{=} 108$ TRUE

The solution is -24.

21.
$$\begin{aligned} -6z - 18 &= -132 \\ -6z - 18 + 18 &= -132 + 18 \\ -6z &= -114 \\ \frac{-6z}{-6} &= \frac{-114}{-6} \\ z &= 19 \end{aligned}$$

Check: $-6z - 18 = -132$

$$\begin{array}{r|l} -6 \cdot 19 - 18 & -132 \\ -114 - 18 & \end{array}$$

$-132 \stackrel{?}{=} -132$ TRUE

The solution is 19.

23.
$$\begin{aligned} 4x + 5x &= 10 \\ 9x &= 10 \quad \text{Combining like terms} \\ \frac{9x}{9} &= \frac{10}{9} \\ x &= \frac{10}{9} \end{aligned}$$

Check: $4x + 5x = 10$

$$\begin{array}{r|l} 4 \cdot \frac{10}{9} + 5 \cdot \frac{10}{9} & 10 \\ \frac{40}{9} + \frac{50}{9} & \\ \frac{90}{9} & \end{array}$$

$10 \stackrel{?}{=} 10$ TRUE

The solution is $\frac{10}{9}$.

25.
$$\begin{aligned} 32 - 7x &= 11 \\ -32 + 32 - 7x &= -32 + 11 \\ -7x &= -21 \\ \frac{-7x}{-7} &= \frac{-21}{-7} \\ x &= 3 \end{aligned}$$

Check: $32 - 7x = 11$

$$\begin{array}{r|l} 32 - 7 \cdot 3 & 11 \\ 32 - 21 & \end{array}$$

$11 \stackrel{?}{=} 11$ TRUE

The solution is 3.

27.
$$\begin{aligned} \frac{3}{5}t - 1 &= 8 \\ \frac{3}{5}t - 1 + 1 &= 8 + 1 \\ \frac{3}{5}t &= 9 \\ \frac{5}{3} \cdot \frac{3}{5}t &= \frac{5}{3} \cdot 9 \\ t &= \frac{5 \cdot \cancel{3} \cdot 3}{\cancel{3} \cdot 1} \\ t &= 15 \end{aligned}$$

Check: $\frac{3}{5}t - 1 = 8$

$$\begin{array}{r|l} \frac{3}{5} \cdot 15 - 1 & 8 \\ 9 - 1 & \end{array}$$

$8 \stackrel{?}{=} 8$ TRUE

The solution is 15.

29.
$$4+\frac{7}{2}x=-10$$
$$-4+4+\frac{7}{2}x=-4-10$$
$$\frac{7}{2}x=-14$$
$$\frac{2}{7}\cdot\frac{7}{2}x=\frac{2}{7}(-14)$$
$$x=-\frac{2\cdot 2\cdot \not{7}}{\not{7}\cdot 1}$$
$$x=-4$$

Check:
$$4+\frac{7}{2}x=-10$$

$4+\frac{7}{2}(-4)$	-10
$4-14$	

$-10 \stackrel{?}{=} -10$ TRUE

The solution is -4.

31.
$$-\frac{3a}{4}-5=2$$
$$-\frac{3a}{4}-5+5=2+5$$
$$-\frac{3a}{4}=7$$
$$-\frac{4}{3}\left(-\frac{3a}{4}\right)=-\frac{4}{3}\cdot 7$$
$$a=-\frac{28}{3}$$

Check:
$$-\frac{3a}{4}-5=2$$

$-\frac{3}{4}\left(-\frac{28}{3}\right)-5$	2
$7-5$	

$2 \stackrel{?}{=} 2$ TRUE

The solution is $-\frac{28}{3}$.

33. $2x=x+x$

$2x=2x$ Adding on the right side

The equation $2x=2x$ is true regardless of the replacement for x, so all real numbers are solutions and the equation is an identity.

35. $4x-6=6x$

$-6=6x-4x$ Subtracting $4x$ from both sides

$-6=2x$ Simplifying

$\frac{-6}{2}=\frac{2x}{2}$ Dividing both sides by 2

$-3=x$

Check:
$$4x-6=6x$$

$4(-3)-6$	$6(-3)$
$-12-6$	-18

$-18 \stackrel{?}{=} -18$ TRUE

The solution is -3.

37. $5y-2=28-y$

$5y-2+y=28-y+y$ Adding y to both sides

$6y-2=28$ Simplifying

$6y-2+2=28+2$ Adding 2 to both sides

$6y=30$ Simplifying

$\frac{6y}{6}=\frac{30}{6}$ Dividing both sides by 6

$y=5$

Check:
$$5y-2=28-y$$

$5\cdot 5-2$	$28-5$
$25-2$	23

$23 \stackrel{?}{=} 23$ TRUE

The solution is 5.

39. $7(2a-1)=21$

$14a-7=21$ Using the distributive law

$14a=21+7$ Adding 7

$14a=28$

$a=2$ Dividing by 14

Check:
$$7(2a-1)=21$$

$7(2\cdot 2-1)$	21
$7(4-1)$	
$7\cdot 3$	

$21 \stackrel{?}{=} 21$ TRUE

The solution is 2.

41. We can write $8=8(x+1)$ as $8\cdot 1=8(x+1)$. Then $1=x+1$, or $x=0$. The solution is 0.

43. $7r-(2r+8)=32$

$7r-2r-8=32$

$5r-8=32$ Combining like terms

$5r=32+8$

$5r=40$

$r=8$

Check:
$$7r-(2r+8)=32$$

$7\cdot 8-(2\cdot 8+8)$	32
$56-(16+8)$	
$56-24$	

$32 \stackrel{?}{=} 32$ TRUE

The solution is 8.

45.
$$6x+3=2x+3$$
$$6x-2x=3-3$$
$$4x=0$$
$$\frac{4x}{4}=\frac{0}{4}$$
$$x=0$$

Check:
$$6x+3=2x+3$$

$6\cdot 0+3$	$2\cdot 0+3$
$0+3$	$0+3$

$3 \stackrel{?}{=} 3$ TRUE

The solution is 0.

47.
$$\begin{aligned} 5-2x &= 3x-7x+25 \\ 5-2x &= -4x+25 \\ 4x-2x &= 25-5 \\ 2x &= 20 \\ \frac{2x}{2} &= \frac{20}{2} \\ x &= 10 \end{aligned}$$

Check:

$5-2x = 3x-7x+25$	
$5-2\cdot 10$	$3\cdot 10-7\cdot 10+25$
$5-20$	$30-70+25$
-15	$-40+25$

$-15 \stackrel{?}{=} -15$ TRUE

The solution is 10.

49.
$$\begin{aligned} 7+3x-6 &= 3x+5-x \\ 3x+1 &= 2x+5 \quad \text{Combining like terms on each side} \\ 3x-2x &= 5-1 \\ x &= 4 \end{aligned}$$

Check:

$7+3x-6 = 3x+5-x$	
$7+3\cdot 4-6$	$3\cdot 4+5-4$
$7+12-6$	$12+5-4$
$19-6$	$17-4$

$13 \stackrel{?}{=} 13$ TRUE

The solution is 4.

51.
$$\begin{aligned} 4y-4+y+24 &= 6y+20-4y \\ 5y+20 &= 2y+20 \\ 5y-2y &= 20-20 \\ 3y &= 0 \\ y &= 0 \end{aligned}$$

Check:

$4y-4+y+24 = 6y+20-4y$	
$4\cdot 0-4+0+24$	$6\cdot 0+20-4\cdot 0$
$0-4+0+24$	$0+20-0$

$20 \stackrel{?}{=} 20$ TRUE

The solution is 0.

53.
$$\begin{aligned} 13-3(2x-1) &= 4 \\ 13-6x+3 &= 4 \\ 16-6x &= 4 \\ -6x &= 4-16 \\ -6x &= -12 \\ x &= 2 \end{aligned}$$

Check:

$13-3(2x-1) = 4$	
$13-3(2\cdot 2-1)$	4
$13-3(4-1)$	
$13-3\cdot 3$	
$13-9$	

$4 \stackrel{?}{=} 4$ TRUE

The solution is 2.

55.
$$\begin{aligned} 7(5x-2) &= 6(6x-1) \\ 35x-14 &= 36x-6 \\ -14+6 &= 36x-35x \\ -8 &= x \end{aligned}$$

Check:

$7(5x-2) = 6(6x-1)$	
$7(5(-8)-2)$	$6(6(-8)-1)$
$7(-40-2)$	$6(-48-1)$
$7(-42)$	$6(-49)$

$-294 \stackrel{?}{=} -294$ TRUE

The solution is -8.

57.
$$\begin{aligned} 19-(2x+3) &= 2(x+3)+x \\ 19-2x-3 &= 2x+6+x \\ 16-2x &= 3x+6 \\ 16-6 &= 3x+2x \\ 10 &= 5x \\ 2 &= x \end{aligned}$$

Check:

$19-(2x+3) = 2(x+3)+x$	
$19-(2\cdot 2+3)$	$2(2+3)+2$
$19-(4+3)$	$2\cdot 5+2$
$19-7$	$10+2$

$12 \stackrel{?}{=} 12$ TRUE

The solution is 2.

59.
$$\begin{aligned} 3(x+4) &= 3(x-1) \\ 3x+12 &= 3x-3 \quad \text{Using the distributive law} \\ -3x+3x+12 &= -3x+3x-3 \quad \text{Using the addition principle} \\ 12 &= -3 \end{aligned}$$

Since the original equation is equivalent to the false equation $12 = -3$, the original equation has no solution. It is a contradiction.

61. $\frac{5}{4}x+\frac{1}{4}x = 2x+\frac{1}{2}+\frac{3}{4}x$

The number 4 is the least common denominator, so we multiply by 4 on both sides.

$$\begin{aligned} 4\left(\frac{5}{4}x+\frac{1}{4}x\right) &= 4\left(2x+\frac{1}{2}+\frac{3}{4}x\right) \\ 4\cdot\frac{5}{4}x+4\cdot\frac{1}{4}x &= 4\cdot 2x+4\cdot\frac{1}{2}+4\cdot\frac{3}{4}x \\ 5x+x &= 8x+2+3x \\ 6x &= 11x+2 \\ 6x-11x &= 2 \\ -5x &= 2 \\ \frac{-5x}{-5} &= \frac{2}{-5} \\ x &= -\frac{2}{5} \end{aligned}$$

Check:

$$\begin{array}{c|c} \frac{5}{4}x+\frac{1}{4}x & 2x+\frac{1}{2}+\frac{3}{4}x \\ \hline \frac{5}{4}\left(-\frac{2}{5}\right)+\frac{1}{4}\left(-\frac{2}{5}\right) & 2\left(-\frac{2}{5}\right)+\frac{1}{2}+\frac{3}{4}\left(-\frac{2}{5}\right) \\ -\frac{1}{2}-\frac{1}{10} & -\frac{4}{5}+\frac{1}{2}-\frac{3}{10} \\ -\frac{5}{10}-\frac{1}{10} & -\frac{8}{10}+\frac{5}{10}-\frac{3}{10} \\ -\frac{6}{10} \stackrel{?}{=} & -\frac{6}{10} \quad \text{TRUE} \end{array}$$

The solution is $-\frac{2}{5}$.

63. $\frac{2}{3}+\frac{1}{4}t=6$

The number 12 is the least common denominator, so we multiply by 12 on both sides.

$$\begin{aligned} 12\left(\frac{2}{3}+\frac{1}{4}t\right) &= 12\cdot 6 \\ 12\cdot\frac{2}{3}+12\cdot\frac{1}{4}t &= 72 \\ 8+3t &= 72 \\ 3t &= 72-8 \\ 3t &= 64 \\ t &= \frac{64}{3} \end{aligned}$$

Check:

$$\begin{array}{c|c} \frac{2}{3}+\frac{1}{4}t & 6 \\ \hline \frac{2}{3}+\frac{1}{4}\left(\frac{64}{3}\right) & 6 \\ \frac{2}{3}+\frac{16}{3} & \\ \frac{18}{3} & \\ 6 \stackrel{?}{=} & 6 \quad \text{TRUE} \end{array}$$

The solution is $\frac{64}{3}$.

65. $\frac{2}{3}+4t=6t-\frac{2}{15}$

The number 15 is the least common denominator, so we multiply by 15 on both sides.

$$\begin{aligned} 15\left(\frac{2}{3}+4t\right) &= 15\left(6t-\frac{2}{15}\right) \\ 15\cdot\frac{2}{3}+15\cdot 4t &= 15\cdot 6t-15\cdot\frac{2}{15} \\ 10+60t &= 90t-2 \\ 10+2 &= 90t-60t \\ 12 &= 30t \\ \frac{12}{30} &= t \\ \frac{2}{5} &= t \end{aligned}$$

Check:

$$\begin{array}{c|c} \frac{2}{3}+4t & 6t-\frac{2}{15} \\ \hline \frac{2}{3}+4\cdot\frac{2}{5} & 6\cdot\frac{2}{5}-\frac{2}{15} \\ \frac{2}{3}+\frac{8}{5} & \frac{12}{5}-\frac{2}{15} \\ \frac{10}{15}+\frac{24}{15} & \frac{36}{15}-\frac{2}{15} \\ \frac{34}{15} \stackrel{?}{=} & \frac{34}{15} \quad \text{TRUE} \end{array}$$

The solution is $\frac{2}{5}$.

67. $\frac{1}{3}x+\frac{2}{5}=\frac{4}{15}+\frac{3}{5}x-\frac{2}{3}$

The number 15 is the least common denominator, so we multiply by 15 on both sides.

$$\begin{aligned} 15\left(\frac{1}{3}x+\frac{2}{5}\right) &= 15\left(\frac{4}{15}+\frac{3}{5}x-\frac{2}{3}\right) \\ 15\cdot\frac{1}{3}x+15\cdot\frac{2}{5} &= 15\cdot\frac{4}{15}+15\cdot\frac{3}{5}x-15\cdot\frac{2}{3} \\ 5x+6 &= 4+9x-10 \\ 5x+6 &= -6+9x \\ 5x-9x &= -6-6 \\ -4x &= -12 \\ \frac{-4x}{-4} &= \frac{-12}{-4} \\ x &= 3 \end{aligned}$$

Check:

$$\begin{array}{c|c} \frac{1}{3}x+\frac{2}{5} & \frac{4}{15}+\frac{3}{5}x-\frac{2}{3} \\ \hline \frac{1}{3}\cdot 3+\frac{2}{5} & \frac{4}{15}+\frac{3}{5}\cdot 3-\frac{2}{3} \\ 1+\frac{2}{5} & \frac{4}{15}+\frac{9}{5}-\frac{2}{3} \\ \frac{5}{5}+\frac{2}{5} & \frac{4}{15}+\frac{27}{15}-\frac{10}{15} \\ \frac{7}{5} & \frac{21}{15} \\ \frac{7}{5} \stackrel{?}{=} & \frac{7}{5} \quad \text{TRUE} \end{array}$$

The solution is 3.

69.

$$\begin{aligned} 2.1x+45.2 &= 3.2-8.4x \\ &\text{Greatest number of decimal places is 1} \\ 10(2.1x+45.2) &= 10(3.2-8.4x) \\ &\text{Multiplying by 10 to clear decimals} \\ 10(2.1x)+10(45.2) &= 10(3.2)-10(8.4x) \\ 21x+452 &= 32-84x \\ 21x+84x &= 32-452 \\ 105x &= -420 \\ x &= \frac{-420}{105} \\ x &= -4 \end{aligned}$$

Check:

$2.1x + 45.2 = 3.2 - 8.4x$	
$2.1(-4) + 45.2$	$3.2 - 8.4(-4)$
$-8.4 + 45.2$	$3.2 + 33.6$
$36.8 \stackrel{?}{=} 36.8$	TRUE

The solution is -4.

71.
$$0.76 + 0.21t = 0.96t - 0.49$$
Greatest number of decimal places is 2
$$100(0.76 + 0.21t) = 100(0.96t - 0.49)$$
Multiplying by 100 to clear decimals
$$100(0.76) + 100(0.21t) = 100(0.96t) - 100(0.49)$$
$$76 + 21t = 96t - 49$$
$$76 + 49 = 96t - 21t$$
$$125 = 75t$$
$$\frac{125}{75} = t$$
$$\frac{5}{3} = t, \text{ or}$$
$$1.\overline{6} = t$$

The answer checks. The solution is $\frac{5}{3}$, or $1.\overline{6}$.

73.
$$\frac{2}{5}x - \frac{3}{2}x = \frac{3}{4}x + 2$$
The least common denominator is 20.
$$20\left(\frac{2}{5}x - \frac{3}{2}x\right) = 20\left(\frac{3}{4}x + 2\right)$$
$$20 \cdot \frac{2}{5}x - 20 \cdot \frac{3}{2}x = 20 \cdot \frac{3}{4}x + 20 \cdot 2$$
$$8x - 30x = 15x + 40$$
$$-22x = 15x + 40$$
$$-22x - 15x = 40$$
$$-37x = 40$$
$$\frac{-37x}{-37} = \frac{40}{-37}$$
$$x = -\frac{40}{37}$$

Check:

$\frac{2}{5}x - \frac{3}{2}x = \frac{3}{4}x + 2$	
$\frac{2}{5}\left(-\frac{40}{37}\right) - \frac{3}{2}\left(-\frac{40}{37}\right)$	$\frac{3}{4}\left(-\frac{40}{37}\right) + 2$
$-\frac{16}{37} + \frac{60}{37}$	$-\frac{30}{37} + \frac{74}{37}$
$\frac{44}{37} \stackrel{?}{=} \frac{44}{37}$	TRUE

The solution is $-\frac{40}{37}$.

75.
$$\frac{1}{3}(2x - 1) = 7$$
$$3 \cdot \frac{1}{3}(2x - 1) = 3 \cdot 7$$
$$2x - 1 = 21$$
$$2x = 22$$
$$x = 11$$

Check:

$\frac{1}{3}(2x - 1) = 7$	
$\frac{1}{3}(2 \cdot 11 - 1)$	7
$\frac{1}{3} \cdot 21$	
$7 \stackrel{?}{=} 7$	TRUE

The solution is 11.

77.
$$\frac{3}{4}(3t - 6) = 9$$
$$\frac{4}{3} \cdot \frac{3}{4}(3t - 6) = \frac{4}{3} \cdot 9$$
$$3t - 6 = 12$$
$$3t = 18$$
$$t = 6$$

Check:

$\frac{3}{4}(3t - 6) = 9$	
$\frac{3}{4}(3 \cdot 6 - 6)$	9
$\frac{3}{4} \cdot 12$	
$9 \stackrel{?}{=} 9$	TRUE

The solution is 6.

79.
$$\frac{1}{6}\left(\frac{3}{4}x - 2\right) = -\frac{1}{5}$$
$$30 \cdot \frac{1}{6}\left(\frac{3}{4}x - 2\right) = 30\left(-\frac{1}{5}\right)$$
$$5\left(\frac{3}{4}x - 2\right) = -6$$
$$\frac{15}{4}x - 10 = -6$$
$$\frac{15}{4}x = 4$$
$$4 \cdot \frac{15}{4}x = 4 \cdot 4$$
$$15x = 16$$
$$x = \frac{16}{15}$$

Check:

$$\begin{array}{c|c} \frac{1}{6}\left(\frac{3}{4}x-2\right) & -\frac{1}{5} \\ \hline \frac{1}{6}\left(\frac{3}{4}\cdot\frac{16}{15}-2\right) & -\frac{1}{5} \\ \frac{1}{6}\left(\frac{4}{5}-2\right) & \\ \frac{1}{6}\left(-\frac{6}{5}\right) & \\ -\frac{1}{5} & -\frac{1}{5} \end{array}$$

$-\frac{1}{5} \stackrel{?}{=} -\frac{1}{5}$ TRUE

The solution is $\frac{16}{15}$.

81.
$$\begin{aligned} 0.7(3x+6) &= 1.1-(x+2) \\ 2.1x+4.2 &= 1.1-x-2 \\ 10(2.1x+4.2) &= 10(1.1-x-2) \quad \text{Clearing decimals} \\ 21x+42 &= 11-10x-20 \\ 21x+42 &= -10x-9 \\ 21x+10x &= -9-42 \\ 31x &= -51 \\ x &= -\frac{51}{31} \end{aligned}$$

The check is left to the student. The solution is $-\frac{51}{31}$.

83.
$$\begin{aligned} a+(a-3) &= (a+2)-(a+1) \\ a+a-3 &= a+2-a-1 \\ 2a-3 &= 1 \\ 2a &= 1+3 \\ 2a &= 4 \\ a &= 2 \end{aligned}$$

Check:

$$\begin{array}{c|c} a+(a-3) & (a+2)-(a+1) \\ \hline 2+(2-3) & (2+2)-(2+1) \\ 2-1 & 4-3 \\ 1 & 1 \end{array}$$

$1 \stackrel{?}{=} 1$ TRUE

The solution is 2.

85. *Writing Exercise*

87. $3-5a = 3-5\cdot 2 = 3-10 = -7$

89. $7x-2x = 7(-3)-2(-3) = -21+6 = -15$

91. *Writing Exercise*

93.
$$\begin{aligned} 8.43x-2.5(3.2-0.7x) &= -3.455x+9.04 \\ 8.43x-8+1.75x &= -3.455x+9.04 \\ 10.18x-8 &= -3.455x+9.04 \\ 10.18x+3.455x &= 9.04+8 \\ 13.635x &= 17.04 \\ x &= 1.\overline{2497}, \text{ or } \frac{1136}{909} \end{aligned}$$

The solution is $1.\overline{2497}$, or $\frac{1136}{909}$.

95.
$$\begin{aligned} -2[3(x-2)+4] &= 4(5-x)-2x \\ -2[3x-6+4] &= 20-4x-2x \\ -2[3x-2] &= 20-6x \\ -6x+4 &= 20-6x \\ 4 &= 20 \quad \text{Adding } 6x \text{ to both sides} \end{aligned}$$

We get a false equation, so there is no solution. The equation is a contradiction.

97.
$$\begin{aligned} 2|x| &= -14 \\ \frac{2|x|}{2} &= -\frac{14}{2} \\ |x| &= -7 \end{aligned}$$

Since the absolute value of a number is always nonnegative, the equation has no solution. It is a contradiction.

99.
$$\begin{aligned} 2x(x+5)-3(x^2+2x-1) &= 9-5x-x^2 \\ 2x^2+10x-3x^2-6x+3 &= 9-5x-x^2 \\ -x^2+4x+3 &= 9-5x-x^2 \\ 4x+3 &= 9-5x \quad \text{Adding } x^2 \\ 4x+5x &= 9-3 \\ 9x &= 6 \\ x &= \frac{2}{3} \end{aligned}$$

The solution is $\frac{2}{3}$.

101.
$$\begin{aligned} 9-3x &= 2(5-2x)-(1-5x) \\ 9-3x &= 10-4x-1+5x \\ 9-3x &= 9+x \\ 9-9 &= x+3x \\ 0 &= 4x \\ 0 &= x \end{aligned}$$

The solution is 0.

103. $[7-2(8\div(-2))]x = 0$

Since $7-2(8\div(-2)) \neq 0$ and the product on the left side of the equation is 0, then x must be 0.

105.
$$\begin{aligned} \frac{5x+3}{4}+\frac{25}{12} &= \frac{5+2x}{3} \\ 12\left(\frac{5x+3}{4}+\frac{25}{12}\right) &= 12\left(\frac{5+2x}{3}\right) \\ 12\left(\frac{5x+3}{4}\right)+12\cdot\frac{25}{12} &= 4(5+2x) \\ 3(5x+3)+25 &= 4(5+2x) \\ 15x+9+25 &= 20+8x \\ 15x+34 &= 20+8x \\ 7x &= -14 \\ x &= -2 \end{aligned}$$

The solution is -2.

Exercise Set 2.3

1. We substitute 10 for t and calculate M.

$$M = \frac{1}{5}t = \frac{1}{5} \cdot 10 = 2$$

The storm is 2 miles away.

3. We substitute 21,345 for n and calculate f.

$$f = \frac{n}{15} = \frac{21,345}{15} = 1423$$

There are 1423 full-time equivalent students.

5. Substitute 1800 for a and calculate B.

$$B = 30a = 30 \cdot 1800 = 54,000$$

The minimum furnace output is 54,000 Btu's.

7. Substitute 1 for t and calculate n.

$$\begin{aligned} n &= 0.5t^4 + 3.45t^3 - 96.65t^2 + 347.7t \\ &= 0.5(1)^4 + 3.45(1)^3 - 96.65(1)^2 + 347.7(1) \\ &= 0.5 + 3.45 - 96.65 + 347.7 \\ &= 255 \end{aligned}$$

255 mg of ibuprofen remains in the bloodstream.

9. $A = bh$

$\frac{A}{h} = \frac{bh}{h}$ Dividing both sides by h

$\frac{A}{h} = b$

11. $d = rt$

$\frac{d}{t} = \frac{rt}{t}$ Dividing both sides by t

$\frac{d}{t} = r$

13. $I = Prt$

$\frac{I}{rt} = \frac{Prt}{rt}$ Dividing both sides by rt

$\frac{I}{rt} = P$

15. $H = 65 - m$

$H + m = 65$ Adding m to both sides

$m = 65 - H$ Subtracting H from both sides

17. $P = 2l + 2w$

$P - 2w = 2l + 2w - 2w$ Subtracting $2w$ from both sides

$P - 2w = 2l$

$\frac{P - 2w}{2} = \frac{2l}{2}$ Dividing both sides by 2

$\frac{P - 2w}{2} = l$, or

$\frac{P}{2} - w = l$

19. $A = \pi r^2$

$\frac{A}{r^2} = \frac{\pi r^2}{r^2}$

$\frac{A}{r^2} = \pi$

21. $A = \frac{1}{2}bh$

$2A = 2 \cdot \frac{1}{2}bh$ Multiplying both sides by 2

$2A = bh$

$\frac{2A}{b} = \frac{bh}{b}$ Dividing both sides by h

$\frac{2A}{b} = h$

23. $E = mc^2$

$\frac{E}{c^2} = \frac{mc^2}{c^2}$ Dividing both sides by c^2

$\frac{E}{c^2} = m$

25. $Q = \frac{c + d}{2}$

$2Q = 2 \cdot \frac{c + d}{2}$ Multiplying both sides by 2

$2Q = c + d$

$2Q - c = c + d - c$ Subtracting c from both sides

$2Q - c = d$

27. $A = \frac{a + b + c}{3}$

$3A = 3 \cdot \frac{a + b + c}{3}$ Multiplying both sides by 3

$3A = a + b + c$

$3A - a - c = a + b + c - a - c$ Subtracting a and c from both sides

$3A - a - c = b$

29. $M = \frac{A}{s}$

$s \cdot M = s \cdot \frac{A}{s}$ Multiplying both sides by s

$sM = A$

31. $F = \frac{9}{5}C + 32$

$F - 32 = \frac{9}{5}C$

$\frac{5}{9}(F - 32) = \frac{5}{9} \cdot \frac{9}{5}C$

$\frac{5}{9}(F - 32) = C$

33. $A = at + bt$

$A = t(a + b)$ Factoring

$\frac{A}{a + b} = t$ Dividing both sides by $a + b$

35.
$$A = \frac{1}{2}ah + \frac{1}{2}bh$$
$$2A = 2\left(\frac{1}{2}ah + \frac{1}{2}bh\right)$$
$$2A = ah + bh$$
$$2A = h(a+b)$$
$$\frac{2A}{a+b} = h$$

37.
$$R = r + \frac{400(W-L)}{N}$$
$$N \cdot R = N\left(r + \frac{400(W-L)}{N}\right)$$
Multiplying both sides by N
$$NR = Nr + 400(W-L)$$
$$NR = Nr + 400W - 400L$$
$$NR + 400L = Nr + 400W \quad \text{Adding } 400L \text{ to both sides}$$
$$400L = Nr + 400W - NR \quad \text{Adding } -NR \text{ to both sides}$$
$$L = \frac{Nr + 400W - NR}{400}$$

39. *Writing Exercise*

41. $0.79(38.4)0$

One factor is 0, so the product is 0.

43.
$20 \div (-4) \cdot 2 - 3$

$= -5 \cdot 2 - 3$ Dividing and

$= -10 - 3$ multiplying from left to right

$= -13$ Subtracting

45. *Writing Exercise*

47.
$$K = 19.18w + 7h - 9.52a + 92.4$$
$$2627 = 19.18(82) + 7(185) - 9.52a + 92.4$$
$$2627 = 1572.76 + 1295 - 9.52a + 92.4$$
$$2627 = 2960.16 - 9.52a$$
$$-333.16 = -9.52a$$
$$35 \approx a$$

The man is about 35 years old.

49. First we substitute 54 for A and solve for s to find the length of a side of the cube.

$A = 6s^2$

$54 = 6s^2$

$9 = s^2$

$3 = s$ Taking the positive square root

Now we substitute 3 for s in the formula for the volume of a cube and compute the volume.

$V = s^3 = 3^3 = 27$

The volume of the cube is 27 in^3.

51.
$$c = \frac{w}{a} \cdot d$$
$$ac = a \cdot \frac{w}{a} \cdot d$$
$$ac = wd$$
$$a = \frac{wd}{c}$$

53.
$$ac = bc + d$$
$$ac - bc = d$$
$$c(a-b) = d$$
$$c = \frac{d}{a-b}$$

55.
$$3a = c - a(b+d)$$
$$3a = c - ab - ad$$
$$3a + ab + ad = c$$
$$a(3+b+d) = c$$
$$a = \frac{c}{3+b+d}$$

57. $K = 917 + 6(2.2046w + 0.3937h - a)$

$K = 917 + 13.2276w + 2.3622h - 6a$

Exercise Set 2.4

1. "What percent of 57 is 23?" can be translated as $n \cdot 57 = 23$, so choice (d) is correct.

3. "23 is 57% of what number?" can be translated as $23 = 0.57y$, so choice (e) is correct.

5. "57 is what percent of 23?" can be translated as $n \cdot 23 = 57$, so choice (c) is correct.

7. "What is 23% of 57?" can be translated as $a = (0.23)57$, so choice (f) is correct.

9. "23% of what number is 57?" can be translated as $57 = 0.23y$, so choice (b) is correct.

11. 30% = 30.0%

30% 0.30.0 ↑⌴

Move the decimal point 2 places to the left.

30% = 0.30, or 0.3

13. 2% = 2.0%

2% 0.02.0 ↑⌴

Move the decimal point 2 places to the left.

2% = 0.02

15. 77% = 77.0%

77% 0.77.0 ↑⌴

Move the decimal point 2 places to the left.

77% = 0.77

17. $9\% = 9.0\%$

9% 0.09.0

Move the decimal point 2 places to the left.

$9\% = 0.09$

19. 62.58% 0.62.58

Move the decimal point 2 places to the left.

$62.58\% = 0.6258$

21. 0.7% 0.00.7

Move the decimal point 2 places to the left.

$0.7\% = 0.007$

23. $125\%=125.0\%$ 1.25.0

Move the decimal point 2 places to the left.

$125\% = 1.25$

25. 0.64

First move the decimal point two places to the right; 0.64.

then write a % symbol: 64%

27. 0.106

First move the decimal point two places to the right; 0.10.6

then write a % symbol: 10.6%

29. 0.42

First move the decimal point two places to the right; 0.42.

then write a % symbol: 42%

31. 0.9

First move the decimal point two places to the right; 0.90.

then write a % symbol: 90%

33. 0.0049

First move the decimal point two places to the right; 0.00.49

then write a % symbol: 0.49%

35. 1.08

First move the decimal point two places to the right; 1.08.

then write a % symbol: 108%

37. 2.3

First move the decimal point two places to the right; 2.30.

then write a % symbol: 230%

39. $\frac{4}{5}$ $\left(\text{Note: } \frac{4}{5} = 0.8\right)$

Move the decimal point two places to the right; 0.80.

then write a % symbol: 80%

41. $\frac{8}{25}$ $\left(\text{Note: } \frac{8}{25} = 0.32\right)$

First move the decimal point two places to the right; 0.32.

then write a % symbol: 32%

43. ***Translate.***

What percent of 68 is 17?

$y \cdot 68 = 17$

We solve the equation and then convert to percent notation.

$$y \cdot 68 = 17$$
$$y = \frac{17}{68}$$
$$y = 0.25 = 25\%$$

The answer is 25%.

45. ***Translate.***

What percent of 125 is 30?

$y \cdot 125 = 30$

We solve the equation and then convert to percent notation.

$$y \cdot 125 = 30$$
$$y = \frac{30}{125}$$
$$y = 0.24 = 24\%$$

The answer is 24%.

47. ***Translate.***

14 is 30% of what number?

$14 = 30\% \cdot y$

We solve the equation.

$$14 = 0.3y \quad (30\% = 0.3)$$
$$\frac{14}{0.3} = y$$
$$46.\overline{6} = y$$

The answer is $46.\overline{6}$, or $46\frac{2}{3}$, or $\frac{140}{3}$.

49. ***Translate.***

0.3 is 12% of what number?

$0.3 = 12\% \cdot y$

We solve the equation.

$$0.3 = 0.12y \quad (12\% = 0.12)$$
$$\frac{0.3}{0.12} = y$$
$$2.5 = y$$

The answer is 2.5.

51. ***Translate.***

What number is 35% of 240?

$$y = 35\% \cdot 240$$

We solve the equation.

$y = 0.35 \cdot 240$ $(35\% = 0.35)$
$y = 84$ Multiplying

The answer is 84.

53. ***Translate.***

What percent of 60 is 75?

$$y \cdot 60 = 75$$

We solve the equation and then convert to percent notation.

$$y \cdot 60 = 75$$
$$y = \frac{75}{60}$$
$$y = 1.25 = 125\%$$

The answer is 125%.

55. ***Translate.***

What is 2% of 40?

$$x = 2\% \cdot 40$$

We solve the equation.

$x = 0.02 \cdot 40$ $(2\% = 0.02)$
$x = 0.8$ Multiplying

The answer is 0.8.

57. Observe that 25 is half of 50. Thus, the answer is 0.5, or 50%. We could also do this exercise by translating to an equation.

Translate.

25 is what percent of 50?

$$25 = y \cdot 50$$

We solve the equation and convert to percent notation.

$$25 = y \cdot 50$$
$$\frac{25}{50} = y$$
$$0.5 = y, \text{ or } 50\% = y$$

The answer is 50%.

59. First we reword and translate, letting p represent the price of a dog.

What is 3% of $6600?

$$p = 0.03 \cdot 6600$$
$$p = 0.03 \cdot 6600 = 198$$

The price of the dog is $198.

61. First we reword and translate, letting v represent the amount spent on veterinary care.

What is 24% of $6600?

$$v = 0.24 \cdot 6600$$
$$v = 0.24 \cdot 6600 = 1584$$

Veterinarian expenses are $1584.

63. First we reword and translate, letting s represent the cost of supplies.

What is 8% of $6600?

$$s = 0.08 \cdot 6600$$
$$s = 0.08 \cdot 6600 = 528$$

The cost of supplies is $528.

65. First we reword and translate, letting c represent the number of credits Frank has completed.

What is 60% of 125?

$$c = 0.6 \cdot 125$$
$$c = 0.6 \cdot 125 = 75$$

Frank has completed 75 credits.

67. First we reword and translate, letting b represent the number of at-bats.

194 is 31% of what number?

$$194 = 0.31 \cdot b$$
$$\frac{194}{0.31} = b$$
$$626 \approx b$$

Ichiro Suzuki had 626 at-bats.

69. a) First we reword and translate, letting p represent the unknown percent.

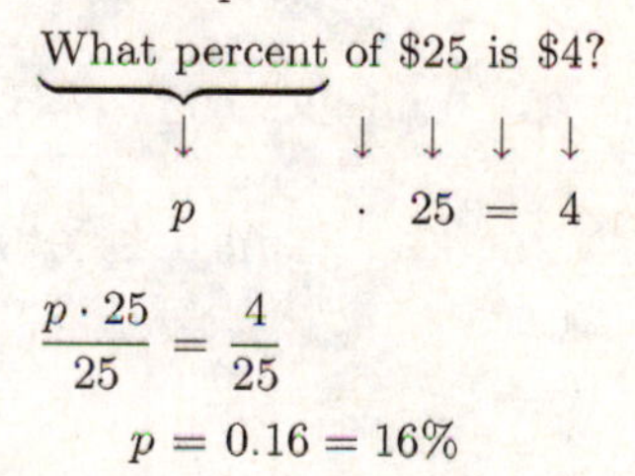

What percent of $25 is $4?

$$p \cdot 25 = 4$$
$$\frac{p \cdot 25}{25} = \frac{4}{25}$$
$$p = 0.16 = 16\%$$

The tip was 16% of the cost of the meal.

b) We add to find the total cost of the meal, including tip:

$$\$25 + \$4 = \$29$$

71. To find the percent of cars manufactured in the U.S., we first reword and translate, letting p represent the unknown percent.

6.0 million is what percent of 8.3 million?

$$6.0 = p \cdot 8.3$$
$$\frac{6.0}{8.3} = p$$
$$0.72 \approx p$$
$$72\% \approx p$$

About 72% of the cars were manufactured in the U.S.

To find the percent of cars manufactured outside the U.S., we subtract:

$$100\% - 72\% = 28\%.$$

About 28% of the cars were manufactured outside the U.S.

73. Let I = the amount of interest Sarah will pay. Then we have:

$$\begin{array}{l} I \text{ is } 8\% \text{ of } \$3500. \\ \downarrow\ \downarrow\quad \downarrow\quad \downarrow\quad\ \downarrow \\ I = 0.08 \cdot \$3500 \\ I = \$280 \end{array}$$

Sarah will pay \$280 interest.

75. If n = the number of women who had babies in good or excellent health, we have:

$$\begin{array}{l} n \text{ is } 95\% \text{ of } 300. \\ \downarrow\ \downarrow\quad \downarrow\quad \downarrow\quad \downarrow \\ n = 0.95 \cdot 300 \\ n = 285 \end{array}$$

285 women had babies in good or excellent health.

77. A self-employed person must earn 120% as much as a non-self-employed person. Let a = the amount Joy would need to earn, in dollars per hour, on her own for a comparable income. Then we have:

$$\begin{array}{l} a \text{ is } 120\% \text{ of } \$15. \\ \downarrow\ \downarrow\quad \downarrow\quad \downarrow\ \downarrow \\ a = 1.2 \cdot 15 \\ a = 18 \end{array}$$

Joy would need to earn \$18 per hour on her own.

79. First we subtract to find the amount of the increase.

$$40,000 - 16,000 = 24,000$$

Then we reword and translate.

$$\begin{array}{l} \underbrace{\text{What percent}}\text{ of } 16,000 \text{ is } 24,000? \\ \qquad\downarrow\qquad\quad \downarrow\quad \downarrow\quad \downarrow\quad \downarrow \\ \qquad p \qquad\qquad \cdot\ 16,000 = 24,000 \end{array}$$

$$\frac{p \cdot 16,000}{16,000} = \frac{24,000}{16,000}$$

$$p = 1.5 = 150\%$$

The number of USA Triathlon members increased by 150% from 1993 to 2002.

81. When the sales tax is 5%, the total amount paid is 105% of the cost of the merchandise. Let c = the cost of the merchandise. Then we have:

$$\begin{array}{l} \$37.80 \text{ is } 105\% \text{ of } c. \\ \quad\downarrow\quad \downarrow\quad \downarrow\quad \downarrow\ \downarrow \\ 37.80 = 1.05 \cdot c \end{array}$$

$$\frac{37.80}{1.05} = c$$

$$36 = c$$

The price of the merchandise was \$36.

83. When the sales tax is 5%, the total amount paid is 105% of the cost of the merchandise. Let c = the amount the school group owes, or the cost of the software without tax. Then we have:

$$\begin{array}{l} \$157.41 \text{ is } 106\% \text{ of } c. \\ \quad\downarrow\quad \downarrow\quad \downarrow\quad \downarrow\ \downarrow \\ 157.41 = 1.06 \cdot c \end{array}$$

$$\frac{157.41}{1.06} = c$$

$$148.5 = c$$

The school group owes \$148.50.

85. First we reword and translate.

$$\begin{array}{l} \text{What is } 16.5\% \text{ of } 191? \\ \downarrow\quad \downarrow\quad \downarrow\quad \downarrow\quad \downarrow \\ a\quad = 0.165 \cdot 191 \end{array}$$

Solve. We convert 16.5% to decimal notation and multiply.

$$a = 0.165 \cdot 191$$

$$a = 31.515 \approx 31.5$$

About 31.5 lb of the author's body weight is fat.

87. Let b = the number of brochures the business can expect to be opened and read. Then we have:

$$\begin{array}{l} b \text{ is } 78\% \text{ of } 9500. \\ \downarrow\ \downarrow\quad \downarrow\quad \downarrow\quad \downarrow \\ b = 0.78 \cdot 9500 \\ b = 7410 \end{array}$$

The business can expect 7410 brochures to be opened and read.

89. The number of calories in a serving of Light Style Bread is 85% of the number of calories in a serving of regular bread. Let c = the number of calories in a serving of regular bread. Then we have:

$$\begin{array}{l} \underbrace{140 \text{ calories}} \text{ is } 85\% \text{ of } c. \\ \qquad\downarrow\qquad\ \downarrow\quad \downarrow\quad \downarrow\ \downarrow \\ \qquad 140 \qquad = 0.85 \cdot c \end{array}$$

$$\frac{140}{0.85} = c$$

$$165 \approx c$$

There are about 165 calories in a serving of regular bread.

91. *Writing Exercise*

93. Let n represent "some number." Then we have $n + 5$, or $5 + n$.

95. $8 \cdot 2a$, or $2a \cdot 8$

97. *Writing Exercise*

99. Let p = the population of Bardville. Then we have:

$$\begin{array}{l} 1332 \text{ is } 15\% \text{ of } 48\% \text{ of } \underbrace{\text{the population.}} \\ \ \downarrow\quad \downarrow\quad \downarrow\quad \downarrow\quad \downarrow\quad \downarrow\qquad\qquad \downarrow \\ 1332 = 0.15 \cdot 0.48 \cdot \qquad\qquad p \end{array}$$

$$\frac{1332}{0.15(0.48)} = p$$

$$18,500 = p$$

The population of Bardville is 18,500.

101. Since 4 ft = 4×1 ft = 4×12 in. = 48 in., we can express 4 ft 8 in. as 48 in. + 8 in., or 56 in. We reword and translate. Let a = Dana's final adult height.

$\underbrace{\text{56 in.}}$ is 84.4% of $\underbrace{\text{adult height}}$.

$$56 = 0.844 \cdot a$$

$$\frac{56}{0.844} = \frac{0.844 \cdot a}{0.844}$$

$$66 \approx a$$

Note that 66 in. = 60 in. + 6 in. = 5 ft 6 in. Dana's final adult height will be about 5 ft 6 in.

103. Using the formula for the area A of a rectangle with length l and width w, $A = l \cdot w$, we first find the area of the photo.

$$A = 8 \text{ in.} \times 6 \text{ in.} = 48 \text{ in}^2$$

Next we find the area of the photo that will be visible using a mat intended for a 5-in. by 7-in. photo.

$$A = 7 \text{ in.} \times 5 \text{ in.} = 35 \text{ in}^2$$

Then the area of the photo that will be hidden by the mat is $48 \text{ in}^2 - 35 \text{ in}^2$, or 13 in^2.

We find what percentage of the area of the photo this represents.

$\underbrace{\text{What percent}}$ of $\underbrace{48 \text{ in}^2}$ is $\underbrace{13 \text{ in}^2}$?

$$p \cdot 48 = 13$$

$$\frac{p \cdot 48}{48} = \frac{13}{48}$$

$$p \approx 0.27$$

$$p \approx 27\%$$

The mat will hide about 27% of the photo.

105. *Writing Exercise*

Exercise Set 2.5

1. ***Familiarize.*** Let n = the number. Then two fewer than ten times the number is $10n - 2$.

Translate.

$\underbrace{\text{Two fewer than ten times a number}}$ is 78.

$$10n - 2 = 78$$

Carry out. We solve the equation.

$10n - 2 = 78$

$10n = 80$ Adding 2

$n = 8$ Dividing by 10

Check. Ten times 8 is 80 and two fewer than 80 is 78. The answer checks.

State. The number is 8.

3. ***Familiarize.*** Let a = the number. Then "five times the sum of 3 and some number" translates to $5(a + 3)$.

Translate.

$\underbrace{\text{Five times the sum of 3 and some number}}$ is 70.

$$5(a + 3) = 70$$

Carry out. We solve the equation.

$5(a + 3) = 70$

$5a + 15 = 70$ Using the distributive law

$5a = 55$ Subtracting 15

$a = 11$ Dividing by 5

Check. The sum of 3 and 11 is 14, and $5 \cdot 14 = 70$. The answer checks.

State. The number is 11.

5. ***Familiarize.*** Let p = the regular price of the shoes. At 15% off, Amy paid (100−15)%, or 85% of the regular price.

Translate.

\$72.25 is 85% of $\underbrace{\text{the regular price.}}$

$$72.25 = 0.85 \cdot p$$

Carry out. We solve the equation.

$72.25 = 0.85p$

$\frac{72.25}{0.85} = p$ Dividing both sides by 0.85

$85 = p$

Check. 85% of \$85, or 0.85(\$85), is \$72.25. The answer checks.

State. The regular price was \$85.

7. ***Familiarize.*** Let c = the price of the graphing calculator itself. When the sales tax rate is 5%, the tax paid on the calculator is 5% of c, or $0.05c$.

Translate.

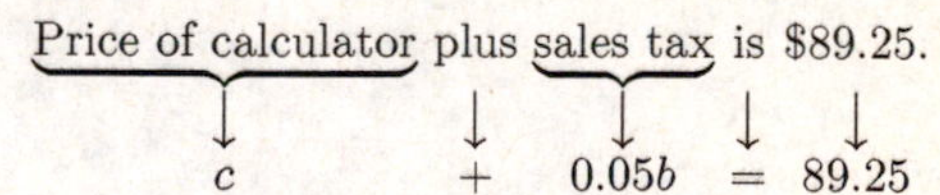

Carry out. We solve the equation.

$c + 0.05c = 89.25$

$1.05c = 89.25$

$c = \frac{89.25}{1.05}$

$c = 85$

Check. 5% of \$85, or 0.05(\$85), is \$4.25 and \$85 + \$4.25 is \$89.25, the total cost. The answer checks.

State. The graphing calculator itself cost \$85.

9. ***Familiarize.*** Let d = Kouros' distance, in miles, from the start after 8 hr. Then the distance from the finish line is $2d$.

Translate.

$\underbrace{\text{Distance from start}}$ plus $\underbrace{\text{distance from finish}}$ is 188 mi.

$$d + 2d = 188$$

Carry out. We solve the equation.

$d + 2d = 188$

$3d = 188$

$d = \frac{188}{3}$, or $62\frac{2}{3}$

Check. If Kouros is $\frac{188}{3}$ mi from the start, then he is $2\cdot\frac{188}{3}$, or $\frac{376}{3}$ mi from the finish. Since $\frac{188}{3}+\frac{376}{3}=\frac{564}{3}=188$, the total distance run, the answer checks.

State. Kouros had run approximately $62\frac{2}{3}$ mi.

11. ***Familiarize.*** Let $d=$ the distance Wheldon had traveled, in miles, at the given point. This is the distance from the start. The corresponding distance from the finish was $300-d$ miles.

Translate. We reword and translate.

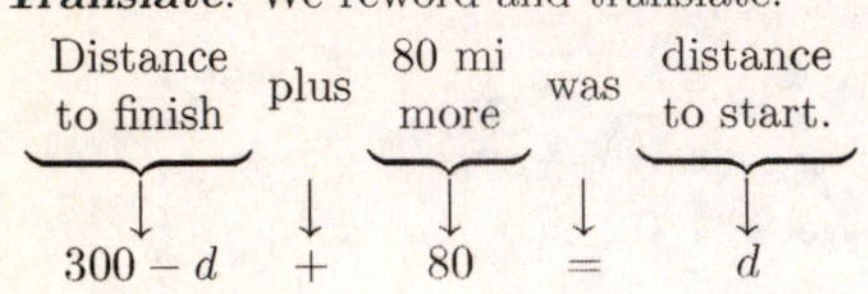

Carry out. We solve the equation.

$$
\begin{aligned}
300-d+80&=d\\
380-d&=d\\
380&=2d\\
190&=d
\end{aligned}
$$

Check. If Wheldon was 190 mi from the start, he was $300-190$, or 110 mi, from the finish. Since 190 is 80 more than 110, the answer checks.

State. Wheldon had traveled 190 mi at the given point.

13. ***Familiarize.*** Let $n=$ the number of Joan's apartment. Then $n+1=$ the number of her next-door neighbor's apartment.

Translate.

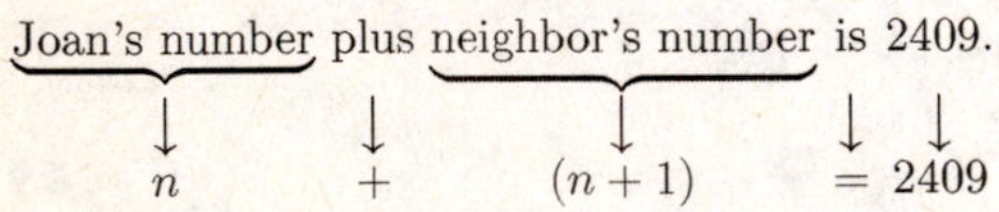

Carry out. We solve the equation.

$$
\begin{aligned}
n+(n+1)&=2409\\
2n+1&=2409\\
2n&=2408\\
n&=1204
\end{aligned}
$$

If Joan's apartment number is 1204, then her next-door neighbor's number is $1204+1$, or 1205.

Check. 1204 and 1205 are consecutive numbers whose sum is 2409. The answer checks.

State. The apartment numbers are 1204 and 1205.

15. ***Familiarize.*** Let $n=$ the smaller house number. Then $n+2=$ the larger number.

Translate.

Smaller number plus larger number is 794.

$$n + (n+2) = 794$$

Carry out. We solve the equation.

$$
\begin{aligned}
n+(n+2)&=794\\
2n+2&=794\\
2n&=792\\
n&=396
\end{aligned}
$$

If the smaller number is 396, then the larger number is $396+2$, or 398.

Check. 396 and 398 are consecutive even numbers and $396+398=794$. The answer checks.

State. The house numbers are 396 and 398.

17. ***Familiarize.*** Let $x=$ the first page number. Then $x+1=$ the second page number, and $x+2=$ the third page number.

Translate.

The sum of three consecutive page numbers is 60.

$$x+(x+1)+(x+2) = 60$$

Carry out. We solve the equation.

$$x+(x+1)+(x+2)=60$$

$3x+3=60$ Combining like terms

$3x=57$ Subtracting 3 from both sides

$x=19$ Dividing both sides by 3

If x is 19, then $x+1$ is 20 and $x+2=21$.

Check. 19, 20, and 21 are consecutive integers, and $19+20+21=60$. The result checks.

State. The page numbers are 19, 20, and 21.

19. ***Familiarize.*** Let $g=$ the groom's age. Then $g+19=$ the bride's age.

Translate.

Groom's age plus bride's age is 185.

$$g + (g+19) = 185$$

Carry out. We solve the equation.

$$
\begin{aligned}
g+(g+19)&=185\\
2g+19&=185\\
2g&=166\\
g&=83
\end{aligned}
$$

If g is 83, then $g+19$ is 102.

Check. 102 is 19 more than 83, and $83+102=185$. The answer checks.

State. The groom was 83 yr old, and the bride was 102 yr old.

21. ***Familiarize.*** Let $a=$ the amount spent to remodel bathrooms, in billions of dollars. Then $2a=$ the amount spent to remodel kitchens. The sum of these two amounts is \$35 billion.

Translate.

$$\underbrace{\text{Amount spent on bathrooms}}_{\displaystyle a} \ \underbrace{\text{plus}}_{\displaystyle +} \ \underbrace{\text{amount spent on kitchens}}_{\displaystyle 2a} \ \underbrace{\text{is}}_{\displaystyle =} \ \underbrace{\text{\$35 billion.}}_{\displaystyle 35}$$

Carry out. We solve the equation.

$$a + 2a = 35$$
$$3a = 35 \quad \text{Combining like terms}$$
$$a = \frac{35}{3}, \text{ or } 11\frac{2}{3}$$

If $a = \frac{35}{3}$, then $2a = 2 \cdot \frac{35}{3} = \frac{70}{3} = 23\frac{1}{3}$.

Check. $\frac{70}{3}$ is twice $\frac{35}{3}$, and $\frac{35}{3} + \frac{70}{3} = \frac{105}{3} = 35$. The answer checks.

State. $\$11\frac{2}{3}$ billion was spent to remodel bathrooms, and $\$23\frac{1}{3}$ billion was spent to remodel kitchens.

23. ***Familiarize.*** The page numbers are consecutive integers. If we let x = the smaller number, then $x + 1$ = the larger number.

Translate. We reword the problem.

$$\underbrace{\text{First integer}}_{\displaystyle x} \ \underbrace{+}_{\displaystyle +} \ \underbrace{\text{Second integer}}_{\displaystyle (x+1)} \ = 281$$

Carry out. We solve the equation.

$$x + (x + 1) = 281$$
$$2x + 1 = 281 \quad \text{Combining like terms}$$
$$2x = 280 \quad \text{Adding } -1 \text{ on both sides}$$
$$x = 140 \quad \text{Dividing on both sides by 2}$$

Check. If $x = 140$, then $x + 1 = 141$. These are consecutive integers, and $140 + 141 = 281$. The answer checks.

State. The page numbers are 140 and 141.

25. ***Familiarize.*** We draw a picture. Let w = the width of the rectangle, in feet. Then $w + 60$ = the length.

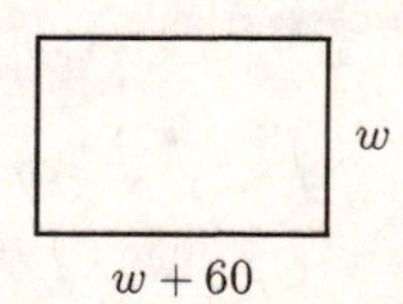

The perimeter is twice the length plus twice the width, and the area is the product of the length and the width.

Translate.

$$\underbrace{\text{Twice the length}}_{\displaystyle 2(w+60)} \ \underbrace{\text{plus}}_{\displaystyle +} \ \underbrace{\text{twice the width}}_{\displaystyle 2w} \ \underbrace{\text{is}}_{\displaystyle =} \ \underbrace{\text{520 ft}}_{\displaystyle 520}.$$

Carry out. We solve the equation.

$$2(w + 60) + 2w = 520$$
$$2w + 120 + 2w = 520$$
$$4w + 120 = 520$$
$$4w = 400$$
$$w = 100$$

Then $w + 60 = 100 + 60 = 160$, and the area is $160 \text{ ft} \cdot 100 \text{ ft} = 16,000 \text{ ft}^2$.

Check. The length, 160 ft, is 60 ft more than the width, 100 ft. The perimeter is $2 \cdot 160 \text{ ft} + 2 \cdot 100 \text{ ft}$, or 320 ft + 200 ft, or 520 ft. We can check the area by doing the calculation again. The answer checks.

State. The length is 160 ft, the width is 100 ft, and the area is 16,000 ft^2.

27. ***Familiarize.*** We draw a picture. Let w = the width of the court, in feet. Then $w + 34$ = the length.

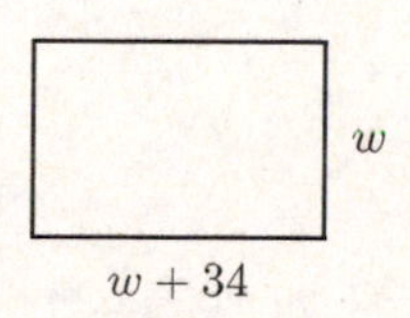

The perimeter is twice the length plus twice the width.

Translate.

$$\underbrace{\text{Twice the length}}_{\displaystyle 2(w+34)} \ \underbrace{\text{plus}}_{\displaystyle +} \ \underbrace{\text{twice the width}}_{\displaystyle 2w} \ \underbrace{\text{is}}_{\displaystyle =} \ \underbrace{\text{268 ft}}_{\displaystyle 268}.$$

Carry out. We solve the equation.

$$2(w + 34) + 2w = 268$$
$$2w + 68 + 2w = 268$$
$$4w + 68 = 268$$
$$4w = 200$$
$$w = 50$$

Then $w + 34 = 50 + 34 = 84$.

Check. The length, 84 ft, is 34 ft more than the width, 50 ft. The perimeter is $2 \cdot 84 \text{ ft} + 2 \cdot 50 \text{ ft} = 168 \text{ ft} + 100 \text{ ft} = 268$ ft. The answer checks.

State. The length of the court is 84 ft, and the width is 50 ft.

29. ***Familiarize.*** Let w = the width, in inches. Then $2w$ = the length. The perimeter is twice the length plus twice the width. We express $10\frac{1}{2}$ as 10.5.

Translate.

$$\underbrace{\text{Twice the length}}_{\displaystyle 2 \cdot 2w} \ \underbrace{\text{plus}}_{\displaystyle +} \ \underbrace{\text{twice the width}}_{\displaystyle 2w} \ \underbrace{\text{is}}_{\displaystyle =} \ \underbrace{\text{10.5 in.}}_{\displaystyle 10.5}$$

Carry out. We solve the equation.

$$2 \cdot 2w + 2w = 10.5$$
$$4w + 2w = 10.5$$
$$6w = 10.5$$
$$w = 1.75, \text{ or } 1\frac{3}{4}$$

Then $2w = 2(1.75) = 3.5$, or $3\frac{1}{2}$.

Check. The length, $3\frac{1}{2}$ in., is twice the width, $1\frac{3}{4}$ in. The perimeter is $2\left(3\frac{1}{2}\text{ in.}\right) + 2\left(1\frac{3}{4}\text{ in.}\right) =$ 7 in. $+ 3\frac{1}{2}$ in. $= 10\frac{1}{2}$ in. The answer checks.

State. The actual dimensions are $3\frac{1}{2}$ in. by $1\frac{3}{4}$ in.

31. ***Familiarize.*** We draw a picture. We let $x =$ the measure of the first angle. Then $3x =$ the measure of the second angle, and $x + 30 =$ the measure of the third angle.

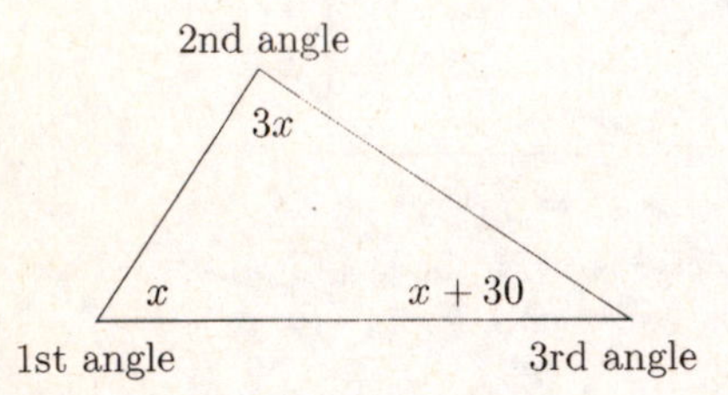

Recall that the measures of the angles of any triangle add up to 180°.

Translate.

Measure of first angle + measure of second angle + measure of third angle is 180°.

$$x + 3x + (x + 30) = 180$$

Carry out. We solve the equation.

$$x + 3x + (x + 30) = 180$$
$$5x + 30 = 180$$
$$5x = 150$$
$$x = 30$$

Possible answers for the angle measures are as follows:

First angle: $x = 30°$

Second angle: $3x = 3(30)° = 90°$

Third angle: $x + 30° = 30° + 30° = 60°$

Check. Consider 30°, 90°, and 60°. The second angle is three times the first, and the third is 30° more than the first. The sum of the measures of the angles is 180°. These numbers check.

State. The measure of the first angle is 30°, the measure of the second angle is 90°, and the measure of the third angle is 60°.

33. ***Familiarize***. Let $x =$ the measure of the first angle. Then $3x =$ the measure of the second angle, and $x + 3x + 10 = 4x + 10 =$ the measure of the third angle. Recall that the sum of the measures of the angles of a triangle is 180°.

Translate.

Measure of first angle + measure of second angle + measure of third angle is 180°.

$$x + 3x + (4x + 10) = 180$$

Carry out. We solve the equation.

$$x + 3x + (4x + 10) = 180$$
$$8x + 10 = 180$$
$$8x = 170$$
$$x = 21.25$$

If x is 21.25, then $3x$ is 63.75, and $4x + 10$ is 95.

Check. Consider 21.25°, 63.75°, and 95°. The second is three times the first, and the third is 10° more than the sum of the other two. The sum of the measures of the angles is 180°. These numbers check.

State. The measure of the third angle is 95°.

35. ***Familiarize***. Let $b =$ the length of the bottom section of the rocket, in feet. Then $\frac{1}{6}b =$ the length of the top section, and $\frac{1}{2}b =$ the length of the middle section.

Translate.

Length of top section + length of middle section + length of bottom section is 240 ft.

$$\frac{1}{6}b + \frac{1}{2}b + b = 240$$

Carry out. We solve the equation. First we multiply by 6 on both sides to clear the fractions.

$$\frac{1}{6}b + \frac{1}{2}b + b = 240$$
$$6\left(\frac{1}{6}b + \frac{1}{2}b + b\right) = 6 \cdot 240$$
$$6 \cdot \frac{1}{6}b + 6 \cdot \frac{1}{2}b + 6 \cdot b = 1440$$
$$b + 3b + 6b = 1440$$
$$10b = 1440$$
$$b = 144$$

Then $\frac{1}{6}b = \frac{1}{6} \cdot 144 = 24$ and $\frac{1}{2}b = \frac{1}{2} \cdot 144 = 72$.

Check. 24 ft is $\frac{1}{6}$ of 144 ft, and 72 ft is $\frac{1}{2}$ of 144 ft. The sum of the lengths of the sections is 24 ft + 72 ft + 144 ft = 240 ft. The answer checks.

State. The length of the top section is 24 ft, the length of the middle section is 72 ft, and the length of the bottom section is 144 ft.

37. ***Familiarize.*** Let m = the number of miles that can be traveled on a \$18 budget. Then the total cost of the taxi ride, in dollars, is $1.90 + 1.60m$, or $1.9 + 1.6m$.

Translate.

$\underbrace{\text{Cost of taxi ride}}$ is \$18.

$$1.9 + 1.6m \quad = \quad 18$$

Carry out. We solve the equation.

$$1.9 + 1.6m = 18$$
$$1.6m = 16.1$$
$$m = \frac{16.1}{1.6} = \frac{161}{16} = 10\frac{1}{16}$$

Check. The mileage change is $\$1.60\left(10\frac{1}{16}\right)$, or \$16.10, and the total cost of the ride is
\$1.90 + \$16.10 = \$18. The answer checks.

State. Debbie and Alex can travel $10\frac{1}{16}$ mi on their budget.

39. ***Familiarize.*** The total cost is the daily charge plus the mileage charge. Let d = the distance that can be traveled, in miles, in one day for \$100. The mileage charge is the cost per mile times the number of miles traveled, or $0.39d$.

Translate.

$\underbrace{\text{Daily rate}}$ plus $\underbrace{\text{mileage charge}}$ is \$100.

$$49.95 \quad + \quad 0.39d \quad = \quad 100$$

Carry out. We solve the equation.

$$49.95 + 0.39d = 100$$
$$0.39d = 50.05$$
$$d = 128.\overline{3}, or 128\frac{1}{3}$$

Check. For a trip of $128\frac{1}{3}$ mi, the mileage charge is $\$0.39\left(128\frac{1}{3}\right)$, or \$50.05, and \$49.95+\$50.05 = \$100. The answer checks.

State. They can travel $128\frac{1}{3}$ mi in one day and stay within their budget.

41. ***Familiarize.*** Let x = the measure of one angle. Then $90 - x$ = the measure of its complement.

Translate.

$\underbrace{\text{Measure of one angle}}$ is 15° $\underbrace{\text{more than}}$ $\underbrace{\text{twice the measure of its complement.}}$

$$x \quad = \quad 15 \quad + \quad 2(90 - x)$$

Carry out. We solve the equation.

$$x = 15 + 2(90 - x)$$
$$x = 15 + 180 - 2x$$
$$x = 195 - 2x$$
$$3x = 195$$
$$x = 65$$

If x is 65, then $90 - x$ is 25.

Check. The sum of the angle measures is 90°. Also, 65° is 15° more than twice its complement, 25°. The answer checks.

State. The angle measures are 65° and 25°.

43. ***Familiarize.*** Let l = the length of the paper, in cm. Then $l - 6.3$ = the width. The perimeter is twice the length plus twice the width.

Translate.

$\underbrace{\text{Twice the length}}$ plus $\underbrace{\text{twice the width}}$ is $\underbrace{\text{99 cm}}$.

$$2l \quad + \quad 2(l - 6.3) \quad = \quad 99$$

Carry out. We solve the equation.

$$2l + 2(l - 6.3) = 99$$
$$2l + 2l - 12.6 = 99$$
$$4l - 12.6 = 99$$
$$4l = 111.6$$
$$l = 27.9$$

Then $l - 6.3 = 27.9 - 6.3 = 21.6$.

Check. The width, 21.6 cm, is 6.3 cm less than the length, 27.9 cm. The perimeter is 2(27.9 cm) + 2(21.6 cm) = 55.8 cm + 43.2 cm = 99 cm. The answer checks.

State. The length of the paper is 27.9 cm, and the width is 21.6 cm.

45. ***Familiarize.*** Let a = the amount Sharon invested. Then the simple interest for one year is $6\% \cdot a$, or $0.06a$.

Translate.

$\underbrace{\text{Amount invested}}$ plus interest is \$6996.

$$a \quad + \quad 0.06a \quad = \quad 6996$$

Carry out. We solve the equation.

$$a + 0.06a = 6996$$
$$1.06a = 6996$$
$$a = 6600$$

Check. An investment of \$6600 at 6% simple interest earns 0.06(\$6600), or \$396, in one year. Since \$6600 + \$396 = \$6996, the answer checks.

State. Sharon invested \$6600.

47. ***Familiarize.*** Let w = the winning score. Then $w - 796$ = the losing score.

Translate.

$\underbrace{\text{Winning score}}$ plus $\underbrace{\text{losing score}}$ was $\underbrace{\text{1302 points}}$.

$$w \quad + \quad w - 796 \quad = \quad 1302$$

Carry out. We solve the equation.

$$\begin{aligned} w + w - 796 &= 1302 \\ 2w - 796 &= 1302 \\ 2w &= 2098 \\ w &= 1049 \end{aligned}$$

Then $w - 796 = 1049 - 796 = 253$.

Check. The winning score, 1049, is 796 points more than the losing score, 253. The total of the two scores is $1049 + 253$, or 1302 points. The answer checks.

State. The winning score was 1049 points.

49. ***Familiarize***. We will use the equation

$$T = \frac{1}{4}N + 40.$$

Translate. We substitute 80 for T.

$$80 = \frac{1}{4}N + 40$$

Carry out. We solve the equation.

$$\begin{aligned} 80 &= \frac{1}{4}N + 40 \\ 40 &= \frac{1}{4}N \\ 160 &= N \quad \text{Multiplying by 4 on both sides} \end{aligned}$$

Check. When $N = 160$, we have $T = \frac{1}{4} \cdot 160 + 40 = 40 + 40 = 80$. The answer checks.

State. A cricket chirps 160 times per minute when the temperature is 80°F.

51. *Writing Exercise*

53. Since -9 is to the left of 5 on the number line, we have $-9 < 5$.

55. Since -4 is to the left of 7 on the number line, we have $-4 < 7$.

57. *Writing Exercise*

59. ***Familiarize***. Let $c =$ the amount the meal originally cost. The 15% tip is calculated on the original cost of the meal, so the tip is $0.15c$.

Translate.

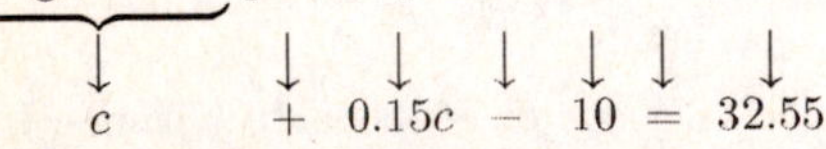

Carry out. We solve the equation.

$$\begin{aligned} c + 0.15c - 10 &= 32.55 \\ 1.15c - 10 &= 32.55 \\ 1.15c &= 42.55 \\ c &= 37 \end{aligned}$$

Check. If the meal originally cost \$37, the tip was 15% of \$37, or 0.15(\$37), or \$5.55. Since \$37 + \$5.55 − \$10 = \$32.55, the answer checks.

State. The meal originally cost \$37.

61. ***Familiarize***. Let $s =$ one score. Then four score $= 4s$ and four score and seven $= 4s + 7$.

Translate. We reword .

$$\underbrace{1776}\ \text{plus}\ \underbrace{\text{four score and seven}}\ \text{is}\ \underbrace{1863}$$

$$1776 + (4s + 7) = 1863$$

Carry out. We solve the equation.

$$\begin{aligned} 1776 + (4s + 7) &= 1863 \\ 4s + 1783 &= 1863 \\ 4s &= 80 \\ s &= 20 \end{aligned}$$

Check. If a score is 20 years, then four score and seven represents 87 years. Adding 87 to 1776 we get 1863. This checks.

State. A score is 20.

63. ***Familiarize***. Let $n =$ the number of half dollars. Then the number of quarters is $2n$; the number of dimes is $2 \cdot 2n$, or $4n$; and the number of nickels is $3 \cdot 4n$, or $12n$. The total value of each type of coin, in dollars, is as follows.

Half dollars: $0.5n$

Quarters: $0.25(2n)$, or $0.5n$

Dimes: $0.1(4n)$, or $0.4n$

Nickels: $0.05(12n)$, or $0.6n$

Then the sum of these amounts is $0.5n + 0.5n + 0.4n + 0.6n$, or $2n$.

Translate.

$$\underbrace{\text{Total amount of change}}\ \text{is}\ \$10.$$

$$2n = 10$$

Carry out. We solve the equation.

$$\begin{aligned} 2n &= 10 \\ n &= 5 \end{aligned}$$

Then $2n = 2 \cdot 5 = 10$, $4n = 4 \cdot 5 = 20$, and $12n = 12 \cdot 5 = 60$.

Check. If there are 5 half dollars, 10 quarters, 20 dimes, and 60 nickels, then there are twice as many quarters as half dollars, twice as many dimes as quarters, and 3 times as many nickels as dimes. The total value of the coins is \$0.5(5) + \$0.25(10) + \$0.1(20) + \$0.05(60) = \$2.50 + \$2.50 + \$2 + \$3 = \$10. The answer checks.

State. The shopkeeper got 5 half dollars, 10 quarters, 20 dimes, and 60 nickels.

65. ***Familiarize***. Let $a =$ the original number of apples in the basket.

Translate.

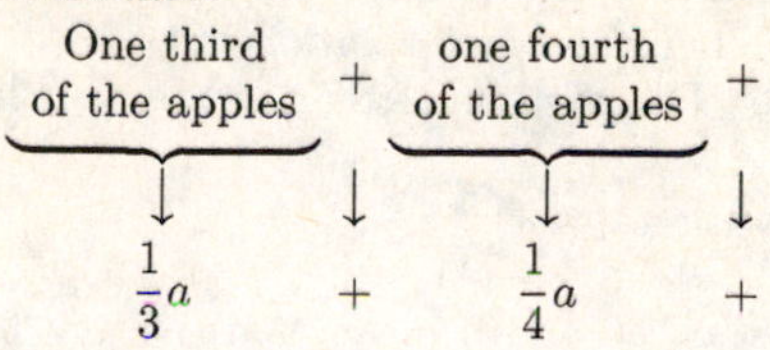

One third of the apples + one fourth of the apples +

$\frac{1}{3}a \quad + \quad \frac{1}{4}a \quad +$

one eighth of the apples + one fifth of the apples + 10 apples +

$\frac{1}{8}a \quad + \quad \frac{1}{5}a \quad + \quad 10 \quad +$

1 apple is the original number of apples.

$1 \quad = \quad a$

Carry out. We solve the equation. Note that the LCD is 120.

$$\frac{1}{3}a+\frac{1}{4}a+\frac{1}{8}a+\frac{1}{5}a+10+1=a$$
$$\frac{1}{3}a+\frac{1}{4}a+\frac{1}{8}a+\frac{1}{5}a+11=a$$
$$120\left(\frac{1}{3}a+\frac{1}{4}a+\frac{1}{8}a+\frac{1}{5}a+11\right)=120\cdot a$$
$$40a+30a+15a+24a+1320=120a$$
$$109a+1320=120a$$
$$1320=11a$$
$$120=a$$

Check. $\frac{1}{3}\cdot 120=40$, $\frac{1}{4}\cdot 120=30$, $\frac{1}{8}\cdot 120=15$, and $\frac{1}{5}\cdot 120=24$. Then $40+30+15+24+10+1=120$. The result checks.

State. There were originally 120 apples in the basket.

67. ***Familiarize.*** Let $x=$ the number of additional games the Falcons will have to play. Then $\frac{x}{2}=$ the number of those games they will win, $15+\frac{x}{2}=$ the total number of games won, and $20+x=$ the total number of games played.

Translate.

Number of games won is 60% of total number of games.

$$15+\frac{x}{2} \quad = \quad 0.6 \quad \cdot \quad 20+x$$

Carry out. We solve the equation.

$$15+\frac{x}{2}=0.6(20+x)$$
$$15+0.5x=12+0.6x \quad \left(\frac{x}{2}=\frac{1}{2}x=0.5x\right)$$
$$15=12+0.1x$$
$$3=0.1x$$
$$30=x$$

Check. If the Falcons play an additional 30 games, then they play a total of $20+30$, or 50, games. If they win half of the 30 additional games, or 15 games, then their wins total $15+15$, or 30. Since 60% of 50 is 30, the answer checks.

State. The Falcons will have to play 30 more games in order to win 60% of the total number of games.

69. ***Familiarize.*** Let $s=$ Ella's score on the third test. Her average score on the first two tests is 85, so she had a total of $2\cdot 85$ points on those tests.

Translate. The average score on the three tests is the sum of the three scores divided by 3.

$$\frac{2\cdot 85+s}{3}=82$$

Carry out. We solve the equation.

$$\frac{2\cdot 85+s}{3}=82$$
$$\frac{170+s}{3}=82$$
$$170+s=246 \quad \text{Multiplying by 3}$$
$$s=76$$

Check. If the score on the third test is 76, Ella's average score is $\frac{2\cdot 85+76}{3}=\frac{246}{3}=82$. The answer checks.

State. Ella's score on the third test was 76.

71. *Writing Exercise*

73. ***Familiarize.*** Let $w=$ the width of the rectangle, in cm. Then $w+4.25=$ the length.

Translate.

The perimeter is 101.74 cm.

$$2(w+4.25)+2w \quad = \quad 101.74$$

Carry out. We solve the equation.

$$2(w+4.25)+2w=101.74$$
$$2w+8.5+2w=101.74$$
$$4w+8.5=101.74$$
$$4w=93.24$$
$$w=23.31$$

Then $w+4.25=23.31+4.25=27.56$.

Check. The length, 27.56 cm, is 4.25 cm more than the width, 23.31 cm. The perimeter is 2(27.56) cm + 2(23.31 cm) = 55.12 cm + 46.62 cm = 101.74 cm. The answer checks.

State. The length of the rectangle is 27.56 cm, and the width is 23.31 cm.

Exercise Set 2.6

1. $-5x\le 30$

$x\ge -6$ Dividing by -5 and reversing the inequality symbol

3. $-2t>-14$

$t<7$ Dividing by -2 and reversing the inequality symbol

5. $x < -2$ and $-2 > x$ are equivalent.

7. If we add 1 to both sides of $-4x - 1 \leq 15$, we get $-4x \leq 16$. The two given inequalities are equivalent.

9. $x > -2$

a) Since $5 > -2$ is true, 5 is a solution.

b) Since $0 > -2$ is true, 0 is a solution.

c) Since $-1.9 > -2$ is true, -1.9 is a solution.

d) Since $-7.3 > -2$ is false, -7.3 is not a solution.

e) Since $1.6 > -2$ is true, 1.6 is a solution.

11. $x \geq 6$

a) Since $-6 \geq 6$ is false, -6 is not a solution.

b) Since $0 \geq 6$ is false, 0 is not a solution.

c) Since $6 \geq 6$ is true, 6 is a solution.

d) Since $6.01 \geq 6$ is true, 6.01 is a solution.

e) Since $-3\frac{1}{2} \geq 6$ is false, $-3\frac{1}{2}$ is not a solution.

13. The solutions of $y < 2$ are those numbers less than 2. They are shown on the graph by shading all points to the left of 2. The open circle at 2 indicates that 2 is not part of the graph.

$y < 2$

−4 −2 0 2 4

15. The solutions of $y > 4$ are those numbers greater than 4. They are shown on the graph by shading all points to the right of 4. The open circle at 4 indicates that 4 is not part of the graph.

$y > 4$

−2 0 2 4 6 8

17. The solutions of $0 \leq t$, or $t \geq 0$, are those numbers greater than or equal to zero. They are shown on the graph by shading the point 0 and all points to the right of 0. The closed circle at 0 indicates that 0 is part of the graph.

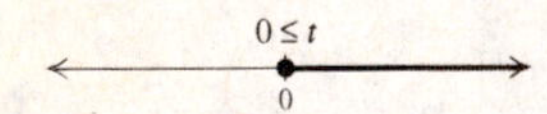

19. In order to be solution of the inequality $-5 \leq x < 2$, a number must be a solution of both $-5 \leq x$ and $x < 2$. The solution set is graphed as follows:

$-5 \leq x < 2$

−6 −5 −4 −3 −2 −1 0 1 2 3 4

The closed circle at -5 means that -5 is part of the graph. The open circle at 2 means that 2 is not part of the graph.

21. In order to be a solution of the inequality $-5 \leq x \leq 0$, a number must be a solution of both $-5 \leq x$ and $x \leq 0$. The solution set is graphed as follows:

$-5 \leq x \leq 0$

−7 −6 −5 −4 −3 −2 −1 0 1 2 3

The closed circles at -5 and 0 mean that -5 and 0 are both part of the graph.

23. All points to the right of -4 are shaded. The open circle at -4 indicates that -4 is not part of the graph. Using set-builder notation we have $\{x|x > -4\}$.

25. The point 2 and all points to the left of 2 are shaded. Using set-builder notation we have $\{x|x \leq 2\}$.

27. All points to the left of -1 are shaded. The open circle at -1 indicates that -1 is not part of the graph. Using set-builder notation we have $\{x|x < -1\}$.

29. The point 0 and all points to the right of 0 are shaded. Using set-builder notation we have $\{x|x \geq 0\}$.

31.
$$\begin{aligned} y + 6 &> 9 \\ y + 6 - 6 &> 9 - 6 \quad \text{Adding } -6 \text{ to both sides} \\ y &> 3 \quad \text{Simplifying} \end{aligned}$$

The solution set is $\{y|y > 3\}$. The graph is as follows:

−3 −2 −1 0 1 2 3 4 5 6 7

33.
$$\begin{aligned} x + 9 &\leq -12 \\ x + 9 - 9 &\leq -12 - 9 \quad \text{Adding } -9 \text{ to both sides} \\ x &\leq -21 \quad \text{Simplifying} \end{aligned}$$

The solution set is $\{x|x \leq -21\}$. The graph is as follows:

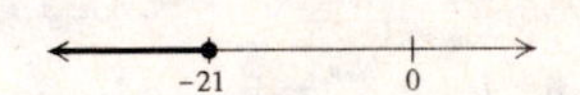

35.
$$\begin{aligned} x - 3 &< 14 \\ x - 3 + 3 &< 14 + 3 \quad \text{Adding 3 to both sides} \\ x &< 17 \quad \text{Simplifying} \end{aligned}$$

The solution set is $\{x|x < 17\}$. The graph is as follows:

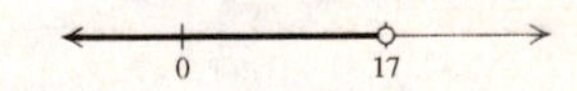

37.
$$\begin{aligned} y - 10 &> -16 \\ y - 10 + 10 &> -16 + 10 \\ y &> -6 \end{aligned}$$

The solution set is $\{y|y > -6\}$. The graph is as follows:

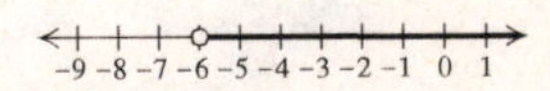

39.
$$\begin{aligned} 2x &\leq x + 9 \\ 2x - x &\leq x + 9 - x \\ x &\leq 9 \end{aligned}$$

The solution set is $\{x|x \leq 9\}$. The graph is as follows:

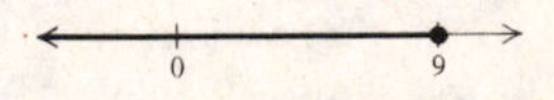

41.
$$y+\frac{1}{3}\le\frac{5}{6}$$
$$y+\frac{1}{3}-\frac{1}{3}\le\frac{5}{6}-\frac{1}{3}$$
$$y\le\frac{5}{6}-\frac{2}{6}$$
$$y\le\frac{3}{6}$$
$$y\le\frac{1}{2}$$

The solution set is $\left\{y\middle|y\le\frac{1}{2}\right\}$. The graph is as follows:

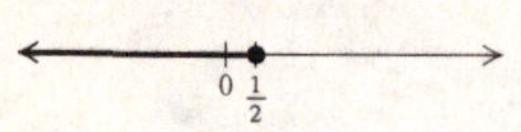

43.
$$t-\frac{1}{8}>\frac{1}{2}$$
$$t-\frac{1}{8}+\frac{1}{8}>\frac{1}{2}+\frac{1}{8}$$
$$t>\frac{4}{8}+\frac{1}{8}$$
$$t>\frac{5}{8}$$

The solution set is $\left\{t\middle|t>\frac{5}{8}\right\}$. The graph is as follows:

45.
$$-9x+17>17-8x$$
$$-9x+17-17>17-8x-17 \quad \text{Adding } -17$$
$$-9x>-8x$$
$$-9x+9x>-8x+9x \quad \text{Adding } 9x$$
$$0>x$$

The solution set is $\{x|x<0\}$. The graph is as follows:

47. $-23<-t$

The inequality states that the opposite of 23 is less than the opposite of t. Thus, t must be less than 23, so the solution set is $\{t|t<23\}$. To solve this inequality using the addition principle, we would proceed as follows:

$$-23<-t$$
$$t-23<0 \quad \text{Adding } t \text{ to both sides}$$
$$t<23 \quad \text{Adding 23 to both sides}$$

The solution set is $\{t|t<23\}$. The graph is as follows:

49.
$$5x<35$$
$$\frac{1}{5}\cdot 5x<\frac{1}{5}\cdot 35 \quad \text{Multiplying by } \frac{1}{5}$$
$$x<7$$

The solution set is $\{x|x<7\}$. The graph is as follows:

51.
$$-7x<13$$
$$-\frac{1}{7}\cdot(-7x)>-\frac{1}{7}\cdot 13 \quad \text{Multiplying by } -\frac{1}{7}$$

↑ The symbol has to be reversed.

$$x>-\frac{13}{7} \quad \text{Simplifying}$$

The solution set is $\left\{x\middle|x>-\frac{13}{7}\right\}$.

53. $-24>8t$

$-3>t$

The solution set is $\{t|t<-3\}$.

55.
$$7y\ge -2$$
$$\frac{1}{7}\cdot 7y\ge\frac{1}{7}(-2) \quad \text{Multiplying by } \frac{1}{7}$$
$$y\ge-\frac{2}{7}$$

The solution set is $\left\{y\middle|y\ge-\frac{2}{7}\right\}$.

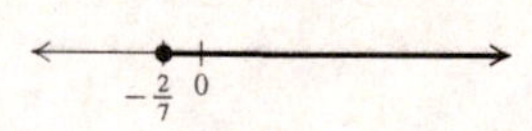

57.
$$-2y\le\frac{1}{5}$$
$$-\frac{1}{2}\cdot(-2y)\ge-\frac{1}{2}\cdot\frac{1}{5}$$

↑ The symbol has to be reversed.

$$y\ge-\frac{1}{10}$$

The solution set is $\left\{y\middle|y\ge-\frac{1}{10}\right\}$.

59.
$$-\frac{8}{5}>-2x$$
$$-\frac{1}{2}\cdot\left(-\frac{8}{5}\right)<-\frac{1}{2}\cdot(-2x)$$
$$\frac{8}{10}<x$$
$$\frac{4}{5}<x, \text{ or } x>\frac{4}{5}$$

The solution set is $\left\{x\middle|\frac{4}{5}<x\right\}$, or $\left\{x\middle|x>\frac{4}{5}\right\}$.

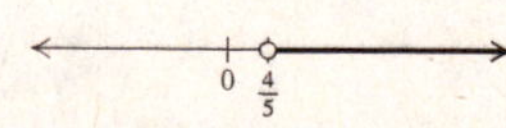

61.
$$\begin{aligned} 7+3x &< 34 \\ 7+3x-7 &< 34-7 && \text{Adding } -7 \text{ to both sides} \\ 3x &< 27 && \text{Simplifying} \\ x &< 9 && \text{Multiplying both sides by } \tfrac{1}{3} \end{aligned}$$
The solution set is $\{x|x<9\}$.

63.
$$\begin{aligned} 6+5y &\geq 26 \\ 6+5y-6 &\geq 26-6 && \text{Adding } -6 \\ 5y &\geq 20 \\ y &\geq 4 && \text{Multiplying by } \tfrac{1}{5} \end{aligned}$$
The solution set is $\{y|y\geq 4\}$.

65.
$$\begin{aligned} 4t-5 &\leq 23 \\ 4t-5+5 &\leq 23+5 && \text{Adding 5 to both sides} \\ 4t &\leq 28 \\ \frac{1}{4}\cdot 4t &\leq \frac{1}{4}\cdot 28 && \text{Multiplying both sides by } \tfrac{1}{4} \\ t &\leq 7 \end{aligned}$$
The solution set is $\{t|t\leq 7\}$.

67.
$$\begin{aligned} 16 &< 4-3y \\ 16-4 &< 4-3y-4 && \text{Adding } -4 \text{ to both sides} \\ 12 &< -3y \\ -\frac{1}{3}\cdot 12 &> -\frac{1}{3}\cdot(-3y) && \text{Multiplying by } -\tfrac{1}{3} \end{aligned}$$
↑ The symbol has to be reversed.
$$-4 > y$$
The solution set is $\{y|-4>y\}$, or $\{y|y<-4\}$.

69.
$$\begin{aligned} 39 &> 3-9x \\ 39-3 &> 3-9x-3 && \text{Adding } -3 \\ 36 &> -9x \\ -\frac{1}{9}\cdot 36 &< -\frac{1}{9}\cdot(-9x) && \text{Multiplying by } -\tfrac{1}{9} \end{aligned}$$
↑ The symbol has to be reversed.
$$-4 < x$$
The solution set is $\{x|-4<x\}$, or $\{x|x>-4\}$.

71.
$$\begin{aligned} 5-6y &> 25 \\ -5+5-6y &> -5+25 \\ -6y &> 20 \\ -\frac{1}{6}\cdot(-6y) &< -\frac{1}{6}\cdot 20 \end{aligned}$$
↑ The symbol has to be reversed.
$$\begin{aligned} y &< -\frac{20}{6} \\ y &< -\frac{10}{3} \end{aligned}$$
The solution set is $\left\{y\middle|y<-\frac{10}{3}\right\}$.

73.
$$\begin{aligned} -3 &< 8x+7-7x \\ -3 &< x+7 && \text{Collecting like terms} \\ -3-7 &< x+7-7 \\ -10 &< x \end{aligned}$$
The solution set is $\{x|-10<x\}$, or $\{x|x>-10\}$.

75.
$$\begin{aligned} 6-4y &> 4-3y \\ 6-4y+4y &> 4-3y+4y && \text{Adding } 4y \\ 6 &> 4+y \\ -4+6 &> -4+4+y && \text{Adding } -4 \\ 2 &> y, \text{ or } y<2 \end{aligned}$$
The solution set is $\{y|2>y\}$, or $\{y|y<2\}$.

77.
$$\begin{aligned} 7-9y &\leq 4-8y \\ 7-9y+9y &\leq 4-8y+9y \\ 7 &\leq 4+y \\ -4+7 &\leq -4+4+y \\ 3 &\leq y, \text{ or } y\geq 3 \end{aligned}$$
The solution set is $\{y|3\leq y\}$, or $\{y|y\geq 3\}$.

79.
$$\begin{aligned} 33-12x &< 4x+97 \\ 33-12x-97 &< 4x+97-97 \\ -64-12x &< 4x \\ -64-12x+12x &< 4x+12x \\ -64 &< 16x \\ -4 &< x \end{aligned}$$
The solution set is $\{x|-4<x\}$, or $\{x|x>-4\}$.

81.
$$\begin{aligned} 2.1x+43.2 &> 1.2-8.4x \\ 10(2.1x+43.2) &> 10(1.2-8.4x) && \text{Multiplying by 10 to clear decimals} \\ 21x+432 &> 12-84x \\ 21x+84x &> 12-432 && \text{Adding } 84x \text{ and } -432 \\ 105x &> -420 \\ x &> -4 && \text{Multiplying by } \tfrac{1}{105} \end{aligned}$$
The solution set is $\{x|x>-4\}$.

83.
$$\begin{aligned} 0.7n-15+n &\geq 2n-8-0.4n \\ 1.7n-15 &\geq 1.6n-8 && \text{Collecting like terms} \\ 10(1.7n-15) &\geq 10(1.6n-8) && \text{Multiplying by 10} \\ 17n-150 &\geq 16n-80 \\ 17n-16n &\geq -80+150 && \text{Adding } -16n \text{ and } 150 \\ n &\geq 70 \end{aligned}$$
The solution set is $\{n|n\geq 70\}$

85.
$$\begin{aligned} \frac{x}{3}-4 &\leq 1 \\ 3\left(\frac{x}{3}-4\right) &\leq 3\cdot 1 && \text{Multiplying by 3 to to clear the fraction} \\ x-12 &\leq 3 && \text{Simplifying} \\ x &\leq 15 && \text{Adding 12} \end{aligned}$$
The solution set is $\{x|x\leq 15\}$.

87.
$$\begin{aligned} 3 &< 5-\frac{t}{7} \\ -2 &< -\frac{t}{7} \\ -7(-2) &> -7\left(-\frac{t}{7}\right) \\ 14 &> t \end{aligned}$$
The solution set is $\{t|t<14\}$.

89. $4(2y-3)<36$

$$\begin{aligned} 8y-12&<36 && \text{Removing parentheses}\\ 8y&<48 && \text{Adding 12}\\ y&<6 && \text{Multiplying by } \frac{1}{8} \end{aligned}$$

The solution set is $\{y|y<6\}$.

91. $3(t-2)\geq 9(t+2)$

$$\begin{aligned} 3t-6&\geq 9t+18\\ 3t-9t&>18+6\\ -6t&\geq 24\\ t&\leq -4 && \text{Multiplying by } -\frac{1}{6} \text{ and reversing the symbol} \end{aligned}$$

The solution set is $\{t|t\leq -4\}$.

93. $3(r-6)+2<4(r+2)-21$

$$\begin{aligned} 3r-18+2&<4r+8-21\\ 3r-16&<4r-13\\ -16+13&<4r-3r\\ -3&<r, \text{ or } r>-3 \end{aligned}$$

The solution set is $\{r|r>-3\}$.

95. $\frac{2}{3}(2x-1)\geq 10$

$$\begin{aligned} \frac{3}{2}\cdot\frac{2}{3}(2x-1)&\geq\frac{3}{2}\cdot 10 && \text{Multiplying by } \frac{3}{2}\\ 2x-1&\geq 15\\ 2x&\geq 16\\ x&\geq 8 \end{aligned}$$

The solution set is $\{x|x\geq 8\}$.

97. $\frac{3}{4}\left(3x-\frac{1}{2}\right)-\frac{2}{3}<\frac{1}{3}$

$$\begin{aligned} \frac{3}{4}\left(3x-\frac{1}{2}\right)&<1 && \text{Adding } \frac{2}{3}\\ \frac{9}{4}x-\frac{3}{8}&<1 && \text{Removing parentheses}\\ 8\cdot\left(\frac{9}{4}x-\frac{3}{8}\right)&<8\cdot 1 && \text{Clearing fractions}\\ 18x-3&<8\\ 18x&<11\\ x&<\frac{11}{18} \end{aligned}$$

The solution set is $\left\{x\middle|x<\frac{11}{18}\right\}$.

99. *Writing Exercise*

101. Let n represent "some number." Then we have $n+3$, or $3+n$.

103. Let x represent "a number." Then we have $2x-3$.

105. *Writing Exercise*

107. $x<x+1$

When any real number is increased by 1, the result is greater than the original number. Thus the solution set is $\{x|x \text{ is a real number}\}$.

109. $27-4[2(4x-3)+7]\geq 2[4-2(3-x)]-3$

$$\begin{aligned} 27-4[8x-6+7]&\geq 2[4-6+2x]-3\\ 27-4[8x+1]&\geq 2[-2+2x]-3\\ 27-32x-4&\geq -4+4x-3\\ 23-32x&\geq -7+4x\\ 23+7&=4x+32x\\ 30&\geq 36x\\ \frac{5}{6}&\geq x \end{aligned}$$

The solution set is $\left\{x\middle|x\leq\frac{5}{6}\right\}$.

111. $\frac{1}{2}(2x+2b)>\frac{1}{3}(21+3b)$

$$\begin{aligned} x+b&>7+b\\ x+b-b&>7+b-b\\ x&>7 \end{aligned}$$

The solution set is $\{x|x>7\}$.

113.

$$\begin{aligned} y&<ax+b && \text{Assume } a<0.\\ y-b&<ax\\ \frac{y-b}{a}&>x && \text{Since } a<0 \text{, the inequality symbol must be reversed.} \end{aligned}$$

The solution set is $\left\{x\middle|x<\frac{y-b}{a}\right\}$.

115. $|x|>-3$

Since absolute value is always nonnegative, the absolute value of any real number will be greater than -3. Thus, the solution set is $\{x|x \text{ is a real number}\}$.

Exercise Set 2.7

1. a is at least b can be translated as $b\leq a$.

3. a is at most b can be translated as $a\leq b$.

5. b is no more than a can be translated as $b\leq a$.

7. b is less than a can be translated as $b<a$.

9. Let n represent the number. Then we have

$$n\geq 8.$$

11. Let t represent the temperature. Then we have

$$t\leq -3.$$

13. Let p represent the price of Pat's PT Cruiser. Then we have

$$p>21,900.$$

15. Let d represent the distance to Normandale Community College. Then we have

$$d\leq 15.$$

17. Let n represent the number. Then we have

$$n>-2.$$

19. Let p represent the number of people attending the Million Man March. Then we have

$$400,000<p<1,200,000.$$

21. ***Familiarize.*** Let s = the length of the service call, in hours. The total charge is \$25 plus \$30 times the number of hours RJ's was there.

Translate.

$$\underbrace{\text{\$25 charge}}_{\downarrow\ 25}\ \ \text{plus}_{\downarrow\ +}\ \ \underbrace{\text{hourly rate}}_{\downarrow\ 30}\ \ \text{times}_{\downarrow\ \cdot}\ \ \underbrace{\text{number of hours}}_{\downarrow\ s}\ \ \underbrace{\text{is greater than}}_{\downarrow\ >}\ \ \underbrace{\$100}_{\downarrow\ 100}.$$

Carry out. We solve the inequality.

$$25 + 30x > 100$$
$$30x > 75$$
$$s > 2.5$$

Check. As a partial check, we show that the cost of a 2.5 hour service call is \$100.

$$\$25 + \$30(2.5) = \$25 + \$75 = \$100$$

State. The length of the service call was more than 2.5 hr.

23. ***Familiarize.*** Let t = the number of one-way trips made per month.

Translate.

$$\underbrace{\text{Cost per trip}}_{\downarrow\ 1\cdot 15}\ \ \text{times}_{\downarrow\ \cdot}\ \ \underbrace{\text{number of trips}}_{\downarrow\ t}\ \ \underbrace{\text{is greater than}}_{\downarrow\ >}\ \ \underbrace{\$21}_{\downarrow\ 21}.$$

Carry out. We solve the inequality.

$$1.15t > 21$$
$$t > 18.3 \quad \text{Rounding}$$

Check. As a partial check we show that the total cost of 18 trips is less than \$21 and the total cost of 19 trips is more than \$21. For 18 trips the cost is \$1.15(18) = \$20.70. For 19 trips the cost is \$1.15(19) = \$21.85.

State. Gail should make more than 18 one-way trips per month in order for the pass to save her money.

25. ***Familiarize.*** The average of the five scores is their sum divided by the number of tests, 5. We let s represent Rod's score on the last test.

Translate. The average of the five scores is given by

$$\frac{73 + 75 + 89 + 91 + s}{5}.$$

Since this average must be at least 85, this means that it must be greater than or equal to 85. Thus, we can translate the problem to the inequality

$$\frac{73 + 75 + 89 + 91 + s}{5} \geq 85.$$

Carry out. We first multiply by 5 to clear the fraction.

$$5\left(\frac{73 + 75 + 89 + 91 + s}{5}\right) \geq 5 \cdot 85$$
$$73 + 75 + 89 + 91 + s \geq 425$$
$$328 + s \geq 425$$
$$s \geq 97$$

Check. As a partial check, we show that Rod can get a score of 97 on the fifth test and have an average of at least 85:

$$\frac{73 + 75 + 89 + 91 + 97}{5} = \frac{425}{5} = 85.$$

State. Scores of 97 and higher will earn Rod an average quiz grade of at least 85.

27. ***Familiarize.*** Let c = the number of credits Millie must complete in the fourth quarter.

Translate.

$$\underbrace{\text{Average number of credits}}_{\downarrow\ \frac{5+7+8+c}{4}}\ \ \underbrace{\text{is at least}}_{\downarrow\ \geq}\ \ 7_{\downarrow\ 7}.$$

Carry out. We solve the inequality.

$$\frac{5 + 7 + 8 + c}{4} \geq 7$$
$$4\left(\frac{5 + 7 + 8 + c}{4}\right) \geq 4 \cdot 7$$
$$5 + 7 + 8 + c \geq 28$$
$$20 + c \geq 28$$
$$c \geq 8$$

Check. As a partial check, we show that Millie can complete 8 credits in the fourth quarter and average 7 credits per quarter.

$$\frac{5 + 7 + 8 + 8}{4} = \frac{28}{4} = 7$$

State. Millie must complete 8 credits or more in the fourth quarter.

29. ***Familiarize.*** The average number of calls per week is the sum of the calls for the three weeks divided by the number of weeks, 3. We let c represent the number of calls made during the third week.

Translate. The average of the three weeks is given by

$$\frac{17 + 22 + c}{3}.$$

Since the average must be at least 20, this means that it must be greater than or equal to 20. Thus, we can translate the problem to the inequality

$$\frac{17 + 22 + c}{3} \geq 20.$$

Carry out. We first multiply by 3 to clear the fraction.

$$3\left(\frac{17 + 22 + c}{3}\right) \geq 3 \cdot 20$$
$$17 + 22 + c \geq 60$$
$$39 + c \geq 60$$
$$c \geq 21$$

Check. Suppose c is a number greater than or equal to 21. Then by adding 17 and 22 on both sides of the inequality we get

$$17 + 22 + c \geq 17 + 22 + 21$$
$$17 + 22 + c \geq 60$$

so

$$\frac{17 + 22 + c}{3} \geq \frac{60}{3}, \text{ or } 20.$$

State. 21 calls or more will maintain an average of at least 20 for the three-week period.

31. ***Familiarize.*** We first make a drawing. We let b represent the length of the base. Then the lengths of the other sides are $b-2$ and $b+3$.

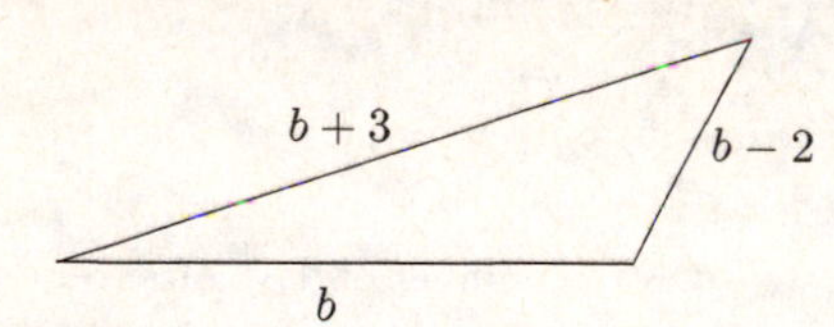

The perimeter is the sum of the lengths of the sides or $b+b-2+b+3$, or $3b+1$.

Translate.

The perimeter (↓ $3b+1$) is greater than (↓ $>$) 19 cm. (↓ 19)

Carry out.

$$3b+1>19$$
$$3b>18$$
$$b>6$$

Check. We check to see if the solution seems reasonable.

When $b=5$, the perimeter is $3\cdot 5+1$, or 16 cm.

When $b=6$, the perimeter is $3\cdot 6+1$, or 19 cm.

When $b=7$, the perimeter is $3\cdot 7+1$, or 22 cm.

From these calculations, it would appear that the solution is correct.

State. For lengths of the base greater than 6 cm the perimeter will be greater than 19 cm.

33. ***Familiarize.*** Let $d=$ the depth of the well, in feet. Then the cost on the pay-as-you-go plan is $\$500+\$8d$. The cost of the guaranteed-water plan is \$4000. We want to find the values of d for which the pay-as-you-go plan costs less than the guaranteed-water plan.

Translate.

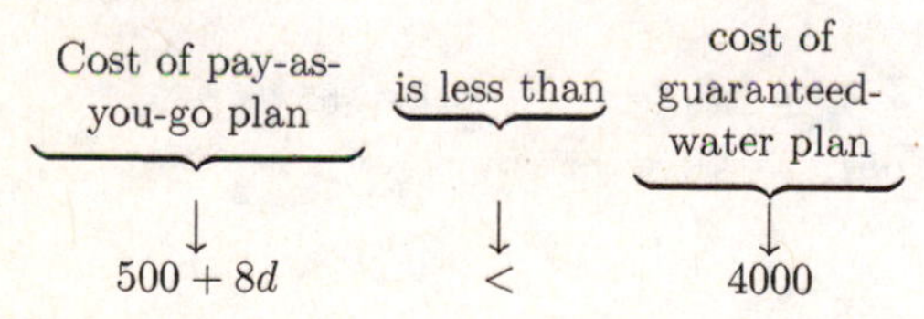

Carry out.

$$500+8d<4000$$
$$8d<3500$$
$$d<437.5$$

Check. We check to see that the solution is reasonable.

When $d=437$, $\$500+\$8\cdot 437=\$3996<\$4000$

When $d=437.5$, $\$500+\$8(437.5)=\$4000$

When $d=438$, $\$500+\$8(438)=\$4004>\$4000$

From these calculations, it appears that the solution is correct.

State. It would save a customer money to use the pay-as-you-go plan for a well of less than 437.5 ft.

35. ***Familiarize.*** Let $v=$ the blue book value of the car. Since the car was repaired, we know that \$8500 does not exceed $0.8v$ or, in other words, $0.8v$ is at least \$8500.

Translate.

80% of the blue book value (↓ $0.8v$) is at least (↓ $\geq$) \$8500. (↓ 8500)

Carry out.

$$0.8v\geq 8500$$
$$v\geq\frac{8500}{0.8}$$
$$v\geq 10,625$$

Check. As a partial check, we show that 80% of \$10,625 is at least \$8500:

$$0.8(\$10,625)=\$8500$$

State. The blue book value of the car was at least \$10,625.

37. ***Familiarize.*** As in the drawing in the text, we let $L=$ the length of the envelope. Recall that the area of a rectangle is the product of the length and the width.

Translate.

Length (↓ L) times (↓ $\cdot$) width (↓ $3\frac{1}{2}$) is at least (↓ $\geq$) $17\frac{1}{2}$ in^2 (↓ $17\frac{1}{2}$)

Carry out.

$$L\cdot 3\frac{1}{2}\geq 17\frac{1}{2}$$
$$L\cdot\frac{7}{2}\geq\frac{35}{2}$$
$$L\cdot\frac{7}{2}\cdot\frac{2}{7}\geq\frac{35}{2}\cdot\frac{2}{7}$$
$$L\geq 5$$

The solution set is $\{L|L\geq 5\}$.

Check. We can obtain a partial check by substituting a number greater than or equal to 5 in the inequality. For example, when $L=6$:

$$L\cdot 3\frac{1}{2}=6\cdot 3\frac{1}{2}=6\cdot\frac{7}{2}=21\geq 17\frac{1}{2}$$

The result appears to be correct.

State. Lengths of 5 in. or more will satisfy the constraints. The solution set is $\{L|L\geq 5 \text{ in.}\}$.

39. ***Familiarize.*** We will use the formula $F=\frac{9}{5}C+32$.

Translate.

Fahrenheit temperature (↓ F) is above (↓ $>$) 98.6°. (↓ 98.6)

Substituting $\frac{9}{5}C+32$ for F, we have

$$\frac{9}{5}C+32>98.6.$$

Carry out. We solve the inequality.

$$\frac{9}{5}C + 32 > 98.6$$
$$\frac{9}{5}C > 66.6$$
$$C > \frac{333}{9}$$
$$C > 37$$

Check. We check to see if the solution seems reasonable.

When $C = 36$, $\frac{9}{5} \cdot 36 + 32 = 96.8$.

When $C = 37$, $\frac{9}{5} \cdot 37 + 32 = 98.6$.

When $C = 38$, $\frac{9}{5} \cdot 38 + 32 = 100.4$.

It would appear that the solution is correct, considering that rounding occurred.

State. The human body is feverish for Celsius temperatures greater than $37°$.

41. ***Familiarize.*** Let h = the height of the triangle, in ft. Recall that the formula for the area of a triangle with base b and height h is $A = \frac{1}{2}bh$.

Translate.

Area is at least 3 ft^2.

$$\frac{1}{2}\left(1\frac{1}{2}\right)h \quad \geq \quad 3$$

Carry out. We solve the inequality.

$$\frac{1}{2}\left(1\frac{1}{2}\right)h \geq 3$$
$$\frac{1}{2} \cdot \frac{3}{2} \cdot h \geq 3$$
$$\frac{3}{4}h \geq 3$$
$$h \geq \frac{4}{3} \cdot 3$$
$$h \geq 3$$

Check. As a partial check, we show that the area of the triangle is 3 ft^2 when the height is 4 ft.

$$\frac{1}{2}\left(1\frac{1}{2}\right)(4) = \frac{1}{2} \cdot \frac{3}{2} \cdot \frac{4}{1} = 3$$

State. The height should be at least 4 ft.

43. ***Familiarize.*** Let r = the amount of fat in a serving of the regular peanut butter, in grams. If reduced fat peanut butter has at least 25% less fat than regular peanut butter, then it has at most 75% as much fat as the regular peanut butter.

Translate.

12 g of fat is at most 75% of the amount of fat in regular peanut butter.

$$12 \quad \leq \quad 0.75 \quad \cdot \quad r$$

Carry out.

$$12 \leq 0.75r$$
$$16 \leq r$$

Check. As a partial check, we show that 12 g of fat does not exceed 75% of 16 g of fat:

$$0.75(16) = 12$$

State. Regular peanut butter contains at least 16 g of fat per serving.

45. ***Familiarize.*** Let d = the number of days after September 5.

Translate.

Weight on September 5 plus 26 lb per day times

$$532 \quad + \quad 26 \quad \cdot$$

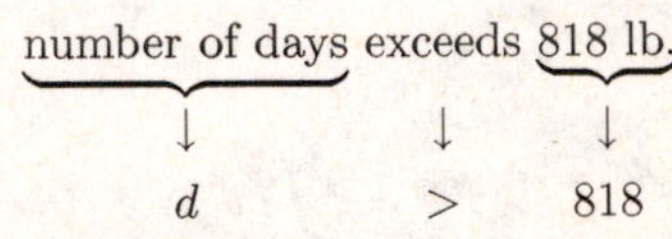

Carry out. We solve the inequality.

$$532 + 26d > 818$$
$$26d > 286$$
$$d > 11$$

Check. As a partial check, we can show that the weight of the pumpkin is 818 lb 11 days after September 5.

$$532 + 26 \cdot 11 = 532 + 286 = 818 \text{ lb}$$

State. The pumpkin's weight will exceed 818 lb more than 11 days after September 5, or on dates after September 16.

47. ***Familiarize.*** Let c = the number of copies Myra has made. The total cost of the copies is the setup fee of \$6 plus \$4 times the number of copies, or $\$4 \cdot c$.

Translate.

Setup fee plus copying cost cannot exceed \$65.

$$6 \quad + \quad 4c \quad \leq \quad 65$$

Carry out. We solve the inequality.

$$6 + 4c \leq 65$$
$$4c \leq 59$$
$$c \leq 14.75$$

Check. As a partial check, we show that Myra can have 14 copies made and not exceed her \$65 budget.

$$\$6 + \$4 \cdot 14 = \$6 + \$56 = \$62$$

State. Myra can have 14 or fewer copies made and stay within her budget.

49. ***Familiarize.*** We will use the formula $R = -0.0065t + 4.3222$.

Translate.

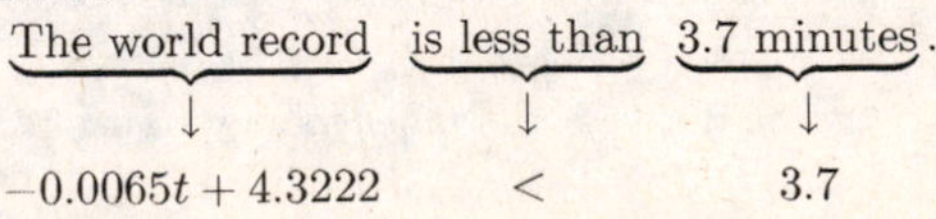

Carry out. We solve the inequality.

$$-0.0065t + 4.3222 < 3.7$$
$$-0.0065t < -0.6222$$
$$t > 95.7$$

Check. As a partial check, we can show that the record is more than 3.7 min 95 yr after 1900 and is less than 3.7 min 96 yr after 1900.

For $t = 95$, $R = -0.0065(95) + 4.3222 = 3.7047$.

For $t = 96$, $R = -0.0065(96) + 4.3222 = 3.6982$.

State. The world record in the mile run is less than 3.7 min more than 95 yr after 1900, or in years after 1995.

51. ***Familiarize.*** We will use the equation $y = 0.03x + 0.21$.

Translate.

$$\underbrace{\text{The cost}}_{0.03x + 0.21}\ \underbrace{\text{is at most}}_{\leq}\ \underbrace{\$6.}_{6}$$

Carry out. We solve the inequality.

$$0.03x + 0.21 \leq 6$$
$$0.03x \leq 5.79$$
$$x \leq 193$$

Check. As a partial check, we show that the cost for driving 193 mi is \$6.

$$0.03(193) + 0.21 = 6$$

State. The cost will be at most \$6 for mileages less than or equal to 193 mi.

53. *Writing Exercise*

55. $\dfrac{9-5}{6-4} = \dfrac{4}{2} = 2$

57. $\dfrac{8-(-2)}{1-4} = \dfrac{10}{-3}$, or $-\dfrac{10}{3}$

59. *Writing Exercise*

61. ***Familiarize.*** We use the formula $F = \dfrac{9}{5}C + 32$.

Translate. We are interested in temperatures such that $5° < F < 15°$. Substituting for F, we have:

$$5 < \frac{9}{5}C + 32 < 15$$

Solve.

$$5 < \frac{9}{5}C + 32 < 15$$
$$5 \cdot 5 < 5\left(\frac{9}{5}C + 32\right) < 5 \cdot 15$$
$$25 < 9C + 160 < 75$$
$$-135 < 9C < -85$$
$$-15 < C < -9\frac{4}{9}$$

Check. The check is left to the student.

State. Green ski wax works best for temperatures between $-15°$C and $-9\frac{4}{9}°$C.

63. Since $8^2 = 64$, the length of a side must be less than or equal to 8 cm (and greater than 0 cm, of course). We can also use the five-step problem-solving procedure.

Familiarize. Let s represent the length of a side of the square. The area s is the square of the length of a side, or s^2.

Translate.

$$\underbrace{\text{The area}}_{s^2}\ \underbrace{\text{is no more than}}_{\leq}\ \underbrace{64\text{ cm}^2.}_{64}$$

Carry out.

$$s^2 \leq 64$$
$$s^2 - 64 \leq 0$$
$$(s+8)(s-8) \leq 0$$

We know that $(s+8)(s-8) = 0$ for $s = -8$ or $s = 8$. Now $(s+8)(s-8) < 0$ when the two factors have opposite signs. That is:

$s+8>0$ *and* $s-8<0$ *or* $s+8<0$ *and* $s-8>0$

$s>-8$ *and* $s<8$ *or* $s<-8$ *and* $s>8$

This can be expressed as $-8 < s < 8$. This is not possible.

Then $(s+8)(s-8) \leq 0$ for $-8 \leq s \leq 8$.

Check. Since the length of a side cannot be negative we only consider positive values of s, or $0 < s \leq 8$. We check to see if this solution seems reasonable.

When $s = 7$, the area is 7^2, or 49 cm^2.

When $s = 8$, the area is 8^2, or 64 cm^2.

When $s = 9$, the area is 9^2, or 81 cm^2.

From these calculations, it appears that the solution is correct.

State. Sides of length 8 cm or less will allow an area of no more than 64 cm^2. (Of course, the length of a side must be greater than 0 also.)

65. ***Familiarize.*** Let f = the fat content of a serving of regular tortilla chips, in grams. A product that contains 60% less fat than another product has 40% of the fat content of that product. If Reduced Fat Tortilla Pops cannot be labeled lowfat, then they contain at least 3 g of fat.

Translate.

$$\underbrace{40\%}_{0.4}\ \underbrace{\text{of}}_{\cdot}\ \underbrace{\text{the fat content of regular tortilla chips}}_{f}\ \underbrace{\text{is at least}}_{\geq}\ \underbrace{\text{3 grams of fat}}_{3}$$

Carry out.

$$0.4f \geq 3$$
$$f \geq 7.5$$

Check. As a partial check, we show that 40% of 7.5 g is not less than 3 g.

$$0.4(7.5) = 3$$

State. A serving of regular tortilla chips contains at least 7.5 g of fat.

67. ***Familiarize.*** Let $p =$ the price of Neoma's tenth book. If the average price of each of the first 9 books is \$12, then the total price of the 9 books is $9 \cdot \$12$, or \$108. The average price of the first 10 books will be $\frac{\$108 + p}{10}$.

Translate.

$$\underbrace{\text{The average price of 10 books}}_{\downarrow} \quad \underbrace{\text{is at least}}_{\downarrow} \quad \underset{\downarrow}{\$15.}$$

$$\frac{108 + p}{10} \quad \geq \quad 15$$

Carry out. We solve the inequality.

$$\frac{108 + p}{10} \geq 15$$
$$108 + p \geq 150$$
$$p \geq 42$$

Check. As a partial check, we show that the average price of the 10 books is \$15 when the price of the tenth book is \$42.

$$\frac{\$108 + \$42}{10} = \frac{\$150}{10} = \$15$$

State. Neoma's tenth book should cost at least \$42 if she wants to select a \$15 book for her free book.

69. *Writing Exercise*

Chapter 3

Introduction to Graphing

Exercise Set 3.1

1. The x-values extend from -9 to 1 and the y-values range from -1 to 5, so (a) is the best choice.

3. The x-values extend from -2 to 4 and the y-values range from -9 to 1, so (b) is the best choice.

5. We go to the top of the bar that is above the body weight 100 lb. Then we move horizontally from the top of the bar to the vertical scale listing numbers of drinks. It appears that consuming approximately 2 drinks in one hour will give a 100 lb person a blood-alcohol level of 0.08%.

7. From 4 on the vertical scale we move horizontally until we reach a bar whose top is above the horizontal line on which we are moving. The first such bar corresponds to a body weight of 220 lb. This means that for body weights represented by bars to the left of this one, consuming 4 drinks will yield a blood-alcohol level of 0.08%. The bar immediately to the left of the 220-pound bar represents 200 pounds. Thus, we can conclude an individual weighs more than 200 lb if 4 drinks are consumed in one hour without reaching a blood-alcohol level of 0.08%.

9. ***Familiarize***. Since there are 292 million Americans and about one-third of them live in the South, there are about $\frac{1}{3}\cdot 292$, or $\frac{292}{3}$ million Southerners. The pie chart indicates that 3% of Americans choose brown as their favorite color. Let $b =$ the number of Southerners, in millions, who choose brown as their favorite color.

Translate. We reword and translate the problem.

$$\begin{array}{ccccc} \text{What} & \text{is} & 3\% & \text{of} & \frac{292}{3} \text{ million?} \\ \downarrow & \downarrow & \downarrow & \downarrow & \downarrow \\ b & = & 3\% & \cdot & \frac{292}{3} \end{array}$$

Carry out. We solve the equation.

$$b = 0.03 \cdot \frac{292}{3} = 2.92$$

Check. We repeat the calculations. The answer checks.

State. About 2.92 million, or 2,920,000 Southerners choose brown as their favorite color.

11. ***Familiarize***. Since there are 292 million Americans and about one-eighth are senior citizens, there are about $\frac{1}{8}\cdot 292$ million, or 36.5 million senior citizens. The pie chart indicates that 4% of Americans choose black as their favorite color. Let $b =$ the number of senior citizens who choose black as their favorite color.

Translate. We reword and translate the problem.

$$\begin{array}{ccccc} \text{What} & \text{is} & 4\% & \text{of} & 36.5 \text{ million?} \\ \downarrow & \downarrow & \downarrow & \downarrow & \downarrow \\ b & = & 4\% & \cdot & 36,500,000 \end{array}$$

Carry out. We solve the equation.

$$b = 0.04 \cdot 36,500,000 = 1,460,000$$

Check. We repeat the calculations. The answer checks.

State. About 1,460,000 senior citizens choose black as their favorite color.

13. ***Familiarize***. From the pie chart we see that 10.7% of solid waste is plastic. We let $x =$ the amount of plastic, in millions of tons, in the waste generated in 2000.

Translate. We reword the problem.

$$\begin{array}{ccccc} \text{What} & \text{is} & 10.7\% & \text{of} & 231.9? \\ \downarrow & \downarrow & \downarrow & \downarrow & \downarrow \\ x & = & 10.7\% & \cdot & 231.9 \end{array}$$

Carry out.

$$x = 0.107 \cdot 231.9 \approx 24.8$$

Check. We can repeat the calculation. The result checks.

State. In 2000, about 24.8 million tons of waste was plastic.

15. ***Familiarize***. From the pie chart we see that 5.5% of solid waste is glass. From Exercise 13 we know that Americans generated 231.9 million tons of waste in 2000. Then the amount of this that is glass is

0.055(231.9), or about 12.8 million tons

We let $x =$ the amount of glass, in millions of tons, that Americans recycled in 2000.

Translate. We reword the problem.

$$\begin{array}{ccccc} \text{What} & \text{is} & 22.7\% & \text{of} & 12.8 \text{ million tons?} \\ \downarrow & \downarrow & \downarrow & \downarrow & \downarrow \\ x & = & 22.7\% & \cdot & 12.8 \end{array}$$

Carry out.

$$x = 0.227(12.8) \approx 2.9$$

Check. We go over the calculations again. The result checks.

State. Americans recycled about 2.9 million tons of glass in 2000.

17. We locate 2003 on the horizontal axis and then move up to the line. From there we move left to the vertical axis and read the number of cell phones, in millions. We estimate that about 120,000,000 cell phones had Internet access in 2003.

19. We locate 150 on the vertical axis and move right to the line. From there we move down to the horizontal scale and read the year. We see that approximately 150 million cell phones had Internet access in 2004.

21. Starting at the origin:

(1,2) is 1 unit right and 2 units up;

$(-2, 3)$ is 2 units left and 3 units up;

$(4, -1)$ is 4 units right and 1 unit down;

$(-5, -3)$ is 5 units left and 3 units down;

(4,0) is 4 units right and 0 units up or down;

$(0, -2)$ is 0 units right or left and 2 units down.

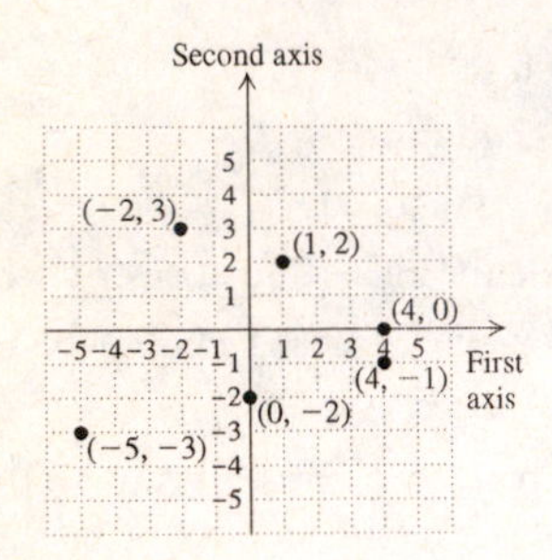

23. Starting at the origin:

(4,4) is 4 units right and 4 units up;

$(-2, 4)$ is 2 units left and 4 units up;

$(5, -3)$ is 5 units right and 3 units down;

$(-5, -5)$ is 5 units left and 5 units down;

(0,4) is 0 units right or left and 4 units up;

$(0, -4)$ is 0 units right or left and 4 units down;

(3,0) is 3 units right and 0 units up or down;

$(-4, 0)$ is 4 units left and 0 units up or down.

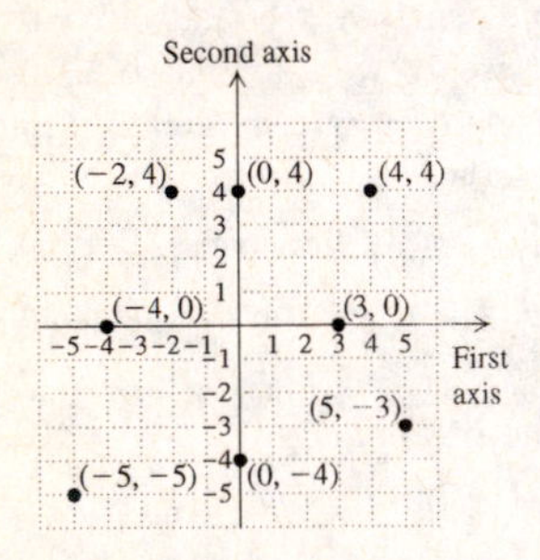

25. We plot the points $(2001, 119.8)$, $(2002, 135.3)$, $(2003, 150.2)$, $(2004, 163.8)$, and $(2005, 176.9)$ and connect adjacent points with line segments.

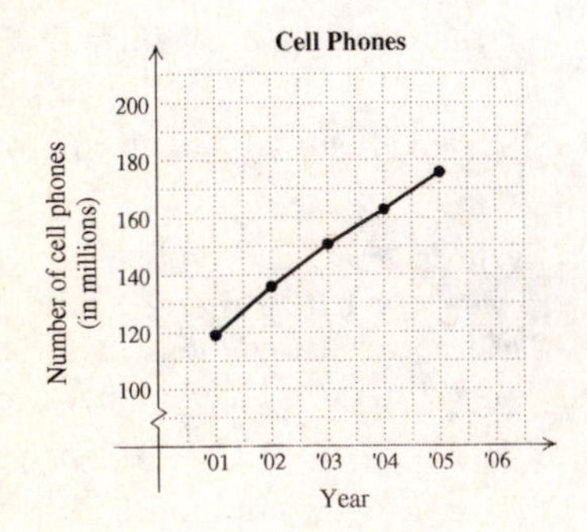

27.

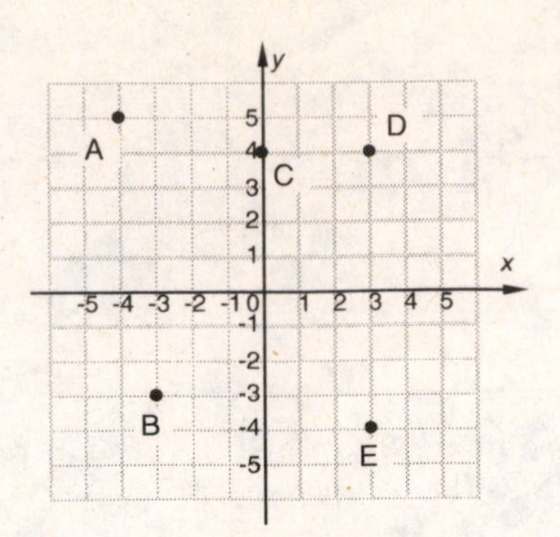

Point A is 4 units left and 5 units up. The coordinates of A are $(-4, 5)$.

Point B is 3 units left and 3 units down. The coordinates of B are $(-3, -3)$.

Point C is 0 units right or left and 4 units up. The coordinates of C are (0,4).

Point D is 3 units right and 4 units up. The coordinates of D are (3,4).

Point E is 3 units right and 4 units down. The coordinates of E are $(3, -4)$.

29.

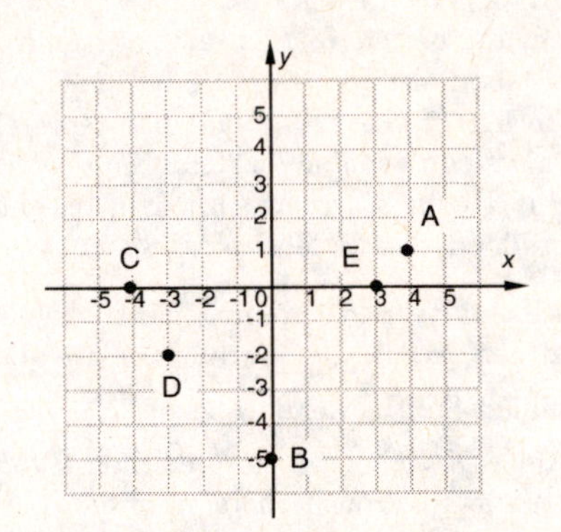

Point A is 4 units right and 1 unit up. The coordinates of A are (4,1).

Point B is 0 units right or left and 5 units down. The coordinates of B are $(0, -5)$.

Point C is 4 units left and 0 units up or down. The coordinates of C are $(-4, 0)$.

Point D is 3 units left and 2 units down. The coordinates of D are $(-3, -2)$.

Point E is 3 units right and 0 units up or down. The coordinates of E are (3,0).

31. Since the x-values range from -75 to 9, the 10 horizontal squares must span $9 - (-75)$, or 84 units. Since 84 is close to 100 and it is convenient to count by 10's, we can count backward from 0 eight squares to -80 and forward from 0 two squares to 20 for a total of $8 + 2$, or 10 squares.

Since the y-values range from -4 to 5, the 10 vertical squares must span $5 - (-4)$, or 9 units. It will be convenient to count by 2's in this case. We count down from 0 five squares to -10 and up from 0 five squares to 10 for a total of $5 + 5$, or 10 squares. (Instead, we might have chosen to count by 1's from -5 to 5.)

Then we plot the points $(-75, 5)$, $(-18, -2)$, and $(9, -4)$.

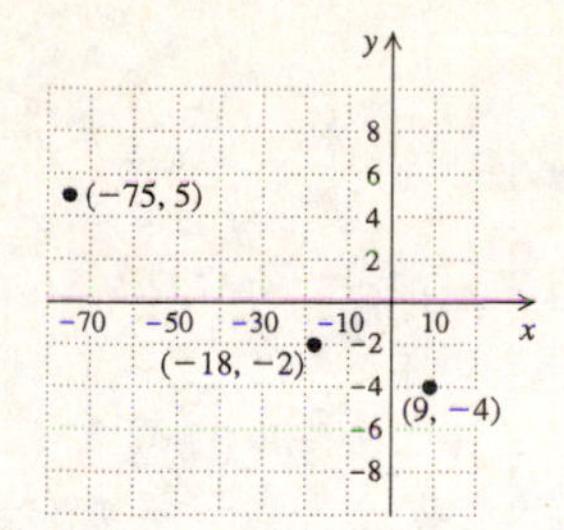

33. Since the x-values range from -5 to 5, the 10 horizontal squares must span $5-(-5)$, or 10 units. It will be convenient to count by 2's in this case. We count backward from 0 five squares to -10 and forward from 0 five squares to 10 for a total of $5+5$, or 10 squares.

Since the y-values range from -14 to 83, the 10 vertical squares must span $83-(-14)$, or 97 units. To include both -14 and 83, the squares should extend from about -20 to 90, or $90-(-20)$, or 110 units. We cannot do this counting by 10's, so we use 20's instead. We count down from 0 four units to -80 and up from 0 six units to 120 for a total of $4+6$, or 10 units. There are other ways to cover the values from -14 to 83 as well.

Then we plot the points $(-1, 83)$, $(-5, -14)$, and $(5, 37)$.

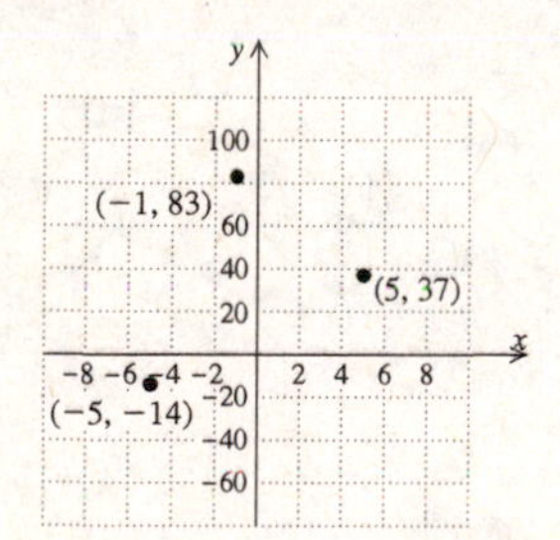

35. Since the x-values range from -16 to 3, the 10 horizontal squares must span $3-(-16)$, or 19 units. We could number by 2's or 3's. We number by 3's, going backward from 0 eight squares to -24 and forward from 0 two squares to 6 for a total of $8+2$, or 10 squares.

Since the y-values range from -4 to 15, the 10 vertical squares must span $15-(-4)$, or 19 units. We will number the vertical axis by 3's as we did the horizontal axis. We go down from 0 four squares to -12 and up from 0 six squares to 18 for a total of $4+6$, or 10 squares.

Then we plot the points $(-10, -4)$, $(-16, 7)$, and $(3, 15)$.

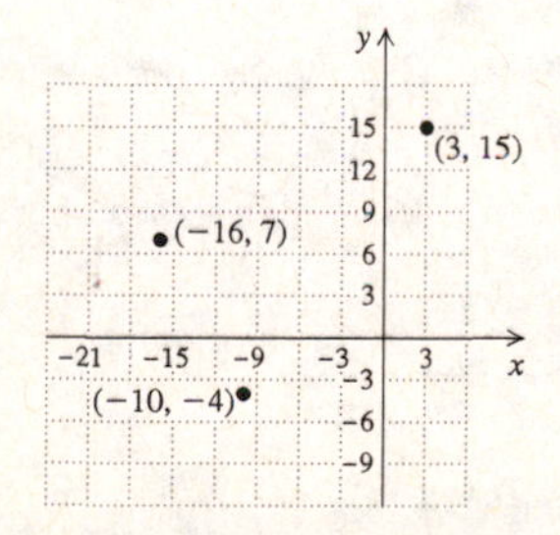

37. Since the x-values range from -100 and 800, the 10 horizontal squares must span $800-(-100)$, or 900 units. Since 900 is close to 1000 we can number by 100's. We go backward from 0 two squares to -200 and forward from 0 eight squares to 800 for a total of $2+8$, or 10 squares. (We could have numbered from -100 to 900 instead.)

Since the y-values range from -5 to 37, the 10 vertical squares must span $37-(-5)$, or 42 units. Since 42 is close to 50, we can count by 5's. We go down from 0 two squares to -10 and up from 0 eight squares to 40 for a total of $2+8$, or 10 squares.

Then we plot the points $(-100, -5)$, $(350, 20)$, and $(800, 37)$.

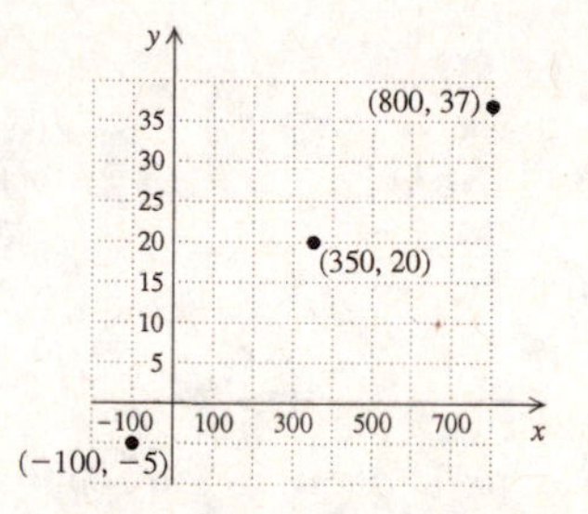

39. Since the x-values range from -124 to 54, the 10 horizontal squares must span $54-(-124)$, or 178 units. We can number by 25's. We go backward from 0 six squares to -150 and forward from 0 four squares to 100 for a total of $6+4$, or 10 squares.

Since the y-values range from -238 to 491, the 10 vertical squares must span $491-(-238)$, or 729 units. We can number by 100's. We go down from 0 four squares to -400 and up from 0 six squares to 600 for a total of $4+6$, or 10 squares.

Then we plot the points $(-83, 491)$, $(-124, -95)$, and $(54, -238)$.

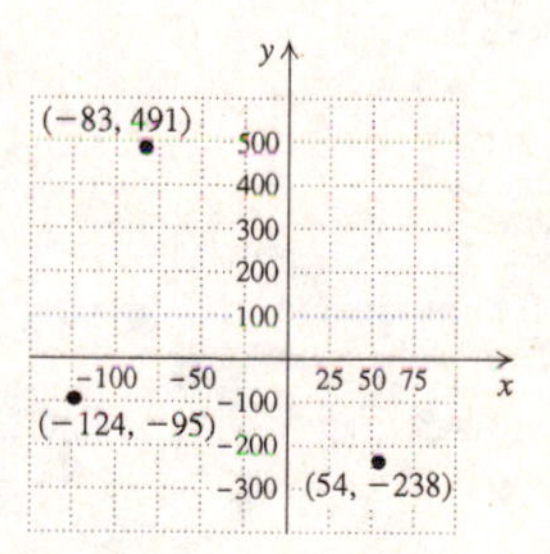

41. Since the first coordinate is positive and the second coordinate negative, the point $(7, -2)$ is located in quadrant IV.

43. Since both coordinates are negative, the point $(-4, -3)$ is in quadrant III.

45. Since both coordinates are positive, the point $(2, 1)$ is in quadrant I.

47. Since the first coordinate is negative and the second coordinate is positive, the point $(-4.9, 8.3)$ is in quadrant II.

49. First coordinates are positive in the quadrants that lie to the right of the origin, or in quadrants I and IV.

51. Points for which both coordinates are positive lie in quadrant I, and points for which both coordinates are negative life in quadrant III. Thus, both coordinates have the same sign in quadrants I and III.

53. *Writing Exercise*

55. $4 \cdot 3 - 6 \cdot 5 = 12 - 30 = -18$

57. $-\frac{1}{2}(-6) + 3 = 3 + 3 = 6$

59.

$$\begin{aligned} 3x - 2y &= 6 \\ -2y &= -3x + 6 \quad \text{Adding } -3x \text{ to both sides} \\ -\frac{1}{2}(-2y) &= -\frac{1}{2}(-3x + 6) \\ y &= -\frac{1}{2}(-3x) - \frac{1}{2}(6) \\ y &= \frac{3}{2}x - 3 \end{aligned}$$

61. *Writing Exercise*

63. The coordinates have opposite signs, so the point could be in quadrant II or quadrant IV.

65.

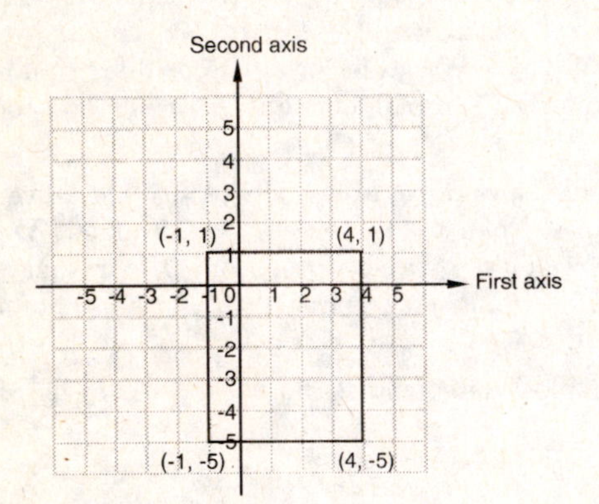

The coordinates of the fourth vertex are $(-1, -5)$.

67. Answers may vary.

We select eight points such that the sum of the coordinates for each point is 7.

$(0,7)$ $0+7=7$
$(1,6)$ $1+6=7$
$(2,5)$ $2+5=7$
$(3,4)$ $3+4=7$
$(4,3)$ $4+3=7$
$(5,2)$ $5+2=7$
$(6,1)$ $6+1=7$
$(7,0)$ $7+0=7$

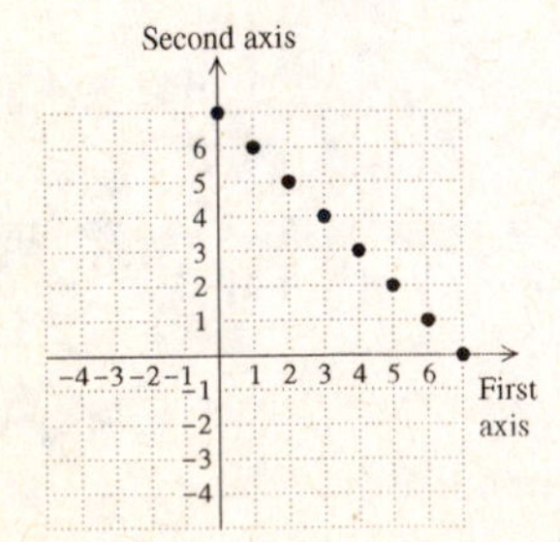

69.

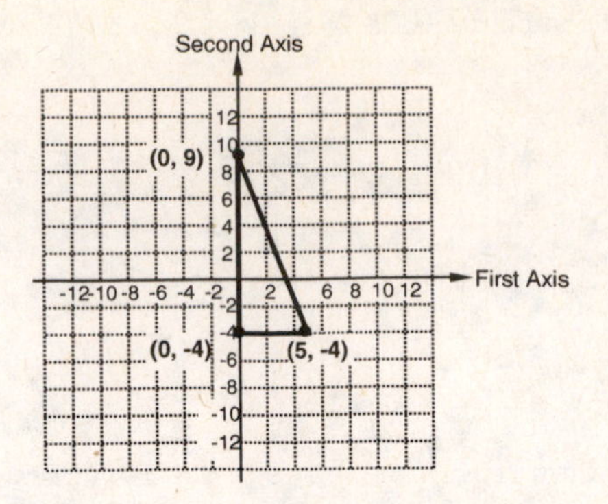

The base is 5 units and the height is 13 units.

$A = \frac{1}{2}bh = \frac{1}{2} \cdot 5 \cdot 13 = \frac{65}{2}$ sq units, or $32\frac{1}{2}$ sq units

71. Latitude 27° North,
Longitude 81° West

73. *Writing Exercise*

Exercise Set 3.2

1. We substitute 0 for x and 2 for y (alphabetical order of variables).

$$\begin{array}{c|l} \multicolumn{2}{c}{y = 5x + 1} \\ \hline 2 & 5 \cdot 0 + 1 \\ & 0 + 1 \end{array}$$

$2 \stackrel{?}{=} 1$ FALSE

Since $2 = 1$ is false, the pair $(0, 2)$ is not a solution.

3. We substitute 4 for x and 2 for y.

$$\begin{array}{c|c} \multicolumn{2}{c}{3y + 2x = 12} \\ \hline 3 \cdot 2 + 2 \cdot 4 & 12 \\ 6 + 8 & \end{array}$$

$14 \stackrel{?}{=} 12$ FALSE

Since $14 = 12$ is false, the pair (4,2) is not a solution.

5. We substitute 2 for a and -1 for b.

$$\begin{array}{c|c} \multicolumn{2}{c}{4a - 3b = 11} \\ \hline 4 \cdot 2 - 3(-1) & 11 \\ 8 + 3 & \end{array}$$

$11 \stackrel{?}{=} 11$ TRUE

Since $11 = 11$ is true, the pair $(2, -1)$ is a solution.

7. To show that a pair is a solution, we substitute, replacing x with the first coordinate and y with the second coordinate in each pair.

$$\begin{array}{c|c} \multicolumn{2}{c}{y = x + 3} \\ \hline 2 & -1 + 3 \end{array} \qquad \begin{array}{c|c} \multicolumn{2}{c}{y = x + 3} \\ \hline 7 & 4 + 3 \end{array}$$

$2 \stackrel{?}{=} 2$ TRUE $\qquad 7 \stackrel{?}{=} 7$ TRUE

In each case the substitution results in a true equation. Thus, $(-1, 2)$ and $(4, 7)$ are both solutions of $y = x + 3$.

We graph these points and sketch the line passing through them.

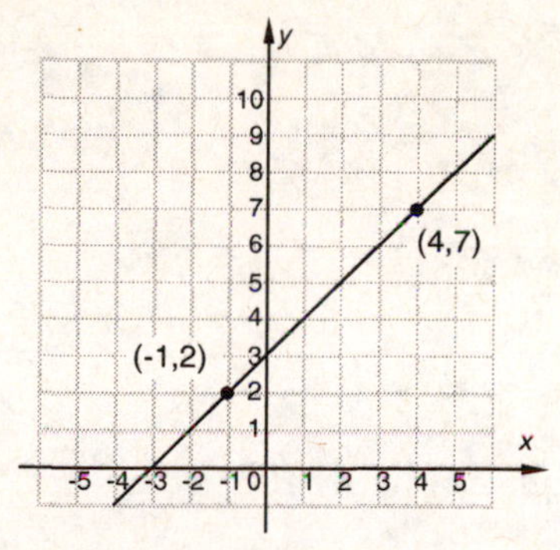

The line appears to pass through $(0,3)$ also. We check to determine if $(0,3)$ is a solution of $y = x + 3$.

$$\begin{array}{c|c} \multicolumn{2}{c}{y = x + 3} \\ \hline 3 & 0+3 \end{array}$$
$$3 \stackrel{?}{=} 3 \qquad \text{TRUE}$$

Thus, $(0,3)$ is another solution. There are other correct answers, including $(-5,-2)$, $(-4,-1)$, $(-3,0)$, $(-2,1)$, $(1,4)$, $(2,5)$, and $(3,6)$.

9. To show that a pair is a solution, we substitute, replacing x with the first coordinate and y with the second coordinate in each pair.

$$\begin{array}{c|c} \multicolumn{2}{c}{y = \frac{1}{2}x + 3} \\ \hline 5 & \frac{1}{2}\cdot 4 + 3 \\ & 2+3 \end{array} \qquad \begin{array}{c|c} \multicolumn{2}{c}{y = \frac{1}{2}x + 3} \\ \hline 2 & \frac{1}{2}(-2) + 3 \\ & -1+3 \end{array}$$
$$5 \stackrel{?}{=} 5 \quad \text{TRUE} \qquad 2 \stackrel{?}{=} 2 \quad \text{TRUE}$$

In each case the substitution results in a true equation. Thus, $(4,5)$ and $(-2,2)$ are both solutions of $y = \frac{1}{2}x + 3$. We graph these points and sketch the line passing through them.

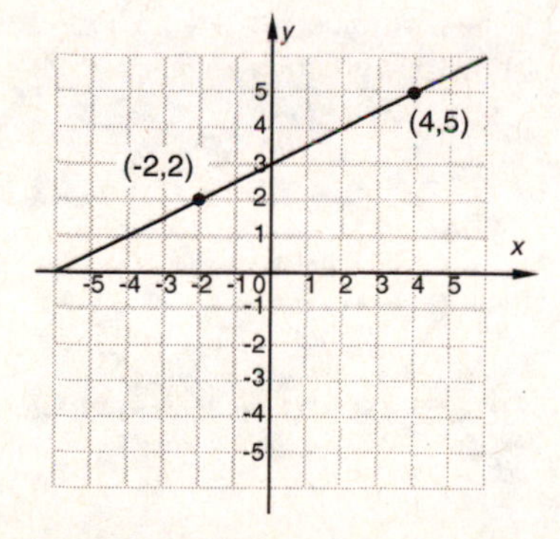

The line appears to pass through $(0,3)$ also. We check to determine if $(0,3)$ is a solution of $y = \frac{1}{2}x + 3$.

$$\begin{array}{c|c} \multicolumn{2}{c}{y = \frac{1}{2}x + 3} \\ \hline 3 & \frac{1}{2}\cdot 0 + 3 \end{array}$$
$$3 \stackrel{?}{=} 3 \qquad \text{TRUE}$$

Thus, $(0,3)$ is another solution. There are other correct answers, including $(-6,0)$, $(-4,1)$, $(2,4)$, and $(6,6)$.

11. To show that a pair is a solution, we substitute, replacing x with the first coordinate and y with the second coordinate in each pair.

$$\begin{array}{c|c} \multicolumn{2}{c}{y + 3x = 7} \\ \hline 1 + 3\cdot 2 & 7 \\ 1+6 & \end{array} \qquad \begin{array}{c|c} \multicolumn{2}{c}{y + 3x = 7} \\ \hline -5 + 3\cdot 4 & 7 \\ -5+12 & \end{array}$$
$$7 \stackrel{?}{=} 7 \quad \text{TRUE} \qquad 7 \stackrel{?}{=} 7 \quad \text{TRUE}$$

In each case the substitution results in a true equation. Thus, $(2,1)$ and $(4,-5)$ are both solutions of $y + 3x = 7$. We graph these points and sketch the line passing through them.

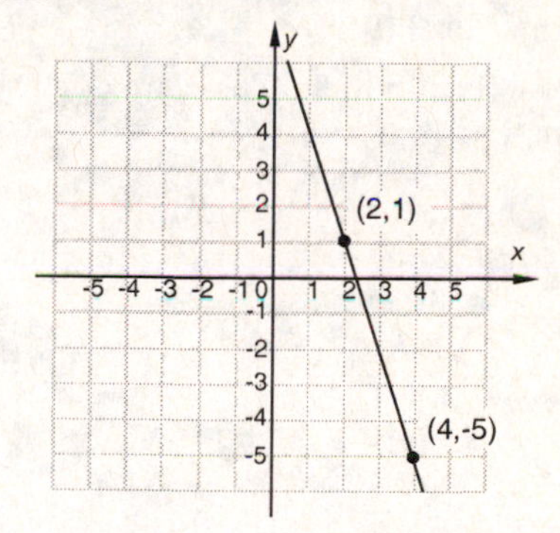

The line appears to pass through $(1,4)$ also. We check to determine if $(1,4)$ is a solution of $y + 3x = 7$.

$$\begin{array}{c|c} \multicolumn{2}{c}{y + 3x = 7} \\ \hline 4 + 3\cdot 1 & 7 \\ 4+3 & \end{array}$$
$$7 \stackrel{?}{=} 7 \quad \text{TRUE}$$

Thus, $(1,4)$ is another solution. There are other correct answers, including $(3,-2)$.

13. To show that a pair is a solution, we substitute, replacing x with the first coordinate and y with the second coordinate in each pair.

$$\begin{array}{c|c} \multicolumn{2}{c}{4x - 2y = 10} \\ \hline 4\cdot 0 - 2(-5) & 10 \end{array}$$
$$10 \stackrel{?}{=} 10 \quad \text{TRUE}$$

$$\begin{array}{c|c} \multicolumn{2}{c}{4x - 2y = 10} \\ \hline 4\cdot 4 - 2\cdot 3 & 10 \\ 16-6 & \end{array}$$
$$10 \stackrel{?}{=} 10 \quad \text{TRUE}$$

In each case the substitution results in a true equation. Thus, $(0,-5)$ and $(4,3)$ are both solutions of $4x - 2y = 10$.

We graph these points and sketch the line passing through them.

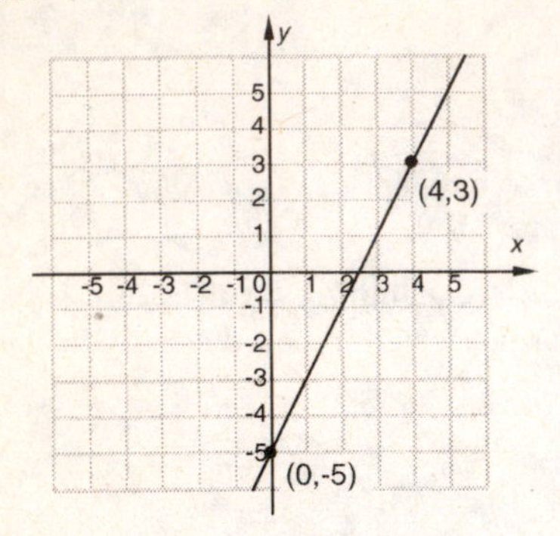

The line appears to pass through $(2,-1)$ also. We check to determine if $(2,-1)$ is a solution of $4x-2y=10$.

$$\begin{array}{c|c} 4x-2y & = 10 \\ \hline 4\cdot 2-2(-1) & 10 \\ 8+2 & \end{array}$$

$$10 \stackrel{?}{=} 10 \quad \text{TRUE}$$

Thus, $(2,-1)$ is another solution. There are other correct answers, including $(1,-3)$, $(2,-1)$, $(3,1)$, and $(5,5)$.

15. $y=x+1$

The equation is in the form $y=mx+b$. The y-intercept is $(0,1)$. We find two other pairs.

When $x=3$, $\quad y=3+1=4$.
When $x=-5$, $\quad y=-5+1=-4$.

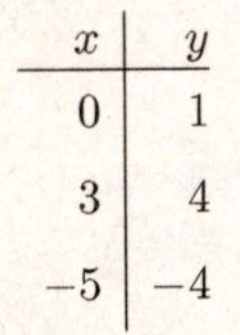

x	y
0	1
3	4
-5	-4

Plot these points, draw the line they determine, and label the graph $y=x+1$.

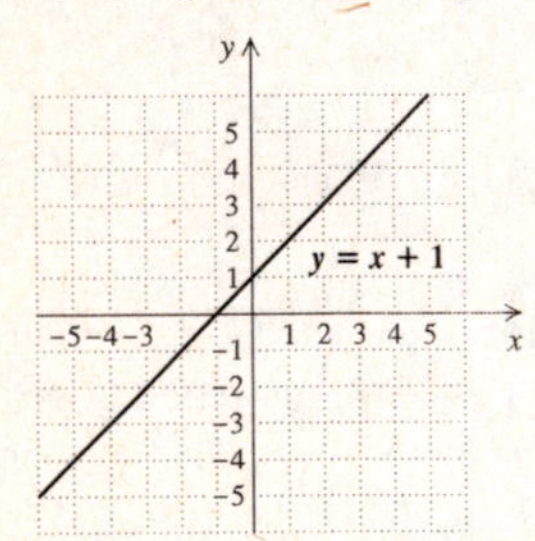

17. $y=-x$

The equation is equivalent to $y=-x+0$. The y-intercept is $(0,0)$. We find two other points.

When $x=-2$, $\quad y=-(-2)=2$.
When $x=3$, $\quad y=-3$.

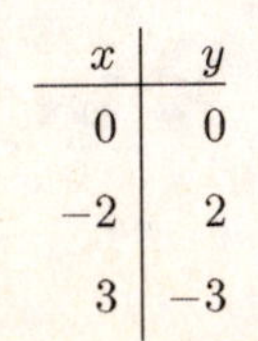

x	y
0	0
-2	2
3	-3

Plot these points, draw the line they determine, and label the graph $y=-x$.

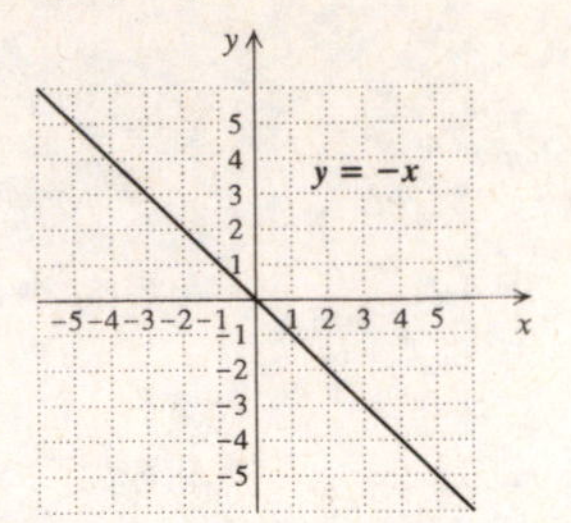

19. $y=\frac{1}{3}x$

The equation is equivalent to $y=\frac{1}{3}x+0$. The y-intercept is $(0,0)$. We find two other points, using multiples of 3 for x to avoid fractions.

When $x=-3$, $y=\frac{1}{3}(-3)=-1$.

When $x=3$, $y=\frac{1}{3}\cdot 3=1$.

x	y
0	0
-3	-1
3	1

Plot these points, draw the line they determine, and label the graph $y=\frac{1}{3}x$.

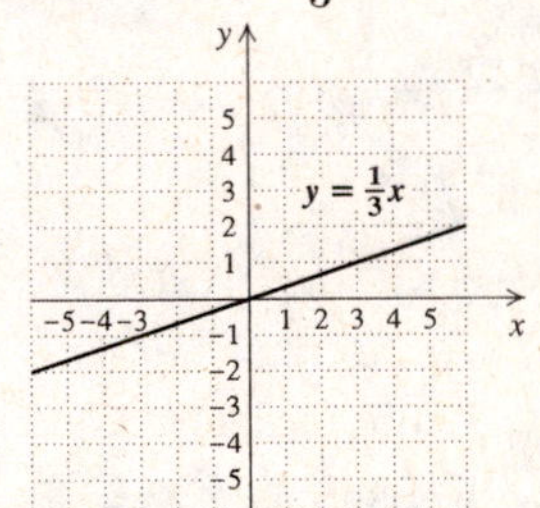

21. $y=x+3$

The equation is in the form $y=mx+b$. The y-intercept is $(0,3)$. We find two other points.

When $x=-4$, $y=-4+3=-1$.

When $x=2$, $y=2+3=5$.

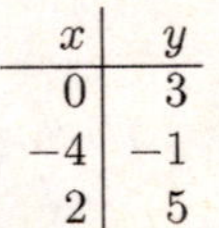

x	y
0	3
-4	-1
2	5

Plot these points, draw the line they determine, and label the graph $y=x+3$.

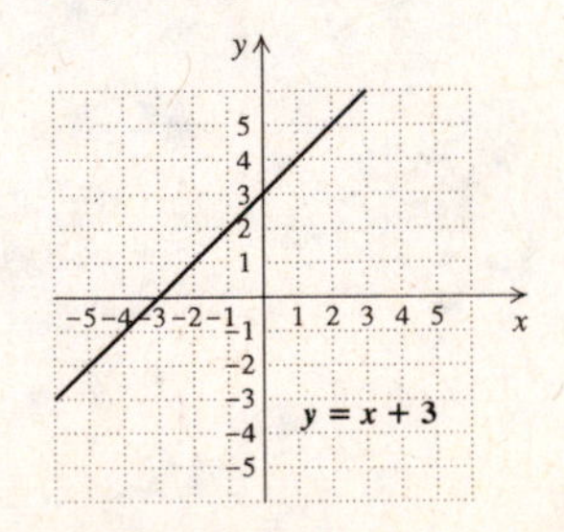

23. $y = 2x + 2$

The y-intercept is $(0, 2)$. We find two other points.

When $x = -3$, $y = 2(-3) + 2 = -6 + 2 = -4$.

When $x = 1$, $y = 2 \cdot 1 + 2 = 2 + 2 = 4$.

x	y
0	2
-3	-4
1	4

Plot these points, draw the line they determine, and label the graph $y = 2x + 2$.

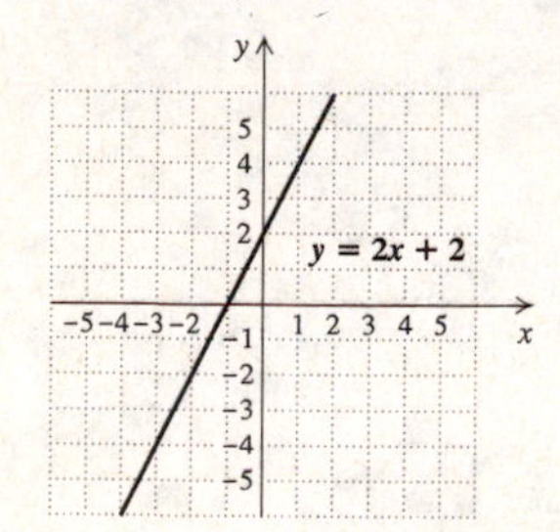

25. $y = \frac{1}{3}x - 4 = \frac{1}{3}x + (-4)$

The y-intercept is $(0, -4)$. We find two other points, using multiples of 3 for x to avoid fractions.

When $x = -3$, $y = \frac{1}{3}(-3) - 4 = -1 - 4 = -5$.

When $x = 3$, $y = \frac{1}{3} \cdot 3 - 4 = 1 - 4 = -3$.

x	y
0	-4
-3	-5
3	-3

Plot these points, draw the line they determine, and label the graph $y = \frac{1}{3}x - 4$.

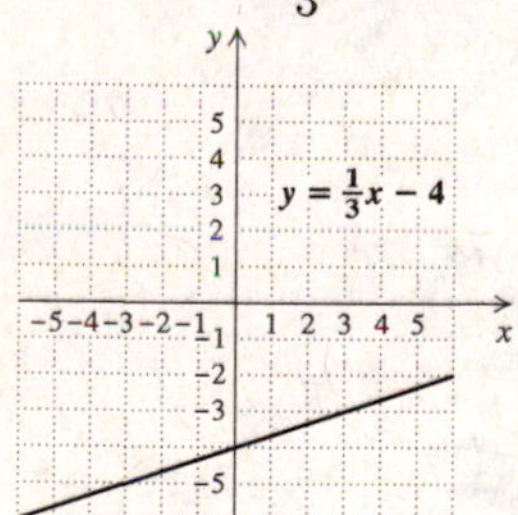

27. $x + y = 4$

$y = -x + 4$

The y-intercept is $(0, 4)$. We find two other points.

When $x = -1$, $y = -(-1) + 4 = 1 + 4 = 5$.

When $x = 2$, $y = -2 + 4 = 2$.

x	y
0	4
-1	5
2	2

Plot these points, draw the line they determine, and label the graph $x + y = 4$.

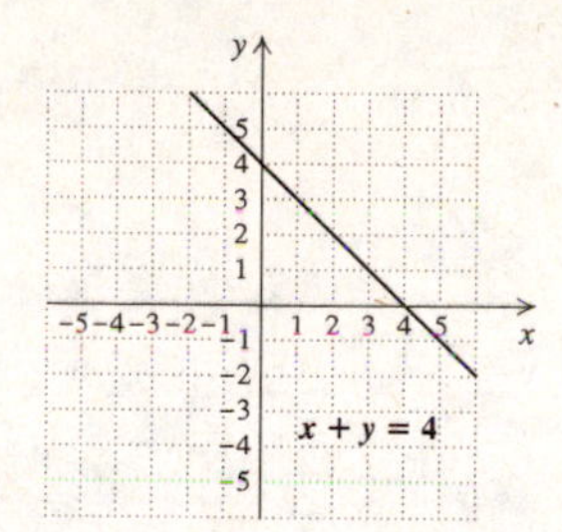

29. $y = \frac{5}{2}x + 3$

The y-intercept is $(0, 3)$. We find two other points, using multiples of 2 for x to avoid fractions.

When $x = -4$, $y = \frac{5}{2}(-4) + 3 = -10 + 3 = -7$.

When $x = -2$, $y = \frac{5}{2}(-2) + 3 = -5 + 3 = -2$.

x	y
0	3
-4	-7
-2	-2

Plot these points, draw the line they determine, and label the graph $y = \frac{5}{2}x + 3$.

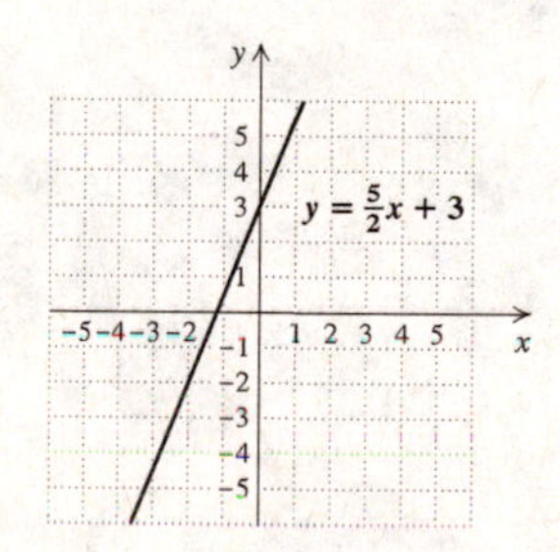

31. $x + 2y = -6$

$$2y = -x - 6$$
$$y = -\frac{1}{2}x - 3$$
$$y = -\frac{1}{2}x + (-3)$$

The y-intercept is $(0, -3)$. We find two other points, using multiples of 2 for x to avoid fractions.

When $x = -4$, $y = -\frac{1}{2}(-4) - 3 = 2 - 3 = -1$.

When $x = 2$, $y = -\frac{1}{2} \cdot 2 - 3 = -1 - 3 = -4$.

x	y
0	-3
-4	-1
2	-4

Plot these points, draw the line they determine, and label the graph $x + 2y = -6$.

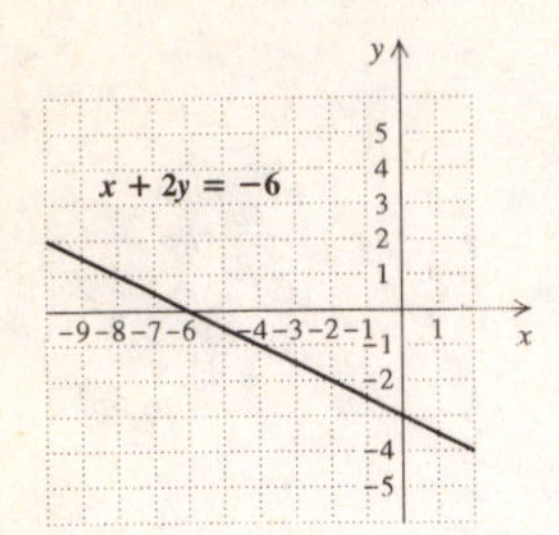

33. $y = -\frac{2}{3} + 4$

The y-intercept is $(0, 4)$. We find two other points, using multiples of 3 for x to avoid fractions.

When $x = 3$, $y = -\frac{2}{3} \cdot 3 + 4 = -2 + 4 = 2$.

When $x = 6$, $y = -\frac{2}{3} \cdot 6 + 4 = -4 + 4 = 0$.

x	y
0	4
3	2
6	0

Plot these points, draw the line they determine, and label the graph $y = -\frac{2}{3}x + 4$.

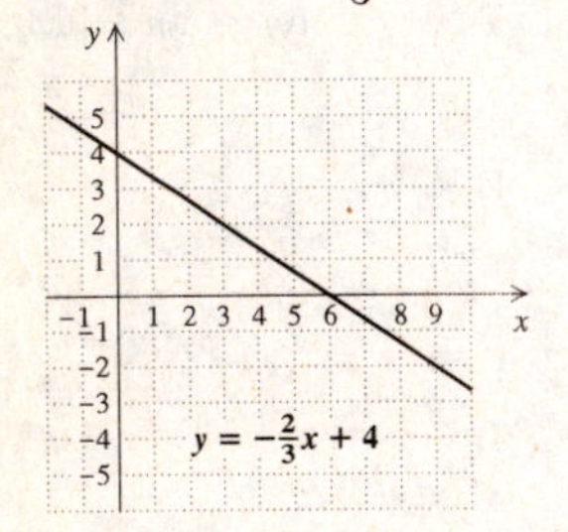

35. $8x - 4y = 12$

$$-4y = -8x + 12$$
$$y = 2x - 3$$
$$y = 2x + (-3)$$

The y-intercept is $(0, -3)$. We find two other points.

When $x = -1$, $y = 2(-1) - 3 = -2 - 3 = -5$.

When $x = 3$, $y = 2 \cdot 3 - 3 = 6 - 3 = 3$.

x	y
0	−3
−1	−5
3	3

Plot these points, draw the line they determine, and label the graph $8x - 4y = 12$.

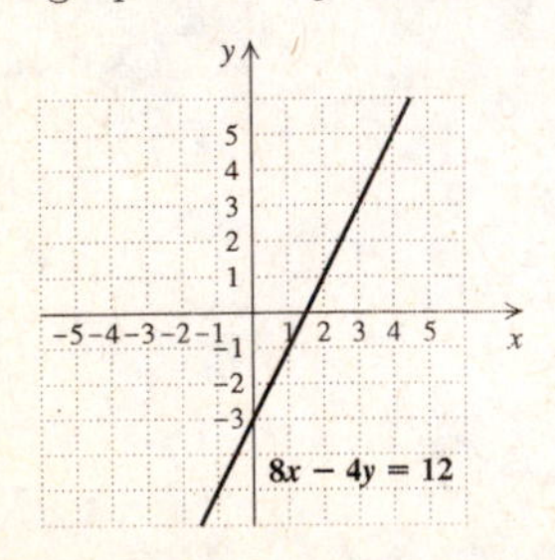

37. $6y + 2x = 8$

$$6y = -2x + 8$$
$$y = -\frac{1}{3}x + \frac{4}{3}$$

The y-intercept is $\left(0, \frac{4}{3}\right)$. We find two other points.

When $x = -2$, $y = -\frac{1}{3}(-2) + \frac{4}{3} = \frac{2}{3} + \frac{4}{3} = 2$.

When $x = 1$, $y = -\frac{1}{3} \cdot 1 + \frac{4}{3} = -\frac{1}{3} + \frac{4}{3} = 1$.

x	y
0	$\frac{4}{3}$
−2	2
1	1

Plot these points, draw the line they determine, and label the graph $6y + 2x = 8$.

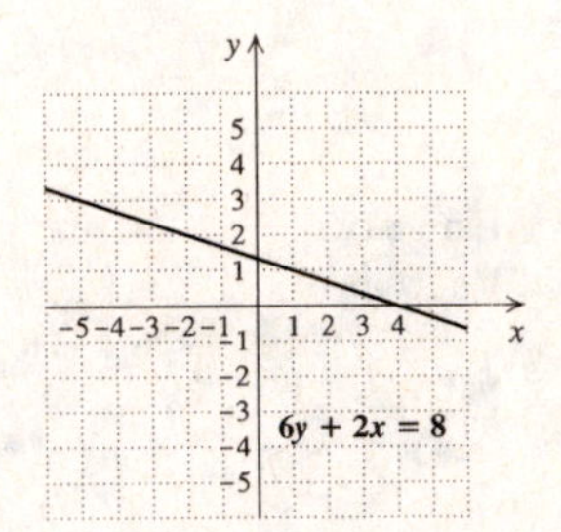

39. We graph $n = 0.9t + 19$ by selecting values for t and then calculating the associated values for n.

If $t = 0$, $n = 0.9(0) + 19 = 19$.

If $t = 3$, $n = 0.9(3) + 19 = 2.7 + 19 = 21.7$.

If $t = 7$, $n = 0.9(7) + 19 = 6.3 + 19 = 25.3$.

t	n
0	19
3	21.7
7	25.3

We plot the points and draw the graph.

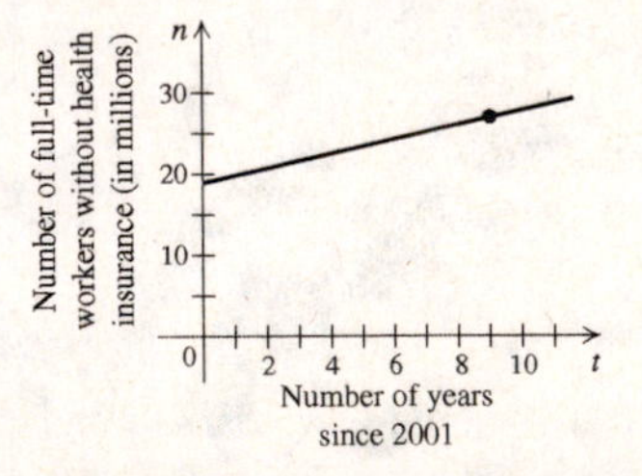

Since $2010 - 2001 = 9$, the year 2010 is 9 years after 2001. Thus, to estimate the number of uninsured full-time workers in 2010, we find the second coordinate associated with 9. Locate the point on the line that is above 9 and then find the value on the vertical axis that corresponds to that point. That value is about 27, so we estimate that there will be about 27 million uninsured full-time workers in 2010.

41. We graph $t + w = 15$, or $w = -t + 15$. Since time cannot be negative in this application, we select only nonnegative values for t.

If $t = 0$, $w = -0 + 15 = 15$.

If $t = 2$, $w = -2 + 15 = 13$.

If $t = 5$, $w = -5 + 15 = 10$.

t	w
0	15
2	13
5	10

We plot the points and draw the graph. Since the likelihood of death cannot be negative, the graph stops at the horizontal axis.

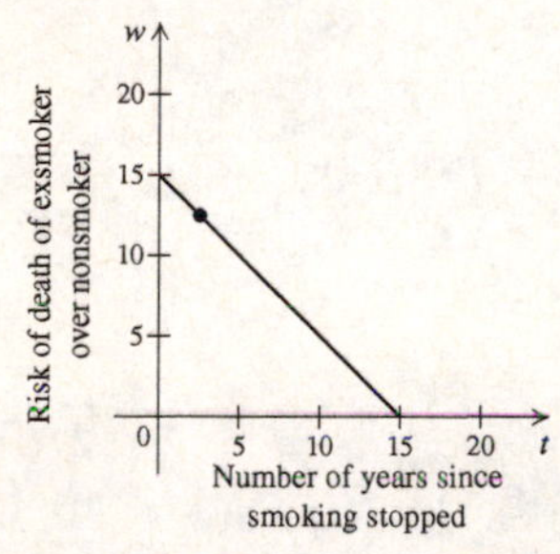

To estimate how much more likely it is for Sandy to die from lung cancer than Polly, we find the second coordinate associated with $2\frac{1}{2}$. Locate the point on the line that is above $2\frac{1}{2}$ and then find the value on the vertical axis that corresponds to that point. That value is about $12\frac{1}{2}$, so it is $12\frac{1}{2}$ times more likely for Sandy to die from lung cancer than Polly.

43. We graph $v = -\frac{3}{4}t + 6$. Since time cannot be negative in this application, we select only nonnegative values for t.

If $t = 0$, $v = -\frac{3}{4} \cdot 0 + 6 = 6$.

If $t = 2$, $v = -\frac{3}{4} \cdot 2 + 6 = -\frac{3}{2} + 6 = \frac{9}{2}$, or $4\frac{1}{2}$.

If $t = 8$, $v = -\frac{3}{4} \cdot 8 + 6 = -6 + 6 = 0$.

t	v
0	6
2	$4\frac{1}{2}$
8	0

We plot the points and draw the graph.

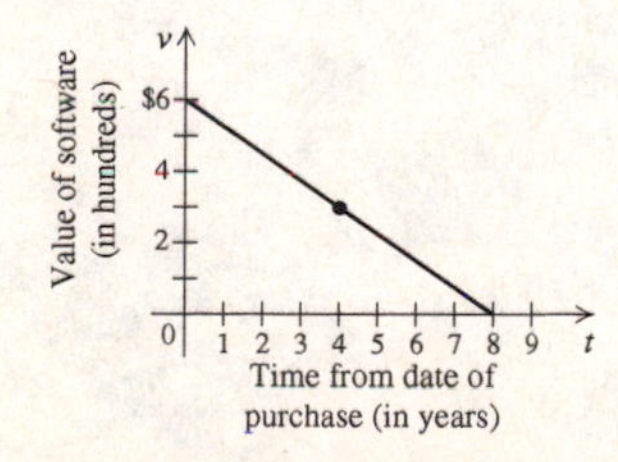

To estimate what the program is worth 4 yr after it was purchased, we find the second coordinate associated with 4. Locate the point on the line that is above 4 and then find the value on the vertical axis that corresponds to that point. That value is 3, so the program is worth $300 after 4 years.

45. We graph $T = \frac{3}{4}c + 1$. Since the number of credits cannot be negative, we select only nonnegative values for c.

If $c = 4$, $T = \frac{3}{4} \cdot 4 + 1 = 3 + 1 = 4$.

If $c = 8$, $T = \frac{3}{4} \cdot 8 + 1 = 6 + 1 = 7$.

If $c = 16$, $T = \frac{3}{4} \cdot 16 = 12 + 1 = 13$.

c	T
4	4
8	7
16	13

We plot the points and draw the graph.

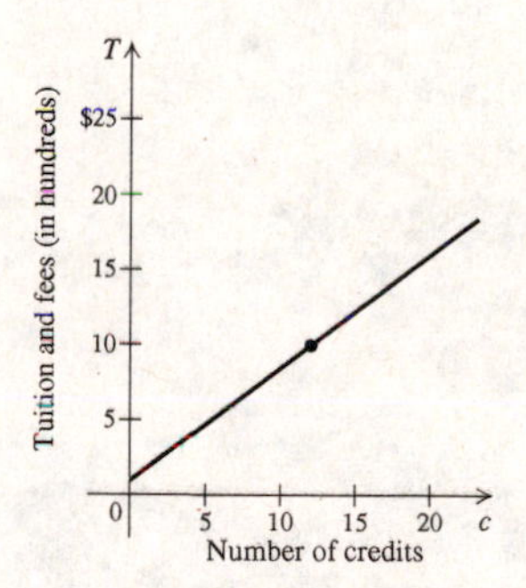

Four three-credit courses total $4 \cdot 3$, or 12, credits. To estimate the cost of tuition and fees for a student who is registered for 12 credits, we find the second coordinate associated with 12. Locate the point on the line that is above 12 and then find the value on the vertical axis that corresponds to that point. That value is about 10, so tuition and fees will cost $10 hundred, or $1000.

47. We graph $T = -2m + 54$.

If $m = 0$, $T = -2 \cdot 0 + 54 = 54$.

If $m = 10$, $T = -2 \cdot 10 + 54 = -20 + 54 = 34$.

If $m = 20$, $T = -2 \cdot 20 + 54 = -40 + 54 = 14$.

m	T
0	54
10	34
20	14

We plot the points and draw the graph.

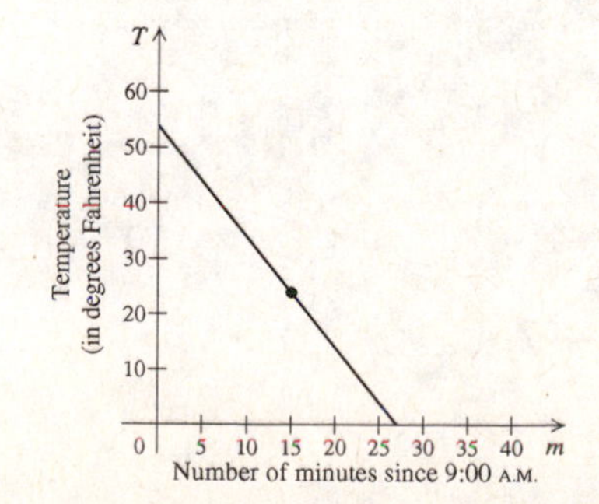

To estimate the temperature at 9:15 A.M., first note that 9:15 is 15 minutes after 9:00 A.M. Then find the second coordinate associated with 15. Locate the point on the line that is above 15 and find the corresponding value on the vertical axis. It appears that the temperature was about 24°F at 9:15 A.M.

49. *Writing Exercise*

51.
$$\begin{aligned} 5x + 3 \cdot 0 &= 12 \\ 5x + 0 &= 12 \\ 5x &= 12 \\ x &= \frac{12}{5} \end{aligned}$$

Check:

$5x + 3 \cdot 0 = 12$	
$5 \cdot \frac{12}{5} + 3 \cdot 0$	12
$12 + 0$	
$12 \stackrel{?}{=} 12$	TRUE

The solution is $\frac{12}{5}$.

53.
$$\begin{aligned} 7 \cdot 0 - 4y &= 10 \\ 0 - 4y &= 10 \\ y &= -\frac{5}{2} \end{aligned}$$

Check:

$7 \cdot 0 - 4y = 10$	
$7 \cdot 0 - 4\left(-\frac{5}{2}\right)$	10
$0 + 10$	
$10 \stackrel{?}{=} 10$	TRUE

The solution is $-\frac{5}{2}$.

55.
$$\begin{aligned} Ax + By &= C \\ By &= C - Ax \quad \text{Subtracting } Ax \\ y &= \frac{C - Ax}{B} \quad \text{Dividing by } B \end{aligned}$$

57. *Writing Exercise*

59. Let s represent the gear that Lauren uses on the southbound portion of her ride and n represent the gear she uses on the northbound portion. Then we have $s + n = 18$. We graph this equation, using only positive integer values for s and n.

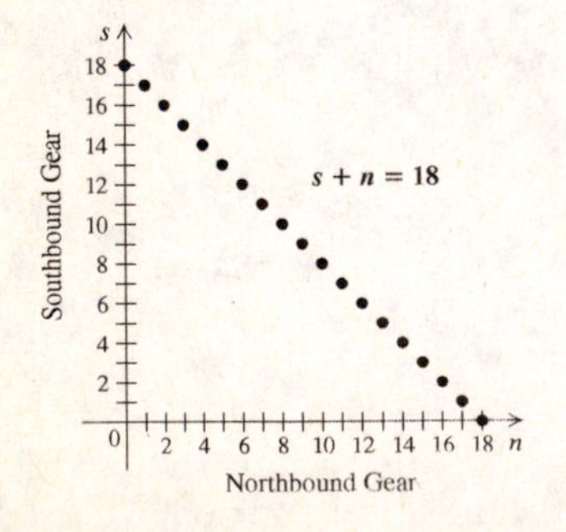

61. Note that the sum of the coordinates of each point on the graph is 2. Thus, we have $x + y = 2$, or $y = -x + 2$.

63. Note that when $x = 0$, $y = -5$ and when $y = 0$, $x = 3$. An equation that fits this situation is $5x - 3y = 15$, or $y = \frac{5}{3}x - 5$.

65. The equation is $25d + 5l = 225$.

Since the number of dinners cannot be negative, we choose only nonnegative values of d when graphing the equation. The graph stops at the horizontal axis since the number of lunches cannot be negative.

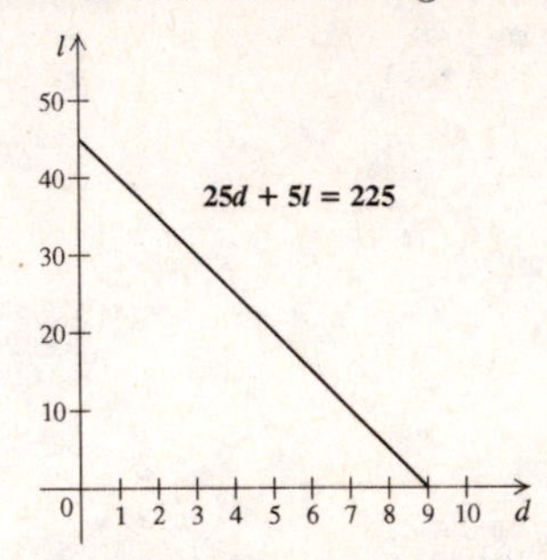

We see that three points on the graph are $(1, 40)$, $(5, 20)$, and $(8, 5)$. Thus, three combinations of dinners and lunches that total \$225 are

1 dinner, 40 lunches,

5 dinners, 20 lunches,

8 dinners, 5 lunches.

67. $y = -|x|$

x	y
-3	-3
-2	-2
-1	-1
0	0
1	-1
2	-2
3	-3

69. $y = -|x| + 2$

x	y
-3	-1
-2	0
-1	1
0	2
1	1
2	0
3	-1

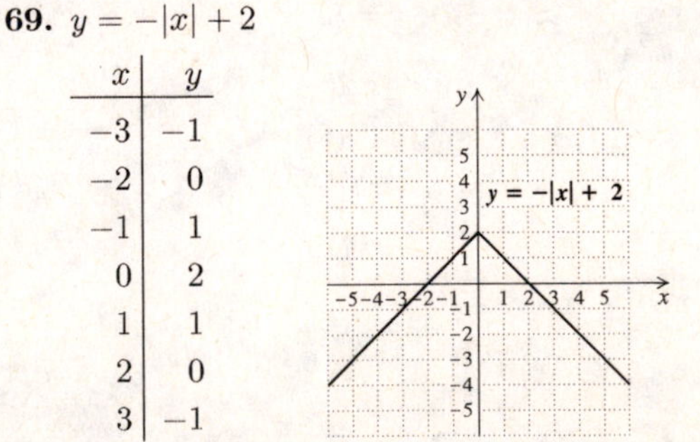

71.

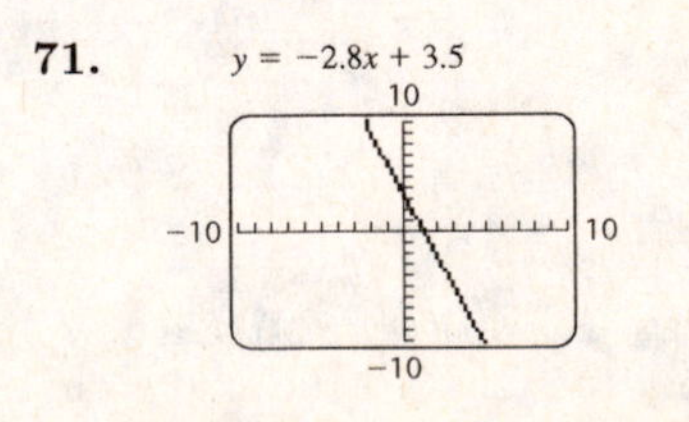

73.

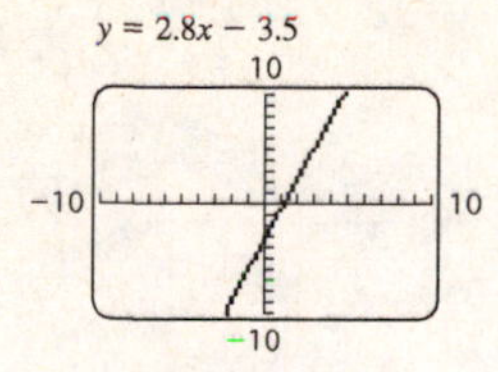

75.

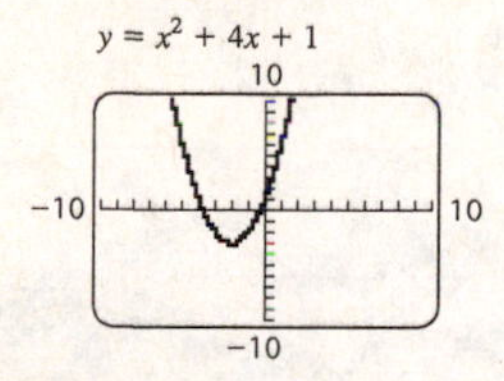

77. *Writing Exercise*

Exercise Set 3.3

1. The graph of $x = -4$ is a vertical line, so (f) is the most appropriate choice.

3. The point $(0, 2)$ lies on the y-axis, so (d) is the most appropriate choice.

5. The point $(3, -2)$ does not lie on an axis, so it could be used as a check when we graph using intercepts. Thus (b) is the most appropriate choice.

7. (a) The graph crosses the y-axis at $(0, 5)$, so the y-intercept is $(0, 5)$.

(b) The graph crosses the x-axis at $(2, 0)$, so the x-intercept is $(2, 0)$.

9. (a) The graph crosses the y-axis at $(0, -4)$, so the y-intercept is $(0, -4)$.

(b) The graph crosses the x-axis at $(3, 0)$, so the x-intercept is $(3, 0)$.

11. (a) The graph crosses the y-axis at $(0, -2)$, so the y-intercept is $(0, -2)$.

(b) The graph crosses the x-axis at $(-3, 0)$, so the x-intercept is $(-3, 0)$.

13. $5x + 3y = 15$

(a) To find the y-intercept, let $x = 0$. This is the same as temporarily ignoring the x-term and then solving.

$$3y = 15$$
$$y = 5$$

The y-intercept is $(0, 5)$.

(b) To find the x-intercept, let $y = 0$. This is the same as temporarily ignoring the y-term and then solving.

$$5x = 15$$
$$x = 3$$

The x-intercept is $(3, 0)$.

15. $7x - 2y = 28$

(a) To find the y-intercept, let $x = 0$. This is the same as temporarily ignoring the x-term and then solving.

$$-2y = 28$$
$$y = -14$$

The y−intercept is $(0, -14)$.

(b) To find the x-intercept, let $y = 0$. This is the same as temporarily ignoring the y-term and then solving.

$$7x = 28$$
$$x = 4$$

The x-intercept is $(4, 0)$.

17. $-4x + 3y = 150$

(a) To find the y-intercept, let $x = 0$. This is the same as temporarily ignoring the x-term and then solving.

$$3y = 150$$
$$y = 50$$

The y-intercept is $(0, 50)$.

(b) To find the x-intercept, let $y = 0$. This is the same as temporarily ignoring the y-term and then solving.

$$-4x = 150$$
$$x = -\frac{75}{2}$$

The x-intercept is $\left(-\frac{75}{2}, 0\right)$.

19. $y = 9$

Observe that this is the equation of a horizontal line 9 units above the x-axis. Thus, (a) the y-intercept is $(0, 9)$ and (b) there is no x-intercept.

21. $x = -7$

Observe that this is the equation of a vertical line 7 units to the left of the y-axis. Thus, (a) there is no y-intercept and (b) the x-intercept is $(-7, 0)$.

23. $3x + 2y = 12$

Find the y-intercept:

$$2y = 12 \quad \text{Ignoring the } x\text{-term}$$
$$y = 6$$

The y-intercept is $(0, 6)$.

Find the x-intercept:

$$3x = 12 \quad \text{Ignoring the } y\text{-term}$$
$$x = 4$$

The x-intercept is $(4, 0)$.

To find a third point we replace x with 2 and solve for y.

$$3 \cdot 2 + 2y = 12$$
$$6 + 2y = 12$$
$$2y = 6$$
$$y = 3$$

The point $(2, 3)$ appears to line up with the intercepts, so we draw the graph.

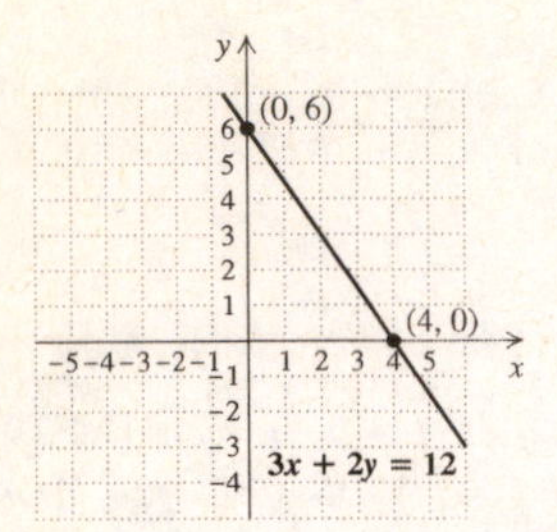

25. $x + 3y = 6$

Find the y-intercept:

$3y = 6$ Ignoring the x-term
$y = 2$

The y-intercept is $(0, 2)$.

Find the x-intercept:

$x = 6$ Ignoring the y-term

The x-intercept is $(6, 0)$.

To find a third point we replace x with 3 and solve for y.

$$3 + 3y = 6$$
$$3y = 3$$
$$y = 1$$

The point $(3, 1)$ appears to line up with the intercepts, so we draw the graph.

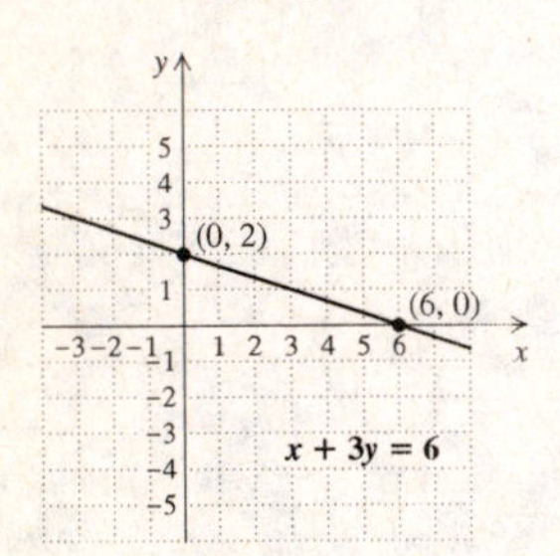

27. $-x + 2y = 8$

Find the y-intercept:

$2y = 8$ Ignoring the x-term
$y = 4$

The y-intercept is $(0, 4)$.

Find the x-intercept:

$-x = 8$ Ignoring the y-term
$x = -8$

The x-intercept is $(-8, 0)$.

To find a third point we replace x with 4 and solve for y.

$$-4 + 2y = 8$$
$$2y = 12$$
$$y = 6$$

The point $(4, 6)$ appears to line up with the intercepts, so we draw the graph.

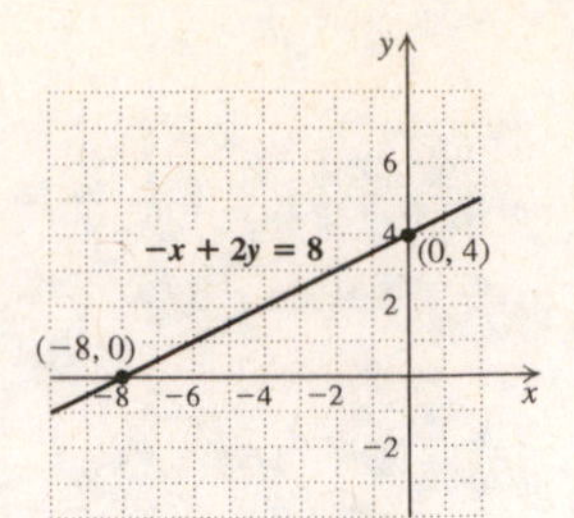

29. $3x + y = 9$

Find the y-intercept:

$y = 9$ Ignoring the x-term

The y-intercept is $(0, 9)$.

Find the $x-$intercept:

$3x = 9$ Ignoring the y-term
$x = 3$

The x-intercept is $(3, 0)$.

To find a third point we replace x with 2 and solve for y.

$$3 \cdot 2 + y = 9$$
$$6 + y = 9$$
$$y = 3$$

The point $(2, 3)$ appears to line up with the intercepts, so we draw the graph.

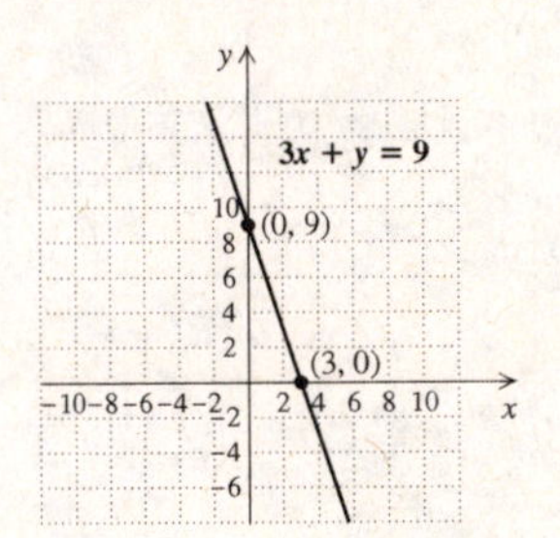

31. $y = 2x - 6$

Find the y-intercept:

$y = -6$ Ignoring the x-term

The y-intercept is $(0, -6)$.

Find the $x-$intercept:

$0 = 2x - 6$ Replacing y with 0
$6 = 2x$
$3 = x$

The x-intercept is $(3, 0)$.

To find a third point we replace x with 2 and find y.

$$y = 2 \cdot 2 - 6 = 4 - 6 = -2$$

The point $(2, -2)$ appears to line up with the intercepts, so we draw the graph.

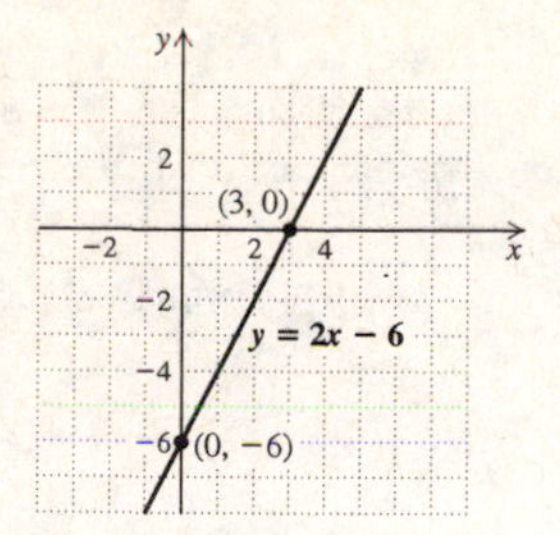

33. $3x - 9 = 3y$

Find the y-intercept:

$$-9 = 3y \quad \text{Ignoring the } x\text{-term}$$
$$-3 = y$$

The y-intercept is $(0, -3)$.

To find the x-intercept, let $y = 0$.

$$3x - 9 = 3 \cdot 0$$
$$3x - 9 = 0$$
$$3x = 9$$
$$x = 3$$

The x-intercept is $(3, 0)$.

To find a third point we replace x with 1 and solve for y.

$$3 \cdot 1 - 9 = 3y$$
$$3 - 9 = 3y$$
$$-6 = 3y$$
$$-2 = y$$

The point $(1, -2)$ appears to line up with the intercepts, so we draw the graph.

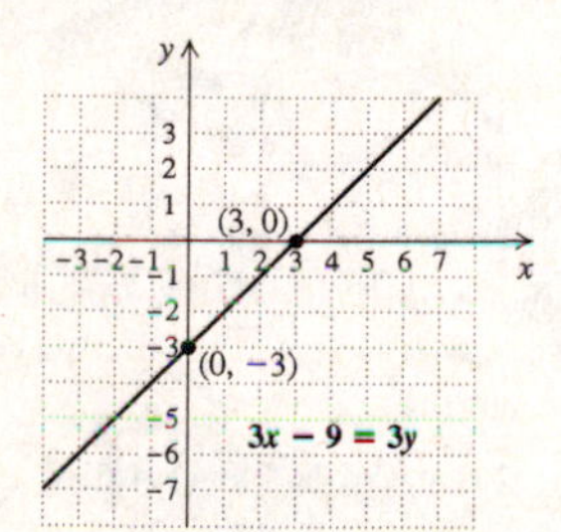

35. $2x - 3y = 6$

Find the y-intercept:

$$-3y = 6 \quad \text{Ignoring the } x\text{-term}$$
$$y = -2$$

The y-intercept is $(0, -2)$.

Find the x-intercept:

$$2x = 6 \quad \text{Ignoring the } y\text{-term}$$
$$x = 3$$

The x-intercept is $(3, 0)$.

To find a third point we replace x with -3 and solve for y.

$$2(-3) - 3y = 6$$
$$-6 - 3y = 6$$
$$-3y = 12$$
$$y = -4$$

The point $(-3, -4)$ appears to line up with the intercepts, so we draw the graph.

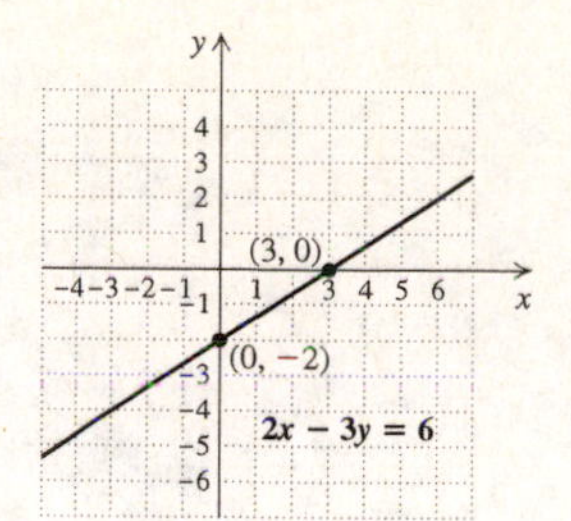

37. $4x + 5y = 20$

Find the y-intercept:

$$5y = 20 \quad \text{Ignoring the } x\text{-term}$$
$$y = 4$$

The y-intercept is $(0, 4)$.

Find the x-intercept:

$$4x = 20 \quad \text{Ignoring the } y\text{-term}$$
$$x = 5$$

The x-intercept is $(5, 0)$.

To find a third point we replace x with 4 and solve for y.

$$4 \cdot 4 + 5y = 20$$
$$16 + 5y = 20$$
$$5y = 4$$
$$y = \frac{4}{5}$$

The point $\left(4, \frac{4}{5}\right)$ appears to line up with the intercepts, so we draw the graph.

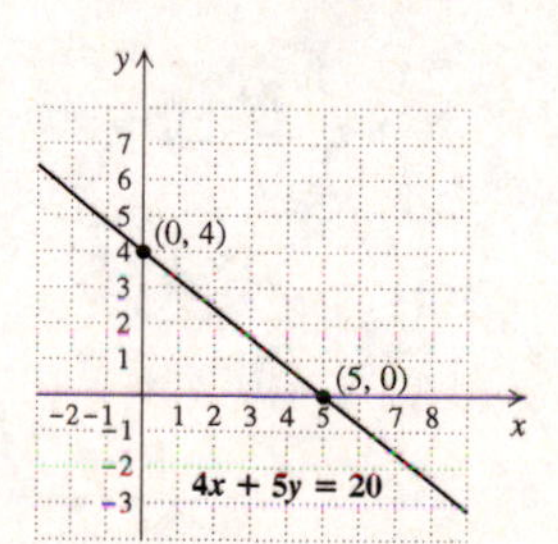

39. $3x + 2y = 8$

Find the y-intercept:

$$2y = 8 \quad \text{Ignoring the } x\text{-term}$$
$$y = 4$$

The y-intercept is $(0, 4)$.

Find the x-intercept:

$$3x = 8 \quad \text{Ignoring the } y\text{-term}$$
$$x = \frac{8}{3}$$

The x-intercept is $\left(\frac{8}{3}, 0\right)$.

To find a third point we replace x with 2 and solve for y.

$$3 \cdot 2 + 2y = 8$$
$$6 + 2y = 8$$
$$2y = 2$$
$$y = 1$$

The point $(2, 1)$ appears to line up with the intercepts, so we draw the graph.

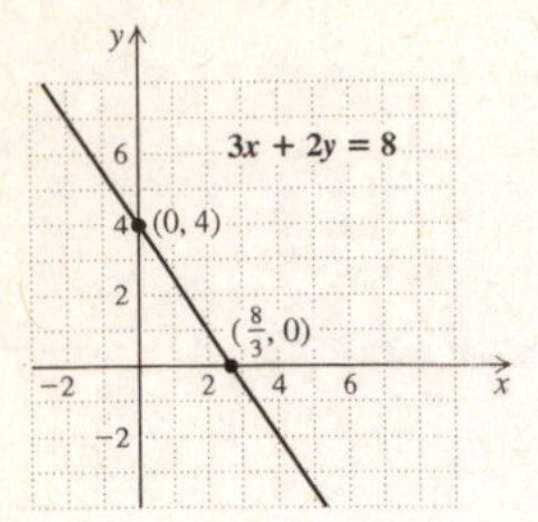

41. $2x + 4y = 6$

Find the y-intercept:

$$4y = 6 \quad \text{Ignoring the } x\text{-term}$$
$$y = \frac{3}{2}$$

The y-intercept is $\left(0, \frac{3}{2}\right)$.

Find the x-intercept:

$$2x = 6 \quad \text{Ignoring the } y\text{-term}$$
$$x = 3$$

The x-intercept is $(3, 0)$.

To find a third point we replace x with -3 and solve for y.

$$2(-3) + 4y = 6$$
$$-6 + 4y = 6$$
$$4y = 12$$
$$y = 3$$

The point $(-3, 3)$ appears to line up with the intercepts, so we draw the graph.

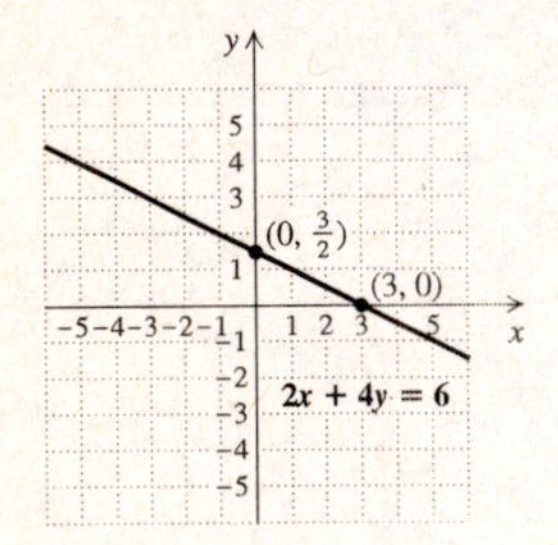

43. $5x + 3y = 180$

Find the y-intercept:

$$3y = 180 \quad \text{Ignoring the } x\text{-term}$$
$$y = 60$$

The y-intercept is $(0, 60)$.

Find the x-intercept:

$$5x = 180 \quad \text{Ignoring the } y\text{-term}$$
$$x = 36$$

The x-intercept is $(36, 0)$.

To find a third point we replace x with 6 and solve for y.

$$5 \cdot 6 + 3y = 180$$
$$30 + 3y = 180$$
$$3y = 150$$
$$y = 50$$

This means that $(6, 50)$ is on the graph.

To graph all three points, the y-axis must go to at least 60 and the x-axis must go to at least 36. Using a scale of 10 units per square allows us to display both intercepts and $(6, 50)$ as well as the origin.

The point $(6, 50)$ appears to line up with the intercepts, so we draw the graph.

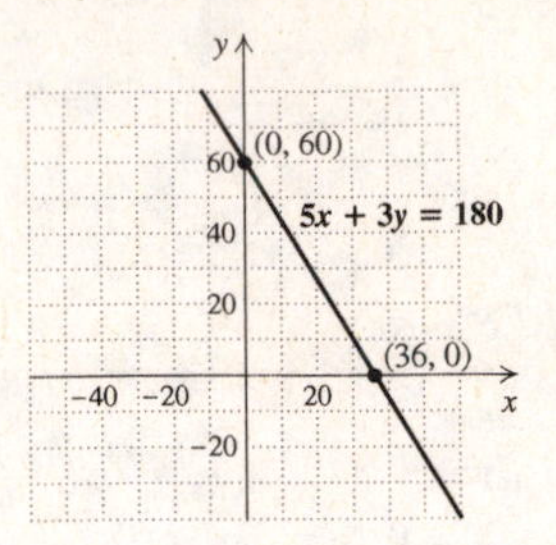

45. $y = -30 + 3x$

Find the y-intercept:

$$y = -30 \quad \text{Ignoring the } x\text{-term}$$

The y-intercept is $(0, -30)$.

To find the x-intercept, let $y = 0$.

$$0 = -30 + 3x$$
$$30 = 3x$$
$$10 = x$$

The x-intercept is $(10, 0)$.

To find a third point we replace x with 5 and solve for y.

$$y = -30 + 3 \cdot 5$$
$$y = -30 + 15$$
$$y = -15$$

This means that $(5, -15)$ is on the graph.

To graph all three points, the y-axis must go to at least -30 and the x-axis must go to at least 10. Using a scale of 5 units per square allows us to display both intercepts and $(5, -15)$ as well as the origin.

The point $(5, -15)$ appears to line up with the intercepts, so we draw the graph.

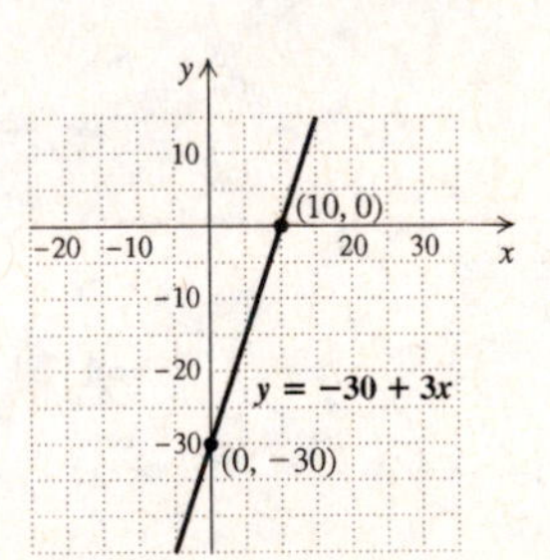

47. $-4x = 20y + 80$

To find the y-intercept, we let $x = 0$.

$$-4 \cdot 0 = 20y + 80$$
$$0 = 20y + 80$$
$$-80 = 20y$$
$$-4 = y$$

The y-intercept is $(0, -4)$.

Find the x-intercept:

$$\begin{aligned} -4x &= 80 \quad \text{Ignoring the } y\text{-term} \\ x &= -20 \end{aligned}$$

The x-intercept is $(-20, 0)$.

To find a third point we replace x with -40 and solve for y.

$$\begin{aligned} -4(-40) &= 20y + 80 \\ 160 &= 20y + 80 \\ 80 &= 20y \\ 4 &= y \end{aligned}$$

This means that $(-40, 4)$ is on the graph.

To graph all three points, the y-axis must go at least from -4 to 4 and the x-axis must go at least from -40 to -20. Since we also want to include the origin we can use a scale of 10 units per square on the x-axis and 1 unit per square on the y-axis.

The point $(-40, 4)$ appears to line up with the intercepts, so we draw the graph.

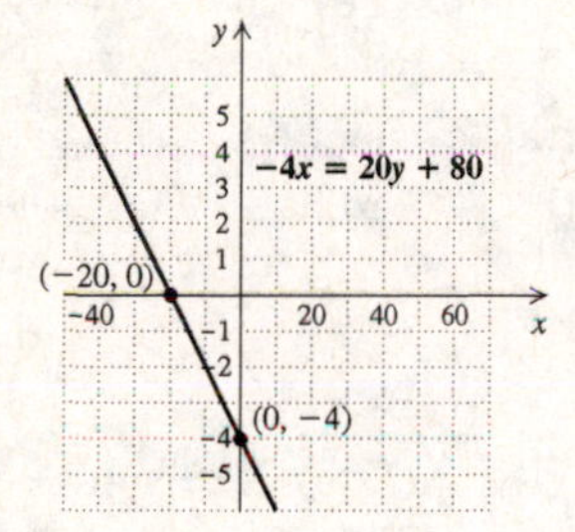

49. $y - 3x = 0$

Find the y-intercept:

$$y = 0 \quad \text{Ignoring the } x\text{-term}$$

The y-intercept is $(0, 0)$. Note that this is also the x-intercept.

In order to graph the line, we will find a second point.

When $x = 1$,
$$\begin{aligned} y - 3 \cdot 1 &= 0 \\ y - 3 &= 0 \\ y &= 3 \end{aligned}$$

To find a third point we replace $x = -1$ and solve for y.

$$\begin{aligned} y - 3(-1) &= 0 \\ y + 3 &= 0 \\ y &= -3 \end{aligned}$$

The point $(-1, -3)$ appears to line up with the other two points, so we draw the graph.

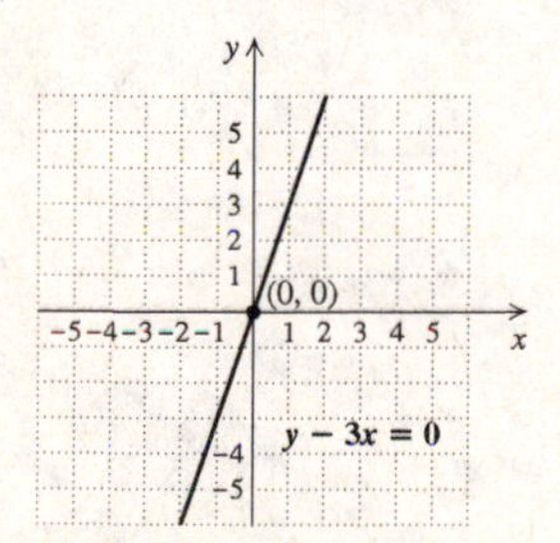

51. $y = 5$

Any ordered pair $(x, 5)$ is a solution. The variable y must be 5, but the x variable can be any number we choose. A few solutions are listed below. Plot these points and draw the line.

x	y
-3	5
0	5
2	5

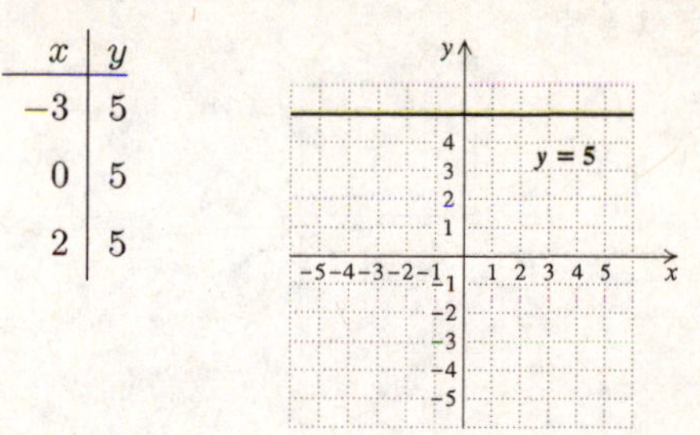

53. $x = 4$

Any ordered pair $(4, y)$ is a solution. The variable x must be 4, but the y variable can be any number we choose. A few solutions are listed below. Plot these points and draw the line.

x	y
4	-2
4	0
4	4

55. $y = -2$

Any ordered pair $(x, -2)$ is a solution. The variable y must be -2, but the x variable can be any number we choose. A few solutions are listed below. Plot these points and draw the line.

x	y
-3	-2
0	-2
4	-2

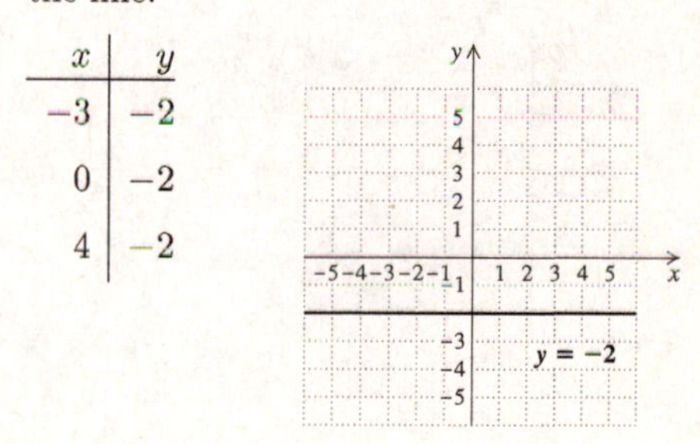

57. $x = -1$

Any ordered pair $(-1, y)$ is a solution. The variable x must be -1, but the y variable can be any number we choose. A few solutions are listed below. Plot these points and draw the line.

x	y
-1	-3
-1	0
-1	2

59. $x = 18$

Any ordered pair $(18, y)$ is a solution. A few solutions are listed below. Plot these points and draw the line choosing an appropriate scale.

x	y
18	-1
18	4
18	5

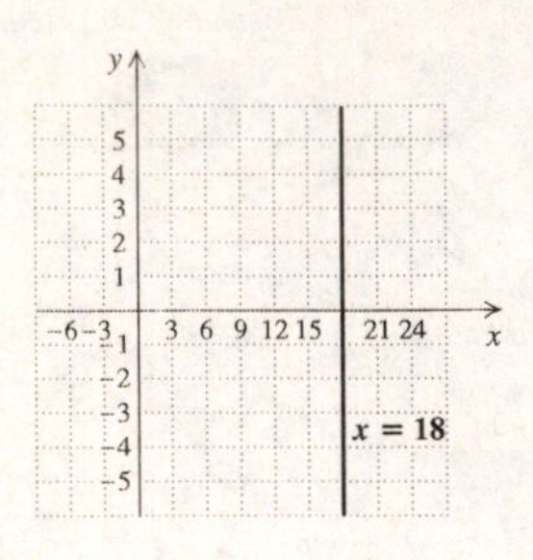

61. $y = 0$

Any ordered pair $(x, 0)$ is a solution. A few solutions are listed below. Plot these points and draw the line.

x	y
-4	0
0	0
2	0

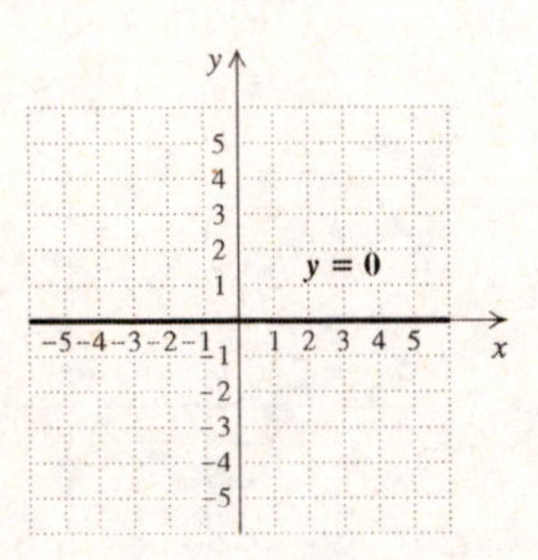

63. $x = -\dfrac{5}{2}$

Any ordered pair $\left(-\dfrac{5}{2}, y\right)$ is a solution. A few solutions are listed below. Plot these points and draw the line.

x	y
$-\dfrac{5}{2}$	-3
$-\dfrac{5}{2}$	0
$-\dfrac{5}{2}$	5

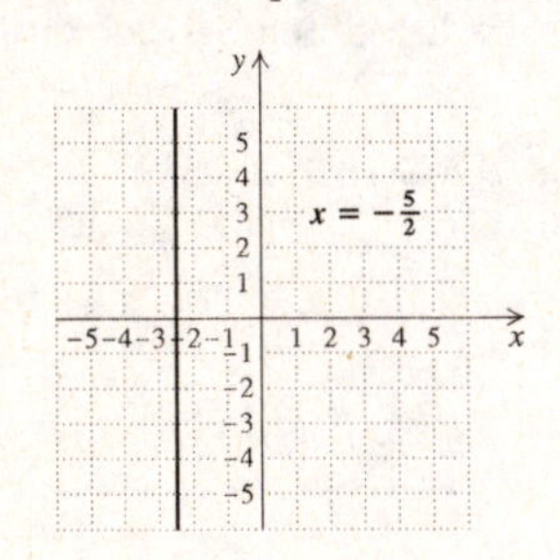

65.
$$-5y = -300$$
$$y = 60 \qquad \text{Dividing by } -5$$

The graph is a horizontal line 60 units above the x-axis.

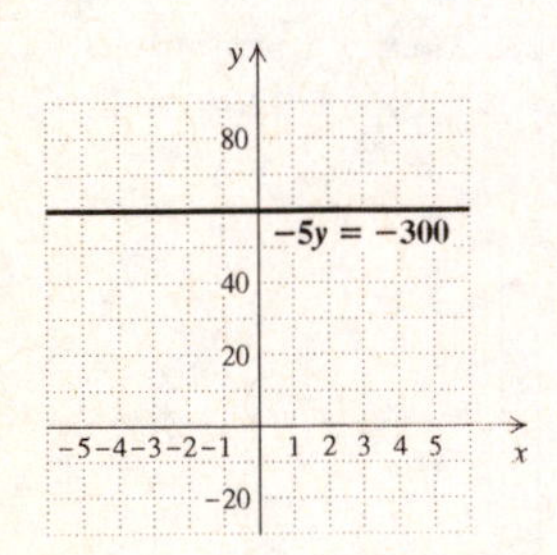

67.
$$35 + 7y = 0$$
$$7y = -35$$
$$y = -5$$

The graph is a horizontal line 5 units below the x-axis.

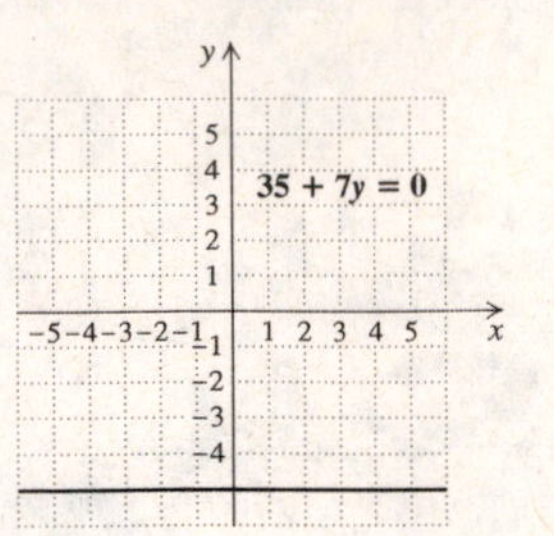

69. Note that every point on the horizontal line passing through $(0, -1)$ has -1 as the y-coordinate. Thus, the equation of the line is $y = -1$.

71. Note that every point on the vertical line passing through $(4, 0)$ has 4 as the x-coordinate. Thus, the equation of the line is $x = 4$.

73. Note that every point on the horizontal line passing through $(0, 0)$ has 0 as the y-coordinate. Thus, the equation of the line is $y = 0$.

75. *Writing Exercise*

77. $d - 7$

79. Let x represent the number. Then we have $2 + x$, or $x + 2$.

81. Let x and y represent the numbers. Then we have $2(x + y)$.

83. *Writing Exercise*

85. The x-axis is a horizontal line, so it is of the form $y = b$. All points on the x-axis are of the form $(x, 0)$, so b must be 0 and the equation is $y = 0$.

87. A line parallel to the y-axis has an equation of the form $x = a$. Since the x-coordinate of one point on the line is -2, then $a = -2$ and the equation is $x = -2$.

89. Since the x-coordinate of the point of intersection must be -3 and y must equal x, the point of intersection is $(-3, -3)$.

91. The y-intercept is $(0, 5)$, so we have $y = mx + 5$. Another point on the line is $(-3, 0)$ so we have
$$0 = m(-3) + 5$$
$$-5 = -3m$$
$$\frac{5}{3} = m$$

The equation is $y = \dfrac{5}{3}x + 5$, or $5x - 3y = -15$, or $-5x + 3y = 15$.

93. Substitute 0 for x and -8 for y.
$$4 \cdot 0 = C - 3(-8)$$
$$0 = C + 24$$
$$-24 = C$$

95. Find the y-intercept:

$2y = 50$ Covering the x-term

$y = 25$

The y-intercept is $(0, 25)$.

Find the x-intercept:

$3x = 50$ Covering the y-term

$x = \frac{50}{3} = 16.\overline{6}$

The x-intercept is $\left(\frac{50}{3}, 0\right)$, or $(16.\overline{6}, 0)$.

97. From the equation we see that the y-intercept is $(0, -9)$.

To find the x-intercept, let $y = 0$.

$0 = 0.2x - 9$

$9 = 0.2x$

$45 = x$

The x-intercept is $(45, 0)$.

99. Find the y-intercept.

$-20y = 1$ Covering the x-term

$y = -\frac{1}{20}$, or -0.05

The y-intercept is $\left(0, -\frac{1}{20}\right)$, or $(0, -0.05)$.

Find the x-intercept:

$25x = 1$ Covering the y-term

$x = \frac{1}{25}$, or 0.04

The x-intercept is $\left(\frac{1}{25}, 0\right)$, or $(0.04, 0)$.

Exercise Set 3.4

1. a) We divide the number of miles traveled by the number of gallons of gas used for that amount of driving.

Rate, in miles per gallon

$= \frac{14,014 \text{ mi} - 13,741 \text{ mi}}{13 \text{ gal}}$

$= \frac{273 \text{ mi}}{13 \text{ gal}}$

$= 21$ mi/gal

$= 21$ miles per gallon

b) We divide the cost of the rental by the number of days. From June 5 to June 8 is $8 - 5$, or 3 days.

Average cost, in dollars per day

$= \frac{118 \text{ dollars}}{3 \text{ days}}$

≈ 39.33 dollars/day

$\approx \$39.33$ per day

c) We divide the number of miles traveled by the number of days. In part (a) we found that the van was driven 273 mi, and in part (b) we found that it was rented for 3 days.

Rate, in miles per day

$= \frac{273 \text{ mi}}{3 \text{ days}}$

$= 91$ mi/day

$= 91$ mi per day

d) Note that $\$118 = 11,800¢$. From part (a) we know that the van was driven 273 mi.

Rate, in cents per mile $= \frac{11,800¢}{273 \text{ mi}}$

$\approx 43¢$ per mi

3. a) From 2:00 to 5:00 is $5 - 2$, or 3 hr.

Average speed, in miles per hour

$= \frac{18 \text{ mi}}{3 \text{ hr}}$

$= 6$ mph

b) From part (a) we know that the bike was rented for 3 hr.

Rate, in dollars per hour $= \frac{\$12}{3 \text{ hr}}$

$= \$4$ per hr

c) Rate, in dollars per mile $= \frac{\$12}{18 \text{ mi}}$

$\approx \$0.67$ per mile

5. a) It is 3 hr from 9:00 A.M. to noon and 5 more hours from noon to 5:00 P.M., so the typist worked $3 + 5$, or 8 hr.

Rate, in dollars per hour $= \frac{\$128}{8 \text{ hr}}$

$= \$16$ per hr

b) The number of pages typed is $48 - 12$, or 36.

In part (a) we found that the typist worked 8 hr.

Rate, in pages per hour $= \frac{36 \text{ pages}}{8 \text{ hr}}$

$= 4.5$ pages per hr

c) In part (b) we found that 36 pages were typed.

Rate, in dollars per page $= \frac{\$128}{36 \text{ pages}}$

$\approx \$3.56$ per page

7. The tuition increased $\$1359 - \$1327$, or $\$32$, in $2001 - 1999$ or 2 yr.

Rate of increase $= \frac{\text{Change in tuition}}{\text{Change in time}}$

$= \frac{\$32}{2 \text{ yr}}$

$= \$16$ per yr

9. a) The elevator traveled $34 - 5$, or 29 floors in 2:40 – 2:38, or 2 min.

Average rate of travel $= \frac{29 \text{ floors}}{2 \text{ min}}$

$= 14.5$ floors per min

b) In part (a) we found that the elevator traveled 29 floors in 2 min. Note that 2 min = 2×1 min = 2×60 sec = 120 sec.

$$\text{Average rate of travel} = \frac{120 \text{ sec}}{29 \text{ floors}}$$
$$\approx 4.14 \text{ sec per floor}$$

11. a) Krakauer ascended 29,028 ft − 27,600 ft, or 1428 ft. From 7:00 A.M. to noon it is 5 hr = 5×1 hr = 5×60 min = 300 min. From noon to 1:25 P.M. is another 1 hr, 25 min, or 1 hr + 25 min = 60 min + 25 min = 85 min. The total time of the ascent is 300 min + 85 min, or 385 min.

$$\text{Rate, in feet per minute} = \frac{1428 \text{ ft}}{385 \text{ min}}$$
$$\approx 3.71 \text{ ft per min}$$

b) We use the information found in part (a).

$$\text{Rate, in minutes per foot} = \frac{385 \text{ min}}{1428 \text{ ft}}$$
$$\approx 0.27 \text{ min per ft}$$

13. The rate of increase of the average copayment is given in dollars per year, so we list Average copayment on the vertical axis and Year on the horizontal axis. If we count by increments of 10 on the vertical axis we can easily reach 19 and beyond. We label the units on the vertical axis in dollars. We list the years on the horizontal axis, beginning with 2003.

We plot the point (2003, \$19). Then, to display the rate of growth, we move from that point to a point that represents a copayment of \$2 more one year later. The coordinates of this point are (2003 + 1, \$19 + \$2), or (2004, \$21). Finally, we draw a line through the two points.

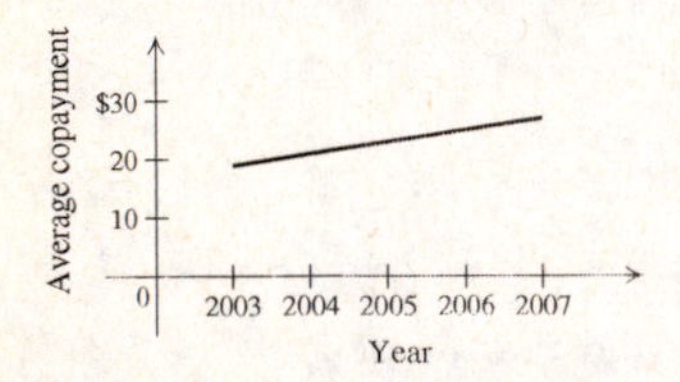

15. The rate is given in millions of crimes per year, so we list Number of crimes, in millions, on the vertical axis and Year on the horizontal axis. If we count by 10's of millions on the vertical axis we can easily reach 26 million and beyond. We plot the point (2000, 26 million). Then, to display the rate of growth, we move from that point to a point that represents 1.2 million fewer crimes 1 year later. The coordinates of this point are (2000 + 1, 26 − 1.2 million), or (2001, 24.8 million). Finally, we draw a line through the two points.

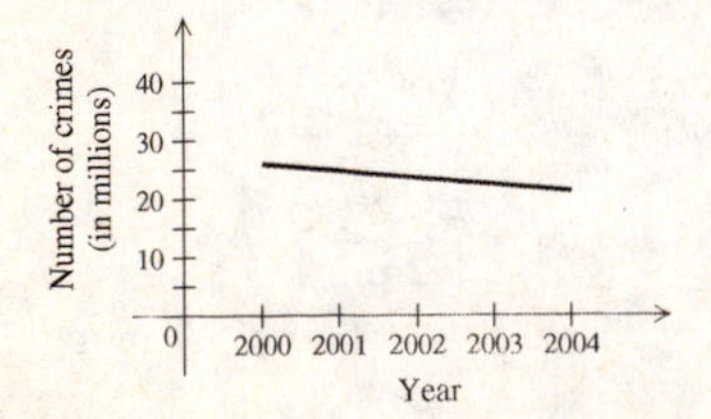

17. The rate is given in miles per hour, so we list the number of miles traveled on the vertical axis and the time of day on the horizontal axis. If we count by 100's of miles on the vertical axis we can easily reach 230 without needing a terribly large graph. We plot the point (3:00, 230). Then to display the rate of travel, we move from that point to a point that represents 90 more miles traveled 1 hour later. The coordinates of this point are (3:00 + 1 hr, 230 + 90), or (4:00, 320). Finally, we draw a line through the two points.

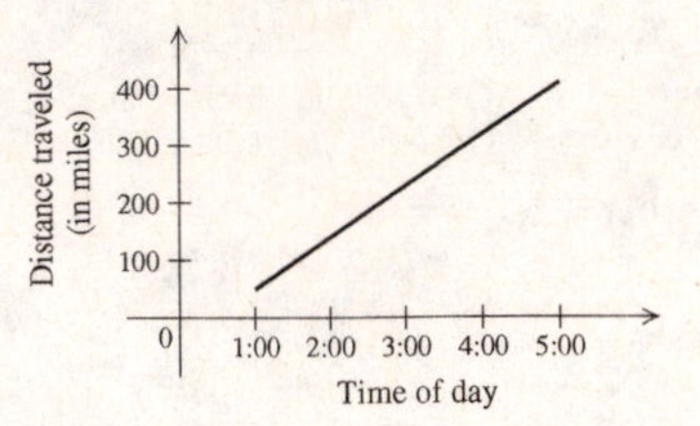

19. The rate is given in dollars per hour so we list money earned on the vertical axis and the time of day on the horizontal axis. We can count by \$20 on the vertical axis and reach \$50 without needing a terribly large graph. Next we plot the point (2:00 P.M., \$50). To display the rate we move from that point to a point that represents \$15 more 1 hour later. The coordinates of this point are (2+1, \$50+ \$15), or (3:00 P.M., \$65). Finally, we draw a line through the two points.

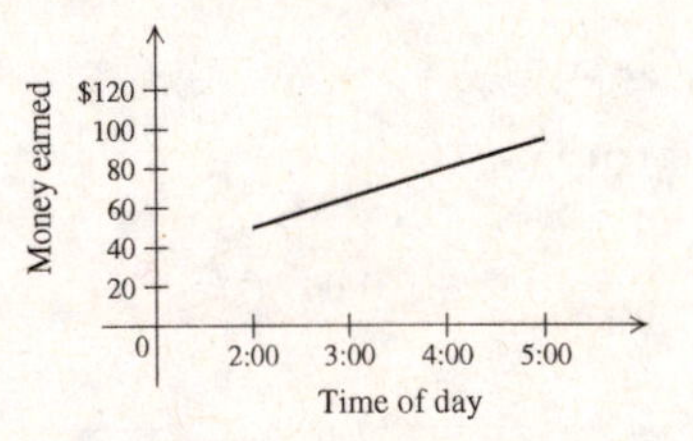

21. The rate is given in cost per minute so we list the amount of the telephone bill on the vertical axis and the number of additional minutes on the horizontal axis. We begin with \$7.50 on the vertical axis and count by \$0.50. A jagged line at the base of the axis indicates that we are not showing amounts smaller than \$7.50. We begin with 0 additional minutes on the horizontal axis and plot the point (0, \$7.50). We move from there to a point that represents \$0.10 more 1 minute later. The coordinates of this point are (0 + 1 min, \$7.50 + \$0.10), or (1 min, \$7.60). Then we draw a line through the two points.

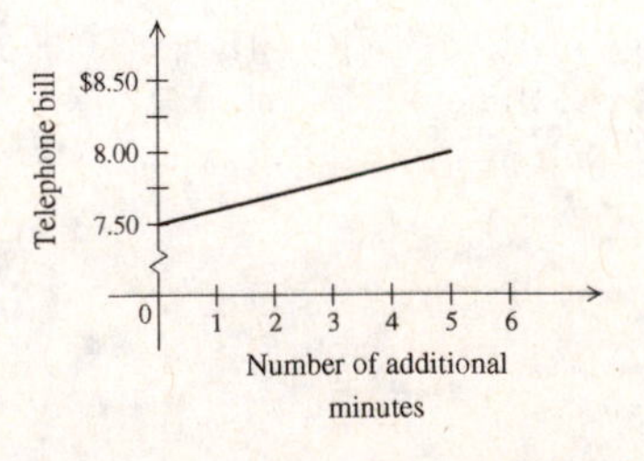

23. The points (2:00, 7 haircuts) and (4:30, 12 haircuts) are on the graph. This tells us that in the 2.5 hr between 2:00 and 4:30 there were 12 − 7 = 5 haircuts completed. The rate is

$\frac{5 \text{ haircuts}}{2.5 \text{ hr}} = 2$ haircuts per hour.

25. The points (12:00, 100 mi) and (2:00, 250 mi) are on the graph. This tells us that in the 2 hr between 12:00 and 2:00 the train traveled $250 - 100 = 150$ mi. The rate is

$\frac{150 \text{ mi}}{2\text{hr}} = 75$ mi per hr.

27. The points (5 min, 35¢) and (10 min, 70¢) are on the graph. This tells us that in $10 - 5 = 5$ min the cost of the call increased 70¢ − 35¢ = 35¢. The rate is

$\frac{35¢}{5 \text{ min}} = 7¢$ per min.

29. The points (2 yr, \$2000) and (4 yr, \$1000) are on the graph. This tells us that in $4 - 2 = 2$ yr the value of the copier changes \$1000 − \$2000 = −\$1000. The rate is

$\frac{-\$1000}{2 \text{ yr}} = -\$500$ per yr.

This means that the value of the copier is decreasing at a rate of \$500 per yr.

31. The points (50 mi, 2 gal) and (200 mi, 8 gal) are on the graph. This tells us that when driven $200 - 50 = 150$ mi the vehicle consumed $8 - 2 = 6$ gal of gas. The rate is

$\frac{6 \text{ gal}}{150 \text{ mi}} = 0.04$ gal per mi.

33. Since swimming is the slowest of the three sports and biking is the fastest, the slope of the line representing swimming speed will be the least steep of the three and that representing biking speed will be the steepest. The second segment of graph (e) rises most steeply and the third segment is the least steep of the three segments. Thus this graph represents running followed by biking and then swimming.

35. Since swimming is the slowest of the three sports and biking is the fastest, the slope of the line representing swimming speed will be the least steep of the three and that representing biking speed will be the steepest. The first segment of graph (d) is the least steep and the second segment is the steepest of the three segments. Thus this graph represents swimming followed by biking and then running.

37. Since swimming is the slowest of the three sports and biking is the fastest, the slope of the line representing swimming speed will be the least steep of the three and that representing biking speed will be the steepest. The first segment of graph (b) is the steepest and the second segment is the least steep of the three segments. Thus this graph represents biking followed by swimming and then running.

39. *Writing Exercise*

41. $-2 - (-7) = -2 + 7 = 5$

43. $\frac{5 - (-4)}{-2 - 7} = \frac{9}{-9} = -1$

45. $\frac{-4 - 8}{7 - (-2)} = \frac{-12}{9} = -\frac{4}{3}$

47. *Writing Exercise*

49. Let t = flight time and a = altitude. While the plane is climbing at a rate of 6500 ft/min, the equation $a = 6500t$ describes the situation. Solving $34,000 = 6500t$, we find that the cruising altitude of 34,000 ft is reached after about 5.23 min. Thus we graph $a = 6500t$ for $0 \le t \le 5.23$.

The plane cruises at 34,000 ft for 3 min, so we graph $a = 34,000$ for $5.23 < t \le 8.23$. After 8.23 min the plane descends at a rate of 3500 ft/min and lands. The equation $a = 34,000 - 3500(t - 8.23)$, or $a = -3500t + 62,805$, describes this situation. Solving $0 = -3500t + 62,805$, we find that the plane lands after about 17.94 min. Thus we graph $a = -3500t + 62,805$ for $8.23 < t \le 17.94$. The entire graph is show below.

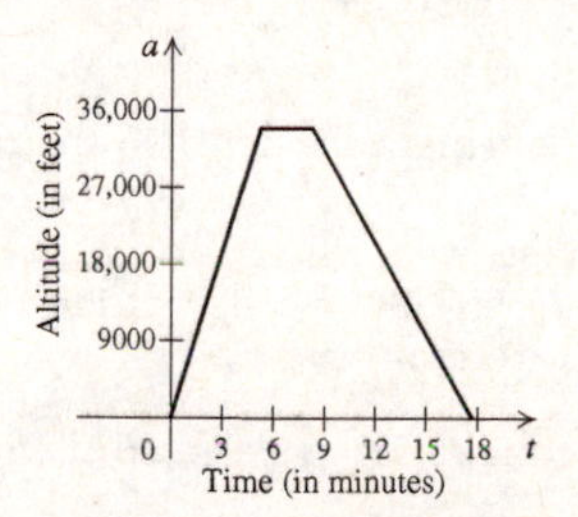

51. Let the horizontal axis represent the distance traveled, in miles, and let the vertical axis represent the fare, in dollars. Use increments of 1/5, or 0.2 mi, on the horizontal axis and of \$1 on the vertical axis. The fare for traveling 0.2 mi is $\$2 + \$0.50 \cdot 1$, or \$2.50 and for 0.4 mi, or 0.2 mi × 2, we have \$2 + \$0.50(2), or \$3. Plot the points (0.2 mi, \$2.50) and (0.4 mi, \$3) and draw the line through them.

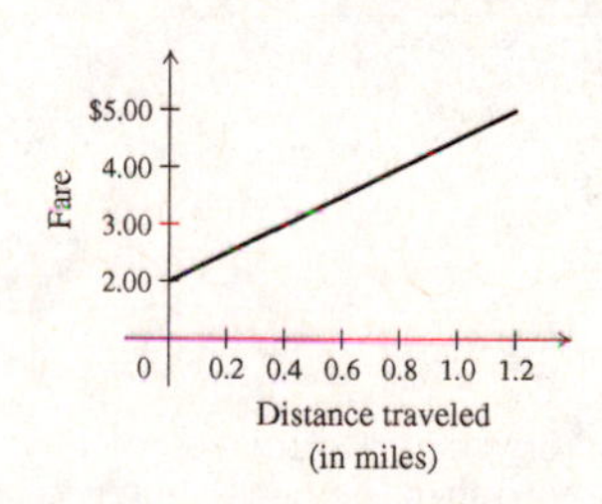

53. Penny walks forward at a rate of $\frac{24 \text{ ft}}{3 \text{ sec}}$, or 8 ft per sec. In addition, the boat is traveling at a rate of 5 ft per sec. Thus, with respect to land, Penny is traveling at a rate of $8 + 5$, or 13 ft per sec.

55. First we find Annette's speed in minutes per kilometer.

$\text{Speed} = \frac{15.5 \text{ min}}{7 \text{ km} - 4 \text{ km}} = \frac{15.5 \text{ min}}{3 \text{ km}}$

Now we convert min/km to min/mi.

$\frac{15.5 \text{ min}}{3 \text{ km}} \approx \frac{15.5 \text{ min}}{3 \text{ km}} \cdot \frac{1 \text{ km}}{0.621 \text{ min}} \approx \frac{15.5 \text{ min}}{1.863 \text{ mi}}$

At a rate of $\frac{15.5 \text{ min}}{1.863 \text{ mi}}$, to run a 5-mi race it would take

$\frac{15.5 \text{ min}}{1.863 \text{ mi}} \cdot 5 \text{ mi} \approx 41.6$ min.

(Answers may vary slightly depending on the conversion factor used.)

57. In the 2 hours from 3 P.M. to 5 P.M., the number of candles made was 100 − 46, or 54. Then candles were being made at the rate of

$$\frac{54 \text{ candles}}{2 \text{ hr}}, \text{ or 27 candles per hour.}$$

Since 82 − 46 = 36, the length of time it took to make the 82nd candle after the 46th candle was made is

$$\frac{36 \text{ candles}}{27 \text{ candles per hour}} = \frac{4}{3} \text{ hr, or 1 hr 20 min.}$$

Thus the 82nd candle was made 1 hr 20 min after 3 P.M., or at 4:20 P.M.

Exercise Set 3.5

1. A teenager's height increases over time, so the rate is positive.

3. The water level decreases during a drought, so the rate is negative.

5. The number of people present increases as the opening tipoff approaches, so the rate is positive.

7. A person's IQ does not change during sleep, so the rate is zero.

9. The inventory decreases as the sale progresses, so the rate is negative.

11. The rate can be found using the coordinates of any two points on the line. We use (2001, 285 million) and (2003, 290 million).

$$\begin{aligned}\text{Rate} &= \frac{\text{change in population}}{\text{corresponding change in time}}\\ &= \frac{290 \text{ million} - 285 \text{ million}}{2003 - 2001}\\ &= \frac{5 \text{ million}}{2 \text{ yr}}\\ &= 2.5 \text{ million people per year}\end{aligned}$$

13. The rate can be found using the coordinates of any two points on the line. We use (1992, 61%) and (2002, 74%).

$$\begin{aligned}\text{Rate} &= \frac{\text{change in percent}}{\text{corresponding change in time}}\\ &= \frac{74\% - 61\%}{2002 - 1992}\\ &= \frac{13\%}{10 \text{ yr}}\\ &= 1.3\% \text{ per year}\end{aligned}$$

15. The rate can be found using the coordinates of any two points on the line. We use (35, 490) and (45, 500), where 35 and 45 are in \$1000's.

$$\begin{aligned}\text{Rate} &= \frac{\text{change in score}}{\text{corresponding change in income}}\\ &= \frac{500 - 490 \text{ points}}{45 - 35}\\ &= \frac{10 \text{ points}}{10}\\ &= 1 \text{ point per \$1000 income}\end{aligned}$$

17. The rate can be found using the coordinates of any two points on the line. We use (0 min, 54°) and (27 min, −4°).

$$\begin{aligned}\text{Rate} &= \frac{\text{change in temperature}}{\text{corresponding change in time}}\\ &= \frac{-4^\circ - 54^\circ}{27 \text{ min} - 0 \text{ min}}\\ &= \frac{-58^\circ}{27 \text{ min}}\\ &\approx -2.1^\circ \text{per min}\end{aligned}$$

19. We can use any two points on the line, such as $(0, 1)$ and $(4, 4)$.

$$\begin{aligned}m &= \frac{\text{change in } y}{\text{change in } x}\\ &= \frac{4-1}{4-0} = \frac{3}{4}\end{aligned}$$

21. We can use any two points on the line, such as $(1, 0)$ and $(3, 3)$.

$$\begin{aligned}m &= \frac{\text{change in } y}{\text{change in } x}\\ &= \frac{3-0}{3-1} = \frac{3}{2}\end{aligned}$$

23. We can use any two points on the line, such as $(-3, -4)$ and $(0, -3)$.

$$\begin{aligned}m &= \frac{\text{change in } y}{\text{change in } x}\\ &= \frac{-3-(-4)}{0-(-3)} = \frac{1}{3}\end{aligned}$$

25. We can use any two points on the line, such as $(0, 2)$ and $(2, 0)$.

$$\begin{aligned}m &= \frac{\text{change in } y}{\text{change in } x}\\ &= \frac{2-0}{0-2} = \frac{2}{-2} = -1\end{aligned}$$

27. This is the graph of a horizontal line. Thus, the slope is 0.

29. We can use any two points on the line, such as $(0, 2)$ and $(3, 1)$.

$$\begin{aligned}m &= \frac{\text{change in } y}{\text{change in } x}\\ &= \frac{1-2}{3-0} = -\frac{1}{3}\end{aligned}$$

31. This is the graph of a vertical line. Thus, the slope is undefined.

33. We can use any two points on the line, such as $(-2, 3)$ and $(2, 2)$.

$$\begin{aligned}m &= \frac{\text{change in } y}{\text{change in } x}\\ &= \frac{2-3}{2-(-2)} = -\frac{1}{4}\end{aligned}$$

35. We can use any two points on the line, such as $(-2,-3)$ and $(2,3)$.

$$m = \frac{\text{change in } y}{\text{change in } x} = \frac{3-(-3)}{2-(-2)} = \frac{6}{4} = \frac{3}{2}$$

37. This is the graph of a horizontal line, so the slope is 0.

39. We can use any two points on the line, such as $(-3,5)$ and $(0,-4)$.

$$m = \frac{\text{change in } y}{\text{change in } x} = \frac{-4-5}{0-(-3)} = \frac{-9}{3} = -3$$

41. $(1,2)$ and $(5,8)$

$$m = \frac{8-2}{5-1} = \frac{6}{4} = \frac{3}{2}$$

43. $(-2,4)$ and $(3,0)$

$$m = \frac{4-0}{-2-3} = \frac{4}{-5} = -\frac{4}{5}$$

45. $(-4,0)$ and $(5,7)$

$$m = \frac{7-0}{5-(-4)} = \frac{7}{9}$$

47. $(0,8)$ and $(-3,10)$

$$m = \frac{8-10}{0-(-3)} = \frac{8-10}{0+3} = \frac{-2}{3} = -\frac{2}{3}$$

49. $(-2,3)$ and $(-6,5)$

$$m = \frac{5-3}{-6-(-2)} = \frac{2}{-6+2} = \frac{2}{-4} = -\frac{1}{2}$$

51. $\left(-2,\frac{1}{2}\right)$ and $\left(-5,\frac{1}{2}\right)$

Observe that the points have the same y-coordinate. Thus, they lie on a horizontal line and its slope is 0. We could also compute the slope.

$$m = \frac{\frac{1}{2}-\frac{1}{2}}{-2-(-5)} = \frac{\frac{1}{2}-\frac{1}{2}}{-2+5} = \frac{0}{3} = 0$$

53. $(3,4)$ and $(9,-7)$

$$m = \frac{-7-4}{9-3} = \frac{-11}{6} = -\frac{11}{6}$$

55. $(6,-4)$ and $(6,5)$

Observe that the points have the same x-coordinate. Thus, they lie on a vertical line and its slope is undefined. We could also compute the slope.

$$m = \frac{-4-5}{6-6} = \frac{-9}{0}, \text{ undefined}$$

57. The line $x=-3$ is a vertical line. The slope is undefined.

59. The line $y=4$ is a horizontal line. A horizontal line has slope 0.

61. The line $x=9$ is a vertical line. The slope is undefined.

63. The line $y=-9$ is a horizontal line. A horizontal line has slope 0.

65. The grade is expressed as a percent.

$$m = \frac{106}{1325} = 0.08 = 8\%$$

67. The grade is expressed as a percent.

$$m = \frac{1}{12} = 0.08\overline{3} = 8.\overline{3}\%$$

69. 2 ft 5 in. $= 2 \cdot 12$ in. $+ 5$ in. $= 24$ in. $+ 5$ in. $= 29$ in.

8 ft 2 in. $= 8 \cdot 12$ in. $+ 2$ in. $= 96$ in. $+ 2$ in. $= 98$ in.

$m = \frac{29}{98}$, or about 30%

71. Longs Peak rises $14,255 - 9600 = 4655$ ft.

$$m = \frac{4655}{15,840} \approx 0.29 \approx 29\%$$

73. *Writing Exercise*

75.

$$ax + by = c$$
$$by = c - ax \quad \text{Adding } -ax \text{ to both sides}$$
$$y = \frac{c-ax}{b} \quad \text{Dividing both sides by } b$$

77.

$$ax - by = c$$
$$-by = c - ax \quad \text{Adding } -ax \text{ to both sides}$$
$$y = \frac{c-ax}{-b} \quad \text{Dividing both sides by } -b$$

We could also express this result as $y = \frac{ax-c}{b}$.

79.

$$\frac{2}{3}x - 5 = \frac{2}{3} \cdot 12 - 5 \quad \text{Substituting}$$
$$= 8 - 5$$
$$= 3$$

81. *Writing Exercise*

83. From the dimensions on the drawing, we see that the ramps labeled A have a rise of 61 cm and a run of 167.6 cm.

$$m = \frac{61 \text{ cm}}{167.6 \text{ cm}} \approx 0.364, \text{ or } 36.4\%$$

85. If the line passes through $(2,5)$ and never enters the second quadrant, then it slants up from left to right or is vertical. This means that its slope is positive. The line slants least steeply if it passes through $(0,0)$. In this case, $m = \frac{5-0}{2-0} = \frac{5}{2}$. Thus, the numbers the line could have for it slope are $\left\{m \middle| m \geq \frac{5}{2}\right\}$.

87. Let $t =$ the number of units each tick mark on the vertical axis represents. Note that the graph drops 4 units for every 3 units of horizontal change. Then we have:

$$\frac{-4t}{3} = -\frac{2}{3}$$
$$-4t = -2 \quad \text{Multiplying by 3}$$
$$t = \frac{1}{2} \quad \text{Dividing by } -4$$

Each tick mark on the vertical axis represents $\frac{1}{2}$ unit.

Exercise Set 3.6

1. $y = 2x - 3 = 2x + (-3)$

The y-intercept is $(0, -3)$, so choice (c) is correct.

3. We can read the slope, 3, directly from the equation. Choice (f) is correct.

5. We can read the slope, $\frac{2}{3}$, directly from the equation. Choice (d) is correct.

7. Slope $\frac{2}{5}$; y-intercept $(0, 1)$

We plot $(0, 1)$ and from there move up 2 units and right 5 units. This locates the point $(5, 3)$. We plot $(5, 3)$ and draw a line passing through $(0, 1)$ and $(5, 3)$.

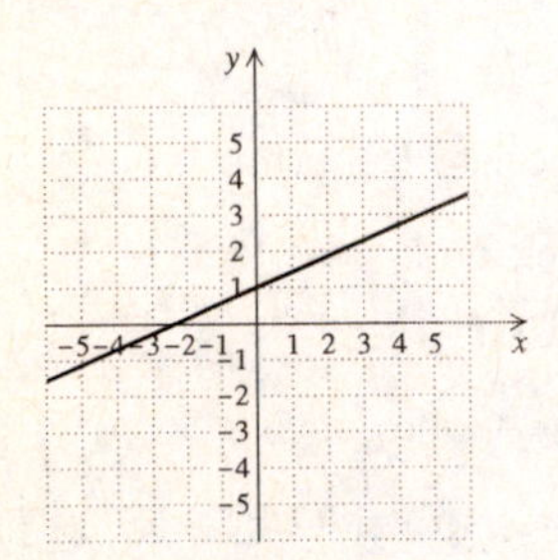

9. Slope $\frac{5}{3}$; y-intercept $(0, -2)$

We plot $(0, -2)$ and from there move up 5 units and right 3 units. This locates the point $(3, 3)$. We plot $(3, 3)$ and draw a line passing through $(0, -2)$ and $(3, 3)$.

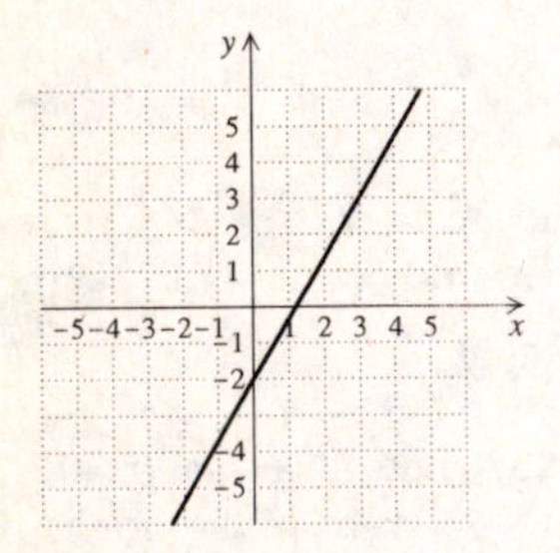

11. Slope $-\frac{3}{4}$; y-intercept $(0, 5)$

We plot $(0, 5)$. We can think of the slope as $\frac{-3}{4}$, so from $(0, 5)$ we move down 3 units and right 4 units. This locates the point $(4, 2)$. We plot $(4, 2)$ and draw a line passing through $(0, 5)$ and $(4, 2)$.

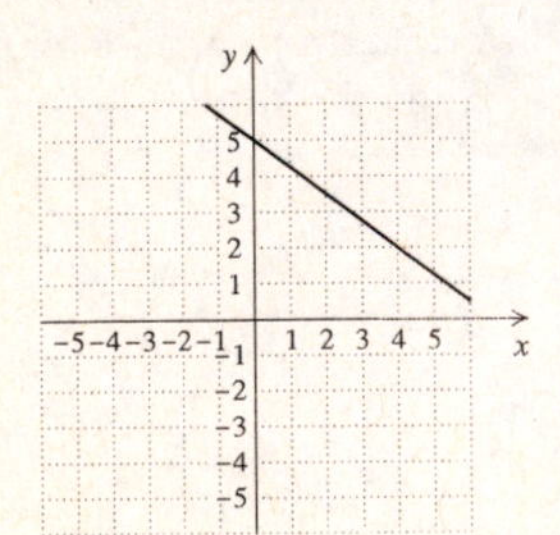

13. Slope 2; y-intercept $(0, -4)$

We plot $(0, -4)$. We can think of the slope as $\frac{2}{1}$, so from $(0, -4)$ we move up 2 units and right 1 unit. This locates the point $(1, -2)$. We plot $(1, -2)$ and draw a line passing through $(0, -4)$ and $(1, -2)$.

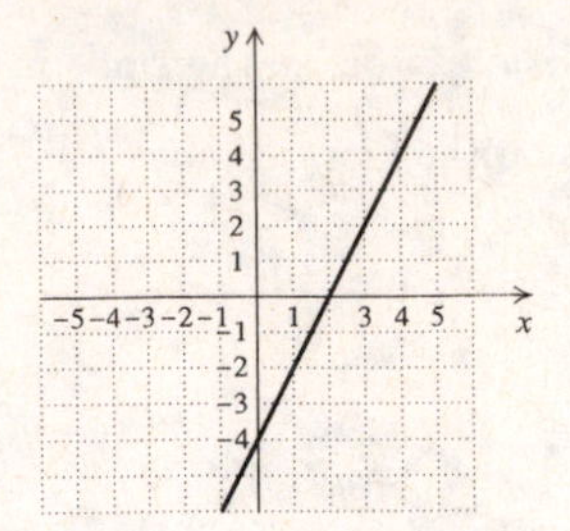

15. Slope -3; y-intercept $(0, 2)$

We plot $(0, 2)$. We can think of the slope as $\frac{-3}{1}$, so from $(0, 2)$ we move down 3 units and right 1 unit. This locates the point $(1, -1)$. We plot $(1, -1)$ and draw a line passing through $(0, 2)$ and $(1, -1)$.

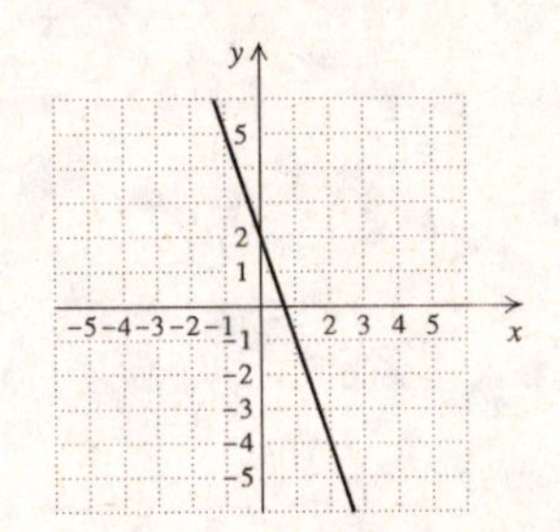

17. We read the slope and y-intercept from the equation.

$$y = -\frac{2}{7}x + 5$$

The slope is $-\frac{2}{7}$. The y-intercept is $(0, 5)$.

19. We read the slope and y-intercept from the equation.

$$y = -\frac{5}{8}x + 3$$

The slope is $-\frac{5}{8}$. The y-intercept is $(0, 3)$.

21. $y = \frac{9}{5}x - 4$

$y = \frac{9}{5}x + (-4)$

The slope is $\frac{9}{5}$, and the y-intercept is $(0, -4)$.

23. We solve for y to rewrite the equation in the form $y = mx + b$.

$-3x + y = 7$

$y = 3x + 7$

The slope is 3, and the y-intercept is $(0, 7)$.

25. $5x + 2y = 8$

$$2y = -5x + 8$$
$$y = \frac{1}{2}(-5x + 8)$$
$$y = -\frac{5}{2}x + 4$$

The slope is $-\frac{5}{2}$, and the y-intercept is $(0, 4)$.

27. Observe that this is the equation of a horizontal line that lies 4 units above the x-axis. Thus, the slope is 0, and the y-intercept is $(0, 4)$. We could also write the equation in slope-intercept form.

$$y = 4$$
$$y = 0x + 4$$

The slope is 0, and the y-intercept is $(0, 4)$.

29. $2x - 5y = -8$

$$-5y = -2x - 8$$
$$y = -\frac{1}{5}(-2x - 8)$$
$$y = \frac{2}{5}x + \frac{8}{5}$$

The slope in $\frac{2}{5}$, and the y-intercept is $\left(0, \frac{8}{5}\right)$.

31. We use the slope-intercept equation, substituting 3 for m and 7 for b:

$$y = mx + b$$
$$y = 3x + 7$$

33. We use the slope-intercept equation, substituting $\frac{7}{8}$ for m and -1 for b:

$$y = mx + b$$
$$y = \frac{7}{8}x - 1$$

35. We use the slope-intercept equation, substituting $-\frac{5}{3}$ for m and -8 for b:

$$y = mx + b$$
$$y = -\frac{5}{3}x - 8$$

37. Since the slope is 0, we know that the line is horizontal. Its y-intercept is $(0, 3)$, so the equation of the line must be $y = 3$.

We could also use the slope-intercept equation, substituting 0 for m and 3 for b.

$$y = mx + b$$
$$y = 0 \cdot x + 3$$
$$y = 3$$

39. From the graph we see that the y-intercept is $(0, 9)$. We also see that the point $(9, 17)$ is on the graph. We find the slope:

$$m = \frac{17 - 9}{9 - 0} = \frac{8}{9}$$

Substituting $\frac{8}{9}$ for m and 9 for b in the slope-intercept equation $y = mx + b$, we have

$$y = \frac{8}{9}x + 9,$$

where y is the number of gallons of bottled water consumed per person and x is the number of years since 1990.

41. From the graph we see that the y-intercept is $(0, 250)$. We also see that the point $(10, 400)$ is on the graph. We find the slope:

$$m = \frac{400 - 250}{10 - 0} = \frac{150}{10} = 15$$

Substituting 15 for m and 250 for b in the slope-intercept equation $y = mx + b$, we have

$$y = 15x + 250,$$

where y is the number of jobs for medical assistants, in thousands, and x is the number of years since 1998.

43. $y = \frac{3}{5}x + 2$

First we plot the y-intercept $(0, 2)$. We can start at the y-intercept and use the slope, $\frac{3}{5}$, to find another point. We move up 3 units and right 5 units to get a new point $(5, 5)$. Thinking of the slope as $\frac{-3}{-5}$ we can start at $(0, 2)$ and move down 3 units and left 5 units to get another point $(-5, -1)$.

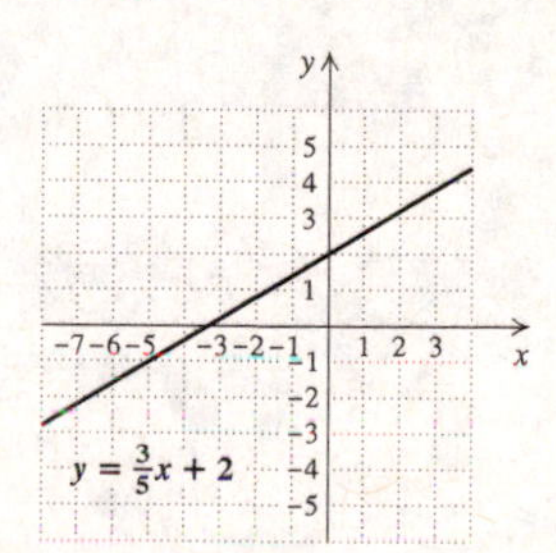

45. $y = -\frac{3}{5}x + 1$

First we plot the y-intercept $(0, 1)$. We can start at the y-intercept and, thinking of the slope as $\frac{-3}{5}$, find another point by moving down 3 units and right 5 units to the point $(5, -2)$. Thinking of the slope as $\frac{3}{-5}$ we can start at $(0, 1)$ and move up 3 units and left 5 units to get another point $(-5, 4)$.

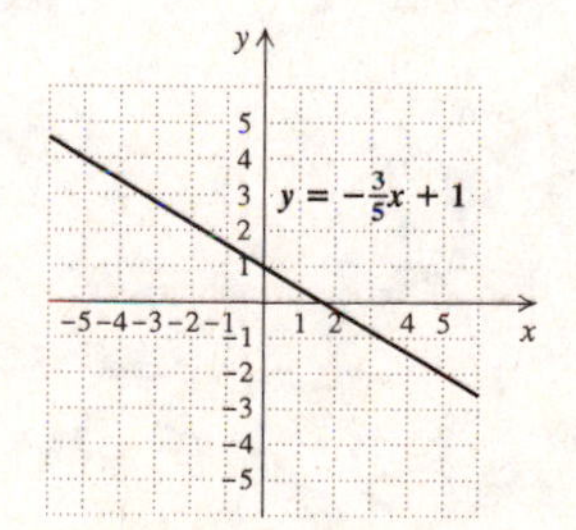

47. $y = \frac{5}{3}x + 3$

First we plot the y-intercept $(0,3)$. We can start at the y-intercept and use the slope, $\frac{5}{3}$, to find another point. We move up 5 units and right 3 units to get a new point $(3,8)$. Thinking of the slope as $\frac{-5}{-3}$ we can start at $(0,3)$ and move down 5 units and left 3 units to get another point $(-3,-2)$.

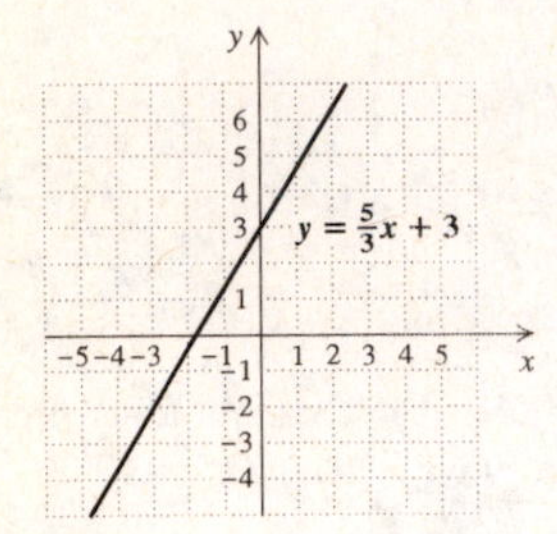

49. $y = -\frac{3}{2}x - 2$

First we plot the y-intercept $(0,-2)$. We can start at the y-intercept and, thinking of the slope as $\frac{-3}{2}$, find another point by moving down 3 units and right 2 units to the point $(2,-5)$. Thinking of the slope as $\frac{3}{-2}$ we can start at $(0,-2)$ and move up 3 units and left 2 units to get another point $(-2,1)$.

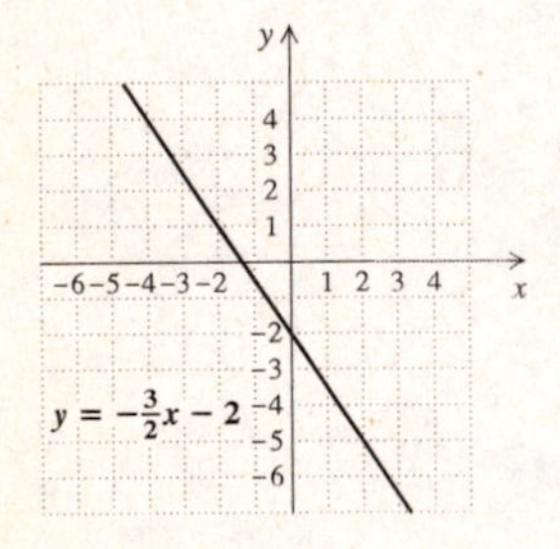

51. We first rewrite the equation in slope-intercept form.

$$2x + y = 1$$
$$y = -2x + 1$$

Now we plot the y-intercept $(0,1)$. We can start at the y-intercept and, thinking of the slope as $\frac{-2}{1}$, find another point by moving down 2 units and right 1 unit to the point $(1,-1)$. In a similar manner, we can move from the point $(1,-1)$ to find a third point $(2,-3)$.

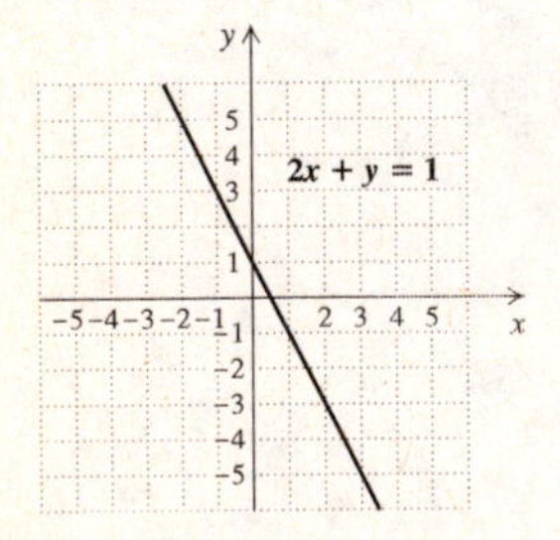

53. We first rewrite the equation in slope-intercept form.

$$3x + y = 0$$
$$y = -3x, \text{ or } y = -3x + 0$$

Now we plot the y-intercept $(0,0)$. We can start at the y-intercept and, thinking of the slope as $\frac{-3}{1}$, find another point by moving down 3 units and right 1 unit to the point $(1,-3)$. Thinking of the slope as $\frac{3}{-1}$ we can start at $(0,0)$ and move up 3 units and left 1 unit to get another point $(-1,3)$.

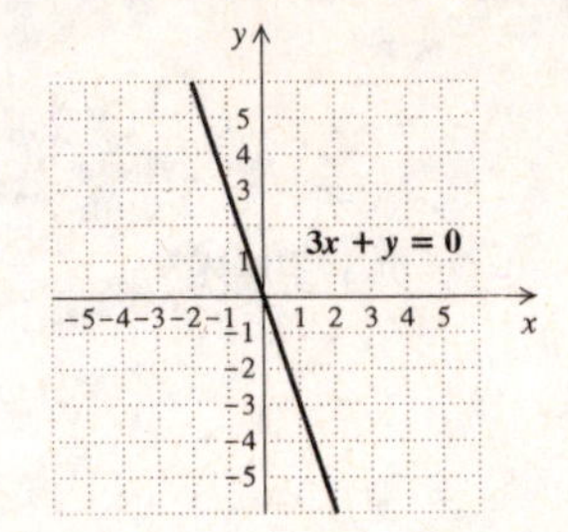

55. We first rewrite the equation in slope-intercept form.

$$2x + 3y = 9$$
$$3y = -2x + 9$$
$$y = \frac{1}{3}(-2x + 9)$$
$$y = -\frac{2}{3}x + 3$$

Now we plot the y-intercept $(0,3)$. We can start at the y-intercept and, thinking of the slope as $\frac{-2}{3}$, find another point by moving down 2 units and right 3 units to the point $(3,1)$. Thinking of the slope as $\frac{2}{-3}$ we can start at $(0,3)$ and move up 2 units and left 3 units to get another point $(-3,5)$.

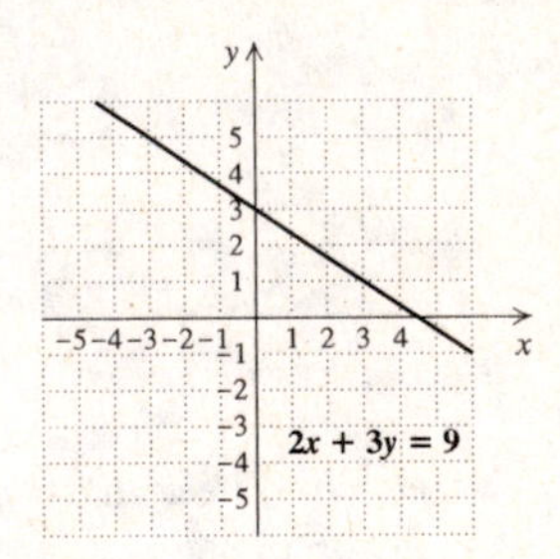

57. We first rewrite the equation in slope-intercept form.

$$x - 4y = 12$$
$$-4y = -x + 12$$
$$y = -\frac{1}{4}(-x + 12)$$
$$y = \frac{1}{4}x - 3$$

Now we plot the y-intercept $(0,-3)$. We can start at the y-intercept and use the slope, $\frac{1}{4}$, to find another point. We move up 1 unit and right 4 units to the point $(4,-2)$. Thinking of the slope as $\frac{-1}{-4}$ we can start at $(0,-3)$ and move down 1 unit and left 4 units to get another point $(-4, -4)$.

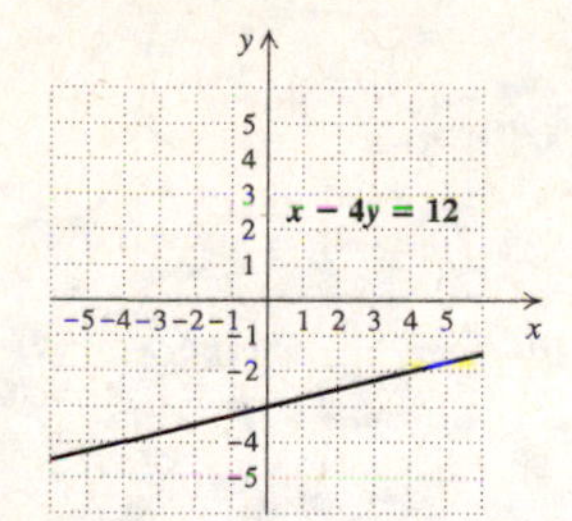

59. $y = \frac{2}{3}x + 7$: The slope is $\frac{2}{3}$, and the y-intercept is $(0, 7)$.

$y = \frac{2}{3}x - 5$: The slope is $\frac{2}{3}$, and the y-intercept is $(0, -5)$.

Since both lines have slope $\frac{2}{3}$ but different y-intercepts, their graphs are parallel.

61. The equation $y = 2x - 5$ represents a line with slope 2 and y-intercept $(0, -5)$. We rewrite the second equation in slope-intercept form.

$$\begin{aligned} 4x + 2y &= 9 \\ 2y &= -4x + 9 \\ y &= \frac{1}{2}(-4x + 9) \\ y &= -2x + \frac{9}{2} \end{aligned}$$

The slope is -2 and the y-intercept is $\left(0, \frac{9}{2}\right)$. Since the lines have different slopes, their graphs are not parallel.

63. Rewrite each equation in slope-intercept form.

$$\begin{aligned} 3x + 4y &= 8 \\ 4y &= -3x + 8 \\ y &= \frac{1}{4}(-3x + 8) \\ y &= -\frac{3}{4}x + 2 \end{aligned}$$

The slope is $-\frac{3}{4}$, and the y-intercept is $(0, 2)$.

$$\begin{aligned} 7 - 12y &= 9x \\ -12y &= 9x - 7 \\ y &= -\frac{1}{12}(9x - 7) \\ y &= -\frac{3}{4}x + \frac{7}{12} \end{aligned}$$

The slope is $-\frac{3}{4}$, and the y-intercept is $\left(0, \frac{7}{12}\right)$.

Since both lines have slope $-\frac{3}{4}$ but different y-intercepts, their graphs are parallel.

65. $y = 4x - 5$,
$4y = 8 - x$

The first equation is in slope-intercept form. It represents a line with slope 4. Now we rewrite the second equation in slope-intercept form.

$$\begin{aligned} 4y &= 8 - x \\ y &= \frac{1}{4}(8 - x) \\ y &= 2 - \frac{1}{4}x \\ y &= -\frac{1}{4}x + 2 \end{aligned}$$

The slope of the line is $-\frac{1}{4}$.

Since $4\left(-\frac{1}{4}\right) = -1$, the equations represent perpendicular lines.

67. $y - 2y = 5$,
$2x + 4y = 8$

We write each equation in slope-intercept form.

$$\begin{aligned} x - 2y &= 5 \\ -2y &= -x + 5 \\ y &= -\frac{1}{2}(-x + 5) \\ y &= \frac{1}{2}x - \frac{5}{2} \end{aligned}$$

The slope is $\frac{1}{2}$.

$$\begin{aligned} 2x + 4y &= 8 \\ 4y &= -2x + 8 \\ y &= \frac{1}{4}(-2x + 8) \\ y &= -\frac{1}{2}x + 2 \end{aligned}$$

The slope is $-\frac{1}{2}$.

Since $\frac{1}{2}\left(-\frac{1}{2}\right) = -\frac{1}{4} \neq -1$, the equations do not represent perpendicular lines.

69. $2x + 3y = 1$,
$3x - 2y = 1$

We write each equation in slope-intercept form.

$$\begin{aligned} 2x + 3y &= 1 \\ 3y &= -2x + 1 \\ y &= \frac{1}{3}(-2x + 1) \\ y &= -\frac{2}{3}x + \frac{1}{3} \end{aligned}$$

The slope is $-\frac{2}{3}$.

$$3x - 2y = 1$$
$$-2y = -3x + 1$$
$$y = -\frac{1}{2}(-3x + 1)$$
$$y = \frac{3}{2}x - \frac{1}{2}$$

The slope is $\frac{3}{2}$.

Since $-\frac{2}{3}\left(\frac{3}{2}\right) = -1$, the equations represent perpendicular lines.

71. The slope of the line represented by $y = 5x - 7$ is 5. Then a line parallel to the graph of $y = 5x - 7$ has slope 5 also. Since the y-intercept is $(0, 11)$, the desired equation is $y = 5x + 11$.

73. First find the slope of the line represented by $2x + y = 0$.

$$2x + y = 0$$
$$y = -2x$$

The slope is -2. Then the slope of a line perpendicular to the graph of $2x + y = 0$ is the negative reciprocal of -2, or $\frac{1}{2}$. Since the y-intercept is $(0, 0)$, the desired equation is $y = \frac{1}{2}x + 0$, or $y = \frac{1}{2}x$.

75. The slope of the line represented by $y = x$ is 1. Then a line parallel to this line also has slope 1. Since the y-intercept is $(0, 3)$, the desired equation is
$y = 1 \cdot x + 3$, or $y = x + 3$.

77. First find the slope of the line represented by $x + y = 3$.

$$x + y = 3$$
$$y = -x + 3, \text{ or } y = -1 \cdot x + 3$$

The slope is -1. Then the slope of a line perpendicular to this line is the negative reciprocal is -1, or 1. Since the y-intercept is -4, the desired equation is $y = 1 \cdot x - 4$, or $y = x - 4$.

79. *Writing Exercise*

81. $y - k = m(x - h)$

$y = m(x - h) + k$ Adding k to both sides

83. $-5 - (-7) = -5 + 7 = 2$

85. $-3 - 6 = -3 + (-6) = -9$

87. *Writing Exercise*

89. See the answer section in the text.

91. Rewrite each equation in slope-intercept form.

$$2x - 6y = 10$$
$$-6y = -2x + 10$$
$$y = \frac{1}{3}x - \frac{5}{3}$$

The slope of the line is $\frac{1}{3}$.

$$9x + 6y = 18$$
$$6y = -9x + 18$$
$$y = -\frac{3}{2}x + 3$$

The y-intercept of the line is $(0, 3)$.

The equation of the line is $y = \frac{1}{3}x + 3$.

93. Rewrite the first equation in slope-intercept form.

$$3x - 5y = 8$$
$$-5y = -3x + 8$$
$$y = -\frac{1}{5}(-3x + 8)$$
$$y = \frac{3}{5}x - \frac{8}{5}$$

The slope is $\frac{3}{5}$.

The slope of a line perpendicular to this line is a number m such that

$$\frac{3}{5}m = -1, \text{ or}$$
$$m = -\frac{5}{3}.$$

Now rewrite the second equation in slope-intercept form.

$$2x + 4y = 12$$
$$4y = -2x + 12$$
$$y = \frac{1}{4}(-2x + 12)$$
$$y = -\frac{1}{2}x + 3$$

The y-intercept of the line is $(0, 3)$.

The equation of the line is $y = -\frac{5}{3}x + 3$.

95. Rewrite the first equation in slope-intercept form.

$$3x - 2y = 9$$
$$-2y = -3x + 9$$
$$y = -\frac{1}{2}(-3x + 9)$$
$$y = \frac{3}{2}x - \frac{9}{2}$$

The slope of the line is $\frac{3}{2}$.

The slope of a line perpendicular to this line is a number m such that

$$\frac{3}{2}m = -1, \text{ or}$$
$$m = -\frac{2}{3}.$$

Now rewrite the second equation in slope-intercept form.

$$2x + 5y = 0$$
$$5y = -2x$$
$$y = -\frac{2}{5}x, \text{ or } y = -\frac{2}{5}x + 0$$

The y-intercept is $(0, 0)$.

The equation of the line is $y = -\frac{2}{3}x + 0$, or $y = -\frac{2}{3}x$.

97. *Writing Exercise*

Exercise Set 3.7

1. Substituting 5 for m, 2 for x_1, and 3 for y_1 in the point-slope equation $y-y_1 = m(x-x_1)$, we have $y-3 = 5(x-2)$. Choice (g) is correct.

3. Substituting -5 for m, -2 for x_1, and -3 for y_1 in the point-slope equation $y-y_1 = m(x-x_1)$, we have $y-(-3) = -5(x-(-2))$, or $y+3 = -5(x+2)$. Choice (e) is correct.

5. Substituting 5 for m, 3 for x_1, and 2 for y_1 in the point-slope equation $y-y_1 = m(x-x_1)$, we have $y-2 = 5(x-3)$. Choice (b) is correct.

7. Substituting -5 for m, -3 for x_1, and -2 for y_1 in the point-slope equation $y-y_1 = m(x-x_1)$, we have $y-(-2) = -5(x-(-3))$, or $y+2 = -5(x+3)$. Choice (f) is correct.

9. We see that the points $(1,-4)$ and $(-3,2)$ are on the line. To go from $(1,-4)$ to $(-3,2)$ we go up 6 units and left 4 units so the slope of the line is $\frac{6}{-4}$, or $-\frac{3}{2}$. Then, substituting $-\frac{3}{2}$ for m, 1 for x_1, and -4 for y_1 in the point-slope equation $y-y_1 = m(x-x_1)$, we have $y-(-4) = -\frac{3}{2}(x-1)$, or $y+4 = -\frac{3}{2}(x-1)$. Choice (c) is correct.

11. We see that the points $(1,-4)$ and $(5,2)$ are on the line. To go from $(1,-4)$ to $(5,2)$ we go up 6 units and right 4 units so the slope of the line is $\frac{6}{4}$, or $\frac{3}{2}$. Then, substituting $\frac{3}{2}$ for m, 1 for x_1, and -4 for y_1 in the point-slope equation $y-y_1 = m(x-x_1)$, we have $y-(-4) = \frac{3}{2}(x-1)$, or $y+4 = \frac{3}{2}(x-1)$. Choice (d) is correct.

13. $y-y_1 = m(x-x_1)$

We substitute 5 for m, 6 for x_1, and 2 for y_1.

$y-2 = 5(x-6)$

15. $y-y_1 = m(x-x_1)$

We substitute -4 for m, 3 for x_1, and 1 for y_1.

$y-1 = -4(x-3)$

17. $y-y_1 = m(x-x_1)$

We substitute $\frac{3}{2}$ for m, 5 for x_1, and -4 for y_1.

$y-(-4) = \frac{3}{2}(x-5)$

19. $y-y_1 = m(x-x_1)$

We substitute $\frac{5}{4}$ for m, -2 for x_1, and 6 for y_1.

$y-6 = \frac{5}{4}(x-(-2))$

21. $y-y_1 = m(x-x_1)$

We substitute -2 for m, -4 for x_1, and -1 for y_1.

$y-(-1) = -2(x-(-4))$

23. $y-y_1 = m(x-x_1)$

We substitute 1 for m, -2 for x_1, and 8 for y_1.

$y-8 = 1(x-(-2))$

25. First we write the equation in point-slope form.

$$y-y_1 = m(x-x_1)$$
$$y-7 = 2(x-5) \quad \text{Substituting}$$

Next we find an equivalent equation of the form $y = mx+b$.

$$y-7 = 2(x-5)$$
$$y-7 = 2x-10$$
$$y = 2x-3$$

27. First we write the equation in point-slope form.

$$y-y_1 = m(x-x_1)$$
$$y-(-2) = \frac{7}{4}(x-4) \quad \text{Substituting}$$

Next we find an equivalent equation of the form $y = mx+b$.

$$y-(-2) = \frac{7}{4}(x-4)$$
$$y+2 = \frac{7}{4}x-7$$
$$y = \frac{7}{4}x-9$$

29. First we write the equation in point-slope form.

$$y-y_1 = m(x-x_1)$$
$$y-6 = -3(x-(-1))$$

Next we find an equivalent equation of the form $y = mx+b$.

$$y-6 = -3(x-(-1))$$
$$y-6 = -3(x+1)$$
$$y-6 = -3x-3$$
$$y = -3x+3$$

31. First we write the equation in point-slope form.

$$y-y_1 = m(x-x_1)$$
$$y-(-1) = -4(x-(-2))$$

Next we find an equivalent equation of the form $y = mx+b$.

$$y-(-1) = -4(x-(-2))$$
$$y+1 = -4(x+2)$$
$$y+1 = -4x-8$$
$$y = -4x-9$$

33. The slope is $-\frac{5}{6}$ and the y-intercept is $(0,4)$. Substituting $-\frac{5}{6}$ for m and 4 for b in the slope-intercept equation $y = mx+b$, we have $y = -\frac{5}{6}x+4$.

35. First solve the equation for y and determine the slope of the given line.

$$x + 2y = 6 \quad \text{Given line}$$
$$2y = -x + 6$$
$$y = -\frac{1}{2}x + 3$$

The slope of the given line is $-\frac{1}{2}$.

The slope of every line parallel to the given line must also be $-\frac{1}{2}$. We find the equation of the line with slope $-\frac{1}{2}$ and containing the point $(4, 7)$.

$$y - y_1 = m(x - x_1) \quad \text{Point-slope equation}$$
$$y - 7 = -\frac{1}{2}(x - 4) \quad \text{Substituting}$$
$$y - 7 = -\frac{1}{2}x + 2$$
$$y = -\frac{1}{2}x + 9$$

37. The slope of $y = 2x + 1$ is 2. The given point, $(0, -7)$, is the y-intercept, so we substitute in the slope-intercept equation.

$$y = 2x - 7$$

39. First solve the equation for y and determine the slope of the given line.

$$5x - 3y = 8 \quad \text{Given line}$$
$$-3y = -5x + 8$$
$$y = \frac{5}{3}x - \frac{8}{3}$$

The slope of the given line is $\frac{5}{3}$.

The slope of every line parallel to the given line must also be $\frac{5}{3}$. We find the equation of the line with slope $\frac{5}{3}$ and containing the point $(2, -6)$.

$$y - y_1 = m(x - x_1) \quad \text{Point-slope equation}$$
$$y - (-6) = \frac{5}{3}(x - 2) \quad \text{Substituting}$$
$$y + 6 = \frac{5}{3}x - \frac{10}{3}$$
$$y = \frac{5}{3}x - \frac{28}{3}$$

41. First solve the equation for y and determine the slope of the given line.

$$3x + 6y = 5 \quad \text{Given line}$$
$$6y = -3x + 5$$
$$y = -\frac{1}{2}x + \frac{5}{6}$$

The slope of the given line is $-\frac{1}{2}$.

The slope of a perpendicular line is given by the opposite of the reciprocal of $-\frac{1}{2}$, 2.

We find the equation of the line with slope 2 and containing the point $(3, -2)$.

$$y - y_1 = m(x - x_1) \quad \text{Point-slope equation}$$
$$y - (-2) = 2(x - 3) \quad \text{Substituting}$$
$$y + 2 = 2x - 6$$
$$y = 2x - 8$$

43. First solve the equation for y and find the slope of the given line.

$$3x - 5y = 6$$
$$-5y = -3x + 6$$
$$y = \frac{3}{5}x - \frac{6}{5}$$

The slope of the given line is $\frac{3}{5}$. The slope of a perpendicular line is given by the opposite of the reciprocal of $\frac{3}{5}$, $-\frac{5}{3}$.

We find the equation of the line with slope $-\frac{5}{3}$ and containing the point $(-4, -7)$.

$$y - y_1 = m(x - x_1) \quad \text{Point-slope equation}$$
$$y - (-7) = -\frac{5}{3}[x - (-4)]$$
$$y + 7 = -\frac{5}{3}(x + 4)$$
$$y + 7 = -\frac{5}{3}x - \frac{20}{3}$$
$$y = -\frac{5}{3}x - \frac{41}{3}$$

45. The slope of a line perpendicular to $2x - 5 = y$ is $-\frac{1}{2}$ and we are given the y-intercept of the desired line, $(0, 6)$. Then we have $y = -\frac{1}{2}x + 6$.

47. We plot $(1, 2)$, move up 4 and to the right 3 to $(4, 6)$ and draw the line.

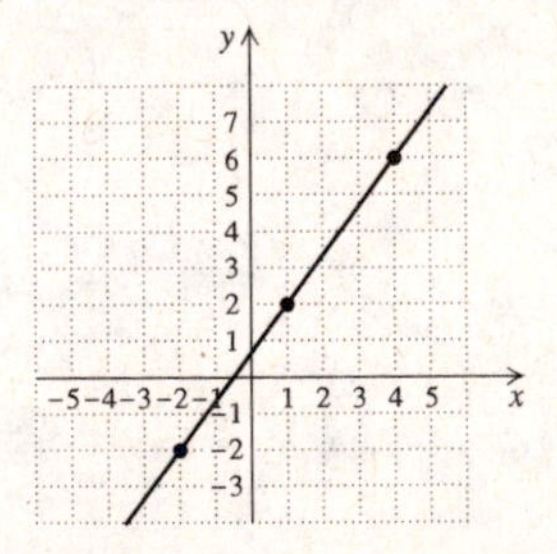

49. We plot $(2, 5)$, move down 3 and to the right 4 to $(6, 2)$ $\left(\text{since } -\frac{3}{4} = \frac{-3}{4}\right)$, and draw the line. We could also think of $-\frac{3}{4}$ and $\frac{3}{-4}$ and move up 3 and to the left 4 from the point $(2, 5)$ to $(-2, 8)$.

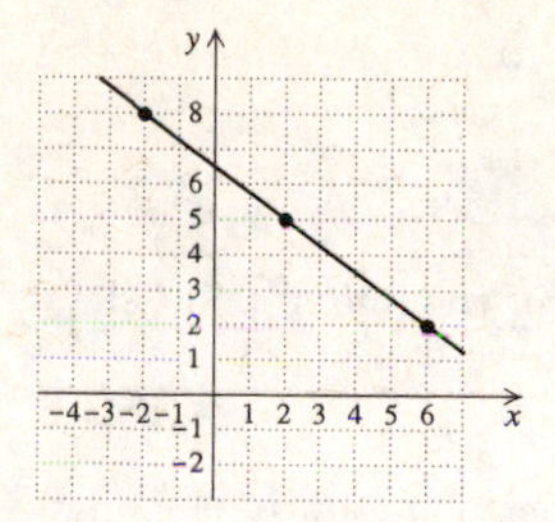

51. $y - 2 = \frac{1}{2}(x - 1)$ Point-slope form

The line has slope $\frac{1}{2}$ and passes through $(1, 2)$. We plot $(1, 2)$ and then find a second point by moving up 1 unit and right 2 units to $(3, 3)$. We draw the line through these points.

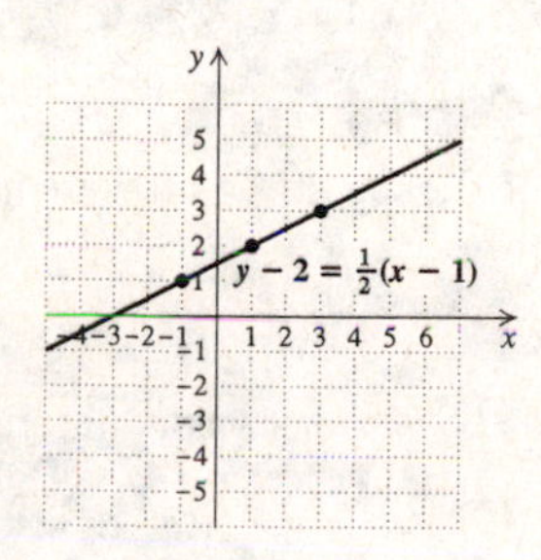

53. $y - 1 = -\frac{1}{2}(x - 3)$ Point-slope form

The line has slope $-\frac{1}{2}$, or $\frac{1}{-2}$ passes through $(3, 1)$. We plot $(3, 1)$ and then find a second point by moving up 1 unit and left 2 units to $(1, 2)$. We draw the line through these points.

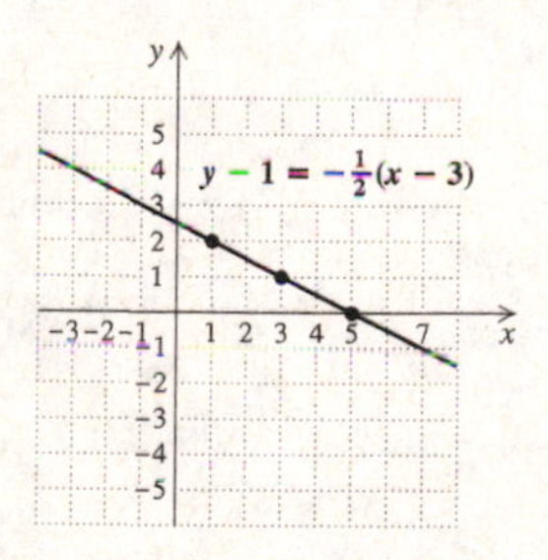

55. $y + 4 = 3(x + 1)$, or $y - (-4) = 3(x - (-1))$

The line has slope 3, or $\frac{3}{1}$, and passes through $(-1, -4)$. We plot $(-1, -4)$ and then find a second point by moving up 3 units and right 1 unit to $(0, -1)$. We draw the line through these points.

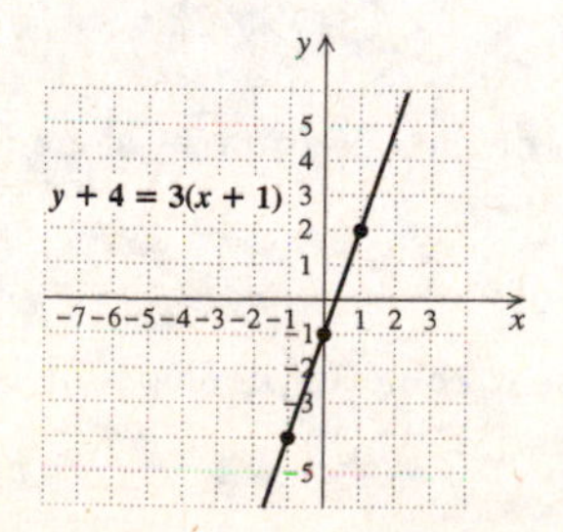

57. $y + 3 = -(x + 2)$, or $y - (-3) = -1(x - (-2))$

The line has slope -1, or $\frac{-1}{1}$, and passes through $(-2, -3)$. We plot $(-2, -3)$ and then find a second point by moving down 1 unit and right 1 unit to $(-1, -4)$. We draw the line through these points.

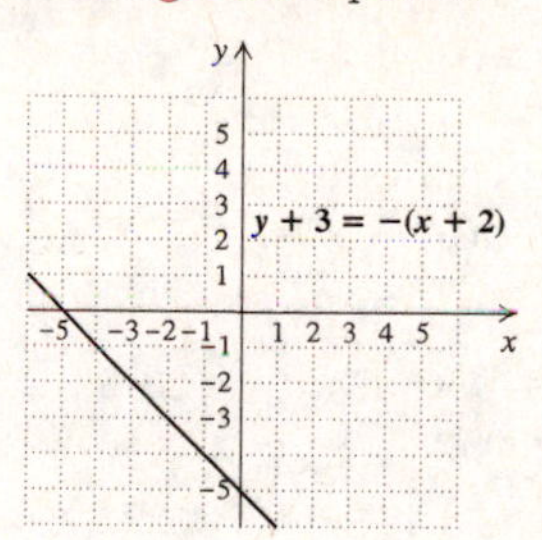

59. $y + 1 = -\frac{3}{5}(x + 2)$, or $y - (-1) = -\frac{3}{5}(x - (-2))$

The line has slope $-\frac{3}{5}$, or $\frac{-3}{5}$ and passes through $(-2, -1)$. We plot $(-2, -1)$ and then find a second point by moving down 3 units and right 5 units to $(3, -4)$, and draw the line.

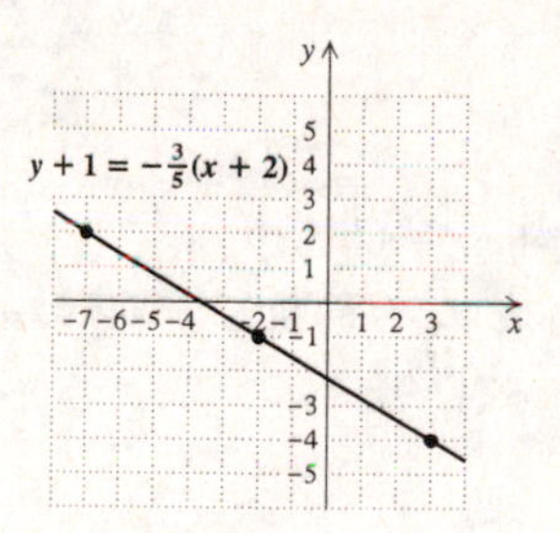

61. a) First find the slope of the line passing through the points $(1, 62.1)$ and $(11, 45.9)$.

$$m = \frac{45.9 - 62.1}{11 - 1} = \frac{-16.2}{10} = -1.62$$

Now write an equation of the line. We use $(1, 62.1)$ in the point-slope equation and then write an equivalent slope-intercept equation.

$$y - y_1 = m(x - x_1)$$
$$y - 62.1 = -1.62(x - 1)$$
$$y - 62.1 = -1.62x + 1.62$$
$$y = -1.62x + 63.72$$

Since 1999 is 9 yr after 1990, we substitute 9 for x to calculate the birth rate in 1999.

$y = -1.62(9) + 63.72 = -14.58 + 63.72 = 49.14$

In 1999, there were 49.14 births per 1000 females age 15 to 19.

b) 2008 is 18 yr after 1990 $(2008 - 1990 = 18)$, so we substitute 18 for x.

$y = -1.62(18) + 63.72 = -29.16 + 63.72 = 34.56$

We predict that the birth rate among teenagers will be 34.56 births per 1000 females in 2008.

63. a) First find the slope of the line passing through the points $(0, 45.7)$ and $(3, 32.5)$. In each case, we let the first coordinate represent the number of years after 1985.

$$m = \frac{32.5 - 45.7}{13 - 0} = \frac{-13.2}{13} \approx -1.02$$

The y-intercept of the line is $(0, 45.7)$. We write the slope-intercept equation: $y = -1.02x + 45.7$.

Since 1990 is 5 yr after 1985, we substitute 5 for x to calculate the percentage in 1990.

$y = -1.02(5) + 45.7 = -5.1 + 45.7 = 40.6\%$

b) Since $2008 - 1985 = 23$, we substitute 23 for x to find the percentage in 2008.

$y = -1.02(23) + 45.7 = -23.46 + 45.7 \approx 22.2\%$

(Answers will vary depending on how the slope was rounded in part (a).)

65. a) First find the slope of the line passing through $(0, 60.3)$ and $(10, 68.3)$. In each case, we let the first coordinate represent the number of years after 1990. The second coordinate represents the enrollment in millions.

$$m = \frac{68.3 - 60.3}{10 - 0} = \frac{8}{10} = 0.8$$

The y-intercept is $(0, 60.3)$. We write the slope-intercept equation: $y = 0.8x + 60.3$.

Since 1996 is 6 yr after 1990, we substitute 6 for x to find the enrollment in 1996.

$y = 0.8(6) + 60.3 = 4.8 + 60.3 = 65.1$ million students

b) Since 2005 is 15 yr after 1990, we substitute 15 for x to find the enrollment in 2005.

$y = 0.8(15) + 60.3 = 12 + 60.3 = 72.3$ million students

67. a) First find the slope of the line through $(0, 31)$ and $(12, 35.6)$. In each case, we let the first coordinate represent the number of years after 1990 and the second millions of residents.

$$m = \frac{35.6 - 31}{12 - 0} = \frac{4.6}{12} \approx 0.38$$

The y-intercept is $(0, 31)$. We write the slope-intercept equation: $y = 0.38x + 31$.

Since 1996 is 6 yr after 1990, we substitute 6 for x to find the number of U.S. residents over the age of 65 in 1996.

$y = 0.38(6) + 31 = 2.28 + 31 = 33.28 \approx 33.3$ million residents

b) Since 2010 is 20 yr after 1990, we substitute 20 for x to find the number of U.S. residents over the age of 65 in 2010.

$y = 0.38(20) + 31 = 7.6 + 31 = 38.6$ million residents

(Answers will vary depending on how the slope is rounded.)

69. $(1, 5)$ and $(4, 2)$

First we find the slope.

$$m = \frac{5 - 2}{1 - 4} = \frac{3}{-3} = -1$$

Then we write an equation of the line in point-slope form using either of the points above.

$$y - 5 = -1(x - 1)$$

Finally, we find an equivalent equation in slope-intercept form.

$$y - 5 = -1(x - 1)$$
$$y - 5 = -x + 1$$
$$y = -x + 6$$

71. $(-3, 1)$ and $(3, 5)$

First we find the slope.

$$m = \frac{1 - 5}{-3 - 3} = \frac{-4}{-6} = \frac{2}{3}$$

Then we write an equation of the line in point-slope form using either of the points above.

$$y - 5 = \frac{2}{3}(x - 3)$$

Finally, we find an equivalent equation in slope-intercept form.

$$y - 5 = \frac{2}{3}(x - 3)$$
$$y - 5 = \frac{2}{3}x - 2$$
$$y = \frac{2}{3}x + 3$$

73. $(5, 0)$ and $(0, -2)$

First we find the slope.

$$m = \frac{0 - (-2)}{5 - 0} = \frac{2}{5}$$

Then we write an equation of the line in point-slope form using either of the points above.

$$y - 0 = \frac{2}{5}(x - 5)$$

Finally, we find an equivalent equation in slope-intercept form.

$$y - 0 = \frac{2}{5}(x - 5)$$
$$y = \frac{2}{5}x - 2$$

75. $(-2, -4)$ and $(2, -1)$

First we find the slope.

$$m = \frac{-4 - (-1)}{-2 - 2} = \frac{-4 + 1}{-2 - 2} = \frac{-3}{-4} = \frac{3}{4}$$

Then we write an equation of the line in point-slope form using either of the points above.

$$y - (-4) = \frac{3}{4}(x - (-2))$$

Finally, we find an equivalent equation in slope-intercept form.

$$y-(-4)=\frac{3}{4}(x-(-2))$$
$$y+4=\frac{3}{4}(x+2)$$
$$y+4=\frac{3}{4}x+\frac{3}{2}$$
$$y=\frac{3}{4}x-\frac{5}{2}$$

77. *Writing Exercise*

79. $(-5)^3=(-5)(-5)(-5)=-125$

81. $3\cdot 2^4-5\cdot 2^3$

$=3\cdot 16-5\cdot 8$ Evaluating the exponential expressions

$=48-40$ Multiplying

$=8$ Subtracting

83. $(-2)^3(-3)^2=-8\cdot 9=-72$

85. *Writing Exercise*

87. $y-3=0(x-52)$

Observe that the slope is 0. Then this is the equation of a horizontal line that passes through $(52,3)$. Thus, its graph is a horizontal line 3 units above the x-axis.

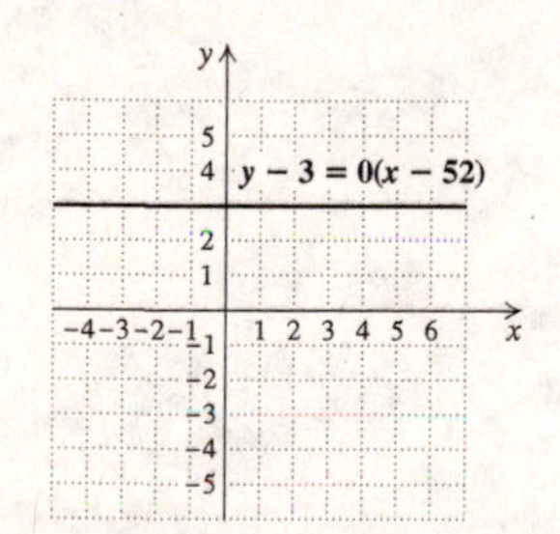

89. First we find the slope of the line using any two points on the line. We will use $(3,-3)$ and $(4,-1)$.

$$m=\frac{-3-(-1)}{3-4}=\frac{-2}{-1}=2$$

Then we write an equation of the line in point-slope form using either of the points above.

$$y-(-3)=2(x-3)$$

Finally, we find an equivalent equation in slope-intercept form.

$$y-(-3)=2(x-3)$$
$$y+3=2x-6$$
$$y=2x-9$$

91. First we find the slope of the line using any two points on the line. We will use $(2,5)$ and $(5,1)$.

$$m=\frac{5-1}{2-5}=\frac{4}{-3}=-\frac{4}{3}$$

Then we write an equation of the line in point-slope form using either of the points above.

$$y-5=-\frac{4}{3}(x-2)$$

Finally, we find an equivalent equation in slope-intercept form.

$$y-5=-\frac{4}{3}(x-2)$$
$$y-5=-\frac{4}{3}x+\frac{8}{3}$$
$$y=-\frac{4}{3}x+\frac{23}{3}$$

93. First find the slope of $2x+3y=11$.

$$2x+3y=11$$
$$3y=-2x+11$$
$$y=-\frac{2}{3}x+\frac{11}{3}$$

The slope is $-\frac{2}{3}$.

Then write a point-slope equation of the line containing $(-4,7)$ and having slope $-\frac{2}{3}$.

$$y-7=-\frac{2}{3}(x-(-4))$$

95. The slope of $y=3-4x$ is -4. We are given the y-intercept of the line, so we use slope-intercept form. The equation is $y=-4x+7$.

97. First find the slope of the line passing through $(2,7)$ and $(-1,-3)$.

$$m=\frac{-3-7}{-1-2}=\frac{-10}{-3}=\frac{10}{3}$$

Now find an equation of the line containing the point $(-1,5)$ and having slope $\frac{10}{3}$.

$$y-5=\frac{10}{3}(x-(-1))$$
$$y-5=\frac{10}{3}(x+1)$$
$$y-5=\frac{10}{3}x+\frac{10}{3}$$
$$y=\frac{10}{3}x+\frac{25}{3}$$

99. *Writing Exercise*

Chapter 4

Polynomials

Exercise Set 4.1

1. By the power rule on 234, choice (b) is correct.

3. By the rule for raising a product to a power on page 234, choice (e) is correct.

5. By the definition of 0 as an exponent on page 233, choice (g) is correct.

7. By the rule for raising a quotient to a power on page 235, choice (c) is correct.

9. $r^4 \cdot r^6 = r^{4+6} = r^{10}$

11. $9^5 \cdot 9^3 = 9^{5+3} = 9^8$

13. $a^6 \cdot a = a^6 \cdot a^1 = a^{6+1} = a^7$

15. $8^4 \cdot 8^7 = 8^{4+7} = 8^{11}$

17. $(3y)^4(3y)^8 = (3y)^{4+8} = (3y)^{12}$

19. $(5t)(5t)^6 = (5t)^1(5t)^6 = (5t)^{1+6} = (5t)^7$

21. $(a^2b^7)(a^3b^2) = a^2b^7a^3b^2$ Using an associative law
$= a^2a^3b^7b^2$ Using a commutative law
$= a^5b^9$ Adding exponents

23. $(x+1)^5(x+1)^7 = (x+1)^{5+7} = (x+1)^{12}$

25. $r^3 \cdot r^7 \cdot r^0 = r^{3+7+0} = r^{10}$

27. $(xy^4)(xy)^3 = (xy^4)(x^3y^3)$
$= x \cdot x^3 \cdot y^4 \cdot y^3$
$= x^{1+3}y^{4+3}$
$= x^4y^7$

29. $\dfrac{7^5}{7^2} = 7^{5-2} = 7^3$ Subtracting exponents

31. $\dfrac{x^{15}}{x^3} = x^{15-3} = x^{12}$ Subtracting exponents

33. $\dfrac{t^5}{t} = \dfrac{t^5}{t^1} = t^{5-1} = t^4$

35. $\dfrac{(5a)^7}{(5a)^6} = (5a)^{7-6} = (5a)^1 = 5a$

37. $\dfrac{(x+y)^8}{(x+y)^8}$

Observe that we have an expression divided by itself. Thus, the result is 1.

We could also do this exercise as follows:

$\dfrac{(x+y)^8}{(x+y)^8} = (x+y)^{8-8} = (x+y)^0 = 1$

39. $\dfrac{6m^5}{8m^2} = \dfrac{6}{8}m^{5-2} = \dfrac{3}{4}m^3$

41. $\dfrac{8a^9b^7}{2a^2b} = \dfrac{8}{2} \cdot \dfrac{a^9}{a^2} \cdot \dfrac{b^7}{b^1} = 4a^{9-2}b^{7-1} = 4a^7b^6$

43. $\dfrac{m^9n^8}{m^0n^4} = \dfrac{m^9}{m^0} \cdot \dfrac{n^8}{n^4} = m^{9-0}n^{8-4} = m^9n^4$

45. When $x = 13$, $x^0 = 13^0 = 1$. (Any nonzero number raised to the 0 power is 1.)

47. When $x = -4$, $5x^0 = 5(-4)^0 = 5 \cdot 1 = 5$.

49. $7^0 + 4^0 = 1 + 1 = 2$

51. $(-3)^1 - (-3)^0 = -3 - 1 = -4$

53. $(x^4)^7 = x^{4 \cdot 7} = x^{28}$ Multiplying exponents

55. $(5^8)^2 = 5^{8 \cdot 2} = 5^{16}$ Multiplying exponents

57. $(m^7)^5 = m^{7 \cdot 5} = m^{35}$

59. $(t^{20})^4 = t^{20 \cdot 4} = t^{80}$

61. $(7x)^2 = 7^2 \cdot x^2 = 49x^2$

63. $(-2a)^3 = (-2)^3a^3 = -8a^3$

65. $(4m^3)^2 = 4^2(m^3)^2 = 16m^6$

67. $(a^2b)^7 = (a^2)^7(b^7) = a^{14}b^7$

69. $(x^3y)^2(x^2y^5) = (x^3)^2y^2x^2y^5 = x^6y^2x^2y^5 = x^8y^7$

71. $(2x^5)^3(3x^4) = 2^3(x^5)^3(3x^4) = 8x^{15} \cdot 3x^4 = 24x^{19}$

73. $\left(\dfrac{a}{4}\right)^3 = \dfrac{a^3}{4^3} = \dfrac{a^3}{64}$ Raising the numerator and the denominator to the third power

75. $\left(\dfrac{7}{5a}\right)^2 = \dfrac{7^2}{(5a)^2} = \dfrac{49}{5^2a^2} = \dfrac{49}{25a^2}$

77. $\left(\dfrac{a^4}{b^3}\right)^5 = \dfrac{(a^4)^5}{(b^3)^5} = \dfrac{a^{20}}{b^{15}}$

79. $\left(\dfrac{y^3}{2}\right)^2 = \dfrac{(y^3)^2}{2^2} = \dfrac{y^6}{4}$

81. $\left(\dfrac{x^2y}{z^3}\right)^4 = \dfrac{(x^2y)^4}{(z^3)^4} = \dfrac{(x^2)^4(y^4)}{z^{12}} = \dfrac{x^8y^4}{z^{12}}$

83. $\left(\dfrac{a^3}{-2b^5}\right)^4 = \dfrac{(a^3)^4}{(-2b^5)^4} = \dfrac{a^{12}}{(-2)^4(b^5)^4} = \dfrac{a^{12}}{16b^{20}}$

85. $\left(\dfrac{5x^7y}{2z^4}\right)^3 = \dfrac{(5x^7y)^3}{(2z^4)^3} = \dfrac{5^3(x^7)^3y^3}{2^3(z^4)^3} =$
$\dfrac{125x^{21}y^3}{8z^{12}}$

87. $\left(\frac{4x^3y^5}{3z^7}\right)^0$

Observe that for $x \neq 0$, $y \neq 0$, and $z \neq 0$, we have a nonzero number raised to the 0 power. Thus, the result is 1.

89. *Writing Exercise*

91. $3s - 3r + 3t = 3 \cdot s - 3 \cdot r + 3 \cdot t = 3(s - r + t)$

93. $9x + 2y - x - 2y = 9x - x + 2y - 2y =$
$(9 - 1)x + (2 - 2)y = 8x + 0y = 8x$

95. $2y + 3x$

97. *Writing Exercise*

99. *Writing Exercise*

101. Choose any number except 0.

For example, let $a = 1$. Then $(a+5)^2 = (1+5)^2 = 6^2 = 36$, but $a^2 + 5^2 = 1^2 + 5^2 = 1 + 25 = 26$.

103. Choose any number except $\frac{7}{6}$. For example let $a = 0$.

Then $\frac{0+7}{7} = \frac{7}{7} = 1$, but $a = 0$.

105. $a^{10k} \div a^{2k} = a^{10k-2k} = a^{8k}$

107. $$\frac{\left(\frac{1}{2}\right)^3\left(\frac{2}{3}\right)^4}{\left(\frac{5}{6}\right)^3} = \frac{\frac{1}{8}\cdot\frac{16}{81}}{\frac{125}{216}} = \frac{1}{8}\cdot\frac{16}{81}\cdot\frac{216}{125} =$$

$$\frac{1 \cdot 2 \cdot \cancel{8} \cdot \cancel{27} \cdot 8}{\cancel{8} \cdot 3 \cdot \cancel{27} \cdot 125} = \frac{16}{375}$$

109.
$$\frac{t^{26}}{t^x} = t^x$$
$$t^{26-x} = t^x$$
$$26 - x = x \quad \text{Equating exponents}$$
$$26 = 2x$$
$$13 = x$$

The solution is 13.

111. Since the bases are the same, the expression with the larger exponent is larger. Thus, $4^2 < 4^3$.

113. $4^3 = 64$, $3^4 = 81$, so $4^3 < 3^4$.

115.
$$25^8 = (5^2)^8 = 5^{16}$$
$$125^5 = (5^3)^5 = 5^{15}$$
$$5^{16} > 5^{15}, \text{ or } 25^8 > 125^5.$$

117. $2^{22} = 2^{10} \cdot 2^{10} \cdot 2^2 \approx 10^3 \cdot 10^3 \cdot 4 \approx 1000 \cdot 1000 \cdot 4 \approx 4,000,000$

Using a calculator, we find that $2^{22} = 4,194,304$. The difference between the exact value and the approximation is $4,194,304 - 4,000,000$, or 194,304.

119. $2^{31} = 2^{10} \cdot 2^{10} \cdot 2^{10} \cdot 2 \approx 10^3 \cdot 10^3 \cdot 10^3 \cdot 2 \approx$
$1000 \cdot 1000 \cdot 1000 \cdot 2 = 2,000,000,000$

Using a calculator, we find that $2^{31} = 2,147,483,648$. The difference between the exact value and the approximation is $2,147,483,648 - 2,000,000,000 = 147,483,648$.

121.
$$\begin{aligned} 1.5 \text{ MB} &= 1.5 \times 1000 \text{ KB} \\ &= 1.5 \times 1000 \times 1 \times 2^{10} \text{ bytes} \\ &= 1,536,000 \text{ bytes} \\ &\approx 1,500,000 \text{ bytes} \end{aligned}$$

Exercise Set 4.2

1. The only expression with 4 terms is (b).

3. Expression (d) is the only expression for which the degree of the leading term is 5.

5. Expression (g) has two terms, and the degree of the leading term is 7.

7. Expression (c) has three terms, but it is not a trinomial because $\frac{3}{x}$ is not a monomial.

9. $7x^4 + x^3 - 5x + 8 = 7x^4 + x^3 + (-5x) + 8$

The terms are $7x^4$, x^3, $-5x$, and 8.

11. $-t^4 + 7t^3 - 3t^2 + 6 = -t^4 + 7t^3 + (-3t^2) + 6$

The terms are $-t^4$, $7t^3$, $-3t^2$, and 6.

13. $4x^5 + 7x$

Term	Coefficient	Degree
$4x^5$	4	5
$7x$	7	1

15. $9t^2 - 3t + 4$

Term	Coefficient	Degree
$9t^2$	9	2
$-3t$	-3	1
4	4	0

17. $7a^4 + 9a + a^3$

Term	Coefficient	Degree
$7a^4$	7	4
$9a$	9	1
a^3	1	3

19. $x^4 - x^3 + 4x - 3$

Term	Coefficient	Degree
x^4	1	4
$-x^3$	-1	3
$4x$	4	1
-3	-3	0

21. $5x - 9x^2 + 3x^6$

a)

Term	$5x$	$-9x^2$	$3x^6$
Degree	1	2	6

b) The term of highest degree is $3x^6$. This is the leading term. Then the leading coefficient is 3.

c) Since the term of highest degree is $3x^6$, the degree of the polynomial is 6.

23. $3a^2 - 7 + 2a^4$

a)

Term	$3a^2$	-7	$2a^4$
Degree	2	0	4

b) The term of highest degree is $2a^4$. This is the leading term. Then the leading coefficient is 2.

c) Since the term of highest degree is $2a^4$, the degree of the polynomial is 4.

25. $8 + 6x^2 - 3x - x^5$

a)

Term	8	$6x^2$	$-3x$	$-x^5$
Degree	0	2	1	5

b) The term of highest degree is $-x^5$. This is the leading term. Then the leading coefficient is -1 since $-x^5 = -1 \cdot x^5$.

c) Since the term of highest degree is $-x^5$, the degree of the polynomial is 5.

27. $7x^2 + 8x^5 - 4x^3 + 6 - \frac{1}{2}x^4$

Term	Coefficient	Degree of Term	Degree of Polynomial
$8x^5$	8	5	
$-\frac{1}{2}x^4$	$-\frac{1}{2}$	4	
$-4x^3$	-4	3	5
$7x^2$	7	2	
6	6	0	

29. Three monomials are added, so $x^2 - 23x + 17$ is a trinomial.

31. The polynomial $x^3 - 7x^2 + 2x - 4$ is none of these because it is composed of four monomials.

33. Two monomials are added, so $8t^2 + 5t$ is a binomial.

35. The polynomial 17 is a monomial because it is the product of a constant and a variable raised to a whole number power. (In this case the variable is raised to the power 0.)

37. $7x^2 + 3x + 4x^2 = (7+4)x^2 + 3x = 11x^2 + 3x$

39. $3a^4 - 2a + 2a + a^4 = (3+1)a^4 + (-2+2)a =$
$4a^4 + 0a = 4a^4$

41. $2x^2 - 6x + 3x + 4x^2 = (2+4)x^2 + (-6+3)x =$
$6x^2 - 3x$

43. $9x^3 + 2x - 4x^3 + 5 - 3x = (9-4)x^3 + (2-3)x + 5 = 5x^3 - x + 5$

45. $10x^2 + 2x^3 - 3x^3 - 4x^2 - 6x^2 - x^4 =$
$-x^4 + (2-3)x^3 + (10-4-6)x^2 = -x^4 - x^3$

47. $\frac{1}{5}x^4 + 7 - 2x^2 + 3 - \frac{2}{15}x^4 + 2x^2 =$
$\left(\frac{1}{5} - \frac{2}{15}\right)x^4 + (-2+2)x^2 + (7+3) =$
$\left(\frac{3}{15} - \frac{2}{15}\right)x^4 + 0x^2 + 10 = \frac{1}{15}x^4 + 10$

49. $5.9x^2 - 2.1x + 6 + 3.4x - 2.5x^2 - 0.5 =$
$(5.9 - 2.5)x^2 + (-2.1 + 3.4)x + (6 - 0.5) =$
$3.4x^2 + 1.3x + 5.5$

51. For $x = 3$: $-7x + 4 = -7 \cdot 3 + 4$
$= -21 + 4$
$= -17$

For $x = -3$: $-7x + 4 = -7(-3) + 4$
$= 21 + 4$
$= 25$

53. For $x = 3$: $2x^2 - 3x + 7 = 2 \cdot 3^2 - 3 \cdot 3 + 7$
$= 2 \cdot 9 - 3 \cdot 3 + 7$
$= 18 - 9 + 7$
$= 16$

For $x = -3$: $2x^2 - 3x + 7 = 2(-3)^2 - 3(-3) + 7$
$= 2 \cdot 9 - 3(-3) + 7$
$= 18 + 9 + 7$
$= 34$

55. For $x = 3$:
$-2x^3 - 3x^2 + 4x + 2 = -2 \cdot 3^3 - 3 \cdot 3^2 + 4 \cdot 3 + 2$
$= -2 \cdot 27 - 3 \cdot 9 + 4 \cdot 3 + 2$
$= -54 - 27 + 12 + 2$
$= -67$

For $x = -3$:
$-2x^3 - 3x^2 + 4x + 2 = -2(-3)^3 - 3(-3)^2 + 4(-3) + 2$
$= -2(-27) - 3 \cdot 9 + 4(-3) + 2$
$= 54 - 27 - 12 + 2$
$= 17$

57. For $x = 3$: $\frac{1}{3}x^4 - 2x^3 = \frac{1}{3} \cdot 3^4 - 2 \cdot 3^3$
$= \frac{1}{3} \cdot 81 - 2 \cdot 27$
$= 27 - 54$
$= -27$

For $x = -3$: $\frac{1}{3}x^4 - 2x^3 = \frac{1}{3}(-3)^4 - 2(-3)^3$
$= \frac{1}{3} \cdot 81 - 2(-27)$
$= 27 + 54$
$= 81$

59. For $x = 3$: $-x^4 - x^3 - x^2 = -3^4 - 3^3 - 3^2$
$= -81 - 27 - 9$
$= -117$

For $x = -3$: $-x^4 - x^3 - x^2 = -(-3)^4 - (-3)^3 - (-3)^2$
$= -81 - (-27) - 9$
$= -81 + 27 - 9$
$= -63$

61. Since 2005 is 5 years after 2000, we evaluate the polynomial for $t = 5$.

$0.15t + 1.42 = 0.15(5) + 1.42$
$= 0.75 + 1.42$
$= 2.17$

The amount of consumer debt in 2005 is about \$2.17 trillion.

63. $11.12t^2 = 11.12(10)^2 = 11.12(100) = 1112$

A skydiver has fallen approximately 1112 ft 10 seconds after jumping from a plane.

65. $2\pi r = 2(3.14)(10)$ Substituting 3.14 for π and 10 for r
$= 62.8$

The circumference is 62.8 cm.

67. $\pi r^2 = 3.14(7)^2$ Substituting 3.14 for π and 7 for r
$= 3.14(49)$
$= 153.86$

The area is 153.86 m^2.

69. Since 2006 is 6 years after 2000, we first locate 6 on the horizontal axis. From there we move vertically to the graph and then horizontally to the E-axis. This locates an E-value of about 5. Thus U.S. electricity consumption in 2006 is about 5 million gigawatt hours.

71. Since 2003 is 3 years after 2000, we first locate 3 on the horizontal axis. From there we move vertically to the graph and then horizontally to the E-axis. This locates an E-value of about 4.5. Thus U.S. electricity consumption in 2003 was about 4.5 million gigawatt hours.

73. Locate 10 on the horizontal axis. From there move vertically to the graph and then horizontally to the M-axis. This locates an M-value of about 9. Thus, about 9 words were memorized in 10 minutes.

75. Locate 8 on the horizontal axis. From there move vertically to the graph and then horizontally to the M-axis. This locates an M-value of about 6. Thus, the value of $-0.001t^3 + 0.1t^2$ for $t = 8$ is approximately 6.

77. We evaluate the polynomial for $x = 20$:

$N = -0.00006(20)^3 + 0.006(20)^2 - 0.1(20) + 1.9$
$= -0.00006(8000) + 0.006(400) - 0.1(20) + 1.9$
$= -0.48 + 2.4 - 2.0 + 1.9$
$= 1.82 \approx 1.8$

There are about 1.8 million hearing-impaired Americans of age 20.

We evaluate the polynomial for $x = 50$:

$N = -0.00006(50)^3 + 0.006(50)^2 - 0.1(50) + 1.9$
$= -0.00006(125,000) + 0.006(2500) - 0.1(50) + 1.9$
$= -7.5 + 15 - 5 + 1.9$
$= 4.4$

There are about 4.4 million hearing-impaired Americans of age 50.

79. *Writing Exercise*

81. $-19 + 24$ A negative and a positive number. We subtract the absolute values: $24 - 19 = 5$. The positive number has the larger absolute value so the answer is positive.

$-19 + 24 = 5$

83. $5x + 15 = 5 \cdot x + 5 \cdot 3 = 5(x + 3)$

85. ***Familiarize.*** Let $x =$ the cost per mile of gasoline in dollars. Then the total cost of the gasoline for the year was $14,800x$.

Translate.

Cost of insurance	+	cost of registration and oil	+	cost of gasoline	=	\$2011.
↓	↓	↓	↓	↓	↓	↓
972	+	114	+	$14,800x$	=	2011

Carry out. We solve the equation.

$972 + 114 + 14,800x = 2011$
$1086 + 14,800x = 2011$
$14,800x = 925$
$x = 0.0625$

Check. If gasoline cost \$0.0625 per mile, then the total cost of the gasoline was 14,800(\$0.0625), or \$925. Then the total auto expense was \$972 + \$114 + \$925, or \$2011. The answer checks.

State. Gasoline cost \$0.0625, or 6.25¢ per mile.

87. *Writing Exercise*

89. Answers may vary. Choose an ax^5-term where a is an even integer. Then choose three other terms with different degrees, each less than degree 5, and coefficients $a+2$, $a+4$, and $a+6$, respectively, when the polynomial is written in descending order. One such polynomial is $2x^5 + 4x^4 + 6x^3 + 8$.

91. Find the total revenue from the sale of 30 monitors:

$250x - 0.5x^2 = 250(30) - 0.5(30)^2$
$= 250(30) - 0.5(900)$
$= 7500 - 450$
$= \$7050$

Find the total cost of producing 30 monitors:

$$\begin{aligned}4000+0.6x^2 &= 4000+0.6(30)^2\\ &= 4000+0.6(900)\\ &= 4000+540\\ &= \$4540\end{aligned}$$

Subtract the cost from the revenue to find the profit: $7050 − $4540 = $2510

93. $(3x^2)^3+4x^2\cdot 4x^4-x^4(2x)^2+[(2x)^2]^3-100x^2(x^2)^2$
$= 27x^6+4x^2\cdot 4x^4-x^4\cdot 4x^2+(2x)^6-100x^2\cdot x^4$
$= 27x^6+16x^6-4x^6+64x^6-100x^6$
$= 3x^6$

95. First locate 1.5 on the vertical axis. Then move horizontally to the graph. We meet the curve at 3 places. At each place move down vertically to the horizontal axis and read the corresponding x-value. We see that the ages at which 1.5 million Americans are hearing impaired are 5, 13, and 80.

97. We first find q, the quiz average, and t, the test average.

$$q=\frac{60+85+72+91}{4}=\frac{308}{4}=77$$

$$t=\frac{89+93+90}{3}=\frac{272}{3}\approx 90.7$$

Now we substitute in the polynomial.

$$\begin{aligned}A &= 0.3q+0.4t+0.2f+0.1h\\ &= 0.3(77)+0.4(90.7)+0.2(84)+0.1(88)\\ &= 23.1+36.28+16.8+8.8\\ &= 84.98\\ &\approx 85.0\end{aligned}$$

99. When $t=3$, $-t^2+10t-18=-3^2+10\cdot 3-18=-9+30-18=3$.
When $t=4$, $-t^2+10t-18=-4^2+10\cdot 4-18=-16+40-18=6$.
When $t=5$, $-t^2+10t-18=-5^2+10\cdot 5-18=-25+50-18=7$.
When $t=6$, $-t^2+10t-18=-6^2+10\cdot 6-18=-36+60-18=6$.
When $t=7$, $-t^2+10t-18=-7^2+10\cdot 7-18=-49+70-18=3$.

We complete the table. Then we plot the points and connect them with a smooth curve.

t	$-t^2+10t-18$
3	3
4	6
5	7
6	6
7	3

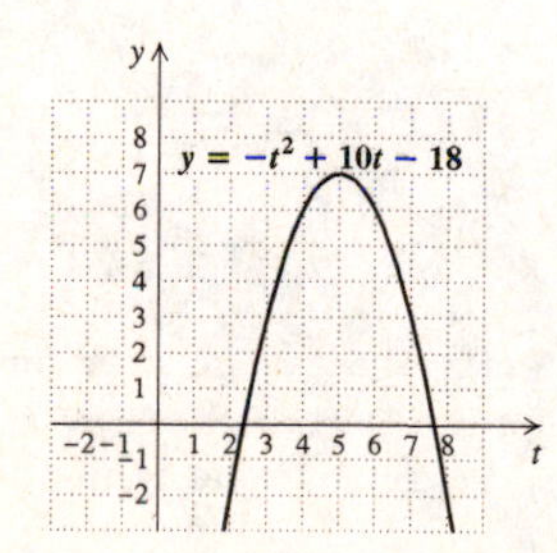

Exercise Set 4.3

1. $(3x+2)+(-5x+4)=(3-5)x+(2+4)=-2x+6$

3. $(-6x+2)+(x^2+x-3)=$
$x^2+(-6+1)x+(2-3)=x^2-5x-1$

5. $(7t^2-3t+6)+(2t^2+8t-9)=$
$(7+2)t^2+(-3+8)t+(6-9)=9t^2+5t-3$

7. $(4m^3-4m^2+m-5)+(4m^3+7m^2-4m-2)=$
$(7+4)m^3+(-4+7)m^2+(1-4)m+(-5-2)=$
$8m^3+3m^2-3m-7$

9. $(3+6a+7a^2+8a^3)+(4+7a-a^2+6a^3)=$
$(3+4)+(6+7)a+(7-1)a^2+(8+6)a^3=$
$7+13a+6a^2+14a^3$

11. $(9x^8-7x^4+2x^2+5)+(8x^7+4x^4-2x)=$
$9x^8+8x^7+(-7+4)x^4+2x^2-2x+5=$
$9x^8+8x^7-3x^4+2x^2-2x+5$

13. $\left(\frac{1}{4}x^4+\frac{2}{3}x^3+\frac{5}{8}x^2+9\right)+\left(-\frac{3}{4}x^4+\frac{3}{8}x^2-7\right)=$
$\left(\frac{1}{4}-\frac{3}{4}\right)x^4+\frac{2}{3}x^3+\left(\frac{5}{8}+\frac{3}{8}\right)x^2+(9-7)=$
$-\frac{2}{4}x^4+\frac{2}{3}x^3+\frac{8}{8}x^2+2=$
$-\frac{1}{2}x^4+\frac{2}{3}x^3+x^2+2$

15. $(5.3t^2-6.4t-9.1)+(4.2t^3-1.8t^2+7.3)=$
$4.2t^3+(5.3-1.8)t^2-6.4t+(-9.1+7.3)=$
$4.2t^3+3.5t^2-6.4t-1.8$

17.
$$\begin{array}{r}-3x^4+6x^2+2x-4\\ -3x^2+2x+4\\ \hline -3x^4+3x^2+4x+0\\ -3x^4+3x^2+4x\end{array}$$

19. Rewrite the problem so the coefficients of like terms have the same number of decimal places.

$$\begin{array}{rrrrr}0.15x^4 & +\,0.10x^3 & -\,0.90x^2 & & \\ & -\,0.01x^3 & +\,0.01x^2 & +\,x & \\ 1.25x^4 & & +\,0.11x^2 & & +\,0.01\\ & 0.27x^3 & & & +\,0.99\\ -0.35x^4 & & +\,15.00x^2 & & -\,0.03\\ \hline 1.05x^4 & +\,0.36x^3 & +\,14.22x^2 & +\,x & +\,0.97\end{array}$$

21. Two forms of the opposite of $-t^3+4t^2-9$ are
i) $-(-t^3+4t^2-9)$ and
ii) t^3-4t^2+9. (Changing the sign of every term)

23. Two forms for the opposite of $12x^4-3x^3+3$ are
i) $-(12x^4-3x^3+3)$ and
ii) $-12x^4+3x^3-3$. (Changing the sign of every term)

25. We change the sign of every term inside parentheses.

$-(8x-9) = -8x+9$

27. We change the sign of every term inside parentheses.

$-(3a^4-5a^2+9) = -3a^4+5a^2-9$

29. We change the sign of every term inside parentheses.

$-\left(-4x^4+6x^2+\frac{3}{4}x-8\right) = 4x^4-6x^2-\frac{3}{4}x+8$

31. $(7x+4)-(-2x+1)$

$= 7x+4+2x-1$ Changing the sign of every term inside parentheses

$= 9x+3$

33. $(-5t+6)-(t^2+3t-1) = -5t+6-t^2-3t+1 = -t^2-8t+7$

35. $(6x^4+3x^3-1)-(4x^2-3x+3)$

$= 6x^4+3x^3-1-4x^2+3x-3$

$= 6x^4+3x^3-4x^2+3x-4$

37. $(1.2x^3+4.5x^2-3.8x)-(-3.4x^3-4.7x^2+23)$

$= 1.2x^3+4.5x^2-3.8x+3.4x^3+4.7x^2-23$

$= 4.6x^3+9.2x^2-3.8x-23$

39. $(7x^3-2x^2+6)-(7x^3-2x^2+6)$

Observe that we are subtracting the polynomial $7x^3-2x^2+6$ from itself. The result is 0.

41. $(3+5a+3a^2-a^3)-(2+3a-4a^2+2a^3) =$

$3+5a+3a^2-a^3-2-3a+4a^2-2a^3 =$

$1+2a+7a^2-3a^3$

43. $\frac{5}{8}x^3-\frac{1}{4}x-\frac{1}{3}-\left(-\frac{1}{8}x^3+\frac{1}{4}x-\frac{1}{3}\right)$

$= \frac{5}{8}x^3-\frac{1}{4}x-\frac{1}{3}+\frac{1}{8}x^3-\frac{1}{4}x+\frac{1}{3}$

$= \frac{6}{8}x^3-\frac{2}{4}x$

$= \frac{3}{4}x^3-\frac{1}{2}x$

45. $(0.07t^3-0.03t^2+0.01t)-(0.02t^3+0.04t^2-1) = 0.07t^3-0.03t^2+0.01t-0.02t^3-0.04t^2+1 = 0.05t^3-0.07t^2+0.01t+1$

47.
$$\begin{array}{r} x^2+5x+6 \\ -(x^2+2x+1) \\ \hline \end{array}$$

$$\begin{array}{rl} x^2+5x+6 & \text{Changing signs and} \\ -x^2-2x-1 & \text{removing parentheses} \\ \hline 3x+5 & \text{Adding} \end{array}$$

49.
$$\begin{array}{r} 5x^4+6x^3-9x^2 \\ -(-6x^4-6x^3+x^2) \\ \hline \end{array}$$

$$\begin{array}{rl} 5x^4+6x^3-9x^2 & \text{Changing signs and} \\ 6x^4+6x^3-x^2 & \text{removing parentheses} \\ \hline 11x^4+12x^3-10x^2 & \text{Adding} \end{array}$$

51. a)

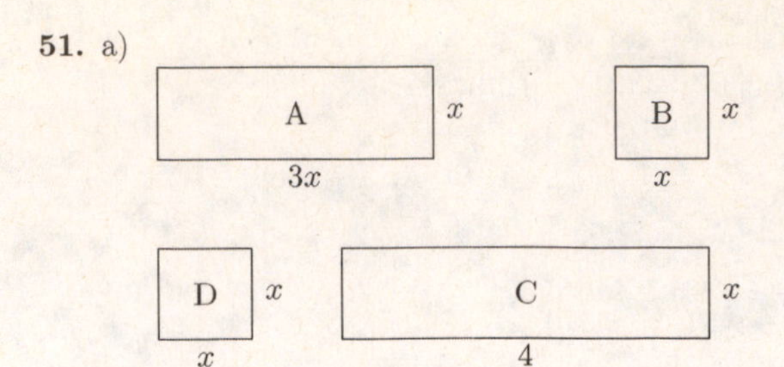

Familiarize. The area of a rectangle is the product of the length and the width.

Translate. The sum of the areas is found as follows:

$$\begin{array}{ccccccc} \text{Area of } A & + & \text{Area of } B & + & \text{Area of } C & + & \text{Area of } D \\ = 3x\cdot x & + & x\cdot x & + & 4\cdot x & + & x\cdot x \end{array}$$

Carry out. We collect like terms.

$3x^2+x^2+4x+x^2 = 5x^2+4x$

Check. We can go over our calculations. We can also assign some value to x, say 2, and carry out the computation of the area in two ways.

Sum of areas: $3\cdot2\cdot2+2\cdot2+4\cdot2+2\cdot2 = 12+4+8+4 = 28$

Substituting in the polynomial: $5(2)^2+4\cdot2 = 20+8 = 28$

Since the results are the same, our solution is probably correct.

State. A polynomial for the sum of the areas is $5x^2+4x$.

b) For $x=5$: $5x^2+4x = 5\cdot5^2+4\cdot5 = 5\cdot25+4\cdot5 = 125+20 = 145$

When $x=5$, the sum of the areas is 145 square units.

For $x=7$: $5x^2+4x = 5\cdot7^2+4\cdot7 = 5\cdot49+4\cdot7 = 245+28 = 273$

When $x=7$, the sum of the areas is 273 square units.

53. ***Familiarize.*** The perimeter is the sum of the lengths of the sides.

Translate. The sum of the lengths is found as follows:

$4a+7+a+\frac{1}{2}a+5+a+2a+3a$

Carry out. We combine like terms.

$4a+7+a+\frac{1}{2}a+5+a+2a+3a$

$= \left(4+1+\frac{1}{2}+1+2+3\right)a+(7+5)$

$= 11\frac{1}{2}a+12$, or $\frac{23}{2}a+12$

Check. We can go over our calculations. We can also perform a partial check by assigning some value to a, say 2, and carry out the computation of the perimeter in two ways.

Sum of lengths of sides:

$4 \cdot 2 + 7 + 2 + \frac{1}{2} \cdot 2 + 5 + 2 + 2 \cdot 2 + 3 \cdot 2 =$

$8 + 7 + 2 + 1 + 5 + 2 + 4 + 6 = 35$

Substituting in the polynomial we found:

$$\frac{23}{2} \cdot 2 + 12 = 23 + 12 = 35$$

Since the results are the same, our solution is probably correct.

State. A polynomial for the perimeter of the figure is $\frac{23}{2}a + 12$.

55. 

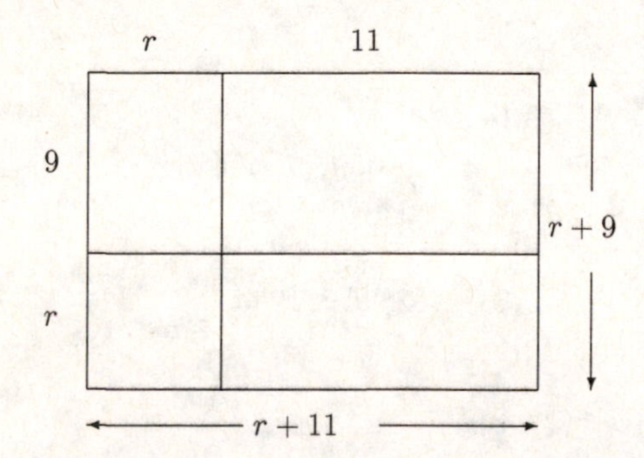

The length and width of the figure can be expressed as $r + 11$ and $r + 9$, respectively. The area of this figure (a rectangle) is the product of the length and width. An algebraic expression for the area is $(r + 11) \cdot (r + 9)$.

The algebraic expressions $9r + 99 + r^2 + 11r$ and $(r + 11) \cdot (r + 9)$ represent the same area.

	r	11	
9	A	B	9
r	C	D	r
	r	11	

The area of the figure can be found by adding the areas of the four rectangles A, B, C, and D. The area of a rectangle is the product of the length and the width.

	Area of A	+	Area of B	+	Area of C	+	Area of D
=	$9 \cdot r$	+	$11 \cdot 9$	+	$r \cdot r$	+	$11 \cdot r$
=	$9r$	+	99	+	r^2	+	$11r$

An algebraic expression for the area of the figure is $9r + 99 + r^2 + 11r$.

57. 

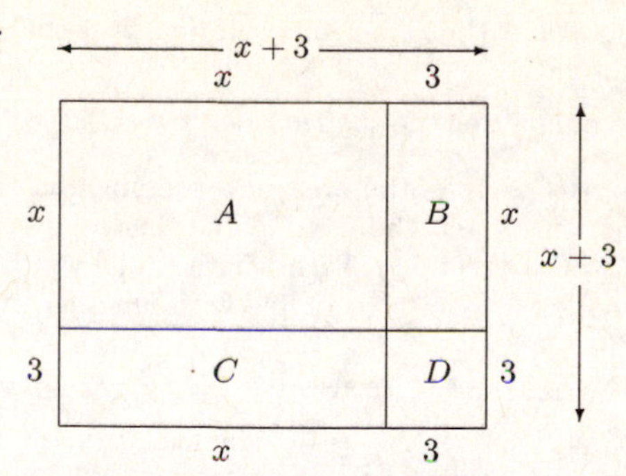

The length and width of the figure can each be expressed as $x + 3$. The area can be expressed as $(x + 3) \cdot (x + 3)$, or $(x + 3)^2$. Another way to express the area is to find an expression for the sum of the areas of the four rectangles A, B, C, and D. The area of each rectangle is the product of its length and width.

	Area of A	+	Area of B	+	Area of C	+	Area of D
=	$x \cdot x$	+	$3 \cdot x$	+	$3 \cdot x$	+	$3 \cdot 3$
=	x^2	+	$3x$	+	$3x$	+	9

The algebraic expressions $(x + 3)^2$ and $x^2 + 3x + 3x + 9$ represent the same area.

$$(x + 3)^2 = x^2 + 3x + 3x + 9$$

59.

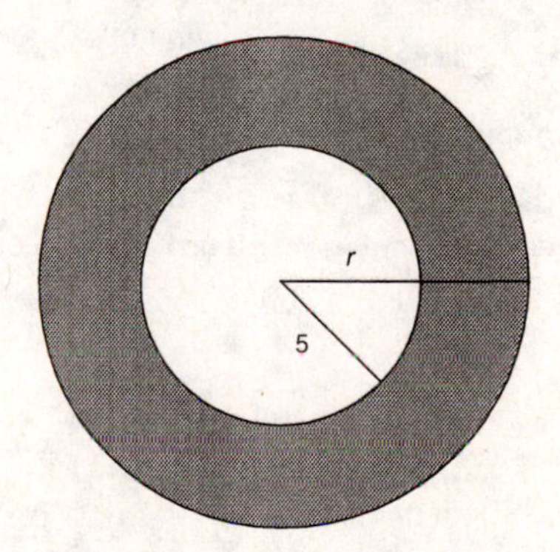

Familiarize. Recall that the area of a circle is the product of π and the square of the radius, r^2.

$$A = \pi r^2$$

Translate.

Area of circle with radius r	−	Area of circle with radius 5	=	Shaded area
$\pi \cdot r^2$	−	$\pi \cdot 5^2$	=	Shaded area

Carry out. We simplify the expression.

$$\pi \cdot r^2 - \pi \cdot 5^2 = \pi r^2 - 25\pi$$

Check. We can go over our calculations. We can also assign some value to r, say 7, and carry out the computation in two ways.

Difference of areas: $\pi \cdot 7^2 - \pi \cdot 5^2 = 49\pi - 25\pi = 24\pi$

Substituting in the polynomial: $\pi \cdot 7^2 - 25\pi = 49\pi - 25\pi = 24\pi$

Since the results are the same, our solution is probably correct.

State. A polynomial for the shaded area is $\pi r^2 - 25\pi$.

61. ***Familiarize.*** Recall that the area of a rectangle is the product of the length and the width. The shaded area is the area of the entire rectangle with length x and width y less the nonshaded area of a rectangle with length 7 and width 3.

Translate.

Area of entire rectangle	$-$	Area not shaded	is	Shaded area
$x \cdot y$	$-$	$7 \cdot 3$	$=$	Shaded area

Carry out. We simplify the expression.

$x \cdot y - 7 \cdot 3 = xy - 21$

Check. We go over the calculation. The answer checks.

State. A polynomial for the shaded area is $xy - 21$.

63. ***Familiarize.*** Recall that the area of a rectangle is the product of the length and the width and that, consequently, the area of a square with side s is s^2. The remaining floor area is the area of the entire floor less the area of the bath enclosure, in square feet.

Translate.

Area of entire floor	$-$	Area of bath enclosure	$=$	Remaining floor area
x^2	$-$	$2 \cdot 6$	$=$	Remaining floor area

Carry out. We simplify the expression.

$x^2 - 2 \cdot 6 = x^2 - 12$

Check. We go over the calculations. The answer checks.

State. A polynomial for the remaining floor area is $(x^2 - 12)$ ft^2.

65. ***Familiarize.*** Recall that the area of a circle with radius r is πr^2 and the area of a square with side s is s^2. The radius of the hot tub is half the width, $\frac{8 \text{ ft}}{2}$, or 4 ft. The remaining area of the patio is the entire area less the area of the hot tub, in square feet.

Translate.

Area of entire patio	$-$	Area of hot tub	is	Remaining patio area
z^2	$-$	$\pi \cdot 4^2$	$=$	Remaining patio area

Carry out. We simplify the expression.

$z^2 - \pi \cdot 4^2 = z^2 - \pi \cdot 16 = z^2 - 16\pi$

Check. We go over the calculations. The answer checks.

State. A polynomial for the remaining area of the patio is $(z^2 - 16\pi)$ ft^2.

67. ***Familiarize.*** Recall that the area of a square with side s is s^2 and the area of a circle with radius r is πr^2. The radius of the circle is half the diameter, or $\frac{d}{2}$ m. The area of the mat outside the circle is the area of the entire mat less the area of the circle, in square meters.

Translate.

Area of mat	$-$	Area of circle	is	Area outside the circle
12^2	$-$	$\pi \cdot \left(\frac{d}{2}\right)^2$	$=$	Area outside the circle

Carry out. We simplify the expression.

$12^2 - \pi \cdot \left(\frac{d}{2}\right)^2 = 144 - \pi \cdot \frac{d^2}{4} = 144 - \frac{d^2}{4}\pi$

Check. We go over the calculations. The answer checks.

State. A polynomial for the area of the mat outside the wrestling circle is $\left(144 - \frac{d^2}{4}\pi\right)$ m^2.

69. *Writing Exercise*

71. $5(4+3) - 5 \cdot 4 - 5 \cdot 3$

$= 5 \cdot 7 - 5 \cdot 4 - 5 \cdot 3$ Adding inside the parentheses

$= 35 - 20 - 15$ Multiplying

$= 0$ Subtracting

73. $2(5t+7) + 3t = 10t + 14 + 3t = 13t + 14$

75. $2(x+3) > 5(x-3) + 7$

$2x + 6 > 5x - 15 + 7$ Removing parentheses

$2x + 6 > 5x - 8$ Collecting like terms

$2x + 14 > 5x$ Adding 8 to both sides

$14 > 3x$ Adding $-2x$ to both sides

$\frac{14}{3} > x$ Dividing both sides by 3

The solution set is $\left\{x \middle| \frac{14}{3} > x\right\}$, or $\left\{x \middle| x < \frac{14}{3}\right\}$.

77. *Writing Exercise*

79. $(6t^2 - 7t) + (3t^2 - 4t + 5) - (9t - 6)$

$= 6t^2 - 7t + 3t^2 - 4t + 5 - 9t + 6$

$= 9t^2 - 20t + 11$

81. $(-8y^2 - 4) - (3y + 6) - (2y^2 - y)$

$= -8y^2 - 4 - 3y - 6 - 2y^2 + y$

$= -10y^2 - 2y - 10$

83. $(345.099x^3 - 6.178x) - (94.508x^3 - 8.99x)$

$= 345.099x^3 - 6.178x - 94.508x^3 + 8.99x$

$= 250.591x^3 + 2.812x$

85. ***Familiarize.*** The surface area is $2lw + 2lh + 2wh$, where l = length, w = width, and h = height of the rectangular solid. Here we have $l = 3$, $w = w$, and $h = 7$.

Translate. We substitute in the formula above.

$2 \cdot 3 \cdot w + 2 \cdot 3 \cdot 7 + 2 \cdot w \cdot 7$

Carry out. We simplify the expression.

$2 \cdot 3 \cdot w + 2 \cdot 3 \cdot 7 + 2 \cdot w \cdot 7$

$= 6w + 42 + 14w$

$= 20w + 42$

Check. We can go over the calculations. We can also assign some value to w, say 6, and carry out the computation in two ways.

Using the formula: $2 \cdot 3 \cdot 6 + 2 \cdot 3 \cdot 7 + 2 \cdot 6 \cdot 7 = 36 + 42 + 84 = 162$

Substituting in the polynomial: $20 \cdot 6 + 42 = 120 + 42 = 162$

Since the results are the same, our solution is probably correct.

State. A polynomial for the surface area is $20w + 42$.

87. *Familiarize*. The surface area is $2lw + 2lh + 2wh$, where l = length, w = width, and h = height of the rectangular solid. Here we have $l = x$, $w = x$, and $h = 5$.

Translate. We substitute in the formula above.

$$2 \cdot x \cdot x + 2 \cdot x \cdot 5 + 2 \cdot x \cdot 5$$

Carry out. We simplify the expression.

$$\begin{aligned} &2 \cdot x \cdot x + 2 \cdot x \cdot 5 + 2 \cdot x \cdot 5 \\ &= 2x^2 + 10x + 10x \\ &= 2x^2 + 20x \end{aligned}$$

Check. We can go over the calculations. We can also assign some value to x, say 3, and carry out the computation in two ways.

Using the formula: $2 \cdot 3 \cdot 3 + 2 \cdot 3 \cdot 5 + 2 \cdot 3 \cdot 5 = 18 + 30 + 30 = 78$

Substituting in the polynomial: $2 \cdot 3^2 + 20 \cdot 3 = 2 \cdot 9 + 60 = 18 + 60 = 78$

Since the results are the same, our solution is probably correct.

State. A polynomial for the surface area is $2x^2 + 20x$.

89. Length of top edges: $x + 6 + x + 6$, or $2x + 12$

Length of bottom edges: $x + 6 + x + 6$, or $2x + 12$

Length of vertical edges: $4 \cdot x$, or $4x$

Total length of edges: $2x + 12 + 2x + 12 + 4x = 8x + 24$

91. *Writing Exercise*

Exercise Set 4.4

1. $4x^4 \cdot 2x^2 = (4 \cdot 2)(x^4 \cdot x^2) = 8x^6$

Choice (b) is correct.

3. $4x^3 \cdot 2x^5 = (4 \cdot 2)(x^3 \cdot x^5) = 8x^8$

Choice (d) is correct.

5. $(4x^3)9 = (4 \cdot 9)x^3 = 36x^3$

7. $(-x^3)(x^4) = (-1 \cdot x^3)(x^4) = -1(x^3 \cdot x^4) = -1 \cdot x^7 = -x^7$

9. $(-x^6)(-x^2) = (-1 \cdot x^6)(-1 \cdot x^2) = (-1)(-1)(x^6 \cdot x^2) = x^8$

11. $(8a^2)(3a^2) = (8 \cdot 3)(a^2 \cdot a^2) = 24a^4$

13. $(0.3x^3)(-0.4x^6) = 0.3(-0.4)(x^3 \cdot x^6) = -0.12x^9$

15. $\left(-\frac{1}{4}x^4\right)\left(\frac{1}{5}x^8\right) = \left(-\frac{1}{4} \cdot \frac{1}{5}\right)(x^4 \cdot x^8) = -\frac{1}{20}x^{12}$

17. $(-5n^3)(-1) = (-5)(-1)n^3 = 5n^3$

19. $(-4y^5)(6y^2)(-3y^3) = -4(6)(-3)(y^5 \cdot y^2 \cdot y^3) = 72y^{10}$

21. $2x(4x - 6) = 2x \cdot 4x + 2x(-6) = 8x^2 - 12x$

23. $3x(x + 2) = 3x \cdot x + 3x \cdot 2 = 3x^2 + 6x$

25. $(a + 9)3a = a \cdot 3a + 9 \cdot 3a = 3a^2 + 27a$

27. $$\begin{aligned} x^2(x^3 + 1) &= x^2(x^3) + x^2(1) \\ &= x^5 + x^2 \end{aligned}$$

29. $$\begin{aligned} 3x(2x^2 - 6x + 1) &= 3x(2x^2) + 3x(-6x) + 3x(1) \\ &= 6x^3 - 18x^2 + 3x \end{aligned}$$

31. $5t^2(3t + 6) = 5t^2(3t) + 5t^2(6) = 15t^3 + 30t^2$

33. $$\begin{aligned} -6x^2(x^2 + x) &= -6x^2(x^2) - 6x^2(x) \\ &= -6x^4 - 6x^3 \end{aligned}$$

35. $$\begin{aligned} &\frac{2}{3}a^4\left(6a^5 - 12a^3 - \frac{5}{8}\right) \\ &= \frac{2}{3}a^4(6a^5) - \frac{2}{3}a^4(12a^3) - \frac{2}{3}a^4\left(\frac{5}{8}\right) \\ &= \frac{12}{3}a^9 - \frac{24}{3}a^7 - \frac{10}{24}a^4 \\ &= 4a^9 - 8a^7 - \frac{5}{12}a^4 \end{aligned}$$

37. $$\begin{aligned} (x + 2)(x + 6) &= (x + 2)x + (x + 2)6 \\ &= x \cdot x + 2 \cdot x + x \cdot 6 + 2 \cdot 6 \\ &= x^2 + 2x + 6x + 12 \\ &= x^2 + 8x + 12 \end{aligned}$$

39. $$\begin{aligned} (x + 5)(x - 2) &= (x + 5)x + (x + 5)(-2) \\ &= x \cdot x + 5 \cdot x + x(-2) + 5(-2) \\ &= x^2 + 5x - 2x - 10 \\ &= x^2 + 3x - 10 \end{aligned}$$

41. $$\begin{aligned} (a - 6)(a - 7) &= (a - 6)a + (a - 6)(-7) \\ &= a \cdot a - 6 \cdot a + a(-7) + (-6)(-7) \\ &= a^2 - 6a - 7a + 42 \\ &= a^2 - 13a + 42 \end{aligned}$$

43. $$\begin{aligned} (x + 3)(x - 3) &= (x + 3)x + (x + 3)(-3) \\ &= x \cdot x + 3 \cdot x + x(-3) + 3(-3) \\ &= x^2 + 3x - 3x - 9 \\ &= x^2 - 9 \end{aligned}$$

45. $$\begin{aligned} (5 - x)(5 - 2x) &= (5 - x)5 + (5 - x)(-2x) \\ &= 5 \cdot 5 - x \cdot 5 + 5(-2x) - x(-2x) \\ &= 25 - 5x - 10x + 2x^2 \\ &= 25 - 15x + 2x^2 \end{aligned}$$

47. $$\begin{aligned} \left(t + \frac{3}{2}\right)\left(t + \frac{4}{3}\right) &= \left(t + \frac{3}{2}\right)t + \left(t + \frac{3}{2}\right)\left(\frac{4}{3}\right) \\ &= t \cdot t + \frac{3}{2} \cdot t + t \cdot \frac{4}{3} + \frac{3}{2} \cdot \frac{4}{3} \\ &= t^2 + \frac{3}{2}t + \frac{4}{3}t + 2 \\ &= t^2 + \frac{9}{6}t + \frac{8}{6}t + 2 \\ &= t^2 + \frac{17}{6}t + 2 \end{aligned}$$

49.
$$\left(\frac{1}{4}a+2\right)\left(\frac{3}{4}a-1\right)$$
$$=\left(\frac{1}{4}a+2\right)\left(\frac{3}{4}a\right)+\left(\frac{1}{4}a+2\right)(-1)$$
$$=\frac{1}{4}a\left(\frac{3}{4}a\right)+2\cdot\frac{3}{4}a+\frac{1}{4}a(-1)+2(-1)$$
$$=\frac{3}{16}a^2+\frac{3}{2}a-\frac{1}{4}a-2$$
$$=\frac{3}{16}a^2+\frac{6}{4}a-\frac{1}{4}a-2$$
$$=\frac{3}{16}a^2+\frac{5}{4}a-2$$

51. Illustrate $x(x+5)$ as the area of a rectangle with width x and length $x+5$.

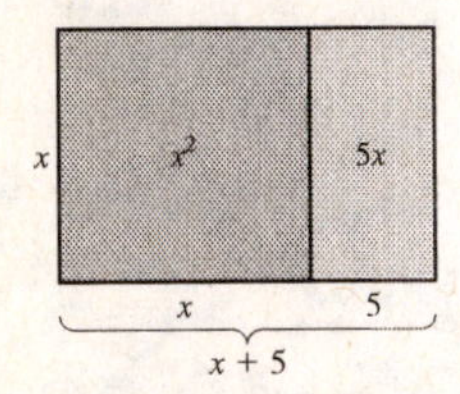

53. Illustrate $(x+1)(x+2)$ as the area of a rectangle with width $x+1$ and length $x+2$.

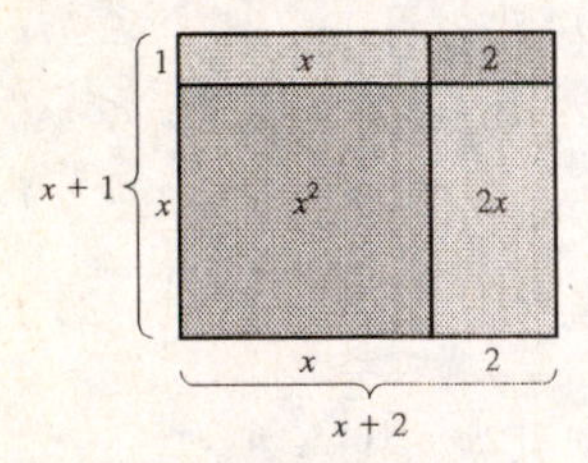

55. Illustrate $(x+5)(x+3)$ as the area of a rectangle with length $x+5$ and width $x+3$.

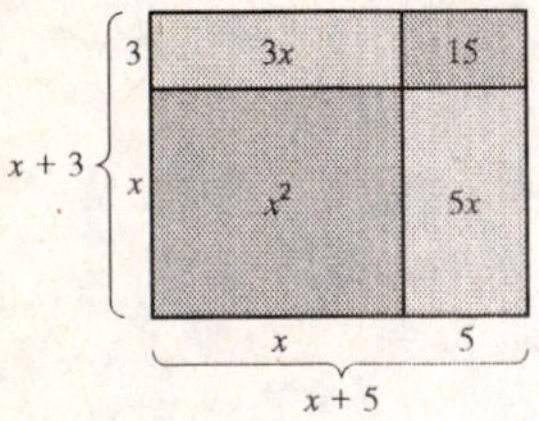

57.
$$(x^2-x+5)(x+1)$$
$$=(x^2-x+5)x+(x^2-x+5)1$$
$$=x^3-x^2+5x+x^2-x+5$$
$$=x^3+4x+5$$

A partial check can be made by selecting a convenient replacement for x, say 1, and comparing the values of the original expression and the result.

$$(1^2-1+5)(1+1) \qquad 1^3+4\cdot 1+5$$
$$=(1-1+5)(1+1) \qquad =1+4+5$$
$$=5\cdot 2 \qquad =10$$
$$=10$$

Since the value of both expressions is 10, the multiplication is very likely correct.

59.
$$(2a+5)(a^2-3a+2)$$
$$=(2a+5)a^2-(2a+5)(3a)+(2a+5)2$$
$$=2a\cdot a^2+5\cdot a^2-2a\cdot 3a-5\cdot 3a+2a\cdot 2+5\cdot 2$$
$$=2a^3+5a^2-6a^2-15a+4a+10$$
$$=2a^3-a^2-11a+10$$

A partial check can be made as in Exercise 57.

61.
$$(y^2-7)(2y^3+y+1)$$
$$=(y^2-7)(2y^3)+(y^2-7)y+(y^2-7)(1)$$
$$=y^2\cdot 2y^3-7\cdot 2y^3+y^2\cdot y-7\cdot y+y^2\cdot 1-7\cdot 1$$
$$=2y^5-14y^3+y^3-7y+y^2-7$$
$$=2y^5-13y^3+y^2-7y-7$$

A partial check can be made as in Exercise 57.

63.
$$(3x+2)(5x+4x+7)$$
$$=(3x+2)(9x+7)$$
$$=(3x+2)(9x)+(3x+2)(7)$$
$$=3x\cdot 9x+2\cdot 9x+3x\cdot 7+2\cdot 7$$
$$=27x^2+18x+21x+14$$
$$=27x^2+39x+14$$

A partial check can be made as in Exercise 57.

65.
$$\begin{array}{rl}
x^2-3x+2 & \text{Line up like terms}\\
x^2+x+1 & \text{in columns}\\ \hline
x^2-3x+2 & \text{Multiplying by } 1\\
x^3-3x^2+2x & \text{Multiplying by } x\\
x^4-3x^3+2x^2 & \text{Multiplying by } x^2\\ \hline
x^4-2x^3-x+2 &
\end{array}$$

A partial check can be made as in Exercise 57.

67.
$$\begin{array}{rl}
2t^2-5t-4 & \\
3t^2-t+\frac{1}{2} & \\ \hline
t^2-\frac{5}{2}t-2 & \text{Multiplying by } \frac{1}{2}\\
-2t^3+5t^2+4t & \text{Multiplying by } -t\\
6t^4-15t^3-12t^2 & \text{Multiplying by } 3t^2\\ \hline
6t^4-17t^3-6t^2+\frac{3}{2}t-2 &
\end{array}$$

A partial check can be made as in Exercise 57.

69. We will multiply horizontally while still aligning like terms.

$$(x+1)(x^3+7x^2+5x+4)$$
$$\begin{array}{rl}
=x^4+7x^3+5x^2+4x & \text{Multiplying by } x\\
+x^3+7x^2+5x+4 & \text{Multiplying by } 1\\ \hline
=x^4+8x^3+12x^2+9x+4 &
\end{array}$$

A partial check can be made as in Exercise 57.

71. *Writing Exercise*

73. $5-3\cdot 2+7=5-6+7=-1+7=6$

75. $(8-2)(8+2)+2^2-8^2$
$= 6\cdot 10+2^2-8^2$
$= 6\cdot 10+4-64$
$= 60+4-64$
$= 64-64$
$= 0$

77. *Writing Exercise*

79. The shaded area is the area of the large rectangle, $6y(14y-5)$ less the area of the unshaded rectangle, $3y(3y+5)$. We have:

$6y(14y-5)-3y(3y+5)$
$= 84y^2-30y-9y^2-15y$
$= 75y^2-45y$

81. Let n = the missing number. Label the figure with the known areas.

Then the area of the figure is $x^2+2x+nx+2n$. This is equivalent to $x^2+7x+10$, so we have $2x+nx=7x$ and $2n=10$. Solving either equation for n, we find that the missing number is 5.

83.

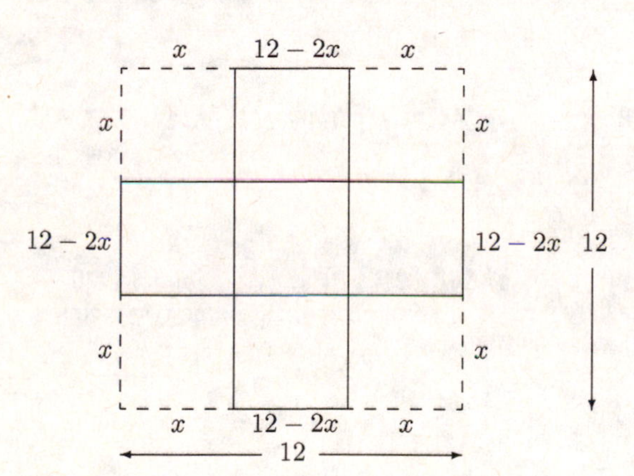

The dimensions, in inches, of the box are $12-2x$ by $12-2x$ by x. The volume is the product of the dimensions (volume = length × width × height):

Volume $= (12-2x)(12-2x)x$
$= (144-48x+4x^2)x$
$= 144x-48x^2+4x^3$ in^3, or
$4x^3-48x^2+144x$ in^3

The outside surface area is the sum of the area of the bottom and the areas of the four sides. The dimensions, in inches, of the bottom are $12-2x$ by $12-2x$, and the dimensions, in inches, of each side are x by $12-2x$.

Surface area = Area of bottom + 4 · Area of each side
$= (12-2x)(12-2x)+4\cdot x(12-2x)$
$= 144-24x-24x+4x^2+48x-8x^2$
$= 144-48x+4x^2+48x-8x^2$
$= 144-4x^2$ in^2, or $-4x^2+144$ in^2

85. We have a rectangular solid with dimensions x m by x m by $x+2$ m with a rectangular solid piece with dimensions 6 m by 5 m by 7 m cut out of it.

Volume = Volume of large solid − Volume of small solid
$= (x\text{ m})(x\text{ m})(x+2\text{ m})-(6\text{ m})(5\text{ m})(7\text{ m})$
$= x^2(x+2)\text{ m}^3-210\text{ m}^3$
$= x^3+2x^2-210\text{ m}^3$

87. Let x = the width of the garden. Then $2x$ = the length of the garden.

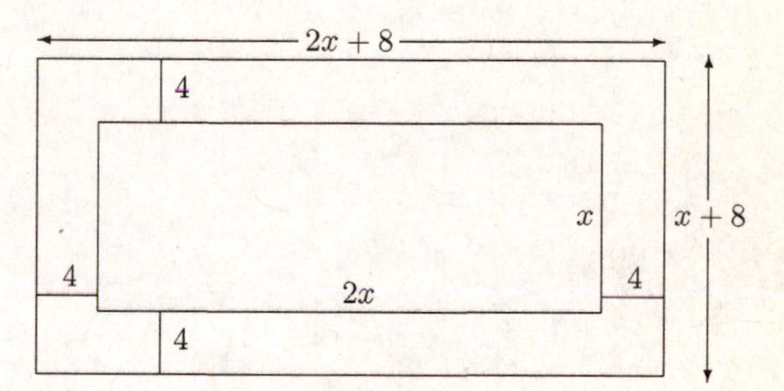

Area of garden and sidewalk together is Area of garden alone + 256 ft^2

$(2x+8)(x+8) = 2x\cdot x + 256$
$2x^2+24x+64 = 2x^2+256$
$24x = 192$
$x = 8$

The dimensions are 8 ft by 16 ft.

89. $(x-2)(x-7)-(x-7)(x-2)$

First observe that, by the commutative law of multiplication, $(x-2)(x-7)$ and $(x-7)(x-2)$ are equivalent expressions. Then when we subtract $(x-7)(x-2)$ from $(x-2)(x-7)$, the result is 0.

91. $(x-a)(x-b)\cdots(x-x)(x-y)(x-z)$
$= (x-a)(x-b)\cdots 0\cdot(x-y)(x-z)$
$= 0$

Exercise Set 4.5

1. It is true that FOIL is simply a memory device for finding the product of two binomials.

3. This statement is false. See the material on squaring binomials at the bottom of page 268 in the text.

5. $(x+3)(x^2+5)$
F O I L
$= x\cdot x^2 + x\cdot 5 + 3\cdot x^2 + 3\cdot 5$
$= x^3+5x+3x^2+15$, or $x^3+3x^2+5x+15$

7. $(x^3+6)(x+2)$
F O I L
$= x^3\cdot x + x^3\cdot 2 + 6\cdot x + 6\cdot 2$
$= x^4+2x^3+6x+12$

9. $(y+2)(y-3)$
F O I L
$= y\cdot y + y\cdot(-3) + 2\cdot y + 2\cdot(-3)$
$= y^2-3y+2y-6$
$= y^2-y-6$

11. $(3x+2)(3x+5)$
F O I L
$= 3x\cdot 3x + 3x\cdot 5 + 2\cdot 3x + 2\cdot 5$
$= 9x^2+15x+6x+10$
$= 9x^2+21x+10$

13. $(5x-4)(x+2)$
F O I L
$= 5x\cdot x + 5x\cdot 2 + (-4)\cdot x + (-4)\cdot 2$
$= 5x^2+10x-4x-8$
$= 5x^2+6x-8$

15. $(1+3t)(2-3t)$
F O I L
$= 1\cdot 2 + 1(-3t) + 3t\cdot 2 + 3t(-3t)$
$= 2-3t+6t-9t^2$
$= 2+3t-9t^2$

17. $(2x-5)(x-4)$
F O I L
$= 2x\cdot x + 2x\cdot(-4) + (-5)\cdot x + (-5)\cdot(-4)$
$= 2x^2-8x-5x+20$
$= 2x^2-13x+20$

19. $\left(p-\frac{1}{4}\right)\left(p+\frac{1}{4}\right)$
F O I L
$= p\cdot p + p\cdot\frac{1}{4} + \left(-\frac{1}{4}\right)\cdot p + \left(-\frac{1}{4}\right)\cdot\frac{1}{4}$
$= p^2+\frac{1}{4}p-\frac{1}{4}p-\frac{1}{16}$
$= p^2-\frac{1}{16}$

21. $(x-0.1)(x+0.1)$
F O I L
$= x\cdot x + x\cdot(0.1) + (-0.1)\cdot x + (-0.1)(0.1)$
$= x^2+0.1x-0.1x-0.01$
$= x^2-0.01$

23. $(-2x+1)(x+6)$
F O I L
$= -2x^2-12x+x+6$
$= -2x^2-11x+6$

25. $(a+9)(a+9)$
F O I L
$= a^2+9a+9a+81$
$= a^2+18a+81$

27. $(1+3t)(1-5t)$
F O I L
$= 1-5t+3t-15t^2$
$= 1-2t-15t^2$

29. $(x^2+3)(x^3-1)$
F O I L
$= x^5-x^2+3x^3-3$, or $x^5+3x^3-x^2-3$

31. $(3x^2-2)(x^4-2)$
F O I L
$= 3x^6-6x^2-2x^4+4$, or $3x^6-2x^4-6x^2+4$

33. $(2t^3+5)(2t^3+3)$
F O I L
$= 4t^6+6t^3+10t^3+15$
$= 4t^6+16t^3+15$

35. $(8x^3+5)(x^2+2)$
F O I L
$= 8x^5+16x^3+5x^2+10$

37. $(4x^2+3)(x-3)$
F O I L
$= 4x^3-12x^2+3x-9$

39. $(x+7)(x-7)$ Product of sum and difference of the same two terms
$= x^2-7^2$
$= x^2-49$

41. $(2x+1)(2x-1)$ Product of sum and difference of the same two terms
$= (2x)^2-1^2$
$= 4x^2-1$

43. $(5m-2)(5m+2)$ Product of sum and difference of the same two terms
$= (5m)^2-2^2$
$= 25m^2-4$

45. $(3x^4-1)(3x^4+1)$
$= (3x^4)^2-1^2$
$= 9x^8-1$

47. $(x^4+7)(x^4-7)$
$= (x^4)^2-7^2$
$= x^8-49$

49. $\left(t-\frac{3}{4}\right)\left(t+\frac{3}{4}\right)$
$= t^2-\left(\frac{3}{4}\right)^2$
$= t^2-\frac{9}{16}$

51. $(x+2)^2$
$= x^2 + 2\cdot x\cdot 2 + 2^2$ Square of a binomial
$= x^2 + 4x + 4$

53. $(7x^3+1)^2$ Square of a binomial
$= (7x^3)^2 + 2\cdot 7x^3\cdot 1 + 1^2$
$= 49x^6 + 14x^3 + 1$

55. $\left(a - \frac{2}{5}\right)^2$ Square of a binomial
$= a^2 - 2\cdot a\cdot \frac{2}{5} + \left(\frac{2}{5}\right)^2$
$= a^2 - \frac{4}{5}a + \frac{4}{25}$

57. $= (t^3+5)^2$ Square of a binomial
$= (t^3)^2 + 2\cdot t^3\cdot 5 + 5^2$
$= t^6 + 10t^3 + 25$

59. $(2-3x^4)^2 = 2^2 - 2\cdot 2\cdot 3x^4 + (3x^4)^2$
$= 4 - 12x^4 + 9x^8$

61. $(5+6t^2)^2 = 5^2 + 2\cdot 5\cdot 6t^2 + (6t^2)^2$
$= 25 + 60t^2 + 36t^4$

63. $(7x-0.3)^2 = (7x)^2 - 2(7x)(0.3) + (0.3)^2$
$= 49x^2 - 4.2x + 0.09$

65. $5a^3(2a^2-1)$
$= 5a^3\cdot 2a^2 - 5a^3\cdot 1$ Multiplying each term of the binomial by the monomial
$= 10a^5 - 5a^3$

67. $(a-3)(a^2+2a-4)$

$$\begin{array}{rl} = a^3 + 2a^2 - 4a & \text{Multiplying horizontally} \\ - 3a^2 - 6a + 12 & \text{and aligning like terms} \\ \hline = a^3 - a^2 - 10a + 12 & \end{array}$$

69. $(3-2x^3)^2$
$= 3^2 - 2\cdot 3\cdot 2x^3 + (2x^3)^2$ Squaring a binomial
$= 9 - 12x^3 + 4x^6$

71. $5x(x^2+6x-2)$
$= 5x\cdot x^2 + 5x\cdot 6x + 5x(-2)$ Multiplying each term of the trinomial by the monomial
$= 5x^3 + 30x^2 - 10x$

73. $(-t^3+1)^2$
$= (-t^3)^2 + 2(-t)^3(1) + 1^2$ Squaring a binomial
$= t^6 - 2t^3 + 1$

75. $3t^2(5t^3 - t^2 + t)$
$= 3t^2\cdot 5t^3 + 3t^2(-t^2) + 3t^2\cdot t$ Multiplying each term of the trinomial by the monomial
$= 15t^5 - 3t^4 + 3t^3$

77. $(6x^4-3)^2$ Squaring a binomial
$= (6x^4)^2 - 2\cdot 6x^4\cdot 3 + 3^2$
$= 36x^8 - 36x^4 + 9$

79. $(3x+2)(4x^2+5)$ Product of two binomials; use FOIL
$= 3x\cdot 4x^2 + 3x\cdot 5 + 2\cdot 4x^2 + 2\cdot 5$
$= 12x^3 + 15x + 8x^2 + 10$, or
$12x^3 + 8x^2 + 15x + 10$

81. $(5-6x^4)^2$ Squaring a binomial
$= 5^2 - 2\cdot 5\cdot 6x^4 + (6x^4)^2$
$= 25 - 60x^4 + 36x^8$

83. $(a+1)(a^2-a+1)$

$$\begin{array}{rl} = a^3 - a^2 + a & \text{Multiplying horizontally} \\ a^2 - a + 1 & \text{and aligning like terms} \\ \hline a^3 \qquad + 1 & \end{array}$$

85.

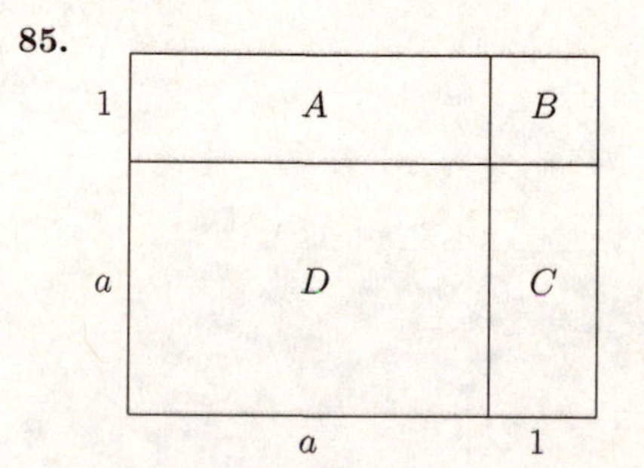

We can find the shaded area in two ways.

Method 1: The figure is a square with side $a+1$, so the area is $(a+1)^2 = a^2 + 2a + 1$.

Method 2: We add the areas of A, B, C, and D.

$1\cdot a + 1\cdot 1 + 1\cdot a + a\cdot a = a + 1 + a + a^2 = a^2 + 2a + 1$.

Either way we find that the total shaded area is $a^2 + 2a + 1$.

87.

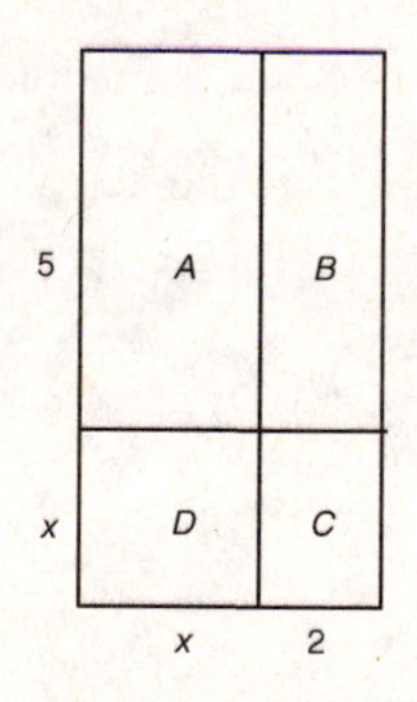

We can find the shaded area in two ways.

Method 1: The figure is a rectangle with dimensions $x+5$ by $x+2$, so the area is

$(x+5)(x+2) = x^2 + 2x + 5x + 10 = x^2 + 7x + 10$.

Method 2: We add the areas of A, B, C, and D.

$5\cdot x + 2\cdot 5 + 2\cdot x + x\cdot x = 5x + 10 + 2x + x^2 = x^2 + 7x + 10$.

Either way, we find that the area is $x^2 + 7x + 10$.

89.

We can find the shaded area in two ways.

Method 1: The figure is a square with side $x+7$, so the area is $(x+7)^2 = x^2+14x+49$.

Method 2: We add the areas of A, B, C, and D.

$x\cdot x+x\cdot 7+7\cdot 7+7\cdot x = x^2+7x+49+7x = x^2+14x+49$.

Either way, we find that the total shaded area is $x^2+14x+49$.

91.

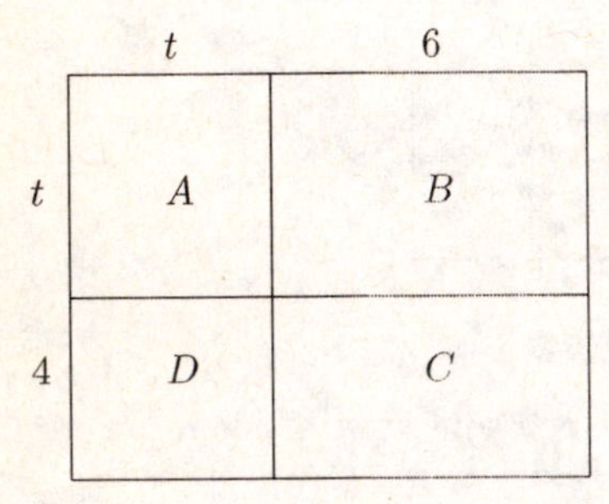

We can find the shaded area in two ways.

Method 1: The figure is a rectangle with dimensions $t+6$ by $t+4$, so the area is $(t+6)(t+4) = t^2+4t+6t+24 = t^2+10t+24$.

Method 2: We add the areas of A, B, C, and D.

$t\cdot t+t\cdot 6+6\cdot 4+4\cdot t = t^2+6t+24+4t = t^2+10t+24$.

Either way, we find that the total shaded area is $t^2+10t+24$.

93.

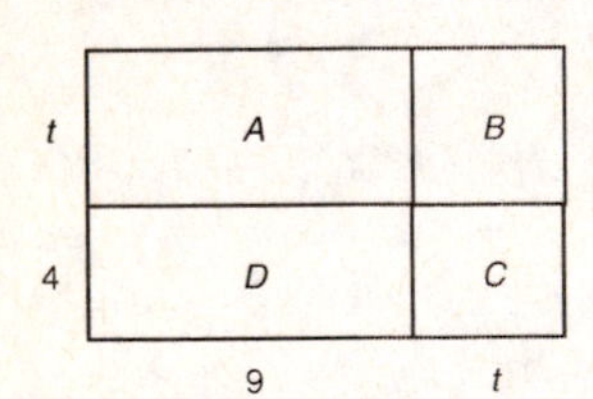

We can find the shaded area in two ways.

Method 1: The figure is a rectangle with dimensions $t+9$ by $t+4$, so the area is

$(t+9)(t+4) = t^2+4t+9t+36 = t^2+13t+36$

Method 2: We add the areas of A, B, C, and D.

$9\cdot t+t\cdot t+4\cdot t+4\cdot 9 = 9t+t^2+4t+36 = t^2+13t+36$.

Either way, we find that the total shaded area is $t^2+13t+36$.

95.

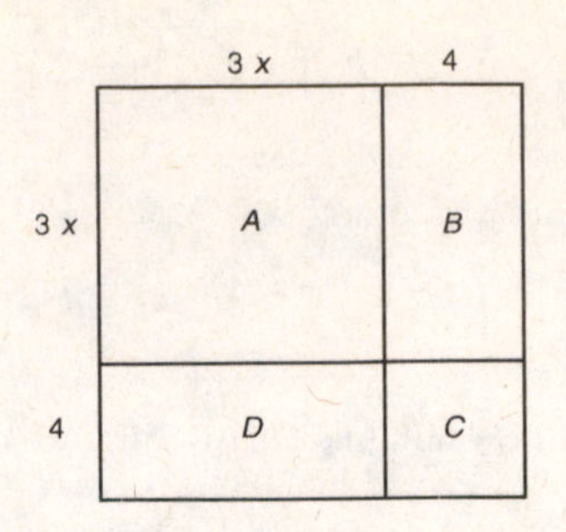

We can find the shaded area in two ways.

Method 1: The figure is a square with side $3x+4$, so the area is $(3x+4)^2 = 9x^2+24x+16$.

Method 2: We add the areas of A, B, C, and D.

$3x\cdot 3x+3x\cdot 4+4\cdot 4+3x\cdot 4 = 9x^2+12x+16+12x = 9x^2+24x+16$.

Either way, we find that the total shaded area is $9x^2+24x+16$.

97. We draw a square with side $x+5$.

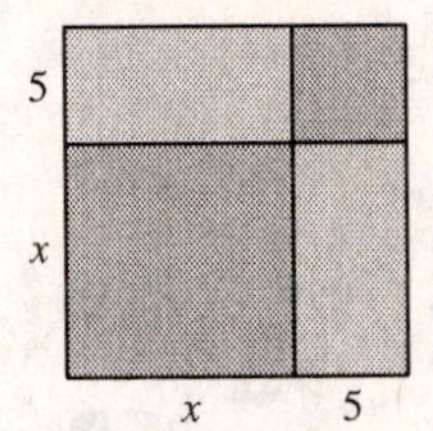

99. We draw a square with side $t+9$.

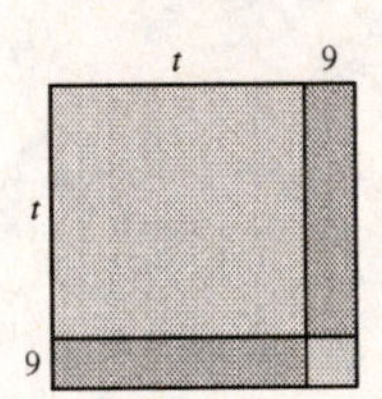

101. We draw a square with side $3+x$.

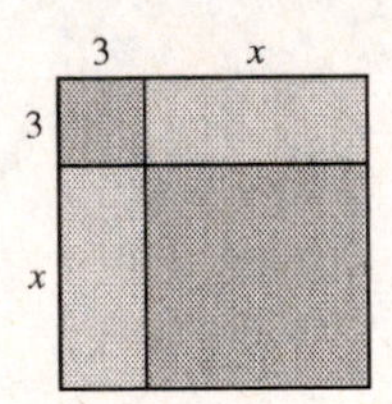

103. *Writing Exercise*

105. ***Familiarize.*** Let t = the number of watts used by the television set. Then $10t$ = the number of watts used by the lamps, and $40t$ = the number of watts used by the air conditioner.

Translate.

$$\underbrace{\text{Lamp watts}} + \underbrace{\text{Air conditioner watts}} + \underbrace{\text{Television watts}} = \underbrace{\text{Total watts}}$$
$$\downarrow \quad \downarrow \quad \downarrow \quad \downarrow \quad \downarrow \quad \downarrow \quad \downarrow$$
$$10t + 40t + t = 2550$$

Solve. We solve the equation.

$$10t + 40t + t = 2550$$
$$51t = 2550$$
$$t = 50$$

The possible solution is:

Television, t: 50 watts

Lamps, $10t$: $10 \cdot 50$, or 500 watts

Air conditioner, $40t$: $40 \cdot 50$, or 2000 watts

Check. The number of watts used by the lamps, 500, is 10 times 50, the number used by the television. The number of watts used by the air conditioner, 2000, is 40 times 50, the number used by the television. Also, $50+500+2000 = 2550$, the total wattage used.

State. The television uses 50 watts, the lamps use 500 watts, and the air conditioner uses 2000 watts.

107. $5xy = 8$

$y = \frac{8}{5x}$ Dividing both sides by $5x$

109. $ax - b = c$

$ax = b + c$ Adding b to both sides

$x = \frac{b+c}{a}$ Dividing both sides by a

111. *Writing Exercise*

113. $(4x^2 + 9)(2x + 3)(2x - 3)$

$= (4x^2 + 9)(4x^2 - 9)$

$= 16x^4 - 81$

115. $(3t - 2)^2(3t + 2)^2$

$= [(3t - 2)(3t + 2)]^2$

$= (9t^2 - 4)^2$

$= 81t^4 - 72t^2 + 16$

117. $(t^3 - 1)^4(t^3 + 1)^4$

$= [(t^3 - 1)(t^3 + 1)]^4$

$= (t^6 - 1)^4$

$= [(t^6 - 1)^2]^2$

$= (t^{12} - 2t^6 + 1)^2$

$= (t^{12} - 2t^6 + 1)(t^{12} - 2t^6 + 1)$

$= t^{24} - 2t^{18} + t^{12} - 2t^{18} + 4t^{12} - 2t^6 +$
$t^{12} - 2t^6 + 1$

$= t^{24} - 4t^{18} + 6t^{12} - 4t^6 + 1$

119. $18 \times 22 = (20 - 2)(20 + 2) = 20^2 - 2^2 =$
$400 - 4 = 396$

121. $(x + 2)(x - 5) = (x + 1)(x - 3)$

$x^2 - 5x + 2x - 10 = x^2 - 3x + x - 3$

$x^2 - 3x - 10 = x^2 - 2x - 3$

$-3x - 10 = -2x - 3$ Adding $-x^2$

$-3x + 2x = 10 - 3$ Adding $2x$ and 10

$-x = 7$

$x = -7$

The solution is -7.

123.

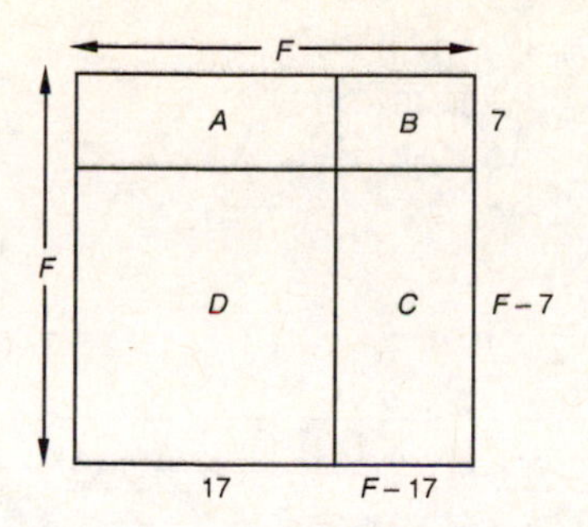

The area of the entire figure is F^2. The area of the unshaded region, C, is $(F - 7)(F - 17)$. Then one expression for the area of the shaded region is
$F^2 - (F - 7)(F - 17)$.

To find a second expression we add the areas of regions A, B, and D. We have:

$17 \cdot 7 + 7(F - 17) + 17(F - 7)$

$= 119 + 7F - 119 + 17F - 119$

$= 24F - 119$

It is possible to find other equivalent expressions also.

125.

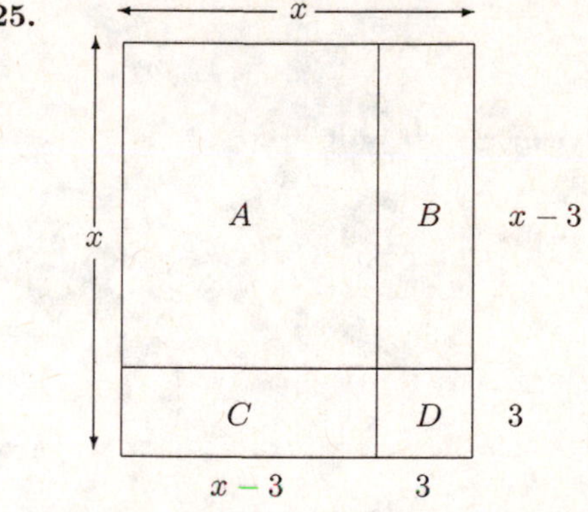

The area of the entire region is x^2. The unshaded area is $3 \cdot 3$, or 9, so an expression for the shaded area is $x^2 - 9$.

To find another expression we add the areas of regions A, B, and C. The dimensions of region A are $x - 3$ by $x - 3$ and regions B and C each have dimensions 3 by $x - 3$, so the sum of the areas is $(x - 3)^2 + 3(x - 3) + 3(x - 3)$. It is possible to find other equivalent expressions also.

127.

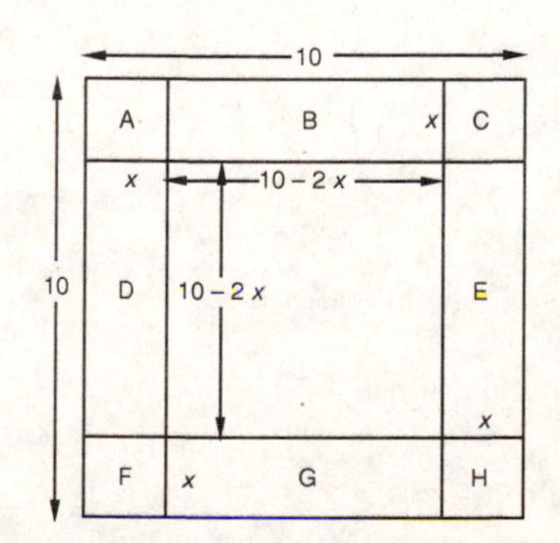

$(10 - 2x)^2$ is the area of the entire square less the areas of A, B, C, D, E, F, G, and H. The areas of A, C, F, and H are each $x \cdot x$, or x^2. The areas of B, D, E, and G are each $x(10 - 2x)$, or $10x - 2x^2$. We have:

$$(10-2x)^2 = 10^2 - 4 \cdot x^2 - 4(10x - 2x^2)$$
$$= 100 - 4x^2 - 40x + 8x^2$$
$$= 100 - 40x + 4x^2$$

129. 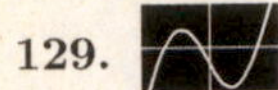

Exercise Set 4.6

1. $(2x-7y)^2$ is the square of a binomial, choice (a).

3. $(5a+6b)(-6b+5a)$, or $(5a+6b)(5a-6b)$ is the product of the sum and difference of the same two terms, choice (b).

5. $(r-3s)(5r+3s)$ is neither the square of a binomial nor the product of the sum and difference of the same two terms, so choice (c) is appropriate.

7. $(4x-9y)(4x-9y)$, or $(4x-9y)^2$ is the square of a binomial, choice (a).

9. We replace x by 5 and y by -2.

$x^2 - 3y^2 + 2xy = 5^2 - 3(-2)^2 + 2 \cdot 5(-2) =$

$25 - 12 - 20 = -7$.

11. We replace x by 2, y by -3, and z by -4.

$xyz^2 - z = 2(-3)(-4)^2 - (-4) = -96 + 4 = -92$

13. Evaluate the polynomial for $h = 160$ and $A = 50$.

$0.041h - 0.018A - 2.69$

$= 0.041(160) - 0.018(50) - 2.69$

$= 6.56 - 0.9 - 2.69$

$= 2.97$

The woman's lung capacity is 2.97 liters.

15. Evaluate the polynomial for $w = 87$, $h = 185$, and $a = 59$.

$19.18w + 7h - 9.52a + 92.4$

$= 19.18(87) + 7(185) - 9.52(59) + 92.4$

$= 1668.66 + 1295 - 561.68 + 92.4$

≈ 2494

The daily caloric needs are 2494 calories.

17. Evaluate the polynomial for $h = 4$, $r = \frac{3}{4}$, and $\pi \approx 3.14$.

$2\pi rh + \pi r^2$

$\approx 2(3.14)\left(\frac{3}{4}\right)(4) + 3.14\left(\frac{3}{4}\right)^2$

$\approx 18.84 + 1.76625$

≈ 20.60625

The surface area is about 20.60625 in^2.

19. Evaluate the polynomial for $h = 50$, $v = 18$, and $t = 2$.

$h + vt - 4.9t^2$

$= 50 + 18 \cdot 2 - 4.9(2)^2$

$= 50 + 36 - 19.6$

$= 66.4$

The ball will be 66.4 m above the ground 2 seconds after it is thrown.

21. $x^3y - 2xy + 3x^2 - 5$

Term	Coefficient	Degree	
x^3y	1	4	(Think: $x^3y = x^3y^1$)
$-2xy$	-2	2	(Think: $-2xy = -2x^1y^1$)
$3x^2$	3	2	
-5	-5	0	(Think: $-5 = -5x^0$)

The degree of the polynomial is the degree of the term of highest degree. The term of highest degree is x^3y. Its degree is 4, so the degree of the polynomial is 4.

23. $17x^2y^3 - 3x^3yz - 7$

Term	Coefficient	Degree	
$17x^2y^3$	17	5	
$-3x^3yz$	-3	5	(Think: $-3x^3yz = -3x^3y^1z^1$)
-7	-7	0	(Think: $-7 = -7x^0$)

The terms of highest degree are $17x^2y^3$ and $-3x^3yz$. Each has degree 5. The degree of the polynomial is 5.

25. $7a + b - 4a - 3b = (7-4)a + (1-3)b = 3a - 2b$

27. $3x^2y - 2xy^2 + x^2 + 5x$

There are no like terms, so none of the terms can be collected.

29.
$$2u^2v - 3uv^2 + 6u^2v - 2uv^2 + 7u^2$$
$$= (2+6)u^2v + (-3-2)uv^2 + 7u^2$$
$$= 8u^2v - 5uv^2 + 7u^2$$

31.
$$5a^2c - 2ab^2 + a^2b - 3ab^2 + a^2c - 2ab^2$$
$$= (5+1)a^2c + (-2-3-2)ab^2 + a^2b$$
$$= 6a^2c - 7ab^2 + a^2b$$

33.
$$(4x^2 - xy + y^2) + (-x^2 - 3xy + 2y^2)$$
$$= (4-1)x^2 + (-1-3)xy + (1+2)y^2$$
$$= 3x^2 - 4xy + 3y^2$$

35.
$$(3a^4 - 5ab + 6ab^2) - (9a^4 + 3ab - ab^2)$$
$$= 3a^4 - 5ab + 6ab^2 - 9a^4 - 3ab + ab^2$$
Adding the opposite
$$= (3-9)a^4 + (-5-3)ab + (6+1)ab^2$$
$$= -6a^4 - 8ab + 7ab^2$$

37. $(5r^2 - 4rt + t^2) + (-6r^2 - 5rt - t^2) + (-5r^2 + 4rt - t^2)$

Observe that the polynomials $5r^2 - 4rt + t^2$ and $-5r^2 + 4rt - t^2$ are opposites. Thus, their sum is 0 and the sum in the exercise is the remaining polynomial, $-6r^2 - 5rt - t^2$.

39.
$$(x^3 - y^3) - (-2x^3 + x^2y - xy^2 + 2y^3)$$
$$= x^3 - y^3 + 2x^3 - x^2y + xy^2 - 2y^3$$
$$= 3x^3 - 3y^3 - x^2y + xy^2, \text{ or}$$
$$3x^3 - x^2y + xy^2 - 3y^3$$

41.
$$(2y^4x^2 - 5y^3x) + (5y^4x^2 - y^3x) + (3y^4x^2 - 2y^3x)$$
$$= (2+5+3)y^4x^2 + (-5-1-2)y^3x$$
$$= 10y^4x^2 - 8y^3x$$

43. $(4x+5y)+(-5x+6y)-(7x+3y)$
$= 4x+5y-5x+6y-7x-3y$
$= (4-5-7)x+(5+6-3)y$
$= -8x+8y$

45. $(3z-u)(2z+3u) = \overset{F}{6z^2} + \overset{O}{9zu} - \overset{I}{2uz} - \overset{L}{3u^2}$
$= 6z^2+7zu-3u^2$

47. $(xy+7)(xy-4) = \overset{F}{x^2y^2} - \overset{O}{4xy} + \overset{I}{7xy} - \overset{L}{28} - 28$
$= x^2y^2+3xy-28$

49. $(2a-b)(2a+b)$ $[(A+B)(A-B)=A^2-B^2]$
$= 4a^2-b^2$

51. $(5rt-2)(3rt+1) = \overset{F}{15r^2t^2} + \overset{O}{5rt} - \overset{I}{6rt} - \overset{L}{2}$
$= 15r^2t^2-rt-2$

53. $(m^3n+8)(m^3n-6)$
$= \overset{F}{m^6n^2} - \overset{O}{6m^3n} + \overset{I}{8m^3n} - \overset{L}{48}$
$= m^6n^2+2m^3n-48$

55. $(6x-2y)(5x-3y)$
$= \overset{F}{30x^2} - \overset{O}{18xy} - \overset{I}{10xy} + \overset{L}{6y^2}$
$= 30x^2-28xy+6y^2$

57. $(pq+0.1)(-pq+0.1)$
$= (0.1+pq)(0.1-pq)$ $[(A+B)(A-B)=A^2-B^2]$
$= 0.01-p^2q^2$

59. $(x+h)^2$
$= x^2+2xh+h^2$ $[(A+B)^2=A^2+2AB+B^2]$

61. $(4a+5b)^2$
$= 16a^2+40ab+25b^2$ $[(A+B)^2=A^2+2AB+B^2]$

63. $(c^2-d)(c^2+d) = (c^2)^2-d^2$
$= c^4-d^2$

65. $(ab+cd^2)(ab-cd^2) = (ab)^2-(cd^2)^2$
$= a^2b^2-c^2d^4$

67. $(a+b-c)(a+b+c)$
$= [(a+b)-c][(a+b)+c]$
$= (a+b)^2-c^2$
$= a^2+2ab+b^2-c^2$

69. $[a+b+c][a-(b+c)]$
$= [a+(b+c)][a-(b+c)]$
$= a^2-(b+c)^2$
$= a^2-(b^2+2bc+c^2)$
$= a^2-b^2-2bc-c^2$

71. The figure is a rectangle with dimensions $a+b$ by $a+c$. Its area is $(a+b)(a+c)=a^2+ac+ab+bc$.

73. The figure is a parallelogram with base $x+z$ and height $x-z$. Thus the area is $(x+z)(x-z)=x^2-z^2$.

75. The figure is a square with side $x+y+z$. Thus the area is
$(x+y+z)^2$
$= [(x+y)+z]^2$
$= (x+y)^2+2(x+y)(z)+z^2$
$= x^2+2xy+y^2+2xz+2yz+z^2$.

77. The figure is a triangle with base $x+2y$ and height $x-y$. Thus the area is $\frac{1}{2}(x+2y)(x-y)=\frac{1}{2}(x^2+xy-2y^2)=\frac{1}{2}x^2+\frac{1}{2}xy-y^2$.

79. We draw a rectangle with dimensions $r+s$ by $u+v$.

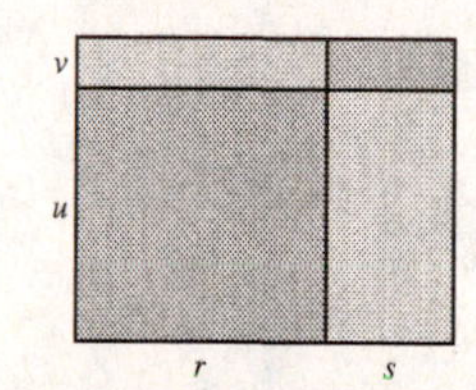

81. We draw a rectangle with dimensions $a+b+c$ by $a+d+f$.

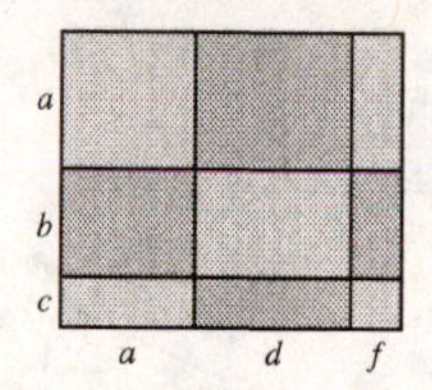

83. *Writing Exercise*

85. $5+\dfrac{7+4+2\cdot 5}{3}$

$=5+\dfrac{7+4+10}{3}$	Multiplying
$=5+\dfrac{21}{3}$	Adding in the numerator
$=5+7$	Dividing
$=12$	Adding

87. $(4+3\cdot5+8)\div3\cdot3$

$=(4+15+8)\div3\cdot3$	Multiplying inside the parentheses
$=27\div3\cdot3$	Adding
$=9\cdot3$	Dividing
$=27$	Multiplying

89. $[3\cdot5-4\cdot2+7(-3)]\div(-2)$

$=(15-8-21)\div(-2)$	Multiplying
$=-14\div(-2)$	Subtracting
$=7$	Dividing

91. *Writing Exercise*

93. It is helpful to add additional labels to the figure.

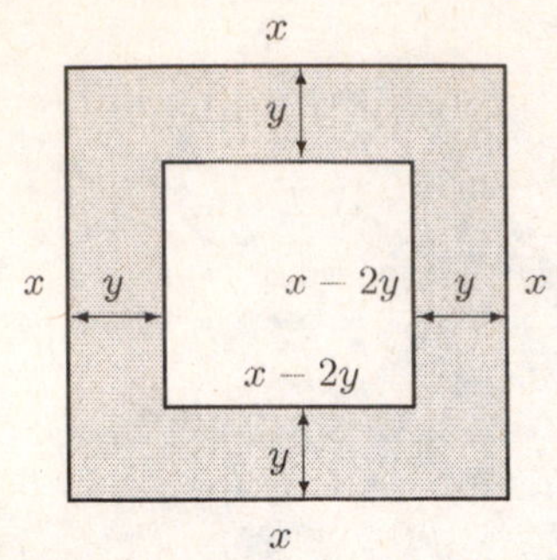

The area of the large square is $x \cdot x$, or x^2. The area of the small square is $(x-2y)(x-2y)$, or $(x-2y)^2$.

Area of shaded region	=	Area of large square	−	Area of small square
Area of shaded region	=	x^2	−	$(x-2y)^2$

$$= x^2 - (x^2 - 4xy + 4y^2)$$
$$= x^2 - x^2 + 4xy - 4y^2$$
$$= 4xy - 4y^2$$

95. It is helpful to add additional labels to the figure.

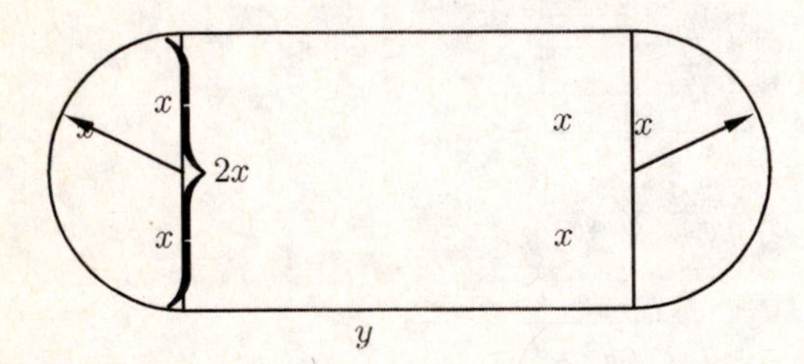

The two semicircles make a circle with radius x. The area of that circle is πx^2. The area of the rectangle is $2x \cdot y$. The sum of the two regions, $\pi x^2 + 2xy$, is the area of the shaded region.

97. The figure can be thought of as a cube with side x, a rectangular solid with dimensions x by x by y, a rectangular solid with dimensions x by y by y, and a rectangular solid with dimensions y by y by $2y$. Thus the volume is

$x^3 + x \cdot x \cdot y + x \cdot y \cdot y + y \cdot y \cdot 2y$, or

$x^3 + x^2y + xy^2 + 2y^3$.

99. The lateral surface area of the outer portion of the solid is the lateral surface area of a right circular cylinder with radius n and height h. The lateral surface area of the inner portion is the lateral surface area of a right circular cylinder with radius m and height h. Recall that the formula for the lateral surface area of a right circular cylinder with radius r and height h is $2\pi rh$.

The surface area of the top is the area of a circle with radius n less the area of a circle with radius m. The surface area of the bottom is the same as the surface area of the top.

Thus, the surface area of the solid is

$$2\pi nh + 2\pi mh + 2\pi n^2 - 2\pi m^2.$$

101. *Writing Exercise*

103. Replace t with 2 and multiply.

$$P(1+r)^2$$
$$= P(1 + 2r + r^2)$$
$$= P + 2Pr + Pr^2$$

105. Substitute \$10,400 for P, 8.5%, or 0.085 for r, and 5 for t.

$$P(1+r)^t$$
$$= \$10,400(1 + 0.085)^5$$
$$\approx \$15,638.03$$

Exercise Set 4.7

1. $\dfrac{32x^5 - 24x}{8} = \dfrac{32x^5}{8} - \dfrac{24x}{8}$

$= \dfrac{32}{8}x^5 - \dfrac{24}{8}x$ Dividing coefficients

$= 4x^5 - 3x$

To check, we multiply the quotient by 8:

$(4x^5 - 3x)8 = 32x^5 - 24x$

The answer checks.

3. $\dfrac{u - 2u^2 + u^7}{u}$

$= \dfrac{u}{u} - \dfrac{2u^2}{u} + \dfrac{u^7}{u}$

$= 1 - 2u + u^6$

Check: We multiply.

$u(1 - 2u + u^6) = u - 2u^2 + u^7$

5. $(18t^3 - 24t^2 + 6t) \div (3t)$

$= \dfrac{18t^3 - 24t^2 + 6t}{3t}$

$= \dfrac{18t^3}{3t} - \dfrac{24t^2}{3t} + \dfrac{6t}{3t}$

$= 6t^2 - 8t + 2$

Check: We multiply.

$3t(6t^2 - 8t + 2) = 18t^3 - 24t^2 + 6t$

7. $(25x^6 - 20x^4 - 5x^2) \div (-5x^2)$

$= \dfrac{25x^6 - 20x^4 - 5x^2}{-5x^2}$

$= \dfrac{25x^6}{-5x^2} - \dfrac{20x^4}{-5x^2} - \dfrac{5x^2}{-5x^2}$

$= -5x^4 - (-4x^2) - (-1)$

$= -5x^4 + 4x^2 + 1$

Check: We multiply.

$-5x^2(-5x^4 + 4x^2 + 1) = 25x^6 - 20x^4 - 5x^2$

9. $(24t^5 - 40t^4 + 6t^3) \div (4t^3)$

$= \dfrac{24t^5 - 40t^4 + 6t^3}{4t^3}$

$= \dfrac{24t^5}{4t^3} - \dfrac{40t^4}{4t^3} + \dfrac{6t^3}{4t^3}$

$= 6t^2 - 10t + \dfrac{3}{2}$

Check: We multiply.

$4t^3\left(6t^2 - 10t + \dfrac{3}{2}\right) = 24t^5 - 40t^4 + 6t^3$

11. $\dfrac{8x^2 - 10x + 1}{2}$

$= \dfrac{8x^2}{2} - \dfrac{10x}{2} + \dfrac{1}{2}$

$= 4x^2 - 5x + \dfrac{1}{2}$

Check: We multiply.

$2\left(4x^2 - 5x + \dfrac{1}{2}\right) = 8x^2 - 10x + 1$

13. $\dfrac{4x^7 + 6x^5 + 4x^2}{4x^3}$

$= \dfrac{4x^7}{4x^3} + \dfrac{6x^5}{4x^3} + \dfrac{4x^2}{4x^3}$

$= x^4 + \dfrac{3}{2}x^2 + \dfrac{1}{x}$

Check: We multiply.

$4x^3\left(x^4 + \dfrac{3}{2}x^2 + \dfrac{1}{x}\right) = 4x^7 + 6x^5 + 4x^2$

15. $\dfrac{9r^2s^2 + 3r^2s - 6rs^2}{-3rs}$

$= \dfrac{9r^2s^2}{-3rs} + \dfrac{3r^2s}{-3rs} - \dfrac{6rs^2}{-3rs}$

$= -3rs - r + 2s$

Check: We multiply.

$-3rs(-3rs - r + 2s) = 9r^2s^2 + 3r^2s - 6rs^2$

17.

$$\begin{array}{r|l} & x+6 \\ \hline x-2 & x^2+4x-12 \\ & \underline{x^2-2x} \\ & \quad 6x-12 \leftarrow (x^2+4x)-(x^2-2x) = 6x \\ & \quad \underline{6x-12} \\ & \qquad 0 \leftarrow (6x-12)-(6x-12) = 0 \end{array}$$

The answer is $x + 6$.

19.

$$\begin{array}{r|l} & t-5 \\ \hline t-5 & t^2-10t-20 \\ & \underline{t^2-5t} \\ & \quad -5t-20 \leftarrow (t^2-10t)-(t^2-5t) = -5t \\ & \quad \underline{-5t+25} \\ & \qquad -45 \leftarrow (-5t-20)-(-5t+25) = -45 \end{array}$$

The answer is $t - 5 + \dfrac{-45}{t-5}$.

21.

$$\begin{array}{r|l} & 2x-1 \\ \hline x+6 & 2x^2+11x-5 \\ & \underline{2x^2+12x} \\ & \quad -x-5 \leftarrow (2x^2+11x)-(2x^2+12x) = -x \\ & \quad \underline{-x-6} \\ & \qquad 1 \leftarrow (-x-5)-(-x-6) = 1 \end{array}$$

The answer is $2x - 1 + \dfrac{1}{x+6}$.

23.

$$\begin{array}{r|l} & a^2-2a+4 \\ \hline a+2 & a^3+0a^2+0a+8 \leftarrow \text{Writing in the missing terms} \\ & \underline{a^3+2a^2} \\ & \quad -2a^2+0a \leftarrow a^3-(a^3+2a^2) = -2a^2 \\ & \quad \underline{-2a^2-4a} \\ & \qquad 4a+8 \leftarrow -2a^2-(-2a^2-4a) = 4a \\ & \qquad \underline{4a+8} \\ & \qquad\quad 0 \leftarrow (4a+8)-(4a+8) = 0 \end{array}$$

The answer is $a^2 - 2a + 4$.

25.

$$\begin{array}{r|l} & t+4 \\ \hline t-4 & t^2+0t-13 \leftarrow \text{Writing in the missing term} \\ & \underline{t^2-4t} \\ & \quad 4t-13 \leftarrow t^2-(t^2-4t) = 4t \\ & \quad \underline{4t-16} \\ & \qquad 3 \leftarrow (4t-13)-(4t-16) = 3 \end{array}$$

The answer is $t + 4 + \dfrac{3}{t-4}$.

27.

$$\begin{array}{r|l} & x+4 \\ \hline 3x-1 & 3x^2+11x-4 \\ & \underline{3x^2-x} \\ & \quad 12x-4 \leftarrow (3x^2+11x)-(3x^2-x) = 12x \\ & \quad \underline{12x-4} \\ & \qquad 0 \leftarrow (12x-4)-(12x-4) = 0 \end{array}$$

The answer is $x + 4$.

29.

$$\begin{array}{r|l} & 3a+1 \\ \hline 2a+5 & 6a^2+17a+8 \\ & \underline{6a^2+15a} \\ & \quad 2a+8 \leftarrow (6a^2+17a)-(6a^2+15a) = 2a+8 \\ & \quad \underline{2a+5} \\ & \qquad 3 \leftarrow (2a+8)-(2a+5) = 3 \end{array}$$

The answer is $3a + 1 + \dfrac{3}{2a+5}$.

31.

$$\begin{array}{r|l} & t^2-3t+1 \\ \hline 2t-3 & 2t^3-9t^2+11t-3 \\ & \underline{2t^3-3t^2} \\ & \quad -6t^2+11t \leftarrow (2t^3-9t^2)-(2t^3-3t^2) = -6t^2 \\ & \quad \underline{-6t^2+9t} \\ & \qquad 2t-3 \leftarrow (-6t^2+11t)-(-6t^2+9t) = 2t \\ & \qquad \underline{2t-3} \\ & \qquad\quad 0 \leftarrow (2t-3)-(2t-3) = 0 \end{array}$$

The answer is $t^2 - 3t + 1$.

33.
$$\begin{array}{rl}
 & t^2-2t+3 \\
t+1 \overline{)} & t^3-t^2+t-1 \\
 & \underline{t^3+t^2} \\
 & -2t^2+t \quad \leftarrow (t^3-t^2)-(t^3+t^2)=-2t^2 \\
 & \underline{-2t^2-2t} \\
 & 3t-1 \quad \leftarrow (-2t^2+t)-(-2t^2-2t)=3t \\
 & \underline{3t+3} \\
 & -4
\end{array}$$

The answer is $t^2-2t+3+\dfrac{-4}{t+1}$.

35.
$$\begin{array}{rl}
 & t^2 \qquad +1 \\
t^2-3 \overline{)} & t^4+0t^3-2t^2+4t-5 \leftarrow \text{Writing in the missing term} \\
 & \underline{t^4 \qquad -3t^2} \\
 & t^2+4t-5 \leftarrow (t^4-2t^2)-(t^4-3t^2)=t^2 \\
 & \underline{t^2 \qquad -3} \\
 & 4t-2 \leftarrow (t^2+4t-5)-(t^2-3)=4t-2
\end{array}$$

The answer is $t^2+1+\dfrac{4t-2}{t^2-3}$.

37.
$$\begin{array}{rl}
 & 2x^2 \qquad +1 \\
2x^2-3 \overline{)} & 4x^4+0x^3-4x^2-x-3 \leftarrow \text{Writing in the missing term} \\
 & \underline{4x^4 \qquad -6x^2} \\
 & 2x^2-x-3 \leftarrow (4x^4-4x^2)-(4x^4-6x^2)=2x^2 \\
 & \underline{2x^2 \qquad -3} \\
 & -x \leftarrow (2x^2-x-3)-(2x^2-3)=-x
\end{array}$$

The answer is $2x^2+1+\dfrac{-x}{2x^2-3}$.

39. *Writing Exercise*

41. $-4+(-13)$ Two negative numbers. Add the absolute values, 4 and 13, to get 17. Make the answer negative.

$-4+(-13)=-17$

43. $-9-(-7)=-9+7=-2$

45. ***Familiarize.*** Let w = the width. Then $w+15$ = the length. We draw a picture.

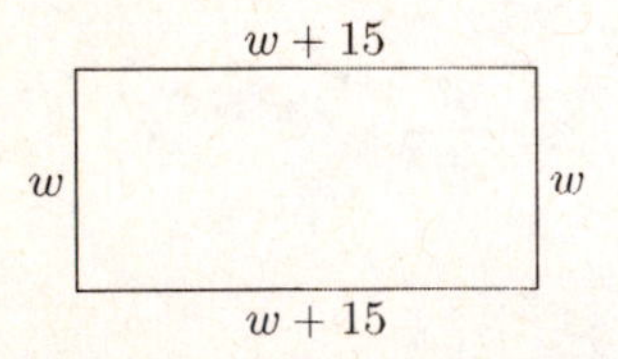

We will use the fact that the perimeter is 640 ft to find w (the width). Then we can find $w+15$ (the length) and multiply the length and the width to find the area.

Translate.

Width+Width+ Length + Length =Perimeter

$w \;+\; w \;+(w+15)+(w+15)= \;640$

Carry out.

$$\begin{aligned}
w+w+(w+15)+(w+15)&=640\\
4w+30&=640\\
4w&=610\\
w&=152.5
\end{aligned}$$

If the width is 152.5, then the length is 152.5+15, or 167.5.

Check. The length, 167.5 ft, is 15 ft greater than the width, 152.5 ft. The perimeter is 152.5 + 152.5 + 167.5 + 167.5, or 640 ft. The answer checks.

State. The length of the rectangle is 167.5 ft.

47. Graph: $3x-2y=12$.

We will graph the equation using intercepts. To find the y-intercept, we let $x=0$.

$$\begin{aligned}
-2y&=12 \quad \text{Ignoring the } x\text{-term}\\
y&=-6
\end{aligned}$$

The y-intercept is $(0,-6)$.

To find the x-intercept, we let $y=0$.

$$\begin{aligned}
3x&=12 \quad \text{Ignoring the } y\text{-term}\\
x&=4
\end{aligned}$$

The x-intercept is $(4,0)$.

To find a third point, replace x with -2 and solve for y:

$$\begin{aligned}
3(-2)-2y&=12\\
-6-2y&=12\\
-2y&=18\\
y&=-9
\end{aligned}$$

The point $(-2,-9)$ appears to line up with the intercepts, so we draw the graph.

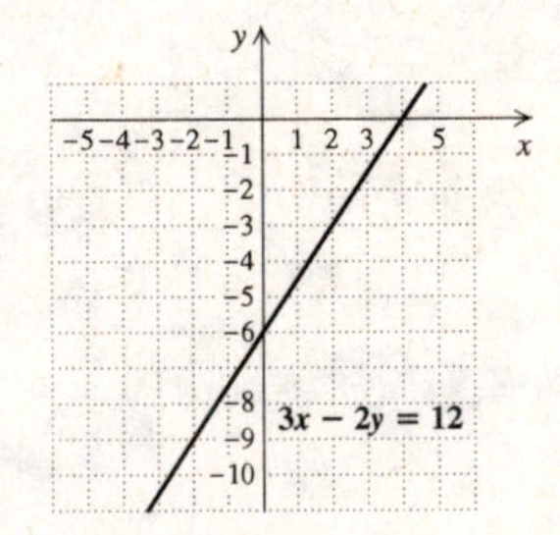

49. *Writing Exercise*

51.
$$\begin{aligned}
&(10x^{9k}-32x^{6k}+28x^{3k})\div(2x^{3k})\\
&=\frac{10x^{9k}-32x^{6k}+28x^{3k}}{2x^{3k}}\\
&=\frac{10x^{9k}}{2x^{3k}}-\frac{32x^{6k}}{2x^{3k}}+\frac{28x^{3k}}{2x^{3k}}\\
&=5x^{9k-3k}-16x^{6k-3k}+14x^{3k-3k}\\
&=5x^{6k}-16x^{3k}+14
\end{aligned}$$

53.

$$\begin{array}{r} 3t^{2h}+2t^h-5 \\ 2t^h+3\,\overline{)\,6t^{3h}+13t^{2h}-4t^h-15} \\ \underline{6t^{3h}+9t^{2h}\qquad\qquad} \\ 4t^{2h}-4t^h\qquad \\ \underline{4t^{2h}+6t^h\qquad} \\ -10t^h-15 \\ \underline{-10t^h-15} \\ 0 \end{array}$$

The answer is $3t^{2h}+2t^h-5$.

55.

$$\begin{array}{r} a+3 \\ 5a^2-7a-2\,\overline{)\,5a^3+8a^2-23a-1} \\ \underline{5a^3-7a^2-2a\qquad} \\ 15a^2-21a-1 \\ \underline{15a^2-21a-6} \\ 5 \end{array}$$

The answer is $a+3+\dfrac{5}{5a^2-7a-2}$.

57.

$$\begin{aligned} &(4x^5-14x^3-x^2+3)+ \\ &\quad(2x^5+3x^4+x^3-3x^2+5x) \\ &=6x^5+3x^4-13x^3-4x^2+5x+3 \end{aligned}$$

$$\begin{array}{r} 2x^2+x-3 \\ 3x^3-2x-1\,\overline{)\,6x^5+3x^4-13x^3-4x^2+5x+3} \\ \underline{6x^5\qquad-4x^3-2x^2\qquad\qquad} \\ 3x^4-9x^3-2x^2+5x\qquad \\ \underline{3x^4\qquad\qquad-2x^2-x\qquad} \\ -9x^3\qquad+6x+3 \\ \underline{-9x^3\qquad+6x+3} \\ 0 \end{array}$$

The answer is $2x^2+x-3$.

59.

$$\begin{array}{r} x-3 \\ x-1\,\overline{)\,x^2-4x+c} \\ \underline{x^2-x\qquad} \\ -3x+c \\ \underline{-3x+3} \\ c-3 \end{array}$$

We set the remainder equal to 0.

$c-3=0$

$c=3$

Thus, c must be 3.

61.

$$\begin{array}{r} c^2x+(2c+c^2) \\ x-1\,\overline{)\,c^2x^2+2cx+1} \\ \underline{c^2x^2-c^2x\qquad} \\ (2c+c^2)x+1 \\ \underline{(2c+c^2)x-(2c+c^2)} \\ 1+(2c+c^2) \end{array}$$

We set the remainder equal to 0.

$c^2+2c+1=0$

$(c+1)^2=0$

$c+1=0$ *or* $c+1=0$

$c=-1$ *or* $c=-1$

Thus, c must be -1.

Exercise Set 4.8

1. $7^{-2}=\frac{1}{7^2}=\frac{1}{49}$

3. $10^{-4}=\frac{1}{10^4}=\frac{1}{10,000}$

5. $(-2)^{-6}=\frac{1}{(-2)^6}=\frac{1}{64}$

7. $x^{-8}=\frac{1}{x^8}$

9. $xy^{-2}=x\cdot\frac{1}{y^2}=\frac{x}{y^2}$

11. $r^{-5}t=\frac{1}{r^5}\cdot t=\frac{t}{r^5}$

13. $\frac{1}{t^{-8}}=t^8$

15. $\frac{1}{h^{-8}}=h^8$

17. $7^{-1}=\frac{1}{7^1}=\frac{1}{7}$

19. $\left(\frac{3}{5}\right)^{-2}=\left(\frac{5}{3}\right)^2=\frac{5^2}{3^2}=\frac{25}{9}$

21. $\left(\frac{a}{2}\right)^{-3}=\left(\frac{2}{a}\right)^3=\frac{2^3}{a^3}=\frac{8}{a^3}$

23. $\left(\frac{s}{t}\right)^{-7}=\left(\frac{t}{s}\right)^7=\frac{t^7}{s^7}$

25. $\frac{1}{6^2}=6^{-2}$

27. $\frac{1}{t^6}=t^{-6}$

29. $\frac{1}{a^4}=a^{-4}$

31. $\frac{1}{p^7}=p^{-7}$

33. $\frac{1}{5}=\frac{1}{5^1}=5^{-1}$

35. $\frac{1}{t}=\frac{1}{t^1}=t^{-1}$

37. $2^{-5}\cdot2^8=2^{-5+8}=2^3$, or 8

39. $x^{-2}\cdot x^{-7}=x^{-2+(-7)}=x^{-9}=\frac{1}{x^9}$

41. $t^{-3}\cdot t=t^{-3}\cdot t^1=t^{-3+1}=t^{-2}=\frac{1}{t^2}$

43. $(a^{-2})^9=a^{-2\cdot9}=a^{-18}=\frac{1}{a^{18}}$

45. $(t^{-3})^{-6}=t^{-3(-6)}=t^{18}$

47. $(t^4)^{-3}=t^{4(-3)}=t^{-12}=\frac{1}{t^{12}}$

49. $(mn)^{-7} = \frac{1}{(mn)^7} = \frac{1}{m^7n^7}$

51. $(3x^{-4})^2 = 3^2(x^{-4})^2 = 9x^{-8} = \frac{9}{x^8}$

53. $(5r^{-4}t^3)^2 = 5^2(r^{-4})^2(t^3)^2 = 25r^{-8}t^6 = \frac{25t^6}{r^8}$

55. $\frac{t^7}{t^{-3}} = t^{7-(-3)} = t^{10}$

57. $\frac{y^{-7}}{y^{-3}} = y^{-7-(-3)} = y^{-4} = \frac{1}{y^4}$

59. $\frac{12y^{-4}}{4y^{-9}} = 3y^{-4-(-9)} = 3y^5$

61. $\frac{2x^6}{x} = 2\frac{x^6}{x^1} = 2x^{6-1} = 2x^5$

63. $\frac{15a^{-7}}{10b^{-9}} = \frac{3b^9}{2a^7}$

65. $\frac{t^{-7}}{t^{-7}}$

Note that we have an expression divided by itself. Thus, the result is 1. We could also find this result as follows:

$$\frac{t^{-7}}{t^{-7}} = t^{-7-(-7)} = t^0 = 1.$$

67. $\frac{3x^{-5}}{y^{-6}z^{-2}} = \frac{3y^6z^2}{x^5}$

69. $(3x^4y^5)^{-3} = 3^{-3}(x^4)^{-3}(y^5)^{-3} = \frac{1}{3^3} \cdot x^{-12}y^{-15} = \frac{1}{27x^{12}y^{15}}$

71. $(x^{-6}y^{-2})^{-4} = (x^{-6})^{-4}(y^{-2})^{-4} = x^{24}y^8$

73. $(a^{-5}b^7c^{-2})(a^{-3}b^{-2}c^6) = a^{-5+(-3)}b^{7+(-2)}c^{-2+6} =$
$a^{-8}b^5c^4 = \frac{b^5c^4}{a^8}$

75. $\left(\frac{a^4}{3}\right)^{-2} = \left(\frac{3}{a^4}\right)^2 = \frac{3^2}{(a^4)^2} = \frac{9}{a^8}$

77. $\left(\frac{m^{-1}}{n^{-4}}\right)^3 = \frac{(m^{-1})^3}{(n^{-4})^3} = \frac{m^{-3}}{n^{-12}} = \frac{n^{12}}{m^3}$

79. $\left(\frac{2a^2}{3b^4}\right)^{-3} = \left(\frac{3b^4}{2a^2}\right)^3 = \frac{(3b^4)^3}{(2a^2)^3} = \frac{3^3(b^4)^3}{2^3(a^2)^3} = \frac{27b^{12}}{8a^6}$

81. $\left(\frac{5x^{-2}}{3y^{-2}z}\right)^0$

Any nonzero expression raised to the 0 power is equal to 1. Thus, the answer is 1.

83. $\frac{-6a^3b^{-5}}{-3a^7b^{-8}} = \frac{-6}{-3} \cdot \frac{a^3}{a^7} \cdot \frac{b^{-5}}{b^{-8}}$

$= 2 \cdot a^{3-7} \cdot b^{-5-(-8)}$

$= 2a^{-4}b^3 = \frac{2b^3}{a^4}$

85. $\frac{10x^{-4}yz^7}{8x^7y^{-3}z^{-3}} = \frac{10}{8} \cdot \frac{x^{-4}}{x^7} \cdot \frac{y}{y^{-3}} \cdot \frac{z^7}{z^{-3}}$

$= \frac{5}{4} \cdot x^{-4-7} \cdot y^{1-(-3)} \cdot z^{7-(-3)}$

$= \frac{5}{4} \cdot x^{-11}y^4z^{10} = \frac{5y^4z^{10}}{4x^{11}}$

87. 7.12×10^4

Since the exponent is positive, the decimal point will move to the right.

7.1200. The decimal point moves right 4 places.

$7.12 \times 10^4 = 71,200$

89. 8.92×10^{-3}

Since the exponent is negative, the decimal point will move to the left.

.008.92 The decimal point moves left 3 places.

$8.92 \times 10^{-3} = 0.00892$

91. 9.04×10^8

Since the exponent is positive, the decimal point will move to the right.

9.04000000.

8 places

$9.04 \times 10^8 = 904,000,000$

93. 2.764×10^{-10}

Since the exponent is negative, the decimal point will move to the left.

0.0000000002.764

10 places

$2.764 \times 10^{-10} = 0.0000000002764$

95. 4.209×10^9

Since the exponent is positive, the decimal point will move to the right.

4.2090000.

7 places

$4.209 \times 10^7 = 42,090,000$

97. $490,000 = 4.9 \times 10^m$

To write 4.9 as 490,000 we move the decimal point 5 places to the right. Thus, m is 5 and

$$490,000 = 4.9 \times 10^5.$$

99. $0.00583 = 5.83 \times 10^m$

To write 5.83 as 0.00583 we move the decimal point 3 places to the left. Thus, m is -3 and

$$0.00583 = 5.83 \times 10^{-3}.$$

101. $78,000,000,000 = 7.8 \times 10^m$

To write 7.8 as 78,000,000,000 we move the decimal point 10 places to the right. Thus, m is 10 and

$$78,000,000,000 = 7.8 \times 10^{10}.$$

103. $0.000000527 = 5.27 \times 10^m$

To write 5.27 as 0.000000527 we move the decimal point 7 places to the left. Thus, m is -7 and

$$0.000000527 = 5.27 \times 10^{-7}.$$

105. $0.000000018 = 1.8 \times 10^m$

To write 1.8 as 0.000000018 we move the decimal point 8 places to the left. Thus, m is -8 and

$$0.000000018 = 1.8 \times 10^{-8}.$$

107. $1,094,000,000,000,000 = 1.094 \times 10^m$

To write 1.094 as 1,094,000,000,000,000 we move the decimal point 15 places to the right. Thus, m is 15 and

$$1,094,000,000,000,000 = 1.094 \times 10^{15}.$$

109. $(4\times 10^7)(2\times 10^5) = (4 \cdot 2) \times (10^7 \cdot 10^5)$
$= 8 \times 10^{7+5}$ Adding exponents
$= 8 \times 10^{12}$

111. $(3.8 \times 10^9)(6.5 \times 10^{-2}) = (3.8 \cdot 6.5) \times (10^9 \cdot 10^{-2})$
$= 24.7 \times 10^7$

The answer is not yet in scientific notation since 24.7 is not a number between 1 and 10. We convert to scientific notation.

$$24.7 \times 10^7 = (2.47 \times 10) \times 10^7 = 2.47 \times 10^8$$

113. $(8.7 \times 10^{-12})(4.5 \times 10^{-5})$
$= (8.7 \cdot 4.5) \times (10^{-12} \cdot 10^{-5})$
$= 39.15 \times 10^{-17}$

The answer is not yet in scientific notation since 39.15 is not a number between 1 and 10. We convert to scientific notation.

$$39.15 \times 10^{-17} = (3.915 \times 10) \times 10^{-17} = 3.915 \times 10^{-16}$$

115. $\dfrac{8.5 \times 10^8}{3.4 \times 10^{-5}} = \dfrac{8.5}{3.4} \times \dfrac{10^8}{10^{-5}}$
$= 2.5 \times 10^{8-(-5)}$
$= 2.5 \times 10^{13}$

117. $(3.0 \times 10^6) \div (6.0 \times 10^9) = \dfrac{3.0 \times 10^6}{6.0 \times 10^9}$
$= \dfrac{3.0}{6.0} \times \dfrac{10^6}{10^9}$
$= 0.5 \times 10^{6-9}$
$= 0.5 \times 10^{-3}$

The answer is not yet in scientific notation because 0.5 is not between 1 and 10. We convert to scientific notation.

$0.5 \times 10^{-3} = (5 \times 10^{-1}) \times 10^{-3} =$
5×10^{-4}

119. $\dfrac{7.5 \times 10^{-9}}{2.5 \times 10^{12}} = \dfrac{7.5}{2.5} \times \dfrac{10^{-9}}{10^{12}}$
$= 3.0 \times 10^{-9-12}$
$= 3.0 \times 10^{-21}$

121. *Writing Exercise*

123. $(3-8)(9-12)$
$= (-5)(-3)$ Subtracting
$= 15$ Multiplying

125. $7 \cdot 2 + 8^2$
$= 7 \cdot 2 + 64$ Evaluating the exponential expression
$= 14 + 64$ Multiplying
$= 78$ Adding

127. $-3(-2)^2 \div 4 \cdot 5$
$= -3(4) \div 4 \cdot 5$ Evaluating the exponential expression
$= -12 \div 4 \cdot 5$ Multiplying and
$= -3 \cdot 5$ dividing from
$= -15$ left to right

129. To plot $(-3, 2)$, we start at the origin and move 3 units to the left and then 2 units up. To plot $(4, -1)$, we start at the origin and move 4 units to the right and then 1 unit down. To plot $(5, 3)$, we start at the origin and move 5 units to the right and 3 units up. To plot $(-5, -2)$, we start at the origin and move 5 units to the left and then 2 units down.

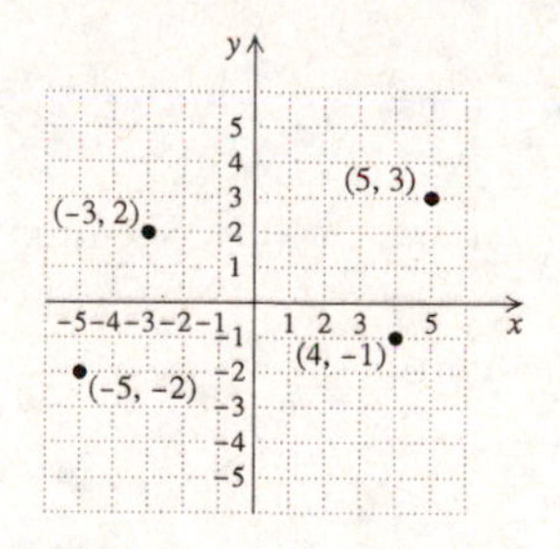

131. *Writing Exercise*

133. $\dfrac{1}{1.25 \times 10^{-6}} = \dfrac{1}{1.25} \times \dfrac{1}{10^{-6}} = 0.8 \times 10^6 =$
$(8 \times 10^{-1}) \times 10^6 = 8 \times 10^5$

135. $8^{-3} \cdot 32 \div 16^2 = (2^3)^{-3} \cdot 2^5 \div (2^4)^2 =$
$2^{-9} \cdot 2^5 \div 2^8 = 2^{-4} \div 2^8 = 2^{-12}$

137. $\dfrac{125^{-4}(25^2)^4}{125} = \dfrac{(5^3)^{-4}((5^2)^2)^4}{5^3} =$
$\dfrac{5^{-12}(5^4)^4}{5^3} = \dfrac{5^{-12} \cdot 5^{16}}{5^3} = \dfrac{5^4}{5^3} = 5^1 = 5$

139. $\dfrac{4.2 \times 10^8[(2.5 \times 10^{-5}) \div (5.0 \times 10^{-9})]}{3.0 \times 10^{-12}}$
$= \dfrac{4.2 \times 10^8[0.5 \times 10^4]}{3.0 \times 10^{-12}}$
$= \dfrac{2.1 \times 10^{12}}{3.0 \times 10^{-12}}$
$= 0.7 \times 10^{24}$
$= (7 \times 10^{-1}) \times 10^{24}$
$= 7 \times 10^{23}$

141. $\dfrac{7.4 \times 10^{29}}{(5.4 \times 10^{-6})(2.8 \times 10^{8})}$

$= \dfrac{7.4}{(5.4 \cdot 2.8)} \times \dfrac{10^{29}}{(10^{-6} \cdot 10^{8})}$

$\approx 0.4894179894 \times 10^{27}$

$\approx (4.894179894 \times 10^{-1}) \times 10^{27}$

$\approx 4.894179894 \times 10^{26}$

143. $\dfrac{(7.8 \times 10^{7})(8.4 \times 10^{23})}{2.1 \times 10^{-12}}$

$= \dfrac{(7.8 \cdot 8.4)}{2.1} \times \dfrac{(10^{7} \cdot 10^{23})}{10^{-12}}$

$= 31.2 \times 10^{42}$

$= (3.12 \times 10) \times 10^{42}$

$= 3.12 \times 10^{43}$

145. a) False; let $x = 2$, $y = 3$, $m = 4$, and $n = 2$:

$2^4 \cdot 3^2 = 16 \cdot 9 = 144$, but

$(2 \cdot 3)^{4 \cdot 2} = 6^8 = 1,679,616$

b) False; let $x = 3$, $y = 4$, and $m = 2$:

$3^2 \cdot 4^2 = 9 \cdot 16 = 144$, but

$(3 \cdot 4)^{2 \cdot 2} = 12^4 = 20,736$

c) False; let $x = 5$, $y = 3$, and $m = 2$:

$(5 - 3)^2 = 2^2 = 4$, but

$5^2 - 3^2 = 25 - 9 = 16$

147. ***Familiarize***. Express 1 billion and 2500 in scientific notation:

1 billion $= 1,000,000,000 = 10^9$

$2500 = 2.5 \times 10^3$

Let $b =$ the number of bytes in the network.

Translate. We reword the problem.

What is 2500 times 1 gigabyte?

$b = 2.5 \times 10^3 \times 10^9$

Carry out. We do the computation.

$b = (2.5 \times 10^3) \times 10^9$

$b = 2.5 \times (10^3 \times 10^9)$

$b = 2.5 \times 10^{12}$

Check. We review the computation. Also, the answer seems reasonable since it is larger than 1 billion.

State. There are 2.5×10^{12} bytes in the network.

149. ***Familiarize***. We must express both dimensions using the same units. Let's choose centimeters. First, convert 1.5 m to centimeters and express the result in scientific notation.

$1.5 \text{ m} = 1.5 \times 1 \text{ m} = 1.5 \times 100 \text{ cm} = 1.5 \times 10^2 \text{ cm}$

Let l represent how many times the DNA is longer than it is wide.

Translate. We reword the problem.

The length is how many times the width.

$1.5 \times 10^2 = l \cdot 1.3 \times 10^{-10}$

Carry out. We solve the equation.

$1.5 \times 10^2 = l \cdot 1.3 \times 10^{-10}$

$\dfrac{1.5 \times 10^2}{1.3 \times 10^{-10}} = l$

$1.15385 \times 10^{12} \approx l$

Check. Since $(1.15385 \times 10^{12}) \times (1.3 \times 10^{-10}) = 1.498705 \times 10^2 \approx 1.5 \times 10^2$, the answer checks.

State. A strand of DNA is about 1.15385×10^{12} times longer than it is wide.

Chapter 5

Polynomials and Factoring

Exercise Set 5.1

1. Since $7a \cdot 5ab = 35a^2b$, choice (h) is most appropriate.

3. $5x + 10 = 5(x+2)$ and $4x + 8 = 4(x+2)$, so $x+2$ is a common factor of $5x+10$ and $4x+8$ and choice (b) is most appropriate.

5. $3x^2(3x^2-1) = 9x^4 - 3x^2$, so choice (c) is most appropriate.

7. $3a + 6a^2 = 3a(1+2a)$, so $1+2a$ is a factor of $3a+6a^2$ and choice (d) is most appropriate.

9. Answers may vary. $10x^3 = (5x)(2x^2) = (10x^2)(x) = (-2)(-5x^3)$

11. Answers may vary. $-15a^4 = (-15)(a^4) = (-5a)(3a^3) = (-3a^2)(5a^2)$

13. Answers may vary. $26x^5 = (2x^4)(13x) = (2x^3)(13x^2) = (-x^2)(-26x^3)$

15. $7x - 14 = 7 \cdot x - 7 \cdot 2$
$= 7(x-2)$

17. $3t^2 + t = t \cdot 3t + t \cdot 1$
$= t(3t+1)$

19. $-4a^2 - 8a = -4a \cdot a - 4a \cdot 2$
$= -4a(a+2)$

We might also factor as follows:

$-4a^2 - 8a = 4a \cdot (-a) + 4a(-2)$
$= 4a(-a-2)$

21. $x^3 + 6x^2 = x^2 \cdot x + x^2 \cdot 6$
$= x^2(x+6)$

23. $8x^4 - 24x^2 = 8x^2 \cdot x^2 - 8x^2 \cdot 3$
$= 8x^2(x^2-3)$

25. $2x^2 + 2x - 8 = 2 \cdot x^2 + 2 \cdot x - 2 \cdot 4$
$= 2(x^2+x-4)$

27. $-7a^6 + 10a^4 - 14a^2 = -a^2 \cdot 7a^4 - a^2 \cdot (-10a^2) - a^2 \cdot 14$
$= -a^2(7a^4 - 10a^2 + 14)$

29. $6x^8 + 12x^6 - 24x^4 + 30x^2$
$= 6x^2 \cdot x^6 + 6x^2 \cdot 2x^4 - 6x^2 \cdot 4x^2 + 6x^2 \cdot 5$
$= 6x^2(x^6 + 2x^4 - 4x^2 + 5)$

31. $x^5y^5 + x^4y^3 + x^3y^3 - x^2y^2$
$= x^2y^2 \cdot x^3y^3 + x^2y^2 \cdot x^2y + x^2y^2 \cdot xy - x^2y^2 \cdot 1$
$= x^2y^2(x^3y^3 + x^2y + xy - 1)$

33. $-5a^3b^4 + 10a^2b^3 - 15a^3b^2$
$= -5a^2b^2 \cdot ab^2 - 5a^2b^2 \cdot (-2b) - 5a^2b^2 \cdot 3a$
$= -5a^2b^2(ab^2 - 2b + 3a)$

35. $y(y-2) + 7(y-2)$
$= (y-2)(y+7)$ Factoring out the common binomial factor $y-2$

37. $x^2(x+3) - 7(x+3)$
$= (x+3)(x^2-7)$ Factoring out the common binomial factor $x+3$

39. $y^2(y+8) + (y+8) = y^2(y+8) + 1(y+8)$
$= (y+8)(y^2+1)$ Factoring out the common factor

41. $x^3 + 3x^2 + 4x + 12$
$= (x^3 + 3x^2) + (4x + 12)$
$= x^2(x+3) + 4(x+3)$ Factoring each binomial
$= (x+3)(x^2+4)$ Factoring out the common factor $x+3$

43. $5a^3 + 15a^2 + 2a + 6$
$= (5a^3 + 15a^2) + (2a + 6)$
$= 5a^2(a+3) + 2(a+3)$ Factoring each binomial
$= (a+3)(5a^2+2)$ Factoring out the common factor $a+3$

45. $9x^3 - 12x^2 + 3x - 4$
$= 3x^2(3x-4) + 1(3x-4)$
$= (3x-4)(3x^2+1)$

47. $4t^3 - 20t^2 + 3t - 15$
$= 4t^2(t-5) + 3(t-5)$
$= (t-5)(4t^2+3)$

49. $7x^3 + 2x^2 - 14x - 4$
$= x^2(7x+2) - 2(7x+2)$
$= (7x+2)(x^2-2)$

51. $6a^3 - 7a^2 + 6a - 7$
$= a^2(6a-7) + 1(6a-7)$
$= (6a-7)(a^2+1)$

53. $x^3 + 8x^2 - 3x - 24 = x^2(x+8) - 3(x+8)$
$= (x+8)(x^2-3)$

55. $2x^3 + 12x^2 - 5x - 30 = 2x^2(x+6) - 5(x+6)$
$= (x+6)(2x^2-5)$

57. We try factoring by grouping.

$p^3 + p^2 - 3p + 10 = p^2(p+1) - (3p - 10)$, or
$p^3 - 3p + p^2 + 10 = p(p^2 - 3) + p^2 + 10$

Because we cannot find a common binomial factor, this polynomial cannot be factored using factoring by grouping.

59. $y^3 + 8y^2 - 2y - 16 = y^2(y+8) - 2(y+8) = (y+8)(y^2-2)$

61. $2x^3 - 8x^2 - 9x + 36 = 2x^2(x-4) - 9(x-4)$
$= (x-4)(2x^2-9)$

63. *Writing Exercise*

65. $(x+3)(x+5)$
F O I L
$= x \cdot x + x \cdot 5 + 3 \cdot x + 3 \cdot 5$
$= x^2 + 5x + 3x + 15$
$= x^2 + 8x + 15$

67. $(a-7)(a+3)$
F O I L
$= a \cdot a + a \cdot 3 - 7 \cdot a - 7 \cdot 3$
$= a^2 + 3a - 7a - 21$
$= a^2 - 4a - 21$

69. $(2x+5)(3x-4)$
F O I L
$= 2x \cdot 3x - 2x \cdot 4 + 5 \cdot 3x - 5 \cdot 4$
$= 6x^2 - 8x + 15x - 20$
$= 6x^2 + 7x - 20$

71. $(3t-5)^2$
$= (3t)^2 - 2 \cdot 3t \cdot 5 + 5^2$
$[(A-B)^2 = A^2 - 2AB + B^2]$
$= 9t^2 - 30t + 25$

73. *Writing Exercise*

75. $4x^5 + 6x^2 + 6x^3 + 9 = 2x^2(2x^3+3) + 3(2x^3+3)$
$= (2x^3+3)(2x^2+3)$

77. $x^{12} + x^7 + x^5 + 1 = x^7(x^5+1) + (x^5+1)$
$= (x^5+1)(x^7+1)$

79. $5x^5 - 5x^4 + x^3 - x^2 + 3x - 3$
$= 5x^4(x-1) + x^2(x-1) + 3(x-1)$
$= (x-1)(5x^4 + x^2 + 3)$

We could also do this exercise as follows:

$5x^5 - 5x^4 + x^3 - x^2 + 3x - 3$
$= (5x^5 + x^3 + 3x) - (5x^4 + x^2 + 3)$
$= x(5x^4 + x^2 + 3) - 1(5x^4 + x^2 + 3)$
$= (5x^4 + x^2 + 3)(x-1)$

81. Answers may vary. $8x^4y^3 - 24x^3y^3 + 16x^2y^4$

Exercise Set 5.2

1. If c is positive, then p and q have the same sign. If both are negative, then b is negative; if both are positive then c is positive. Thus we replace each blank with "positive."

3. If p is negative and q is negative, then b is negative because it is the sum of two negative numbers and c is positive because it is the product of two negative numbers.

5. Since c is negative, it is the product of a negative and a positive number. Then because c is the product of p and q and we know that p is negative, q must be positive.

7. $x^2 + 7x + 6$

Since the constant term and the coefficient of the middle term are both positive, we look for a factorization of 6 in which both factors are positive. Their sum must be 7.

Pairs of factors	Sums of factors
1, 6	7
2, 3	5

The numbers we want are 1 and 6.

$x^2 + 7x + 6 = (x+1)(x+6)$

9. $x^2 + 7x + 12$

Since the constant term is positive and the coefficient of the middle term is positive, we look for a factorization of 12 in which both factors are positive. Their sum must be 7.

Pairs of factors	Sums of factors
1, 2	11
2, 6	7
3, 4	7

The numbers we want are 3 and 4.

$x^2 + 7x + 12 = (x+3)(x+4)$

11. $x^2 - 6x + 9$

Since the constant term is positive and the middle term is negative, we look for a factorization of 9 in which both factors are negative.

Pairs of factors	Sums of factors
−1, −9	−10
−3, −3	−6

The numbers we want are −3 and −3.

$x^2 - 6x + 9 = (x-3)(x-3)$, or $(x-3)^2$

13. $x^2 + 9x + 14$

Since the constant term is positive and the coefficient of the middle term is positive, we look for a factorization of 14 in which both factors are positive. Their sum must be 9.

Pairs of factors	Sums of factors
1, 14	15
2, 7	9

The numbers we want are 2 and 7.

$x^2 + 9x + 14 = (x+2)(x+7)$

15. $b^2 - 5b + 4$

Since the constant term is positive and the coefficient of the middle term is negative, we look for a factorization of 4 in which both factors are negative. Their sum must be -5.

Pairs of factors	Sums of factors
$-1, -4$	-5
$-2, -2$	-4

The numbers we want are -1 and -4.

$b^2 - 5b + 4 = (b - 1)(b - 4)$.

17. $a^2 + 4a - 12$

Since the constant term is negative, we look for a factorization of -12 in which one factor is positive and one factor is negative. Their sum must be 4, the coefficient of the middle term, so the positive factor must have the larger absolute value. Thus we consider only pairs of factors in which the positive factor has the larger absolute value.

Pairs of factors	Sums of factors
$-1, \quad 12$	11
$-2, \quad 6$	4
$-3, \quad 4$	1

The numbers we need are -2 and 6.

$a^2 + 4a - 12 = (a - 2)(a + 6)$.

19. $d^2 - 7d + 10$

Since the constant term is positive and the coefficient of the middle term is negative, we look for a factorization of 10 in which both factors are negative. Their sum must be -7.

Pairs of factors	Sums of factors
$-1, -10$	-11
$-2, \quad -5$	-7

The numbers we want are -2 and -5.

$d^2 - 7d + 10 = (d - 2)(d - 5)$.

21. $x^2 - 2x - 15$

The constant term, -15, must be expressed as the product of a negative number and a positive number. Since the sum of those two numbers must be negative, the negative number must have the greater absolute value.

Pairs of factors	Sums of factors
$1, \quad -15$	-14
$3, \quad -5$	-2

The numbers we need are 3 and -5.

$x^2 - 2x - 15 = (x + 3)(x - 5)$.

23. $x^2 + 2x - 15$

The constant term, -15, must be expressed as the product of a negative number and a positive number. Since the sum of those two numbers must be positive, the positive number must have the greater absolute value.

Pairs of factors	Sums of factors
$-1, \quad 15$	14
$-3, \quad 5$	2

The numbers we need are -3 and 5.

$x^2 + 2x - 15 = (x - 3)(x + 5)$.

25. $3y^2 - 9y - 84 = 3(y^2 - 3y - 28)$

After factoring out the common factor, 3, we consider $y^2-3y+28$. The constant term, -28, must be expressed as the product of a negative number and a positive number. Since the sum of those two numbers must be negative, the negative number must have the greater absolute value.

Pairs of factors	Sums of factors
$1, \quad -28$	-27
$2, \quad -14$	-12
$4, \quad -7$	-3

The numbers we need are 4 and -7. The factorization of $y^2 - 3y - 28$ is $(y + 4)(y - 7)$. We must not forget the common factor, 3. Thus, $3y^2 - 9y - 84 = 3(y^2 - 3y - 28) = 3(y + 4)(y - 7)$.

27. $-x^3 + x^2 + 42x = -x(x^2 - x - 42)$

After factoring out the common factor, $-x$, we consider $x^2 - x - 42$. The constant term, -42, must be expressed as the product of a negative number and a positive number. Since the sum of those two numbers must be negative, the negative number must have the greater absolute value.

Pairs of factors	Sums of factors
$1, \quad -42$	-41
$2, \quad -21$	-19
$3, \quad -14$	-11
$6, \quad -7$	-1

The numbers we need are 6 and -7. The factorization of x^2-x-42 is $(x+6)(x-7)$. We must not forget the common factor, $-x$. Thus, $-x^3 + x^2 + 42x = -x(x^2 - x - 42) = -x(x + 6)(x - 7)$.

29. $7x - 60 + x^2 = x^2 + 7x - 60$

The constant term, -60, must be expressed as the product of a negative number and a positive number. Since the sum of those two numbers must be positive, the positive number must have the greater absolute value.

Pairs of factors	Sums of factors
$-1, \quad 60$	59
$-2, \quad 30$	28
$-3, \quad 20$	17
$-4, \quad 15$	11
$-5, \quad 12$	7

The numbers we need are -5 and 12.

$7x - 60 + x^2 = (x - 5)(x + 12)$

31. $x^2 - 72 + 6x = x^2 + 6x - 72$

The constant term, -72, must be expressed as the product of a negative number and a positive number. Since the sum of those two numbers must be positive, the positive number must have the greater absolute value.

Pairs of factors	Sums of factors
−1, 72	71
−2, 36	34
−3, 24	21
−4, 18	14
−6, 12	6

The numbers we need are −6 and 12.

$x^2 - 72 + 6x = (x - 6)(x + 12)$

33. $-5b^2 - 25b + 120 = -5(b^2 + 5b - 24)$

After factoring out the common factor, −5, we consider $b^2 + 5b - 24$. The constant term, −24, must be expressed as the product of a negative number and a positive number. Since the sum of those two numbers must be positive, the positive number must have the greater absolute value.

Pairs of factors	Sums of factors
−1, 24	23
−2, 12	10
−3, 8	5

The numbers we need are −3 and 8. The factorization of $b^2+5b-24$ is $(b-3)(b+8)$. We must not forget the common factor. Thus, $-5b^2 - 25b + 120 = -5(b^2 + 5b - 24) = -5(b - 3)(b + 8)$.

35. $x^5 - x^4 - 2x^3 = x^3(x^2 - x - 2)$

After factoring out the common factor, x^3, we consider $x^2 - x - 2$. The constant term, −2, must be expressed as the product of a negative number and a positive number. Since the sum of those two numbers must be negative, the negative number must have the greater absolute value. The only possible factors that fill these requirements are 1 and −2. These are the numbers we need. The factorization of $x^2 - x - 2$ is $(x + 1)(x - 2)$. We must not forget the common factor, x^3. Thus, $x^5 - x^4 - 2x^3 = x^3(x^2 - x - 2) = x^3(x + 1)(x - 2)$.

37. $x^2 + 2x + 3$

Since the constant term and the coefficient of the middle term are both positive, we look for a factorization of 3 in which both factors are positive. Their sum must be 2. The only possible pair of positive factors is 1 and 3, but their sum is not 2. Thus, this polynomial is not factorable into polynomials with integer coefficients. It is prime.

39. $50 + 15t + t^2 = t^2 + 15t + 50$

Since the constant term is positive and the coefficient of the middle term is positive, we look for a factorization of 50 in which both factors are positive. Their sum must be 15.

Pairs of factors	Sums of factors
1, 50	51
2, 25	27
5, 10	15

The numbers we want are 5 and 10.

$50 + 15t + t^2 = (t + 5)(t + 10)$.

41. $x^2 + 20x + 99$

We look for two factors, both positive, whose product is 99 and whose sum is 20.

They are 9 and 11: $9 \cdot 11 = 99$ and $9 + 11 = 20$.

$x^2 + 20 + 99 = (x + 9)(x + 11)$

43. $2x^3 - 40x^2 + 192x = 2x(x^2 - 20x + 96)$

After factoring out the common factor, $2x$, we consider $x^2 - 20x + 96$. We look for two factors, both negative, whose product is 96 and whose sum is −20.

They are −8 and −12: $-8(-12) = 96$ and $-8 + (-12) = -20$.

$x^2 - 20x + 96 = (x - 8)(x - 12)$, so $2x^3 - 40x^2 + 192x = 2x(x - 8)(x - 12)$.

45. $-4x^2 - 40x - 100 = -4(x^2 + 10x + 25)$

After factoring out the common factor, −4, we consider $x^2 + 10x + 25$. We look for two factors, both positive, whose product is 25 and whose sum is 10. They are 5 and 5.

$x^2 + 10x + 25 = (x + 5)(x + 5)$, so $-4x^2 - 40x - 100 = -4(x + 5)(x + 5)$, or $-4(x + 5)^2$.

47. $y^2 - 21y + 108$

We look for two factors, both negative, whose product is 108 and whose sum is −21. They are −9 and −12.

$y^2 - 21y + 108 = (y - 9)(y - 12)$

49. $-a^6 - 9a^5 + 90a^4 = -a^4(a^2 + 9a - 90)$

After factoring out the common factor, $-a^4$, we consider $a^2 + 9a - 90$. We look for two factors, one positive and one negative, whose product is −90 and whose sum is 9. They are −6 and 15.

$a^2 + 9a - 90 = (a - 6)(a + 15)$, so $-a^6 - 9a^5 + 90a^4 = -a^4(a - 6)(a + 15)$.

51. $t^2 + \frac{2}{3}t + \frac{1}{9}$

We look for two factors, both positive, whose product is $\frac{1}{9}$ and whose sum is $\frac{2}{3}$. They are $\frac{1}{3}$ and $\frac{1}{3}$.

$t^2 + \frac{2}{3}t + \frac{1}{9} = \left(t + \frac{1}{3}\right)\left(t + \frac{1}{3}\right)$, or $\left(t + \frac{1}{3}\right)^2$

53. $11 + w^2 - 4w = w^2 - 4w + 11$

Since the constant term is positive and the coefficient of the middle term is negative, we look for a factorization of 11 in which both factors are negative. Their sum must be −4. The only possible pair of factors is −1 and −11, but their sum is not −4. Thus, this polynomial is not factorable into polynomials with integer coefficients. It is prime.

55. $p^2 + 3pq - 10q^2 = p^2 + 3qp - 10q^2$

Think of $3q$ as a "coefficient" of p. Then we look for factors of $-10q^2$ whose sum is $3q$. They are $5q$ and $-2q$.

$p^2 + 3pq - 10q^2 = (p + 5q)(p - 2q)$.

57. $m^2 + 5mn + 5n^2 = m^2 + 5nm + 5n^2$

We look for factors of $5n^2$ whose sum is $5n$. The only reasonable possibilities are shown below.

Pairs of factors	Sums of factors
$5n,\ n$	$6n$
$-5n,\ -n$	$-6n$

There are no factors whose sum is $5n$. Thus, the polynomial is not factorable into polynomials with integer coefficients. It is prime.

59. $s^2 - 2st - 15t^2 = s^2 - 2ts - 15t^2$

We look for factors of $-15t^2$ whose sum is $-2t$. They are $-5t$ and $3t$.

$s^2 - 2st - 15t^2 = (s - 5t)(s + 3t)$

61. $6a^{10} - 30a^9 - 84a^8 = 6a^8(a^2 - 5a - 14)$

After factoring out the common factor, $6a^8$, we consider $a^2 - 5a - 14$. We look for two factors, one positive and one negative, whose product is -14 and whose sum is -5. They are 2 and -7.

$a^2 - 5a - 14 = (a + 2)(a - 7)$, so $6a^{10} - 30a^9 - 84a^8 = 6a^8(a + 2)(a - 7)$.

63. *Writing Exercise*

65. $3x - 8 = 0$

$3x = 8$ Adding 8 on both sides

$x = \frac{8}{3}$ Dividing by 3 on both sides

The solution is $\frac{8}{3}$.

67. $(x + 6)(3x + 4)$

$= 3x^2 + 4x + 18x + 24$ Using FOIL

$= 3x^2 + 22x + 24$

69. ***Familiarize***. Let n = the number of people arrested the year before.

Translate. We reword the problem.

$\underbrace{\text{Number arrested the year before}}$ less 1.2% of $\underbrace{\text{that number}}$ is 29,090.

$n \quad - \quad 1.2\% \quad \cdot \quad n \quad = \quad 29,090$

Carry out. We solve the equation.

$n - 1.2\% \cdot n = 29,090$

$1 \cdot n - 0.012n = 29,090$

$0.988n = 29,090$

$n \approx 29,443$ Rounding

Check. 1.2% of 29,443 is $0.012(29,443) \approx 353$ and $29,443 - 353 = 29,090$. The answer checks.

State. Approximately 29,443 people were arrested the year before.

71. *Writing Exercise*

73. $a^2 + ba - 50$

We look for all pairs of integer factors whose product is -50. The sum of each pair is represented by b.

Pairs of factors whose product is -50	Sums of factors
$-1,\ 50$	49
$1,\ -50$	-49
$-2,\ 25$	23
$2,\ -25$	-23
$-5,\ 10$	5
$5,\ -10$	-5

The polynomial $a^2 + ba - 50$ can be factored if b is 49, -49, 23, -23, 5, or -5.

75. $y^2 - 0.2y - 0.08$

We look for two factors, one positive and one negative, whose product is -0.08 and whose sum is -0.2. They are -0.4 and 0.2.

$y^2 - 0.2y - 0.08 = (y - 0.4)(y + 0.2)$

77. $-\frac{1}{3}a^3 + \frac{1}{3}a^2 + 2a = -\frac{1}{3}a(a^2 - a - 6)$

After factoring out the common factor, $-\frac{1}{3}a$, we consider $a^2 - a - 6$. We look for two factors, one positive and one negative, whose product is -6 and whose sum is -1. They are 2 and -3.

$a^2 - a - 6 = (a + 2)(a - 3)$, so
$-\frac{1}{3}a^3 + \frac{1}{3}a^2 + 2a = -\frac{1}{3}a(a + 2)(a - 3)$.

79. $x^{2m} + 11x^m + 28 = (x^m)^2 + 11x^m + 28$

We look for numbers p and q such that $x^{2m} + 11x^m + 28 = (x^m + p)(x^m + q)$. We find two factors, both positive, whose product is 28 and whose sum is 11. They are 4 and 7.

$x^{2m} + 11x^m + 28 = (x^m + 4)(x^m + 7)$

81. $(a + 1)x^2 + (a + 1)3x + (a + 1)2$

$= (a + 1)(x^2 + 3x + 2)$

After factoring out the common factor $a + 1$, we consider $x^2 + 3x + 2$. We look for two factors, whose product is 2 and whose sum is 3. They are 1 and 2.

$x^2 + 3x + 2 = (x + 1)(x + 2)$, so
$(a + 1)x^2 + (a + 1)3x + (a + 1)2 =$
$(a + 1)(x + 1)(x + 2)$.

83. $6x^2 + 36x + 54 = 6(x^2 + 6x + 9) = 6(x+3)(x+3) = 6(x+3)^2$

Since the surface area of a cube with sides is given by $6s^2$, we know that this cube has side $x + 3$. The volume of a cube with side s is given by s^3, so the volume of this cube is $(x + 3)^3$, or $x^3 + 9x^2 + 27x + 27$.

85. Shaded area = Area of circle − Area of triangle =

$\pi x^2 - \frac{1}{2}(2x)(x) = \pi x^2 - x^2 = x^2(\pi - 1)$

87. The shaded area consists of the area of a square with side $x + x + x$, or $3x$, less the area of a semicircle with radius x. It can be expressed as follows:

$3x \cdot 3x - \frac{1}{2}\pi x^2 = 9x^2 - \frac{1}{2}\pi x^2 = x^2\left(9 - \frac{1}{2}\pi\right)$

89. $x^2 + 5x + 7x + 35 = x^2 + 12x + 35 = (x+5)(x+7)$

Exercise Set 5.3

1. Since $(6x-1)(2x+3) = 12x^2+16x-3$, choice (c) is correct.

3. Since $(7x + 1)(2x - 3) = 14x^2 - 19x - 3$, choice (d) is correct.

5. $2x^2 + 7x - 4$

(1) There is no common factor (other than 1 or −1).

(2) Because $2x^2$ can be factored as $2x \cdot x$, we have this possibility:

$(2x + \quad)(x + \quad)$

(3) There are 3 pairs of factors of −4 and they can be listed two ways:

$$\begin{array}{llll} & -4,1 & 4,-1 & 2,-2 \\ \text{and} & 1,-4 & -1,4 & -2,2 \end{array}$$

(4) Look for Outer and Inner products resulting from steps (2) and (3) for which the sum is $7x$. We can immediately reject all possibilities in which a factor has a common factor, such as $(2x - 4)$ or $(2x + 2)$, because we determined at the outset that there is no common factor other than 1 and −1. We try some possibilities:

$(2x+1)(x-4) = 2x^2 - 7x - 4$

$(2x-1)(x+4) = 2x^2 + 7x - 4$

The factorization is $(2x - 1)(x + 4)$.

7. $3t^2 + 4t - 15$

(1) There is no common factor (other than 1 or −1).

(2) Because $3t^2$ can be factored as $3t \cdot t$, we have this possibility:

$(3t + \quad)(t + \quad)$

(3) There are 4 pairs of factors of −15 and they can be listed two ways:

$$\begin{array}{lllll} & -15,1 & 15,-1 & -5,3 & 5,-3 \\ \text{and} & 1,-15 & -1,15 & 3,-5 & -3,5 \end{array}$$

(4) Look for Outer and Inner products resulting from steps (2) and (3) for which the sum is $4t$. We can immediately reject all possibilities in which either factor has a common factor, such as $(3t - 15)$ or $(3t + 3)$, because at the outset we determined that there is no common factor other than 1 or −1. We try some possibilities:

$(3t+1)(t-15) = 3t^2 - 44t - 15$

$(3t-5)(t+3) = 3t^2 + 4t - 15$

The factorization is $(3t - 5)(t + 3)$.

9. $6x^2 - 23x + 7$

(1) There is no common factor (other than 1 or −1).

(2) Because $6x^2$ can be factored as $6x \cdot x$ or $3x \cdot 2x$, we have these possibilities:

$(6x + \quad)(x + \quad)$ and $(3x + \quad)(2x + \quad)$

(3) There are 2 pairs of factors of 7 and they can be listed two ways:

$$\begin{array}{lll} & 7,1 & -7,-1 \\ \text{and} & 1,7 & -1,-7 \end{array}$$

(4) Look for Outer and Inner products resulting from steps (2) and (3) for which the sum is $-23x$. Since the sign of the middle term is negative and the sign of the last term is positive, the factors of 7 must both be negative. We try some possibilities:

$(6x-7)(x-1) = 6x^2 - 13x + 7$

$(3x-7)(2x-1) = 6x^2 - 17x + 7$

$(6x-1)(x-7) = 6x^2 - 43x + 7$

$(3x-1)(2x-7) = 6x^2 - 23x + 7$

The factorization is $(3x - 1)(2x - 7)$.

11. $7x^2 + 15x + 2$

(1) There is no common factor (other than 1 or −1).

(2) Because $7x^2$ can be factored as $7x \cdot x$, we have this possibility:

$(7x + \quad)(x + \quad)$

(3) There are 2 pairs of factors of 2 and they can be listed two ways:

$$\begin{array}{lll} & 2,1 & -2,-1 \\ \text{and} & 1,2 & -1,-2 \end{array}$$

(4) Look for Outer and Inner products resulting from steps (2) and (3) for which the sum is $15x$. Since all coefficients are positive, we need consider only positive factors of 2. We try some possibilities:

$(7x+2)(x+1) = 7x^2 + 9x + 2$

$(7x+1)(x+2) = 7x^2 + 15x + 2$

The factorization is $(7x + 1)(x + 2)$.

13. $9a^2 - 6a - 8$

(1) There is no common factor (other than 1 or −1).

(2) Because $9a^2$ can be factored as $9a \cdot a$ and $3a \cdot 3a$, we have these possibilities:

$(9a + \quad)(a + \quad)$ and $(3a + \quad)(3a + \quad)$

(3) There are 4 pairs of factors of −8 and they can be listed two ways:

$$\begin{array}{lllll} & -8,1 & 8,-1 & -4,2 & 4,-2 \\ \text{and} & 1,-8 & -1,8 & 2,-4 & -2,4 \end{array}$$

(4) Look for Outer and Inner products resulting from steps (2) and (3) for which the sum is $-6a$. We try some possibilities:

$(9a-8)(a+1) = 9a^2 + a - 8$

$(9a-4)(a+2) = 9a^2 + 14a - 8$

$(3a+8)(3a-1) = 9a^2 + 21a - 8$

$(3a - 4)(3a + 2) = 9a^2 - 6a - 8$

The factorization is $(3a - 4)(3a + 2)$.

15. $6x^2 - 10x - 4$

(1) We factor out the largest common factor, 2:

$2(3x^2 - 5x - 2)$.

Then we factor the trinomial $3x^2 - 5x - 2$.

(2) Because $3x^2$ can be factored as $3x \cdot x$, we have this possibility:

$(3x + \quad)(x + \quad)$

(3) There are 2 pairs of factors of -2 and they can be listed two ways:

$-2, 1 \quad 2, -1$

and $\quad 1, -2 \quad -1, 2$

(4) Look for Outer and Inner products resulting from steps (2) and (3) for which the sum is $-5x$. We try some possibilities:

$(3x - 2)(x + 1) = 3x^2 + x - 2$

$(3x + 2)(x - 1) = 3x^2 - x - 2$

$(3x + 1)(x - 2) = 3x^2 - 5x - 2$

The factorization of $3x^2 - 5x - 2$ is $(3x + 1)(x - 2)$. We must include the common factor in order to get a factorization of the original trinomial.

$6x^2 - 10x - 4 = 2(3x + 1)(x - 2)$

17. $12t^2 - 6t - 6$

(1) We factor out the common factor, 6:

$6(2t^2 - t - 1)$.

Then we factor the trinomial $2t^2 - t - 1$.

(2) Because $2t^2$ can be factored as $2t \cdot t$, we have this possibility:

$(2t + \quad)(t + \quad)$

(3) There are 2 pairs of factors of -1. In this case they can be listed in only one way:

$-1, 1 \quad 1, -1$

(4) Look for Outer and Inner products resulting from steps (2) and (3) for which the sum is $-t$. We try some possibilities:

$(2t - 1)(t + 1) = 2t^2 + t - 1$

$(2t + 1)(t - 1) = 2t^2 - t - 1$

The factorization of $2t^2 - t - 1$ is $(2t + 1)(t - 1)$. We must include the common factor in order to get a factorization of the original trinomial.

$12t^2 - 6t - 6 = 6(2t + 1)(t - 1)$

19. $6 + 19x + 15x^2 = 15x^2 + 19x + 6$

(1) There is no common factor (other than 1 or -1).

(2) Because $15x^2$ can be factored as $15x \cdot x$ and $5x \cdot 3x$, we have these possibilities:

$(15x + \quad)(x + \quad)$ and $(5x + \quad)(3x + \quad)$

(3) Since all coefficients are positive, we need consider only positive pairs of factors of 6. There are 2 such pairs and they can be listed two ways:

$6, 1 \quad 3, 2$

and $\quad 1, 6 \quad 2, 3$

(4) We can immediately reject all possibilities in which either factor has a common factor, such as $(15x + 6)$ or $(3x + 3)$, because we determined at the outset that there is no common factor other than 1 or -1. We try some possibilities:

$(15x + 2)(x + 3) = 15x^2 + 47x + 6$

$(5x + 6)(3x + 1) = 15x^2 + 23x + 6$

$(5x + 3)(3x + 2) = 15x^2 + 19x + 6$

The factorization is $(5x + 3)(3x + 2)$.

21. $-35x^2 - 34x - 8$

(1) We factor out -1 in order to have a trinomial with a positive leading coefficient.

$-35x^2 - 34x - 8 = -1(35x^2 + 34x + 8)$

Now we factor $35x^2 + 34x + 8$.

(2) Because $35x^2$ can be factored as $35x \cdot x$ or $7x \cdot 5x$, we have these possibilities:

$(35x + \quad)(x + \quad)$ and $(7x + \quad)(5x + \quad)$

(3) Since all coefficients are positive, we need consider only positive pairs of factors of 8. There are 2 such pairs and they can be listed two ways:

$8, 1 \quad 4, 2$

and $\quad 1, 8 \quad 2, 4$

(4) We try some possibilities:

$(35x + 8)(x + 1) = 35x^2 + 43x + 8$

$(7x + 8)(5x + 1) = 35x^2 + 47x + 8$

$(7x + 4)(5x + 2) = 35x^2 + 34x + 8$

The factorization of $35x^2 + 34x + 8$ is $(7x + 4)(5x + 2)$.

We must include the factor of -1 in order to get a factorization of the original trinomial.

$-35x^2 - 34x - 8 = -1(7x + 4)(5x + 2)$, or $-(7x + 4)(5x + 2)$.

23. $4 + 6t^2 - 13t = 6t^2 - 13t + 4$

(1) There is no common factor (other than 1 or -1).

(2) Because $6t^2$ can be factored as $6t \cdot t$ or $3t \cdot 2t$, we have these possibilities:

$(6t + \quad)(t + \quad)$ and $(3t + \quad)(2t + \quad)$

(3) Since the sign of the middle term is negative but the sign of the last term is positive, we need to consider only negative factors of 4. There is only 1 such pair and it can be listed two ways:

$-4, -1$ and $-1, -4$

(4) We can immediately reject all possibilities in which either factor has a common factor, such as $(6t - 4)$ or $(2t - 4)$, because we determined at the outset that there is no common factor other than 1 or -1. We try some possibilities:

$$(6t-1)(t-4) = 6t^2 - 25t + 4$$
$$(3t-4)(2t-1) = 6t^2 - 11t + 4$$

These are the only possibilities that do not contain a common factor. Since neither is the desired factorization, we must conclude that $4 + 6t^2 - 13t$ is prime.

25. $25x^2 + 40x + 16$

(1) There is no common factor (other than 1 or -1).

(2) Because $25x^2$ can be factored as $25x \cdot x$ or $5x \cdot 5x$, we have these possibilities:

$$(25x + \quad)(x + \quad) \text{ and } (5x + \quad)(5x + \quad)$$

(3) Since all coefficients are positive, we need consider only positive pairs of factors of 16. There are 3 such pairs and two of them can be listed two ways:

$16, 1 \quad 8, 2 \quad 4, 4$

and $\quad 1, 16 \quad 2, 8$

(4) We try some possibilities:

$$(25x+16)(x+1) = 25x^2 + 41x + 16$$
$$(5x+8)(5x+2) = 25x^2 + 50x + 16$$
$$(5x+4)(5x+4) = 25x^2 + 40x + 16$$

The factorization is $(5x+4)(5x+4)$, or $(5x+4)^2$.

27. $16a^2 + 78a + 27$

(1) There is no common factor (other than 1 or -1).

(2) Because $16a^2$ can be factored as $16a \cdot a$, $8a \cdot 2a$, or $4a \cdot 4a$, we have these possibilities:

$(16a + \quad)(a + \quad)$ and $(8a + \quad)(2a + \quad)$ and $(4a + \quad)(4a + \quad)$

(3) Since all coefficients are positive, we need consider only positive pairs of factors of 27. There are 2 such pairs and two of them can be listed two ways:

$27, 1 \quad 3, 9$

and $\quad 1, 27 \quad 9, 3$

(4) We try some possibilities:

$$(16a+27)(a+1) = 16a^2 + 43a + 27$$
$$(8a+3)(2a+9) = 16a^2 + 78a + 27$$

The factorization is $(8a+3)(2a+9)$.

29. $18t^2 + 24t - 10$

(1) Factor out the common factor, 2:

$$2(9t^2 + 12t - 5)$$

Then we factor the trinomial $9t^2 + 12t - 5$.

(2) Because $9t^2$ can be factored as $9t \cdot t$ or $3t \cdot 3t$, we have these possibilities:

$$(9t + \quad)(t + \quad) \text{ and } (3t + \quad)(3t + \quad)$$

(3) There are 2 pairs of factors of -5 and they can be listed two ways:

$-5, 1 \quad 5, -1$

and $\quad 1, -5 \quad -1, 5$

(4) We try some possibilities:

$$(9t-5)(t+1) = 9t^2 + 4t - 5$$
$$(9t+1)(t-5) = 9t^2 - 44t - 5$$
$$(3t+1)(3t-5) = 9t^2 - 12t - 5$$
$$(3t-1)(3t+5) = 9t^2 + 12t - 5$$

The factorization of $9t^2+12t-5$ is $(3t-1)(3t+5)$. We must include the common factor in order to get a factorization of the original trinomial.

$18t^2 + 24t - 10 = 2(3t-1)(3t+5)$

31. $-2x^2 + 15 + x = -2x^2 + x + 15$

(1) We factor out -1 in order to have a trinomial with a positive leading coefficient.

$$-2x^2 + x + 15 = -1(2x^2 - x - 15)$$

Now we factor $2x^2 - x - 15$.

(2) Because $2x^2$ can be factored as $2x \cdot x$ we have this possibility:

$$(2x + \quad)(x + \quad)$$

(3) There are 4 pairs of factors of -15 and they can be listed two ways:

$-15, 1 \quad 15, -1 \quad -5, 3 \quad 5, -3$

and $\quad 1, -15 \quad -1, 15 \quad 3, -5 \quad -3, 5$

(4) We try some possibilities:

$$(2x-15)(x+1) = 2x^2 - 13x - 15$$
$$(2x-5)(x+3) = 2x^2 + x - 15$$
$$(2x+5)(x-3) = 2x^2 - x - 15$$

The factorization of $2x^2-x-15$ is $(2x+5)(x-3)$. We must include the factor of -1 in order to get a factorization of the original trinomial.

$-2x^2 + 15 + x = -1(2x+5)(x-3)$, or $-(2x+5)(x-3)$

33. $-6x^2 - 33x - 15$

(1) Factor out -3. This not only removes the largest common factor, 3. It also produces a trinomial with a positive leading coefficient.

$$-3(2x^2 + 11x + 5)$$

Then we factor the trinomial $2x^2 + 11x + 5$.

(2) Because $2x^2$ can be factored as $2x \cdot x$ we have this possibility:

$$(2x + \quad)(x + \quad)$$

(3) Since all coefficients are positive, we need consider only positive pairs of factors of 5. There is one such pair and it can be listed two ways:

$5, 1 \quad$ and $\quad 1, 5$

(4) We try some possibilities:

$$(2x+5)(x+1) = 2x^2 + 7x + 5$$
$$(2x+1)(x+5) = 2x^2 + 11x + 5$$

The factorization of $2x^2+11x+5$ is $(2x+1)(x+5)$. We must include the common factor in order to get a factorization of the original trinomial.

$-6x^2 - 33x - 15 = -3(2x+1)(x+5)$

35. $20x^2 - 25x + 5$

(1) Factor out the common factor, 5:

$5(4x^2 - 5x + 1)$

Then we factor the trinomial $4x^2 - 5x + 1$.

(2) Because $4x^2$ can be factored as $4x \cdot x$ or $2x \cdot 2x$, we have these possibilities:

$(4x + \quad)(x + \quad)$ and $(2x + \quad)(2x + \quad)$

(3) Since the sign of the middle term is negative but the sign of the last term is positive, we need to consider only negative factors of 1. There is only 1 such pair, $-1, -1$.

(4) We try the possibilities:

$(4x - 1)(x - 1) = 4x^2 - 5x + 1$

The factorization of $4x^2 - 5x + 1$ is $(4x - 1)(x - 1)$. We must include the common factor in order to get a factorization of the original trinomial.

$20x^2 - 25x + 5 = 5(4x - 1)(x - 1)$

37. $12x^2 + 68x - 24$

(1) Factor out the common factor, 4:

$4(3x^2 + 17x - 6)$

Then we factor the trinomial $3x^2 + 17x - 6$.

(2) Because $3x^2$ can be factored as $3x \cdot x$ we have this possibility:

$(3x + \quad)(x + \quad)$

(3) There are 4 pairs of factors of -6 and they can be listed two ways:

	$6, -1$	$-6, 1$	$3, -2$	$-3, 2$
and	$-1, 6$	$1, -6$	$-2, 3$	$2, -3$

(4) We can immediately reject all possibilities in which either factor has a common factor, such as $(3x + 6)$ or $(3x - 3)$, because we determined at the outset that there is no common factor other than 1 or -1. We try some possibilities:

$(3x - 1)(x + 6) = 3x^2 + 17x - 6$

The factorization of $3x^2 + 17x - 6$ is $(3x - 1)(x + 6)$. We must include the common factor in order to get a factorization of the original trinomial.

$12x^2 + 68x - 24 = 4(3x - 1)(x + 6)$

39. $4x + 1 + 3x^2 = 3x^2 + 4x + 1$

(1) There is no common factor (other than 1 or -1).

(2) Because $3x^2$ can be factored as $3x \cdot x$ we have this possibility:

$(3x + \quad)(x + \quad)$

(3) Since all coefficients are positive, we need consider only positive pairs of factors of 1. There is one such pair: 1,1.

(4) We try the possible factorization:

$(3x + 1)(x + 1) = 3x^2 + 4x + 1$

The factorization is $(3x + 1)(x + 1)$.

41. $y^2 + 4y - 2y - 8 = y(y + 4) - 2(y + 4)$
$= (y + 4)(y - 2)$

43. $8t^2 - 6t - 28t + 21 = 2t(4t - 3) - 7(4t - 3)$
$= (4t - 3)(2t - 7)$

45. $6x^2 + 4x + 9x + 6 = 2x(3x + 2) + 3(3x + 2)$
$= (3x + 2)(2x + 3)$

47. $2t^2 + 6t - t - 3 = 2t(t + 3) - 1(t + 3)$
$= (t + 3)(2t - 1)$

49. $3a^2 - 12a - a + 4 = 3a(a - 4) - 1(a - 4)$
$= (a - 4)(3a - 1)$

51. $9t^2 + 14t + 5$

(1) First note that there is no common factor (other than 1 or -1).

(2) Multiply the leading coefficient, 9, and the constant, 5:

$9 \cdot 5 = 45$

(3) We look for factors of 45 that add to 14. Since all coefficients are positive, we need to consider only positive factors.

Pairs of factors	Sums of factors
1, 45	46
3, 15	18
5, 9	14

The numbers we need are 5 and 9.

(4) Rewrite the middle term:

$14t = 5t + 9t$

(5) Factor by grouping:

$$9t^2 + 14t + 5 = 9t^2 + 5t + 9t + 5$$
$$= t(9t + 5) + 1(9t + 5)$$
$$= (9t + 5)(t + 1)$$

53. $-16x^2 - 32x - 7$

(1) We factor out -1 in order to have a trinomial with a positive leading coefficient.

$-16x^2 - 32x - 7 = -1(16x^2 + 32x + 7)$

Now we factor $16x^2 + 32x + 7$.

(2) Multiply the leading coefficient, 16, and the constant, 7:

$16 \cdot 7 = 112$

(3) We look for factors of 112 that add to 32. Since all coefficients are positive, we need to consider only positive factors.

Pairs of factors	Sums of factors
1, 112	113
2, 56	58
4, 28	32
7, 16	23
8, 14	22

The numbers we need are 4 and 28.

(4) Rewrite the middle term:

$32x = 4x + 28x$

(5) Factor by grouping:

$$\begin{aligned}16x^2 + 32x + 7 &= 16x^2 + 4x + 28x + 7\\ &= 4x(4x+1) + 7(4x+1)\\ &= (4x+1)(4x+7)\end{aligned}$$

We must include the factor of -1 in order to get a factorization of the original trinomial.

$-16x^2 - 32x - 7 = -1(4x+1)(4x+7)$, or
$-(4x+1)(4x+7)$

55. $10a^2 + 25a - 15$

(1) Factor out the largest common factor, 5:

$$10a^2 + 25a - 15 = 5(2a^2 + 5a - 3)$$

(2) To factor $2a^2 + 5a - 3$ by grouping we first multiply the leading coefficient, 2, and the constant, -3:

$$2(-3) = -6$$

(3) We look for factors of -6 that add to 5.

Pairs of factors	Sums of factors
−1, 6	5
−6, 1	−5
−2, 3	1
2, −3	−1

The numbers we need are -1 and 6.

(4) Rewrite the middle term:

$$5a = -a + 6a$$

(5) Factor by grouping:

$$\begin{aligned}2a^2 + 5a - 3 &= 2a^2 - a + 6a - 3\\ &= a(2a-1) + 3(2a-1)\\ &= (2a-1)(a+3)\end{aligned}$$

The factorization of $2a^2+5a-3$ is $(2a-1)(a+3)$. We must include the common factor in order to get a factorization of the original trinomial:

$$10a^2 + 25a - 15 = 5(2a-1)(a+3)$$

57. $18x^3 + 21x^2 - 9x$

(1) Factor out the largest common factor, $3x$:

$$18x^3 + 21x^2 - 9x = 3x(6x^2 + 7x - 3)$$

(2) To factor $6x^2 + 7x - 3$ by grouping we first multiply the leading coefficient, 6, and the constant, -3:

$$6(-3) = -18$$

(3) We look for factors of -18 that add to 7.

Pairs of factors	Sums of factors
−1, 18	17
1, −18	−17
−2, 9	7
2, −9	−7
−3, 6	3
3, −6	−3

The numbers we need are -2 and 9.

(4) Rewrite the middle term:

$$7x = -2x + 9x$$

(5) Factor by grouping:

$$\begin{aligned}6x^2 + 7x - 3 &= 6x^2 - 2x + 9x - 3\\ &= 2x(3x-1) + 3(3x-1)\\ &= (3x-1)(2x+3)\end{aligned}$$

The factorization of $6x^2+7x-3$ is $(3x-1)(2x+3)$. We must include the common factor in order to get a factorization of the original trinomial:

$$18x^3 + 21x^2 - 9x = 3x(3x-1)(2x+3)$$

59. $89x + 64 + 25x^2 = 25x^2 + 89x + 64$

(1) First note that there is no common factor (other than 1 or -1).

(2) Multiply the leading coefficient, 25, and the constant, 64:

$$25 \cdot 64 = 1600$$

(3) We look for factors of 1600 that add to 89. Since all coefficients are positive, we need to consider only positive factors. The numbers we need are 25 and 64.

(4) Rewrite the middle term:

$$89x = 25x + 64x$$

(5) Factor by grouping:

$$\begin{aligned}25x^2 + 89x + 64 &= 25x^2 + 25x + 64x + 64\\ &= 25x(x+1) + 64(x+1)\\ &= (x+1)(25x+64)\end{aligned}$$

61. $168x^3 + 45x^2 + 3x$

(1) Factor out the largest common factor, $3x$:

$$168x^3 + 45x^2 + 3x = 3x(56x^2 + 15x + 1)$$

(2) To factor $56x^2 + 15x + 1$ we first multiply the leading coefficient, 56, and the constant, 1:

$$56 \cdot 1 = 56$$

(3) We look for factors of 56 that add to 15. Since all coefficients are positive, we need to consider only positive factors. The numbers we need are 7 and 8.

(4) Rewrite the middle term:

$$15x = 7x + 8x$$

(5) Factor by grouping:

$$\begin{aligned}56x^2 + 15x + 1 &= 56x^2 + 7x + 8x + 1\\ &= 7x(8x+1) + 1(8x+1)\\ &= (8x+1)(7x+1)\end{aligned}$$

The factorization of $56x^2+15x+1$ is $(8x+1)(7x+1)$. We must include the common factor in order to get a factorization of the original trinomial:

$$168x^3 + 45x^2 + 3x = 3x(8x+1)(7x+1)$$

63. $-14t^4 + 19t^3 + 3t^2$

(1) Factor out $-t^2$. This not only removes the largest common factor, t^2. It also produces a trinomial with a positive leading coefficient.

$$-14t^4 + 19t^3 + 3t^2 = -t^2(14t^2 - 19t - 3)$$

(2) To factor $14t^2 - 19t - 3$ we first multiply the leading coefficient, 14, and the constant, -3:

$14(-3) = -42$

(3) We look for factors of -42 that add to -19. The numbers we need are -21 and 2.

(4) Rewrite the middle term:

$-19t = -21t + 2t$

(5) Factor by grouping:

$$\begin{aligned} 14t^2 - 19t - 3 &= 14t^2 - 21t + 2t - 3 \\ &= 7t(2t-3) + 1(2t-3) \\ &= (2t-3)(7t+1) \end{aligned}$$

The factorization of $14t^2 - 19t - 3$ is $(2t-3)(7t+1)$. We must include the common factor in order to get a factorization of the original trinomial:

$-14t^4 + 19t^3 + 3t^2 = -t^2(2t-3)(7t+1)$

65. $3x + 45x^2 - 18 = 45x^2 + 3x - 18$

(1) Factor out the largest common factor, 3:

$45x^2 + 3x - 18 = 3(15x^2 + x - 6)$

(2) To factor $15x^2 + x - 6$ we first multiply the leading coefficient, 15, and the constant, -6:

$15(-6) = -90$

(3) We look for factors of -90 that add to 1. The numbers we need are 10 and -9.

(4) Rewrite the middle term:

$x = 10x - 9x$

(5) Factor by grouping:

$$\begin{aligned} 15x^2 + x - 6 &= 15x^2 + 10x - 9x - 6 \\ &= 5x(3x+2) - 3(3x+2) \\ &= (3x+2)(5x-3) \end{aligned}$$

The factorization of $15x^2+x-6$ is $(3x+2)(5x-3)$. We must include the common factor in order to get a factorization of the original trinomial:

$3x + 45x^2 - 18 = 3(3x+2)(5x-3)$

67. $2a^2 + 5ab + 2b^2$

(1) There is no common factor (other than 1 or -1).

(2) Multiply the leading coefficient, 2, and the constant, 2:

$2 \cdot 2 = 4$

(3) We look for factors of 4 that add to 5. The numbers we need are 1 and 4.

(4) Rewrite the middle term:

$5ab = ab + 4ab$

(5) Factor by grouping:

$$\begin{aligned} 2a^2 + 5ab + 2b^2 &= 2a^2 + ab + 4ab + 2b^2 \\ &= a(2a+b) + 2b(2a+b) \\ &= (2a+b)(a+2b) \end{aligned}$$

69. $8s^2 + 18st + 9t^2$

(1) There is no common factor (other than 1 or -1).

(2) Multiply the leading coefficient, 8, and the constant, 9:

$8 \cdot 9 = 72$

(3) We look for factors of 72 that add to 18. The numbers we need are 6 and 12.

(4) Rewrite the middle term:

$18st = 6st + 12st$

(5) Factor by grouping:

$$\begin{aligned} 8s^2 + 18st + 9t^2 &= 8s^2 + 6st + 12st + 9t^2 \\ &= 2s(4s+3t) + 3t(4s+3t) \\ &= (4s+3t)(2s+3t) \end{aligned}$$

71. $18x^2 - 6xy - 24y^2$

(1) Factor out the largest common factor, 6:

$18x^2 - 6xy - 24y^2 = 6(3x^2 - xy - 4y^2)$

(2) To factor $3x^2 - xy - 4y^2$, we first multiply the leading coefficient, 3, and the constant, -4:

$3(-4) = -12$

(3) We look for factors of -12 that add to -1. The numbers we need are -4 and 3.

(4) Rewrite the middle term:

$-xy = -4xy + 3xy$

(5) Factor by grouping:

$$\begin{aligned} 3x^2 - xy - 4y^2 &= 3x^2 - 4xy + 3xy - 4y^2 \\ &= x(3x-4y) + y(3x-4y) \\ &= (3x-4y)(x+y) \end{aligned}$$

The factorization of $3x^2 - xy - 4y^2$ is $(3x-4y)(x+y)$. We must include the common factor in order to get a factorization of the original trinomial:

$18x^2 - 6xy - 24y^2 = 6(3x-4y)(x+y)$

73. $-24a^2 + 34ab - 12b^2$

(1) Factor out -2. This not only removes the largest common factor, 2. It also produces a trinomial with a positive leading coefficient.

$-24a^2 + 34ab - 12b^2 = -2(12a^2 - 17ab + 6b^2)$

(2) To factor $12a^2-17ab+6b^2$, we first multiply the leading coefficient, 12, and the constant, 6:

$12 \cdot 6 = 72$

(3) We look for factors of 72 that add to -17. The numbers we need are -8 and -9.

(4) Rewrite the middle term:

$-17ab = -8ab - 9ab$

(5) Factor by grouping:

$$\begin{aligned} 12a^2 - 17ab + 6b^2 &= 12a^2 - 8ab - 9ab + 6b^2 \\ &= 4a(3a-2b) - 3b(3a-2b) \\ &= (3a-2b)(4a-3b) \end{aligned}$$

The factorization of $12a^2 - 17ab + 6b^2$ is $(3a-2b)(4a-3b)$. We must include the common factor in order to get a factorization of the original trinomial:

$-24a^2 + 34ab - 12b^2 = -2(3a-2b)(4a-3b)$

75. $35x^2 + 34x^3 + 8x^4 = 8x^4 + 34x^3 + 35x^2$

(1) Factor out the largest common factor, x^2:

$x^2(8x^2 + 34x + 35)$

(2) To factor $8x^2 + 34x + 35$ by grouping we first multiply the leading coefficient, 8, and the constant, 35:

$8 \cdot 35 = 280$

(3) We look for factors of 280 that add to 34. The numbers we need are 14 and 20.

(4) Rewrite the middle term:

$34x = 14x + 20x$

(5) Factor by grouping:

$$\begin{aligned} 8x^2 + 34x + 35 &= 8x^2 + 14x + 20x + 35 \\ &= 2x(4x+7) + 5(4x+7) \\ &= (4x+7)(2x+5) \end{aligned}$$

The factorization of $8x^2 + 34x + 35$ is $(4x+7)(2x+5)$. We must include the common factor in order to get a factorization of the original trinomial:

$35x^2 + 34x^3 + 8x^4 = x^2(4x+7)(2x+5)$

77. $18a^7 + 8a^6 + 9a^8 = 9a^8 + 18a^7 + 8a^6$

(1) Factor out the largest common factor, a^6:

$9a^8 + 18a^7 + 8a^6 = a^6(9a^2 + 18a + 8)$

(2) To factor $9a^2 + 18a + 8$ we first multiply the leading coefficient, 9, and the constant, 8:

$9 \cdot 8 = 72$

(3) Look for factors of 72 that add to 18. The numbers we need are 6 and 12.

(4) Rewrite the middle term:

$18a = 6a + 12a$

(5) Factor by grouping:

$$\begin{aligned} 9a^2 + 18a + 8 &= 9a^2 + 6a + 12a + 8 \\ &= 3a(3a+2) + 4(3a+2) \\ &= (3a+2)(3a+4) \end{aligned}$$

The factorization of $9a^2 + 18a + 8$ is $(3a+2)(3a+4)$. We must include the common factor in order to get a factorization of the original trinomial:

$18a^7 + 8a^6 + 9a^8 = a^6(3a+2)(3a+4)$

79. *Writing Exercise*

81. ***Familiarize***. We will use the formula $C = 2\pi r$, where C is circumference and r is radius, to find the radius in kilometers. Then we will multiply that number by 0.62 to find the radius in miles.

Translate.

$$\underbrace{\text{Circumference}} = \underbrace{2 \cdot \pi \cdot \text{radius}}$$
$$\downarrow \qquad \downarrow \qquad \downarrow$$
$$40,000 \approx 2(3.14)r$$

Carry out. First we solve the equation.

$$\begin{aligned} 40,000 &\approx 2(3.14)r \\ 40,000 &\approx 6.28r \\ 6369 &\approx r \end{aligned}$$

Then we multiply to find the radius in miles:

$6369(0.62) \approx 3949$

Check. If $r = 6369$, then $2\pi r = 2(3.14)(6369) \approx 40,000$. We should also recheck the multiplication we did to find the radius in miles. Both values check.

State. The radius of the earth is about 6369 km or 3949 mi. (These values may differ slightly if a different approximation is used for π.)

83. $$\begin{aligned} (3x+1)^2 &= (3x)^2 + 2 \cdot 3x \cdot 1 + 1^2 \\ &\qquad [(A+B)^2 = A^2 + 2AB + B^2] \\ &= 9x^2 + 6x + 1 \end{aligned}$$

85. $$\begin{aligned} (4t-5)^2 &= (4t)^2 - 2 \cdot 4t \cdot 5 + 5^2 \\ &\qquad [(A-B)^2 = A^2 - 2AB + B^2] \\ &= 16t^2 - 40t + 25 \end{aligned}$$

87. $$\begin{aligned} (5x-2)(5x+2) &= (5x)^2 - 2^2 \\ &\qquad [(A+B)(A-B) = A^2 - B^2] \\ &= 25x^2 - 4 \end{aligned}$$

89. $$\begin{aligned} (2t+7)(2t-7) &= (2t)^2 - 7^2 \\ &\qquad [(A+B)(A-B) = A^2 - B^2] \\ &= 4t^2 - 49 \end{aligned}$$

91. *Writing Exercise*

93. $18x^2y^2 - 3xy - 10$

We will factor by grouping.

(1) There is no common factor (other than 1 or -1).

(2) Multiply the leading coefficient, 18, and the constant, -10:

$18(-10) = -180$

(3) We look for factors of -180 that add to -3. The numbers we want are -15 and 12.

(4) Rewrite the middle term:

$-3xy = -15xy + 12xy$

(5) Factor by grouping:

$$\begin{aligned} 18x^2y^2 - 3xy - 10 &= 18x^2y^2 - 15xy + 12xy - 10 \\ &= 3xy(6xy-5) + 2(6xy-5) \\ &= (6xy-5)(3xy+2) \end{aligned}$$

95. $9a^2b^3 + 25ab^2 + 16$

We cannot factor the leading term, $9a^2b^3$, in a way that will produce a middle term with variable factors ab^2, so this trinomial is prime.

97. $16t^{10} - 8t^5 + 1 = 16(t^5)^2 - 8t^5 + 1$

(1) There is no common factor (other than 1 or -1).

(2) Because $16t^{10}$ can be factored as $16t^5 \cdot t^5$ or $8t^5 \cdot 2t^5$ or $4t^5 \cdot 4t^5$, we have these possibilities:

$(16t^5 + \quad)(t^5 + \quad)$ and $(8t^5 + \quad)(2t^5 + \quad)$

and $(4t^5 + \quad)(4t^5 + \quad)$

(3) Since the last term is positive and the middle term is negative we need consider only negative factors of 1. The only negative pair of factors is $-1, -1$.

(4) We try some possibilities:

$(16t^5 - 1)(t^5 - 1) = 16t^{10} - 17t^5 + 1$

$(8t^5 - 1)(2t^5 - 1) = 16t^{10} - 10t^5 + 1$

$(4t^5 - 1)(4t^5 - 1) = 16t^{10} - 8t^5 + 1$

The factorization is $(4t^5 - 1)(4t^5 - 1)$, or $(4t^5 - 1)^2$.

99. $-20x^{2n} - 16x^n - 3 = -20(x^n)^2 - 16x^n - 3$

(1) Factor out -1 in order to have a trinomial with a positive leading coefficient.

$-20x^{2n} - 16x^n - 3 = -1(20x^{2n} + 16x^n + 3)$

(2) Because $20x^{2n}$ can be factored as $20x^n \cdot x^n$, $10x^n \cdot 2x^n$, or $5x^n \cdot 4x^n$, we have these possibilities:

$(20x^n + \quad)(x^n + \quad)$ and $(10x^n + \quad)(2x^n + \quad)$ and $(5x^n + \quad)(4x^n + \quad)$

(3) Since all the signs are positive, we need consider only the positive factor pair 3,1 when factoring 3. This pair can also be listed as 1,3.

(4) We try some possibilities:

$(20x^n + 3)(x^n + 1) = 20x^{2n} + 23x^n + 3$

$(10x^n + 3)(2x^n + 1) = 20x^{2n} + 16x^n + 3$

The factorization of $20x^{2n} + 16x^n + 3$ is $(10x^n + 3)(2x^n + 1)$. We must include the common factor to get a factorization of the original trinomial.

$-20x^{2n} - 16x^n - 3 = -1(10x^n + 3)(2x^n + 1)$, or $-(10x^n + 3)(2x^n + 1)$

101. $a^{2n+1} - 2a^{n+1} + a$

(1) Factor out the largest common factor, a:

$a^{2n+1} - 2a^{n+1} + a = a(a^{2n} - 2a^n + 1)$

(2) Multiply the leading coefficient, 1, and the constant, 1:

$1 \cdot 1 = 1$

(3) Look for factors of 1 that add to -2. The numbers we need are -1 and -1.

(4) Rewrite the middle term.

$-2a^n = -a^n - a^n$

(5) Factor by grouping:

$$\begin{aligned} a^{2n} - 2a^n + 1 &= a^{2n} - a^n - a^n + 1 \\ &= a^n(a^n - 1) - 1(a^n - 1) \\ &= (a^n - 1)(a^n - 1), \text{ or } (a^n - 1)^2 \end{aligned}$$

The factorization of $a^{2n} - 2a^n + 1$ is $(a^n - 1)^2$. We must include the common factor in order to get a factorization of the original trinomial:

$a^{2n+1} - 2a^{n+1} + a = a(a^n - 1)^2$

103.

$$\begin{aligned} & 3(a+1)^{n+1}(a+3)^2 - 5(a+1)^n(a+3)^3 \\ &= (a+1)^n(a+3)^2[3(a+1) - 5(a+3)] \quad \text{Removing the common factors} \\ &= (a+1)^n(a+3)^2[3a + 3 - 5a - 15] \quad \text{Simplifying inside the brackets} \\ &= (a+1)^n(a+3)^2(-2a - 12) \\ &= (a+1)^n(a+3)^2(-2)(a+6) \quad \text{Removing the common factor} \\ &= -2(a+1)^n(a+3)^2(a+6) \quad \text{Rearranging} \end{aligned}$$

Exercise Set 5.4

1. $x^2 - 64 = x^2 - 8^2$, so $x^2 - 64$ is a difference of squares.

3. Two terms of $x^2 - 5x + 4$ are squares (x^2 and 4) and neither is being subtracted. However, the remaining term is neither $2 \cdot x \cdot 2$ nor $-2 \cdot x \cdot 2$, so the polynomial is not a perfect-square trinomial. It is not a binomial so it cannot be a difference of squares. It is not prime since it can be factored: $x^2 - 5x + 4 = (x - 1)(x - 4)$. Thus it is none of the given possibilities.

5. $a^2 - 8a + 16 = a^2 - 2 \cdot a \cdot 4 + 4^2$, so this is a perfect-square trinomial.

7. $-25x^2 - 9$ is not a trinomial. It is not a difference of squares because the terms do not have different signs. There is no common factor (other than 1 or -1), so $-25x^2 - 9$ is a prime polynomial.

9. $4t^2 + 20t + 25 = (2t)^2 + 2 \cdot 2t \cdot 5 + 5^2$, so this is a perfect square trinomial.

11. $x^2 + 18x + 81$

(1) Two terms, x^2 and 81, are squares.

(2) Neither x^2 nor 81 is being subtracted.

(3) Twice the product of the square roots, $2 \cdot x \cdot 9$, is $18x$, the remaining term.

Thus, $x^2 + 18x + 81$ is a perfect-square trinomial.

13. $x^2 - 16x - 64$

(1) Two terms, x^2 and 64, are squares.

(2) There is a minus sign before 64, so $x^2 + 16x - 64$ is not a perfect-square trinomial.

15. $x^2 - 3x + 9$

(1) Two terms, x^2 and 9, are squares.

(2) There is no minus sign before x^2 or 9.

(3) Twice the product of the square roots, $2 \cdot x \cdot 3$, is $6x$. This is neither the remaining term nor its opposite, so $x^2 - 3x + 9$ is not a perfect-square trinomial.

17. $9x^2 - 36x + 24$

(1) Only one term, $9x^2$, is a square. Thus, $9x^2 - 36x + 24$ is not a perfect-square trinomial.

19.

$$\begin{aligned} & x^2 + 16x + 64 \\ &= x^2 + 2 \cdot x \cdot 8 + 8^2 = (x+8)^2 \\ & \;\uparrow \quad\;\; \uparrow \;\; \uparrow \;\; \uparrow \quad \uparrow \\ &= A^2 + 2 \;\; A \;\; B + B^2 = (A+B)^2 \end{aligned}$$

21.

$$\begin{aligned} & x^2 - 14x + 49 \\ &= x^2 - 2 \cdot x \cdot 7 + 7^2 = (x-7)^2 \\ & \;\uparrow \quad\;\; \uparrow \;\; \uparrow \;\; \uparrow \quad \uparrow \\ &= A^2 - 2 \;\; A \;\; B + B^2 = (A-B)^2 \end{aligned}$$

23.

$$\begin{aligned} 3x^2 + 6x + 3 &= 3(x^2 + 2x + 1) \\ &= 3(x^2 + 2 \cdot x \cdot 1 + 1^2) \\ &= 3(x+1)^2 \end{aligned}$$

25. $4 - 4x + x^2 = 2^2 - 2 \cdot 2 \cdot x + x^2$
$= (2 - x)^2$

We could also factor as follows:

$4 - 4x + x^2 = x^2 - 4x + 4$
$= x^2 - 2 \cdot x \cdot 2 + 2^2$
$= (x - 2)^2$

27. $18x^2 + 12x + 2 = 2(9x^2 + 6x + 1)$
$= 2[(3x)^2 + 2 \cdot 3x \cdot 1 + 1^2]$
$= 2(3x + 1)^2$

29. $49 - 56y + 16y^2 = 16y^2 - 56y + 49$
$= (4y)^2 - 2 \cdot 4y \cdot 7 + 7^2$
$= (4y - 7)^2$

We could also factor as follows:

$49 - 56y + 16y^2 = 7^2 - 2 \cdot 7 \cdot 4y + (4y)^2$
$= (7 - 4y)^2$

31. $-x^5 + 18x^4 - 81x^3 = -x^3(x^2 - 18x + 81)$
$= -x^3(x^2 - 2 \cdot x \cdot 9 + 9^2)$
$= -x^3(x - 9)^2$

33. $2x^3 - 4x^2 + 2x = 2x(x^2 - 2x + 1)$
$= 2x(x^2 - 2 \cdot x \cdot 1 + 1^2)$
$= 2x(x - 1)^2$

35. $20x^2 + 100x + 125 = 5(4x^2 + 20x + 25)$
$= 5[(2x)^2 + 2 \cdot 2x \cdot 5 + 5^2]$
$= 5(2x + 5)^2$

37. $49 - 42x + 9x^2 = 7^2 - 2 \cdot 7 \cdot 3x + (3x)^2 = (7 - 3x)^2$, or $(3x - 7)^2$

39. $16x^2 + 24x + 9 = (4x)^2 + 2 \cdot 4x \cdot 3 + 3^2 = (4x + 3)^2$

41. $2 + 20x + 50x^2 = 2(1 + 10x + 25x^2)$
$= 2[1^2 + 2 \cdot 1 \cdot 5x + (5x)^2]$
$= 2(1 + 5x)^2$, or $2(5x + 1)^2$

43. $4p^2 + 12pq + 9q^2 = (2p)^2 + 2 \cdot 2p \cdot 3q + (3q)^2$
$= (2p + 3q)^2$

45. $a^2 - 12ab + 49b^2$

This is not a perfect square trinomial because $-2 \cdot a \cdot 7b = -14ab \neq -12ab$. Nor can it be factored using the methods of Sections 5.2 and 5.3. Thus, it is prime.

47. $-64m^2 - 16mn - n^2 = -1(64m^2 + 16mn + n^2)$
$= -1[(8m)^2 + 2 \cdot 8m \cdot n + n^2]$
$= -1(8m + n)^2$, or $-(8m + n)^2$

49. $-32s^2 + 80st - 50t^2 = -2(16s^2 - 40st + 25t^2)$
$= -2[(4s)^2 - 2 \cdot 4s \cdot 5t + (5t)^2]$
$= -2(4s - 5t)^2$

51. $x^2 + 100$

(1) The first expression is a square: x^2

The second expression is a square: $100 = 10^2$

(2) The terms do not have different signs.

Thus, $x^2 + 100$ is not a difference of squares.

53. $x^2 - 81$

(1) The first expression is a square: x^2

The second expression is a square: $81 = 9^2$

(2) The terms have different signs.

Thus, $x^2 - 81$ is a difference of squares, $x^2 - 9^2$.

55. $-26 + 4t^2$

(1) The expression 26 is not a square. Thus, $-26 + 4t^2$ is not a difference of squares.

57. $y^2 - 4 = y^2 - 2^2 = (y + 2)(y - 2)$

59. $p^2 - 9 = p^2 - 3^2 = (p + 3)(p - 3)$

61. $-49 + t^2 = t^2 - 49 = t^2 - 7^2 = (t + 7)(t - 7)$, or $(7 + t)(-7 + t)$

63. $6a^2 - 54 = 6(a^2 - 9) = 6(a^2 - 3^2) = 6(a + 3)(a - 3)$

65. $49x^2 - 14x + 1 = (7x)^2 - 2 \cdot 7x \cdot 1 + 1^2 = (7x - 1)^2$

67. $200 - 2t^2 = 2(100 - t^2) = 2(10^2 - t^2) = 2(10 + t)(10 - t)$

69. $-80a^2 + 45 = -5(16a^2 - 9) = -5[(4a^2) - 3^2] = -5(4a + 3)(4a - 3)$

71. $5t^2 - 80 = 5(t^2 - 16) = 5(t^2 - 4^2) = 5(t + 4)(t - 4)$

73. $8x^2 - 98 = 2(4x^2 - 49) = 2[(2x)^2 - 7^2] = 2(2x + 7)(2x - 7)$

75. $36x - 49x^3 = x(36 - 49x^2) = x[6^2 - (7x)^2] = x(6 + 7x)(6 - 7x)$

77. $49a^4 - 20$

There is no common factor (other than 1 or -1). Since 20 is not a square, this is not a difference of squares. Thus, the polynomial is prime.

79. $t^4 - 1$
$= (t^2)^2 - 1^2$
$= (t^2 + 1)(t^2 - 1)$
$= (t^2 + 1)(t + 1)(t - 1)$ Factoring further; $t^2 - 1$ is a difference of squares

81. $-3x^3 + 24x^2 - 48x = -3x(x^2 - 8x + 16)$
$= -3x(x^2 - 2 \cdot x \cdot 4 + 4^2)$
$= -3x(x - 4)^2$

83. $48t^2 - 27 = 3(16t^2 - 9)$
$= 3[(4t)^2 - 3^2]$
$= 3(4t + 3)(4t - 3)$

85. $a^8 - 2a^7 + a^6 = a^6(a^2 - 2a + 1)$

$= a^6(a^2 - 2 \cdot a \cdot 1 + 1^2)$

$= a^6(a-1)^2$

87. $7a^2 - 7b^2 = 7(a^2 - b^2)$

$= 7(a+b)(a-b)$

89. $25x^2 - 4y^2 = (5x)^2 - (2y)^2$

$= (5x + 2y)(5x - 2y)$

91. $18t^2 - 8s^2 = 2(9t^2 - 4s^2)$

$= 2[(3t)^2 - (2s)^2]$

$= 2(3t + 2s)(3t - 2s)$

93. *Writing Exercise*

95. ***Familiarize.*** Let $a =$ the amount of oxygen, in liters, that can be dissolved in 100 L of water at 20° C.

Translate. We reword the problem.

$\underbrace{5\text{ L}}$ is 1.6 times $\underbrace{\text{amount } a}$.

$\downarrow \quad \downarrow \quad \downarrow \quad \downarrow \qquad \downarrow$

$5 \quad = 1.6 \quad \cdot \quad a$

Carry out. We solve the equation.

$5 = 1.6a$

$3.125 = a$ Dividing both sides by 1.6

Check. Since 1.6 times 3.125 is 5, the answer checks.

State. 3.125 L of oxygen can be dissolved in 100 L of water at 20° C.

97. $(x^3y^5)(x^9y^7) = x^{3+9}y^{5+7} = x^{12}y^{12}$

99. Graph: $y = \frac{3}{2}x - 3$

Because the equation is in the form $y = mx + b$, we know the y-intercept is $(0, -3)$. We find two other points on the line, substituting multiples of 2 for x to avoid fractions.

When $x = -2$, $y = \frac{3}{2}(-2) - 3 = -3 - 3 = -6$.

When $x = 4$, $y = \frac{3}{2} \cdot 4 - 3 = 6 - 3 = 3$.

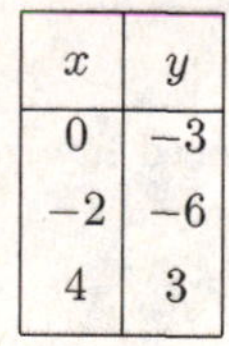

x	y
0	−3
−2	−6
4	3

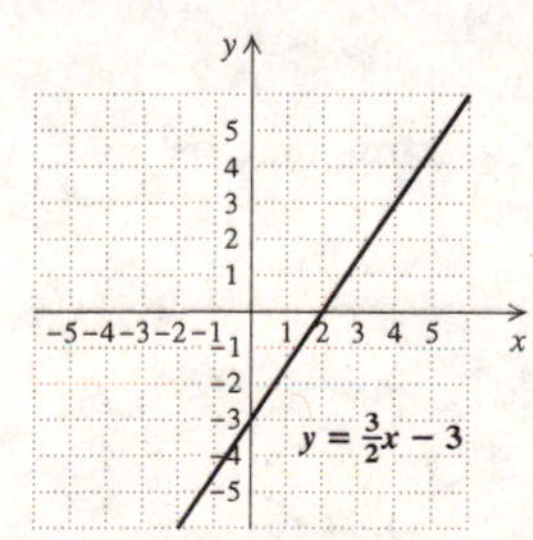

101. *Writing Exercise*

103. $x^8 - 2^8 = (x^4 + 2^4)(x^4 - 2^4)$

$= (x^4 + 2^4)(x^2 + 2^2)(x^2 - 2^2)$

$= (x^4 + 2^4)(x^2 + 2^2)(x + 2)(x - 2)$, or

$(x^4 + 16)(x^2 + 4)(x + 2)(x - 2)$

105. $18x^3 - \frac{8}{25}x = 2x\left(9x^2 - \frac{4}{25}\right) =$

$2x\left(3x + \frac{2}{5}\right)\left(3x - \frac{2}{5}\right)$

107. $(y-5)^4 - z^8$

$= [(y-5)^2 + z^4][(y-5)^2 - z^4]$

$= [(y-5)^2 + z^4][y - 5 + z^2][y - 5 - z^2]$

$= (y^2 - 10y + 25 + z^4)(y - 5 + z^2)(y - 5 - z^2)$

109. $-x^4 + 8x^2 + 9 = -1(x^4 - 8x^2 - 9)$

$= -1(x^2 - 9)(x^2 + 1)$

$= -1(x + 3)(x - 3)(x^2 + 1)$

111. $(y+3)^2 + 2(y+3) + 1$

$= (y+3)^2 + 2 \cdot (y+3) \cdot 1 + 1^2$

$= [(y+3) + 1]^2$

$= (y+4)^2$

113. $27x^3 - 63x^2 - 147x + 343$

$= 9x^2(3x - 7) - 49(3x - 7)$

$= (3x - 7)(9x^2 - 49)$

$= (3x - 7)(3x + 7)(3x - 7)$, or

$(3x - 7)^2(3x + 7)$

115. $81 - b^{4k} = 9^2 - (b^{2k})^2$

$= (9 + b^{2k})(9 - b^{2k})$

$= (9 + b^{2k})[3^2 - (b^k)^2]$

$= (9 + b^{2k})(3 + b^k)(3 - b^k)$

117. $x^2(x+1)^2 - (x^2+1)^2$

$= x^2(x^2 + 2x + 1) - (x^4 + 2x^2 + 1)$

$= x^4 + 2x^3 + x^2 - x^4 - 2x^2 - 1$

$= 2x^3 + x^2 - 2x^2 - 1$

$= (2x^3 - 2x^2) + (x^2 - 1)$

$= 2x^3 - x^2 - 1$

119. $y^2 + 6y + 9 - x^2 - 8x - 16$

$= (y^2 + 6y + 9) - (x^2 + 8x + 16)$

$= (y+3)^2 - (x+4)^2$

$= [(y+3) + (x+4)][(y+3) - (x+4)]$

$= (y + 3 + x + 4)(y + 3 - x - 4)$

$= (y + x + 7)(y - x - 1)$

121. For $c = a^2$, $2 \cdot a \cdot 3 = 24$. Then $a = 4$, so $c = 4^2 = 16$.

123. $(x+1)^2 - x^2$

$= [(x+1) + x][(x+1) - x]$

$= 2x + 1$

$= (x+1) + x$

Exercise Set 5.5

1. $t^3 + 8 = t^3 + 2^3$
$= (t + 2)(t^2 - t \cdot 2 + 2^2)$
$= (t + 2)(t^2 - 2t + 4)$

3. $a^3 - 64 = a^3 - 4^3$
$= (a - 4)(a^2 + a \cdot 4 + 4^2)$
$= (a - 4)(a^2 + 4a + 16)$

5. $z^3 + 125 = z^3 + 5^3$
$= (z + 5)(z^2 - x \cdot 5 + 5^2)$
$= (z + 5)(x^2 - 5z + 25)$

7. $8a^3 - 1 = (2a)^3 - 1^3$
$= (2a - 1)[(2a)^2 + 2a \cdot 1 + 1^2]$
$= (2a - 1)(4a^2 + 2a + 1)$

9. $y^3 - 27 = y^3 - 3^3$
$= (y - 3)(y^3 + y \cdot 3 + 3^2)$
$= (y - 3)(y^2 + 3y + 9)$

11. $64 + 125x^3 = 4^3 + (5x)^3$
$= (4 + 5x)[4^2 - 4 \cdot 5x + (5x)^2]$
$= (4 + 5x)(16 - 20x + 25x^2)$

13. $125p^3 - 1 = (5p)^3 - 1^3$
$= (5p - 1)[(5p)^2 + 5p \cdot 1 + 1^2]$
$= (5p - 1)(25p^2 + 5p + 1)$

15. $27m^3 + 64 = (3m)^3 + 4^3$
$= (3m + 4)[(3m)^2 - 3m \cdot 4 + 4^2]$
$= (3m + 4)(9m^2 - 12m + 16)$

17. $p^3 - q^3 = (p - q)(p^2 + pq + q^2)$

19. $x^3 + \frac{1}{8} = x^3 + \left(\frac{1}{2}\right)^3$
$= \left(x + \frac{1}{2}\right)\left[x^2 - x \cdot \frac{1}{2} + \left(\frac{1}{2}\right)^2\right]$
$= \left(x + \frac{1}{2}\right)\left(x^2 - \frac{1}{2}x + \frac{1}{4}\right)$

21. $2y^3 - 128 = 2(y^3 - 64)$
$= 2(y^3 - 4^3)$
$= 2(y - 4)(y^2 + 4 \cdot y + 4^2)$
$= 2(y - 4)(y^2 + 4y + 16)$

23. $24a^3 + 3 = 3(8a^3 + 1)$
$= 3[(2a)^3 + 1^3]$
$= 3(2a + 1)[(2a)^2 - 2a \cdot 1 + 1^2]$
$= 3(2a + 1)(4a^2 - 2a + 1)$

25. $rs^3 - 64r = r(s^3 - 64)$
$= r(s^3 - 4^3)$
$= r(s - 4)(s^2 + s \cdot 4 + 4^2)$
$= r(s - 4)(s^2 + 4s + 16)$

27. $5x^3 + 40z^3 = 5(x^3 + 8z^3)$
$= 5[x^3 + (2z)^3]$
$= 5(x + 2z)[x^2 - x \cdot 2z + (2z)^2]$
$= 5(x + 2z)(x^2 - 2xz + 4z^2)$

29. $x^3 + 0.001 = x^3 + (0.1)^3$
$= (x + 0.1)[x^2 - x(0.1) + (0.1)^2]$
$= (x + 0.1)(x^2 - 0.1x + 0.01)$

31. $3z^5 - 3z^2 = 3z^2(z^3 - 1)$
$= 3z^2(z^3 - 1^3)$
$= 3z^2(z - 1)(z^2 + z \cdot 1 + 1^2)$
$= 3z^2(z - 1)(z^2 + z + 1)$

33. $t^6 + 1 = (t^2)^3 + 1^3$
$= (t^2 + 1)[(t^2)^2 - t^2 \cdot 1 + 1^2]$
$= (t^2 + 1)(t^4 - t^2 + 1)$

35. $p^6 - q^6$
$= (p^3)^2 - (q^3)^2$
$= (p^3 + q^3)(p^3 - q^3)$
$= (p + q)(p^2 - pq + q^2)(p - q)(p^2 + pq + q^2)$

37. *Writing Exercise*

39. $(3x + 5)(3x - 5) = (3x)^2 - 5^2 = 9x^2 - 25$

41. $(x - 7)(x + 4) = x^2 + 4x - 7x - 28 = x^2 - 3x - 28$

43. ***Familiarize.*** Let n = the number of people who had high-speed Internet access at work in 2000.

Translate.

Number with high speed access in 2005 | is 2.3 times | number with high-speed access in 2000.

$55.2 = 2.3 \times n$

Carry out. We solve the equation.

$55.2 = 2.3n$

$24 = n$ Dividing both sides by 2.3

Check. Since $2.3(24) = 55.2$, the answer checks.

State. In 2000, 24 million people had high-speed Internet access at work.

45. *Writing Exercise*

47. $125c^6 + 8d^6$
$= (5c^2)^3 + (2d^2)^3$
$= (5c^2 + 2d^2)[(5c^2)^2 - 5c^2 \cdot 2d^2 + (2d^2)^2]$
$= (5c^2 + 2d^2)(25c^4 - 10c^2d^2 + 4d^4)$

49. $3x^{3a} - 24y^{3b}$
$= 3(x^{3a} - 8y^{3b})$
$= 3[(x^a)^3 - (2y^b)^3]$
$= 3(x^a - 2y^b)[(x^a)^2 + x^a \cdot 2y^b + (2y^b)^2]$
$= 3(x^a - 2y^b)(x^{2a} + 2x^ay^b + 4y^{2b})$

51. $\frac{1}{24}x^3y^3 + \frac{1}{3}z^3$

$= \frac{1}{3}\left(\frac{1}{8}x^3y^3 + z^3\right)$

$= \frac{1}{3}\left[\left(\frac{1}{2}xy\right)^3 + z^3\right]$

$= \frac{1}{3}\left(\frac{1}{2}xy + z\right)\left[\left(\frac{1}{2}xy\right)^2 - \frac{1}{2}xy \cdot z + z^2\right]$

$= \frac{1}{3}\left(\frac{1}{2}xy + 1\right)\left(\frac{1}{4}x^2y^2 - \frac{1}{2}xyz + z^2\right)$

53. $(x+5)^3 + (x-5)^3$

$= (x+5+x-5)[(x+5)^2-(x+5)(x-5)+(x-5)^2]$

$= 2x(x^2+10x+25-(x^2-25)+x^2-10x+25)$

$= 2x(x^2+10x+25-x^2+25+x^2-10x+25)$

$= 2x(x^2+75)$, or $2x^3+150x$

55. $t^4 - 8t^3 - t + 8$

$= t^4 - t - 8t^3 + 8$

$= t(t^3 - 1) - 8(t^3 - 1)$

$= (t^3 - 1)(t - 8)$

$= (t - 1)(t^2 + t + 1)(t - 8)$

Exercise Set 5.6

1. common factor

3. grouping

5. $10a^2 - 640$

$= 10(a^2 - 64)$ 10 is a common factor.

$= 10(a+8)(a-8)$ Factoring the difference of squares

7. $y^2 + 49 - 14y$

$= y^2 - 14y + 49$ Perfect-square trinomial

$= (y-7)^2$

9. $2t^2 + 11t + 12$

There is no common factor (other than 1 of -1). This trinomial has three terms, but it is not a perfect-square trinomial. Multiply the leading coefficient and the constant, 2 and 12: $2 \cdot 12 = 24$. Try to factor 24 so that the sum of the factors is 11. The numbers we want are 3 and 8: $3 \cdot 8 = 24$ and $3 + 8 = 11$. Split the middle term and factor by grouping.

$2t^2 + 11t + 12 = 2t^2 + 3t + 8t + 12$

$= t(2t+3) + 4(2t+3)$

$= (2t+3)(t+4)$

11. $x^3 - 18x^2 + 81x$

$= x(x^2 - 18x + 81)$ x is a common factor.

$= x(x^2 - 2 \cdot x \cdot 9 + 9^2)$ Perfect-square trinomial

$= x(x-9)^2$

13. $x^3 - 5x^2 - 25x + 125$

$= x^2(x-5) - 25(x-5)$ Factoring by grouping

$= (x-5)(x^2-25)$

$= (x-5)(x+5)(x-5)$ Factoring the difference of squares

$= (x-5)^2(x+5)$

15. $27t^3 - 3t$

$= 3t(9t^2 - 1)$ $3t$ is a common factor.

$= 3t[(3t)^2 - 1^2]$ Difference of squares

$= 3t(3t+1)(3t-1)$

17. $9x^3 + 12x^2 - 45x$

$= 3x(3x^2 + 4x - 15)$ $3x$ is a common factor.

$= 3x(x+3)(3x-5)$ Factoring the trinomial

19. $t^2 + 25$

The polynomial has no common factor and is not a difference of squares. It is prime.

21. $6x^2 + 3x - 45$

$= 3(2x^2 + x - 15)$ 3 is a common factor.

$= 3(2x-5)(x+3)$ Factoring the trinomial

23. $-2a^6 + 8a^5 - 8a^4$

$= -2a^4(a^2 - 4a + 4)$ Factoring out $-2a^4$

$= -2a^4(a-2)^2$ Factoring the perfect-square trinomial

25. $5x^5 - 80x$

$= 5x(x^4 - 16)$ $5x$ is a common factor.

$= 5x[(x^2)^2 - 4^2]$ Difference of squares

$= 5x(x^2+4)(x^2-4)$ Difference of squares

$= 5x(x^2+4)(x+2)(x-2)$

27. $t^4 - 9$ Difference of squares

$= (t^2+3)(t^2-3)$

29. $-x^6 + 2x^5 - 7x^4$

$= -x^4(x^2 - 2x + 7)$

The trinomial is prime, so this is the complete factorization.

31. $x^3 - y^3$ Difference of cubes

$= (x-y)(x^2+xy+y^2)$

33. $ax^2 + ay^2 = a(x^2+y^2)$

35. $36mn - 9m^2n^2 = 9mn(4 - mn)$

37. $2\pi rh + 2\pi r^2 = 2\pi r(h+r)$

39. $(a+b)(5a) + (a+b)(3b)$

$= (a+b)(5a+3b)$ $(a+b)$ is a common factor.

41. $x^2 + x + xy + y$

$= x(x+1) + y(x+1)$ Factoring by grouping

$= (x+1)(x+y)$

43. $a^2 - 3a + ay - 3y$
$= a(a-3) + y(a-3)$ Factoring by grouping
$= (a-3)(a+y)$

45. $3x^2 + 13xy - 10y^2 = (3x - 2y)(x + 5y)$

47. $4b^2 + a^2 - 4ab$
$= 4b^2 - 4ab + a^2$
$= (2b)^2 - 2 \cdot 2b \cdot a + a^2$ Perfect-square trinomial
$= (2b - a)^2$

This result can also be expressed as $(a - 2b)^2$.

49. $16x^2 + 24xy + 9y^2$
$= (4x)^2 + 2 \cdot 4x \cdot 3y + (3y)^2$ Perfect-square trinomial
$= (4x + 3y)^2$

51. $t^2 - 8t + 10$

There is no common factor, this is not a perfect-square trinomial, and we cannot find a pair of factors whose product is 10 and whose sum is -8, so this polynomial is prime.

53. $64t^6 - 1$
$= (8t^3)^2 - 1^2$ Difference of squares
$= (8t^3 + 1)(8t^3 - 1)$ Sum of cubes; difference of cubes
$= (2t+1)(4t^2 - 2t + 1)(2t - 1)(4t^2 + 2t + 1)$

55. $4p^2q - pq^2 + 4p^3$
$= p(4pq - q^2 + 4p^2)$
$= p(4p^2 + 4pq - q^2)$

The trinomial cannot be factored further, so this is the complete factorization.

57. $3b^2 + 17ab - 6a^2 = (3b - a)(b + 6a)$

59. $-12 - x^2y^2 - 8xy$
$= -x^2y^2 - 8xy - 12$
$= -1(x^2y^2 + 8xy + 12)$
$= -1(xy + 2)(xy + 6)$, or $-(xy + 2)(xy + 6)$

61. $p^2q^2 + 7pq + 6 = (pq + 1)(pq + 6)$

63. $54a^4 + 16ab^3$
$= 2a(27a^3 + 8b^3)$ $2a$ is a common factor.
$= 2a[(3a)^3 + (2b)^3]$ Sum of cubes
$= 2a(3a + 2b)[(3a)^2 - 3a \cdot 2b + (2b)^2]$
$= 2a(3a + 2b)(9a^2 - 6ab + 4b^2)$

65. $x^6 + x^5y - 2x^4y^2$
$= x^4(x^2 + xy - 2y^2)$ x^4 is a common factor.
$= x^4(x + 2y)(x - y)$ Factoring the trinomial

67. $36a^2 - 15a + \frac{25}{16}$
$= (6a)^2 - 2 \cdot 6a \cdot \frac{5}{4} + \left(\frac{5}{4}\right)^2$ Perfect-square trinomial
$= \left(6a - \frac{5}{4}\right)^2$

69. $\frac{1}{81}x^2 - \frac{8}{27}x + \frac{16}{9}$
$= \left(\frac{1}{9}x\right)^2 - 2 \cdot \frac{1}{9}x \cdot \frac{4}{3} + \left(\frac{4}{3}\right)^2$ Perfect-square trinomial
$= \left(\frac{1}{9}x - \frac{4}{3}\right)^2$

If we had factored out $\frac{1}{9}$ at the outset, the final result would have been $\frac{1}{9}\left(\frac{1}{3}x - 4\right)^2$.

71. $1 - 16x^{12}y^{12}$
$= (1 + 4x^6y^6)(1 - 4x^6y^6)$ Difference of squares
$= (1 + 4x^6y^6)(1 + 2x^3y^3)(1 - 2x^3y^3)$ Difference of squares

73. $4a^2b^2 + 12ab + 9$
$= (2ab)^2 + 2 \cdot 2ab \cdot 3 + 3^2$ Perfect-square trinomial
$= (2ab + 3)^2$

75. $a^4 + 8a^2 + 8a^3 + 64a$
$= a(a^3 + 8a + 8a^2 + 64)$ a is a common factor.
$= a[a(a^2 + 8) + 8(a^2 + 8)]$ Factoring by grouping
$= a(a^2 + 8)(a + 8)$

77. *Writing Exercise*

79.

$y = -4x + 7$	
11	$-4(-1) + 7$
	$4 + 7$

$11 \stackrel{?}{=} 11$ TRUE

Since $11 = 11$ is true, $(-1, 11)$ is a solution.

$y = -4x + 7$	
7	$-4 \cdot 0 + 7$
	$0 + 7$

$7 \stackrel{?}{=} 7$ TRUE

Since $7 = 7$ is true, $(0, 7)$ is a solution.

$y = -4x + 7$	
-5	$-4 \cdot 3 + 7$
	$-12 + 7$

$-5 \stackrel{?}{=} -5$ TRUE

Since $-5 = -5$ is true, $(3, -5)$ is a solution.

81. $3x + 7 = 0$
$3x = -7$ Subtracting 7 from both sides
$x = -\frac{7}{3}$ Dividing both sides by 3

The solution is $-\frac{7}{3}$.

83. $4x - 9 = 0$

$4x = 9$ Adding 9 to both sides

$x = \frac{9}{4}$ Dividing both sides by 4

The solution is $\frac{9}{4}$.

85. *Writing Exercise*

87.
$$\begin{aligned} & -(x^5 + 7x^3 - 18x) \\ = & -x(x^4 + 7x^2 - 18) \\ = & -x(x^2 + 9)(x^2 - 2) \end{aligned}$$

89.
$$\begin{aligned} & -3a^4 + 15a^2 - 12 \\ = & -3(a^4 - 5a^2 + 4) \\ = & -3(a^2 - 1)(a^2 - 4) \\ = & -3(a + 1)(a - 1)(a + 2)(a - 2) \end{aligned}$$

91.
$$\begin{aligned} & y^2(y + 1) - 4y(y + 1) - 21(y + 1) \\ = & (y + 1)(y^2 - 4y - 21) \\ = & (y + 1)(y - 7)(y + 3) \end{aligned}$$

93.
$$\begin{aligned} & 6(x - 1)^2 + 7y(x - 1) - 3y^2 \\ = & [2(x - 1) + 3y][3(x - 1) - y] \\ = & (2x + 3y - 2)(3x - y - 3) \end{aligned}$$

95.
$$\begin{aligned} & 2(a + 3)^4 - (a + 3)^3(b - 2) - (a + 3)^2(b - 2)^2 \\ = & (a + 3)^2[2(a + 3)^2 - (a + 3)(b - 2) - (b - 2)^2] \\ = & (a + 3)^2[2(a + 3) + (b - 2)][(a + 3) - (b - 2)] \\ = & (a + 3)^2(2a + 6 + b - 2)(a + 3 - b + 2) \\ = & (a + 3)^2(2a + b + 4)(a - b + 5) \end{aligned}$$

Exercise Set 5.7

1. Equations of the type $ax^2 + bx + c = 0$, with $a \neq 0$, are quadratic, so choice (c) is correct.

3. Most quadratic equations have 2 solutions, so choice (a) is correct.

5. $(x + 5)(x + 7) = 0$

We use the principle of zero products.

$x + 5 = 0$ *or* $x + 7 = 0$

$x = -5$ *or* $x = -7$

Check:

For -5:

$(x + 5)(x + 7) = 0$	
$(-5 + 5)(-5 + 7)$	0
$0 \cdot 2$	

$0 \stackrel{?}{=} 0$ TRUE

For -7:

$(x + 5)(x + 7) = 0$	
$(-7 + 5)(-7 + 7)$	0
$-2 \cdot 0$	

$0 \stackrel{?}{=} 0$ TRUE

The solutions are -5 and -7.

7. $(2x - 9)(x + 4) = 0$

$2x - 9 = 0$ *or* $x + 4 = 0$

$2x = 9$ *or* $x = -4$

$x = \frac{9}{2}$ *or* $x = -4$

The solutions are $\frac{9}{2}$ and -4.

9. $(10x - 9)(4x + 7) = 0$

$10x - 9 = 0$ *or* $4x + 7 = 0$

$10x = 9$ *or* $4x = -7$

$x = \frac{9}{10}$ *or* $x = -\frac{7}{4}$

The solutions are $\frac{9}{10}$ and $-\frac{7}{4}$.

11. $x(x + 2) = 0$

$x = 0$ *or* $x + 2 = 0$

$x = 0$ *or* $x = -2$

The solutions are 0 and -2.

13. $\left(\frac{2}{3}x - \frac{12}{11}\right)\left(\frac{7}{4}x - \frac{1}{12}\right) = 0$

$\frac{2}{3}x - \frac{12}{11} = 0$ *or* $\frac{7}{4}x - \frac{1}{12} = 0$

$\frac{2}{3}x = \frac{12}{11}$ *or* $\frac{7}{4}x = \frac{1}{12}$

$x = \frac{3}{2} \cdot \frac{12}{11}$ *or* $x = \frac{4}{7} \cdot \frac{1}{12}$

$x = \frac{18}{11}$ *or* $x = \frac{1}{21}$

The solutions are $\frac{18}{11}$ and $\frac{1}{21}$.

15. $5x(2x + 9) = 0$

$5x = 0$ *or* $2x + 9 = 0$

$x = 0$ *or* $2x = -9$

$x = 0$ *or* $x = -\frac{9}{2}$

The solutions are 0 and $-\frac{9}{2}$.

17. $(20x - 0.4x)(7 - 0.1x) = 0$

$20 - 0.4x = 0$ *or* $7 - 0.1x = 0$

$-0.4x = -20$ *or* $-0.1x = -7$

$x = 50$ *or* $x = 70$

The solutions are 50 and 70.

19. $(3x - 2)(x + 5)(x - 1) = 0$

$3x - 2 = 0$ *or* $x + 5 = 0$ *or* $x - 1 = 0$

$3x = 2$ *or* $x = -5$ *or* $x = 1$

$x = \frac{2}{3}$ *or* $x = -5$ *or* $x = 1$

The solutions are $\frac{2}{3}$, -5, and 1.

21. $x^2 - 7x + 6 = 0$

$(x-6)(x-1) = 0$ Factoring

$x - 6 = 0$ *or* $x - 1 = 0$ Using the principle of zero products

$x = 6$ *or* $x = 1$

The solutions are 6 and 1.

23. $x^2 + 4x - 21 = 0$

$(x-3)(x+7) = 0$ Factoring

$x - 3 = 0$ *or* $x + 7 = 0$ Using the principle of zero products

$x = 3$ *or* $x = -7$

The solutions are 3 and -7.

25. $x^3 - 3x^2 + 2x = 0$

$x(x^2 - 3x + 2) = 0$

$x(x-1)(x-2) = 0$

$x = 0$ *or* $x - 1 = 0$ *or* $x - 2 = 0$

$x = 0$ *or* $x = 1$ *or* $x = 2$

The solutions are 0, 1, and 2.

27. $x^2 - 6x = 0$

$x(x-6) = 0$

$x = 0$ *or* $x - 6 = 0$

$x = 0$ *or* $x = 6$

The solutions are 0 and 6.

29. $6t + t^2 = 0$

$t(6+t) = 0$

$t = 0$ *or* $6 + t = 0$

$t = 0$ *or* $t = -6$

The solutions are 0 and -6.

31. $9x^2 = 4$

$9x^2 - 4 = 0$

$(3x+2)(3x-2) = 0$

$3x + 2 = 0$ *or* $3x - 2 = 0$

$3x = -2$ *or* $3x = 2$

$x = -\frac{2}{3}$ *or* $x = \frac{2}{3}$

The solutions are $-\frac{2}{3}$ and $\frac{2}{3}$.

33. $0 = 25 + x^2 + 10x$

$0 = x^2 + 10x + 25$ Writing in descending order

$0 = (x+5)(x+5)$

$x + 5 = 0$ *or* $x + 5 = 0$

$x = -5$ *or* $x = -5$

The solution is -5.

35. $1 + x^2 = 2x$

$x^2 - 2x + 1 = 0$

$(x-1)(x-1) = 0$

$x - 1 = 0$ *or* $x - 1 = 0$

$x = 1$ *or* $x = 1$

The solution is 1.

37. $4t^2 = 8t$

$4t^2 - 8t = 0$

$4t(t-2) = 0$

$t = 0$ *or* $t - 2 = 0$

$t = 0$ *or* $t = 2$

The solutions are 0 and 2.

39. $3x^2 - 7x = 20$

$3x^2 - 7x - 20 = 0$

$(3x+5)(x-4) = 0$

$3x + 5 = 0$ *or* $x - 4 = 0$

$3x = -5$ *or* $x = 4$

$x = -\frac{5}{3}$ *or* $x = 4$

The solutions are $-\frac{5}{3}$ and 4.

41. $2y^2 + 12y = -10$

$2y^2 + 12y + 10 = 0$

$2(y^2 + 6y + 5) = 0$

$2(y+5)(y+1) = 0$

$y + 5 = 0$ *or* $y + 1 = 0$

$y = -5$ *or* $y = -1$

The solutions are -5 and -1.

43. $(x-7)(x+1) = -16$

$x^2 - 6x - 7 = -16$

$x^2 - 6x + 9 = 0$

$(x-3)(x-3) = 0$

$x - 3 = 0$ *or* $x - 3 = 0$

$x = 3$ *or* $x = 3$

The solution is 3.

45. $14z^2 - 3 = 21z - 3$

$14z^2 - 21z - 3 = -3$

$14z^2 - 21z = 0$

$7z(2z-3) = 0$

$z = 0$ *or* $2z - 3 = 0$

$z = 0$ *or* $2z = 3$

$z = 0$ *or* $z = \frac{3}{2}$

The solutions are 0 and $\frac{3}{2}$.

47.
$$81x^2 - 5 = 20$$
$$81x^2 - 25 = 0$$
$$(9x+5)(9x-5) = 0$$
$$9x+5 = 0 \quad or \quad 9x-5=0$$
$$9x = -5 \quad or \quad 9x = 5$$
$$x = -\frac{5}{9} \quad or \quad x = \frac{5}{9}$$

The solutions are $-\frac{5}{9}$ and $\frac{5}{9}$.

49. $(x-1)(5x+4) = 2$
$$5x^2 - x - 4 = 2$$
$$5x^2 - x - 6 = 0$$
$$(5x-6)(x+1) = 0$$
$$5x-6 = 0 \quad or \quad x+1=0$$
$$5x = 6 \quad or \quad x = -1$$
$$x = \frac{6}{5} \quad or \quad x = -1$$

The solutions are $\frac{6}{5}$ and -1.

51.
$$x^2 - 2x = 18 + 5x$$
$$x^2 - 7x - 18 = 0 \quad \text{Subtracting 18 and } 5x$$
$$(x-9)(x+2) = 0$$
$$x-9=0 \quad or \quad x+2=0$$
$$x = 9 \quad or \quad x = -2$$

The solutions are 9 and -2.

53.
$$x^2(2x-1) = 3x$$
$$2x^3 - x^2 = 3x$$
$$2x^3 - x^2 - 3x = 0$$
$$x(2x^2 - x - 3) = 0$$
$$x(2x-3)(x+1) = 0$$
$$x = 0 \quad or \quad 2x-3=0 \quad or \quad x+1=0$$
$$x = 0 \quad or \quad 2x = 3 \quad or \quad x = -1$$
$$x = 0 \quad or \quad x = \frac{3}{2} \quad or \quad x = -1$$

The solutions are 0, $\frac{3}{2}$, and -1.

55. $(2x-5)(3x^2+29x+56) = 0$
$$(2x-5)(x+7)(3x+8) = 0$$
$$2x-5=0 \quad or \quad x+7=0 \quad or \quad 3x+8=0$$
$$2x = 5 \quad or \quad x = -7 \quad or \quad 3x = -8$$
$$x = \frac{5}{2} \quad or \quad x = -7 \quad or \quad x = -\frac{8}{3}$$

The solutions are $\frac{5}{2}$, -7, and $-\frac{8}{3}$.

57. The solutions of the equation are the first coordinates of the x-intercepts of the graph. From the graph we see that the x-intercepts are $(-3,0)$ and $(2,0)$, so the solutions of the equation are -3 and 2.

59. The solutions of the equation are the first coordinates of the x-intercepts of the graph. From the graph we see that the x-intercepts are $(-1,0)$ and $(3,0)$, so the solutions of the equation are -1 and 3.

61. We let $y = 0$ and solve for x.
$$0 = x^2 + 3x - 4$$
$$0 = (x+4)(x-1)$$
$$x+4=0 \quad or \quad x-1=0$$
$$x = -4 \quad or \quad x = 1$$

The x-intercepts are $(-4,0)$ and (1,0).

63. We let $y = 0$ and solve for x.
$$0 = x^2 - 2x - 15$$
$$0 = (x-5)(x+3)$$
$$x-5=0 \quad or \quad x+3=0$$
$$x = 5 \quad or \quad x = -3$$

The x-intercepts are $(5,0)$ and $(-3,0)$.

65. We let $y = 0$ and solve for x
$$0 = 2x^2 + x - 10$$
$$0 = (2x+5)(x-2)$$
$$2x+5=0 \quad or \quad x-2=0$$
$$2x = -5 \quad or \quad x = 2$$
$$x = -\frac{5}{2} \quad or \quad x = 2$$

The x-intercepts are $\left(-\frac{5}{2}, 0\right)$ and $(2,0)$.

67. *Writing Exercise*

69. $(a+b)^2$

71. Let x represent the smaller integer; $x + (x+1)$

73. Let x represent the number; $\frac{1}{2}x - 7 > 24$

75. *Writing Exercise*

77. a)
$$x = -4 \quad or \quad x = 5$$
$$x+4=0 \quad or \quad x-5=0$$
$$(x+4)(x-5) = 0 \quad \text{Principle of zero products}$$
$$x^2 - x - 20 = 0 \quad \text{Multiplying}$$

b)
$$x = -1 \quad or \quad x = 7$$
$$x+1=0 \quad or \quad x-7=0$$
$$(x+1)(x-7) = 0$$
$$x^2 - 6x + -7 = 0$$

c)
$$x = \frac{1}{4} \quad or \quad x = 3$$
$$x - \frac{1}{4} = 0 \quad or \quad x - 3 = 0$$

$$\left(x-\frac{1}{4}\right)(x-3)=0$$
$$x^2-\frac{13}{4}x+\frac{3}{4}=0$$
$$4\left(x^2-\frac{13}{4}x+\frac{3}{4}\right)=4\cdot 0 \quad \text{Multiplying both sides by 4}$$
$$4x^2-13x+3=0$$

d) $x=\frac{1}{2}$ *or* $x=\frac{1}{3}$
$$x-\frac{1}{2}=0 \quad or \quad x-\frac{1}{3}=0$$
$$\left(x-\frac{1}{2}\right)\left(x-\frac{1}{3}\right)=0$$
$$x^2-\frac{5}{6}x+\frac{1}{6}=0$$
$$6x^2-5x+1=0 \quad \text{Multiplying by 6}$$

e) $x=\frac{2}{3}$ *or* $x=\frac{3}{4}$
$$x-\frac{2}{3}=0 \quad or \quad x-\frac{3}{4}=0$$
$$\left(x-\frac{2}{3}\right)\left(x-\frac{3}{4}\right)=0$$
$$x^2-\frac{17}{12}x+\frac{1}{2}=0$$
$$12x^2-17x+6=0 \quad \text{Multiplying by 12}$$

f) $x=-1$ *or* $x=2$ *or* $x=3$
$$x+1=0 \quad or \quad x-2=0 \quad or \quad x-3=0$$
$$(x+1)(x-2)(x-3)=0$$
$$(x^2-x-2)(x-3)=0$$
$$x^3-4x^2+x+6=0$$

79.
$$a(9+a)=4(2a+5)$$
$$9a+a^2=8a+20$$
$$a^2+a-20=0 \quad \text{Subtracting } 8a \text{ and } 20$$
$$(a+5)(a-4)=0$$
$$a+5=0 \quad or \quad a-4=0$$
$$a=-5 \quad or \quad a=4$$

The solutions are -5 and 4.

81.
$$-x^2+\frac{9}{25}=0$$
$$x^2-\frac{9}{25}=0 \quad \text{Multiplying by } -1$$
$$\left(x-\frac{3}{5}\right)\left(x+\frac{3}{5}\right)=0$$
$$x-\frac{3}{5}=0 \quad or \quad x+\frac{3}{5}=0$$
$$x=\frac{3}{5} \quad or \quad x=-\frac{3}{5}$$

The solutions are $\frac{3}{5}$ and $-\frac{3}{5}$.

83. $(t+1)^2=9$

Observe that $t+1$ is a number which yields 9 when it is squared. Thus, we have
$$t+1=-3 \quad or \quad t+1=3$$
$$t=-4 \quad or \quad t=2$$

The solutions are -4 and 2.

We could also do this exercise as follows:
$$(t+1)^2=9$$
$$t^2+2t+1=9$$
$$t^2+2t-8=0$$
$$(t+4)(t-2)=0$$
$$t+4=0 \quad or \quad t-2=0$$
$$t=-4 \quad or \quad t=2$$

Again we see that the solutions are -4 and 2.

85. a) $2(x^2+10x-2)=2\cdot 0$ Multiplying (a) by 2
$$2x^2+20x-4=0$$
(a) and $2x^2+20x-4=0$ are equivalent.

b) $(x-6)(x+3)=x^2-3x-18$ Multiplying
(b) and $x^2-3x-18=0$ are equivalent.

c) $5x^2-5=5(x^2-1)=5(x+1)(x-1)=$
$(x+1)5(x-1)=(x+1)(5x-5)$
(c) and $(x+1)(5x-5)=0$ are equivalent.

d) $2(2x-5)(x+4)=2\cdot 0$ Multiplying (d) by 2
$$2(x+4)(2x-5)=0$$
$$(2x+8)(2x-5)=0$$
(d) and $(2x+8)(2x-5)=0$ are equivalent.

e) $4(x^2+2x+9)=4\cdot 0$ Multiplying (e) by 4
$$4x^2+8x+36=0$$
(e) and $4x^2+8x+36=0$ are equivalent.

f) $3(3x^2-4x+8)=3\cdot 0$ Multiplying (f) by 3
$$9x^2-12x+24=0$$
(f) and $9x^2-12x+24=0$ are equivalent.

87. *Writing Exercise*

89. $-0.25, 0.88$

91. $4.55, -3.23$

93. $-3.76, 0$

Exercise Set 5.8

1. ***Familiarize.*** Let $x=$ the number (or numbers).

Translate. We reword the problem.

The square of a number minus the number is 20.
$$\underbrace{\text{The square of a number}}_{x^2}\ \underset{-}{\text{minus}}\ \underbrace{\text{the number}}_{x}\ \underset{=}{\text{is}}\ \underset{20}{20.}$$

Carry out. We solve the equation.

$$x^2 - x = 20$$
$$x^2 - x - 20 = 0$$
$$(x-5)(x+4) = 0$$

$$x - 5 = 0 \quad or \quad x + 4 = 0$$
$$x = 5 \quad or \quad x = -4$$

Check. For 5: The square of 5 is 5^2, or 25, and $25-5 = 20$.

For -4: The square of -4 is $(-4)^2$, or 16, and $16-(-4) = 16+4 = 20$. Both numbers check.

State. The numbers are 5 and -4.

3. ***Familiarize.*** Let x = the length of the shorter leg, in cm. Then $x+3$ = the length of the longer leg.

Translate. we use the Pythagorean theorem.

$$a^2 + b^2 = c^2$$
$$x^2 + (x+3)^2 = 15^2$$

Carry out. We solve the equation.

$$x^2 + (x+3)^2 = 15^2$$
$$x^2 + x^2 + 6x + 9 = 225$$
$$2x^2 + 6x + 9 = 225$$
$$2x^2 + 6x - 216 = 0$$
$$2(x^2 + 3x - 108) = 0$$
$$2(x+12)(x-9) = 0$$

$$x + 12 = 0 \quad or \quad x - 9 = 0$$
$$x = -12 \quad or \quad x = 9$$

Check. The number -12 cannot be the length of a side because it is negative. When $x = 9$, then $x + 3 = 12$, and $9^2 + 12^2 = 81 + 144 = 225 = 15^2$, so the number 9 checks.

State.The lengths of the sides are 9 cm, 12 cm, and 15 cm

5. ***Familiarize.*** The locker numbers are consecutive integers. Let x = the smaller integer. Then $x + 1$ = the larger integer.

Translate. We reword the problem.

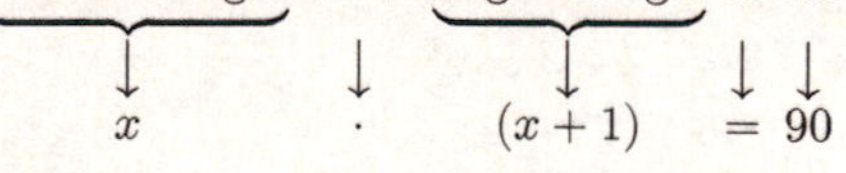

Carry out. We solve the equation.

$$x(x+1) = 90$$
$$x^2 + x = 90$$
$$x^2 + x - 90 = 0$$
$$(x+10)(x-9) = 0$$
$$x + 10 = 0 \quad or \quad x - 9 = 0$$
$$x = -10 \quad or \quad x = 9$$

Check. The solutions of the equation are -10 and 9. Since a locker number cannot be negative, -10 cannot be a solution of the original problem. We only need to check 9. When $x = 9$, then $x + 1 = 10$, and $9 \cdot 10 = 90$. This checks.

State. The locker numbers are 9 and 10.

7. ***Familiarize.*** Let x = the smaller odd integer. Then $x + 2$ = the larger odd integer.

Translate. We reword the problem.

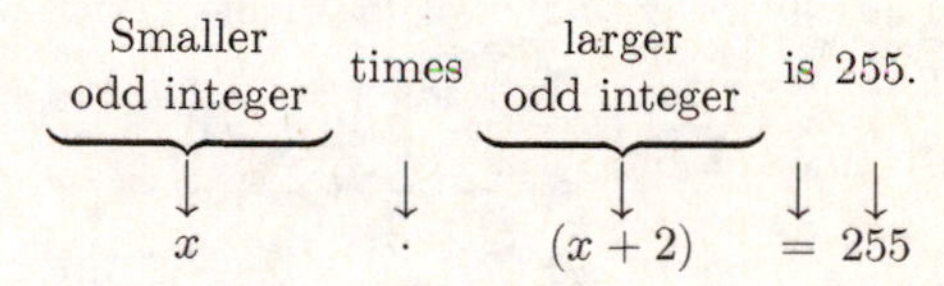

Carry out.

$$x(x+2) = 255$$
$$x^2 + 2x = 255$$
$$x^2 + 2x - 255 = 0$$
$$(x+17)(x-15) = 0$$
$$x + 17 = 0 \quad or \quad x - 15 = 0$$
$$x = -17 \quad or \quad x = 15$$

Check. The solutions of the equation are -17 and 15. When x is -17, then $x + 2$ is -15 and $-17(-15) = 225$. The numbers -17 and -15 are consecutive odd integers which are solutions of the problem. When x is 15, then $x + 2 = 17$ and $15 \cdot 17 = 255$. The numbers 15 and 17 are also consecutive odd integers which are solutions of the problem.

State. We have two solutions, each of which consists of a pair of numbers: -17 and -15 or 15 and 17.

9. ***Familiarize.*** Let w = the width of the table, in feet. Then $6w$ = the width. Recall that the area of a rectangle is Length · Width.

Translate.

$$\underbrace{\text{The area of the rectangle}}_{6w \cdot w} \ \underbrace{\text{is}}_{=} \ \underbrace{24 \text{ ft}^2.}_{24}$$

Carry out. We solve the equation.

$$6w \cdot w = 24$$
$$6w^2 = 24$$
$$6w^2 - 24 = 0$$
$$6(w^2 - 4) = 0$$
$$2(w+2)(w-2) = 0$$
$$w + 2 = 0 \quad or \quad w - 2 = 0$$
$$w = -2 \quad or \quad w = 2$$

Check. Since the width must be positive, -2 cannot be a solution. If the width is 2 ft, then the length is $6 \cdot 2$ ft, or 12 ft, and the area is 2 ft · 12 ft = 24 ft^2. Thus, 2 checks.

State. The table is 12 ft long and 2 ft wide.

11. ***Familiarize***. We make a drawing. Let w = the width, in cm. Then $w + 2$ = the length, in cm.

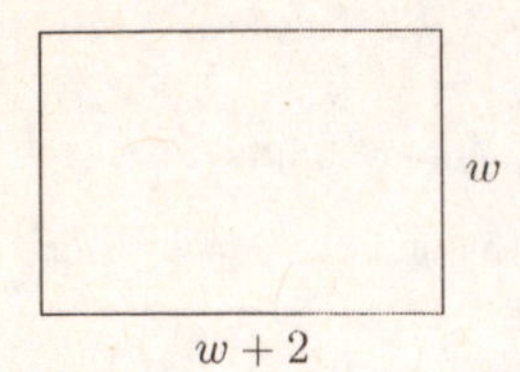

Recall that the area of a rectangle is length times width.

Translate. We reword the problem.

$$\underbrace{\text{Length}}_{\downarrow\ (w+2)}\ \underbrace{\text{times}}_{\downarrow\ \cdot}\ \underbrace{\text{width}}_{\downarrow\ w}\ \underbrace{\text{is}}_{\downarrow\ =}\ \underbrace{24\text{ cm}^2.}_{\downarrow\ 24}$$

Carry out. We solve the equation.

$$(w + 2)w = 24$$
$$w^2 + 2w = 24$$
$$w^2 + 2w - 24 = 0$$
$$(w + 6)(w - 4) = 0$$
$$w + 6 = 0 \quad \text{or} \quad w - 4 = 0$$
$$w = -6 \quad \text{or} \quad w = 4$$

Check. Since the width must be positive, -6 cannot be a solution. If the width is 4 cm, then the length is $4 + 2$, or 6 cm, and the area is $6 \cdot 4$, or 24 cm^2. Thus, 4 checks.

State. The width is 4 cm, and the length is 6 cm.

13. ***Familiarize***. Using the labels shown on the drawing in the text, we let h = the height, in cm, and $h + 10$ = the base, in cm. Recall that the formula for the area of a triangle is $\frac{1}{2} \cdot \text{(base)} \cdot \text{(height)}$.

Translate.

$$\underbrace{\tfrac{1}{2}}_{\downarrow\ \frac{1}{2}}\ \underbrace{\text{times}}_{\downarrow\ \cdot}\ \underbrace{\text{base}}_{\downarrow\ (h+10)}\ \underbrace{\text{times}}_{\downarrow\ \cdot}\ \underbrace{\text{height}}_{\downarrow\ h}\ \underbrace{\text{is}}_{\downarrow\ =}\ \underbrace{28\text{ cm}^2.}_{\downarrow\ 28}$$

Carry out.

$$\frac{1}{2}(h + 10)h = 28$$
$$(h + 10)h = 56 \quad \text{Multiplying by 2}$$
$$h^2 + 10h = 56$$
$$h^2 + 10h - 56 = 0$$
$$(h + 14)(h - 4) = 0$$
$$h + 14 = 0 \quad \text{or} \quad h - 4 = 0$$
$$h = -14 \quad \text{or} \quad h = 4$$

Check. Since the height of the triangle must be positive, -14 cannot be a solution. If the height is 4 cm, then the base is $4+10$, or 14 cm, and the area is $\frac{1}{2} \cdot 14 \cdot 4$, or 28 cm^2. Thus, 4 checks.

State. The height of the triangle is 4 cm, and the base is 14 cm.

15. ***Familiarize***. Using the labels show on the drawing in the text, we let x = the length of the foot of the sail, in ft, and $x + 5$ = the height of the sail, in ft. Recall that the formula for the area of a triangle is $\frac{1}{2} \cdot \text{(base)} \cdot \text{(height)}$.

Translate.

$$\underbrace{\tfrac{1}{2}}_{\downarrow\ \frac{1}{2}}\ \underbrace{\text{times}}_{\downarrow\ \cdot}\ \underbrace{\text{base}}_{\downarrow\ x}\ \underbrace{\text{times}}_{\downarrow\ \cdot}\ \underbrace{\text{height}}_{\downarrow\ (x+5)}\ \underbrace{\text{is}}_{\downarrow\ =}\ \underbrace{42\text{ ft}^2.}_{\downarrow\ 42}$$

Carry out.

$$\frac{1}{2}x(x + 5) = 42$$
$$x(x + 5) = 84 \quad \text{Multiplying by 2}$$
$$x^2 + 5x = 84$$
$$x^2 + 5x - 84 = 0$$
$$(x + 12)(x - 7) = 0$$
$$x + 12 = 0 \quad \text{or} \quad x - 7 = 0$$
$$x = -12 \quad \text{or} \quad x = 7$$

Check. The solutions of the equation are -12 and 7. The length of the base of a triangle cannot be negative, so -12 cannot be a solution. Suppose the length of the foot of the sail is 7 ft. Then the height is $7 + 5$, or 12 ft, and the area is $\frac{1}{2} \cdot 7 \cdot 12$, or 42 ft^2. These numbers check.

State. The length of the foot of the sail is 7 ft, and the height is 12 ft.

17. ***Familiarize and Translate***. We substitute 150 for A in the formula.

$$A = -50t^2 + 200t$$
$$150 = -50t^2 + 200t$$

Carry out. We solve the equation.

$$150 = -50t^2 + 200t$$
$$0 = -50t^2 + 200t - 150$$
$$0 = -50(t^2 - 4t + 3)$$
$$0 = -50(t - 1)(t - 3)$$
$$t - 1 = 0 \quad \text{or} \quad t - 3 = 0$$
$$t = 1 \quad \text{or} \quad t = 3$$

Check. Since $-50 \cdot 1^2 + 200 \cdot 1 = -50 + 200 = 150$, the number 1 checks. Since $-50 \cdot 3^2 + 200 \cdot 3 = -450 + 600 = 150$, the number 3 checks also.

State. There will be about 150 micrograms of Albuterol in the bloodstream 1 minute and 3 minutes after an inhalation.

19. ***Familiarize***. We will use the formula $x^2 - x = N$.

Translate. Substitute 240 for N.

$$x^2 - x = 240$$

Carry out.

$$x^2 - x = 240$$
$$x^2 - x - 240 = 0$$
$$(x - 16)(x + 15) = 0$$
$$x - 16 = 0 \quad \text{or} \quad x + 15 = 0$$
$$x = 16 \quad \text{or} \quad x = -15$$

Check. The solutions of the equation are 16 and -15. Since the number of teams cannot be negative, -15 cannot be a solution. But 16 checks since $16^2 - 16 = 256 - 16 = 240$.

State. There are 16 teams in the league.

21. *Familiarize.* We will use the formula $H = \frac{1}{2}(n^2 - n)$.

Translate. Substitute 15 for n.

$$H = \frac{1}{2}(15^2 - 15)$$

Carry out. We do the computation on the right.

$$H = \frac{1}{2}(15^2 - 15)$$
$$H = \frac{1}{2}(225 - 15)$$
$$H = \frac{1}{2}(210)$$
$$H = 105$$

Check. We can recheck the computation, or we can solve the equation $105 = \frac{1}{2}(n^2 - n)$. The answer checks.

State. 105 handshakes are possible.

23. *Familiarize.* We will use the formula $H = \frac{1}{2}(n^2 - n)$, since "high fives" can be substituted for handshakes.

Translate. Substitute 66 for H.

$$66 = \frac{1}{2}(n^2 - n)$$

Carry out.

$$66 = \frac{1}{2}(n^2 - n)$$
$$132 = n^2 - n \qquad \text{Multiplying by 2}$$
$$0 = n^2 - n - 132$$
$$0 = (n - 12)(n + 11)$$
$$n - 12 = 0 \quad or \quad n + 11 = 0$$
$$n = 12 \quad or \quad n = -11$$

Check. The solutions of the equation are 12 and -11. Since the number of players cannot be negative, -11 cannot be a solution. However, 12 checks since $\frac{1}{2}(12^2 - 12) = \frac{1}{2}(144 - 12) = \frac{1}{2}(132) = 66$.

State. 12 players were on the team.

25. *Familiarize.* Let h = the vertical height to which each brace reaches, in feet. We have a right triangle with hypotenuse 15 ft and legs 12 ft and h.

Translate. We use the Pythagorean theorem.

$$a^2 + b^2 = c^2$$
$$12^2 + h^2 = 15^2$$

Carry out. We solve the equation.

$$12^2 + h^2 = 15^2$$
$$144 + h^2 = 225$$
$$h^2 - 81 = 0$$
$$(h + 9)(h - 9) = 0$$
$$h + 9 = 0 \quad \text{or} \quad h - 9 = 0$$
$$h = -9 \quad \text{or} \quad h = 9$$

Check. Since the vertical height must be positive, -9 cannot be a solution. If the height is 9 ft, then we have $12^2 + 9^2 = 144 + 81 = 225 = 15^2$. The number 9 checks.

State. Each brace reaches 9 ft vertically.

27. *Familiarize.* We make a drawing. Let l = the length of the cable, in ft.

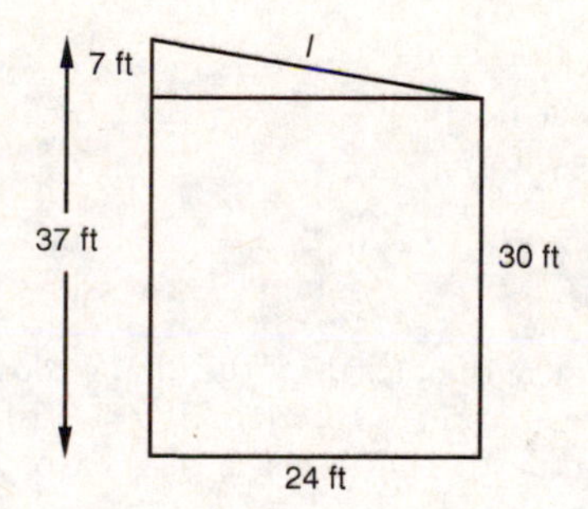

Note that we have a right triangle with hypotenuse l and legs of 24 ft and $37 - 30$, or 7 ft.

Translate. We use the Pythagorean theorem.

$$a^2 + b^2 = c^2$$
$$7^2 + 24^2 = l^2 \qquad \text{Substituting}$$

Carry out.

$$7^2 + 24^2 = l^2$$
$$49 + 576 = l^2$$
$$625 = l^2$$
$$0 = l^2 - 625$$
$$0 = (l + 25)(l - 25)$$
$$l + 25 = 0 \quad or \quad l - 25 = 0$$
$$l = -25 \quad or \quad l = 25$$

Check. The integer -25 cannot be the length of the cable, because it is negative. When $l = 25$, we have $7^2 + 24^2 = 25^2$. This checks.

State. The cable is 25 ft long.

29. *Familiarize.* We label the drawing. Let x = the length of a side of the dining room, in ft. Then the dining room has dimensions x by x and the kitchen has dimensions x by 10. The entire rectangular space has dimension x by $x + 10$. Recall that we multiply these dimensions to find the area of the rectangle.

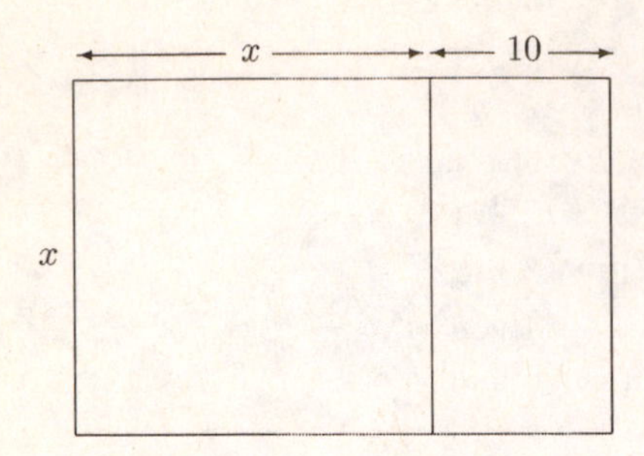

Translate.

The area of the rectangular space is 264 ft^2.

$$x(x+10) = 264$$

Carry out. We solve the equation.

$$x(x+10) = 264$$
$$x^2 + 10x = 264$$
$$x^2 + 10x - 264 = 0$$
$$(x+22)(x-12) = 0$$
$$x + 22 = 0 \quad or \quad x - 12 = 0$$
$$x = -22 \quad or \quad x = 12$$

Check. Since the length of a side of the dining room must be positive, -22 cannot be a solution. If x is 12 ft, then $x+10$ is 22 ft, and the area of the space is $12 \cdot 22$, or 264 ft^2. The number 12 checks.

State. The dining room is 12 ft by 12 ft, and the kitchen is 12 ft by 10 ft.

31. ***Familiarize.*** We will use the formula $h = 48t - 16t^2$.

Translate. Substitute $\frac{1}{2}$ for t.

$$h = 48 \cdot \frac{1}{2} - 16\left(\frac{1}{2}\right)^2$$

Carry out. We do the computation on the right.

$$h = 48 \cdot \frac{1}{2} - 16\left(\frac{1}{2}\right)^2$$
$$h = 48 \cdot \frac{1}{2} - 16 \cdot \frac{1}{4}$$
$$h = 24 - 4$$
$$h = 20$$

Check. We can recheck the computation, or we can solve the equation $20 = 48t - 16t^2$. The answer checks.

State. The rocket is 20 ft high $\frac{1}{2}$ sec after it is launched.

33. ***Familiarize.*** We will use the formula $h = 48t - 16t^2$.

Translate. Substitute 32 for h.

$$32 = 48t - 16t^2$$

Carry out. We solve the equation.

$$32 = 48t - 16t^2$$
$$0 = -16t^2 + 48t - 32$$
$$0 = -16(t^2 - 3t + 2)$$
$$0 = -16(t-1)(t-2)$$
$$t - 1 = 0 \quad or \quad t - 2 = 0$$
$$t = 1 \quad or \quad t = 2$$

Check. When $t = 1$, $h = 48 \cdot 1 - 16 \cdot 1^2 = 48 - 16 = 32$. When $t = 2$, $h = 48 \cdot 2 - 16 \cdot 2^2 = 96 - 64 = 32$. Both numbers check.

State. The rocket will be exactly 32 ft above the ground at 1 sec and at 2 sec after it is launched.

35. *Writing Exercise*

37. $-\frac{2}{3} \cdot \frac{4}{7} = -\frac{2 \cdot 4}{3 \cdot 7} = -\frac{8}{21}$

39. $\frac{5}{6}\left(\frac{-7}{9}\right) = \frac{5(-7)}{6 \cdot 9} = \frac{-35}{54}$, or $-\frac{35}{54}$

41.
$$-\frac{2}{3} + \frac{4}{7} = -\frac{2}{3} \cdot \frac{7}{7} + \frac{4}{7} \cdot \frac{3}{3}$$
$$= -\frac{14}{21} + \frac{12}{21}$$
$$= -\frac{2}{21}$$

43.
$$\frac{5}{6} + \frac{-7}{9} = \frac{5}{6} \cdot \frac{3}{3} + \frac{-7}{9} \cdot \frac{2}{2}$$
$$= \frac{15}{18} + \frac{-14}{18}$$
$$= \frac{1}{18}$$

45. *Writing Exercise*

47. ***Familiarize.*** First we find the length of the other leg of the right triangle. Then we find the area of the triangle, and finally we multiply by the cost per square foot of the sailcloth. Let x = the length of the other leg of the right triangle, in feet.

Translate. We use the Pythagorean theorem to find x.

$$a^2 + b^2 = c^2$$
$$x^2 + 24^2 = 26^2 \quad \text{Substituting}$$

Carry out.

$$x^2 + 24^2 = 26^2$$
$$x^2 + 576 = 676$$
$$x^2 - 100 = 0$$
$$(x+10)(x-10) = 0$$
$$x + 10 = 0 \quad or \quad x - 10 = 0$$
$$x = -10 \quad or \quad x = 10$$

Since the length of the leg must be positive, -10 cannot be a solution. We use the number 10. Find the area of the triangle:

$$\frac{1}{2}bh = \frac{1}{2} \cdot 10 \text{ ft} \cdot 24 \text{ ft} = 120 \text{ ft}^2$$

Finally, we multiply the area, 120 ft^2, by the price per square foot of the sailcloth, \$10:

$$120 \cdot 10 = 1200$$

Check. Recheck the calculations. The answer checks.

State. A new main sail costs \$1200.

49. ***Familiarize.*** We add labels to the drawing in the text.

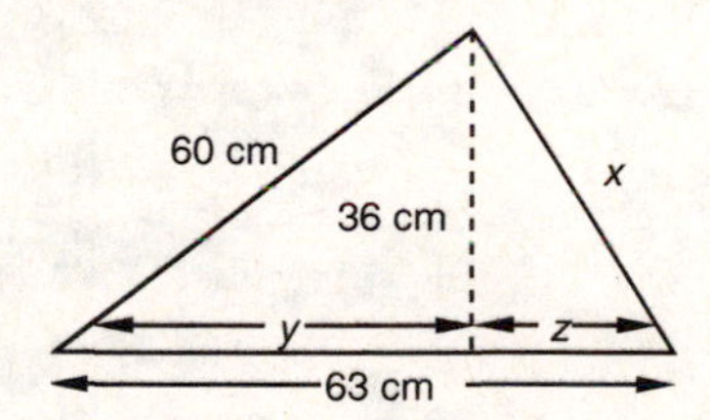

First we will use the Pythagorean theorem to find y. Then we will subtract to find z and, finally, we will use the Pythagorean theorem again to find x.

Translate. We use the Pythagorean theorem to find y.

$$a^2 + b^2 = c^2$$
$$y^2 + 36^2 = 60^2 \quad \text{Substituting}$$

Carry out.

$$y^2 + 36^2 = 60^2$$
$$y^2 + 1296 = 3600$$
$$y^2 - 2304 = 0$$
$$(y+48)(y-48) = 0$$
$$y + 48 = 0 \quad or \quad y - 48 = 0$$
$$y = -48 \quad or \quad y = 48$$

Since the length y cannot be negative, we use 48 cm. Then $z = 63 - 48 = 15$ cm.

Now we find x. We use the Pythagorean theorem again.

$$15^2 + 36^2 = x^2$$
$$225 + 1296 = x^2$$
$$1521 = x^2$$
$$0 = x^2 - 1521$$
$$0 = (x+39)(x-39)$$
$$x + 39 = 0 \quad or \quad x - 39 = 0$$
$$x = -39 \quad or \quad x = 39$$

Since the length x cannot be negative, we use 39 cm.

Check. We repeat all of the calculations. The answer checks.

State. The value of x is 39 cm.

51. ***Familiarize.*** Let w = the width of the piece of cardboard, in cm. Then $2w$ = the length, in cm. The length and width of the base of the box are $2x - 8$ and $x - 8$, respectively, and its height is 4.

Recall that the formula for the volume of a rectangular solid is given by length · width · height.

Translate.

$$\underbrace{\text{The volume}}_{\downarrow} \quad \underset{\downarrow}{\text{is}} \quad \underbrace{616 \text{ cm}^3}_{\downarrow}.$$
$$(2w-8)(w-8)(4) = 616$$

Carry out. We solve the equation.

$$(2w-8)(w-8)(4) = 616$$
$$(2w^2 - 24w + 64)(4) = 616$$
$$8w^2 - 96 + 256 = 616$$
$$8w^2 - 96w - 360 = 0$$
$$8(w^2 - 12w - 45) = 0$$
$$w^2 - 12w - 45 = 0 \quad \text{Dividing by 8}$$
$$(w-15)(w+3) = 0$$
$$w - 15 = 0 \quad or \quad w + 3 = 0$$
$$w = 15 \quad or \quad w = -3$$

Check. The width cannot be negative, so we only need to check 15. When $w = 15$, then $2w = 30$ and the dimensions of the box are $30 - 8$ by $15 - 8$ by 4, or 22 by 7 by 4. The volume is $22 \cdot 7 \cdot 4$, or 616.

State. The original dimension of the cardboard are 15 cm by 30 cm.

53. ***Familiarize.*** First we can use the Pythagorean theorem to find x, in ft. Then the height of the telephone pole is $x + 5$.

Translate. We use the Pythagorean theorem.

$$a^2 + b^2 = c^2$$
$$\left(\frac{1}{2}x + 1\right)^2 + x^2 = 34^2$$

Carry out. We solve the equation.

$$\left(\frac{1}{2}x + 1\right)^2 + x^2 = 34^2$$
$$\frac{1}{4}x^2 + x + 1 + x^2 = 1156$$
$$x^2 + 4x + 4 + 4x^2 = 4624 \quad \text{Multiplying by 4}$$
$$5x^2 + 4 + 4 = 4624$$
$$5x^2 + 4x - 4620 = 0$$
$$(5x + 154)(x - 30) = 0$$
$$5x + 154 = 0 \quad or \quad x - 30 = 0$$
$$5x = -154 \quad or \quad x = 30$$
$$x = -30.8 \quad or \quad x = 30$$

Check. Since the length x must be positive, -30.8 cannot be a solution. If x is 30 ft, then $\frac{1}{2}x + 1$ is $\frac{1}{2} \cdot 30 + 1$, or 16 ft. Since $16^2 + 30^2 = 1156 = 34^2$, the number 30 checks. When x is 30 ft, then $x + 5$ is 35 ft.

State. The height of the telephone pole is 35 ft.

55. First substitute 18 for N in the given formula.

$$18 = -0.009t(t-12)^3$$

Graph $y_1 = 18$ and $y_2 = -0.009x(x-12)^3$ in the given window and use the TRACE feature to find the first coordinates of the points of intersection of the graphs. We find $x \approx 2$ hr and $x \approx 4.2$ hr.

57. Graph $y = -0.009x(x-12)^3$ and use the TRACE feature to find the first coordinate of the highest point on the graph. We find $x = 3$ hr.

Chapter 6

Rational Expressions and Equations

Exercise Set 6.1

1. $x-2=0$ when $x=2$ and $x+3=0$ when $x=-3$, so choice (e) is correct.

3. $a^2-a-12=(a-4)(a+3)$; $a-4=0$ when $a=4$ and $a+3=0$ when $a=-3$, so choice (d) is correct.

5. $2t-1=0$ when $t=\frac{1}{2}$ and $3t+4=0$ when $t=-\frac{4}{3}$, so choice (c) is correct.

7. $\frac{12}{-7x}$

We find the real number(s) that make the denominator 0. To do so we set the denominator equal to 0 and solve for x:

$$-7x=0$$
$$x=0$$

The expression is undefined for $x=0$.

9. $\frac{t-6}{t+8}$

Set the denominator equal to 0 and solve for t:

$$t+8=0$$
$$t=-8$$

The expression is undefined for $t=-8$.

11. $\frac{a-4}{3a-12}$

Set the denominator equal to 0 and solve for a:

$$3a-12=0$$
$$3a=12$$
$$a=4$$

The expression is undefined for $a=4$.

13. $\frac{x^2-25}{x^2-3x-28}$

Set the denominator equal to 0 and solve for x:

$$x^2-3x-28=0$$
$$(x-7)(x+4)=0$$
$$x-7=0 \quad or \quad x+4=0$$
$$x=7 \quad or \quad x=-4$$

The expression is undefined for $x=7$ and $x=-4$.

15. $\frac{t^2-4}{2t^3+11t^2-6t}$

Set the denominator equal to 0 and solve for t:

$$2t^3+11t^2-6t=0$$
$$t(2t^2+11t-6)=0$$
$$t(2t-1)(t+6)=0$$
$$t==0 \quad or \quad 2t-1=0 \quad or \quad t+6=0$$
$$t==0 \quad or \quad 2t=1 \quad or \quad t=-6$$
$$t==0 \quad or \quad t=\frac{1}{2} \quad or \quad t=-6$$

The expression is undefined for $t=0, t=\frac{1}{2}$, and $t=-6$.

17. $\frac{50a^2b}{40ab^3}$

$=\frac{5a\cdot 10ab}{4b^2\cdot 10ab}$ — Factoring the numerator and denominator. Note the common factor of $10ab$.

$=\frac{5a}{4b^2}\cdot\frac{10ab}{10ab}$ — Rewriting as a product of two rational expressions

$=\frac{5a}{4b^2}\cdot 1$ — $\frac{10ab}{10ab}=1$

$=\frac{5a}{4b^2}$ — Removing the factor 1

19.
$$\frac{28x^2y}{21x^3y^5}=\frac{4\cdot 7x^2y}{3xy^4\cdot 7x^2y}=\frac{4}{3xy^4}\cdot\frac{7x^2y}{7x^2y}=\frac{4}{3xy^4}\cdot 1=\frac{4}{3xy^4}$$

21.
$$\frac{9x+15}{12x+20}=\frac{3(3x+5)}{4(3x+5)}=\frac{3}{4}\cdot\frac{3x+5}{3x+5}=\frac{3}{4}\cdot 1=\frac{3}{4}$$

23.
$$\frac{a^2-9}{a^2+4a+3}=\frac{(a+3)(a-3)}{(a+3)(a+1)}=\frac{a+3}{a+3}\cdot\frac{a-3}{a+1}=1\cdot\frac{a-3}{a+1}=\frac{a-3}{a+1}$$

25. $\dfrac{36x^6}{24x^9} = \dfrac{3 \cdot 12x^6}{2x^3 \cdot 12x^6}$
$= \dfrac{3}{2x^3} \cdot \dfrac{12x^6}{12x^6}$
$= \dfrac{3}{2x^3} \cdot 1$
$= \dfrac{3}{2x^3}$

Check: Let $x = 1$.
$\dfrac{36x^6}{24x^9} = \dfrac{36 \cdot 1^6}{24 \cdot 1^9} = \dfrac{36}{24} = \dfrac{3}{2}$
$\dfrac{3}{2x^3} = \dfrac{3}{2 \cdot 1^3} = \dfrac{3}{2}$
The answer is probably correct.

27. $\dfrac{-2y+6}{-8y} = \dfrac{-2(y-3)}{-2 \cdot 4y}$
$= \dfrac{-2}{-2} \cdot \dfrac{y-3}{4y}$
$= 1 \cdot \dfrac{y-3}{4y}$
$= \dfrac{y-3}{4y}$

Check: Let $x = 2$.
$\dfrac{-2y+6}{-8y} = \dfrac{-2 \cdot 2 + 6}{-8 \cdot 2} = \dfrac{2}{-16} = -\dfrac{1}{8}$
$\dfrac{y-3}{4y} = \dfrac{2-3}{4 \cdot 2} = \dfrac{-1}{8} = -\dfrac{1}{8}$
The answer is probably correct.

29. $\dfrac{t^2-16}{t^2+t-20} = \dfrac{(t+4)(t-4)}{(t+5)(t-4)}$
$= \dfrac{t+4}{t+5} \cdot \dfrac{t-4}{t-4}$
$= \dfrac{t+4}{t+5} \cdot 1$
$= \dfrac{t+4}{t+5}$

Check: Let $t = 1$.
$\dfrac{t^2-16}{t^2+t-20} = \dfrac{1^2-16}{1^2+1-20} = \dfrac{-15}{-18} = \dfrac{5}{6}$
$\dfrac{t+4}{t+5} = \dfrac{1+4}{1+5} = \dfrac{5}{6}$
The answer is probably correct.

31. $\dfrac{3a^2+9a-12}{6a^2-30a+24} = \dfrac{3(a^2+3a-4)}{6(a^2-5a+4)}$
$= \dfrac{3(a+4)(a-1)}{3 \cdot 2(a-4)(a-1)}$
$= \dfrac{3(a-1)}{3(a-1)} \cdot \dfrac{a+4}{2(a-4)}$
$= 1 \cdot \dfrac{a+4}{2(a-4)}$
$= \dfrac{a+4}{2(a-4)}$

Check: Let $a = 2$.
$\dfrac{3a^2+9a-12}{6a^2-30a+24} = \dfrac{3 \cdot 2^2 + 9 \cdot 2 - 12}{6 \cdot 2^2 - 30 \cdot 2 + 24} = \dfrac{18}{-12} = -\dfrac{3}{2}$
$\dfrac{a+4}{2(a-4)} = \dfrac{2+4}{2(2-4)} = \dfrac{6}{-4} = -\dfrac{3}{2}$
The answer is probably correct.

33. $\dfrac{x^2+8x+16}{x^2-16} = \dfrac{(x+4)(x+4)}{(x+4)(x-4)}$
$= \dfrac{x+4}{x+4} \cdot \dfrac{x+4}{x-4}$
$= 1 \cdot \dfrac{x+4}{x-4}$
$= \dfrac{x+4}{x-4}$

Check: Let $x = 1$.
$\dfrac{x^2+8x+16}{x^2-16} = \dfrac{1^2+8 \cdot 1+16}{1^2-16} = \dfrac{25}{-15} = -\dfrac{5}{3}$
$\dfrac{x+4}{x-4} = \dfrac{1+4}{1-4} = \dfrac{5}{-3} = -\dfrac{5}{3}$
The answer is probably correct.

35. $\dfrac{t^2-1}{t+1} = \dfrac{(t+1)(t-1)}{t+1}$
$= \dfrac{t+1}{t+1} \cdot \dfrac{t-1}{1}$
$= 1 \cdot \dfrac{t-1}{1}$
$= t-1$

Check: Let $t = 2$.
$\dfrac{t^2-1}{t+1} = \dfrac{2^2-1}{2+1} = \dfrac{3}{3} = 1$
$t-1 = 2-1 = 1$

The answer is probably correct.

37. $\dfrac{a-2}{a^3-8} = \dfrac{a-2}{(a-2)(a^2+2a+4)}$
$= \dfrac{a-2}{a-2} \cdot \dfrac{1}{a^2+2a+4}$
$= 1 \cdot \dfrac{1}{a^2+2a+4}$
$= \dfrac{1}{a^2+2a+4}$

Check: Let $a = 1$.
$\dfrac{a-2}{a^3-8} = \dfrac{1-2}{1^3-8} = \dfrac{-1}{-7} = \dfrac{1}{7}$
$\dfrac{1}{a^2+2a+4} = \dfrac{1}{1^2+2 \cdot 1+4} = \dfrac{1}{7}$
The answer is probably correct.

39. $\dfrac{y^2+4}{y+2}$ cannot be simplified.

Neither the numerator nor the denominator can be factored.

41. $\frac{5x^2+20}{10x^2+40} = \frac{5(x^2+4)}{10(x^2+4)}$

$= \frac{1 \cdot \cancel{5} \cdot \cancel{(x^2+4)}}{2 \cdot \cancel{5} \cdot \cancel{(x^2+4)}}$

$= \frac{1}{2}$

Check: Let $x = 1$.

$\frac{5x^2+20}{10x^2+40} = \frac{5 \cdot 1^2+20}{10 \cdot 1^2+40} = \frac{25}{50} = \frac{1}{2}$

$\frac{1}{2} = \frac{1}{2}$

The answer is probably correct.

43. $\frac{y^2+6y}{2y^2+13y+6} = \frac{y(y+6)}{(2y+1)(y+6)}$

$= \frac{y}{2y+1} \cdot \frac{y+6}{y+6}$

$= \frac{y}{2y+1} \cdot 1$

$= \frac{y}{2y+1}$

Check: Let $y = 1$.

$\frac{y^2+6y}{2y^2+13y+6} = \frac{1^2+6 \cdot 1}{2 \cdot 1^2+13 \cdot 1+6} = \frac{7}{21} = \frac{1}{3}$

$\frac{y}{2y+1} = \frac{1}{2 \cdot 1+1} = \frac{1}{3}$

The answer is probably correct.

45. $\frac{4x^2-12x+9}{10x^2-11x-6} = \frac{(2x-3)(2x-3)}{(2x-3)(5x+2)}$

$= \frac{2x-3}{2x-3} \cdot \frac{2x-3}{5x+2}$

$= 1 \cdot \frac{2x-3}{5x+2}$

$= \frac{2x-3}{5x+2}$

Check: Let $t = 1$.

$\frac{4x^2-12x+9}{10x^2-11x-6} = \frac{4 \cdot 1^2-12 \cdot 1+9}{10 \cdot 1^2-11 \cdot 1-6} = \frac{1}{-7} = -\frac{1}{7}$

$\frac{2x-3}{5x+2} = \frac{2 \cdot 1-3}{5 \cdot 1+2} = \frac{-1}{7} = -\frac{1}{7}$

The answer is probably correct.

47. $\frac{x-9}{9-x} = \frac{x-9}{-(x-9)}$

$= \frac{1}{-1} \cdot \frac{x-9}{x-9}$

$= \frac{1}{-1} \cdot 1$

$= -1$

Check: Let $x = 2$.

$\frac{x-9}{9-x} = \frac{2-9}{9-2} = \frac{-7}{7} = -1$

The answer is probably correct.

49. $\frac{7t-14}{2-t} = \frac{7(t-2)}{-(t-2)}$

$= \frac{7}{-1} \cdot \frac{t-2}{t-2}$

$= \frac{7}{-1} \cdot 1$

$= -7$

Check: Let $t = 1$.

$\frac{7t-14}{2-t} = \frac{7 \cdot 1-14}{2-1} = \frac{-7}{1} = -7$

The answer is probably correct.

51. $\frac{a-b}{3b-3a} = \frac{a-b}{-3(a-b)}$

$= \frac{1}{-3} \cdot \frac{a-b}{a-b}$

$= \frac{1}{-3} \cdot 1$

$= -\frac{1}{3}$

Check: Let $a = 2$ and $b = 1$.

$\frac{a-b}{3b-3a} = \frac{2-1}{3 \cdot 1-3 \cdot 2} = \frac{1}{-3} = -\frac{1}{3}$

The answer is probably correct.

53. $\frac{3x^2-3y^2}{2y^2-2x^2} = \frac{3(x^2-y^2)}{2(y^2-x^2)}$

$= \frac{3(x^2-y^2)}{2(-1)(x^2-y^2)}$

$= \frac{3}{2(-1)} \cdot \frac{x^2-y^2}{x^2-y^2}$

$= \frac{3}{2(-1)} \cdot 1$

$= -\frac{3}{2}$

Check: Let $x = 1$ and $y = 2$.

$\frac{3x^2-3y^2}{2y^2-2x^2} = \frac{3 \cdot 1^2-3 \cdot 2^2}{2 \cdot 2^2-2 \cdot 1^2} = \frac{-9}{6} = -\frac{3}{2}$

$-\frac{3}{2} = -\frac{3}{2}$

The answer is probably correct.

55. $\frac{7s^2-28t^2}{28t^2-7s^2}$

Note that the numerator and denominator are opposites. Thus, we have an expression divided by its opposite, so the result is -1.

57. *Writing Exercise*

59. $-\frac{2}{3} \cdot \frac{6}{7} = -\frac{2 \cdot 6}{3 \cdot 7}$

$= -\frac{2 \cdot 2 \cdot \cancel{3}}{\cancel{3} \cdot 7}$

$= -\frac{4}{7}$

61. $\frac{5}{8} \div \left(-\frac{1}{6}\right) = \frac{5}{8}\cdot(-6)$
$= -\frac{5\cdot 6}{8}$
$= -\frac{5\cdot \cancel{2}\cdot 3}{\cancel{2}\cdot 4}$
$= -\frac{15}{4}$

63. $\frac{7}{9} - \frac{2}{3}\cdot\frac{6}{7} = \frac{7}{9} - \frac{4}{7} = \frac{7}{9}\cdot\frac{7}{7} - \frac{4}{7}\cdot\frac{9}{2} =$
$\frac{49}{63} - \frac{36}{63} = \frac{13}{63}$

65. *Writing Exercise*

67. $\frac{16y^4 - x^4}{(x^2+4y^2)(x-2y)}$
$= \frac{(4y^2+x^2)(4y^2-x^2)}{(x^2+4y^2)(x-2y)}$
$= \frac{(4y^2+x^2)(2y+x)(2y-x)}{(x^2+4y^2)(x-2y)}$
$= \frac{(x^2+4y^2)(2y+x)(-1)(x-2y)}{(x^2+4y^2)(x-2y)}$
$= \frac{(x^2+4y^2)(x-2y)}{(x^2+4y^2)(x-2y)}\cdot\frac{(2y+x)(-1)}{1}$
$= -2y - x$, or $-x-2y$, or $-(2y+x)$

69. $\frac{x^5 - 2x^3 + 4x^2 - 8}{x^7 + 2x^4 - 4x^3 - 8}$
$= \frac{x^3(x^2-2) + 4(x^2-2)}{x^4(x^3+2) - 4(x^3+2)}$
$= \frac{(x^2-2)(x^3+4)}{(x^3+2)(x^4-4)}$
$= \frac{(x^2-2)(x^3+4)}{(x^3+2)(x^2+2)(x^2-2)}$
$= \frac{\cancel{(x^2-2)}(x^3+4)}{(x^3+2)(x^2+2)\cancel{(x^2-2)}}$
$= \frac{x^3+4}{(x^3+2)(x^2+2)}$

71. $\frac{(t^4-1)(t^2-9)(t-9)^2}{(t^4-81)(t^2+1)(t+1)^2}$
$= \frac{(t^2+1)(t+1)(t-1)(t+3)(t-3)(t-9)(t-9)}{(t^2+9)(t+3)(t-3)(t^2+1)(t+1)(t+1)}$
$= \frac{\cancel{(t^2+1)}\cancel{(t+1)}(t-1)\cancel{(t+3)}\cancel{(t-3)}(t-9)(t-9)}{(t^2+9)\cancel{(t+3)}\cancel{(t-3)}\cancel{(t^2+1)}\cancel{(t+1)}(t+1)}$
$= \frac{(t-1)(t-9)(t-9)}{(t^2+9)(t+1)}$, or $\frac{(t-1)(t-9)^2}{(t^2+9)(t+1)}$

73. $\frac{(x^2-y^2)(x^2-2xy+y^2)}{(x+y)^2(x^2-4xy-5y^2)}$
$= \frac{(x+y)(x-y)(x-y)(x-y)}{(x+y)(x+y)(x-5y)(x+y)}$
$= \frac{\cancel{(x+y)}(x-y)(x-y)(x-y)}{\cancel{(x+y)}(x+y)(x-5y)(x+y)}$
$= \frac{(x-y)^3}{(x+y)^2(x-5y)}$

75. *Writing Exercise*

Exercise Set 6.2

1. $\frac{7x}{5}\cdot\frac{x-5}{2x+1} = \frac{7x(x-5)}{5(2x+1)}$

3. $\frac{a-4}{a+6}\cdot\frac{a+2}{a+6} = \frac{(a-4)(a+2)}{(a+6)(a+6)}$, or $\frac{(a-4)(a+2)}{(a+6)^2}$

5. $\frac{2x+3}{4}\cdot\frac{x+1}{x-5} = \frac{(2x+3)(x+1)}{4(x-5)}$

7. $\frac{a-4}{a^2+4}\cdot\frac{a+4}{a^2-4} = \frac{(a-4)(a+4)}{(a^2+4)(a^2-4)}$

9. $\frac{x+6}{3+x}\cdot\frac{x-1}{x+1} = \frac{(x+6)(x-1)}{(3+x)(x+1)}$

11. $\frac{5a^4}{6a}\cdot\frac{2}{a}$
$= \frac{5a^4\cdot 2}{6a\cdot a}$ Multiplying the numerators and the denominators
$= \frac{5\cdot a\cdot a\cdot a\cdot a\cdot 2}{2\cdot 3\cdot a\cdot a}$ Factoring the numerator and the denominator
$= \frac{5\cdot \cancel{a}\cdot\cancel{a}\cdot a\cdot a\cdot\cancel{2}}{\cancel{2}\cdot 3\cdot\cancel{a}\cdot\cancel{a}}$ Removing a factor equal to 1
$= \frac{5a^2}{3}$ Simplifying

13. $\frac{3c}{d^2}\cdot\frac{8d}{6c^3}$
$= \frac{3c\cdot 8d}{d^2\cdot 6c^3}$ Multiplying the numerators and the denominators
$= \frac{3\cdot c\cdot 2\cdot 4\cdot d}{d\cdot d\cdot 3\cdot 2\cdot c\cdot c\cdot c}$ Factoring the numerator and the denominator
$= \frac{\cancel{3}\cdot\cancel{c}\cdot\cancel{2}\cdot 4\cdot\cancel{d}}{\cancel{d}\cdot d\cdot\cancel{3}\cdot\cancel{2}\cdot\cancel{c}\cdot c\cdot c}$
$= \frac{4}{dc^2}$

15. $\frac{x^2-3x-10}{(x-2)^2}\cdot\frac{x-2}{x-5} = \frac{(x^2-3x-10)(x-2)}{(x-2)^2(x-5)}$
$= \frac{(x-5)(x+2)(x-2)}{(x-2)(x-2)(x-5)}$
$= \frac{\cancel{(x-5)}(x+2)\cancel{(x-2)}}{\cancel{(x-2)}(x-2)\cancel{(x-5)}}$
$= \frac{x+2}{x-2}$

17. $\frac{a^2+25}{a^2-4a+3}\cdot\frac{a-5}{a+5} = \frac{(a^2+25)(a-5)}{(a^2-4a+3)(a+5)}$
$= \frac{(a^2+25)(a-5)}{(a-3)(a-1)(a+5)}$

(No simplification is possible.)

19. $$\frac{a^2-9}{a^2}\cdot\frac{7a}{a^2+a-12}=\frac{(a+3)(a-3)\cdot 7\cdot a}{a\cdot a(a+4)(a-3)}$$
$$=\frac{(a+3)\cancel{(a-3)}\cdot 7\cdot \cancel{a}}{\cancel{a}\cdot a(a+4)\cancel{(a-3)}}$$
$$=\frac{7(a+3)}{a(a+4)}$$

21. $$\frac{4a^2}{3a^2-12a+12}\cdot\frac{3a-6}{2a}$$
$$=\frac{4a^2(3a-6)}{(3a^2-12a+12)2a}$$
$$=\frac{2\cdot 2\cdot a\cdot a\cdot 3\cdot(a-2)}{3\cdot(a-2)\cdot(a-2)\cdot 2\cdot a}$$
$$=\frac{\cancel{2}\cdot 2\cdot \cancel{a}\cdot a\cdot \cancel{3}\cdot\cancel{(a-2)}}{\cancel{3}\cdot\cancel{(a-2)}\cdot(a-2)\cdot\cancel{2}\cdot\cancel{a}}$$
$$=\frac{2a}{a-2}$$

23. $$\frac{t^2+2t-3}{t^2+4t-5}\cdot\frac{t^2-3t-10}{t^2+5t+6}$$
$$=\frac{(t^2+2t-3)(t^2-3t-10)}{(t^2+4t-5)(t^2+5t+6)}$$
$$=\frac{(t+3)(t-1)(t-5)(t+2)}{(t+5)(t-1)(t+3)(t+2)}$$
$$=\frac{\cancel{(t+3)}\cancel{(t-1)}(t-5)\cancel{(t+2)}}{(t+5)\cancel{(t-1)}\cancel{(t+3)}\cancel{(t+2)}}$$
$$=\frac{t-5}{t+5}$$

25. $$\frac{5a^2-180}{10a^2-10}\cdot\frac{20a+20}{2a-12}$$
$$=\frac{(5a^2-180)(20a+20)}{(10a^2-10)(2a-12)}$$
$$=\frac{5(a+6)(a-6)(2)(10)(a+1)}{10(a+1)(a-1)(2)(a-6)}$$
$$=\frac{5(a+6)\cancel{(a-6)}\cancel{(2)}\cancel{(10)}\cancel{(a+1)}}{\cancel{10}\cancel{(a+1)}(a-1)\cancel{(2)}\cancel{(a-6)}}$$
$$=\frac{5(a+6)}{a-1}$$

27. $$\frac{x^2+4x+4}{(x-1)^2}\cdot\frac{x^2-2x+1}{(x+2)^2}=\frac{(x+2)^2(x-1)^2}{(x-1)^2(x+2)^2}=1$$

29. $$\frac{5t^2+12t+4}{t^2+4t+4}\cdot\frac{t^2+8t+16}{5t^2+22t+8}$$
$$=\frac{(5t^2+12t+4)(t^2+8t+16)}{(t^2+4t+4)(5t^2+22t+8)}$$
$$=\frac{(5t+2)(t+2)(t+4)(t+4)}{(t+2)(t+2)(5t+2)(t+4)}$$
$$=\frac{\cancel{(5t+2)}\cancel{(t+2)}\cancel{(t+4)}(t+4)}{\cancel{(t+2)}(t+2)\cancel{(5t+2)}\cancel{(t+4)}}$$
$$=\frac{t+4}{t+2}$$

31. $$\frac{2x^2-5x+3}{6x^2-5x-1}\cdot\frac{6x^2+13x+2}{2x^2+3x-9}$$
$$=\frac{(2x^2-5x+3)(6x^2+13x+2)}{(6x^2-5x-1)(2x^2+3x-9)}$$
$$=\frac{(2x-3)(x-1)(6x+1)(x+2)}{(6x+1)(x-1)(2x-3)(x+3)}$$
$$=\frac{\cancel{(2x-3)}\cancel{(x-1)}\cancel{(6x+1)}(x+2)}{\cancel{(6x+1)}\cancel{(x-1)}\cancel{(2x-3)}(x+3)}$$
$$=\frac{x+2}{x+3}$$

33. $$\frac{c^3+8}{c^5-4c^3}\cdot\frac{c^6-4c^5+4c^4}{c^2-2c+4}$$
$$=\frac{(c^3+8)(c^6-4c^5+4c^4)}{(c^5-4c^3)(c^2-2c+4)}$$
$$=\frac{(c+2)(c^2-2c+4)(c^4)(c-2)(c-2)}{c^3(c+2)(c-2)(c^2-2c+4)}$$
$$=\frac{c^3(c+2)(c^2-2c+4)(c-2)}{c^3(c+2)(c^2-2c+4)(c-2)}\cdot\frac{c(c-2)}{1}$$
$$=c(c-2)$$

35. The reciprocal of $\frac{3x}{7}$ is $\frac{7}{3x}$ because $\frac{3x}{7}\cdot\frac{7}{3x}=1$.

37. The reciprocal of a^3-8a is $\frac{1}{a^3-8a}$ because $\frac{a^3-8a}{1}\cdot\frac{1}{a^3-8a}=1$.

39. $$\frac{5}{8}\div\frac{3}{7}$$
$=\frac{5}{8}\cdot\frac{7}{3}$ Multiplying by the reciprocal of the divisor
$$=\frac{5\cdot 7}{8\cdot 3}$$
$$=\frac{35}{24}$$

No simplification is possible.

41. $$\frac{x}{4}\div\frac{5}{x}$$
$=\frac{x}{4}\cdot\frac{x}{5}$ Multiplying by the reciprocal of the divisor
$$=\frac{x\cdot x}{4\cdot 5}$$
$$=\frac{x^2}{20}$$

43. $$\frac{a^5}{b^4}\div\frac{a^2}{b}=\frac{a^5}{b^4}\cdot\frac{b}{a^2}$$
$$=\frac{a^5\cdot b}{b^4\cdot a^2}$$
$$=\frac{a^2\cdot a^3\cdot b}{b\cdot b^3\cdot a^2}$$
$$=\frac{a^2b}{a^2b}\cdot\frac{a^3}{b^3}$$
$$=\frac{a^3}{b^3}$$

45. $\dfrac{y+5}{4} \div \dfrac{y}{2} = \dfrac{y+5}{4} \cdot \dfrac{2}{y}$

$= \dfrac{(y+5)(2)}{4 \cdot y}$

$= \dfrac{(y+5)(\cancel{2})}{\cancel{2} \cdot 2y}$

$= \dfrac{y+5}{2y}$

47. $\dfrac{4y-8}{y+2} \div \dfrac{y-2}{y^2-4} = \dfrac{4y-8}{y+2} \cdot \dfrac{y^2-4}{y-2}$

$= \dfrac{(4y-8)(y^2-4)}{(y+2)(y-2)}$

$= \dfrac{4\cancel{(y-2)}\cancel{(y+2)}(y-2)}{\cancel{(y+2)}\cancel{(y-2)}(1)}$

$= 4(y-2)$

49. $\dfrac{a}{a-b} \div \dfrac{b}{b-a} = \dfrac{a}{a-b} \cdot \dfrac{b-a}{b}$

$= \dfrac{a(b-a)}{(a-b)(b)}$

$= \dfrac{a(-1)\cancel{(a-b)}}{\cancel{(a-b)}(b)}$

$= \dfrac{-a}{b} = -\dfrac{a}{b}$

51. $(y^2-9) \div \dfrac{y^2-2y-3}{y^2+1} = \dfrac{(y^2-9)}{1} \cdot \dfrac{y^2+1}{y^2-2y-3}$

$= \dfrac{(y^2-9)(y^2+1)}{y^2-2y-3}$

$= \dfrac{(y+3)(y-3)(y^2+1)}{(y-3)(y+1)}$

$= \dfrac{(y+3)\cancel{(y-3)}(y^2+1)}{\cancel{(y-3)}(y+1)}$

$= \dfrac{(y+3)(y^2+1)}{y+1}$

53. $\dfrac{-3+3x}{16} \div \dfrac{x-1}{5} = \dfrac{3x-3}{16} \cdot \dfrac{5}{x-1}$

$= \dfrac{(3x-3) \cdot 5}{16(x-1)}$

$= \dfrac{3(x-1) \cdot 5}{16(x-1)}$

$= \dfrac{3\cancel{(x-1)} \cdot 5}{16\cancel{(x-1)}}$

$= \dfrac{15}{16}$

55. $\dfrac{a+2}{a-1} \div \dfrac{3a+6}{a-5} = \dfrac{a+2}{a-1} \cdot \dfrac{a-5}{3a+6}$

$= \dfrac{(a+2)(a-5)}{(a-1)(3a+6)}$

$= \dfrac{(a+2)(a-5)}{(a-1) \cdot 3 \cdot (a+2)}$

$= \dfrac{\cancel{(a+2)}(a-5)}{(a-1) \cdot 3 \cdot \cancel{(a+2)}}$

$= \dfrac{a-5}{3(a-1)}$

57. $(2x-1) \div \dfrac{2x^2-11x+5}{4x^2-1}$

$= \dfrac{2x-1}{1} \cdot \dfrac{4x^2-1}{2x^2-11x+5}$

$= \dfrac{(2x-1)(4x^2-1)}{1 \cdot (2x^2-11x+5)}$

$= \dfrac{(2x-1)(2x+1)(2x-1)}{1 \cdot (2x-1)(x-5)}$

$= \dfrac{\cancel{(2x-1)}(2x+1)(2x-1)}{1 \cdot \cancel{(2x-1)}(x-5)}$

$= \dfrac{(2x-1)(2x+1)}{x-5}$

59. $\dfrac{a^2-10a+25}{2a^2-a-21} \div \dfrac{a^2-a-20}{3a^2+5a-12}$

$= \dfrac{a^2-10a+25}{2a^2-a-21} \cdot \dfrac{3a^2+5a-12}{a^2-a-20}$

$= \dfrac{\cancel{(a-5)}(a-5)\cancel{(a+3)}(3a-4)}{\cancel{(a+3)}(2a-7)\cancel{(a-5)}(a+4)}$

$= \dfrac{(a-5)(3a-4)}{(2a-7)(a+4)}$

61. $\dfrac{c^2+10c+21}{c^2-2c-15} \div (5c^2+32c-21)$

$= \dfrac{c^2+10c+21}{c^2-2c-15} \cdot \dfrac{1}{5c^2+32c-21}$

$= \dfrac{(c^2+10c+21) \cdot 1}{(c^2-2c-15)(5c^2+32c-21)}$

$= \dfrac{(c+7)(c+3)}{(c-5)(c+3)(5c-3)(c+7)}$

$= \dfrac{(c+7)(c+3)}{(c+7)(c+3)} \cdot \dfrac{1}{(c-5)(5c-3)}$

$= \dfrac{1}{(c-5)(5c-3)}$

63. $\dfrac{x-y}{x^2+2xy+y^2} \div \dfrac{x^2-y^2}{x^2-5xy+4y^2}$

$= \dfrac{x-y}{x^2+2xy+y^2} \cdot \dfrac{x^2-5xy+4y^2}{x^2-y^2}$

$= \dfrac{\cancel{(x-y)}(x-y)(x-4y)}{(x+y)(x+y)(x+y)\cancel{(x-y)}}$

$= \dfrac{(x-y)(x-4y)}{(x+y)^3}$

65. $\dfrac{x^3-64}{x^3+64} \div \dfrac{x^2-16}{x^2-4x+16}$

$= \dfrac{x^3-64}{x^3+64} \cdot \dfrac{x^2-4x+16}{x^2-16}$

$= \dfrac{(x^3-64)(x^2-4x+16)}{(x^3+64)(x^2-16)}$

$= \dfrac{(x-4)(x^2+4x+16)(x^2-4x+16)}{(x+4)(x^2-4x+16)(x+4)(x-4)}$

$= \dfrac{(x-4)(x^2-4x+16)}{(x-4)(x^2-4x+16)} \cdot \dfrac{x^2+4x+16}{(x+4)(x+4)}$

$= \dfrac{x^2+4x+16}{(x+4)(x+4)}$, or $\dfrac{x^2+4x+16}{(x+4)^2}$

67. $$\frac{8a^3+b^3}{2a^2+3ab+b^2} \div \frac{8a^2-4ab+2b^2}{4a^2+4ab+b^2}$$
$$= \frac{8a^3+b^3}{2a^2+3ab+b^2} \cdot \frac{4a^2+4ab+b^2}{8a^2-4ab+2b^2}$$
$$= \frac{(8a^3+b^3)(4a^2+4ab+b^2)}{(2a^2+3ab+b^2)(8a^2-4ab+2b^2)}$$
$$= \frac{(2a+b)(4a^2-2ab+b^2)(2a+b)(2a+b)}{(2a+b)(a+b)(2)(4a^2-2ab+b^2)}$$
$$= \frac{(2a+b)(4a^2-2ab+b^2)}{(2a+b)(4a^2-2ab+b^2)} \cdot \frac{(2a+b)(2a+b)}{(a+b)(2)}$$
$$= \frac{(2a+b)(2a+b)}{2(a+b)}, \text{ or } \frac{(2a+b)^2}{2(a+b)}$$

69. *Writing Exercise*

71. $$\frac{3}{4}+\frac{5}{6} = \frac{3}{4}\cdot\frac{3}{3}+\frac{5}{6}\cdot\frac{2}{2}$$
$$= \frac{9}{12}+\frac{10}{12}$$
$$= \frac{19}{12}$$

73. $$\frac{2}{9}-\frac{1}{6} = \frac{2}{9}\cdot\frac{2}{2}-\frac{1}{6}\cdot\frac{3}{3}$$
$$= \frac{4}{18}-\frac{3}{18}$$
$$= \frac{1}{18}$$

75. $$\frac{2}{5}-\left(\frac{3}{2}\right)^2 = \frac{2}{5}-\frac{9}{4} = \frac{8}{20}-\frac{45}{20} = -\frac{37}{20}$$

77. *Writing Exercise*

79. $$\frac{2a^2-5ab}{c-3d} \div (4a^2-25b^2)$$
$$= \frac{2a^2-5ab}{c-3d} \cdot \frac{1}{4a^2-25b^2}$$
$$= \frac{a(2a-5b)}{(c-3d)(2a+5b)(2a-5b)}$$
$$= \frac{2a-5b}{2a-5b} \cdot \frac{a}{(c-3d)(2a+5b)}$$
$$= \frac{a}{(c-3d)(2a+5b)}$$

81. $$\frac{a^2-3b}{a^2+2b} \cdot \frac{a^2-2b}{a^2+3b} \cdot \frac{a^2+2b}{a^2-3b}$$

Note that $\frac{a^2-3b}{a^2+2b} \cdot \frac{a^2+2b}{a^2-3b}$ is the product of reciprocals and thus is equal to 1. Then the product in the original exercise is the remaining factor, $\frac{a^2-2b}{a^2+3b}$.

83. $$\frac{z^2-8z+16}{z^2+8z+16} \div \frac{(z-4)^5}{(z+4)^5} \div \frac{3z+12}{z^2-16}$$
$$= \frac{(z-4)^2}{(z+4)^2} \cdot \frac{(z+4)^5}{(z-4)^5} \cdot \frac{(z+4)(z-4)}{3(z+4)}$$
$$= \frac{(z-4)^2(z+4)^2(z+4)^3(z+4)(z-4)}{(z+4)^2(z-4)^2(z-4)(z-4)^2(3)(z+4)}$$
$$= \frac{(z+4)^3}{3(z-4)^2}$$

85. $$\frac{3x+3y+3}{9x} \div \frac{x^2+2xy+y^2-1}{x^4+x^2}$$
$$= \frac{3x+3y+3}{9x} \cdot \frac{x^4+x^2}{x^2+2xy+y^2-1}$$
$$= \frac{3(x+y+1)(x^2)(x^2+1)}{9x[(x+y)+1][(x+y)-1]}$$
$$= \frac{3(x+y+1)(x)(x)(x^2+1)}{3\cdot 3\cdot x(x+y+1)(x+y-1)}$$
$$= \frac{3x(x+y+1)}{3x(x+y+1)} \cdot \frac{x(x^2+1)}{3(x+y-1)}$$
$$= \frac{x(x^2+1)}{3(x+y-1)}$$

87. $$\frac{3y^3+6y^2}{y^2-y-12} \div \frac{y^2-y}{y^2-2y-8} \cdot \frac{y^2+5y+6}{y^2}$$
$$= \frac{3y^3+6y^2}{y^2-y-12} \cdot \frac{y^2-2y-8}{y^2-y} \cdot \frac{y^2+5y+6}{y^2}$$
$$= \frac{3\cdot y^2(y+2)(y-4)(y+2)(y+3)(y+2)}{(y-4)(y+3)(y)(y-1)(y^2)}$$
$$= \frac{3(y+2)^3}{y(y-1)}$$

89. Enter $y_1 = \frac{x-1}{x^2+2x+1} \div \frac{x^2-1}{x^2-5x+4}$ and $y_2 = \frac{x^2-5x+4}{(x+1)^3}$, display the values of y_1 and y_2 in a table, and compare the values. (See the Technology Connection on page 384 in the text.)

Exercise Set 6.3

1. To add two rational expressions when the denominators are the same, add numerators and keep the common denominator. (See page 388 in the text.)

3. The least common multiple of two denominators is usually referred to as the least common denominator and is abbreviated LCD. (See page 390 in the text.)

5. $\frac{6}{x}+\frac{4}{x} = \frac{10}{x}$ Adding numerators

7. $\frac{x}{12}+\frac{2x+5}{12} = \frac{3x+5}{12}$ Adding numerators

9. $\frac{4}{a+3}+\frac{5}{a+3} = \frac{9}{a+3}$

11. $\frac{8}{a+2}-\frac{2}{a+2} = \frac{6}{a+2}$ Subtracting numerators

13. $$\frac{3y+8}{2y}-\frac{y+1}{2y}$$
$$= \frac{3y+8-(y+1)}{2y}$$
$$= \frac{3y+8-y-1}{2y} \quad \text{Removing parentheses}$$
$$= \frac{2y+7}{2y}$$

15. $\dfrac{7x+8}{x+1}+\dfrac{4x+3}{x+1}$

$=\dfrac{11x+11}{x+1}$ Adding numerators

$=\dfrac{11(x+1)}{x+1}$ Factoring

$=\dfrac{11\cancel{(x+1)}}{\cancel{x+1}}$ Removing a factor equal to 1

$=11$

17. $\dfrac{7x+8}{x+1}-\dfrac{4x+3}{x+1}=\dfrac{7x+8-(4x+3)}{x+1}$

$=\dfrac{7x+8-4x-3}{x+1}$

$=\dfrac{3x+5}{x+1}$

19. $\dfrac{a^2}{a-4}+\dfrac{a-20}{a-4}=\dfrac{a^2+a-20}{a-4}$

$=\dfrac{(a+5)(a-4)}{a-4}$

$=\dfrac{(a+5)\cancel{(a-4)}}{\cancel{a-4}}$

$=a+5$

21. $\dfrac{x^2}{x-2}-\dfrac{6x-8}{x-2}=\dfrac{x^2-(6x-8)}{x-2}$

$=\dfrac{x^2-6x+8}{x-2}$

$=\dfrac{(x-4)(x-2)}{x-2}$

$=\dfrac{(x-4)\cancel{(x-2)}}{\cancel{x-2}}$

$=x-4$

23. $\dfrac{t^2-5t}{t-1}+\dfrac{5t-t^2}{t-1}$

Note that the numerators are opposites, so their sum is 0. Then we have $\dfrac{0}{t-1}$, or 0.

25. $\dfrac{x-6}{x^2+5x+6}+\dfrac{9}{x^2+5x+6}=\dfrac{x+3}{x^2+5x+6}$

$=\dfrac{x+3}{(x+3)(x+2)}$

$=\dfrac{\cancel{x+3}}{\cancel{(x+3)}(x+2)}$

$=\dfrac{1}{x+2}$

27. $\dfrac{t^2-5t}{t^2+6t+9}+\dfrac{4t-12}{t^2+6t+9}=\dfrac{t^2-t-12}{t^2+6t+9}$

$=\dfrac{(t-4)(t+3)}{(t+3)^2}$

$=\dfrac{(t-4)\cancel{(t+3)}}{(t+3)\cancel{(t+3)}}$

$=\dfrac{t-4}{t+3}$

29. $\dfrac{2x^2+x}{x^2-8x+12}-\dfrac{x^2-2x+10}{x^2-8x+12}$

$=\dfrac{2x^2+x-(x^2-2x+10)}{x^2-8x+12}$

$=\dfrac{2x^2+x-x^2+2x-10}{x^2-8x+12}$

$=\dfrac{x^2+3x-10}{x^2-8x+12}$

$=\dfrac{(x+5)(x-2)}{(x-6)(x-2)}$

$=\dfrac{(x+5)\cancel{(x-2)}}{(x-6)\cancel{(x-2)}}$

$=\dfrac{x+5}{x-6}$

31. $\dfrac{3-2x}{x^2-6x+8}+\dfrac{7-3x}{x^2-6x+8}$

$=\dfrac{10-5x}{x^2-6x+8}$

$=\dfrac{5(2-x)}{(x-4)(x-2)}$

$=\dfrac{5(-1)(x-2)}{(x-4)(x-2)}$

$=\dfrac{5(-1)\cancel{(x-2)}}{(x-4)\cancel{(x-2)}}$

$=\dfrac{-5}{x-4}$, or $-\dfrac{5}{x-4}$, or $\dfrac{5}{4-x}$

33. $\dfrac{x-9}{x^2+3x-4}-\dfrac{2x-5}{x^2+3x-4}$

$=\dfrac{x-9-(2x-5)}{x^2+3x-4}$

$=\dfrac{x-9-2x+5}{x^2+3x-4}$

$=\dfrac{-x-4}{x^2+3x-4}$

$=\dfrac{-(x+4)}{(x+4)(x-1)}$

$=\dfrac{-1\cancel{(x+4)}}{\cancel{(x+4)}(x-1)}$

$=\dfrac{-1}{x-1}$, or $-\dfrac{1}{x-1}$, or $\dfrac{1}{1-x}$

35. $15=3\cdot 5$

$27=3\cdot 3\cdot 3$

$\text{LCM}=3\cdot 3\cdot 3\cdot 5$, or 135

37. $8=2\cdot 2\cdot 2$

$9=3\cdot 3$

$\text{LCM}=2\cdot 2\cdot 2\cdot 3\cdot 3$, or 72

39. $6=2\cdot 3$

$9=3\cdot 3$

$21=3\cdot 7$

$\text{LCM}=2\cdot 3\cdot 3\cdot 7$, or 126

41. $12x^2 = 2 \cdot 2 \cdot 3 \cdot x \cdot x$

$6x^3 = 2 \cdot 3 \cdot x \cdot x \cdot x$

$\text{LCM} = 2 \cdot 2 \cdot 3 \cdot x \cdot x \cdot x$, or $12x^3$

43. $15a^4b^7 = 3 \cdot 5 \cdot a \cdot a \cdot a \cdot a \cdot b \cdot b \cdot b \cdot b \cdot b \cdot b \cdot b$

$10a^2b^8 = 2 \cdot 5 \cdot a \cdot a \cdot b \cdot b \cdot b \cdot b \cdot b \cdot b \cdot b \cdot b$

$\text{LCM} = 2 \cdot 3 \cdot 5 \cdot a \cdot a \cdot a \cdot a \cdot b \cdot b \cdot b \cdot b \cdot b \cdot b \cdot b \cdot b$, or $30a^4b^8$

45. $2(y-3) = 2 \cdot (y-3)$

$6(y-3) = 2 \cdot 3 \cdot (y-3)$

$\text{LCM} = 2 \cdot 3 \cdot (y-3)$, or $6(y-3)$

47. $x^2 - 4 = (x+2)(x-2)$

$x^2 + 5x + 6 = (x+3)(x+2)$

$\text{LCM} = (x+2)(x-2)(x+3)$

49. $t^3 + 4t^2 + 4t = t(t^2 + 4t + 4) = t(t+2)(t+2)$

$t^2 - 4t = t(t-4)$

$\text{LCM} = t(t+2)(t+2)(t-4) = t(t+2)^2(t-4)$

51. $10x^2y = 2 \cdot 5 \cdot x \cdot x \cdot y$

$6y^2z = 2 \cdot 3 \cdot y \cdot y \cdot z$

$5xz^3 = 5 \cdot x \cdot z \cdot z \cdot z$

$\text{LCM} = 2 \cdot 3 \cdot 5 \cdot x \cdot x \cdot y \cdot y \cdot z \cdot z \cdot z = 30x^2y^2z^3$

53. $a + 1 = a + 1$

$(a-1)^2 = (a-1)(a-1)$

$a^2 - 1 = (a+1)(a-1)$

$\text{LCM} = (a+1)(a-1)(a-1) = (a+1)(a-1)^2$

55. $m^2 - 5m + 6 = (m-3)(m-2)$

$m^2 - 4m + 4 = (m-2)(m-2)$

$\text{LCM} = (m-3)(m-2)(m-2) = (m-3)(m-2)^2$

57. $6x^3 - 24x^2 + 18x = 6x(x^2 - 4x + 3) =$
$2 \cdot 3 \cdot x(x-1)(x-3)$

$4x^5 - 24x^4 + 20x^3 = 4x^3(x^2 - 6x + 5) =$
$2 \cdot 2 \cdot x \cdot x \cdot x(x-1)(x-5)$

$\text{LCM} = 2 \cdot 2 \cdot 3 \cdot x \cdot x \cdot x(x-1)(x-3)(x-5) =$
$12x^3(x-1)(x-3)(x-5)$

59. $2x^3 - 2 = 2(x^3 - 1) = 2(x-1)(x^2 + x + 1)$

$x^2 - 1 = (x+1)(x-1)$

$\text{LCM} = 2(x-1)(x^2 + x + 1)(x+1)$, or
$2(x-1)(x+1)(x^2 + x + 1)$

61. $10a^3 = 2 \cdot 5 \cdot a \cdot a \cdot a$

$5a^6 = 5 \cdot a \cdot a \cdot a \cdot a \cdot a \cdot a$

The LCD is $2 \cdot 5 \cdot a \cdot a \cdot a \cdot a \cdot a \cdot a$, or $10a^6$.

$$\frac{3}{10a^3} \cdot \frac{a^3}{a^3} = \frac{3a^3}{10a^6} \text{ and}$$
$$\frac{b}{5a^6} \cdot \frac{2}{2} = \frac{2b}{10a^6}$$

63. $3x^4y^2 = 3 \cdot x \cdot x \cdot x \cdot x \cdot y \cdot y$

$9xy^3 = 3 \cdot 3 \cdot x \cdot y \cdot y \cdot y$

The LCD is $3 \cdot 3 \cdot x \cdot x \cdot x \cdot x \cdot y \cdot y \cdot y$, or $9x^4y^3$.

$$\frac{7}{3x^4y^2} \cdot \frac{3y}{3y} = \frac{21y}{9x^4y^3} \text{ and}$$
$$\frac{4}{9xy^3} \cdot \frac{x^3}{x^3} = \frac{4x^3}{9x^4y^3}$$

65. The LCD is $(x+2)(x-2)(x+3)$. (See Exercise 47.)

$$\frac{2x}{x^2-4} = \frac{2x}{(x+2)(x-2)} \cdot \frac{x+3}{x+3} = \frac{2x(x+3)}{(x+2)(x-2)(x+3)}$$

$$\frac{4x}{x^2+5x+6} = \frac{4x}{(x+3)(x+2)} \cdot \frac{x-2}{x-2} = \frac{4x(x-2)}{(x+3)(x+2)(x-2)}$$

67. *Writing Exercise*

69. $\dfrac{7}{-9} = -\dfrac{7}{9} = \dfrac{-7}{9}$

71.
$$\frac{5}{18} - \frac{7}{12} = \frac{5}{18} \cdot \frac{2}{2} - \frac{7}{12} \cdot \frac{3}{3} = \frac{10}{36} - \frac{21}{36} = -\frac{11}{36}$$

73. The shaded area has dimensions $x - 6$ by $x - 3$. Then the area is $(x-6)(x-3)$, or $x^2 - 9x + 18$.

75. *Writing Exercise*

77.
$$\frac{6x-1}{x-1} + \frac{3(2x+5)}{x-1} + \frac{3(2x-3)}{x-1} = \frac{6x-1+6x+15+6x-9}{x-1} = \frac{18x+5}{x-1}$$

79.
$$\frac{x^2}{3x^2-5x-2} - \frac{2x}{3x+1} \cdot \frac{1}{x-2}$$
$$= \frac{x^2}{(3x+1)(x-2)} - \frac{2x}{(3x+1)(x-2)}$$
$$= \frac{x^2-2x}{(3x+1)(x-2)}$$
$$= \frac{x(x-2)}{(3x+1)(x-2)}$$
$$= \frac{x}{3x+1}$$

81. The smallest number of strands that can be used is the LCM of 10 and 3.

$10 = 2 \cdot 5$

$3 = 3$

$\text{LCM} = 2 \cdot 5 \cdot 3 = 30$

The smallest number of stands that can be used is 30.

83. If the number of strands must also be a multiple of 4, we find the smallest multiple of 30 that is also a multiple of 4.

$1 \cdot 30 = 30$, not a multiple of 4

$2 \cdot 30 = 60 = 15 \cdot 4$, a multiple of 4

The smallest number of strands that can be used is 60.

85. $6t^2 - 6 = 6(t^2 - 1) = 2 \cdot 3(t+1)(t-1)$

$(3t^2 - 6t + 3)^2 = [3(t^2 - 2t + 1)]^2 = [3(t-1)^2]^2 =$

$3^2(t-1)^4 = 3 \cdot 3(t-1)(t-1)(t-1)(t-1)$

$8t - 8 = 8(t-1) = 2 \cdot 2 \cdot 2(t-1)$

$\text{LCM} = 2 \cdot 2 \cdot 2 \cdot 3 \cdot 3(t+1)(t-1)^4 = 72(t+1)(t-1)^4$

87. The time it takes until the machines begin copying a page at exactly the same time again is the LCM of their copying rates.

$10 = 2 \cdot 5$

$14 = 2 \cdot 7$

$\text{LCM} = 2 \cdot 5 \cdot 7 = 70$

It takes 70 min.

89. The number of minutes after 5:00 A.M. when the shuttles will first leave at the same time again is the LCM of their departure intervals, 25 minutes and 35 minutes.

$25 = 5 \cdot 5$

$35 = 5 \cdot 7$

$\text{LCM} = 5 \cdot 5 \cdot 7$, or 175

Thus, the shuttles will leave at the same time 175 minutes after 5:00 A.M., or at 7:55 A.M.

91. *Writing Exercise*

Exercise Set 6.4

1. To add or subtract when denominators are different, first find the LCD.

3. Add or subtract the numerators, as indicated. Write the sum or difference over the LCD.

5. $\frac{4}{x} + \frac{9}{x^2} = \frac{4}{x} + \frac{9}{x \cdot x}$ $\quad$ LCD $= x \cdot x$, or x^2

$$= \frac{4}{x} \cdot \frac{x}{x} + \frac{9}{x \cdot x}$$

$$= \frac{4x+9}{x^2}$$

7. $\left.\begin{array}{l} 6r = 2 \cdot 3 \cdot r \\ 8r = 2 \cdot 2 \cdot 2 \cdot r \end{array}\right\}$ LCD $= 2 \cdot 2 \cdot 2 \cdot 3 \cdot r$, or $24r$

$$\frac{1}{6r} - \frac{3}{8r} = \frac{1}{6r} \cdot \frac{4}{4} - \frac{3}{8r} \cdot \frac{3}{3}$$

$$= \frac{4-9}{24r}$$

$$= \frac{-5}{24r}, \text{ or } -\frac{5}{24r}$$

9. $\left.\begin{array}{l} c^2d = c \cdot c \cdot d \\ cd^3 = c \cdot d \cdot d \cdot d \end{array}\right\}$ LCD $= c \cdot c \cdot d \cdot d \cdot d$, or c^2d^3

$$\frac{2}{c^2d} + \frac{7}{cd^3} = \frac{2}{c^2d} \cdot \frac{d^2}{d^2} + \frac{7}{cd^3} \cdot \frac{c}{c} = \frac{2d^2 + 7c}{c^2d^3}$$

11. $\left.\begin{array}{l} 3xy^2 = 3 \cdot x \cdot y \cdot y \\ x^2y^3 = x \cdot x \cdot y \cdot y \cdot y \end{array}\right\}$ LCD $= 3 \cdot x \cdot x \cdot y \cdot y \cdot y$, or $3x^2y^3$

$$\frac{-2}{3xy^2} - \frac{6}{x^2y^3} = \frac{-2}{3xy^2} \cdot \frac{xy}{xy} - \frac{6}{x^2y^3} \cdot \frac{3}{3} = \frac{-2xy - 18}{3x^2y^3}$$

13. $\left.\begin{array}{l} 9 = 3 \cdot 3 \\ 6 = 2 \cdot 3 \end{array}\right\}$ LCD $= 3 \cdot 3 \cdot 2$, or 18

$$\frac{x-4}{9} + \frac{x+5}{6} = \frac{x-4}{9} \cdot \frac{2}{2} + \frac{x+5}{6} \cdot \frac{3}{3}$$

$$= \frac{2(x-4) + 3(x+5)}{18}$$

$$= \frac{2x - 8 + 3x + 15}{18}$$

$$= \frac{5x+7}{18}$$

15. $\left.\begin{array}{l} 6 = 2 \cdot 3 \\ 3 = 3 \end{array}\right\}$ LCD $= 2 \cdot 3$, or 6

$$\frac{x-2}{6} - \frac{x+1}{3} = \frac{x-2}{6} - \frac{x+1}{3} \cdot \frac{2}{2}$$

$$= \frac{x-2}{6} - \frac{2x+2}{6}$$

$$= \frac{x - 2 - (2x+2)}{6}$$

$$= \frac{x - 2 - 2x - 2}{6}$$

$$= \frac{-x-4}{6}, \text{ or } \frac{-(x+4)}{6}$$

17. $\left.\begin{array}{l} 16a = 2 \cdot 2 \cdot 2 \cdot 2 \cdot a \\ 4a^2 = 2 \cdot 2 \cdot a \cdot a \end{array}\right\}$ LCD $= 2 \cdot 2 \cdot 2 \cdot 2 \cdot a \cdot a$, or $16a^2$

$$\frac{a+4}{16a} + \frac{3a+4}{4a^2} = \frac{a+4}{16a} \cdot \frac{a}{a} + \frac{3a+4}{4a^2} \cdot \frac{4}{4}$$

$$= \frac{a^2 + 4a}{16a^2} + \frac{12a + 16}{16a^2}$$

$$= \frac{a^2 + 16a + 16}{16a^2}$$

19. $\left.\begin{array}{l} 3z = 3 \cdot z \\ 4z = 2 \cdot 2 \cdot z \end{array}\right\}$ LCD $= 2 \cdot 2 \cdot 3 \cdot z$, or $12z$

$$\frac{4z-9}{3z} - \frac{3z-8}{4z} = \frac{4z-9}{3z} \cdot \frac{4}{4} - \frac{3z-8}{4z} \cdot \frac{3}{3}$$

$$= \frac{16z - 36}{12z} - \frac{9z - 24}{12z}$$

$$= \frac{16z - 36 - (9z - 24)}{12z}$$

$$= \frac{16z - 36 - 9z + 24}{12z}$$

$$= \frac{7z - 12}{12z}$$

21. $\left.\begin{array}{l} xy^2 = x\cdot y\cdot y \\ x^2y = x\cdot x\cdot y \end{array}\right\}\text{LCD} = x\cdot x\cdot y\cdot y, \text{ or } x^2y^2$

$$\frac{x+y}{xy^2}+\frac{3x+y}{x^2y} = \frac{x+y}{xy^2}\cdot\frac{x}{x}+\frac{3x+y}{x^2y}\cdot\frac{y}{y}$$
$$= \frac{x(x+y)+y(3x+y)}{x^2y^2}$$
$$= \frac{x^2+xy+3xy+y^2}{x^2y^2}$$
$$= \frac{x^2+4xy+y^2}{x^2y^2}$$

23. $\left.\begin{array}{l} 3xt^2 = 3\cdot x\cdot t\cdot t \\ x^2t = x\cdot x\cdot t \end{array}\right\}\text{LCD} = 3\cdot x\cdot x\cdot t\cdot t, \text{ or } 3x^2t^2$

$$\frac{4x+2t}{3xt^2}-\frac{5x-3t}{x^2t}$$
$$= \frac{4x+2t}{3xt^2}\cdot\frac{x}{x}-\frac{5x-3t}{x^2t}\cdot\frac{3t}{3t}$$
$$= \frac{4x^2+2tx}{3x^2t^2}-\frac{15xt-9t^2}{3x^2t^2}$$
$$= \frac{4x^2+2tx-(15xt-9t^2)}{3x^2t^2}$$
$$= \frac{4x^2+2tx-15xt+9t^2}{3x^2t^2}$$
$$= \frac{4x^2-13xt+9t^2}{3x^2t^2}$$

(Although $4x^2-13xt+9t^2$ can be factored, doing so will not enable us to simplify the result further.)

25. The denominators cannot be factored, so the LCD is their product, $(x-2)(x+2)$.

$$\frac{3}{x-2}+\frac{3}{x+2} = \frac{3}{x-2}\cdot\frac{x+2}{x+2}+\frac{3}{x+2}\cdot\frac{x-2}{x-2}$$
$$= \frac{3(x+2)+3(x-2)}{(x-2)(x+2)}$$
$$= \frac{3x+6+3x-6}{(x-2)(x+2)}$$
$$= \frac{6x}{(x-2)(x+2)}$$

27. $\dfrac{5}{x+5}-\dfrac{3}{x-5}$ LCD $=(x+5)(x-5)$

$$= \frac{5}{x+5}\cdot\frac{x-5}{x-5}-\frac{3}{x-5}\cdot\frac{x+5}{x+5}$$
$$= \frac{5x-25}{(x+5)(x-5)}-\frac{3x+15}{(x+5)(x-5)}$$
$$= \frac{5x-25-(3x+15)}{(x+5)(x-5)}$$
$$= \frac{5x-25-3x-15}{(x+5)(x-5)}$$
$$= \frac{2x-40}{(x+5)(x-5)}$$

(Although $2x-40$ can be factored, doing so will not enable us to simplify the result further.)

29. $\left.\begin{array}{l} 3x = 3\cdot x \\ x+1 = x+1 \end{array}\right\}\text{LCD} = 3x(x+1)$

$$\frac{3}{x+1}+\frac{2}{3x} = \frac{3}{x+1}\cdot\frac{3x}{3x}+\frac{2}{3x}\cdot\frac{x+1}{x+1}$$
$$= \frac{9x+2(x+1)}{3x(x+1)}$$
$$= \frac{9x+2x+2}{3x(x+1)}$$
$$= \frac{11x+2}{3x(x+1)}$$

31. $\dfrac{3}{2t^2-2t}-\dfrac{5}{2t-2}$

$$= \frac{3}{2t(t-1)}-\frac{5}{2(t-1)} \quad \text{LCD} = 2t(t-1)$$
$$= \frac{3}{2t(t-1)}-\frac{5}{2(t-1)}\cdot\frac{t}{t}$$
$$= \frac{3-5t}{2t(t-1)}$$

33. $\dfrac{2x}{x^2-16}+\dfrac{x}{x-4}$

$$= \frac{2x}{(x+4)(x-4)}+\frac{x}{x-4}$$

LCD $=(x+4)(x-4)$

$$= \frac{2x}{(x+4)(x-4)}+\frac{x}{x-4}\cdot\frac{x+4}{x+4}$$
$$= \frac{2x+x(x+4)}{(x+4)(x-4)}$$
$$= \frac{2x+x^2+4x}{(x+4)(x-4)}$$
$$= \frac{x^2+6x}{(x+4)(x-4)}$$

(Although x^2+6x can be factored, doing so will not enable us to simplify the result further.)

35. $\dfrac{6}{z+4}-\dfrac{2}{3z+12} = \dfrac{6}{z+4}-\dfrac{2}{3(z+4)}$

LCD $=3(z+4)$

$$= \frac{6}{z+4}\cdot\frac{3}{3}-\frac{2}{3(z+4)}$$
$$= \frac{18}{3(z+4)}-\frac{2}{3(z+4)}$$
$$= \frac{16}{3(z+4)}$$

37. $\dfrac{3}{x-1}+\dfrac{2}{(x-1)^2}$ LCD $=(x-1)^2$

$$= \frac{3}{x-1}\cdot\frac{x-1}{x-1}+\frac{2}{(x-1)^2}$$
$$= \frac{3(x-1)+2}{(x-1)^2}$$
$$= \frac{3x-3+2}{(x-1)^2}$$
$$= \frac{3x-1}{(x-1)^2}$$

39. $$\frac{m-3}{m^3-1}-\frac{2}{1-m^3}=\frac{m-3}{m^3-1}-\frac{2}{1-m^3}\cdot\frac{-1}{-1}$$
$$=\frac{m-3}{m^3-1}-\frac{-2}{m^3-1}$$
$$=\frac{m-3-(-2)}{m^3-1}$$
$$=\frac{m-3+2}{m^3-1}$$
$$=\frac{m-1}{m^3-1}$$
$$=\frac{(\cancel{m-1})\cdot 1}{(\cancel{m-1})(m^2+m+1)}$$
$$=\frac{1}{m^2+m+1}$$

41. $$\frac{3a}{4a-20}+\frac{9a}{6a-30}$$
$$=\frac{3a}{2\cdot 2(a-5)}+\frac{9a}{2\cdot 3(a-5)}$$
$$\text{LCD }=2\cdot 2\cdot 3(a-5)$$
$$=\frac{3a}{2\cdot 2(a-5)}\cdot\frac{3}{3}+\frac{9a}{2\cdot 3(a-5)}\cdot\frac{2}{2}$$
$$=\frac{9a+18a}{2\cdot 2\cdot 3(a-5)}$$
$$=\frac{27a}{2\cdot 2\cdot 3(a-5)}$$
$$=\frac{\cancel{3}\cdot 9\cdot a}{2\cdot 3\cdot \cancel{3}(a-5)}$$
$$=\frac{9a}{4(a-5)}$$

43. $$\frac{x}{x-5}+\frac{x}{5-x}=\frac{x}{x-5}+\frac{x}{5-x}\cdot\frac{-1}{-1}$$
$$=\frac{x}{x-5}+\frac{-x}{x-5}$$
$$=0$$

45. $$\frac{6}{a^2+a-2}+\frac{4}{a^2-4a+3}$$
$$=\frac{6}{(a+2)(a-1)}+\frac{4}{(a-3)(a-1)}$$
$$\text{LCD }=(a+2)(a-1)(a-3)$$
$$=\frac{6}{(a+2)(a-1)}\cdot\frac{a-3}{a-3}+\frac{4}{(a-3)(a-1)}\cdot\frac{a+2}{a+2}$$
$$=\frac{6(a-3)+4(a+2)}{(a+2)(a-1)(a-3)}$$
$$=\frac{6a-18+4a+8}{(a+2)(a-1)(a-3)}$$
$$=\frac{10a-10}{(a+2)(a-1)(a-3)}$$
$$=\frac{10(\cancel{a-1})}{(a+2)(\cancel{a-1})(a-3)}$$
$$=\frac{10}{(a+2)(a-3)}$$

47. $$\frac{x}{x^2+9x+20}-\frac{4}{x^2+7x+12}$$
$$=\frac{x}{(x+4)(x+5)}-\frac{4}{(x+3)(x+4)}$$
$$\text{LCD }=(x+3)(x+4)(x+5)$$
$$=\frac{x}{(x+4)(x+5)}\cdot\frac{x+3}{x+3}-\frac{4}{(x+3)(x+4)}\cdot\frac{x+5}{x+5}$$
$$=\frac{x(x+3)-4(x+5)}{(x+3)(x+4)(x+5)}$$
$$=\frac{x^2+3x-4x-20}{(x+3)(x+4)(x+5)}$$
$$=\frac{x^2-x-20}{(x+3)(x+4)(x+5)}$$
$$=\frac{(\cancel{x+4})(x-5)}{(x+3)(\cancel{x+4})(x+5)}$$
$$=\frac{x-5}{(x+3)(x+5)}$$

49. $$\frac{3z}{z^2-4x+4}+\frac{10}{z^2+z-6}$$
$$=\frac{3z}{(z-2)^2}+\frac{10}{(z-2)(z+3)},$$
$$\text{LCD }=(z-2)^2(z+3)$$
$$=\frac{3z}{(z-2)^2}\cdot\frac{z+3}{z+3}+\frac{10}{(z-2)(z+3)}\cdot\frac{z-2}{z-2}$$
$$=\frac{3z(z+3)+10(z-2)}{(x-2)^2(z+3)}$$
$$=\frac{3z^2+9z+10z-20}{(z-2)^2(z+3)}$$
$$=\frac{3z^2+19z-20}{(z-2)^2(z+3)}$$

51. $$\frac{-5}{x^2+17x+16}-\frac{0}{x^2+9x+8}$$

Note that $\frac{0}{x^2+9x+8}=0$, so the difference is $\frac{-5}{x^2+17x+16}$.

53. $$\frac{4x}{5}-\frac{x-3}{-5}=\frac{4x}{5}-\frac{x-3}{-5}\cdot\frac{-1}{-1}$$
$$=\frac{4x}{5}-\frac{3-x}{5}$$
$$=\frac{4x-(3-x)}{5}$$
$$=\frac{4x-3+x}{5}$$
$$=\frac{5x-3}{5}$$

55. $$\frac{y^2}{y-3}+\frac{9}{3-y}=\frac{y^2}{y-3}+\frac{9}{3-y}\cdot\frac{-1}{-1}$$
$$=\frac{y^2}{y-3}+\frac{-9}{-3+y}$$
$$=\frac{y^2-9}{y-3}$$
$$=\frac{(y+3)(y-3)}{y-3}$$
$$=y+3$$

57. $$\frac{b-7}{b^2-16}+\frac{7-b}{16-b^2}=\frac{b-7}{b^2-16}+\frac{7-b}{16-b^2}\cdot\frac{-1}{-1}$$
$$=\frac{b-7}{b^2-16}+\frac{b-7}{b^2-16}$$
$$=\frac{2b-14}{b^2-16}$$

(Although both $2b-14$ and b^2-16 can be factored, doing so will not enable us to simplify the result further.)

59. $$\frac{4-p}{25-p^2}+\frac{p+1}{p-5}$$
$$=\frac{4-p}{(5+p)(5-p)}+\frac{p+1}{p-5}$$
$$=\frac{4-p}{(5+p)(5-p)}\cdot\frac{-1}{-1}+\frac{p+1}{p-5}$$
$$=\frac{p-4}{(p+5)(p-5)}+\frac{p+1}{p-5}\quad \text{LCD}=(p+5)(p-5)$$
$$=\frac{p-4}{(p+5)(p-5)}+\frac{p+1}{p-5}\cdot\frac{p+5}{p+5}$$
$$=\frac{p-4+p^2+6p+5}{(p+5)(p-5)}$$
$$=\frac{p^2+7p+1}{(p+5)(p-5)}$$

61. $$\frac{8x}{16-x^2}-\frac{5}{x-4}$$
$$=\frac{8x}{(4+x)(4-x)}-\frac{5}{x-4}$$
$$=\frac{8x}{(4+x)(4-x)}-\frac{5}{x-4}\cdot\frac{-1}{-1}$$
$$=\frac{8x}{(4+x)(4-x)}-\frac{-5}{4-x}\quad \text{LCD}=(4+x)(4-x)$$
$$=\frac{8x}{(4+x)(4-x)}-\frac{-5}{4-x}\cdot\frac{4+x}{4+x}$$
$$=\frac{8x-(-5)(4+x)}{(4+x)(4-x)}$$
$$=\frac{8x+20+5x}{(4+x)(4-x)}$$
$$=\frac{13x+20}{(4+x)(4-x)},\text{ or }\frac{-13x-20}{(x+4)(x-4)}$$

63. $$\frac{a}{a^2-1}+\frac{2a}{a-a^2}=\frac{a}{a^2-1}+\frac{2\cdot a}{a(1-a)}$$
$$=\frac{a}{(a+1)(a-1)}+\frac{2}{1-a}$$
$$=\frac{a}{(a+1)(a-1)}+\frac{2}{1-a}\cdot\frac{-1}{-1}$$
$$=\frac{a}{(a+1)(a-1)}+\frac{-2}{a-1}$$
$$\text{LCD}=(a+1)(a-1)$$
$$=\frac{a}{(a+1)(a-1)}+\frac{-2}{a-1}\cdot\frac{a+1}{a+1}$$
$$=\frac{a-2a-2}{(a+1)(a-1)}$$
$$=\frac{-a-2}{(a+1)(a-1)},\text{ or}$$
$$=\frac{a+2}{(1+a)(1-a)}$$

65. $$\frac{4x}{x^2-y^2}-\frac{6}{y-x}$$
$$=\frac{4x}{(x+y)(x-y)}-\frac{6}{y-x}$$
$$=\frac{4x}{(x+y)(x-y)}-\frac{6}{y-x}\cdot\frac{-1}{-1}$$
$$=\frac{4x}{(x+y)(x-y)}-\frac{-6}{x-y}\quad \text{LCD}=(x+y)(x-y)$$
$$=\frac{4x}{(x+y)(x-y)}-\frac{-6}{x-y}\cdot\frac{x+y}{x+y}$$
$$=\frac{4x-(-6)(x+y)}{(x+y)(x-y)}$$
$$=\frac{4x+6x+6y}{(x+y)(x-y)}$$
$$=\frac{10x+6y}{(x+y)(x-y)}$$

(Although $10x+6y$ can be factored, doing so will not enable us to simplify the result further.)

67. $$\frac{x-3}{2-x}-\frac{x+3}{x+2}+\frac{x+6}{4-x^2}$$
$$=\frac{x-3}{2-x}-\frac{x+3}{x+2}+\frac{x+6}{(2+x)(2-x)}$$
$$\text{LCD}=(2+x)(2-x)$$
$$=\frac{x-3}{2-x}\cdot\frac{2+x}{2+x}-\frac{x+3}{x+2}\cdot\frac{2-x}{2-x}+\frac{x+6}{(2+x)(2-x)}$$
$$=\frac{(x-3)(2+x)-(x+3)(2-x)+(x+6)}{(2+x)(2-x)}$$
$$=\frac{x^2-x-6-(-x^2-x+6)+x+6}{(2+x)(2-x)}$$
$$=\frac{x^2-x-6+x^2+x-6+x+6}{(2+x)(2-x)}$$
$$=\frac{2x^2+x-6}{(2+x)(2-x)}$$
$$=\frac{(2x-3)(x+2)}{(2+x)(2-x)}$$
$$=\frac{2x-3}{2-x}$$

69. $\frac{x+5}{x+3}+\frac{x+7}{x+2}-\frac{7x+19}{(x+3)(x+2)}$

LCD is $(x+3)(x+2)$

$=\frac{x+5}{x+3}\cdot\frac{x+2}{x+2}+\frac{x+7}{x+2}\cdot\frac{x+3}{x+3}-\frac{7x+19}{(x+3)(x+2)}$

$=\frac{(x+5)(x+2)+(x+7)(x+3)-(7x+19)}{(x+3)(x+2)}$

$=\frac{x^2+7x+10+x^2+10x+21-7x-19}{(x+3)(x+2)}$

$=\frac{2x^2+10x+12}{(x+3)(x+2)}$

$=\frac{2(x^2+5x+6)}{(x+3)(x+2)}$

$=\frac{2\cancel{(x+3)}\cancel{(x+2)}}{\cancel{(x+3)}\cancel{(x+2)}}$

$=2$

71. $\frac{1}{x+y}+\frac{1}{x-y}-\frac{2x}{x^2-y^2}$

LCD $=(x+y)(x-y)$

$=\frac{1}{x+y}\cdot\frac{x-y}{x-y}+\frac{1}{x-y}\cdot\frac{x+y}{x+y}-\frac{2x}{(x+y)(x-y)}$

$=\frac{(x-y)+(x+y)-2x}{(x+y)(x-y)}$

$=0$

73. *Writing Exercise*

75. $-\frac{3}{7}\div\frac{6}{13}=-\frac{3}{7}\cdot\frac{13}{6}$

$=-\frac{\cancel{3}\cdot 13}{7\cdot 2\cdot\cancel{3}}$

$=-\frac{13}{14}$

77. $\frac{\frac{2}{9}}{\frac{5}{3}}=\frac{2}{9}\div\frac{5}{3}$

$=\frac{2}{9}\cdot\frac{3}{5}$

$=\frac{2\cdot\cancel{3}}{\cancel{3}\cdot 3\cdot 5}$

$=\frac{2}{15}$

79. Graph: $y=-\frac{1}{2}x-5$

Since the equation is in the form $y=mx+b$, we know the y–intercept is $(0,-5)$. We find two other solutions, substituting multiples of 2 for x to avoid fractions.

When $x=-2$, $y=-\frac{1}{2}(-2)-5=1-5=-4$.

When $x=-4$, $y=-\frac{1}{2}(-4)-5=2-5=-3$.

x	y
0	−5
−2	−4
−4	−3

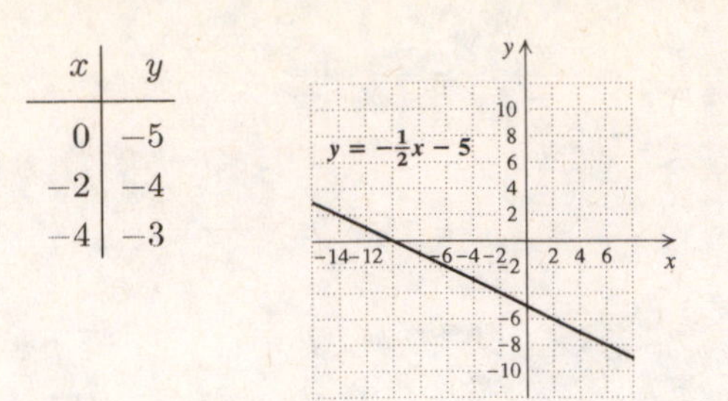

81. *Writing Exercise*

83. $P=2\left(\frac{3}{x+4}\right)+2\left(\frac{2}{x-5}\right)$

$=\frac{6}{x+4}+\frac{4}{x-5}$ LCD $=(x+4)(x-5)$

$=\frac{6}{x+4}\cdot\frac{x-5}{x-5}+\frac{4}{x-5}\cdot\frac{x+4}{x+4}$

$=\frac{6x-30+4x+16}{(x+4)(x-5)}$

$=\frac{10x-14}{(x+4)(x-5)}$, or $\frac{10x-14}{x^2-x-20}$

$A=\left(\frac{3}{x+4}\right)\left(\frac{2}{x-5}\right)=\frac{6}{(x+4)(x-5)}$

85. $\frac{2x+11}{x-3}\cdot\frac{3}{x+4}+\frac{2x+1}{4+x}\cdot\frac{3}{3-x}$

$=\frac{6x+33}{(x-3)(x+4)}+\frac{6x+3}{(4+x)(3-x)}$

$=\frac{6x+33}{(x-3)(x+4)}+\frac{6x+3}{(4+x)(3-x)}\cdot\frac{-1}{-1}$

$=\frac{6x+33}{(x-3)(x+4)}+\frac{-6x-3}{(x+4)(x-3)}$

$=\frac{6x+33-6x-3}{(x-3)(x+4)}$

$=\frac{30}{(x-3)(x+4)}$

87. We recognize that this is the product of the sum and difference of two terms: $(A+B)(A-B)=A^2-B^2$.

$\left(\frac{x}{x+7}-\frac{3}{x+2}\right)\left(\frac{x}{x+7}+\frac{3}{x+2}\right)$

$=\frac{x^2}{(x+7)^2}-\frac{9}{(x+2)^2}$ LCD $=(x+7)^2(x+2)^2$

$=\frac{x^2}{(x+7)^2}\cdot\frac{(x+2)^2}{(x+2)^2}-\frac{9}{(x+2)^2}\cdot\frac{(x+7)^2}{(x+7)^2}$

$=\frac{x^2(x+2)^2-9(x+7)^2}{(x+7)^2(x+2)^2}$

$=\frac{x^2(x^2+4x+4)-9(x^2+14x+49)}{(x+7)^2(x+2)^2}$

$=\frac{x^4+4x^3+4x^2-9x^2-126x-441}{(x+7)^2(x+2)^2}$

$=\frac{x^4+4x^3-5x^2-126x-441}{(x+7)^2(x+2)^2}$

89. $\dfrac{2x^2+5x-3}{2x^2-9x+9}+\dfrac{x+1}{3-2x}+\dfrac{4x^2+8x+3}{x-3}\cdot\dfrac{x+3}{9-4x^2}$

$=\dfrac{2x^2+5x-3}{(2x-3)(x-3)}+\dfrac{x+1}{3-2x}+\dfrac{(4x^2+8x+3)(x+3)}{(x-3)(3+2x)(3-2x)}$

$=\dfrac{2x^2+5x-3}{(2x-3)(x-3)}\cdot\dfrac{-1}{-1}+\dfrac{x+1}{3-2x}+\dfrac{4x^3+20x^2+27x+9}{(x-3)(3+2x)(3-2x)}$

$=\dfrac{-2x^2-5x+3}{(3-2x)(x-3)}+\dfrac{x+1}{3-2x}+\dfrac{4x^3+20x^2+27x+9}{(x-3)(3+2x)(3-2x)}$

LCD $=(x-3)(3+2x)(3-2x)$

$=\dfrac{-2x^2-5x+3}{(3-2x)(x-3)}\cdot\dfrac{3+2x}{3+2x}+\dfrac{x+1}{3-2x}\cdot\dfrac{(x-3)(3+2x)}{(x-3)(3+2x)}+\dfrac{4x^3+20x^2+27x+9}{(x-3)(3+2x)(3-2x)}$

$=[(-4x^3-16x^2-9x+9+2x^3-x^2-12x-9+4x^3+20x^2+27x+9)]/[(x-3)(3+2x)(3-2x)]$

$=\dfrac{2x^3+3x^2+6x+9}{(x-3)(3+2x)(3-2x)}$

$=\dfrac{x^2(2x+3)+3(2x+3)}{(x-3)(3+2x)(3-2x)}$

$=\dfrac{(2x+3)(x^2+3)}{(x-3)(3+2x)(3-2x)}$

$=\dfrac{x^2+3}{(x-3)(3-2x)}$, or $\dfrac{-x^2-3}{(x-3)(2x-3)}$

91. Answers may vary. $\dfrac{a}{a-b}+\dfrac{3b}{b-a}$

93. *Writing Exercise*

Exercise Set 6.5

1. The LCD is the LCM of x^2, x, 2, and $4x$. It is $4x^2$.

$\dfrac{\frac{5}{x^2}+\frac{1}{x}}{\frac{7}{2}-\frac{3}{4x}}\cdot\dfrac{4x^2}{4x^2}=\dfrac{\frac{5}{x^2}\cdot 4x^2+\frac{1}{x}\cdot 4x^2}{\frac{7}{2}\cdot 4x^2-\frac{3}{4x}\cdot 4x^2}$

Choice (d) is correct.

3. We subtract to get a single rational expression in the numerator and add to get a single rational expression in the denominator.

$\dfrac{\frac{4}{5x}-\frac{1}{10}}{\frac{8}{x^2}+\frac{7}{2}}=\dfrac{\frac{4}{5x}\cdot\frac{2}{2}-\frac{1}{10}\cdot\frac{x}{x}}{\frac{8}{x^2}\cdot\frac{2}{2}+\frac{7}{2}\cdot\frac{x^2}{x^2}}$

$=\dfrac{\frac{8}{10x}-\frac{x}{10x}}{\frac{16}{2x^2}+\frac{7x^2}{2x^2}}$

$=\dfrac{\frac{8-x}{10x}}{\frac{16+7x^2}{2x^2}}$

Choice (b) is correct.

5. $\dfrac{1+\frac{1}{2}}{1+\frac{1}{4}}$ LCD is 4

$=\dfrac{1+\frac{1}{2}}{1+\frac{1}{4}}\cdot\dfrac{4}{4}$ Multiplying by $\frac{4}{4}$

$=\dfrac{\left(1+\frac{1}{2}\right)4}{\left(1+\frac{1}{4}\right)4}$ Multiplying numerator and denominator by 4

$=\dfrac{1\cdot 4+\frac{1}{2}\cdot 4}{1\cdot 4+\frac{1}{4}\cdot 4}$

$=\dfrac{4+2}{4+1}$

$=\dfrac{6}{5}$

7. $\dfrac{4+\frac{1}{3}}{1-\frac{5}{27}}$

$=\dfrac{4\cdot\frac{3}{3}+\frac{1}{3}}{1\cdot\frac{27}{27}-\frac{5}{27}}$ Getting a common denominator in numerator and in denominator

$=\dfrac{\frac{12}{3}+\frac{1}{3}}{\frac{27}{27}-\frac{5}{27}}$

$=\dfrac{\frac{13}{3}}{\frac{22}{27}}$ Adding in the numerator; subtracting in the denominator

$=\dfrac{13}{3}\cdot\dfrac{27}{22}$ Multiplying by the reciprocal of the divisor

$=\dfrac{13\cdot 3\cdot 9}{3\cdot 22}$

$=\dfrac{13\cdot \cancel{3}\cdot 9}{\cancel{3}\cdot 22}$

$=\dfrac{117}{22}$

9. $\dfrac{\dfrac{s}{3}+s}{\dfrac{3}{s}+s}$ LCD is $3s$

$$= \frac{\dfrac{s}{3}+s}{\dfrac{3}{s}+s}\cdot\frac{3s}{3s}$$

$$= \frac{\left(\dfrac{s}{3}+s\right)(3s)}{\left(\dfrac{3}{s}+s\right)(3s)}$$

$$= \frac{\dfrac{5}{3}\cdot 3s+s\cdot 3s}{\dfrac{3}{s}\cdot 3s+s\cdot 3s}$$

$$= \frac{s^2+3s^2}{9+3s^2}$$

$$= \frac{4s^2}{9+3s^2}$$

11. $\dfrac{\dfrac{4}{x}}{\dfrac{3}{x}+\dfrac{2}{x^2}}$ LCD is x^2

$$= \frac{\dfrac{4}{x}}{\dfrac{3}{x}+\dfrac{2}{x^2}}\cdot\frac{x^2}{x^2}$$

$$= \frac{\dfrac{4}{x}\cdot x^2}{\left(\dfrac{3}{x}+\dfrac{2}{x^2}\right)x^2}$$

$$= \frac{4x}{\dfrac{3}{x}\cdot x^2+\dfrac{2}{x^2}\cdot x^2}$$

$$= \frac{4x}{3x+2}$$

13. $\dfrac{\dfrac{2a-5}{3a}}{\dfrac{a-7}{6a}}$

$$= \frac{2a-5}{3a}\cdot\frac{6a}{a-7}$$ Multiplying by the reciprocal of the divisor

$$= \frac{(2a-5)\cdot 2\cdot 3a}{3a\cdot(a-7)}$$

$$= \frac{(2a-5)\cdot 2\cdot \cancel{3a}}{\cancel{3a}\cdot(a-7)}$$

$$= \frac{2(2a-5)}{a-7}$$

$$= \frac{4a-10}{a-7}$$

15. $\dfrac{\dfrac{x}{4}-\dfrac{4}{x}}{\dfrac{1}{4}+\dfrac{1}{x}}$ LCD is $4x$

$$= \frac{\dfrac{x}{4}-\dfrac{4}{x}}{\dfrac{1}{4}+\dfrac{1}{x}}\cdot\frac{4x}{4x}$$

$$= \frac{\dfrac{x}{4}\cdot 4x-\dfrac{4}{x}\cdot 4x}{\dfrac{1}{4}\cdot 4x+\dfrac{1}{x}\cdot 4x}$$

$$= \frac{x^2-16}{x+4}$$

$$= \frac{(x+4)(x-4)}{x+4}$$

$$= \frac{\cancel{(x+4)}(x-4)}{\cancel{(x+4)}\cdot 1}$$

$$= x-4$$

17. $\dfrac{\dfrac{1}{6}-\dfrac{1}{x}}{\dfrac{6-x}{6}}$ LCD is $6x$

$$= \frac{\dfrac{1}{6}-\dfrac{1}{x}}{\dfrac{6-x}{6}}\cdot\frac{6x}{6x}$$

$$= \frac{\dfrac{1}{6}\cdot 6x-\dfrac{1}{x}\cdot 6x}{\left(\dfrac{6-x}{6}\right)(6x)}$$

$$= \frac{x-6}{(6-x)(x)}$$

$$= \frac{x-6}{-(x-6)(x)}$$ $(6-x=-1(-6+x)=-(x-6))$

$$= \frac{\cancel{(x-6)}\cdot 1}{-\cancel{(x-6)}(x)}$$

$$= \frac{1}{-x} = -\frac{1}{x}$$

19. $\dfrac{\dfrac{1}{t^2}+1}{\dfrac{1}{t}-1}$ LCD is t^2

$$= \frac{\dfrac{1}{t^2}+1}{\dfrac{1}{t}-1}\cdot\frac{t^2}{t^2}$$

$$= \frac{\dfrac{1}{t^2}\cdot t^2+1\cdot t^2}{\dfrac{1}{t}\cdot t^2-1\cdot t^2}$$

$$= \frac{1+t^2}{t-t^2}$$

(Although the denominator can be factored, doing so will not enable us to simplify further.)

21. $\dfrac{\frac{x^2}{x^2-y^2}}{\frac{x}{x+y}}$

$= \dfrac{x^2}{x^2-y^2}\cdot\dfrac{x+y}{x}$ Multiplying by the reciprocal of the divisor

$= \dfrac{x^2(x+y)}{(x^2-y^2)(x)}$

$= \dfrac{x\cdot x\cdot(x+y)}{(x+y)(x-y)(x)}$

$= \dfrac{\cancel{x}\cdot x\cdot\cancel{(x+y)}}{\cancel{(x+y)}(x-y)(\cancel{x})}$

$= \dfrac{x}{x-y}$

23. $\dfrac{\frac{7}{a^2}+\frac{2}{a}}{\frac{5}{a^3}-\frac{3}{a}}$ LCD is a^3

$= \dfrac{\frac{7}{a^2}+\frac{2}{a}}{\frac{5}{a^3}-\frac{3}{a}}\cdot\dfrac{a^3}{a^3}$

$= \dfrac{\frac{7}{a^2}\cdot a^3+\frac{2}{a}\cdot a^3}{\frac{5}{a^3}\cdot a^3-\frac{3}{a}\cdot a^3}$

$= \dfrac{7a+2a^2}{5-3a^2}$

(Although the numerator can be factored, doing so will not enable us to simplify further.)

25. $\dfrac{\frac{x}{5y^3}+\frac{3}{10y}}{\frac{3}{10y}+\frac{x}{5y^3}}$

Observe that, by the commutative law of addition, the numerator and denominator are equivalent, so the result is 1.

27. $\dfrac{\frac{3}{ab^4}+\frac{4}{a^3b}}{\frac{5}{a^3b}-\frac{3}{ab}} = \dfrac{\frac{3}{ab^4}\cdot\frac{a^2}{a^2}+\frac{4}{a^3b}\cdot\frac{b^3}{b^3}}{\frac{5}{a^3b}-\frac{3}{ab}\cdot\frac{a^2}{a^2}}$

$= \dfrac{\frac{3a^2+4b^3}{a^3b^4}}{\frac{5-3a^2}{a^3b}}$

$= \dfrac{3a^2+4b^3}{a^3b^4}\cdot\dfrac{a^3b}{5-3a^2}$

$= \dfrac{\cancel{a^3b}(3a^2+4b^3)}{\cancel{a^3b}\cdot b^3(5-3a^2)}$

$= \dfrac{3a^2+4b^3}{b^3(5-3a^2)}$, or $\dfrac{3a^2+4b^3}{5b^3-3a^2b^3}$

29. $\dfrac{2-\frac{3}{x^2}}{2+\frac{3}{x^4}} = \dfrac{2-\frac{3}{x^2}}{2+\frac{3}{x^4}}\cdot\dfrac{x^4}{x^4}$

$= \dfrac{2\cdot x^4-\frac{3}{x^2}\cdot x^4}{2\cdot x^4+\frac{3}{x^4}\cdot x^4}$

$= \dfrac{2x^4-3x^2}{2x^4+3}$

31. $\dfrac{t-\frac{2}{t}}{t+\frac{5}{t}} = \dfrac{t\cdot\frac{t}{t}-\frac{2}{t}}{t\cdot\frac{t}{t}+\frac{5}{t}}$

$= \dfrac{\frac{t^2-2}{t}}{\frac{t^2+5}{t}}$

$= \dfrac{t^2-2}{t}\cdot\dfrac{t}{t^2+5}$

$= \dfrac{\cancel{t}(t^2-2)}{\cancel{t}(t^2+5)}$

$= \dfrac{t^2-2}{t^2+5}$

33. $\dfrac{\frac{1}{a}+\frac{1}{b}}{\frac{1}{a^3}+\frac{1}{b^3}}$ LCD is a^3b^3

$= \dfrac{\frac{1}{a}+\frac{1}{b}}{\frac{1}{a^3}+\frac{1}{b^3}}\cdot\dfrac{a^3b^3}{a^3b^3}$

$= \dfrac{\frac{1}{a}\cdot a^3b^3+\frac{1}{b}\cdot a^3b^3}{\frac{1}{a^3}\cdot a^3b^3+\frac{1}{b^3}\cdot a^3b^3}$

$= \dfrac{a^2b^3+a^3b^2}{b^3+a^3}$

$= \dfrac{a^2b^2\cancel{(b+a)}}{\cancel{(b+a)}(b^2-ba+a^2)}$

$= \dfrac{a^2b^2}{b^2-ba+a^2}$

35. $\dfrac{3+\frac{4}{ab^3}}{\frac{3+a}{a^2b}} = \dfrac{3+\frac{4}{ab^3}}{\frac{3+a}{a^2b}}\cdot\dfrac{a^2b^3}{a^2b^3}$

$= \dfrac{3\cdot a^2b^3+\frac{4}{ab^3}\cdot a^2b^3}{\frac{3+a}{a^2b}\cdot a^2b^3}$

$= \dfrac{3a^2b^3+4a}{b^2(3+a)}$, or $\dfrac{3a^2b^3+4a}{3b^2+ab^2}$

37. $\dfrac{t+5+\dfrac{3}{t}}{t+2+\dfrac{1}{t}}$ LCD is t

$$= \dfrac{t+5+\dfrac{3}{t}}{t+2+\dfrac{1}{t}} \cdot \dfrac{t}{t}$$

$$= \dfrac{t\cdot t+5\cdot t+\dfrac{3}{t}\cdot t}{t\cdot t+2\cdot t+\dfrac{1}{t}\cdot t}$$

$$= \dfrac{t^2+5t+3}{t^2+2t+1}$$

$$= \dfrac{t^2+5t+3}{(t+1)^2}$$

39. $\dfrac{x-2-\dfrac{1}{x}}{x-5-\dfrac{4}{x}} = \dfrac{x-2-\dfrac{1}{x}}{x-5-\dfrac{4}{x}} \cdot \dfrac{x}{x}$

$$= \dfrac{x\cdot x-2\cdot x-\dfrac{1}{x}\cdot x}{x\cdot x-5\cdot x-\dfrac{4}{x}\cdot x}$$

$$= \dfrac{x^2-2x-1}{x^2-5x-4}$$

41. *Writing Exercise*

43. $3x-5+2(4x-1) = 12x-3$

$$3x-5+8x-2 = 12x-3$$
$$11x-7 = 12x-3$$
$$-7 = x-3$$
$$-4 = x$$

The solution is -4.

45. $\dfrac{3}{4}x - \dfrac{5}{8} = \dfrac{3}{8}x + \dfrac{7}{4}$ LCD is 8

$$8\left(\frac{3}{4}x - \frac{5}{8}\right) = 8\left(\frac{3}{8}x + \frac{7}{4}\right)$$
$$8\cdot\frac{3}{4}x - 8\cdot\frac{5}{8} = 8\cdot\frac{3}{8}x + 8\cdot\frac{7}{4}$$
$$6x-5 = 3x+14$$
$$3x-5 = 14$$
$$3x = 19$$
$$x = \frac{19}{3}$$

The solution is $\dfrac{19}{3}$.

47. $x^2-7x-30 = 0$

$$(x-10)(x+3) = 0$$
$$x-10 = 0 \quad or \quad x+3 = 0$$
$$x = 10 \quad or \quad x = -3$$

The solutions are 10 and -3.

49. *Writing Exercise*

51. $\dfrac{\dfrac{x-5}{x-6}}{\dfrac{x-7}{x-8}}$

This expression is undefined for any value of x that makes a denominator 0. We see that $x-6=0$ when $x=6$, $x-7=0$ when $x=7$, and $x-8=0$ when $x=8$, so the expression is undefined for the x-values 6, 7, and 8.

53. $\dfrac{\dfrac{2x+3}{5x+4}}{\dfrac{3}{7}-\dfrac{2x}{21}}$

This expression is undefined for any value of x that makes a denominator 0. First we find the value of x for which $5x+4=0$.

$$5x+4 = 0$$
$$5x = -4$$
$$x = -\frac{4}{5}$$

Then we find the value of x for which $\dfrac{3}{7}-\dfrac{2x}{21}=0$:

$$\frac{3}{7}-\frac{2x}{21} = 0$$
$$21\left(\frac{3}{7}-\frac{2x}{21}\right) = 21\cdot 0$$
$$21\cdot\frac{3}{7} - 21\cdot\frac{2x}{21} = 0$$
$$9-2x = 0$$
$$9 = 2x$$
$$\frac{9}{2} = x$$

The expression is undefined for the x-values $-\dfrac{4}{5}$ and $\dfrac{9}{2}$.

55. $$\frac{\dfrac{P\left(1+\dfrac{i}{12}\right)^2}{\left(1+\dfrac{1}{12}\right)^2-1}}{\dfrac{i}{12}} = \frac{\dfrac{P\left(1+\dfrac{i}{6}+\dfrac{i^2}{144}\right)}{\left(1+\dfrac{i}{6}+\dfrac{i^2}{144}\right)-1}}{\dfrac{i}{12}}$$

$$= \frac{\dfrac{P\left(1+\dfrac{i}{6}+\dfrac{i^2}{144}\right)}{\dfrac{i}{6}+\dfrac{i^2}{144}}}{\dfrac{i}{12}}$$

$$= \frac{P\left(1+\dfrac{i}{6}+\dfrac{i^2}{144}\right)}{\left(\dfrac{i}{6}+\dfrac{i^2}{144}\right)\left(\dfrac{12}{i}\right)}$$

$$= \frac{P\left(1+\dfrac{i}{6}+\dfrac{i^2}{144}\right)}{2+\dfrac{i}{12}}$$

$$= \frac{P\left(1+\dfrac{i}{6}+\dfrac{i^2}{144}\right)}{2+\dfrac{i}{12}}\cdot\frac{144}{144}$$

$$= \frac{144P\left(1+\dfrac{i}{6}+\dfrac{i^2}{144}\right)}{144\left(2+\dfrac{i}{12}\right)}$$

$$= \frac{P(144+24i+i^2)}{288+12i}$$

$$= \frac{P(12+i)^2}{12(24+i)}, \text{ or}$$

$$\frac{P(i+12)^2}{12(i+24)}$$

57. $$\frac{\dfrac{5}{x+2}-\dfrac{3}{x-2}}{\dfrac{x}{x-1}+\dfrac{x}{x+1}} = \frac{\dfrac{5}{x+2}\cdot\dfrac{x-2}{x-2}-\dfrac{3}{x-2}\cdot\dfrac{x+2}{x+2}}{\dfrac{x}{x-1}\cdot\dfrac{x+1}{x+1}+\dfrac{x}{x+1}\cdot\dfrac{x-1}{x-1}}$$

$$= \frac{\dfrac{5(x-2)-3(x+2)}{(x+2)(x-2)}}{\dfrac{x(x+1)+x(x-1)}{(x+1)(x-1)}}$$

$$= \frac{\dfrac{5x-10-3x-6}{(x+2)(x-2)}}{\dfrac{x^2+x+x^2-x}{(x+1)(x-1)}}$$

$$= \frac{\dfrac{2x-16}{(x+2)(x-2)}}{\dfrac{2x^2}{(x+1)(x-1)}}$$

$$= \frac{2x-16}{(x+2)(x-2)}\cdot\frac{(x+1)(x-1)}{2x^2}$$

$$= \frac{\cancel{2}(x-8)(x+1)(x-1)}{\cancel{2}\cdot x^2(x+2)(x-2)}$$

$$= \frac{(x-8)(x+1)(x-1)}{x^2(x+2)(x-2)}$$

59. $$\left[\frac{\dfrac{x-1}{x-1}-1}{\dfrac{x+1}{x-1}+1}\right]^5$$

Consider the numerator of the complex rational expression:

$$\frac{x-1}{x-1}-1 = 1-1 = 0$$

Since the denominator, $\dfrac{x+1}{x-1}+1$ is not equal to 0, the simplified form of the original expression is 0.

61. $$\frac{\dfrac{z}{1-\dfrac{z}{2+2z}}-2z}{\dfrac{2z}{5z-2}-3} = \frac{\dfrac{z}{\dfrac{2+2z-z}{2+2z}}-2z}{\dfrac{2z-15z+6}{5z-2}}$$

$$= \frac{\dfrac{z}{\dfrac{2+z}{2+2z}}-2z}{\dfrac{-13z+6}{5z-2}}$$

$$= \frac{z\cdot\dfrac{2+2z}{2+z}-2z}{\dfrac{-13z+6}{5z-2}}$$

$$= \frac{\dfrac{z(2+2z)-2z(2+z)}{2+z}}{\dfrac{-13z+6}{5z-2}}$$

$$= \frac{\dfrac{2z+2z^2-4z-2z^2}{2+z}}{\dfrac{-13z+6}{5z-2}}$$

$$= \frac{\dfrac{-2z}{2+z}}{\dfrac{-13z+6}{5z-2}}$$

$$= \frac{-2z}{2+z}\cdot\frac{5z-2}{-13z+6}$$

$$= \frac{-2z(5z-2)}{(2+z)(-13z+6)}, \text{ or}$$

$$\frac{2z(5z-2)}{(2+z)(13z-6)}$$

63. 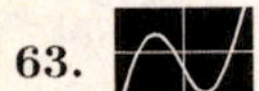

Exercise Set 6.6

1. The statement is false. See Example 2(c).

3. The statement is true. See page 414 in the text.

5. Because no variable appears in a denominator, no restrictions exist.

$$\frac{3}{5}-\frac{5}{8}=\frac{x}{20}, \text{ LCD} = 40$$

$$40\left(\frac{3}{5}-\frac{5}{8}\right) = 40\cdot\frac{x}{20}$$

$$40\cdot\frac{3}{5}-40\cdot\frac{5}{8} = 40\cdot\frac{x}{20}$$

$$24-25 = 2x$$

$$-1 = 2x$$

$$-\frac{1}{2} = x$$

Check:

$\frac{3}{5}-\frac{5}{8}=\frac{x}{20}$	
$\frac{3}{5}-\frac{5}{8}$	$\dfrac{-\frac{1}{2}}{20}$
$\frac{24}{40}-\frac{25}{40}$	$-\frac{1}{2}\cdot\frac{1}{20}$
$-\frac{1}{40}$	$\overset{?}{=}\ -\frac{1}{40}$ TRUE

This checks, so the solution is $-\frac{1}{2}$.

7. Note that x cannot be 0.

$$\frac{1}{3}+\frac{5}{6}=\frac{1}{x}, \text{ LCD} = 6x$$

$$6x\left(\frac{1}{3}+\frac{5}{6}\right) = 6x\cdot\frac{1}{x}$$

$$6x\cdot\frac{1}{3}+6x\cdot\frac{5}{6} = 6x\cdot\frac{1}{x}$$

$$2x+5x = 6$$

$$7x = 6$$

$$x = \frac{6}{7}$$

Check:

$\frac{1}{3}+\frac{5}{6}=\frac{1}{x}$	
$\frac{1}{3}+\frac{5}{6}$	$\dfrac{1}{\frac{6}{7}}$
$\frac{2}{6}+\frac{5}{6}$	$1\cdot\frac{7}{6}$
$\frac{7}{6}$	$\overset{?}{=}\ \frac{7}{6}$ TRUE

This checks, so the solution is $\frac{6}{7}$.

9. Note that t cannot be 0.

$$\frac{1}{6}+\frac{1}{8}=\frac{1}{t}, \text{ LCD} = 24t$$

$$24t\left(\frac{1}{6}+\frac{1}{8}\right) = 24t\cdot\frac{1}{t}$$

$$24t\cdot\frac{1}{6}+24t\cdot\frac{1}{8} = 24t\cdot\frac{1}{t}$$

$$4t+3t = 24$$

$$7t = 24$$

$$t = \frac{24}{7}$$

Check:

$$\begin{array}{c|c} \multicolumn{2}{c}{\frac{1}{6}+\frac{1}{8}=\frac{1}{t}} \\ \hline \frac{1}{6}+\frac{1}{8} & \dfrac{1}{\frac{24}{7}} \\ \frac{4}{24}+\frac{3}{24} & 1\cdot\frac{7}{24} \\ \end{array}$$

$$\frac{7}{24} \stackrel{?}{=} \frac{7}{24} \quad \text{TRUE}$$

This checks, so the solution is $\frac{24}{7}$.

11. Note that x cannot be 0.

$$x+\frac{5}{x}=-6, \text{ LCD}=x$$
$$x\left(x+\frac{5}{x}\right)=-6\cdot x$$
$$x\cdot x+x\cdot\frac{5}{x}=-6\cdot x$$
$$x^2+5=-6x$$
$$x^2+6x+5=0$$
$$(x+5)(x+1)=0$$
$$x+5=0 \quad \text{or} \quad x+1=0$$
$$x=-5 \quad \text{or} \quad x=-1$$

Check:

$$\begin{array}{c|c} \multicolumn{2}{c}{x+\frac{5}{x}=-6} \\ \hline -5+\frac{5}{-5} & -6 \\ -5-1 & \\ \end{array} \qquad \begin{array}{c|c} \multicolumn{2}{c}{x+\frac{5}{x}=-6} \\ \hline -1+\frac{5}{-1} & -6 \\ -1-5 & \\ \end{array}$$

$$-6 \stackrel{?}{=} -6 \quad \text{TRUE} \qquad -6 \stackrel{?}{=} -6 \quad \text{TRUE}$$

Both of these check, so the two solutions are -5 and -1.

13. Note that x cannot be 0.

$$\frac{x}{6}-\frac{6}{x}=0, \text{ LCD}=6x$$
$$6x\left(\frac{x}{6}-\frac{6}{x}\right)=6x\cdot 0$$
$$6x\cdot\frac{x}{6}-6x\cdot\frac{6}{x}=6x\cdot 0$$
$$x^2-36=0$$
$$(x+6)(x-6)=0$$
$$x+6=0 \quad \text{or} \quad x-6=0$$
$$x=-6 \quad \text{or} \quad x=6$$

Check:

$$\begin{array}{c|c} \multicolumn{2}{c}{\frac{x}{6}-\frac{6}{x}=0} \\ \hline \frac{-6}{6}-\frac{6}{-6} & 0 \\ -1+1 & \\ \end{array} \qquad \begin{array}{c|c} \multicolumn{2}{c}{\frac{x}{6}-\frac{6}{x}=0} \\ \hline \frac{6}{6}-\frac{6}{6} & 0 \\ 1-1 & \\ \end{array}$$

$$0 \stackrel{?}{=} 0 \quad \text{TRUE} \qquad 0 \stackrel{?}{=} 0 \quad \text{TRUE}$$

Both of these check, so the two solutions are -6 and 6

15. Note that x cannot be 0.

$$\frac{5}{x}=\frac{6}{x}-\frac{1}{3}, \text{ LCD}=3x$$
$$3x\cdot\frac{5}{x}=3x\left(\frac{6}{x}-\frac{1}{3}\right)$$
$$3x\cdot\frac{5}{x}=3x\cdot\frac{6}{x}-3x\cdot\frac{1}{3}$$
$$15=18-x$$
$$-3=-x$$
$$3=x$$

Check:

$$\begin{array}{c|c} \multicolumn{2}{c}{\frac{5}{x}=\frac{6}{x}-\frac{1}{3}} \\ \hline \frac{5}{3} & \frac{6}{3}-\frac{1}{3} \\ \end{array}$$

$$\frac{5}{3} \stackrel{?}{=} \frac{5}{3} \quad \text{TRUE}$$

This checks, so the solution is 3.

17. Note that t cannot be 0.

$$\frac{5}{3t}+\frac{3}{t}=1, \text{ LCD}=3t$$
$$3t\left(\frac{5}{3t}+\frac{3}{t}\right)=3t\cdot 1$$
$$3t\cdot\frac{5}{3t}+3t\cdot\frac{3}{t}=3t\cdot 1$$
$$5+9=3t$$
$$14=3t$$
$$\frac{14}{3}=t$$

Check:

$$\begin{array}{c|c} \multicolumn{2}{c}{\frac{5}{3t}+\frac{3}{t}=1} \\ \hline \dfrac{5}{3\cdot\frac{14}{3}}+\dfrac{3}{\frac{14}{3}} & 1 \\ \frac{5}{14}+\frac{9}{14} & \\ \frac{14}{14} & \\ \end{array}$$

$$1 \stackrel{?}{=} 1 \quad \text{TRUE}$$

This checks, so the solution is $\frac{14}{3}$.

19. To avoid division by 0, we must have $x+3 \neq 0$, or $x \neq -3$.

$$\frac{x-8}{x+3} = \frac{1}{4}, \text{ LCD} = 4(x+3)$$
$$4(x+3)\cdot\frac{x-8}{x+3} = 4(x+3)\cdot\frac{1}{4}$$
$$4(x-8) = x+3$$
$$4x-32 = x+3$$
$$3x = 35$$
$$x = \frac{35}{3}$$

Check:

$$\frac{x-8}{x+3} = \frac{1}{4}$$
$$\frac{\frac{35}{3}-8}{\frac{35}{3}+3} \;\Big|\; \frac{1}{4}$$
$$\frac{\frac{35}{3}-\frac{24}{3}}{\frac{35}{3}+\frac{9}{3}}$$
$$\frac{\frac{11}{3}}{\frac{44}{3}}$$
$$\frac{11}{3}\cdot\frac{3}{44}$$
$$\frac{1}{4} \stackrel{?}{=} \frac{1}{4} \quad \text{TRUE}$$

This checks, so the solution is $\frac{35}{3}$.

21. Note that x cannot be 0.

$$x+\frac{12}{x} = -7, \text{ LCD is } x$$
$$x\left(x+\frac{12}{x}\right) = x\cdot(-7)$$
$$x\cdot x + x\cdot\frac{12}{x} = -7x$$
$$x^2+12 = -7x$$
$$x^2+7x+12 = 0$$
$$(x+3)(x+4) = 0$$
$$x+3 = 0 \quad \text{or} \quad x+4 = 0$$
$$x = -3 \quad \text{or} \quad x = -4$$

Both numbers check, so the solutions are -3 and -4.

23. To avoid division by 0, we must have $x+1 \neq 0$ and $x-2 \neq 0$, or $x \neq -1$ and $x \neq 2$.

$$\frac{2}{x+1} = \frac{1}{x-2},$$
$$\text{LCD} = (x+1)(x-2)$$
$$(x+1)(x-2)\cdot\frac{2}{x+1} = (x+1)(x-2)\cdot\frac{1}{x-2}$$
$$2(x-2) = x+1$$
$$2x-4 = x+1$$
$$x = 5$$

This checks, so the solution is 5.

25. Because no variable appears in a denominator, no restrictions exist.

$$\frac{a}{6}-\frac{a}{10} = \frac{1}{6}, \text{ LCD} = 30$$
$$30\left(\frac{a}{6}-\frac{a}{10}\right) = 30\cdot\frac{1}{6}$$
$$30\cdot\frac{a}{6} - 30\cdot\frac{a}{10} = 30\cdot\frac{1}{6}$$
$$5a-3a = 5$$
$$2a = 5$$
$$a = \frac{5}{2}$$

This checks, so the solution is $\frac{5}{2}$.

27. Because no variable appears in a denominator, no restrictions exist.

$$\frac{x+1}{3}-1 = \frac{x-1}{2}, \text{ LCD} = 6$$
$$6\left(\frac{x+1}{3}-1\right) = 6\cdot\frac{x-1}{2}$$
$$6\cdot\frac{x+1}{3} - 6\cdot 1 = 6\cdot\frac{x-1}{2}$$
$$2(x+1)-6 = 3(x-1)$$
$$2x+2-6 = 3x-3$$
$$2x-4 = 3x-3$$
$$-1 = x$$

This checks, so the solution is -1.

29. To avoid division by 0, we must have $t-5 \neq 0$, or $t \neq 5$.

$$\frac{4}{t-5} = \frac{t-1}{t-5}, \text{ LCD} = t-5$$
$$(t-5)\cdot\frac{4}{t-5} = (t-5)\cdot\frac{t-1}{t-5}$$
$$4 = t-1$$
$$5 = t$$

Because of the restriction $t \neq 5$, the number 5 must be rejected as a solution. The equation has no solution.

31. To avoid division by 0, we must have $x+4 \neq 0$ and $x \neq 0$, or $x \neq -4$ and $x \neq 0$.

$$\frac{3}{x+4} = \frac{5}{x}, \text{ LCD} = x(x+4)$$
$$x(x+4) \cdot \frac{3}{x+4} = x(x+4) \cdot \frac{5}{x}$$
$$3x = 5(x+4)$$
$$3x = 5x + 20$$
$$-2x = 20$$
$$x = -10$$

This checks, so the solution is -10.

33. To avoid division by 0, we must have $a-1 \neq 0$ and $a-2 \neq 0$, or $a \neq 1$ and $a \neq 2$.

$$\frac{a-4}{a-1} = \frac{a+2}{a-2}, \text{ LCD} = (a-1)(a-2)$$
$$(a-1)(a-2) \cdot \frac{a-4}{a-1} = (a-1)(a-2) \cdot \frac{a+2}{a-2}$$
$$(a-2)(a-4) = (a-1)(a+2)$$
$$a^2 - 6a + 8 = a^2 + a - 2$$
$$-6a + 8 = a - 2$$
$$10 = 7a$$
$$\frac{10}{7} = a$$

This checks, so the solution is $\frac{10}{7}$.

35. To avoid division by 0, we must have $t-2 \neq 0$, or $t \neq 2$.

$$\frac{5}{t-2} + \frac{3t}{t-2} = \frac{4}{t^2-4t+4}, \text{ LCD is } (t-2)^2$$
$$(t-2)^2\left(\frac{5}{t-2} + \frac{3t}{t-2}\right) = (t-2)^2 \cdot \frac{4}{(t-2)^2}$$
$$5(t-2) + 3t(t-2) = 4$$
$$5t - 10 + 3t^2 - 6t = 4$$
$$3t^2 - t - 10 = 4$$
$$3t^2 - t - 14 = 0$$
$$(3t-7)(t+2) = 0$$
$$3t - 7 = 0 \quad or \quad t + 2 = 0$$
$$3t = 7 \quad or \quad t = -2$$
$$t = \frac{7}{3} \quad or \quad t = -2$$

Both numbers check. The solutions are $\frac{7}{3}$ and -2.

37. To avoid division by 0, we must have $x-3 \neq 0$ and $x+3 \neq 0$, or $x \neq 3$ and $x \neq -3$.

$$\frac{4}{x-3} + \frac{2x}{x^2-9} = \frac{1}{x+3},$$
$$\text{LCD} = (x-3)(x+3)$$
$$(x-3)(x+3)\left(\frac{4}{x-3} + \frac{2x}{(x+3)(x-3)}\right) = (x-3)(x+3) \cdot \frac{1}{x+3}$$
$$4(x+3) + 2x = x - 3$$
$$4x + 12 + 2x = x - 3$$
$$6x + 12 = x - 3$$
$$5x = -15$$
$$x = -3$$

Because of the restriction of $x \neq -3$, we must reject the number -3 as a solution. The equation has no solution.

39. To avoid division by 0, we must have $y-3 \neq 0$ and $y+3 \neq 0$, or $y \neq 3$ and $y \neq -3$.

$$\frac{5}{y-3} - \frac{30}{y^2-9} = 1$$
$$\frac{5}{y-3} - \frac{30}{(y+3)(y-3)} = 1,$$
$$\text{LCD} = (y-3)(y+3)$$
$$(y-3)(y+3)\left(\frac{5}{y-3} - \frac{30}{(y+3)(y-3)}\right) = (y-3)(y+3) \cdot 1$$
$$5(y+3) - 30 = (y+3)(y-3)$$
$$5y + 15 - 30 = y^2 - 9$$
$$0 = y^2 - 5y + 6$$
$$0 = (y-3)(y-2)$$
$$y - 3 = 0 \quad or \quad y - 2 = 0$$
$$y = 3 \quad or \quad y = 2$$

Because of the restriction $y \neq 3$, we must reject the number 3 as a solution. The number 2 checks, so it is the solution.

41. To avoid division by 0, we must have $8 - a \neq 0$ (or equivalently $a - 8 \neq 0$), or $a \neq 8$.

$$\frac{4}{8-a} = \frac{4-a}{a-8}$$
$$\frac{-1}{-1} \cdot \frac{4}{8-a} = \frac{4-a}{a-8}$$
$$\frac{-4}{a-8} = \frac{4-a}{a-8}, \text{ LCD} = a - 8$$
$$(a-8) \cdot \frac{-4}{a-8} = (a-8) \cdot \frac{4-a}{a-8}$$
$$-4 = 4 - a$$
$$-8 = -a$$
$$8 = a$$

Because of the restriction $a \neq 8$, we must reject the number 8 as a solution. The equation has no solution.

43. $\frac{-2}{x+2} = \frac{x}{x+2}$

To avoid division by 0, we must have $x+2 \neq 0$, or $x \neq -2$. Now observe that the denominators are the same, so the numerators must be the same. Thus, we have $-2 = x$, but because of the restriction $x \neq -2$ this cannot be a solution. The equation has no solution.

45. *Writing Exercise*

47. ***Familiarize***. Let $x =$ the first odd integer. Then $x+2 =$ the next odd integer.

Translate.

The sum of two consecutive odd integers is 276.

$$\underbrace{x + (x+2)} = 276$$

Carry out. We solve the equation.

$$x + (x+2) = 276$$
$$2x + 2 = 276$$
$$2x = 274$$
$$x = 137$$

When $x = 137$, then $x + 2 = 137 + 2 = 139$.

Check. The numbers 137 and 139 are consecutive odd integers and $137 + 139 = 276$. These numbers check.

State. The integers are 137 and 139.

49. ***Familiarize***. Let $b =$ the base of the triangle, in cm. Then $b + 3 =$ the height. Recall that the area of a triangle is given by $\frac{1}{2} \times$ base $\times$ height.

Translate.

The area of the triangle is 54 cm^2.

$$\frac{1}{2} \cdot b \cdot (b+3) = 54$$

Carry out. We solve the equation.

$$\frac{1}{2}b(b+3) = 54$$
$$2 \cdot \frac{1}{2}b(b+3) = 2 \cdot 54$$
$$b(b+3) = 108$$
$$b^2 + 3b = 108$$
$$b^2 + 3b - 108 = 0$$
$$(b-9)(b+12) = 0$$
$$b - 9 = 0 \quad or \quad b + 12 = 0$$
$$b = 9 \quad or \quad b = -12$$

Check. The length of the base cannot be negative so we need to check only 9. If the base is 9 cm, then the height is 9+3, or 12 cm, and the area is $\frac{1}{2} \cdot 9 \cdot 12$, or 54 cm^2. The answer checks.

State. The base measures 9 cm, and the height measures 12 cm.

51. To find the rate, in centimeters per day, we divide the amount of growth by the number of days. From June 9 to June 24 is $24 - 9 = 15$ days.

$$\text{Rate, in cm per day} = \frac{0.9 \text{ cm}}{15 \text{ days}} = 0.06 \text{ cm/day} = 0.06 \text{ cm per day}$$

53. *Writing Exercise*

55. To avoid division by 0, we must have $x - 3 \neq 0$, or $x \neq 3$.

$$1 + \frac{x-1}{x-3} = \frac{2}{x-3} - x, \text{ LCD} = x-3$$
$$(x-3)\left(1 + \frac{x-1}{x-3}\right) = (x-3)\left(\frac{2}{x-3} - x\right)$$
$$(x-3) \cdot 1 + (x-3) \cdot \frac{x-1}{x-3} = (x-3) \cdot \frac{2}{x-3} - (x-3)x$$
$$x - 3 + x - 1 = 2 - x^2 + 3x$$
$$2x - 4 = 2 - x^2 + 3x$$
$$x^2 - x - 6 = 0$$
$$(x-3)(x+2) = 0$$
$$x - 3 = 0 \quad or \quad x + 2 = 0$$
$$x = 3 \quad or \quad x = -2$$

Because of the restriction $x \neq 3$, we must reject the number 3 as a solution. The number -2 checks, so it is the solution.

57. To avoid division by 0, we must have $x + 4 \neq 0$ and $x - 1 \neq 0$ and $x + 2 \neq 0$, or $x \neq -4$ and $x \neq 1$ and $x \neq -2$.

$$\frac{x}{x^2+3x-4} + \frac{x+1}{x^2+6x+8} = \frac{2x}{x^2+x-2}$$
$$\frac{x}{(x+4)(x-1)} + \frac{x+1}{(x+2)(x+4)} = \frac{2x}{(x+2)(x-1)},$$
$$\text{LCD} = (x+4)(x-1)(x+2)$$
$$(x+4)(x-1)(x+2)\left(\frac{x}{(x+4)(x-1)} + \frac{x+1}{(x+2)(x+4)}\right) =$$
$$(x+4)(x-1)(x+2)\left(\frac{2x}{(x+2)(x-1)}\right)$$
$$x(x+2) + (x-1)(x+1) = 2x(x+4)$$
$$x^2 + 2x + x^2 - 1 = 2x^2 + 8x$$
$$2x^2 + 2x - 1 = 2x^2 + 8x$$
$$-1 = 6x$$
$$-\frac{1}{6} = x$$

This checks, so the solution is $-\frac{1}{6}$.

59. To avoid division by 0, we must have $x + 2 \neq 0$ and $x - 2 \neq 0$, or $x \neq -2$ and $x \neq 2$.

$$\frac{x^2}{x^2-4} = \frac{x}{x+2} - \frac{2x}{2-x}$$
$$\frac{x^2}{x^2-4} = \frac{x}{x+2} - \frac{2x}{2-x} \cdot \frac{-1}{-1}$$
$$\frac{x^2}{(x+2)(x-2)} = \frac{x}{x+2} - \frac{-2x}{x-2},$$
$$\text{LCD} = (x+2)(x-2)$$
$$(x+2)(x-2) \cdot \frac{x^2}{(x+2)(x-2)} =$$
$$(x+2)(x-2)\left(\frac{x}{x+2} - \frac{-2x}{x-2}\right)$$
$$x^2 = x(x-2) - (-2x)(x+2)$$
$$x^2 = x^2 - 2x + 2x^2 + 4x$$
$$x^2 = 3x^2 + 2x$$
$$0 = 2x^2 + 2x$$
$$0 = 2x(x+1)$$

$2x = 0$ *or* $x + 1 = 0$

$x = 0$ *or* $x = -1$

Both of these check, so the solutions are -1 and 0.

61. To avoid division by 0, we must have $x - 1 \neq 0$, or $x \neq 1$.

$$\frac{1}{x-1} + x - 5 = \frac{5x-4}{x-1} - 6, \text{ LCD} = x-1$$

$$(x-1)\left(\frac{1}{x-1} + x - 5\right) = (x-1)\left(\frac{5x-4}{x-1} - 6\right)$$

$$1 + x(x-1) - 5(x-1) = 5x - 4 - 6(x-1)$$

$$1 + x^2 - x - 5x + 5 = 5x - 4 - 6x + 6$$

$$x^2 - 6x + 6 = -x + 2$$

$$x^2 - 5x + 4 = 0$$

$$(x-1)(x-4) = 0$$

$x - 1 = 0$ *or* $x - 4 = 0$

$x = 1$ *or* $x = 4$

Because of the restriction $x \neq 1$, we must reject the number 1 as a solution. The number 4 checks, so it is the solution.

63. 

Exercise Set 6.7

1. ***Familiarize.*** The job takes Ned 20 min working alone and Linda 30 min working alone. Then in 1 min Ned does $\frac{1}{20}$ of the job and Linda does $\frac{1}{30}$ of the job. Working together they can do $\frac{1}{20} + \frac{1}{30}$, or $\frac{5}{60}$, or $\frac{1}{12}$ of the job in 1 min. In 10 min, Ned does $10 \cdot \frac{1}{20}$ or the job and Linda does $10 \cdot \frac{1}{30}$ of the job. Working together they can do $10 \cdot \frac{1}{20} + 10 \cdot \frac{1}{30}$, or $\frac{5}{6}$, of the job in 10 min. In 15 min, Ned does $15 \cdot \frac{1}{20}$ of the job and Linda does $15 \cdot \frac{1}{30}$ of the job. Working together they can do $15 \cdot \frac{1}{20} + 15 \cdot \frac{1}{30}$, or $1\frac{1}{4}$ of the job which is more of the job than needs to be done. The answer is somewhere between 10 min and 15 min. (When we determined that Ned and Linda could do $\frac{1}{12}$ of the job working together for 1 min, we could have observed that it would take them 12 min to do the entire job, but we will continue with the full solution here.)

Translate. If they work together t minutes, then Ned does $t \cdot \frac{1}{20}$ of the job and Linda does $t \cdot \frac{1}{30}$ of the job. We want a number t such that

$$\left(\frac{1}{20} + \frac{1}{30}\right)t = 1, \text{ or } \frac{1}{12} \cdot t = 1.$$

Carry out. We solve the equation.

$$\frac{1}{12} \cdot t = 1$$

$$12 \cdot \frac{1}{12} \cdot t = 12 \cdot 1$$

$$t = 12$$

Check. We can repeat the computations. We also expected the result to be between 10 min and 15 min as it is.

State. Working together, it takes Ned and Linda 12 min to do the job.

3. ***Familiarize.*** The job takes Juanita 12 hours working alone and Anton 16 hours working alone. Then in 1 hour Juanita does $\frac{1}{12}$ of the job and Anton does $\frac{1}{16}$ of the job. Working together, they can do $\frac{1}{12} + \frac{1}{16}$, or $\frac{7}{48}$ of the job in 1 hour. In four hours, Juanita does $4 \cdot \frac{1}{12}$ of the job and Anton does $4 \cdot \frac{1}{16}$ of the job. Working together they can do $4 \cdot \frac{1}{12} + 4 \cdot \frac{1}{16}$, or $\frac{7}{12}$ of the job in 4 hours. In 7 hours they can do $7 \cdot \frac{1}{12} + 7 \cdot \frac{1}{16}$, or $\frac{49}{48}$ or $1\frac{1}{48}$ of the job which is more of the job than needs to be done. The answer is somewhere between 4 hr and 7 hr.

Translate. If they work together t hours, then Juanita does $t\left(\frac{1}{12}\right)$ of the job and Anton does $t\left(\frac{1}{16}\right)$ of the job. We want some number t such that

$$\left(\frac{1}{12} + \frac{1}{16}\right)t = 1, \text{ or } \frac{7}{48} \cdot t = 1.$$

Carry out. We solve the equation.

$$\frac{7}{48} \cdot t = 1$$

$$\frac{48}{7} \cdot \frac{7}{48} \cdot t = \frac{48}{7} \cdot 1$$

$$t = \frac{48}{7}, \text{ or } 6\frac{6}{7}$$

Check. The check can be done by repeating the computations. We also have a partial check in that we expected from our familiarization step that the answer would be between 4 hr and 7 hr.

State. Working together, it takes them $\frac{48}{7}$ hr, or 6 hr, $51\frac{3}{7}$ min, to build the shed.

5. ***Familiarize.*** Let t = the number of minutes it would take the two machines to copy the dissertation, working together.

Translate. We use the work principle.

$$\left(\frac{1}{12} + \frac{1}{18}\right)t = 1, \text{ or } \frac{5}{36} \cdot t = 1$$

Carry out. We solve the equation.

$$\frac{5}{36} \cdot 1 = 1$$

$$\frac{36}{5} \cdot \frac{5}{36} \cdot t = \frac{36}{5} \cdot 1$$

$$t = \frac{36}{5}$$

Check. In $\frac{36}{5}$ min, the portion of the job done is $\frac{1}{12} \cdot \frac{36}{5} + \frac{1}{18} \cdot \frac{36}{5} = \frac{3}{5} + \frac{2}{5} = 1$. The answer checks.

State. It would take the two machines $\frac{36}{5}$ min, or $7\frac{1}{5}$ min to copy the dissertation, working together.

7. ***Familiarize***. The pool can be filled in 12 hours with only the pipe and in 30 hours with only the hose. Then in 1 hour, the pipe fills $\frac{1}{12}$ of the pool, and the hose fills $\frac{1}{30}$ of the pool. Using both the pipe and the hose, $\frac{1}{12}+\frac{1}{30}$ of the pool can be filled in 1 hour.

Suppose that it takes t hours to fill the pool using both the pipe and hose.

Translate. We use the work principle.

$$\left(\frac{1}{12}+\frac{1}{30}\right)t=1, \text{ or } \frac{7}{60}t=1$$

Carry out. We solve the equation.

$$\frac{7}{60}\cdot t=1$$
$$\frac{60}{7}\cdot\frac{7}{60}\cdot t=\frac{60}{7}\cdot 1$$
$$t=\frac{60}{7}$$

Check. The possible solution is $\frac{60}{7}$ hours. If the pipe is used $\frac{60}{7}$ hours, it fills $\frac{1}{12}\cdot\frac{60}{7}$, or $\frac{5}{7}$ of the pool. If the hose is used $\frac{60}{7}$ hours, it fills $\frac{1}{30}\cdot\frac{60}{7}$, or $\frac{2}{7}$ of the pool. Using both, $\frac{5}{7}+\frac{2}{7}$ of the pool, or all of it, will be filled in $\frac{60}{7}$ hours.

State. Using both the pipe and the hose, it will take $\frac{60}{7}$, or $8\frac{4}{7}$ hours, to fill the pool.

9. ***Familiarize***. Let t represent the time, in minutes, that it takes the HP copier to do the job, working alone. Then $2t$ represents the time, in minutes, it takes the Canon copier to do the same job, working alone. In 1 minute, the HP copier does $\frac{1}{t}$ of the job and the Canon copier does $\frac{1}{2t}$ of the job.

Translate. Working together, they can do the entire job in 10 min, so we want to find t such that

$$\left(\frac{1}{t}+\frac{1}{2t}\right)10=1, \text{ or } \frac{3}{2t}\cdot 10=1.$$

Carry out. We solve the equation.

$$\frac{3}{2t}\cdot 10=1$$
$$\frac{15}{t}=1$$
$$t\cdot\frac{15}{t}=t\cdot 1$$
$$15=t$$

Check. If the HP copier does the job in 15 min, then in 10 min it does $10\cdot\frac{1}{15}$, or $\frac{2}{3}$ of the job. If it takes the Canon copier $2\cdot 15$, or 30 min, to do the job, then in 10 min it does $10\left(\frac{1}{30}\right)$, or $\frac{1}{3}$ of the job. Working together they do $\frac{2}{3}+\frac{1}{3}$, or 1 entire job in 10 min. The answer checks.

State. Working alone, it takes the HP copier 15 min and the Canon copier 30 min to do the job.

11. ***Familiarize***. Let t represent the number of hours it takes the Erickson helicopter to douse the fire, working alone. Then $4t$ represents the time it takes the S-58T helicopter to douse the fire. In 1 hour the Erickson does $\frac{1}{t}$ of the job and the S-58T does $\frac{1}{4t}$ of the job.

Translate. Working together, the two helicopters can douse the fire in 8 hours, so we want to find t such that

$$8\left(\frac{1}{t}\right)+8\left(\frac{1}{4t}\right)=1.$$

Carry out. We solve the equation.

$$8\left(\frac{1}{t}\right)+8\left(\frac{1}{4t}\right)=1$$
$$\frac{8}{t}+\frac{2}{t}=1$$
$$\frac{10}{t}=1, \text{ LCD is } t$$
$$t\cdot\frac{10}{t}=t\cdot 1$$
$$10=t$$

Then $4t=4\cdot 10=40$.

Check. In 8 hr the Erickson does $8\cdot\frac{1}{10}$, or $\frac{4}{5}$ of the job, working alone, and the S-58T does $8\cdot\frac{1}{40}$, or $\frac{1}{5}$ of the job. Working together, they do $\frac{4}{5}+\frac{1}{5}$, or 1 entire job. The answer checks.

State. The Erickson helicopter can douse the fire in 10 hr working alone and the S-58T helicopter can do the same job in 40 hr.

13. ***Familiarize***. Let t represent the number of hours it takes Mariah to deliver the papers alone. Then $3t$ represents the number of hours it takes Stan to deliver the papers alone.

Translate. In 1 hr Mariah and Stan will do one entire job, so we have

$$\left(\frac{1}{t}+\frac{1}{3t}\right)\cdot 1=1, \text{ or } \frac{4}{3t}=1.$$

Carry out. We solve the equation.

$$\frac{4}{3t}=1$$
$$t\cdot\frac{4}{3t}=t\cdot 1$$
$$\frac{4}{3}=t$$

Check. If Mariah does the job alone in $\frac{4}{3}$ hr, then in 1 hr she does $\frac{1}{4/3}$, or $\frac{3}{4}$ of the job. If Stan does the job alone in

$3 \cdot \frac{4}{3}$, or 4 hr, then in 1 hr he does $\frac{1}{4}$ of the job. Together, they do $\frac{3}{4}+\frac{1}{4}$, or 1 entire job, in 1 hr. The result checks.

State. It would take Mariah $\frac{4}{3}$ hours and it would take Stan 4 hours to deliver the papers alone.

15. ***Familiarize***. Let t represent the number of hours it would take Bill to pave the driveway, working alone. Then $t+4$ represents the time it would take Larry to do the job. In 1 hr Bill does $\frac{1}{t}$ of the job and Larry does $\frac{1}{t+4}$.

We convert minutes to hours:

$$48 \text{ min} = 48 \cancel{\text{min}} \cdot \frac{1 \text{ hr}}{60 \cancel{\text{min}}} = \frac{48}{60} \text{ hr} = 0.8 \text{ hr}$$

Then 4 hr 48 min is 4.8 hr.

Translate. In 4.8 hr they do 1 entire job working together, so we have

$$4.8\left(\frac{1}{t}\right) + 4.8\left(\frac{1}{t+4}\right) = 1, \text{ or } \frac{4.8}{t} + \frac{4.8}{t+4} = 1.$$

Carry out. We solve the equation. First we multiply both sides by the LCD, $t(t+4)$.

$$t(t+4)\left(\frac{4.8}{t} + \frac{4.8}{t+4}\right) = t(t+4)\cdot 1$$
$$4.8(t+4) + 4.8t = t(t+4)$$
$$4.8t + 19.2 + 4.8t = t^2 + 4t$$
$$9.6t + 19.2 = t^2 + 4t$$
$$0 = t^2 - 5.6t - 19.2$$
$$0 = 10t^2 - 56t - 192$$

Multiplying by 10

$$0 = 2(5t^2 - 28t - 96)$$
$$0 = 2(t-8)(5t+12)$$

$$t = 8 \quad or \quad t = -\frac{12}{5}$$

Check. Since negative time has no meaning in this application, $-\frac{12}{5}$ cannot be a solution. In 8 hr, Bill does $4.8\left(\frac{1}{8}\right)$, or 0.6 of the job working alone, and Larry does $4.8\left(\frac{1}{8+4}\right)$, or $4.8\left(\frac{1}{12}\right)$, or 0.4 of the job. Together they do $0.6 + 0.4$, or 1 entire job. The answer checks.

State. It takes Bill 8 hr to pave the driveway working alone.

17. ***Familiarize***. We complete the table shown in the text.

$$d = r \cdot t$$

	Distance	Speed	Time
B & M	330	$r-14$	$\frac{330}{r-14}$
AMTRAK	400	r	$\frac{400}{r}$

Translate. Since the time must be the same for both trains, we have the equation

$$\frac{330}{r-14} = \frac{400}{r}.$$

Carry out. We first multiply by the LCD, $r(r-14)$.

$$r(r-14)\cdot\frac{330}{r-14} = r(r-14)\cdot\frac{400}{r}$$
$$330r = 400(r-14)$$
$$330r = 400r - 5600$$
$$-70r = -5600$$
$$r = 80$$

If the speed of the AMTRAK train is 80 km/h, then the speed of the B & M train is $80 - 14$, or 66 km/h.

Check. The speed of the B&M train is 14 km/h slower than the speed of the AMTRAK train. At 66 km/h the B&M train travels 330 km in 330/66, or 5 hr. At 80 km/h the AMTRAK train travels 400 km in 400/80, or 5 hr. The times are the same, so the answer checks.

State. The speed of the AMTRAK train is 80 km/h, and the speed of the B&M freight train is 66 km/h.

19. ***Familiarize***. Let r = the speed of Bill's Harley, in mph. Then $r+30$ = the speed of Hillary's Lexus. We organize the information in a table using the formula time = distance/rate to fill in the last column.

	Distance	Speed	Time
Harley	75	r	$\frac{75}{r}$
Lexus	120	$r+30$	$\frac{120}{r+30}$

Translate. Since the times must be the same, we have the equation

$$\frac{75}{r} = \frac{120}{r+30}.$$

Carry out. We first multiply by the LCD, $r(r+30)$.

$$r(r+30)\cdot\frac{75}{r} = r(r+30)\cdot\frac{120}{r+30}$$
$$75(r+30) = 120r$$
$$75r + 2250 = 120r$$
$$2250 = 45r$$
$$50 = r$$

Then $r + 30 = 50 + 30 = 80$.

Check. The speed of the Lexus is 30 mph faster than the speed of the Harley. At 50 mph, the Harley travels 75 mi in 75/50, or 1.5 hr. At 80 mph, the Lexus travels 120 mi in 120/80, or 1.5 hr. The times are the same, so the answer checks.

State. The speed of Bill's Harley is 50 mph, and the speed of Hillary's Lexus is 80 mph.

21. ***Familiarize***. Let t = the time it takes Caledonia to drive to town and organize the given information in a table.

	Distance	Speed	Time
Caledonia	15	r	t
Manley	20	r	$t+1$

Translate. We can replace the r's in the table above using the formula $r = d/t$.

	Distance	Speed	Time
Caledonia	15	$\frac{15}{t}$	t
Manley	20	$\frac{20}{t+1}$	$t+1$

Since the speeds are the same for both riders, we have the equation

$$\frac{15}{t} = \frac{20}{t+1}.$$

Carry out. We multiply by the LCD, $t(t+1)$.

$$t(t+1) \cdot \frac{15}{t} = t(t+1) \cdot \frac{20}{t+1}$$
$$15(t+1) = 20t$$
$$15t + 15 = 20t$$
$$15 = 5t$$
$$3 = t$$

If $t = 3$, then $t + 1 = 3 + 1$, or 4.

Check. If Caledonia's time is 3 hr and Manley's time is 4 hr, then Manley's time is 1 hr more than Caledonia's. Caledonia's speed is 15/3, or 5 mph. Manley's speed is 20/4, or 5 mph. Since the speeds are the same, the answer checks.

State. It takes Caledonia 3 hr to drive to town.

23. ***Familiarize***. We first make a drawing. Let r = the kayak's speed in still water in mph. Then $r - 3$ = the speed upstream and $r + 3$ = the speed downstream.

Upstream 4 miles $r - 3$ mph

10 miles $r + 3$ mph Downstream

We organize the information in a table. The time is the same both upstream and downstream so we use t for each time.

	Distance	Speed	Time
Upstream	4	$r-3$	t
Downstream	10	$r+3$	t

Translate. Using the formula Time = Distance/Rate in each row of the table and the fact that the times are the same, we can write an equation.

$$\frac{4}{r-3} = \frac{10}{r+3}$$

Carry out. We solve the equation.

$$\frac{4}{r-3} = \frac{10}{r+3}, \text{ LCD is } (r-3)(r+3)$$
$$(r-3)(r+3) \cdot \frac{4}{r-3} = (r-3)(r+3) \cdot \frac{10}{r+3}$$
$$4(r+3) = 10(r-3)$$
$$4r + 12 = 10r - 30$$
$$42 = 6r$$
$$7 = r$$

Check. If $r = 7$ mph, then $r - 3$ is 4 mph and $r + 3$ is 10 mph. The time upstream is $\frac{4}{4}$, or 1 hour. The time downstream is $\frac{10}{10}$, or 1 hour. Since the times are the same, the answer checks.

State. The speed of the kayak in still water is 7 mph.

25. ***Familiarize***. We first make a drawing. Let r = Benny's speed on a nonmoving sidewalk in ft/sec. Then his speed moving forward on the moving sidewalk is $r + 1.7$, and his speed in the opposite direction is $r - 1.7$.

Forward $r + 1.7$ 120 ft

52 ft $r - 1.7$ Opposite direction

We organize the information in a table. The time is the same both forward and in the opposite direction so we use t for each time.

	Distance	Speed	Time
Forward	120	$r+1.7$	t
Opposite direction	52	$r-1.7$	t

Translate. Using the formula Time = Distance/Rate in each row of the table and the fact that the times are the same, we can write an equation.

$$\frac{120}{r+1.7} = \frac{52}{r-1.7}$$

Carry out. We solve the equation.

$$\frac{120}{r+1.7} = \frac{52}{r-1.7},$$
$$\text{LCD is } (r+1.7)(r-1.7)$$
$$(r+1.7)(r-1.7) \cdot \frac{120}{r+1.7} = (r+1.7)(r-1.7) \cdot \frac{52}{r-1.7}$$
$$120(r-1.7) = 52(r+1.7)$$
$$120r - 204 = 52r + 88.4$$
$$68r = 292.4$$
$$r = 4.3$$

Check. If Benny's speed on a nonmoving sidewalk is 4.3 ft/sec, then his speed moving forward on the moving sidewalk is 4.3 + 1.7, or 6 ft/sec, and his speed moving in the opposite direction on the sidewalk is 4.3 − 1.7, or 2.6 ft/sec. Moving 120 ft at 6 ft/sec takes $\frac{120}{6} = 20$ sec. Moving 52 ft at 2.6 ft/sec takes $\frac{52}{2.6}$, or 20 sec. Since the times are the same, the answer checks.

State. Benny would be walking 4.3 ft/sec on a nonmoving sidewalk.

27. ***Familiarize***. Let r = the speed of the passenger train in mph. Then $r - 14$ = the speed of the freight train in mph. We organize the information in a table. The time is the same for both trains so we use t for each time.

	Distance	Speed	Time
Passenger train	400	r	t
Freight train	330	$r-14$	t

Translate. Using the formula Time = Distance/Rate in each row of the table and the fact that the times are the same, we can write an equation.

$$\frac{400}{r} = \frac{330}{r-14}$$

Carry out. We solve the equation.

$$\frac{400}{r} = \frac{330}{r-14}, \text{ LCD is } r(r-14)$$
$$r(r-14)\cdot\frac{400}{r} = r(r-14)\cdot\frac{330}{r-14}$$
$$400(r-14) = 330r$$
$$400r - 5600 = 330r$$
$$-5600 = -70r$$
$$80 = r$$

Check. If the passenger train's speed is 80 mph, then the freight train's speed is 80 − 14, or 66 mph. Traveling 400 mi at 80 mph takes $\frac{400}{80} = 5$ hr. Traveling 330 mi at 66 mph takes $\frac{330}{66} = 5$ hr. Since the times are the same, the answer checks.

State. The speed of the passenger train is 80 mph; the speed of the freight train is 66 mph.

29. ***Familiarize.*** We let r = the speed of the river. Then $15+r$ = Laverne's speed downstream in km/h and $15-r$ = her speed upstream in km/h. The times are the same. Let t represent the time. We organize the information in a table.

	Distance	Speed	Time
Downstream	140	$15+r$	t
Upstream	35	$15-r$	t

Translate. Using the formula Time = Distance/Rate in each row of the table and the fact that the times are the same, we can write an equation.

$$\frac{140}{15+r} = \frac{35}{15-r}$$

Carry out. We solve the equation.

$$\frac{140}{15+r} = \frac{35}{15-r},$$
$$\text{LCD is } (15+r)(15-r)$$
$$(15+r)(15-r)\cdot\frac{140}{15+r} = (15+r)(15-r)\cdot\frac{35}{15-r}$$
$$140(15-r) = 35(15+r)$$
$$2100 - 140r = 525 + 35r$$
$$1575 = 175r$$
$$9 = r$$

Check. If $r = 9$, then the speed downstream is 15 + 9, or 24 km/h and the speed upstream is 15 − 9, or 6 km/h. The time for the trip is downstream is $\frac{140}{24}$, or $5\frac{5}{6}$ hours. The time for the trip upstream is $\frac{35}{6}$, or $5\frac{5}{6}$ hours. The times are the same. The values check.

State. The speed of the river is 9 km/h.

31. ***Familiarize.*** Let c = the speed of the current, in km/h. Then $7+c$ = the speed downriver and $7-c$ = the speed upriver. We organize the information in a table.

	Distance	Speed	Time
Downriver	45	$7+c$	t_1
Upriver	45	$7-c$	t_2

Translate. Using the formula Time = Distance/Rate we see that $t_1 = \frac{45}{7+c}$ and $t_2 = \frac{45}{7-c}$. The total time upriver and back is 14 hr, so $t_1 + t_2 = 14$, or

$$\frac{45}{7+c} + \frac{45}{7-c} = 14.$$

Carry out. We solve the equation. Multiply both sides by the LCD, $(7+c)(7-c)$.

$$(7+c)(7-c)\left(\frac{45}{7+c} + \frac{45}{7-c}\right) = (7+c)(7-c)14$$
$$45(7-c) + 45(7+c) = 14(49-c^2)$$
$$315 - 45c + 315 + 45c = 686 - 14c^2$$
$$14c^2 - 56 = 0$$
$$14(c+2)(c-2) = 0$$
$$c+2 = 0 \quad or \quad c-2 = 0$$
$$c = -2 \quad or \quad c = 2$$

Check. Since speed cannot be negative in this problem, −2 cannot be a solution of the original problem. If the speed of the current is 2 km/h, the barge travels upriver at 7 − 2, or 5 km/h. At this rate it takes $\frac{45}{5}$, or 9 hr, to travel 45 km. The barge travels downriver at 7 + 2, or 9 km/h. At this rate it takes $\frac{45}{9}$, or 5 hr, to travel 45 km. The total travel time is 9+5, or 14 hr. The answer checks.

State. The speed of the current is 2 km/h.

33. We write a proportion and then solve it.

$$\frac{b}{6} = \frac{7}{4}$$
$$b = \frac{7}{4}\cdot 6$$
$$b = \frac{42}{4}, \text{ or } 10.5$$

(Note that the proportions $\frac{6}{b} = \frac{4}{7}$, $\frac{b}{7} = \frac{6}{4}$, or $\frac{7}{b} = \frac{4}{6}$ could also be used.)

35. We write a proportion and then solve it.

$$\frac{4}{f} = \frac{6}{4}$$
$$4f \cdot \frac{4}{f} = 4f \cdot \frac{6}{4}$$
$$16 = 6f$$
$$\frac{8}{3} = f \quad \text{Simplifying}$$

(One of the following proportions could also be used: $\frac{f}{4} = \frac{4}{6}, \frac{4}{f} = \frac{9}{6}, \frac{f}{4} = \frac{6}{9}, \frac{4}{9} = \frac{f}{6}, \frac{9}{4} = \frac{6}{f}$)

37. From the blueprint we see that 9 in. represents 36 ft and that p in. represent 15 ft. We use a proportion to find p.

$$\frac{9}{36} = \frac{p}{15}$$
$$180 \cdot \frac{9}{36} = 180 \cdot \frac{p}{15}$$
$$45 = 12p$$
$$\frac{15}{4} = p, \text{ or}$$
$$3\frac{3}{4} = p$$

The length of p is $3\frac{3}{4}$ in.

39. From the blueprint we see that 9 in. represents 36 ft and that 5 in. represents r ft. We use a proportion to find r.

$$\frac{9}{36} = \frac{5}{r}$$
$$36r \cdot \frac{9}{36} = 36r \cdot \frac{5}{r}$$
$$9r = 180$$
$$r = 20$$

The length of r is 20 ft.

41. Consider the two similar right triangles in the drawing. One has legs 1.5 ft and 18 ft. The other has legs h ft and 32 ft. We use a proportion to find h.

$$\frac{1.5}{18} = \frac{h}{32}$$
$$288 \cdot \frac{1.5}{18} = 288 \cdot \frac{h}{32}$$
$$24 = 9h$$
$$\frac{8}{3} = h, \text{ or}$$
$$2\frac{2}{3} = h$$

The length of h is $2\frac{2}{3}$ ft.

43. Consider the two similar right triangles in the drawing. One has legs 5 and 7. The other has legs 9 and r. We use a proportion to find r.

$$\frac{5}{7} = \frac{9}{r}$$
$$7r \cdot \frac{5}{7} = 7r \cdot \frac{9}{r}$$
$$5r = 63$$
$$r = \frac{63}{5}, \text{ or } 12.6$$

45. ***Familiarize.*** A rate of 140 steps per minute corresponds to a speed of 4 mph, and we wish to find the number of steps per minute S that correspond to a speed of 3 mph. We can use a proportion.

Translate.

$$\begin{array}{r} \text{Steps per min} \rightarrow \\ \text{Speed} \rightarrow \end{array} \frac{140}{4} = \frac{S}{3} \begin{array}{l} \leftarrow \text{Steps per min} \\ \leftarrow \text{Speed} \end{array}$$

Carry out. We solve the proportion.

$$12 \cdot \frac{140}{4} = 12 \cdot \frac{S}{3}$$
$$420 = 4S$$
$$105 = S$$

Check. $\frac{140}{4} = 35, \ \frac{105}{3} = 35$

The ratios are the same so the answer checks.

State. 105 steps per minute corresponds to a speed of 3 mph.

47. Observe that 42 days = $3 \cdot 14$ days. Then it follows that the number of photos taken in 42 days is $3 \cdot 234$ photos = 702 photos.

49. ***Familiarize.*** U.S. women earn 77 cents for each dollar earned by a man. This gives us one ratio, expressed in dollars: $\frac{0.77}{1}$. If a male sales manager earns $42,000, we want to find how much a female would earn for comparable work. This gives us a second ratio, also expressed in dollars: $\frac{F}{42,000}$.

Translate. We translate to a proportion.

$$\begin{array}{r} \text{Female's} \\ \text{earnings} \rightarrow \\ \text{Male's earnings} \rightarrow \end{array} \frac{0.77}{1} = \frac{F}{42,000} \begin{array}{l} \text{Female's} \\ \leftarrow \text{earnings} \\ \leftarrow \text{Male's earnings} \end{array}$$

Carry out. We solve the proportion.

$$42,000 \cdot \frac{0.77}{1} = 42,000 \cdot \frac{F}{42,000}$$
$$32,340 = F$$

Check.

$$\frac{0.77}{1} = 0.77, \ \frac{32,340}{42,000} = 0.77$$

The ratios are the same, so the answer checks.

State. If a male sales manager earns $42,000, a female would earn $32,340 for comparable work.

51. ***Familiarize.*** Let D = the number of duds you would expect in a sample of 320 firecrackers. We can use a proportion to find D.

Translate.

$$\begin{matrix} \text{Duds} \rightarrow \\ \text{Firecrackers} \rightarrow \end{matrix} \frac{9}{144} = \frac{D}{320} \begin{matrix} \leftarrow \text{Duds} \\ \leftarrow \text{Firecrackers} \end{matrix}$$

Carry out. We solve the proportion. We multiply by the LCD, 2880.

$$2880 \cdot \frac{9}{144} = 2880 \cdot \frac{D}{320}$$
$$180 = 9D$$
$$20 = D$$

Check. $\frac{9}{144} = 0.0625$, $\frac{20}{320} = 0.0625$

The ratios are the same, so the answer checks.

State. You would expect 20 duds in a sample of 320 fireworks.

53. ***Familiarize.*** The ratio of moose tagged to the total number of moose in the park, M, is $\frac{69}{M}$. Of the 40 moose caught later, 15 are tagged. The ratio of tagged moose to moose caught is $\frac{15}{40}$.

Translate. We translate to a proportion.

$$\begin{matrix} \text{Moose originally tagged} \rightarrow \\ \text{Moose in forest} \rightarrow \end{matrix} \frac{69}{M} = \frac{15}{40} \begin{matrix} \leftarrow \text{Tagged moose caught later} \\ \leftarrow \text{Moose caught later} \end{matrix}$$

Carry out. We solve the proportion. We multiply by the LCD, $40M$.

$$40M \cdot \frac{69}{M} = 40M \cdot \frac{15}{40}$$
$$40 \cdot 69 = M \cdot 15$$
$$2760 = 15M$$
$$\frac{2760}{15} = M$$
$$184 = M$$

Check.

$\frac{69}{184} = 0.375$, $\frac{15}{40} = 0.375$

The ratios are the same, so the answer checks.

State. We estimate that there are 184 moose in the park.

55. ***Familiarize.*** The ratio of the weight of an object on the moon to the weight of an object on the earth is 0.16 to 1.

a) We wish to find how much a 12-ton rocket would weigh on the moon.

b) We wish to find how much a 180-lb astronaut would weigh on the moon.

Translate. We translate to proportions.

a)
$$\begin{matrix} \text{Weight on the moon} \rightarrow \\ \text{Weight on earth} \rightarrow \end{matrix} \frac{0.16}{1} = \frac{T}{12} \begin{matrix} \leftarrow \text{Weight on the moon} \\ \leftarrow \text{Weight on earth} \end{matrix}$$

b)
$$\begin{matrix} \text{Weight on the moon} \rightarrow \\ \text{Weight on earth} \rightarrow \end{matrix} \frac{0.16}{1} = \frac{P}{180} \begin{matrix} \leftarrow \text{Weight on the moon} \\ \leftarrow \text{Weight on earth} \end{matrix}$$

Carry out. We solve each proportion.

a)
$$\frac{0.16}{1} = \frac{T}{12}$$
$$12(0.16) = T$$
$$1.92 = T$$

b)
$$\frac{0.16}{1} = \frac{P}{180}$$
$$180(0.16) = P$$
$$28.8 = P$$

Check. $\frac{0.16}{1} = 0.16$, $\frac{1.92}{12} = 0.16$, $\frac{28.8}{180} = 0.16$

The ratios are the same, so the answer checks.

State.

a) A 12-ton rocket would weigh 1.92 tons on the moon.

b) A 180-lb astronaut would weigh 28.8 lb on the moon.

57. *Writing Exercise*

59. Graph: $y = 2x - 6$.

We select some x-values and compute y-values.

If $x = 1$, then $y = 2 \cdot 1 - 6 = -4$.

If $x = 3$, then $y = 2 \cdot 3 - 6 = 0$.

If $x = 5$, then $y = 2 \cdot 5 - 6 = 4$.

x	y	(x, y)
1	-4	$(1, -4)$
3	0	$(3, 0)$
5	4	$(5, 4)$

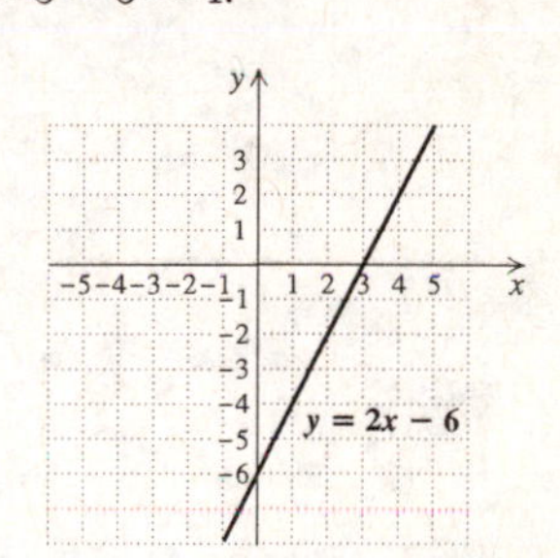

61. Graph: $3x + 2y = 12$.

We can replace either variable with a number and then calculate the other coordinate. We will find the intercepts and one other point.

If $y = 0$, we have:

$$3x + 2 \cdot 0 = 12$$
$$3x = 12$$
$$x = 4$$

The x-intercept is $(4, 0)$.

If $x = 0$, we have:

$$3 \cdot 0 + 2y = 12$$
$$2y = 12$$
$$y = 6$$

The y-intercept is $(0, 6)$.

If $y = -3$, we have:

$$3x + 2(-3) = 12$$
$$3x - 6 = 12$$
$$3x = 18$$
$$x = 6$$

The point $(6, -3)$ is on the graph.

We plot these points and draw a line through them.

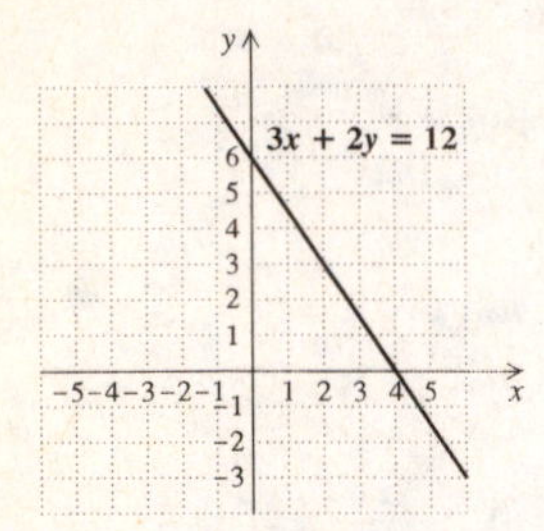

63. Graph: $y = -\frac{3}{4}x + 2$

We select some x-values and compute y-values. We use multiples of 4 to avoid fractions.

If $x = -4$, then $y = -\frac{3}{4}(-4) + 2 = 5$.

If $x = 0$, then $y = -\frac{3}{4} \cdot 0 + 2 = 2$.

If $x = 4$, then $y = -\frac{3}{4} \cdot 4 + 2 = -1$.

x	y	(x, y)
-4	5	$(-4, 5)$
0	2	$(0, 2)$
4	-1	$(4, -1)$

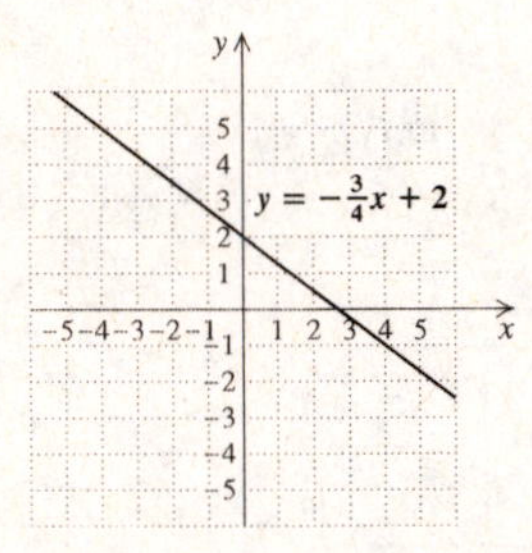

65. *Writing Exercise*

67. ***Familiarize.*** Let t = the time, in hours, it takes Michelle to wax the car alone. Then $\frac{t}{2}$ = Sal's time alone, and $t-2$ = Kristen's time alone. In 1 hr they do $\frac{1}{t} + \frac{1}{\frac{t}{2}} + \frac{1}{t-2}$, or $\frac{1}{t} + \frac{2}{t} + \frac{1}{t-2}$ of the job working together. The entire job takes 1 hr and 20 min, or $\frac{4}{3}$ hr.

Translate. To get an entire job, we multiply the amount of work done in 1 hr by the number of hours required to complete the job.

$$\frac{4}{3}\left(\frac{1}{t} + \frac{2}{t} + \frac{1}{t-2}\right) = 1$$

Carry out. We solve the equation.

$$3 \cdot \frac{4}{3}\left(\frac{1}{t} + \frac{2}{t} + \frac{1}{t-2}\right) = 3 \cdot 1$$

$$4\left(\frac{1}{t} + \frac{2}{t} + \frac{1}{t-2}\right) = 3$$

$$4\left(\frac{3}{t} + \frac{1}{t-2}\right) = 3 \quad \text{Adding: } \frac{1}{t} + \frac{2}{t} = \frac{3}{t}$$

$$\frac{12}{t} + \frac{4}{t-2} = 3$$

$$t(t-2)\left(\frac{12}{t} + \frac{4}{t-2}\right) = t(t-2)(3)$$

$$12(t-2) + t \cdot 4 = 3t(t-2)$$

$$12t - 24 + 4t = 3t^2 - 6t$$

$$16t - 24 = 3t^2 - 6t$$

$$0 = 3t^2 - 22t + 24$$

$$0 = (3t - 4)(t - 6)$$

$$3t - 4 = 0 \quad or \quad t - 6 = 0$$

$$3t = 4 \quad or \quad t = 6$$

$$t = \frac{4}{3} \quad or \quad t = 6$$

Check. If $t = \frac{4}{3}$, then $t - 2 = -\frac{2}{3}$. Since time cannot be negative in this problem, $\frac{4}{3}$ cannot be a solution. If $t = 6$, then $t/2 = 3$ and $t - 2 = 4$. If Michelle, Sal, and Kristen can do the job in 6 hr, 3 hr, and 4 hr, respectively, then in one hour they do $\frac{1}{6} + \frac{1}{3} + \frac{1}{4}$, or $\frac{3}{4}$ of the job working together and in $\frac{4}{3}$ hr they do $\frac{4}{3} \cdot \frac{3}{4}$, or 1 entire job. The answer checks.

State. Working alone, the job would take Michelle 6 hr, Sal 3 hr, and Kristen 4 hr.

69. ***Familiarize.*** Let t = the number of hours it takes to wire one house, working together. We want to find the number of hours it takes to wire two houses, working together.

Translate. We write an equation.

$$\frac{t}{28} + \frac{t}{34} = 2$$

Carry out. We solve the equation.

$$\frac{t}{28} + \frac{t}{34} = 2, \text{ LCD} = 476$$

$$476\left(\frac{t}{28} + \frac{t}{34}\right) = 476 \cdot 2$$

$$17t + 14t = 952$$

$$31t = 952$$

$$t = \frac{952}{31}, \text{ or } 32\frac{22}{31}$$

Check. If $30\frac{22}{31}$ hr, Janet does $\frac{952}{31} \cdot \frac{1}{28} = \frac{34}{31}$ of one complete job and Linus does $\frac{952}{31} \cdot \frac{1}{34}$, or $\frac{28}{31}$ of one complete job. Together they do $\frac{34}{31} + \frac{28}{31}$, or $\frac{62}{31}$, or 2 complete jobs. The answer checks.

State. It will take Janet and Linus $30\frac{22}{31}$ hr to wire two houses, working together.

71. ***Familiarize***. The correct ratio of oil to gasoline is 3.2/160, or 0.02. The ratio in Gus' original mixture is 5.6/200, or 0.028. Since this is a larger number than 0.02, Gus needs to add more gasoline to make the ratio lower. Let $g =$ the number of ounces of gasoline Gus should add.

Translate. We translate to an equation.

$$\begin{array}{r} \text{Oil} \rightarrow \\ \text{Gasoline} \rightarrow \end{array} \frac{5.6}{200+g} = 0.02$$

Carry out. We solve the equation.

$$(200+g)\cdot\frac{5.6}{200+g} = (200+g)(0.02)$$
$$5.6 = 4 + 0.02g$$
$$1.6 = 0.02g$$
$$80 = g$$

Check. If Gus adds an additional 80 oz of gasoline, the ratio of oil to gasoline is $\frac{5.6}{200+80} = \frac{5.6}{280} = 0.02$, the correct ratio. The answer checks.

State. Gus should add 80 oz of gasoline.

73. ***Familiarize***. Let $x =$ the numerator in the equivalent ratio. Then $104 - x =$ the denominator.

Translate. The ratios $\frac{9}{17}$ and $\frac{x}{104-x}$ are equivalent, so we write a proportion.

$$\frac{9}{17} = \frac{x}{104-x}$$

Carry out. We solve the proportion. We multiply by the LCD, $17(104-x)$.

$$17(104-x)\cdot\frac{9}{17} = 17(104-x)\cdot\frac{x}{104-x}$$
$$9(104-x) = 17x$$
$$936 - 9x = 17x$$
$$936 = 26x$$
$$36 = x$$

Then $104 - x = 104 - 36 = 68$.

Check. $\frac{36}{68} = \frac{4\cdot 9}{4\cdot 17} = \frac{9}{17}$, so the ratios are equivalent.

State. The numerator will be 36, and the denominator will be 68.

75. Find a second proportion:

$$\frac{A}{B} = \frac{C}{D} \quad \text{Given}$$
$$\frac{D}{A}\cdot\frac{A}{B} = \frac{D}{A}\cdot\frac{C}{D} \quad \text{Multiplying by } \frac{D}{A}$$
$$\frac{D}{B} = \frac{C}{A}$$

Find a third proportion:

$$\frac{A}{B} = \frac{C}{D} \quad \text{Given}$$
$$\frac{B}{C}\cdot\frac{A}{B} = \frac{B}{C}\cdot\frac{C}{D} \quad \text{Multiplying by } \frac{B}{C}$$
$$\frac{A}{C} = \frac{B}{D}$$

Find a fourth proportion:

$$\frac{A}{B} = \frac{C}{D} \quad \text{Given}$$
$$\frac{DB}{AC}\cdot\frac{A}{B} = \frac{DB}{AC}\cdot\frac{C}{D} \quad \text{Multiplying by } \frac{DB}{AC}$$
$$\frac{D}{C} = \frac{B}{A}$$

77. *Writing Exercise*

Chapter 7

Functions and Graphs

Exercise Set 7.1

1. For any function, the set of all inputs, or first values, is called the domain.

3. In any function, each member of the domain is paired with exactly one member of the range.

5. In the notation $f(5) = 8$, the input is 5.

7. In the notation $f(x) = c$, the independent variable is x.

9. The correspondence is a function, because each member of the domain corresponds to exactly one member of the range.

11. The correspondence is a function, because each member of the domain corresponds to exactly one member of the range.

13. The correspondence is a function, because each member of the domain corresponds to exactly one member of the range.

15. This correspondence is not a function because a member of the domain (July 24) corresponds to more than one member of the range. (July 18 also corresponds to more than one member of the range.)

17. This correspondence is a function, because each pumpkin has only one price.

The correspondence is a relation, since it is reasonable to assume that each member of a rock band plays at least one instrument.

19. This correspondence is a function, because each player has only one uniform number.

21. a) Locate 1 on the horizontal axis and then find the point on the graph for which 1 is the first coordinate. From that point, look to the vertical axis to find the corresponding y-coordinate, -1. Thus, $f(1) = -1$.

b) The domain is the set of all x-values in the graph. It is $\{x| -4 \leq x \leq 3\}$.

c) To determine which member(s) of the domain are paired with 2, locate 2 on the vertical axis. From there look left and right to the graph to find any points for which 2 is the second coordinate. One such point exists. Its first coordinate is -3. Thus, the x-value for which $f(x) = 2$ is -3.

d) The range is the set of all y-values in the graph. It is $\{y| -2 \leq y \leq 5\}$.

23. a) Locate 1 on the horizontal axis and then find the point on the graph for which 1 is the first coordinate. From that point, look to the vertical axis to find the corresponding y-coordinate, 3. Thus, $f(1) = 3$.

b) The set of all x-values in the graph extends from -1 to 4, so the domain is $\{x| -1 \leq x \leq 4\}$.

c) To determine which member(s) of the domain are paired with 2, locate 2 on the vertical axis. From there look left and right to the graph to find any points for which 2 is the second coordinate. One such point exists. Its first coordinate is 3. Thus, the x-value for which $f(x) = 2$ is 3.

d) The set of all y-values in the graph extends from 1 to 4, so the range is $\{y|1 \leq y \leq 4\}$.

25. a) Locate 1 on the horizontal axis and then find the point on the graph for which 1 is the first coordinate. From that point, look to the vertical axis to find the corresponding y-coordinate. It appears to be 3. Thus, $f(1) = 3$.

b) The set of all x-values in the graph extends from -4 to 3 so the domain is $\{x| -4 \leq x \leq 3\}$.

c) To determine which member(s) of the domain are paired with 2, locate 2 on the vertical axis. From there look left and right to the graph to find any points for which 2 is the second coordinate. One such point exists. Its first coordinate is 0, so the x-value for which $f(x) = 2$ is 0.

d) The set of all y-values in the graph extends from -5 to 4, so the range is $\{y| -5 \leq y \leq 4\}$.

27. a) Locate 1 on the horizontal axis and then find the point on the graph for which 1 is the first coordinate. From that point, look to the vertical axis to find the corresponding y-coordinate, 3. Thus, $f(1) = 3$.

b) The set of all x-values in the graph extends from -4 to 3, so the domain is $\{x| -4 \leq x \leq 3\}$.

c) To determine which member(s) of the domain are paired with 2, locate 2 on the vertical axis. From there look left and right to the graph to find any points for which 2 is the second coordinate. One such point exists. Its first coordinate is -3. Thus, the x-value for which $f(x) = 2$ is -3.

d) The set of all y-values in the graph extends from -2 to 5, so the range is $\{y| -2 \leq y \leq 5\}$.

29. a) Locate 1 on the horizontal axis and then find the point on the graph for which 1 is the first coordinate. From that point, look to the vertical axis to find the corresponding y-coordinate, 1. Thus, $f(1) = 1$.

b) The domain is the set of all x-values in the graph. It is $\{-3, -1, 1, 3, 5\}$.

c) To determine which member(s) of the domain are paired with 2, locate 2 on the vertical axis. From there look left and right to the graph to find any points for which 2 is the second coordinate. One such point exists. Its first coordinate is 3. Thus, the x-value for which $f(x) = 2$ is 3.

d) The range is the set of all y-values in the graph. It is $\{-1, 0, 1, 2, 3\}$.

31. a) Locate 1 on the horizontal axis and then find the point on the graph for which 1 is the first coordinate. From that point, look to the vertical axis to find the corresponding y-coordinate, 4. Thus, $f(1) = 4$.

b) The set of all x-values in the graph extends from -3 to 4, so the domain is $\{x| -3 \leq x \leq 4\}$.

c) To determine which member(s) of the domain are paired with 2, locate 2 on the vertical axis. From there look left and right to the graph to find any points for which 2 is the second coordinate. There are two such points, $(-1, 2)$ and $(3, 2)$. Thus, the x-values for which $f(x) = 2$ are -1 and 3.

d) The set of all y-values in the graph extends from -4 to 5, so the range is $\{y| -4 \leq y \leq 5\}$.

33. a) Locate 1 on the horizontal axis and then find the point on the graph for which 1 is the first coordinate. From that point, look to the vertical axis to find the corresponding y-coordinate, 2. Thus, $f(1) = 2$.

b) The set of all x-values in the graph extends from -4 to 4, so the domain is $\{x| -4 \leq x \leq 4\}$.

c) To determine which member(s) of the domain are paired with 2, locate 2 on the vertical axis. From there look left and right to the graph to find any points for which 2 is the second coordinate. All points in the set $\{x|0 < x \leq 2\}$ satisfy this condition. These are the x-values for which $f(x) = 2$.

d) The range is the set of all y-values in the graph. It is $\{1, 2, 3, 4\}$.

35. $g(x) = x + 5$

a) $g(0) = 0 + 5 = 5$

b) $g(-4) = -4 + 5 = 1$

c) $g(-7) = -7 + 5 = -2$

d) $g(8) = 8 + 5 = 13$

e) $g(a + 2) = a + 2 + 5 = a + 7$

f) $g(a) + 2 = a + 5 + 2 = a + 7$

37. $f(n) = 5n^2 + 4n$

a) $f(0) = 5 \cdot 0^2 + 4 \cdot 0 = 0 + 0 = 0$

b) $f(-1) = 5(-1)^2 + 4(-1) = 5 - 4 = 1$

c) $f(3) = 5 \cdot 3^2 + 4 \cdot 3 = 45 + 12 = 57$

d) $f(t) = 5t^2 + 4t$

e) $f(2a) = 5(2a)^2 + 4 \cdot 2a = 5 \cdot 4a^2 + 8a = 20a^2 + 8a$

f) $f(3) - 9 = 5 \cdot 3^2 + 4 \cdot 3 - 9 = 5 \cdot 9 + 4 \cdot 3 - 9 = 45 + 12 - 9 = 48$

39. $f(x) = \dfrac{x - 3}{2x - 5}$

a) $f(0) = \dfrac{0 - 3}{2 \cdot 0 - 5} = \dfrac{-3}{0 - 5} = \dfrac{-3}{-5} = \dfrac{3}{5}$

b) $f(4) = \dfrac{4 - 3}{2 \cdot 4 - 5} = \dfrac{1}{8 - 5} = \dfrac{1}{3}$

c) $f(-1) = \dfrac{-1 - 3}{2(-1) - 5} = \dfrac{-4}{-2 - 5} = \dfrac{-4}{-7} = \dfrac{4}{7}$

d) $f(3) = \dfrac{3 - 3}{2 \cdot 3 - 5} = \dfrac{0}{6 - 5} = \dfrac{0}{1} = 0$

e) $f(x + 2) = \dfrac{x + 2 - 3}{2(x + 2) - 5} = \dfrac{x - 1}{2x + 4 - 5} = \dfrac{x - 1}{2x - 1}$

41. $A(s) = s^2\dfrac{\sqrt{3}}{4}$

$A(4) = 4^2\dfrac{\sqrt{3}}{4} = 4\sqrt{3} \approx 6.93$

The area is $4\sqrt{3}$ cm$^2 \approx 6.93$ cm^2.

43. $V(r) = 4\pi r^2$

$V(3) = 4\pi(3)^2 = 36\pi$

The area is 36π in$^2 \approx 113.10$ in^2.

45. $P(d) = 1 + \dfrac{d}{33}$

$P(20) = 1 + \dfrac{20}{33} = 1\dfrac{20}{33}$ atm

$P(30) = 1 + \dfrac{30}{33} = 1 + \dfrac{10}{11} = 1\dfrac{10}{11}$ atm

$P(100) = 1 + \dfrac{100}{33} = 1 + 3\dfrac{1}{33} = 4\dfrac{1}{33}$ atm

47. $H(x) = 2.75x + 71.48$

$H(32) = 2.75(32) + 71.48 = 159.48$

The predicted height is 159.48 cm.

49. $F(C) = \dfrac{9}{5}C + 32$

$F(-10) = \dfrac{9}{5}(-10) + 32 = -18 + 32 = 14$

The equivalent temperature is 14°F.

51. Locate the point that is directly above 225. Then estimate its second coordinate by moving horizontally from the point to the vertical axis. The rate is about 75 per 10,000 men.

53. Locate the point on the graph that is directly above '60. Then estimate its second coordinate by moving horizontally from the point to the vertical axis. In 1960, about 56% of Americans were willing to vote for a woman for president. That is, $P(1960) \approx 56\%$.

55. Plot and connect the points, using CFL wattage as the first coordinate and the wattage of the incandescent equivalent as the second coordinate.

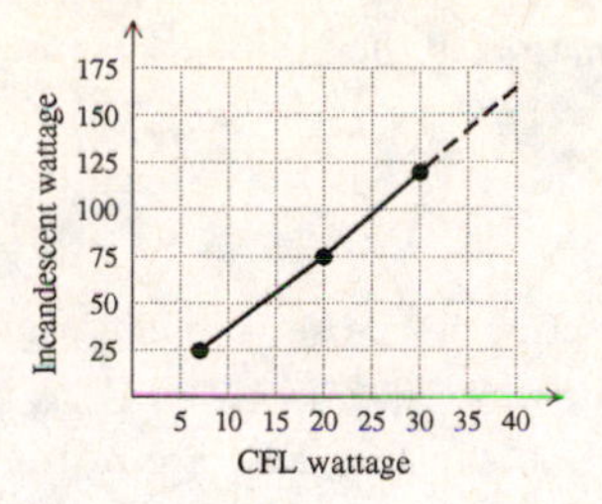

To estimate the wattage of an incandescent bulb that creates light equivalent to a 15-watt CFL bulb, first locate the point that is directly above 15. Then estimate the second coordinate by moving horizontally from the point to the vertical axis. Read the approximate function value there. The wattage is about 60 watts.

To predict the wattage of an incandescent bulb that creates light equivalent to a 35-watt CFL bulb, extend the graph and extrapolate. The wattage is about 140 watts.

57. Plot and connect the points, using body weight as the first coordinate and the corresponding number of drinks as the second coordinate.

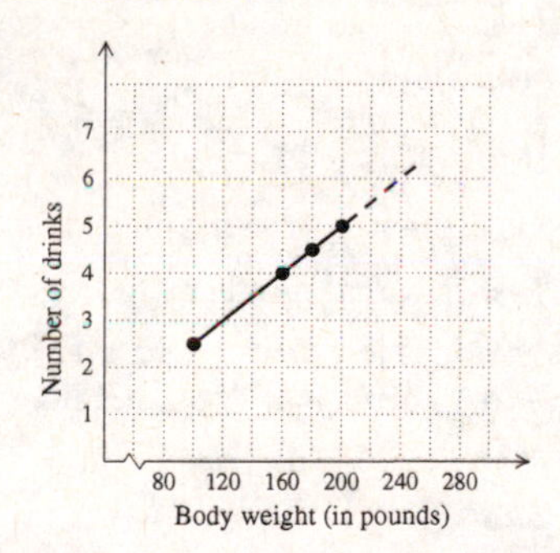

To estimate the number of drinks that a 140-lb person would have to drink to be considered intoxicated, first locate the point that is directly above 140. Then estimate its second coordinate by moving horizontally from the point to the vertical axis. Read the approximate function value there. The estimated number of drinks is 3.5.

To predict the number of drinks it would take for a 250-lb person to be considered intoxicated, extend the graph and extrapolate. It appears that it would take about 6 drinks.

59. Plot and connect the points, using the year as the first coordinate and the corresponding number of reported cases of AIDS as the second coordinate.

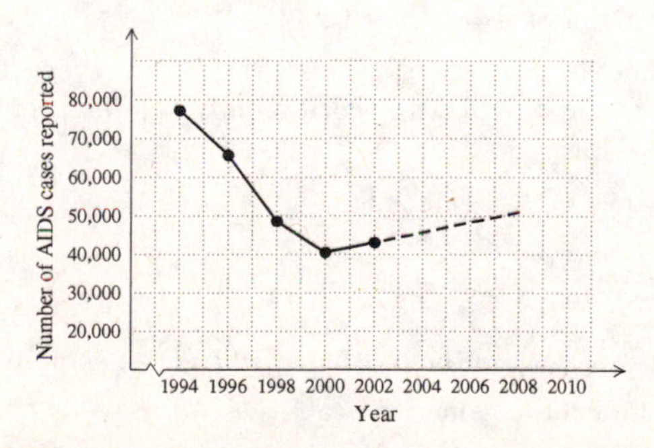

To estimate the number of cases of AIDS reported in 1997, first locate the point that is directly above 1997. Then move horizontally from the point to the vertical axis and read the approximate function value there. We estimate that about 57,000 cases of AIDS were reported in 1997.

To estimate the number of cases of AIDS that will be reported in 2007, extend the graph and extrapolate. It appears that about 50,000 cases of AIDS will be reported in 2007.

61. Plot and connect the points, using the year as the first coordinate and the total sales as the second coordinate.

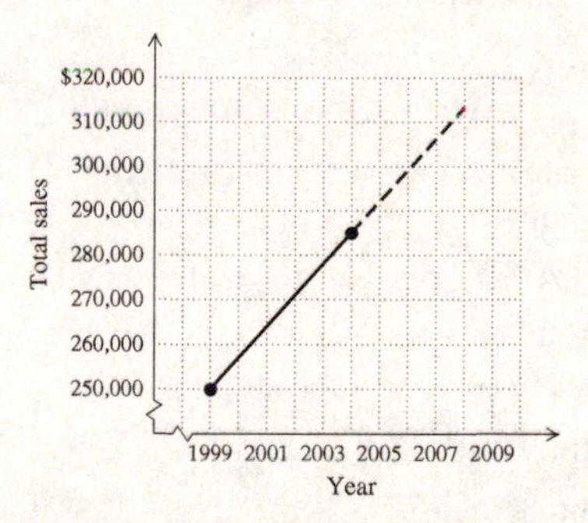

To estimate the total sales for 2000, first locate the point directly above 2000. Then estimate its second coordinate by moving horizontally to the vertical axis. Read the approximate function value there. The estimated 2000 sales total is about $257,000.

To estimate the sales for 2007, extend the graph and extrapolate. We estimate the sales for 2007 to be about $306,000.

63. *Writing Exercise*

65. $\dfrac{10-3^2}{9-2\cdot 3} = \dfrac{10-9}{9-6} = \dfrac{1}{3}$

67.
$$\begin{aligned} S &= 2lh + 2lw + 2wh \\ S - 2wh &= 2lh + 2lw \\ S - 2wh &= l(2h + 2w) \\ \frac{S-2wh}{2h+2w} &= l \end{aligned}$$

69.
$$\begin{aligned} 2x + 3y &= 6 \\ 3y &= 6 - 2x \\ y &= \frac{6-2x}{3}, \text{ or } 2 - \frac{2}{3}x, \text{ or } -\frac{2}{3}x + 2 \end{aligned}$$

71. *Writing Exercise*

73. To find $f(g(-4))$, we first find $g(-4)$:

$g(-4) = 2(-4) + 5 = -8 + 5 = -3$.

Then $f(g(-4)) = f(-3) = 3(-3)^2 - 1 = 3\cdot 9 - 1 = 27 - 1 = 26$.

To find $g(f(-4))$, we first find $f(-4)$:

$f(-4) = 3(-4)^2 - 1 = 3\cdot 16 - 1 = 48 - 1 = 47$.

Then $g(f(-4)) = g(47) = 2\cdot 47 + 5 = 94 + 5 = 99$.

75. $f(\text{tiger}) = \text{dog}$

$f(\text{dog}) = f(f(\text{tiger})) = \text{cat}$

$f(\text{cat}) = f(f(f(\text{tiger}))) = \text{fish}$

$f(\text{fish}) = f(f(f(f(\text{tiger})))) = \text{worm}$

77. Locate the highest point on the graph. Then move vertically to the horizontal axis and read the corresponding time. It is about 2 min, 50 sec.

79. The two largest contractions occurred at about 2 minutes, 50 seconds and 5 minutes, 40 seconds. The difference in these times, is 2 minutes, 50 seconds, so the frequency is about 1 every 3 minutes.

81. We know that $(-1,-7)$ and $(3,8)$ are both solutions of $g(x)=mx+b$. Substituting, we have

$$-7=m(-1)+b, \text{ or } -7=-m+b,$$
$$\text{and } 8=m(3)+b, \text{ or } 8=3m+b.$$

Solve the first equation for b and substitute that expression into the second equation.

$-7=-m+b$ First equation

$m-7=b$ Solving for b

$8=3m+b$ Second equation

$8=3m+(m-7)$ Substituting

$8=3m+m-7$

$8=4m-7$

$15=4m$

$\frac{15}{4}=m$

We know that $m-7=b$, so $\frac{15}{4}-7=b$, or $-\frac{13}{4}=b$.

We have $m=\frac{15}{4}$ and $b=-\frac{13}{4}$, so $g(x)=\frac{15}{4}x-\frac{13}{4}$.

Exercise Set 7.2

1. Since $-1\le 0<10$, function (c) would be used.

3. Since $10\ge 10$, function (d) would be used.

5. Since $-1\le -1<10$, function (c) would be used.

7. The domain is the set of all first coordinates, $\{2,9,-2,-4\}$.

The range is the set of all second coordinates, $\{8,3,10,4\}$.

9. The domain is the set of all first coordinates, $\{0,4,-5,-1\}$.

The range is the set of all second coordinates, $\{0,-2\}$.

11. The function f can be written as $\{(-4,-2),(-2,-1),(0,0),(2,1),(4,2)\}$. The domain is the set of all first coordinates, $\{-4,-2,0,2,4\}$ and the range is the set of all second coordinates, $\{-2,-1,0,1,2\}$.

Domain of $f=\{-4,-2,0,3,5\}$; range of $f=\{4,1,3,-2,0\}$.

13. The function f can be written as $\{(-5,-1),(-3,-1),(-1,-1),(0,1),(2,1),(4,1)\}$. The domain is the set of all first coordinates, $\{-5,-3,-1,0,2,4\}$ and the range is the set of all second coordinates, $\{-1,1\}$.

15. The domain of the function is the set of all x-values that are in the graph, $\{x|-4\le x\le 3\}$.

The range is the set of all y-values that are in the graph, $\{y|-3\le y\le 4\}$.

17. The domain of the function is the set of all x-values that are in the graph, $\{x|-4\le x\le 5\}$.

The range is the set of all y-values that are in the graph, $\{y|-2\le y\le 4\}$.

19. The domain of the function is the set of all x-values that are in the graph, $\{x|-4\le x\le 4\}$.

The range is the set of all y-values that are in the graph, $\{-3,-1,1\}$.

21. For any x-value and for any y-value there is a point on the graph. Thus,

Domain of $f=\{x|x \text{ is a real number}\}$ and

Range of $f=\{y|y \text{ is a real number}\}$.

23. For any x-value there is a point on the graph. Thus,

Domain of $f=\{x|x \text{ is a real number}\}$.

The only y-value on the graph is 4. Thus,

Range of $f=\{4\}$.

25. For an x-value there is a point on the graph. Thus,

Domain of $f=\{x|x \text{ is a real number}\}$.

The function has no y-values less than 1 and every y-value greater than or equal to 1 corresponds to a member of the domain. Thus,

Range of $f=\{y|y\ge 1\}$.

27. The hole in the graph at $(-2,-4)$ indicates that the function is not defined for $x=-2$. For any other x-value there is a point on the graph. Thus,

Domain of $f=\{x|x \text{ is a real number } and\ x\ne -2\}$.

There is no function value at $(-2,-4)$, so -4 is not in the range of the function. For any other y-value there is a point on the graph. Thus,

Range of $f=\{y|y \text{ is a real number } and\ y\ne -4\}$.

29. The function has no x-values less than 0 and every x-value greater than or equal to 0 corresponds to a member of the domain. Thus,

Domain of $f=\{x|x\ge 0\}$.

The function has no y-values less than 0 and every y-value greater than or equal to 0 corresponds to a member of the range. Thus,

Range of $f=\{y|y\ge 0\}$.

31. $f(x)=\frac{5}{x-3}$

Since $\frac{5}{x-3}$ cannot be computed when the denominator is 0, we find the x-value that causes $x-3$ to be 0:

$x-3=0$

$x=3$ Adding 3 to both sides

Thus, 3 is not in the domain of f, while all other real numbers are. The domain of f is $\{x|x \text{ is a real number } and\ x \neq 3\}$.

33. $f(x) = \dfrac{3}{2x-1}$

Since $\dfrac{3}{2x-1}$ cannot be computed when the denominator is 0, we find the x-value that causes $2x-1$ to be 0:

$$2x - 1 = 0$$
$$2x = 1$$
$$x = \frac{1}{2}$$

Thus, $\dfrac{1}{2}$ is not in the domain of f, while all other real numbers are. The domain of f is $\left\{x|x \text{ is a real number } and\ x \neq \dfrac{1}{2}\right\}$.

35. $f(x) = 2x + 1$

Since we can compute $2x+1$ for any real number x, the domain is the set of all real numbers.

37. $g(x) = |5 - x|$

Since we can compute $|5-x|$ for any real number x, the domain is the set of all real numbers.

39. $f(x) = \dfrac{5}{x^2-9}$

The expression $\dfrac{5}{x^2-9}$ is undefined when $x^2 - 9 = 0$.

$$x^2 - 9 = 0$$
$$(x+3)(x-3) = 0$$
$$x + 3 = 0 \quad or \quad x - 3 = 0$$
$$x = -3 \quad or \quad x = 3$$

Thus, Domain of $f = \{x|x \text{ is a real number } and\ x \neq -3\ and\ x \neq 3\}$.

41. $f(x) = x^2 - 9$

Since we can compute $x^2 - 9$ for any real number x, the domain is the set of all real numbers.

43. $f(x) = \dfrac{2x-7}{x^2+8x+7}$

The expression $\dfrac{2x-7}{x^2+8x+7}$ is undefined when $x^2 + 8x + 7 = 0$.

$$x^2 + 8x + 7 = 0$$
$$(x+1)(x+7) = 0$$
$$x + 1 = 0 \quad or \quad x + 7 = 0$$
$$x = -1 \quad or \quad x = -7$$

Thus, Domain of $f = \{x|x \text{ is a real number } and\ x \neq -1\ and\ x \neq -7\}$.

45. $R(t) = 46.8 - 0.075t$

If we assume the function is not valid for years before 1930, we must have $t \geq 0$. In addition, $R(t)$ must be positive, so we have:

$$46.8 - 0.075t > 0$$
$$-0.075t > -46.8$$
$$t < 624$$

Then the domain of the function is $\{t|0 \leq t < 624\}$.

47. $A(p) = -2.5p + 26.5$

The price must be positive, so we have $p > \$0$. In addition $A(p)$ must be nonnegative, so we have:

$$-2.5p + 26.5 \geq 0$$
$$26.5 \geq 2.5p$$
$$10.6 \geq p$$

Then the domain of the function is $\{p|\$0 < p \leq \$10.60\}$.

49. $P(d) = 0.03d + 1$

The depth must be nonnegative, so we have $d \geq 0$. In addition, $P(d)$ must be nonnegative, so we have:

$$0.03d + 1 \geq 0$$
$$0.03d \geq -1$$
$$d \geq -33.\overline{3}$$

Then we have $d \geq 0$ *and* $d \geq -33.\overline{3}$, so the domain of the function is $\{d|d \geq 0\}$.

51. $h(t) = -16t^2 + 64t + 80$

The time cannot be negative, so we have $t \geq 0$. The height cannot be negative either, so an upper limit for t will be the positive value of t for which $h(t) = 0$.

$$-16t^2 + 64t + 80 = 0$$
$$-16(t^2 - 4t - 5) = 0$$
$$-16(t-5)(t+1) = 0$$
$$t - 5 = 0 \quad or \quad t + 1 = 0$$
$$t = 5 \quad or \quad t = -1$$

We know that -1 is not in the domain of the function. We also see that 5 is an upper limit for t. Then the domain of the function is $\{t|0 \leq t \leq 5\}$.

53. $f(x) = \begin{cases} x, & \text{if } x < 0 \\ 2x+1, & \text{if } x \geq 0 \end{cases}$

a) Since $-5 < 0$, we use the equation $f(x) = x$.

Thus, $f(-5) = -5$.

b) Since $0 \geq 0$, we use the equation $f(x) = 2x + 1$.

$f(0) = 2 \cdot 0 + 1 = 0 + 1 = 1$

c) Since $10 \geq 0$, we use the equation $f(x) = 2x + 1$.

$f(10) = 2 \cdot 10 + 1 = 20 + 1 = 21$

55. $G(x) = \begin{cases} x - 5, & \text{if } x < -1 \\ x, & \text{if } -1 \leq x \leq 2 \\ x + 2, & \text{if } x > 2 \end{cases}$

a) Since $-1 \leq 0 \leq 2$, we use the equation $G(x) = x$.

$G(0) = 0$

b) Since $-1 \leq 2 \leq 2$, we use the equation $G(x) = x$.

$G(2) = 2$

c) Since $5 > 2$, we use the equation $G(x) = x + 2$.

$G(5) = 5 + 2 = 7$

57. $f(x) = \begin{cases} x^2 - 10, \text{ if } x < -10 \\ x^2, \text{ if } -10 \le x \le 10 \\ x^2 + 10, \text{ if } x > 10 \end{cases}$

a) Since $-10 \le -10 \le 10$, we use the equation $f(x) = x^2$.

$f(-10) = (-10)^2 = 100$

b) Since $-10 \le 10 \le 10$, we use the equation $f(x) = x^2$.

$f(10) = 10^2 = 100$

c) Since $11 > 10$, we use the equation $f(x) = x^2 + 10$.

$f(11) = 11^2 + 10 = 121 + 10 = 131$

59. *Writing Exercise*

61. $y = 2x - 3$

First we plot the y-intercept, $(0, -3)$. We can think of the slope as $\frac{2}{1}$. Starting at $(0, -3)$, find a second point by moving up 2 units and to the right 1 unit to the point $(1, -1)$. In a similar manner we can move from $(1, -1)$ to $(2, 1)$. Then we connect the points.

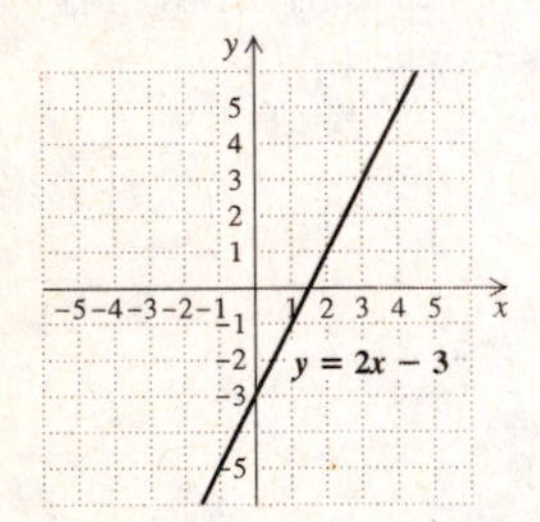

63. $\frac{2}{3}x - 4$

The slope is $\frac{2}{3}$, and the y-intercept is $(0, -4)$.

65. $y = \frac{4}{3}x$, or $y = \frac{4}{3}x + 0$

The slope is $\frac{4}{3}$ and the y-intercept is $(0, 0)$.

67. *Writing Exercise*

69.

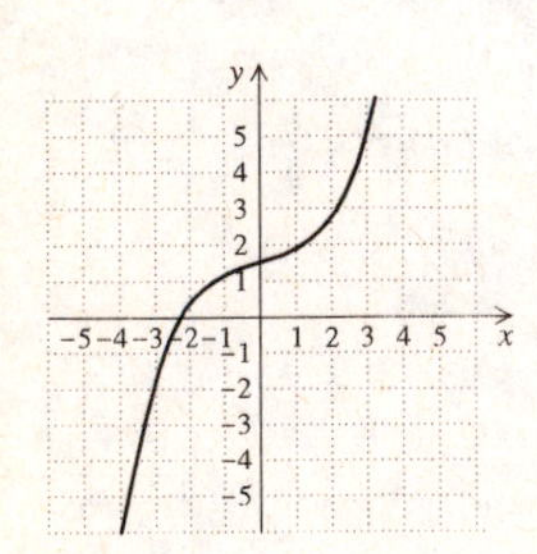

71.

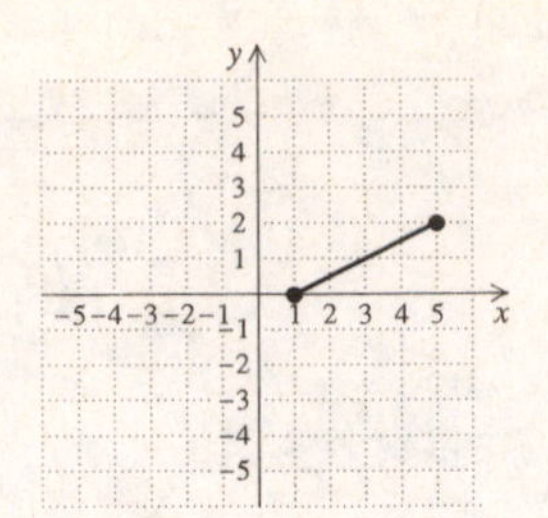

73. The graph indicates that the function is not defined for $x = 0$. For any other x-value there is a point on the graph. Thus,

Domain of $f = \{x|x \text{ is a real number } and\ x \neq 0\}$.

The graph also indicates that the function is not defined for $y = 0$. For any other y-value there is a point on the graph. Thus,

Domain of $f = \{y|y \text{ is a real number } and\ y \neq 0\}$.

75. The function has no x-values for $-2 \le x \le 0$. For any other x-value there is a point on the graph. Thus, the domain of the function is $\{x|x < -2 \text{ } or \text{ } x > 0\}$.

The function has no y-values for $-2 \le y \le 3$. Every other y-value corresponds to a member of the range. Then the range is $\{y|y < -2 \text{ } or \text{ } y > 3\}$.

77.

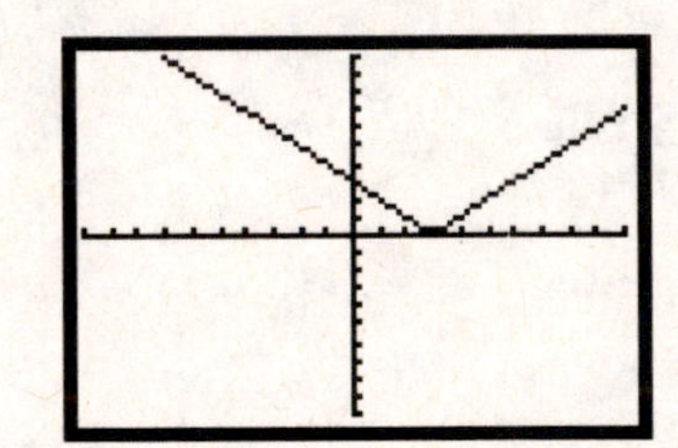

From the graph we see that the domain of f is $\{x|x \text{ is a real number}\}$ and the range is $\{y|y \ge 0\}$.

79.

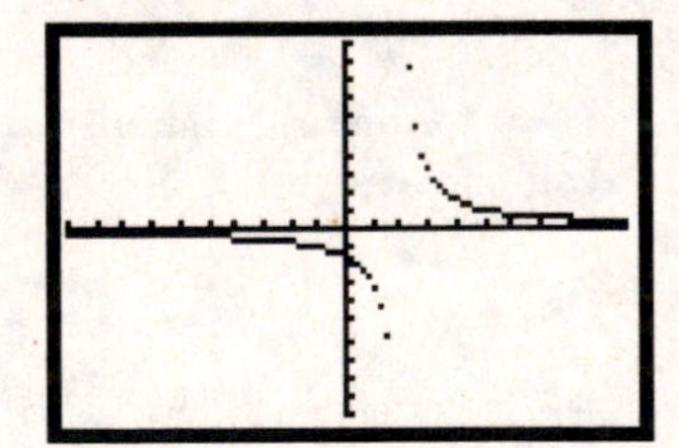

From the graph we see that the domain of f is $\{x|x \text{ is a real number } and\ y \neq 2\}$ and the range is $\{y|y \text{ is a real number } and\ y \neq 0\}$.

81. We graph the function $h(t) = -16t^2 + 64t + 80$ in the window $[0, 5, -5, 150]$ with Xscl = 1 and Yscl = 15.

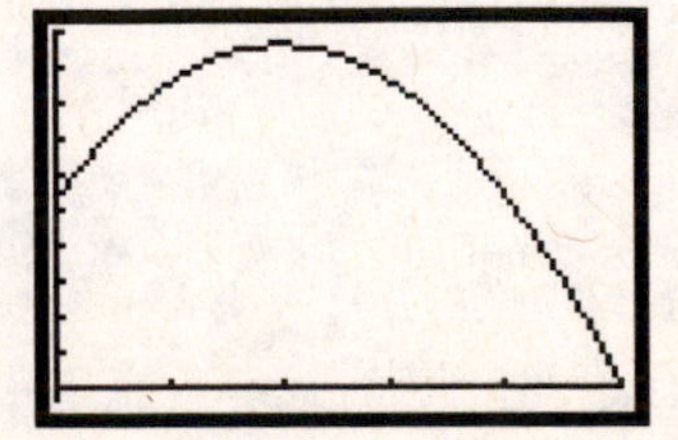

From the graph we estimate that the range of the function is $\{h|0 \leq h \leq 144\}$.

83.

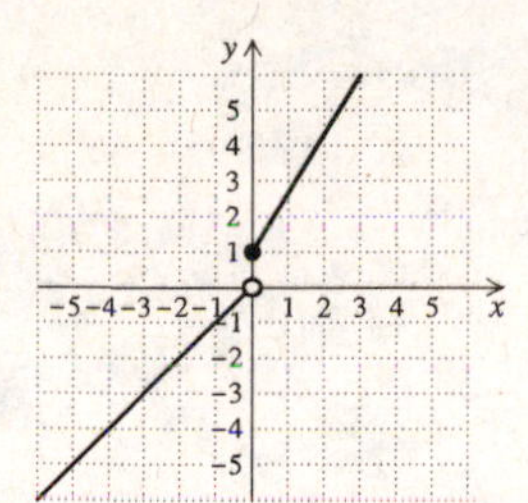

85.

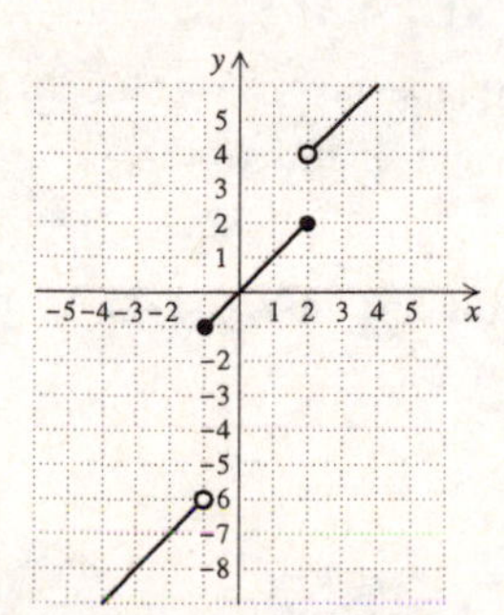

87.

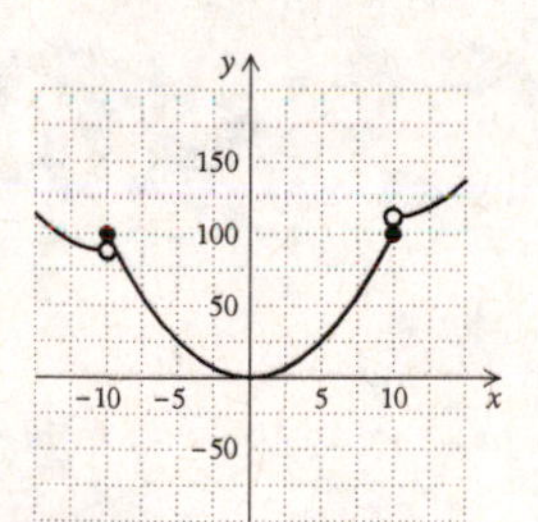

89. 

Exercise Set 7.3

1. False; see page 464 in the text.

3. False; unless restricted, the domain of a constant function is the set of all real numbers.

5. True; see page 467 in the text.

7. We can use the vertical-line test:

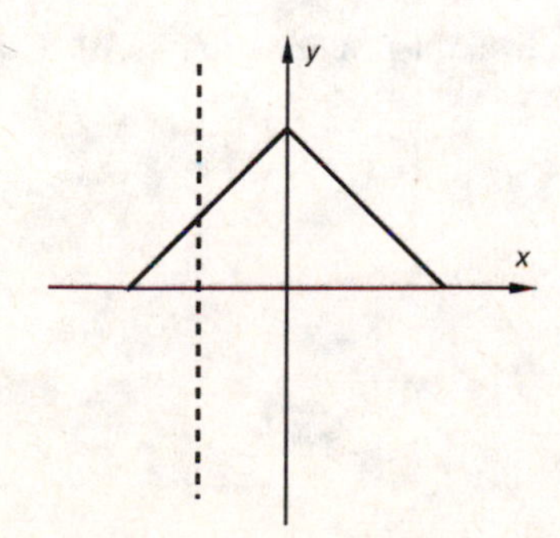

Visualize moving this vertical line across the graph. No vertical line will intersect the graph more than once. Thus, the graph is a graph of a function.

9. We can use the vertical-line test:

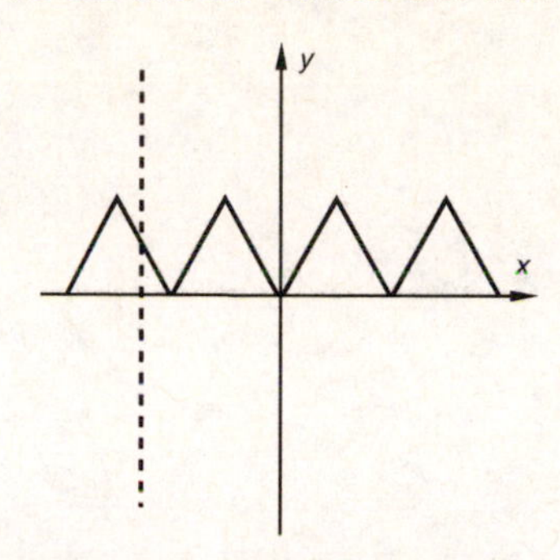

Visualize moving this vertical line across the graph. No vertical line will intersect the graph more than once. Thus, the graph is a graph of a function.

11. We can use the vertical-line test.

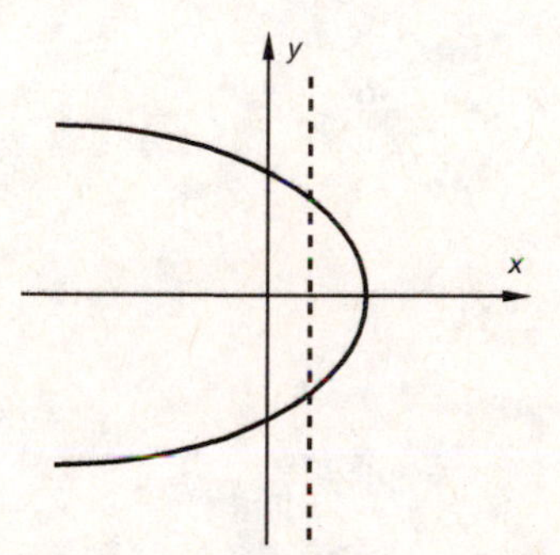

It is possible for a vertical line to intersect the graph more than once. Thus this is not the graph of a function.

13. We can use the vertical-line test.

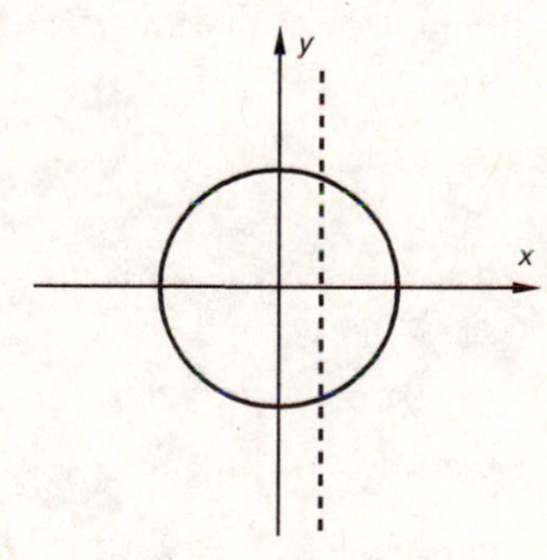

It is possible for a vertical line to intersect the graph more than once. Thus this is not a graph of a function.

15. This is a straight line that is not vertical, so it is the graph of a function.

17. The vertical line that is the graph itself intersects the graph more than once, so this is not the graph of a function.

19. Graph $x - y = 4$.

First we determine the intercepts.

Let $y = 0$ to find the x-intercept.

$$x - 0 = 4$$
$$x = 4$$

The x-intercept is (4,0).

Let $x = 0$ to find the y-intercept.

$$0 - y = 4$$
$$-y = 4$$
$$y = -4$$

The y-intercept is $(0, -4)$.

We plot (4,0) and $(0, -4)$ and draw the line.

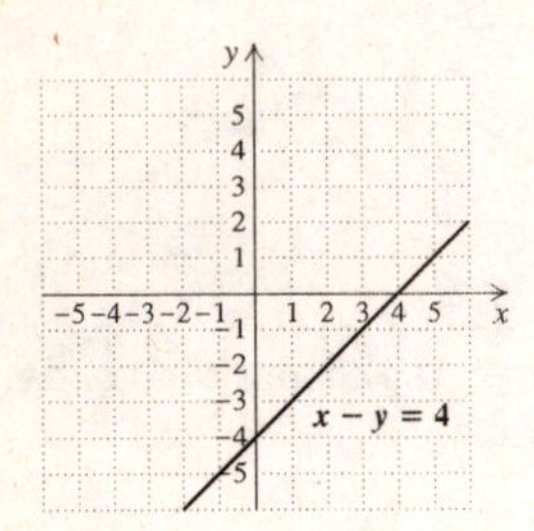

21. Graph $4x - 5y = 20$.

First we determine the intercepts.

Let $y = 0$ to find the x-intercept.

$$4x - 5 \cdot 0 = 20$$
$$4x = 20$$
$$x = 5$$

The x-intercept is (5,0).

Let $x = 0$ to find the y-intercept.

$$4 \cdot 0 - 5y = 20$$
$$-5y = 20$$
$$y = -4$$

The y-intercept is $(0, -4)$.

We plot (5,0) and $(0, -4)$ and draw the line.

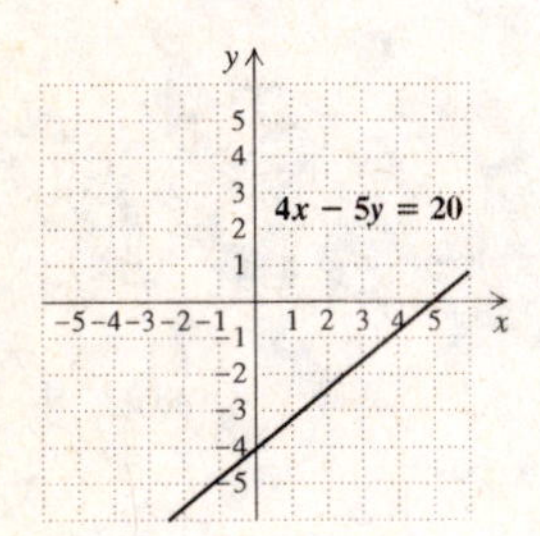

23. Graph $y = 2x - 1$.

The slope is 2 and the y-intercept is $(0, -1)$. We plot $(0, -1)$. Then, thinking of the slope as $\frac{2}{1}$, we start at $(0, -1)$ and move up 2 units and right 1 unit to the point (1,1). Alternatively, we can think of the slope as $\frac{-2}{-1}$ and, starting at $(0, -1)$, move down 2 units and left 1 unit to $(-1, -3)$. Using the points found we draw the graph.

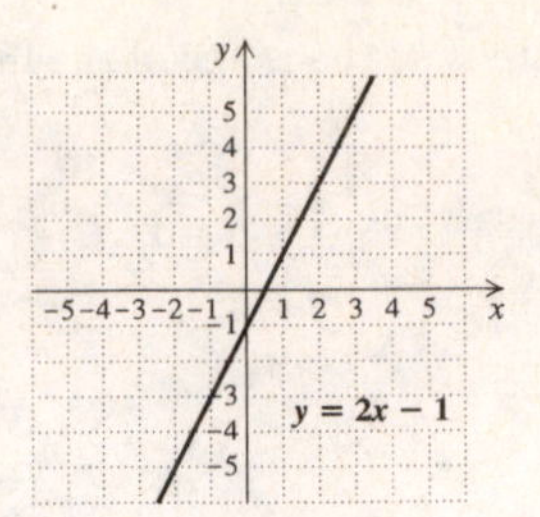

25. Graph $y = -\frac{2}{3}x + 3$.

The slope is $-\frac{2}{3}$ and the y-intercept is $(0, 3)$. We plot $(0, 3)$. Then, thinking of the slope as $\frac{-2}{3}$, we start at $(0, 3)$ and move down 2 units and right 3 units to the point (3,1). Alternatively, we can think of the slope as $\frac{2}{-3}$ and, starting at $(0, 3)$, move up 2 units and left 3 units to $(-3, 5)$. Using the points found we draw the graph.

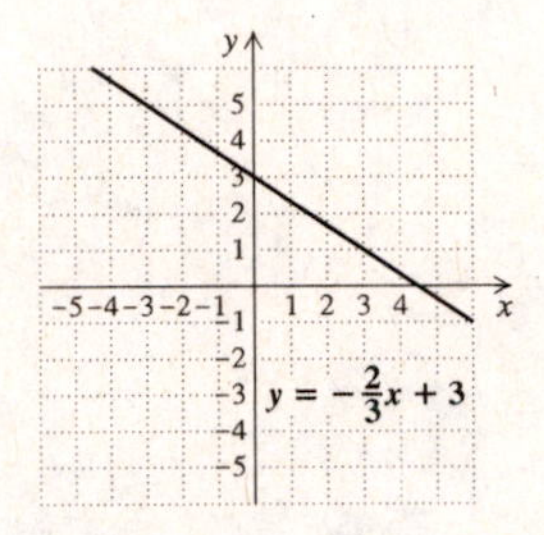

27. Graph $3y = 6 - 4x$.

First we write the equation in slope-intercept form.

$$3y = 6 - 4x$$
$$y = \frac{1}{3}(6 - 4x)$$
$$y = 2 - \frac{4}{3}x$$
$$y = -\frac{4}{3}x + 2$$

The slope is $-\frac{4}{3}$ and the y-intercept is $(0, 2)$. We plot $(0, 2)$. Then, thinking of the slope as $\frac{-4}{3}$, we start at $(0, 2)$ and move down 4 units and right 3 units to the point $(3, -2)$. Alternatively, we can think of the slope as $\frac{4}{-3}$ and, starting at $(0, 2)$, move up 4 units and left 3 units to $(-3, 6)$. Using the points found we draw the graph.

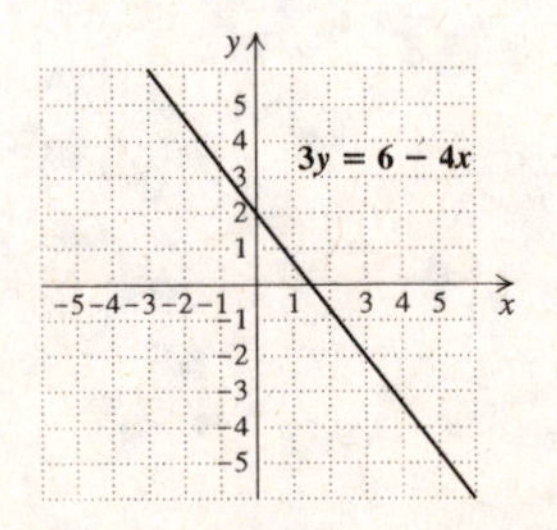

29. Graph $y = -2$.

For every value of x, the value of y is -2. The graph is a horizontal line.

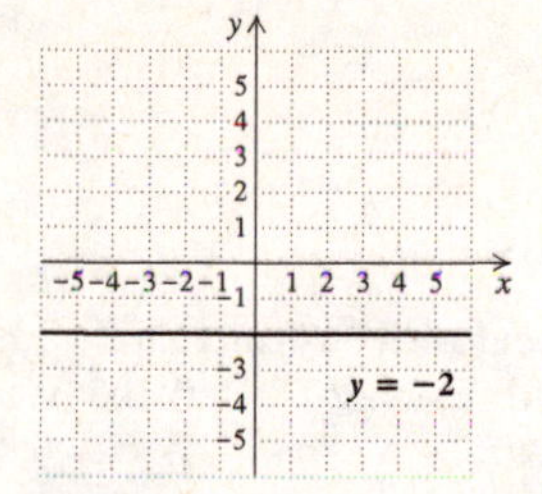

31. Graph $x = 4$.

The value of x is 4 for any value of y. The graph is a vertical line.

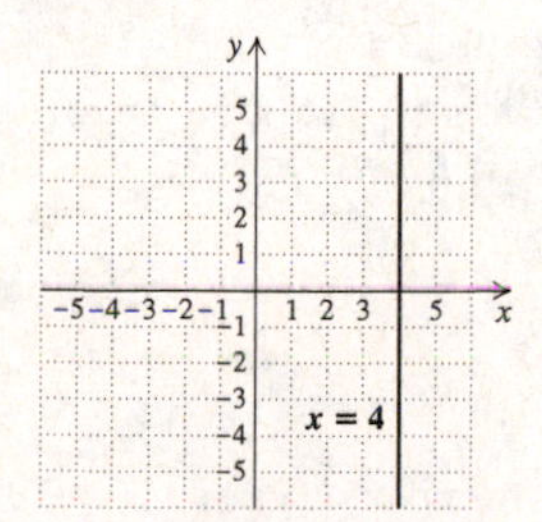

33. Graph $f(x) = x + 3$.

The slope is 1, or $\dfrac{1}{1}$ and the y-intercept is $(0, 3)$. We plot $(0, 3)$ and move up 1 unit and right 1 unit to the point (1,4). After we have sketched the line, a third point can be calculated as a check.

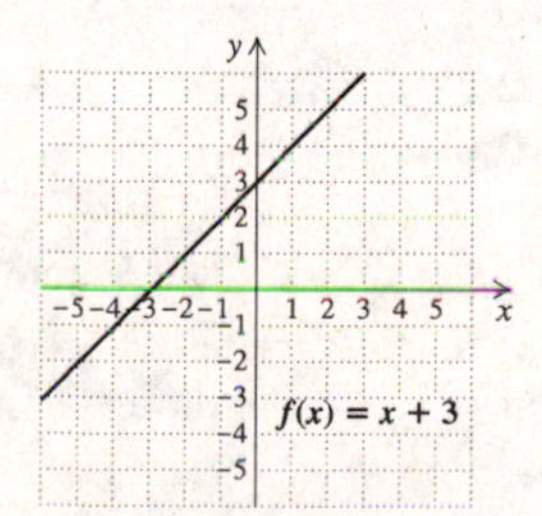

35. Graph $f(x) = \dfrac{3}{4}x + 1$.

The slope is $\dfrac{3}{4}$ and the y-intercept is $(0, 1)$. We plot $(0, 1)$ and move up 3 units and right 4 units to the point (4,4). After we have sketched the line, a third point can be calculated as a check.

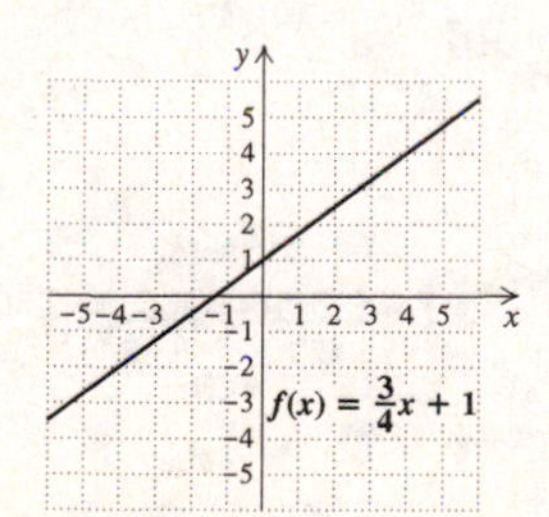

37. Graph $g(x) = 4$.

For every input x, the output is 4. The graph is a horizontal line.

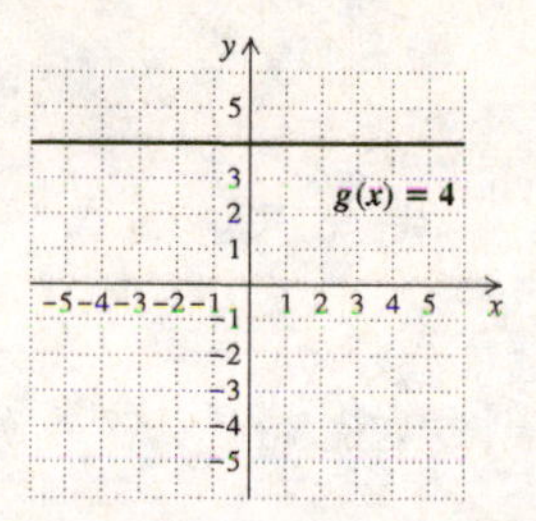

39. ***Familiarize***. A monthly fee is charged after the purchase of the phone. After one month of service, the total cost will be $\$50 + \$25 = \$75$. After two months, the total cost will be $\$50 + 2 \cdot \$25 = \$100$. We can generalize this with a model, letting $C(t)$ represent the total cost, in dollars, for t months of service.

Translate. We reword the problem and translate.

Total cost	is	cost of phone	plus	\$25 per month.
↓	↓	↓	↓	↓
$C(t)$	$=$	50	$+$	$25 \cdot t$

Carry out. The model can be written $C(t) = 25t + 50$. To find the time required for the total cost to reach \$150, substitute 150 for $C(t)$ and solve for t.

$$C(t) = 25t + 50$$
$$150 = 25t + 50$$
$$100 = 25t$$
$$4 = t$$

Check. We evaluate.

$$C(4) = 25 \cdot 4 + 50$$
$$= 100 + 50$$
$$= 150$$

The answer checks.

State. It takes 4 months for the total cost to reach \$150.

41. ***Familiarize***. Oscar's hair is initially 1 in. long, and it grows at a rate of $\dfrac{1}{2}$ in. per month. After 1 month the length of her hair will be $1 + \dfrac{1}{2}$, or $1\dfrac{1}{2}$ in. After 2 months the length will be $1 + 2 \cdot \dfrac{1}{2}$, or 2 in. This can be generalized with a model, letting $L(t)$ represent the length of Oscar's hair, in inches, after t months.

Translate.

Total length	is	initial length	plus	$\frac{1}{2}$ in. per month.
↓	↓	↓	↓	↓
$L(t)$	$=$	1	$+$	$\frac{1}{2} \cdot t$

Carry out. The model can be written $L(t) = \dfrac{1}{2}t + 1$. To find the time required for the hair to be 3 in. long, substitute 3 for $L(t)$ and solve for t.

$$L(t) = \frac{1}{2}t + 1$$
$$3 = \frac{1}{2}t + 1$$
$$2 = \frac{1}{2}t$$
$$4 = t \quad \text{Multiplying by 2}$$

Check. We evaluate.

$$L(4) = \frac{1}{2} \cdot 4 + 1$$
$$= 2 + 1$$
$$= 3$$

The answer checks.

State. Oscar's hair will be 3 in. long after 4 months.

43. ***Familiarize***. A charge for each mile is applied after an initial charge. For a 1 mile taxi ride, the total cost is \$2 + \$0.75 = \$2.75. For a two mile ride the cost is \$2 + 2(\$0.75) = \$3.50. We can generalize this with a model, letting $C(d)$ represent the total cost, in dollars, for a taxi ride of d miles.

Translate.

Total cost	is	initial charge	plus	\$0.75 per mile.
↓	↓	↓	↓	↓
$C(d)$	$=$	2	$+$	$0.75 \cdot d$

Carry out. The model can be written $C(d) = 0.75d + 2$. To find the length of a taxi ride when the cost is \$4.25, substitute 4.25 for $C(d)$ and solve for d.

$$C(d) = 0.75d + 2$$
$$4.25 = 0.75d + 2$$
$$2.25 = 0.75d$$
$$3 = d$$

Check. We evaluate.

$$C(3) = 0.75(3) + 2$$
$$= 2.25 + 2$$
$$= 4.25$$

State. The length of a taxi ride is 3 mi when the cost is \$4.25.

45. a) Letting t represent the number of years after 1971, we form the pairs $(0, 1542)$ and $(29, 1877)$. First we find the slope of the function that fits the data:

$$m = \frac{1877 - 1542}{29 - 0} = \frac{335}{29}.$$

We know that the y-intercept is $(0, 1542)$, so we write a function in slope-intercept form.

$$C(t) = \frac{335}{29}t + 1542.$$

b) In 2009, $t = 2009 - 1971 = 38$.

$$C(38) = \frac{335}{29} \cdot 38 + 1542 \approx 1981 \text{ calories}$$

c) We substitute 2000 for $C(t)$ and solve for t.

$$2000 = \frac{335}{29}t + 1542$$
$$458 = \frac{335}{29}t$$
$$\frac{29}{335} \cdot 458 = t$$
$$40 \approx t$$

The average number of calories an American woman will consume per day will reach 2000 approximately 40 yr after 1971, or in 2011.

47. a) Letting t represent the number of years since 1990, we form the pairs $(0, 71.8)$ and $(10, 74.1)$. First we find the slope of the function that fits the data:

$$m = \frac{74.1 - 71.8}{10 - 0} = \frac{2.3}{10} = 0.23.$$

We know the y-intercept is $(0, 71.8)$, so we write a function in slope-intercept form.

$$E(t) = 0.23t + 71.8$$

b) In 2009, $t = 2009 - 1990 = 19$.

$$E(19) = 0.23(19) + 71.8 \approx 76.2 \text{ yr}$$

49. a) Letting t represent the number of years since 1992, we form the pairs $(0, 178.6)$ and $(10, 282)$. First we find the slope of the function that fits the data:

$$m = \frac{282 - 178.6}{10 - 0} = \frac{103.4}{10} = 10.34$$

We know the y-intercept is $(0, 178.6)$, so we write a function in slope-intercept form.

$$A(t) = 10.34t + 178.6$$

b) In 2008, $t = 2008 - 1992 = 16$.

$$A(16) = 10.34(16) + 178.6 \approx \$344 \text{ million}$$

51. a) Letting t represent the number of years after 1999, we form the pairs (0,48) and (4,64). First we find the slope of the function that fits the data:

$$m = \frac{64 - 48}{4 - 0} = \frac{16}{4} = 4$$

We know the y-intercept is (0,48), so we write a function in slope-intercept form.

$$N(t) = 4t + 48$$

b) In 2009, $t = 2009 - 1999 = 10$.

$$N(10) = 4 \cdot 10 + 48 = 88 \text{ million Americans}$$

c) Substitute 104 for $N(t)$ and solve for t.

$$104 = 4t + 48$$
$$56 = 4t$$
$$14 = t$$

104 million Americans will find travel information on the Internet 14 yr after 1999, or in 2013.

53. a) Letting t represent the number of years after 1994, we form the pairs $(0, 74.9)$ and $(6, 78.2)$. First we find the slope of the function that fits the data:

$$m = \frac{78.2 - 74.9}{6 - 0} = \frac{3.3}{6} = 0.55$$

We know the y-intercept is $(0, 74.9)$, so we write a function in slope-intercept form.

$$A(t) = 0.55t + 74.9$$

b) In 2009, $t = 2009 - 1994 = 15$.

$$A(15) = 0.55(15) + 74.9 = 83.15 \text{ million acres}$$

55. $f(x) = \frac{1}{3}x - 7$

The function is in the form $f(x) = mx + b$, so it is a linear function. We can compute $\frac{1}{3}x - 7$ for any value of x, so the domain is the set of all real numbers.

57. $p(x) = x^2 + x + 1$

The function is in the form $(x) = ax^2 + bx + c$, $a \neq 0$, so it is a quadratic function. We can compute $x^2 + x + 1$ for any value of x, so the domain is the set of all real numbers.

59. $f(t) = \dfrac{12}{3t + 4}$

The function is described by a rational equation, so it is a rational function. The expression $\dfrac{12}{3t + 4}$ is undefined when $t = -\dfrac{4}{3}$, so the domain is $\left\{t \middle| t \text{ is a real number } \textit{and } t \neq -\dfrac{4}{3}\right\}$.

61. $f(x) = 0.02x^4 - 0.1x + 1.7$

The function is described by a polynomial equation that is neither linear nor quadratic, so it is a polynomial function. We can compute $0.02x^4 - 0.1x + 1.7$ for any value of x, so the domain is the set of all real numbers.

63. $f(x) = \dfrac{x}{2x - 5}$

The function is described by a rational equation, so it is a rational function. The expression $\dfrac{x}{2x - 5}$ is undefined when $x = \dfrac{5}{2}$, so the domain is $\left\{x \middle| x \text{ is a real number } \textit{and } x \neq \dfrac{5}{2}\right\}$.

65. $f(n) = \dfrac{4n - 7}{n^2 + 3n + 2}$

The function is described by a rational equation, so it is a rational function. The expression $\dfrac{4n - 7}{n^2 + 3n + 2}$ is undefined for values of n that make the denominator 0. We find those values:

$$n^2 + 3n + 2 = 0$$
$$(n + 1)(n + 2) = 0$$
$$n + 1 = 0 \quad \text{or} \quad n + 2 = 0$$
$$n = -1 \quad \text{or} \quad n = -2$$

Then the domain is $\{n | n \text{ is a real number } \textit{and } n \neq -1 \textit{ and } n \neq -2\}$.

67. $f(n) = 200 - 0.1n$

The function can be written in the form $f(n) = mn + b$, so it is a linear function. We can compute $200 - 0.1n$ for any value of n, so the domain is the set of all real numbers.

69. The function has no y-values less than 0 and every y-value greater than or equal to 0 corresponds to a member of the domain. Thus, the range is $\{y | y \geq 0\}$.

71. Every y-value corresponds to a member of the domain, so the range is the set of all real numbers.

73. There is no y-value greater than 0 and every y-value less than or equal to 0 corresponds to a member of the domain. Thus, the range is $\{y | y \leq 0\}$.

75.

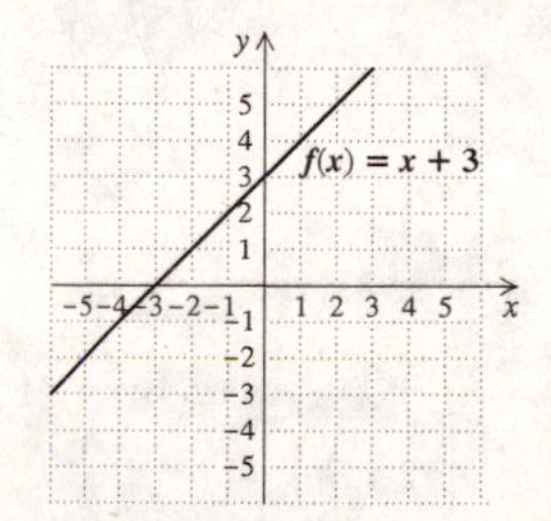

For any x-value and for any y-value there is a point on the graph. Thus,

Domain of $f = \{x | x \text{ is a real number}\}$ and

Range of $f = \{y | y \text{ is a real number}\}$

77.

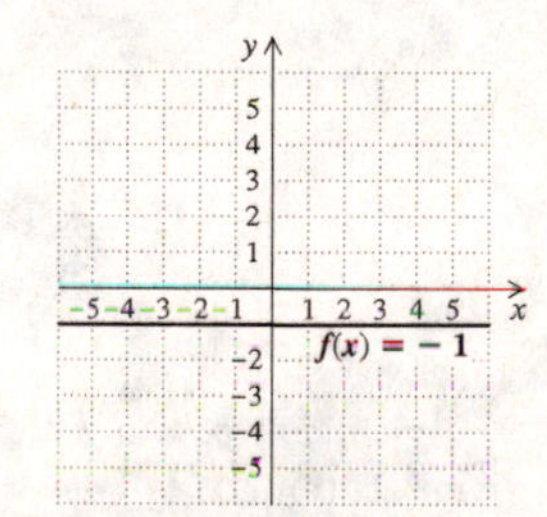

For any x-value there is a point on the graph, so

Domain of $f = \{x | x \text{ is a real number}\}$.

The only y-value on the graph is -1, so

Range of $f = \{-1\}$.

79.

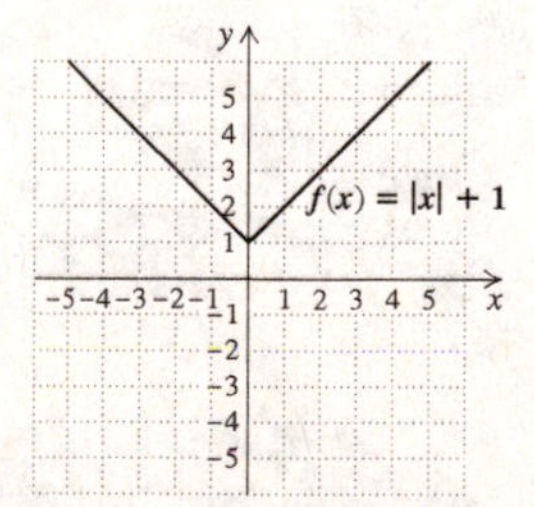

For any x-value there is a point on the graph, so

Domain of $f = \{x | x \text{ is a real number}\}$.

There is no y-value less than 1 and every y-value greater than or equal to 1 corresponds to a member of the domain. Thus,

Range of $f = \{y | y \geq 1\}$.

81.

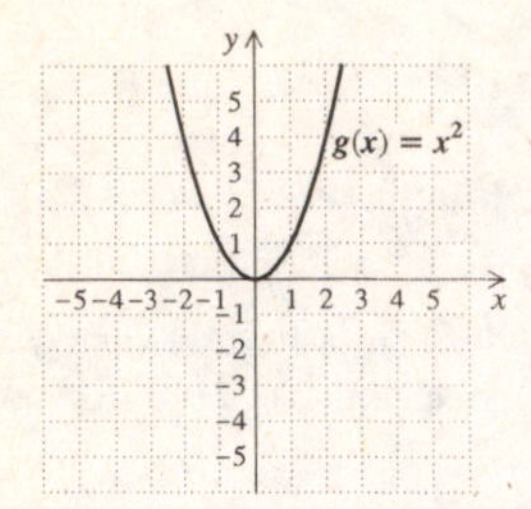

For any x-value there is a point on the graph, so

Domain of $g = \{x|x \text{ is a real number}\}$.

There is no y-value less than 0 and every y-value greater than or equal to 0 corresponds to a member of the domain. Thus,

Range of $g = \{y|y \geq 0\}$.

83. *Writing Exercise*

85.
$$\begin{aligned}&(x^2 + 2x + 7) + (3x^2 - 8)\\ &= (x^2 + 3x^2) + 2x + (7 - 8)\\ &= 4x^2 + 2x - 1\end{aligned}$$

87.
$$\begin{aligned}(2x + 1)(x - 7) &= 2x^2 - 14x + x - 7\\ &= 2x^2 - 13x - 7\end{aligned}$$

89.
$$\begin{aligned}&(x^3 + x^2 - 4x + 7) - (3x^2 - x + 2)\\ &= x^3 + x^2 - 4x + 7 - 3x^2 + x - 2\\ &= x^3 + (x^2 - 3x^2) + (-4x + x) + (7 - 2)\\ &= x^3 - 2x^2 - 3x + 5\end{aligned}$$

91. *Writing Exercise*

93. Let $c = 1$ and $d = 2$. Then $f(c + d) = f(1 + 2) = f(3) = 3m + b$, but $f(c) + f(d) = (m + b) + (2m + b) = 3m + 2b$. The given statement is false.

95. Let $k = 2$. Then $f(kx) = f(2x) = 2mx + b$, but $kf(x) = 2(mx + b) = 2mx + 2b$. The given statement is false.

97. ***Familiarize***. Celsius temperature C corresponding to a Fahrenheit temperature F can be modeled by a line that contains the points $(32, 0)$ and $(212, 100)$.

Translate. We find an equation relating C and F.

$$m = \frac{100 - 0}{212 - 32} = \frac{100}{180} = \frac{5}{9}$$

$$C - 0 = \frac{5}{9}(F - 32)$$

$$C = \frac{5}{9}(F - 32)$$

Carry out. Using function notation we have $C(F) = \frac{5}{9}(F - 32)$. Now we find $C(70)$:

$$C(70) = \frac{5}{9}(70 - 32) = \frac{5}{9}(38) \approx 21.1.$$

Check. We can repeat the calculations. We could also graph the function and determine that $(70, 21.1)$ is on the graph.

State. A temperature of about 21.1° C corresponds to a temperature of 70° F.

99. ***Familiarize***. The total cost C of the phone, in dollars, after t months, can be modeled by a line that contains the points $(5, 230)$ and $(9, 390)$.

Translate. We find an equation relating C and t.

$$m = \frac{390 - 230}{9 - 5} = \frac{160}{4} = 40$$

$$\begin{aligned}C - 230 &= 40(t - 5)\\ C - 230 &= 40t - 200\\ C &= 40t + 30\end{aligned}$$

Carry out. Using function notation we have $C(t) = 40t + 30$. To find the costs already incurred when the service began we find $C(0)$:

$$C(0) = 40 \cdot 0 + 30 = 30$$

Check. We can repeat the calculations. We could also graph the function and determine that $(0, 30)$ is on the graph.

State. Mel had already incurred \$30 in costs when his service just began.

101. a) We have two pairs, $(3, -5)$ and $(7, -1)$. Use the point-slope form:

$$m = \frac{-1 - (-5)}{7 - 3} = \frac{-1 + 5}{4} = \frac{4}{4} = 1$$

$$\begin{aligned}y - (-5) &= 1(x - 3)\\ y + 5 &= x - 3\\ y &= x - 8\\ g(x) &= x - 8 \quad \text{Using function notation}\end{aligned}$$

b) $g(-2) = -2 - 8 = -10$

c) $g(a) = a - 8$

If $g(a) = 75$, we have

$$\begin{aligned}a - 8 &= 75\\ a &= 83.\end{aligned}$$

103. Following the instructions for entering data and using the linear regression option on a graphing calculator, we find the following function:

$f(x) = 0.229x - 515.3622222$, where x is the year

Then we have $f(2010) \approx 85.6$ yr.

The result we found in Exercise 48 was 80.2 yr. The answer found using linear regression seems more reliable because it was found using a function that is based on more data points than the function in Exercise 48.

Exercise Set 7.4

1. domain; see page 479 in the text.

3. evaluate; see page 479 in the text.

5. excluding; see page 482 in the text.

7. Since $f(2) = -3 \cdot 2 + 1 = -5$, and $g(2) = 2^2 + 2 = 6$, we have $f(2) + g(2) = -5 + 6 = 1$.

9. Since $f(5) = -3 \cdot 5 + 1 = -14$ and $g(5) = 5^2 + 2 = 27$, we have $f(5) - g(5) = -14 - 27 = -41$.

11. Since $f(-1) = -3(-1) + 1 = 4$ and

$g(-1) = (-1)^2 + 2 = 3$, we have

$f(-1) \cdot g(-1) = 4 \cdot 3 = 12$.

13. Since $f(-4) = -3(-4) + 1 = 13$ and

$g(-4) = (-4)^2 + 2 = 18$, we have

$f(-4)/g(-4) = 13/18$.

15. Since $g(1) = 1^2 + 2 = 3$ and

$f(1) = -3 \cdot 1 + 1 = -2$, we have

$g(1) - f(1) = 3 - (-2) = 3 + 2 = 5$.

17. $(f+g)(x) = f(x)+g(x) = (-3x+1)+(x^2+2) = x^2-3x+3$

19.
$$\begin{aligned}(F+G)(x) &= F(x) + G(x)\\ &= x^2 - 2 + 5 - x\\ &= x^2 - x + 3\end{aligned}$$

21. Using our work in Exercise 19, we have
$$\begin{aligned}(F+G)(-4) &= (-4)^2 - (-4) + 3\\ &= 16 + 4 + 3\\ &= 23.\end{aligned}$$

23.
$$\begin{aligned}(F-G)(x) &= F(x) - G(x)\\ &= x^2 - 2 - (5 - x)\\ &= x^2 - 2 - 5 + x\\ &= x^2 + x - 7\end{aligned}$$

Then we have
$$\begin{aligned}(F-G)(3) &= 3^2 + 3 - 7\\ &= 9 + 3 - 7\\ &= 5.\end{aligned}$$

25.
$$\begin{aligned}(F \cdot G)(x) &= F(x) \cdot G(x)\\ &= (x^2 - 2)(5 - x)\\ &= 5x^2 - x^3 - 10 + 2x\end{aligned}$$

Then we have
$$\begin{aligned}(F \cdot G)(-3) &= 5(-3)^2 - (-3)^3 - 10 + 2(-3)\\ &= 5 \cdot 9 - (-27) - 10 - 6\\ &= 45 + 27 - 10 - 6\\ &= 56.\end{aligned}$$

27.
$$\begin{aligned}(F/G)(x) &= F(x)/G(x)\\ &= \frac{x^2 - 2}{5 - x},\ x \neq 5\end{aligned}$$

29. Using our work in Exercise 27, we have
$$(F/G)(-2) = \frac{(-2)^2 - 2}{5 - (-2)} = \frac{4 - 2}{5 + 2} = \frac{2}{7}.$$

31. $(P - L)(2) = P(2) - L(2) \approx 26.5\% - 22.5\% \approx 4\%$

33.
$$\begin{aligned}N(2000) &= (R + W)(2000)\\ &= R(2000) + W(2000)\\ &\approx 1.3 + 2.2\\ &\approx 3.5\end{aligned}$$

We estimate that 3.5 million U.S. women had children in 2000.

35. $(n + l)('98) = n('98) + l('98)$

From the middle line of the graph, we can see that $n('98) + l('98) \approx 50$ million.

This represents the total number of passengers serviced by Newark Liberty and LaGuardia airports in 1998.

37. From the top line of the graph, we see that $F('02) \approx$ 81 million. This represents the number of passengers using the three airports in 2002.

39.
$$\begin{aligned}(F - k)('02) &= F('02) - k('02)\\ &\approx 81 - 30\\ &\approx 51 \text{ million}\end{aligned}$$

This represents the number of passengers using LaGuardia and Newark Liberty in 2002.

41. The domain of f and of g is all real numbers. Thus, Domain of $f + g$ = Domain of $f - g$ = Domain of $f \cdot g$ = $\{x|x \text{ is a real number}\}$.

43. Because division by 0 is undefined, we have

Domain of $f = \{x|x \text{ is a real number } \textit{and } x \neq 3\}$,

and

Domain of $g = \{x|x \text{ is a real number}\}$.

Thus, Domain of $f+g$ = Domain of $f-g$ = Domain of $f \cdot g = \{x|x \text{ is a real number } \textit{and } x \neq 3\}$.

45. Because division by 0 is undefined, we have

Domain of $f = \{x|x \text{ is a real number } \textit{and } x \neq 0\}$,

and

Domain of $g = \{x|x \text{ is a real number}\}$.

Thus, Domain of $f+g$ = Domain of $f-g$ = Domain of $f \cdot g = \{x|x \text{ is a real number } \textit{and } x \neq 0\}$.

47. Because division by 0 is undefined, we have

Domain of $f = \{x|x \text{ is a real number } \textit{and } x \neq 1\}$,

and

Domain of $g = \{x|x \text{ is a real number}\}$.

Thus, Domain of $f+g$ = Domain of $f-g$ = Domain of $f \cdot g = \{x|x \text{ is a real number } \textit{and } x \neq 1\}$.

49. Because division by 0 is undefined, we have

Domain of $f = \{x|x \text{ is a real number } \textit{and } x \neq 2\}$,

and

Domain of $g = \{x|x \text{ is a real number } \textit{and } x \neq 4\}$.

Thus, Domain of $f+g$ = Domain of $f-g$ = Domain of $f \cdot g = \{x|x \text{ is a real number } \textit{and } x \neq 2 \textit{ and } x \neq 4\}$.

51. Domain of f = Domain of g = $\{x|x$ is a real number$\}$.

Since $g(x) = 0$ when $x - 3 = 0$, we have $g(x) = 0$ when $x = 3$. We conclude that Domain of f/g = $\{x|x$ is a real number *and* $x \neq 3\}$.

53. Domain of f = Domain of g = $\{x|x$ is a real number$\}$.

Since $g(x) = 0$ when $2x - 8 = 0$, we have $g(x) = 0$ when $x = 4$. We conclude that Domain of f/g = $\{x|x$ is a real number *and* $x \neq 4\}$.

55. Domain of $f = \{x|x$ is a real number *and* $x \neq 4\}$.

Domain of $g = \{x|x$ is a real number$\}$.

Since $g(x) = 0$ when $5 - x = 0$, we have $g(x) = 0$ when $x = 5$. We conclude that Domain of f/g = $\{x|x$ is a real number *and* $x \neq 4$ *and* $x \neq 5\}$.

57. Domain of $f = \{x|x$ is a real number *and* $x \neq -1\}$.

Domain of $g = \{x|x$ is a real number$\}$.

Since $g(x) = 0$ when $2x + 5 = 0$, we have $g(x) = 0$ when $x = -\frac{5}{2}$. We conclude that Domain of f/g = $\left\{x \middle| x \text{ is a real number } and\ x \neq -1\ and\ x \neq -\frac{5}{2}\right\}$.

59. $(F + G)(5) = F(5) + G(5) = 1 + 3 = 4$

$(F + G)(7) = F(7) + G(7) = -1 + 4 = 3$

61. $(G - F)(7) = G(7) - F(7) = 4 - (-1) = 4 + 1 = 5$

$(G - F)(3) = G(3) - F(3) = 1 - 2 = -1$

63. From the graph we see that Domain of $F = \{x|0 \leq x \leq 9\}$ and Domain of $G = \{x|3 \leq x \leq 10\}$. Then Domain of $F + G = \{x|3 \leq x \leq 9\}$. Since $G(x)$ is never 0, Domain of $F/G = \{x|3 \leq x \leq 9\}$.

65. We use $(F + G)(x) = F(x) + G(x)$.

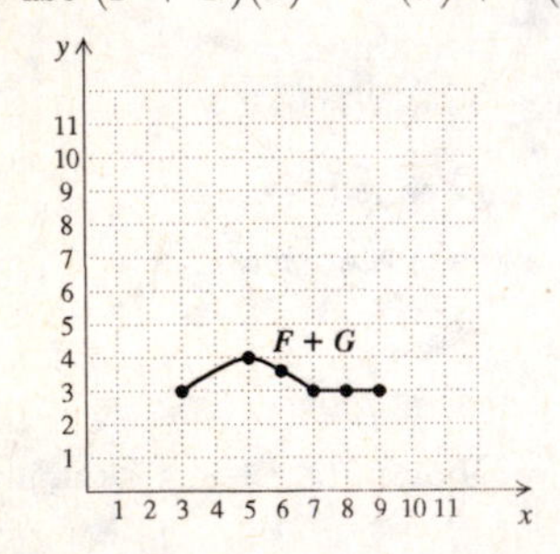

67. *Writing Exercise*

69.
$$4x - 7y = 8$$
$$4x = 7y + 8 \quad \text{Adding } 7y$$
$$\frac{1}{4} \cdot 4x = \frac{1}{4}(7y + 8) \quad \text{Multiplying by } \frac{1}{4}$$
$$x = \frac{7}{4}y + 2$$

71.
$$5x + 2y = -3$$
$$2y = -5x - 3 \quad \text{Subtracting } 5x$$
$$\frac{1}{2} \cdot 2y = \frac{1}{2}(-5x - 3) \quad \text{Multiplying by } \frac{1}{2}$$
$$y = -\frac{5}{2}x - \frac{3}{2}$$

73. Let n represent the number; $2n + 5 = 49$.

75. Let n represent the first integer; $x + (x + 1) = 145$.

77. *Writing Exercise*

79. Domain of $f = \left\{x \middle| x \text{ is a real number } and\ x \neq -\frac{5}{2}\right\}$;

domain of $g = \{x|x$ is a real number *and* $x \neq -3\}$;

$g(x) = 0$ when $x^4 - 1 = 0$, or when $x = 1$ or $x = -1$.

Then domain of $f/g = \left\{x \middle| x \text{ is a real number } and\ x \neq -\frac{5}{2}\ and\ x \neq -3\ and\ x \neq 1\ and\ x \neq -1\right\}$.

81. Answers may vary.

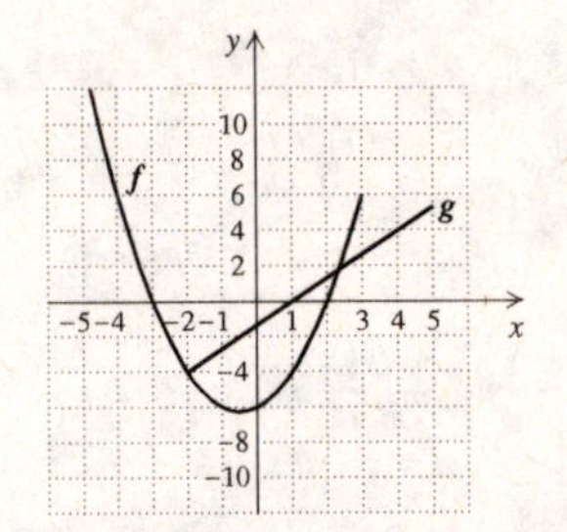

83. The problem states that Domain of $m = \{x| -1 < x < 5\}$. Since $n(x) = 0$ when $2x - 3 = 0$, we have $n(x) = 0$ when $x = \frac{3}{2}$. We conclude that Domain of m/n = $\left\{x \middle| x \text{ is a real number } and\ -1 < x < 5\ and\ x \neq \frac{3}{2}\right\}$.

85. Answers may vary. $f(x) = \frac{1}{x + 2}$, $g(x) = \frac{1}{x - 5}$

87.

Exercise Set 7.5

1. LCD; see Examples 1, 2, and 3.

3. factor; see Examples 1 and 3.

5. As the number of painters increases, the time required to scrape the house decreases, so we have inverse variation.

7. As the number of laps increases, the time required to swim them increases, so we have direct variation.

9. As the number of volunteers increases, the time required to wrap the toys decreases, so we have inverse variation.

11.
$$f = \frac{L}{d}$$
$$df = L \quad \text{Multiplying by } d$$
$$d = \frac{L}{f} \quad \text{Dividing by } f$$

13.
$$s = \frac{(v_1 + v_2)t}{2}$$
$$2s = (v_1 + v_2)t \quad \text{Multiplying by 2}$$
$$\frac{2s}{t} = v_1 + v_2 \quad \text{Dividing by } t$$
$$\frac{2s}{t} - v_2 = v_1$$

This result can also be expressed as $v_1 = \frac{2s - tv_2}{t}$.

15.
$$\frac{t}{a} + \frac{t}{b} = 1$$
$$ab\left(\frac{t}{a} + \frac{t}{b}\right) = ab \cdot 1 \quad \text{Multiplying by the LCD}$$
$$ab \cdot \frac{t}{a} + ab \cdot \frac{t}{b} = ab$$
$$bt + at = ab$$
$$at = ab - bt$$
$$at = b(a - t) \quad \text{Factoring}$$
$$\frac{at}{a - t} = b$$

17.
$$I = \frac{2V}{R + 2r}$$
$$I(R + 2r) = \frac{2V}{R + 2r} \cdot (R + 2r) \quad \text{Multiplying by the LCD}$$
$$I(R + 2r) = 2V$$
$$R + 2r = \frac{2V}{I}$$
$$R = \frac{2V}{I} - 2r, \text{ or } \frac{2V - 2Ir}{I}$$

19.
$$R = \frac{gs}{g + s}$$
$$(g + s) \cdot R = (g + s) \cdot \frac{gs}{g + s} \quad \text{Multiplying by the LCD}$$
$$Rg + Rs = gs$$
$$Rs = gs - Rg$$
$$Rs = g(s - R) \quad \text{Factoring out } g$$
$$\frac{Rs}{s - R} = g \quad \text{Multiplying by } \frac{1}{s - R}$$

21.
$$I = \frac{nE}{R + nr}$$
$$I(R + nr) = \frac{nE}{R + nr} \cdot (R + nr) \quad \text{Multiplying by the LCD}$$
$$IR + Inr = nE$$
$$IR = nE - Inr$$
$$IR = n(E - Ir)$$
$$\frac{IR}{E - Ir} = n$$

23.
$$\frac{1}{p} + \frac{1}{q} = \frac{1}{f}$$
$$pqf\left(\frac{1}{p} + \frac{1}{q}\right) = pqf \cdot \frac{1}{f} \quad \text{Multiplying by the LCD}$$
$$qf + pf = pq$$
$$pf = pq - qf$$
$$pf = q(p - f)$$
$$\frac{pf}{p - f} = q$$

25.
$$S = \frac{H}{m(t_1 - t_2)}$$
$$(t_1 - t_2)S = \frac{H}{m} \quad \text{Multiplying by } t_1 - t_2$$
$$t_1 - t_2 = \frac{H}{Sm} \quad \text{Dividing by } S$$
$$t_1 = \frac{H}{Sm} + t_2, \text{ or } \frac{H + Smt_2}{Sm}$$

27.
$$\frac{E}{e} = \frac{R + r}{r}$$
$$er \cdot \frac{E}{e} = er \cdot \frac{R + r}{r} \quad \text{Multiplying by the LCD}$$
$$Er = e(R + r)$$
$$Er = eR + er$$
$$Er - er = eR$$
$$r(E - e) = eR$$
$$r = \frac{eR}{E - e}$$

29.
$$S = \frac{a}{1 - r}$$
$$(1 - r)S = a \quad \text{Multiplying by the LCD, } 1 - r$$
$$1 - r = \frac{a}{S} \quad \text{Dividing by } S$$
$$1 - \frac{a}{S} = r \quad \text{Adding } r \text{ and } -\frac{a}{S}$$

This result can also be expressed as $r = \frac{S - a}{S}$.

31.
$$c = \frac{f}{(a + b)c}$$
$$\frac{a + b}{c} \cdot c = \frac{a + b}{c} \cdot \frac{f}{(a + b)c}$$
$$a + b = \frac{f}{c^2}$$

33.
$$P = \frac{A}{1 + r}$$
$$P(1 + r) = \frac{A}{1 + r} \cdot (1 + r)$$
$$P(1 + r) = A$$
$$1 + r = \frac{A}{P}$$
$$r = \frac{A}{P} - 1, \text{ or } \frac{A - P}{P}$$

35.
$$v = \frac{d_2 - d_1}{t_2 - t_1}$$
$$(t_2 - t_1)v = (t_2 - t_1) \cdot \frac{d_2 - d_1}{t_2 - t_1}$$
$$(t_2 - t_1)v = d_2 - d_1$$
$$t_2 - t_1 = \frac{d_2 - d_1}{v}$$
$$t_2 = \frac{d_2 - d_1}{v} + t_1, \text{ or } \frac{d_2 - d_1 + t_1 v}{v}$$

37.
$$\frac{x^2}{a^2} + \frac{y^2}{b^2} = 1$$
$$a^2b^2\left(\frac{x^2}{a^2} + \frac{y^2}{b^2}\right) = a^2b^2 \cdot 1$$
$$a^2b^2 \cdot \frac{x^2}{a^2} + a^2b^2 \cdot \frac{y^2}{b^2} = a^2b^2$$
$$b^2x^2 + a^2y^2 = a^2b^2$$
$$a^2y^2 = a^2b^2 - b^2x^2$$
$$a^2y^2 = b^2(a^2 - x^2)$$
$$\frac{a^2y^2}{a^2 - x^2} = b^2$$

39.
$$A = \frac{2Tt + Qq}{2T + Q}$$
$$(2T + Q) \cdot A = (2T + Q) \cdot \frac{2Tt + Qq}{2T + Q}$$
$$2AT + AQ = 2Tt + Qq$$
$$AQ - Qq = 2Tt - 2AT \quad \text{Adding } -2AT \text{ and } -Qq$$
$$Q(A - q) = 2Tt - 2AT$$
$$Q = \frac{2Tt - 2AT}{A - q}$$

41. $y = kx$

$28 = k \cdot 4$ Substituting

$7 = k$

The variation constant is 7.
The equation of variation is $y = 7x$.

43. $y = kx$

$3.4 = k \cdot 2$ Substituting

$1.7 = k$

The variation constant is 1.7.
The equation of variation is $y = 1.7x$.

45. $y = kx$

$2 = k \cdot \frac{1}{3}$ Substituting

$6 = k$ Multiplying by 3

The variation constant is 6.

The equation of variation is $y = 6x$.

47. $y = \frac{k}{x}$

$3 = \frac{k}{20}$ Substituting

$60 = k$

The variation constant is 60.

The equation of variation is $y = \frac{60}{x}$.

49. $y = \frac{k}{x}$

$28 = \frac{k}{4}$ Substituting

$112 = k$

The variation constant is 112.

The equation of variation is $y = \frac{112}{x}$.

51. $y = \frac{k}{x}$

$27 = \frac{k}{\frac{1}{3}}$ Substituting

$9 = k$

The variation constant is 9.

The equation of variation is $y = \frac{9}{x}$.

53. ***Familiarize***. Because N varies directly as the number of people P using the cans, we write N as a function of P: $N(P) = kP$. We know that $N(250) = 60,000$.

Translate.
$$N(P) = kP$$
$$N(250) = k \cdot 250 \quad \text{Replacing } P \text{ with } 250$$
$$60,000 = k \cdot 250 \quad \text{Replacing } N(250) \text{ with } 60,000$$
$$\frac{60,000}{250} = k$$
$$240 = k \quad \text{Variation constant}$$
$$N(P) = 240P \quad \text{Equation of variation}$$

Carry out. Find $N(1,008,000)$.
$$N(P) = 240P$$
$$N(1,008,000) = 240 \cdot 1,008,000$$
$$= 241,920,000$$

Check. Reexamine the calculation.

State. 241,920,000 aluminum cans are used each year in Dallas.

55. ***Familiarize***. Because of the phrase "I ... varies directly as ... V," we express the current as a function of the voltage. Thus we have $I(V) = kV$. We know that $I(15) = 5$.

Translate. We find the variation constant and then find the equation of variation.

$I(V) = kV$

$I(15) = k \cdot 15$ Replacing V with 15

$5 = k \cdot 15$ Replacing $I(15)$ with 5

$\frac{5}{15} = k$

$\frac{1}{3} = k$ Variation constant

The equation of variation is $I(V) = \frac{1}{3}V$.

Carry out. We compute $I(18)$.

$I(V) = \frac{1}{3}V$

$I(18) = \frac{1}{3} \cdot 18$ Replacing V with 18

$= 6$

Check. Reexamine the calculations. Note that the answer seems reasonable since $15/5 = 18/6$.

State. The current is 6 amperes when 18 volts is applied.

57. ***Familiarize***. Because T varies inversely as P, we write $T(p) = k/p$. We know that $T(7) = 5$.

Translate. We find the variation constant and the equation of variation.

$T(P) = \frac{k}{p}$

$T(7) = \frac{k}{7}$ Replacing P with 7

$5 = \frac{k}{7}$ Replacing $T(P)$ with 5

$35 = k$ Variation constant

$T(P) = \frac{35}{P}$ Equation of variation

Carry out. We find $T(10)$.

$T(10) = \frac{35}{10}$

$= 3.5$

Check. Reexamine the calculations.

State. It would take 3.5 hr for 10 volunteers to complete the job.

59. Since we have direct variation and $48 = \frac{1}{2} \cdot 96$, then the result is $\frac{1}{2} \cdot 64$ kg, or 32 kg. We could also do this problem as follows.

Familiarize. Because W varies directly as the total mass, we write $W(m) = km$. We know that $W(96) = 64$.

Translate.

$W(m) = km$

$W(96) = k \cdot 96$ Replacing m with 96

$64 = k \cdot 96$ Replacing $W(96)$ with 64

$\frac{2}{3} = k$ Variation constant

$W(m) = \frac{2}{3}m$ Equation of variation

Carry out. Find $W(48)$.

$W(m) = \frac{2}{3}m$

$W(48) = \frac{2}{3} \cdot 48$

$= 32$

Check. Reexamine the calculations.

State. There are 32 kg of water in a 64 kg person.

61. ***Familiarize***. Because the number of calories burned C varies directly as the time t spent bicycling, we write $C(t) = kt$. We know that $C(30) = 150$.

Translate.

$C(t) = kt$

$C(30) = k \cdot 30$ Replacing t with 30

$150 = k \cdot 30$ Replacing $C(30)$ with 150

$5 = k$ Variation constant

$C(t) = 5t$ Equation of variation

Carry out. We find t when $C(t)$ is 250.

$C(t) = 5t$

$250 = 5t$

$50 = t$

Check. Reexamine the calculations.

State. It would take 50 min to burn 250 calories when biking 10 mph.

63. ***Familiarize***. Because of the phrase "t varies inversely as ... u," we write $t(u) = k/u$. We know that $t(4) = 75$.

Translate. We find the variation constant and then we find the equation of variation.

$t(u) = \frac{k}{u}$

$t(4) = \frac{k}{4}$ Replacing u with 4

$75 = \frac{k}{4}$ Replacing $t(4)$ with 70

$300 = k$ Variation constant

$t(u) = \frac{300}{u}$ Equation of variation

Carry out. We find $t(14)$.

$t(14) = \frac{300}{14} \approx 21$

Check. Reexamine the calculations. Note that, as expected, as the UV rating increases, the time it takes to burn goes down.

State. It will take about 21 min to burn when the UV rating is 14.

65. ***Familiarize***. The amount A of carbon monoxide released, in tons, varies directly as the population P. We write A as a function of P: $A(P) = kP$. We know that $A(2.6) = 1.1$.

Translate.

$$A(P) = kP$$

$$A(2.6) = k \cdot 2.6 \quad \text{Replacing } P \text{ with } 2.6$$

$$1.1 = k \cdot 2.6 \quad \text{Replacing } A(2.6) \text{ with } 1.1$$

$$\frac{11}{26} = k \quad \text{Variation constant}$$

$$A(P) = \frac{11}{26}P \quad \text{Equation of variation}$$

Carry out. Find $A(289,000,000)$.

$$A(P) = \frac{11}{26}P$$

$$A(289,000,000) = \frac{11}{26}(289,000,000)$$

$$\approx 122,269,231$$

Check. Reexamine the calculations. Answers may vary slightly due to rounding differences.

State. About 122,269,231 tons of carbon monoxide were released nationally.

67. $y = kx^2$

$$6 = k \cdot 3^2 \quad \text{Substituting}$$

$$6 = 9k$$

$$\frac{6}{9} = k$$

$$\frac{2}{3} = k \quad \text{Variation constant}$$

The equation of variation is $y = \frac{2}{3}x^2$.

69. $y = \frac{k}{x^2}$

$$6 = \frac{k}{3^2} \quad \text{Substituting}$$

$$6 = \frac{k}{9}$$

$$6 \cdot 9 = k$$

$$54 = k \quad \text{Variation constant}$$

The equation of variation is $y = \frac{54}{x^2}$.

71. $y = kxz^2$

$$105 = k \cdot 14 \cdot 5^2 \quad \text{Substituting 105 for } y, \text{ 14 for } x, \text{ and 5 for } z$$

$$105 = 350k$$

$$\frac{105}{350} = k$$

$$0.3 = k$$

The equation of variation is $y = 0.3xz^2$.

73. $y = k \cdot \frac{wx^2}{z}$

$$49 = k \cdot \frac{3 \cdot 7^2}{12} \quad \text{Substituting}$$

$$4 = k \quad \text{Variation constant}$$

The equation of variation is $y = \frac{4wx^2}{z}$.

75. ***Familiarize.*** Because time t, in seconds, varies inversely as square of the current c, in amperes, we write $t = k/c^2$. We know that $t = 3.4$ when $c = 0.089$.

Translate. Find k and the equation of variation.

$$t = \frac{k}{c^2}$$

$$3.4 = \frac{k}{(0.089)^2}$$

$$3.4(0.089)^2 = k$$

$$0.027 \approx k$$

$$t = \frac{0.027}{c^2}$$

Carry out. We find the value of t when c is 0.096.

$$t = \frac{0.027}{(0.096)^2} \approx 2.9$$

Check. Reexamine the calculations.

State. It would take about 2.9 sec for a 0.096-amp current to stop a 150-lb person's heart from beating.

77. ***Familiarize.*** Because V varies directly as T and inversely as P, we write $V = kT/P$. We know that $V = 231$ when $T = 300$ and $P = 20$.

Translate. Find k and the equation of variation.

$$V = \frac{kT}{P}$$

$$231 = \frac{k \cdot 300}{20}$$

$$\frac{20}{300} \cdot 231 = k$$

$$15.4 = k$$

$$V = \frac{15.4T}{P} \quad \text{Equation of variation}$$

Carry out. Substitute 320 for T and 16 for P and find V.

$$V = \frac{15.4(320)}{16} = 308$$

Check. Reexamine the calculations.

State. The volume is 308 cm^3 when $T = 320°\text{K}$ and $P = 16\ \text{lb/cm}^2$.

79. ***Familiarize.*** The drag W varies jointly as the surface area A and velocity v, so we write $W = kAv$. We know that $W = 222$ when $A = 37.8$ and $v = 40$.

Translate. Find k.

$$W = kAv$$

$$222 = k(37.8)(40)$$

$$\frac{222}{37.8(40)} = k$$

$$\frac{37}{252} = k$$

$$W = \frac{37}{252}Av \quad \text{Equation of variation}$$

Carry out. Substitute 51 for A and 430 for W and solve for v.

$$430 = \frac{37}{252} \cdot 51 \cdot v$$

$$57.42 \text{ mph} \approx v$$

(If we had used the rounded value 0.1468 for k, the resulting speed would have been approximately 57.43 mph.)

Check. Reexamine the calculations.

State. The car must travel about 57.42 mph.

81. *Writing Exercise*

83. $2x - 5 = 8$

$$2x = 13$$

$$x = \frac{13}{2}$$

The solution is $\frac{13}{2}$.

85. $\frac{1}{x+1} = \frac{3}{x}$, LCD is $x(x+1)$

$$x(x+1) \cdot \frac{1}{x+1} = x(x+1) \cdot \frac{3}{x}$$

$$x = 3(x+1)$$

$$x = 3x + 3$$

$$-2x = 3$$

$$x = -\frac{3}{2}$$

The solution is $-\frac{3}{2}$.

87. $3a + 1 = 3(a+1)$

$$3a + 1 = 3a + 3$$

$$1 = 3 \quad \text{Subtracting } 3a$$

We get a false equation, so there is no solution. The equation is a contradiction.

89. *Writing Exercise*

91. Use the result of Example 2.

$$h = \frac{2R^2 g}{V^2} - R$$

We have $V = 6.5$ mi/sec, $R = 3960$ mi, and $g = 32.2$ ft/sec^2. We must convert 32.2 ft/sec^2 to mi/sec^2 so all units of length are the same.

$$32.2 \frac{\text{ft}}{\text{sec}^2} \cdot \frac{1 \text{ mi}}{5280 \text{ ft}} \approx 0.0060984 \frac{\text{mi}}{\text{sec}^2}$$

Now we substitute and compute.

$$h = \frac{2(3960)^2(0.0060984)}{(6.5)^2} - 3960$$

$$h \approx 567$$

The satellite is about 567 mi from the surface of the earth.

93. $c = \frac{a}{a+12} \cdot d$

$$c = \frac{2a}{2a+12} \cdot d \quad \text{Doubling } a$$

$$= \frac{2a}{2(a+6)} \cdot d$$

$$= \frac{a}{a+6} \cdot d \quad \text{Simplifying}$$

The ratio of the larger dose to the smaller dose is

$$\frac{\frac{a}{a+6} \cdot d}{\frac{a}{a+12} \cdot d} = \frac{\frac{ad}{a+6}}{\frac{ad}{a+12}}$$

$$= \frac{ad}{a+6} \cdot \frac{a+12}{ad}$$

$$= \frac{ad(a+12)}{(a+6)ad}$$

$$= \frac{a+12}{a+6}.$$

The amount by which the dosage increases is

$$\frac{a}{a+6} \cdot d - \frac{a}{a+12} \cdot d$$

$$\frac{ad}{a+6} - \frac{ad}{a+12}$$

$$= \frac{ad}{a+6} \cdot \frac{a+12}{a+12} - \frac{ad}{a+12} \cdot \frac{a+6}{a+6}$$

$$= \frac{ad(a+12) - ad(a+6)}{(a+6)(a+12)}$$

$$= \frac{a^2d + 12ad - a^2d - 6ad}{(a+6)(a+12)}$$

$$= \frac{6ad}{(a+6)(a+12)}.$$

Then the percent by which the dosage increases is

$$\frac{\frac{6ad}{(a+6)(a+12)}}{\frac{a}{a+12} \cdot d} = \frac{\frac{6ad}{(a+6)(a+12)}}{\frac{ad}{a+12}}$$

$$= \frac{6ad}{(a+6)(a+12)} \cdot \frac{a+12}{ad}$$

$$= \frac{6 \cdot ad \cdot (a+12)}{(a+6)(a+12) \cdot ad}$$

$$= \frac{6}{a+6}.$$

This is a decimal representation for the percent of increase. To give the result in percent notation we multiply by 100 and use a percent symbol. We have

$$\frac{6}{a+6} \cdot 100\%, \text{ or } \frac{600}{a+6}\%.$$

95.

$$a = \frac{\frac{d_4 - d_3}{t_4 - t_3} - \frac{d_2 - d_1}{t_2 - t_1}}{t_4 - t_2}$$

$$a(t_4 - t_2) = \frac{d_4 - d_3}{t_4 - t_3} - \frac{d_2 - d_1}{t_2 - t_1} \quad \text{Multiplying by } t_4 - t_2$$

$$a(t_4-t_2)(t_4-t_3)(t_2-t_1) = (d_4-d_3)(t_2-t_1) - (d_2-d_1)(t_4-t_3)$$

Multiplying by $(t_4 - t_3)(t_2 - t_1)$

$$a(t_4 - t_2)(t_4 - t_3)(t_2 - t_1) - (d_4 - d_3)(t_2 - t_1) = -(d_2 - d_1)(t_4 - t_3)$$

$$(t_2 - t_1)[a(t_4 - t_2)(t_4 - t_3) - (d_4 - d_3)] = -(d_2 - d_1)(t_4 - t_3)$$

$$t_2 - t_1 = \frac{-(d_2 - d_1)(t_4 - t_3)}{a(t_4 - t_2)(t_4 - t_3) - (d_4 - d_3)}$$

$$t_2 + \frac{(d_2 - d_1)(t_4 - t_3)}{a(t_4 - t_2)(t_4 - t_3) + d_3 - d_4} = t_1$$

97. Let w = the wattage of the bulb. Then we have $I = \dfrac{kw}{d^2}$.
Now substitute $2w$ for w and $2d$ for d.

$$I = \frac{k(2w)}{(2d)^2} = \frac{2kw}{4d^2} = \frac{kw}{2d^2} = \frac{1}{2} \cdot \frac{kw}{d^2}$$

We see that the intensity is halved.

99. ***Familiarize***. We write $T = kml^2f^2$. We know that $T = 100$ when $m = 5$, $l = 2$, and $f = 80$.

Translate. Find k.

$$T = kml^2f^2$$
$$100 = k(5)(2)^2(80)^2$$
$$0.00078125 = k$$
$$T = 0.00078125ml^2f^2$$

Carry out. Substitute 72 for T, 5 for m, and 80 for f and solve for l.

$$72 = 0.00078125(5)(l^2)(80)^2$$
$$2.88 = l^2$$
$$1.697 \approx l$$

Check. Recheck the calculations.

State. The string should be about 1.697 m long.

101. ***Familiarize***. Because d varies inversely as s, we write $d(s) = k/s$. We know that $d(0.56) = 50$.

Translate.

$$d(s) = \frac{k}{s}$$
$$d(0.56) = \frac{k}{0.56} \quad \text{Replacing } s \text{ with } 0.56$$
$$50 = \frac{k}{0.56} \quad \text{Replacing } d(0.56) \text{ with } 50$$
$$28 = k$$
$$d(s) = \frac{28}{s} \quad \text{Equation of variation}$$

Carry out. Find $d(0.40)$.

$$d(0.40) = \frac{28}{0.40}$$
$$= 70$$

Check. Reexamine the calculations. Also observe that, as expected, when d decreases, then s increases.

State. The equation of variation is $d(s) = \dfrac{28}{s}$. The distance is 70 yd.

Chapter 8

Systems of Linear Equations and Problem Solving

Exercise Set 8.1

1. True; see page 510 in the text.

3. True; see Example 4(c).

5. True; see Example 4(b).

7. False; see page 513 in the text.

9. We use alphabetical order for the variables. We replace x by 1 and y by 2.

$$\begin{array}{c|c} \multicolumn{2}{c}{4x-y=2} \\ \hline 4\cdot 1-2 & 2 \\ 4-2 & \\ \multicolumn{2}{c}{2 \stackrel{?}{=} 2 \quad \text{TRUE}} \end{array} \qquad \begin{array}{c|c} \multicolumn{2}{c}{10x-3y=4} \\ \hline 10\cdot 1-3\cdot 2 & 4 \\ 10-6 & \\ \multicolumn{2}{c}{4 \stackrel{?}{=} 4 \quad \text{TRUE}} \end{array}$$

The pair $(1, 2)$ makes both equations true, so it is a solution of the system.

11. We use alphabetical order for the variables. We replace x by 2 and y by 5.

$$\begin{array}{c|c} \multicolumn{2}{c}{y=3x-1} \\ \hline 5 & 3\cdot 2-1 \\ & 6-1 \\ \multicolumn{2}{c}{5 \stackrel{?}{=} 5 \quad \text{TRUE}} \end{array} \qquad \begin{array}{c|c} \multicolumn{2}{c}{2x+y=4} \\ \hline 2\cdot 2+5 & 4 \\ 4+5 & \\ \multicolumn{2}{c}{9 \stackrel{?}{=} 4 \quad \text{FALSE}} \end{array}$$

The pair $(2, 5)$ is not a solution of $2x + y = 4$. Therefore, it is not a solution of the system of equations.

13. We replace x by 1 and y by 5.

$$\begin{array}{c|c} \multicolumn{2}{c}{x+y=6} \\ \hline 1+5 & 6 \\ \multicolumn{2}{c}{6 \stackrel{?}{=} 6 \quad \text{TRUE}} \end{array} \qquad \begin{array}{c|c} \multicolumn{2}{c}{y=2x+3} \\ \hline 5 & 2\cdot 1+3 \\ & 2+3 \\ \multicolumn{2}{c}{5 \stackrel{?}{=} 5 \quad \text{TRUE}} \end{array}$$

The pair $(1, 5)$ makes both equations true, so it is a solution of the system.

15. Observe that if we multiply both sides of the first equation by 2, we get the second equation. Thus, if we find that the given point makes the one equation true, we will also know that it makes the other equation true. We replace x by 3 and y by 1 in the first equation.

$$\begin{array}{c|c} \multicolumn{2}{c}{3x+4y=13} \\ \hline 3\cdot 3+4\cdot 1 & 13 \\ 9+4 & \\ \multicolumn{2}{c}{13 \stackrel{?}{=} 13 \quad \text{TRUE}} \end{array}$$

The pair $(3, 1)$ makes both equations true, so it is a solution of the system.

17. Graph both equations.

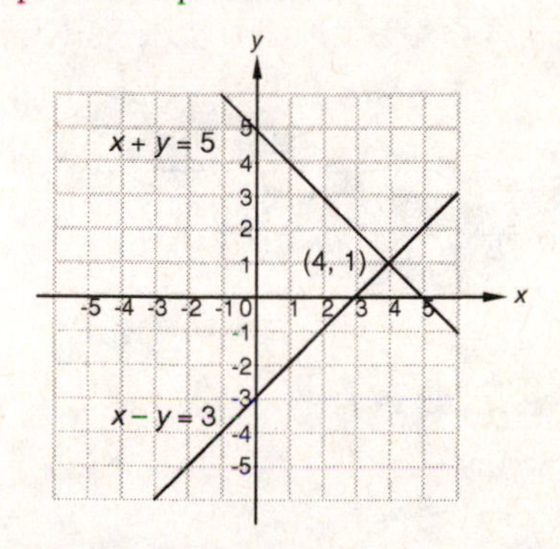

The solution (point of intersection) is apparently $(4, 1)$.

Check:

$$\begin{array}{c|c} \multicolumn{2}{c}{x-y=3} \\ \hline 4-1 & 3 \\ \multicolumn{2}{c}{3 \stackrel{?}{=} 3 \quad \text{TRUE}} \end{array} \qquad \begin{array}{c|c} \multicolumn{2}{c}{x+y=5} \\ \hline 4+1 & 5 \\ \multicolumn{2}{c}{5 \stackrel{?}{=} 5 \quad \text{TRUE}} \end{array}$$

The solution is $(4, 1)$.

19. Graph the equations.

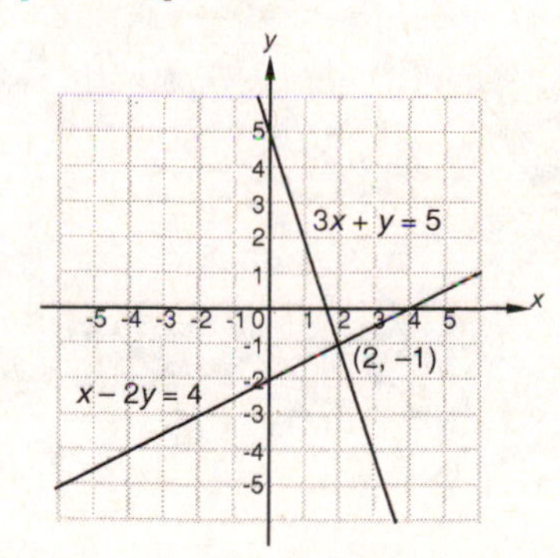

The solution (point of intersection) is apparently $(2, -1)$.

Check:

$$\begin{array}{c|c} \multicolumn{2}{c}{3x+y=5} \\ \hline 3\cdot 2+(-1) & 5 \\ 6-1 & \\ \multicolumn{2}{c}{5 \stackrel{?}{=} 5 \quad \text{TRUE}} \end{array} \qquad \begin{array}{c|c} \multicolumn{2}{c}{x-2y=4} \\ \hline 2-2(-1) & 4 \\ 2+2 & \\ \multicolumn{2}{c}{4 \stackrel{?}{=} 4 \quad \text{TRUE}} \end{array}$$

The solution is $(2, -1)$.

21. Graph both equations.

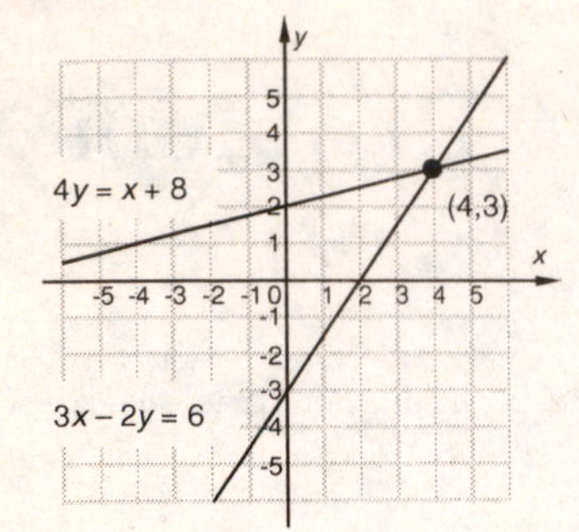

The solution (point of intersection) is apparently $(4, 3)$.

Check:

$$\begin{array}{c|c} \multicolumn{2}{c}{4y = x + 8} \\ \hline 4 \cdot 3 & 4 + 8 \\ \end{array}$$

$12 \stackrel{?}{=} 12$ TRUE

$$\begin{array}{c|c} \multicolumn{2}{c}{3x - 2y = 6} \\ \hline 3 \cdot 4 - 2 \cdot 3 & 6 \\ 12 - 6 & \end{array}$$

$6 \stackrel{?}{=} 6$ TRUE

The solution is $(4, 3)$.

23. Graph both equations.

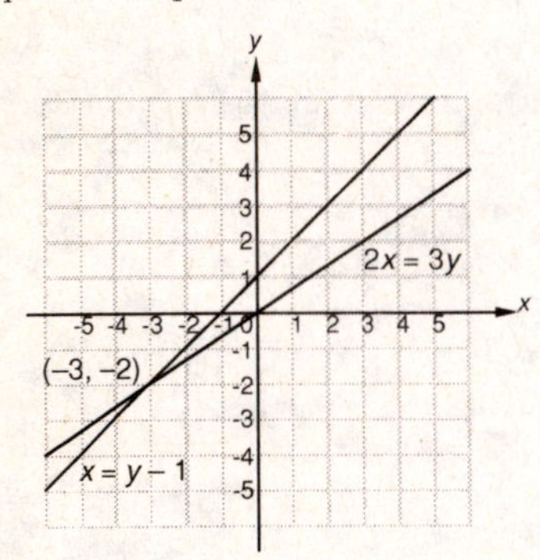

The solution (point of intersection) is apparently $(-3, -2)$.

Check:

$$\begin{array}{c|c} \multicolumn{2}{c}{x = y - 1} \\ \hline -3 & -2 - 1 \end{array}$$

$-3 \stackrel{?}{=} -3$ TRUE

$$\begin{array}{c|c} \multicolumn{2}{c}{2x = 3y} \\ \hline 2(-3) & 3(-2) \end{array}$$

$-6 \stackrel{?}{=} -6$ TRUE

The solution is $(-3, -2)$.

25. Graph both equations.

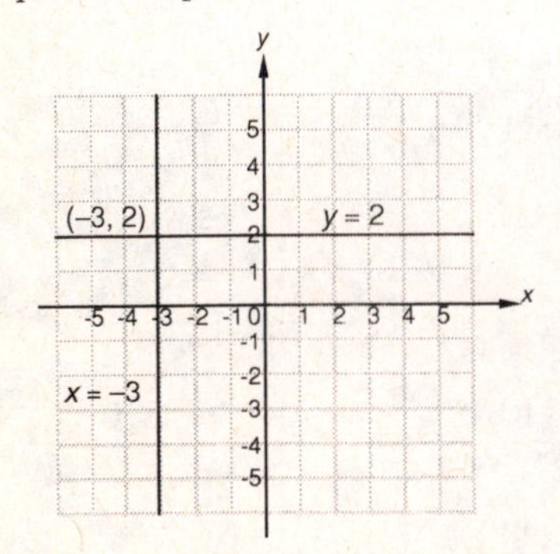

The ordered pair $(-3, 2)$ checks in both equations. It is the solution.

27. Graph both equations.

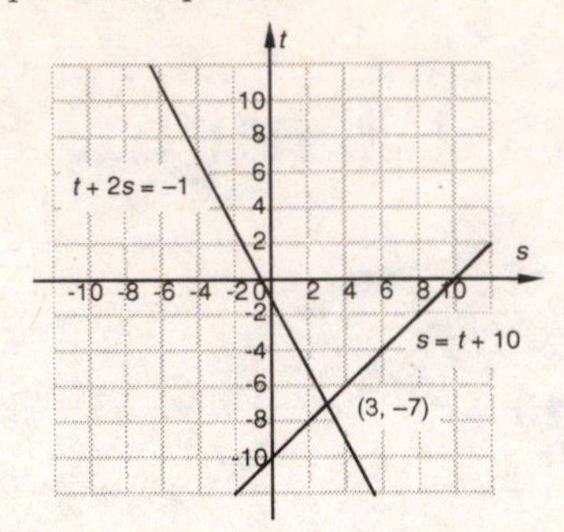

The solution (point of intersection) is apparently $(3, -7)$.

Check:

$$\begin{array}{c|c} \multicolumn{2}{c}{t + 2s = -1} \\ \hline -7 + 2 \cdot 3 & -1 \\ -7 + 6 & \end{array}$$

$-1 \stackrel{?}{=} -1$ TRUE

$$\begin{array}{c|c} \multicolumn{2}{c}{s = t + 10} \\ \hline 3 & -7 + 10 \end{array}$$

$3 \stackrel{?}{=} 3$ TRUE

The solution is (3,-7).

29. Graph both equations.

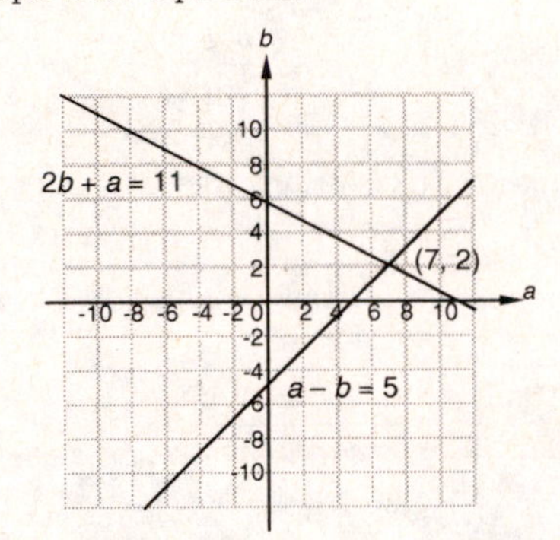

The solution (point of intersection) is apparently $(7, 2)$.

Check:

$$\begin{array}{c|c} \multicolumn{2}{c}{2b + a = 11} \\ \hline 2 \cdot 2 + 7 & 11 \\ 4 + 7 & \end{array}$$

$11 \stackrel{?}{=} 11$ TRUE

$$\begin{array}{c|c} \multicolumn{2}{c}{a - b = 5} \\ \hline 7 - 2 & 5 \end{array}$$

$5 \stackrel{?}{=} 5$ TRUE

The solution is (7,2).

31. Graph both equations.

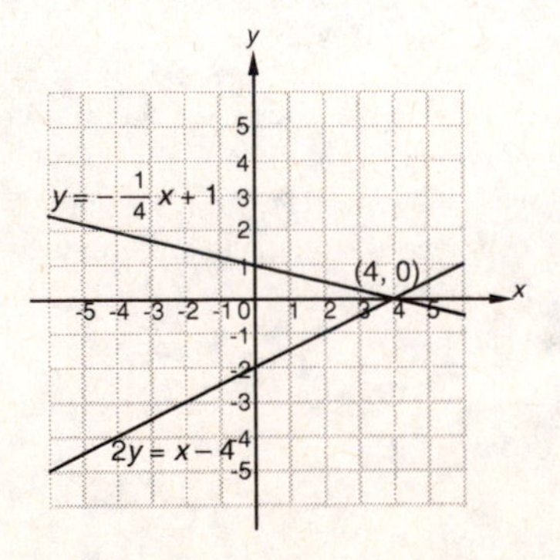

The solution (point of intersection) is apparently $(4, 0)$.

Check:

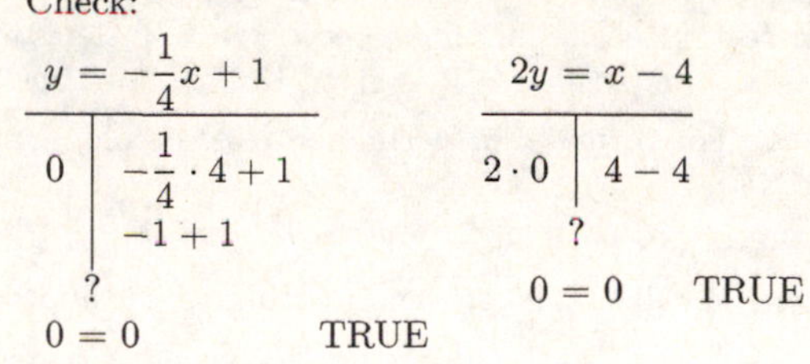

The solution is (4,0).

33. Graph both equations.

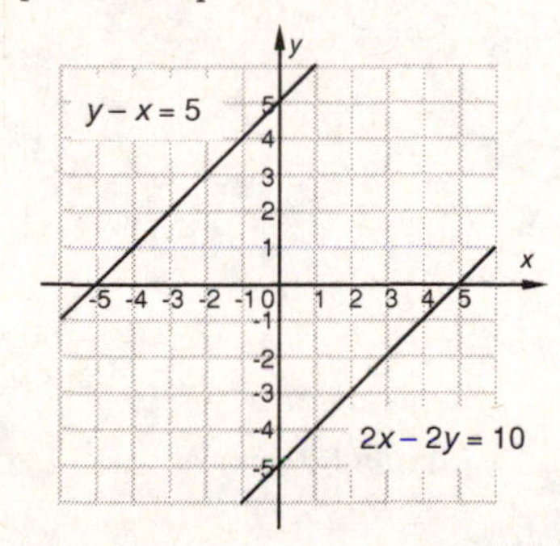

The lines are parallel. The system has no solution.

35. Graph both equations.

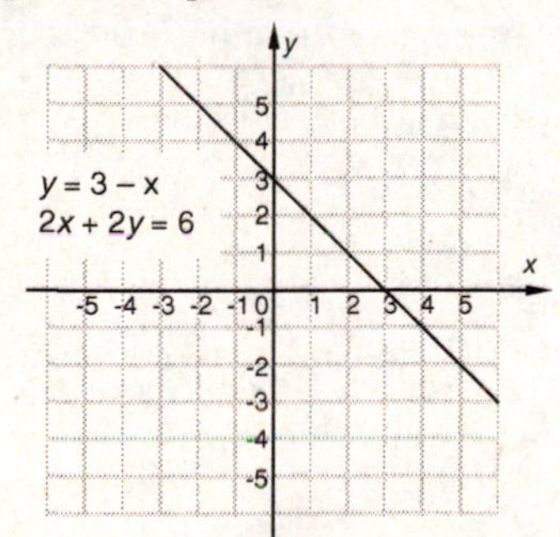

The graphs are the same. Any solution of one equation is a solution of the other. Each equation has infinitely many solutions. The solution set is the set of all pairs (x, y) for which $y = 3 - x$, or $\{(x, y) | y = 3 - x\}$. (In place of $y = 3 - x$ we could have used $2x + 2y = 6$ since the two equations are equivalent.)

37. A system of equations is consistent if it has at least one solution. Of the systems under consideration, only the one in Exercise 33 has no solution. Therefore, all except the system in Exercise 33 are consistent.

39. A system of two equations in two variables is dependent if it has infinitely many solutions. Only the system in Exercise 35 is dependent.

41. ***Familiarize***. Let x = the first number and y = the second number.

Translate.

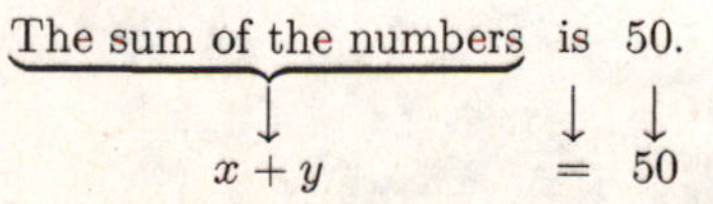

The first number is 25% of the second number.

$x = 0.25 \cdot y$

We have a system of equations:

$x + y = 50,$

$x = 0.25y$

43. ***Familiarize***. Let m = the number of ounces of mineral oil to be used and v = the number of ounces of vinegar.

Translate.

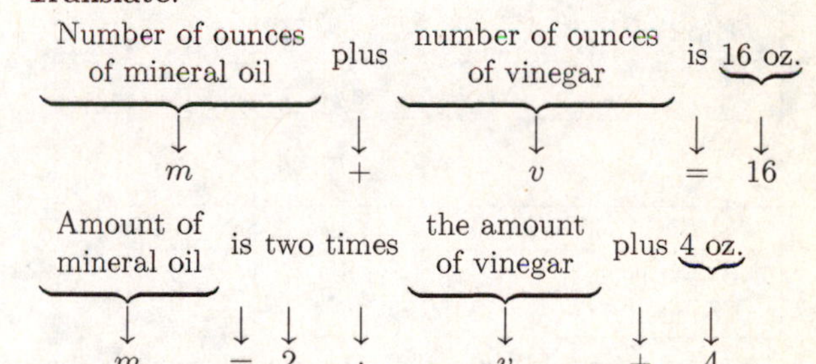

We have a system of equations:

$m + v = 16,$

$m = 2v + 4$

45. ***Familiarize***. Let x = the measure of one angle and y = the measure of the other angle.

Translate.

Two angles are supplementary.

Rewording: The sum of the measures is 180°.

$x + y = 180$

One angle is 3° less than twice the other.

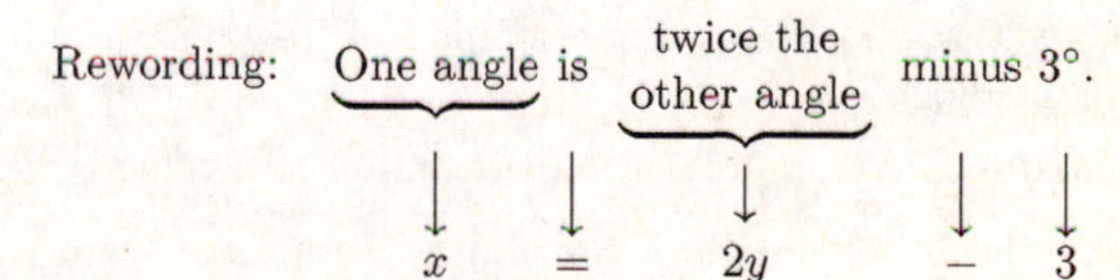

We have a system of equations:

$x + y = 180,$

$x = 2y - 3$

47. ***Familiarize***. Let g = the number of field goals and t = the number of foul shots made.

Translate. We organize the information in a table.

Kind of shot	Field goal	Foul shot	Total
Number scored	g	t	64
Points per score	2	1	
Points scored	$2g$	t	100

From the "Number scored" row of the table we get one equation:

$g + t = 64$

The "Points scored" row gives us another equation:

$2g + t = 100$

We have a system of equations:

$g + t = 64,$

$2g + t = 100$

49. ***Familiarize.*** Let x = the number of less expensive brushes sold and y = the number of more expensive brushes sold.

Translate. We organize the information in a table.

Kind of brush	Less expen-sive	More expen-sive	Total
Number sold	x	y	45
Price	\$8.50	\$9.75	
Amount taken in	$8.50x$	$9.75y$	398.75

The "Number sold" row of the table gives us one equation:

$x + y = 45$

The "Amount taken in" row gives us a second equation:

$8.50x + 9.75y = 398.75$

We have a system of equations:

$x + y = 45,$

$8.50x + 9.75y = 398.75$

We can multiply both sides of the second equation by 100 to clear the decimals:

$x + y = 45,$

$850x + 975y = 39,875$

51. ***Familiarize.*** Let h = the number of vials of Humulin Insulin sold and n = the number of vials of Novolin Velosulin Insulin sold.

Translate. We organize the information in a table.

Brand	Humulin	Novolin	Total
Number sold	h	n	50
Price	\$27.06	\$34.39	
Amount taken in	$27.06h$	$34.39n$	1565.57

The "Number sold" row of the table gives us one equation:

$h + n = 50$

The "Amount taken in" row gives us a second equation:

$27.06h + 34.39n = 1565.57$

We have a system of equations:

$h + n = 50,$

$27.06h + 34.39n = 1565.57$

We can multiply both sides of the second equation by 100 to clear the decimals:

$h + n = 50,$

$2706h + 3439n = 156,557$

53. ***Familiarize.*** The basketball court is a rectangle with perimeter 288 ft. Let l = the length, in feet, and w = width, in feet. Recall that for a rectangle with length l and width w, the perimeter P is given by $P = 2l + 2w$.

Translate. The formula for perimeter gives us one equation:

$2l + 2w = 288$

The statement relating length and width gives us another equation:

The length is 44 ft longer than the width.

$l = w + 44$

We have a system of equations:

$2l + 2w = 288,$

$l = w + 44$

55. *Writing Exercise*

57. $2(4x - 3) - 7x = 9$

$8x - 6 - 7x = 9$ Removing parentheses

$x - 6 = 9$ Collecting like terms

$x = 15$ Adding 6 to both sides

The solution is 15.

59. $4x - 5x = 8x - 9 + 11x$

$-x = 19x - 9$ Collecting like terms

$-20x = -9$ Adding $-19x$ to both sides

$x = \frac{9}{20}$ Multiplying both sides by $-\frac{1}{20}$

The solution is $\frac{9}{20}$.

61. $3x + 4y = 7$

$4y = -3y + 7$ Adding $-3x$ to both sides

$y = \frac{1}{4}(-3x + 7)$ Multiplying both sides by $\frac{1}{4}$

$y = -\frac{3}{4}x + \frac{7}{4}$

63. *Writing Exercise*

65. a) There are many correct answers. One can be found by expressing the sum and difference of the two numbers:

$x + y = 6,$

$x - y = 4$

b) There are many correct answers. For example, write an equation in two variables. Then write a second equation by multiplying the left side of the first equation by one nonzero constant and multiplying the right side by another nonzero constant.

$x + y = 1,$

$2x + 2y = 3$

c) There are many correct answers. One can be found by writing an equation in two variables and then writing a nonzero constant multiple of that equation:

$x + y = 1,$

$2x + 2y = 2$

67. Substitute 4 for x and -5 for y in the first equation:

$$A(4) - 6(-5) = 13$$
$$4A + 30 = 13$$
$$4A = -17$$
$$A = -\frac{17}{4}$$

Substitute 4 for x and -5 for y in the second equation:

$$4 - B(-5) = -8$$
$$4 + 5B = -8$$
$$5B = -12$$
$$B = -\frac{12}{5}$$

We have $A = -\frac{17}{4}$, $B = -\frac{12}{5}$.

69. ***Familiarize***. Let x = the number of years Lou has taught and y = the number of years Juanita has taught. Two years ago, Lou and Juanita had taught $x - 2$ and $y - 2$ years, respectively.

Translate.

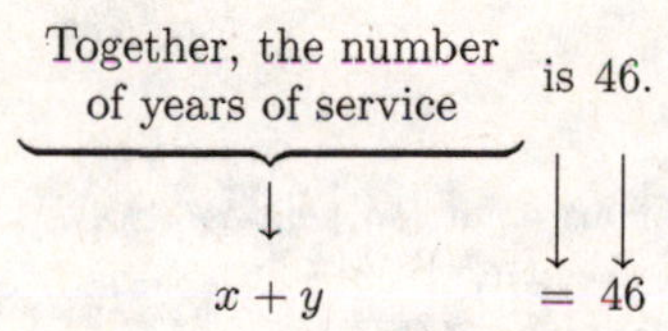

Two years ago
Lou had taught 2.5 times as many years as Juanita.

$$x - 2 = 2.5(y - 2)$$

We have a system of equations:

$$x + y = 46,$$
$$x - 2 = 2.5(y - 2)$$

71. ***Familiarize***. Let b = the number of ounces of baking soda and v = the number of ounces of vinegar to be used. The amount of baking soda in the mixture will be four times the amount of vinegar.

Translate.

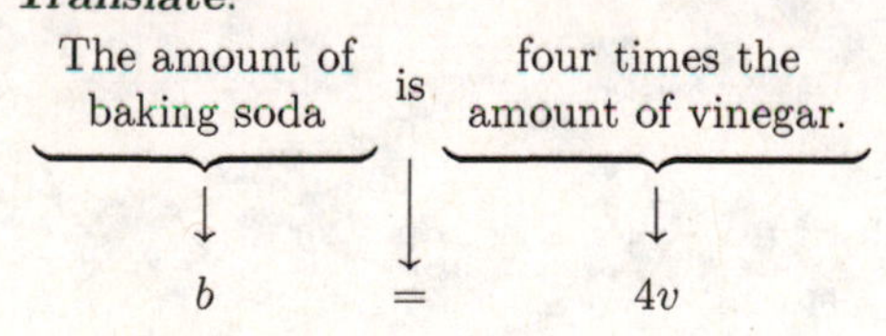

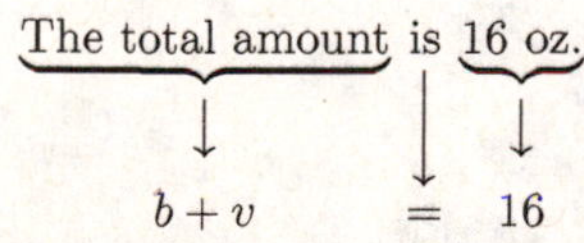

We have a system of equations.

$$b = 4v,$$
$$b + v = 16$$

73. Graph both equations.

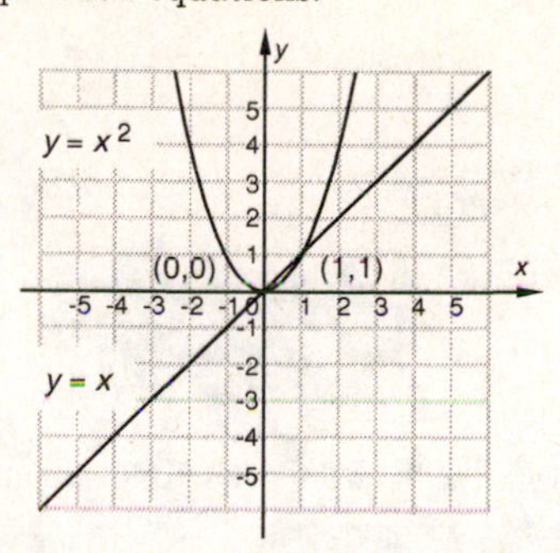

The solutions are apparently $(0, 0)$ and $(1, 1)$. Both pairs check.

75. $(0.07, -7.95)$

77. $(0.02, 1.25)$

Exercise Set 8.2

1. Adding the equations, we get $8y = -1$, so choice (d) is correct.

3. Multiplying the second equation by 2 gives us the system of equations in (a), so choice (a) is correct.

5. Substituting $4x - 7$ for y in the second equation gives us $6x + 3(4x - 7) = 19$, so choice (c) is correct.

7. $y = 5 - 4x$, (1)
$2x - 3y = 13$ (2)

We substitute $5 - 4x$ for y in the second equation and solve for x.

$$2x - 3y = 13 \quad (2)$$
$$2x - 3(5 - 4x) = 13 \quad \text{Substituting}$$
$$2x - 15 + 12x = 13$$
$$14x - 15 = 13$$
$$14x = 28$$
$$x = 2$$

Next we substitute 2 for x in either equation of the original system and solve for y.

$$y = 5 - 4x \quad (1)$$
$$y = 5 - 4 \cdot 2 \quad \text{Substituting}$$
$$y = 5 - 8$$
$$y = -3$$

We check the ordered pair $(2, -3)$.

$y = 5 - 4x$	
-3	$5 - 4 \cdot 2$
	$5 - 8$
$-3 \stackrel{?}{=} -3$	TRUE

$$\begin{array}{r|l} \multicolumn{2}{c}{2x - 3y = 13} \\ \hline 2 \cdot 2 - 3(-3) & 13 \\ 4 + 9 & \\ \end{array}$$

$$13 \stackrel{?}{=} 13 \quad \text{TRUE}$$

Since $(2, -3)$ checks, it is the solution.

9. $x = 8 - 4y, \quad (1)$

$3x + 5y = 3 \quad (2)$

We substitute $8 - 4y$ for x in the second equation and solve for y.

$$\begin{aligned} 3x + 5y &= 3 \quad (2) \\ 3(8 - 4y) + 5y &= 3 \quad \text{Substituting} \\ 24 - 12y + 5y &= 3 \\ 24 - 7y &= 3 \\ -7y &= -21 \\ y &= 3 \end{aligned}$$

Next we substitute 3 for y in either equation of the original system and solve for x.

$$x = 8 - 4y \quad (1)$$

$$x = 8 - 4 \cdot 3 = 8 - 12 = -4$$

We check the ordered pair $(-4, 3)$.

$$\begin{array}{r|l} \multicolumn{2}{c}{x = 8 - 4y} \\ \hline -4 & 8 - 4 \cdot 3 \\ & 8 - 12 \\ \end{array}$$

$$-4 \stackrel{?}{=} -4 \quad \text{TRUE}$$

$$\begin{array}{r|l} \multicolumn{2}{c}{3x + 5y = 3} \\ \hline 3(-4) + 5 \cdot 3 & 3 \\ -12 + 15 & \\ \end{array}$$

$$3 \stackrel{?}{=} 3 \quad \text{TRUE}$$

Since $(-4, 3)$ checks, it is the solution.

11. $3s - 4t = 14, \quad (1)$

$5s + t = 8 \quad (2)$

We solve the second equation for t.

$$5s + t = 8 \quad (2)$$

$$t = 8 - 5s \quad (3)$$

We substitute $8 - 5s$ for t in the first equation and solve for s.

$$\begin{aligned} 3s - 4t &= 14 \quad (1) \\ 3s - 4(8 - 5s) &= 14 \quad \text{Substituting} \\ 3s - 32 + 20s &= 14 \\ 23s - 32 &= 14 \\ 23s &= 46 \\ s &= 2 \end{aligned}$$

Next we substitute 2 for s in Equation (1), (2), or (3). It is easiest to use Equation (3) since it is already solved for t.

$$t = 8 - 5 \cdot 2 = 8 - 10 = -2$$

We check the ordered pair $(2, -2)$.

$$\begin{array}{r|l} \multicolumn{2}{c}{3s - 4t = 14} \\ \hline 3 \cdot 2 - 4(-2) & 14 \\ 6 + 8 & \\ \end{array}$$

$$14 \stackrel{?}{=} 14 \quad \text{TRUE}$$

$$\begin{array}{r|l} \multicolumn{2}{c}{5s + t = 8} \\ \hline 5 \cdot 2 + (-2) & 8 \\ 10 - 2 & \\ \end{array}$$

$$8 \stackrel{?}{=} 8 \quad \text{TRUE}$$

Since $(2, -2)$ checks, it is the solution.

13. $4x - 2y = 6, \quad (1)$

$2x - 3 = y \quad (2)$

We substitute $2x - 3$ for y in the first equation and solve for x.

$$\begin{aligned} 4x - 2y &= 6 \quad (1) \\ 4x - 2(2x - 3) &= 6 \\ 4x - 4x + 6 &= 6 \\ 6 &= 6 \end{aligned}$$

We have an identity, or an equation that is always true. The equations are dependent and the solution set is infinite: $\{(x, y) | 2x - 3 = y\}$.

15. $-5s + t = 11, \quad (1)$

$4s + 12t = 4 \quad (2)$

We solve the first equation for t.

$$-5s + t = 11 \quad (1)$$

$$t = 5s + 11 \quad (3)$$

We substitute $5s + 11$ for t in the second equation and solve for s.

$$\begin{aligned} 4s + 12t &= 4 \quad (2) \\ 4s + 12(5s + 11) &= 4 \\ 4s + 60s + 132 &= 4 \\ 64s + 132 &= 4 \\ 64s &= -128 \\ s &= -2 \end{aligned}$$

Next we substitute -2 for s in Equation (3).

$$t = 5s + 11 = 5(-2) + 11 = -10 + 11 = 1$$

We check the ordered pair $(-2, 1)$.

$$\begin{array}{r|l} \multicolumn{2}{c}{-5s + t = 11} \\ \hline -5(-2) + 1 & 11 \\ 10 + 1 & \\ \end{array}$$

$$11 \stackrel{?}{=} 11 \quad \text{TRUE}$$

$$\begin{array}{r|l} \multicolumn{2}{c}{4s + 12t = 4} \\ \hline 4(-2) + 12 \cdot 1 & 4 \\ -8 + 12 & \\ \end{array}$$

$$4 \stackrel{?}{=} 4 \quad \text{TRUE}$$

Since $(-2, 1)$ checks, it is the solution.

17. $2x + 2y = 2,\quad (1)$

$3x - y = 1 \quad (2)$

We solve the second equation for y.

$$\begin{aligned} 3x - y &= 1 \quad (2) \\ -y &= -3x + 1 \\ y &= 3x - 1 \quad (3) \end{aligned}$$

We substitute $3x - 1$ for y in the first equation and solve for x.

$$\begin{aligned} 2x + 2y &= 2 \quad (1) \\ 2x + 2(3x - 1) &= 2 \\ 2x + 6x - 2 &= 2 \\ 8x - 2 &= 2 \\ 8x &= 4 \\ x &= \frac{1}{2} \end{aligned}$$

Next we substitute $\frac{1}{2}$ for x in Equation (3).

$$y = 3x - 1 = 3 \cdot \frac{1}{2} - 1 = \frac{3}{2} - 1 = \frac{1}{2}$$

The ordered pair $\left(\frac{1}{2}, \frac{1}{2}\right)$ checks in both equations. It is the solution.

19. $x - 4y = 3,\quad (1)$

$5x + 3y = 4 \quad (2)$

We solve the first equation for x.

$$\begin{aligned} x - 4y &= 3 \quad (1) \\ x &= 4y + 3 \quad (3) \end{aligned}$$

Substitute $4y + 3$ for x in the second equation and solve for y.

$$\begin{aligned} 5x + 3y &= 4 \quad (2) \\ 5(4y + 3) + 3y &= 4 \\ 20y + 15 + 3y &= 4 \\ 23y + 15 &= 4 \\ 23y &= -11 \\ y &= -\frac{11}{23} \end{aligned}$$

Substitute $-\frac{11}{23}$ for y in Equation (3).

$$x = 4\left(-\frac{11}{23}\right) + 3 = -\frac{44}{23} + \frac{69}{25} = \frac{25}{23}$$

The ordered pair is $\left(\frac{25}{23}, -\frac{11}{23}\right)$ checks in both equations. It is the solution.

21. $2x - 3 = y \quad (1)$

$y - 2x = 1,\quad (2)$

We substitute $2x - 3$ for y in the second equation and solve for x.

$$\begin{aligned} y - 2x &= 1 \quad (2) \\ 2x - 3 - 2x &= 1 \quad \text{Substituting} \\ -3 &= 1 \quad \text{Collecting like terms} \end{aligned}$$

We have a contradiction, or an equation that is always false. Therefore, there is no solution.

23.

$$\begin{aligned} x + 3y &= 7 \quad (1) \\ -x + 4y &= 7 \quad (2) \\ \hline 0 + 7y &= 14 \quad \text{Adding} \\ 7y &= 14 \\ y &= 2 \end{aligned}$$

Substitute 2 for y in one of the original equations and solve for x.

$$\begin{aligned} x + 3y &= 7 \quad (1) \\ x + 3 \cdot 2 &= 7 \quad \text{Substituting} \\ x + 6 &= 7 \\ x &= 1 \end{aligned}$$

Check:

$x + 3y = 7$	
$1 + 3 \cdot 2$	7
$1 + 6$	
$7 \stackrel{?}{=} 7$ TRUE	

$-x + 4y = 7$	
$-1 + 4 \cdot 2$	7
$-1 + 8$	
$7 \stackrel{?}{=} 7$ TRUE	

Since $(1, 2)$ checks, it is the solution.

25.

$$\begin{aligned} x + y &= 9 \quad (1) \\ 2x - y &= -3 \quad (2) \\ \hline 3x &= 6 \quad \text{Adding} \\ x &= 2 \end{aligned}$$

Substitute 2 for x in Equation (1) and solve for y.

$$\begin{aligned} x + y &= 9 \quad (1) \\ 2 + y &= 9 \quad \text{Substituting} \\ y &= 7 \end{aligned}$$

We obtain $(2, 7)$. This checks, so it is the solution.

27.

$$\begin{aligned} 9x + 3y &= -3 \quad (1) \\ 2x - 3y &= -8 \quad (2) \\ \hline 11x + 0 &= -11 \quad \text{Adding} \\ 11x &= -11 \\ x &= -1 \end{aligned}$$

Substitute -1 for x in Equation (1) and solve for y.

$$\begin{aligned} 9x + 3y &= -3 \\ 9(-1) + 3y &= -3 \quad \text{Substituting} \\ -9 + 3y &= -3 \\ 3y &= 6 \\ y &= 2 \end{aligned}$$

We obtain $(-1, 2)$. This checks, so it is the solution.

29. $5x + 3y = 19,$ (1)

$2x - 5y = 11$ (2)

We multiply twice to make two terms become opposites.

From (1): $25x + 15y = 95$ Multiplying by 5

From (2): $6x - 15y = 33$ Multiplying by 3

$31x + 0 = 128$ Adding

$x = \frac{128}{31}$

Substitute $\frac{128}{31}$ for x in Equation (1) and solve for y.

$5x + 3y = 19$

$5 \cdot \frac{128}{31} + 3y = 19$ Substituting

$\frac{640}{31} + 3y = \frac{589}{31}$

$3y = -\frac{51}{31}$

$\frac{1}{3} \cdot 3y = \frac{1}{3} \cdot \left(-\frac{51}{31}\right)$

$y = -\frac{17}{31}$

We obtain $\left(\frac{128}{31}, -\frac{17}{31}\right)$. This checks, so it is the solution.

31. $5r - 3s = 24,$ (1)

$3r + 5s = 28$ (2)

We multiply twice to make two terms become additive inverses.

From (1): $25r - 15s = 120$ Multiplying by 5

From (2): $9r + 15s = 84$ Multiplying by 3

$34r + 0 = 204$ Adding

$r = 6$

Substitute 6 for r in Equation (2) and solve for s.

$3r + 5s = 28$

$3 \cdot 6 + 5s = 28$ Substituting

$18 + 5s = 28$

$5s = 10$

$s = 2$

We obtain $(6, 2)$. This checks, so it is the solution.

33. $6s + 9t = 12,$ (1)

$4s + 6t = 5$ (2)

We multiply twice to make two terms become opposites.

From (1): $12s + 18t = 24$ Multiplying by 2

From (2): $-12s - 18t = -15$ Multiplying by -3

$0 = 9$

We get a contradiction, or an equation that is always false. The system has no solution.

35. $\frac{1}{2}x - \frac{1}{6}y = 3$ (1)

$\frac{2}{5}x + \frac{1}{2}y = 2,$ (2)

We first multiply each equation by the LCM of the denominators to clear fractions.

$3x - y = 18$ (3) Multiplying (1) by 6

$4x + 5y = 20$ (4) Multiplying (2) by 10

We multiply by 5 on both sides of Equation (3) and then add.

$15x - 5y = 90$ Multiplying (3) by 5

$4x + 5y = 20$ (4)

$19x + 0 = 110$ Adding

$x = \frac{110}{19}$

Substitute $\frac{110}{19}$ for x in one of the equations in which the fractions were cleared and solve for y.

$3x - y = 18$ (3)

$3\left(\frac{110}{19}\right) - y = 18$ Substituting

$\frac{330}{19} - y = \frac{342}{19}$

$-y = \frac{12}{19}$

$y = -\frac{12}{19}$

We obtain $\left(\frac{110}{19}, -\frac{12}{19}\right)$. This checks, so it is the solution.

37. $\frac{x}{2} + \frac{y}{3} = \frac{7}{6},$ (1)

$\frac{2x}{3} + \frac{3y}{4} = \frac{5}{4}$ (2)

We first multiply each equation by the LCM of the denominators to clear fractions.

$3x + 2y = 7$ (3) Multiplying (1) by 6

$8x + 9y = 15$ (4) Multiplying (2) by 12

We multiply twice to make two terms become opposites.

From (3): $27x + 18y = 63$ Multiplying by 9

From (4): $-16x - 18y = -30$ Multiplying by -2

$11x = 33$ Adding

$x = 3$

Substitute 3 for x in one of the equations in which the fractions were cleared and solve for y.

$3x + 2y = 7$ (3)

$3 \cdot 3 + 2y = 7$ Substituting

$9 + 2y = 7$

$2y = -2$

$y = -1$

We obtain $(3, -1)$. This checks, so it is the solution.

39. $12x - 6y = -15,$ (1)

$-4x + 2y = 5$ (2)

Observe that, if we multiply Equation (1) by $-\frac{1}{3}$, we obtain Equation (2). Thus, any pair that is a solution of Equation (1) is also a solution of Equation (2). The equations are dependent and the solution set is infinite: $\{(x, y)| - 4x + 2y = 5\}$.

41. $0.2a + 0.3b = 1,$

$0.3a - 0.2b = 4,$

We first multiply each equation by 10 to clear decimals.

$2a + 3b = 10 \quad (1)$

$3a - 2b = 40 \quad (2)$

We multiply so that the b-terms can be eliminated.

$$\begin{aligned} \text{From (1):} \quad 4a + 6b &= 20 \quad \text{Multiplying by 2} \\ \text{From (2):} \quad 9a - 6b &= 120 \quad \text{Multiplying by 3} \\ \hline 13a + 0 &= 140 \quad \text{Adding} \\ a &= \frac{140}{13} \end{aligned}$$

Substitute $\frac{140}{13}$ for a in Equation (1) and solve for b.

$$\begin{aligned} 2a + 3b &= 10 \\ 2 \cdot \frac{140}{13} + 3b &= 10 \quad \text{Substituting} \\ \frac{280}{13} + 3b &= \frac{130}{13} \\ 3b &= -\frac{150}{13} \\ b &= -\frac{50}{13} \end{aligned}$$

We obtain $\left(\frac{140}{13}, -\frac{50}{13}\right)$. This checks, so it is the solution.

43. $a - 2b = 16, \quad (1)$

$b + 3 = 3a \quad (2)$

We will use the substitution method. First solve Equation (1) for a.

$a - 2b = 16$

$a = 2b + 16 \quad (3)$

Now substitute $2b + 16$ for a in Equation (2) and solve for b.

$$\begin{aligned} b + 3 &= 3a \quad (2) \\ b + 3 &= 3(2b + 16) \quad \text{Substituting} \\ b + 3 &= 6b + 48 \\ -45 &= 5b \\ -9 &= b \end{aligned}$$

Substitute -9 for b in Equation (3).

$a = 2(-9) + 16 = -2$

We obtain $(-2, -9)$. This checks, so it is the solution.

45. $10x + y = 306, \quad (1)$

$10y + x = 90 \quad (2)$

We will use the substitution method. First solve Equation (1) for y.

$10x + y = 306$

$y = -10x + 306 \quad (3)$

Now substitute $-10x + 306$ for y in Equation (2) and solve for y.

$$\begin{aligned} 10y + x &= 90 \quad (2) \\ 10(-10x + 306) + x &= 90 \quad \text{Substituting} \\ -100x + 3060 + x &= 90 \\ -99x + 3060 &= 90 \\ -99x &= -2970 \\ x &= 30 \end{aligned}$$

Substitute 30 for x in Equation (3).

$y = -10 \cdot 30 + 306 = 6$

We obtain $(30, 6)$. This checks, so it is the solution.

47. $3y = x - 2, \quad (1)$

$x = 2 + 3y \quad (2)$

We will use the substitution method. Substitute $2 + 3y$ for x in the first equation and solve for y.

$$\begin{aligned} 3y &= x - 2 \quad (1) \\ 3y &= 2 + 3y - 2 \quad \text{Substituting} \\ 3y &= 3y \quad \text{Collecting like terms} \end{aligned}$$

We get an identity. The system is dependent and the solution set is infinite: $\{(x, y)|x = 2 + 3y\}$.

49. $3s - 7t = 5,$

$7t - 3s = 8$

First we rewrite the second equation with the variables in a different order. Then we use the elimination method.

$$\begin{aligned} 3s - 7t &= 5, \quad (1) \\ -3s + 7t &= 8 \quad (2) \\ \hline 0 &= 13 \end{aligned}$$

We get a contradiction, so the system has no solution.

51. $0.05x + 0.25y = 22, \quad (1)$

$0.15x + 0.05y = 24 \quad (2)$

We first multiply each equation by 100 to clear decimals.

$5x + 25y = 2200$

$15x + 5y = 2400$

We multiply by -5 on both sides of the second equation and add.

$$\begin{aligned} 5x + 25y &= 2200 \\ -75x - 25y &= -12,000 \quad \text{Multiplying (2) by } -5 \\ \hline -70x &= -9800 \quad \text{Adding} \\ x &= \frac{-9800}{-70} \\ x &= 140 \end{aligned}$$

Substitute 140 for x in one of the equations in which the decimals were cleared and solve for y.

$$\begin{aligned} 5x + 25y &= 2200 \quad (1) \\ 5 \cdot 140 + 25y &= 2200 \quad \text{Substituting} \\ 700 + 25y &= 2200 \\ 25y &= 1500 \\ y &= 60 \end{aligned}$$

We obtain $(140, 60)$. This checks, so it is the solution.

53. $13a - 7b = 9$, (1)

$2a - 8b = 6$ (2)

We will use the elimination method. First we multiply so that the b-terms can be eliminated.

$$\begin{aligned} &\text{From (1):} \quad 104a - 56b = 72 \quad \text{Multiplying by 8} \\ &\text{From (2):} \quad \underline{-14a + 56b = -42} \quad \text{Multiplying by } -7 \\ &\qquad\qquad\quad 90a \qquad\;\; = 30 \quad \text{Adding} \\ &\qquad\qquad\qquad\quad a = \frac{1}{3} \end{aligned}$$

Substitute $\frac{1}{3}$ for a in one of the equations and solve for b.

$$\begin{aligned} 2a - 8b &= 6 \quad (2) \\ 2 \cdot \frac{1}{3} - 8b &= 6 \\ \frac{2}{3} - 8b &= 6 \\ -8b &= \frac{16}{3} \\ b &= -\frac{2}{3} \end{aligned}$$

We obtain $\left(\frac{1}{3}, -\frac{2}{3}\right)$. This checks, so it is the solution.

55. *Writing Exercise*

57. ***Familiarize.*** Let m = the number of $\frac{1}{4}$-mi units traveled after the first $\frac{1}{2}$ mi. The total distance traveled will be $\frac{1}{2}$ mi $+ m \cdot \frac{1}{4}$ mi.

Translate.

Fare for first $\frac{1}{2}$ mi	plus	fare for additional $\frac{1}{4}$-mi units	is $5.20.
↓	↓	↓	↓ ↓
1	+	$0.3m$	= 5.20

Carry out. We solve the equation.

$$\begin{aligned} 1 + 0.3m &= 5.20 \\ 0.3m &= 4.20 \\ m &= 14 \end{aligned}$$

If the taxi travels the first $\frac{1}{2}$ mi plus 14 additional $\frac{1}{4}$-mi units, then it travels a total of $\frac{1}{2} + 14 \cdot \frac{1}{4}$, or $\frac{1}{2} + \frac{7}{2}$, or 4 mi.

Check. We have 4 mi $= \frac{1}{2}$ mi $+ \frac{7}{2}$ mi $= \frac{1}{2}$ mi $+ 14 \cdot \frac{1}{4}$ mi. The fare for traveling this distance is $\$1.00 + \$0.30(14) = \$1.00 + \$4.20 = \$5.20$. The answer checks.

State. It is 4 mi from Johnson Street to Elm Street.

59. ***Familiarize.*** Let a = the amount spent to remodel bathrooms, in billions of dollars. Then $2a$ = the amount spent to remodel kitchens. The sum of these two amounts is $35 billion.

Translate.

Amount spent on bathrooms	plus	amount spent on kitchens	is	$35 billion.
↓	↓	↓	↓	↓
a	+	$2a$	=	35

Carry out. We solve the equation.

$$\begin{aligned} a + 2a &= 35 \\ 3a &= 35 \quad \text{Combining like terms} \\ a &= \frac{35}{3}, \text{ or } 11\frac{2}{3} \end{aligned}$$

If $a = \frac{35}{3}$, then $2a = 2 \cdot \frac{35}{3} = \frac{70}{3} = 23\frac{1}{3}$.

Check. $\frac{70}{3}$ is twice $\frac{35}{3}$, and $\frac{35}{3} + \frac{70}{3} = \frac{105}{3} = 35$. The answer checks.

State. $\$11\frac{2}{3}$ billion was spent to remodel bathrooms, and $\$23\frac{1}{3}$ billion was spent to remodel kitchens.

61. ***Familiarize.*** The total cost is the daily charge plus the mileage charge. The mileage charge is the cost per mile times the number of miles driven. Let m = the number of miles that can be driven for $80.

Translate. We reword the problem.

Daily rate	plus	Cost per mile	times	Number of miles driven	is	Amount.
↓	↓	↓	↓	↓	↓	↓
34.95	+	0.10	·	m	=	80

Carry out. We solve the equation.

$$\begin{aligned} 34.95 + 0.10m &= 80 \\ 100(34.95 + 0.10m) &= 100(80) \quad \text{Clearing decimals} \\ 3495 + 10m &= 8000 \\ 10m &= 4505 \\ m &= 450.5 \end{aligned}$$

Check. The mileage cost is found by multiplying 450.5 by $0.10 obtaining $45.05. Then we add $45.05 to $34.95, the daily rate, and get $80.

State. The businessperson can drive 450.5 mi on the car-rental allotment.

63. *Writing Exercise*

65. First write $f(x) = mx + b$ as $y = mx + b$. Then substitute 1 for x and 2 for y to get one equation and also substitute -3 for x and 4 for y to get a second equation:

$$\begin{aligned} 2 &= m \cdot 1 + b \\ 4 &= m(-3) + b \end{aligned}$$

Solve the resulting system of equations.

$$2 = m + b$$
$$4 = -3m + b$$

Multiply the second equation by -1 and add.

$$\begin{aligned} 2 &= m + b \\ -4 &= 3m - b \\ \hline -2 &= 4m \\ -\frac{1}{2} &= m \end{aligned}$$

Substitute $-\frac{1}{2}$ for m in the first equation and solve for b.

$$2 = -\frac{1}{2} + b$$
$$\frac{5}{2} = b$$

Thus, $m = -\frac{1}{2}$ and $b = \frac{5}{2}$.

67. Substitute -4 for x and -3 for y in both equations and solve for a and b.

$-4a - 3b = -26$, (1)

$-4b + 3a = 7$ (2)

$$\begin{aligned} -12a - 9b &= -78 \quad \text{Multiplying (1) by 3} \\ 12a - 16b &= 28 \quad \text{Multiplying (2) by 4} \\ \hline -25b &= -50 \\ b &= 2 \end{aligned}$$

Substitute 2 for b in Equation (2).

$$-4 \cdot 2 + 3a = 7$$
$$3a = 15$$
$$a = 5$$

Thus, $a = 5$ and $b = 2$.

69. $\frac{x+y}{2} - \frac{x-y}{5} = 1$,

$\frac{x-y}{2} + \frac{x+y}{6} = -2$

After clearing fractions we have:

$3x + 7y = 10$, (1)

$4x - 2y = -12$ (2)

$$\begin{aligned} 6x + 14y &= 20 \quad \text{Multiplying (1) by 2} \\ 28x - 14y &= -84 \quad \text{Multiplying (2) by 7} \\ \hline 34x &= -64 \\ x &= -\frac{32}{17} \end{aligned}$$

Substitute $-\frac{32}{17}$ for x in Equation (1).

$$3\left(-\frac{32}{17}\right) + 7y = 10$$
$$7y = \frac{266}{17}$$
$$y = \frac{38}{17}$$

The solution is $\left(-\frac{32}{17}, \frac{38}{17}\right)$.

71. $\frac{2}{x} + \frac{1}{y} = 0$, $\qquad 2 \cdot \frac{1}{x} + \frac{1}{y} = 0$,

or

$\frac{5}{x} + \frac{2}{y} = -5 \qquad 5 \cdot \frac{1}{x} + 2 \cdot \frac{1}{y} = -5$

Substitute u for $\frac{1}{x}$ and v for $\frac{1}{y}$.

$2u + v = 0$, (1)

$5u + 2v = -5$ (2)

$$\begin{aligned} -4u - 2v &= 0 \quad \text{Multiplying (1) by } -2 \\ 5u + 2v &= -5 \quad (2) \\ \hline u &= -5 \end{aligned}$$

Substitute -5 for u in Equation (1).

$$2(-5) + v = 0$$
$$-10 + v = 0$$
$$v = 10$$

If $u = -5$, then $\frac{1}{x} = -5$. Thus $x = -\frac{1}{5}$.

If $v = 10$, then $\frac{1}{y} = 10$. Thus $y = \frac{1}{10}$.

The solution is $\left(-\frac{1}{5}, \frac{1}{10}\right)$.

73. *Writing Exercise*

Exercise Set 8.3

1. The Familiarize and Translate steps were done in Exercise 41 of Exercise Set 8.1.

Carry out. We solve the system of equations

$x + y = 50$, (1)

$x = 0.25y$, (2)

where x is the first number and y is the second number. We use substitution.

Substitute $0.25y$ for x in (1) and solve for y.

$$0.25y + y = 50$$
$$1.25y = 50$$
$$y = 40$$

Now substitute 40 for y in (2).

$$x = 0.25(40) = 10$$

Check. The sum of the numbers is $10 + 40$, or 50, and 0.25 times the second number, 40, is the first number, 10. The answer checks.

State. The first number is 10, and the second number is 40.

3. The Familiarize and Translate steps were done in Exercise 43 of Exercise Set 8.1.

Carry out. We solve the system of equations

$m + v = 16$, (1)

$m = 2v + 4$, (2)

where m and v represent the number of ounces of mineral oil and vinegar to be used, respectively. We use substitution.

Substitute $2v + 4$ for m in (1) and solve for v.

$$(2v + 4) + v = 16$$
$$3v + 4 = 16$$
$$3v = 12$$
$$v = 4$$

Now substitute 4 for v in (2).

$$m = 2 \cdot 4 + 4 = 8 + 4 = 12$$

Check. The mixture contains 12+4, or 16 oz. The amount of mineral oil, 12 oz, is 4 oz more than twice the amount of vinegar, 4 oz: $2 \cdot 4 + 4 = 12$. The answer checks.

State. 12 oz of mineral oil and 4 oz of vinegar should be used.

5. The Familiarize and Translate steps were done in Exercise 45 of Exercise Set 8.1

Carry out. We solve the system of equations

$$x + y = 180, \quad (1)$$
$$x = 2y - 3 \quad (2)$$

where x = the measure of one angle and y = the measure of the other angle. We use substitution.

Substitute $2y - 3$ for x in (1) and solve for y.

$$2y - 3 + y = 180$$
$$3y - 3 = 180$$
$$3y = 183$$
$$y = 61$$

Now substitute 61 for y in (2).

$$x = 2 \cdot 61 - 3 = 122 - 3 = 119$$

Check. The sum of the angle measures is $119° + 61°$, or $180°$, so the angles are supplementary. Also $2 \cdot 61° - 3° = 122° - 3° = 119°$. The answer checks.

State. The measures of the angles are 119° and 61°.

7. The Familiarize and Translate steps were done in Exercise 47 of Exercise Set 8.1

Carry out. We solve the system of equations

$$g + t = 64, \quad (1)$$
$$2g + t = 100 \quad (2)$$

where g = the number of field goals and t = the number of foul shots Chamberlain made. We use elimination.

$$-g - t = -64 \quad \text{Multiplying (1) by } -1$$
$$\underline{2g + t = 100}$$
$$g = 36$$

Substitute 36 for g in (1) and solve for t.

$$36 + t = 64$$
$$t = 28$$

Check. The total number of scores was 36+28, or 64. The total number of points was $2 \cdot 36 + 28 = 72 + 28 = 100$. The answer checks.

State. Chamberlain made 36 field goals and 28 foul shots.

9. The Familiarize and Translate steps were done in Exercise 49 of Exercise Set 8.1

Carry out. We solve the system of equations

$$x + y = 45, \quad (1)$$
$$850x + 975y = 39,875 \quad (2)$$

where x = the number of less expensive brushes sold and y = the number of more expensive brushes sold. We use elimination. Begin by multiplying Equation (1) by -850.

$$-850x - 850y = -38,250 \quad \text{Multiplying (1)}$$
$$\underline{850x + 975y = 39,875}$$
$$125y = 1625$$
$$y = 13$$

Substitute 13 for y in (1) and solve for x.

$$x + 13 = 45$$
$$x = 32$$

Check. The number of brushes sold is 32 + 13, or 45. The amount taken in was $8.50(32) + $9.75(13) = $272 + $126.75 = $398.75. The answer checks.

State. 32 of the less expensive brushes were sold, and 13 of the more expensive brushes were sold.

11. The Familiarize and Translate steps were done in Exercise 51 of Exercise Set 8.1

Carry out. We solve the system of equations

$$h + n = 50, \quad (1)$$
$$2706h + 3439n = 156,557 \quad (2)$$

where h = the number of vials of Humulin Insulin sold and n = the number of vials of Novolin Velosulin Insulin sold. We use elimination.

$$-2706h - 2706n = -135,300 \quad \text{Multiplying (1) by } -2706$$
$$\underline{2706h + 3439n = 156,557}$$
$$733n = 21,257$$
$$n = 29$$

Substitute 29 for n in (1) and solve for h.

$$h + 29 = 50$$
$$h = 21$$

Check. A total of 21 + 29, or 50 vials, were sold. The amount collected was $27.06(21) + $34.39(29) = $568.26 + $997.31 = $1565.57. The answer checks.

State. 21 vials of Humulin Insulin and 29 vials of Novolin Velosulin Insulin were sold.

13. The Familiarize and Translate steps were done in Exercise 53 of Exercise Set 8.1

Carry out. We solve the system of equations

$$2l + 2w = 288, \quad (1)$$
$$l = w + 44 \quad (2)$$

where l = the length, in feet, and w = the width, in feet, of the basketball court. We use substitution.

Substitute $w + 44$ for l in (1) and solve for w.

$$2(w+44)+2w=288$$
$$2w+88+2w=288$$
$$4w+88=288$$
$$4w=200$$
$$w=50$$

Now substitute 50 for w in (2).

$$l=50+44=94$$

Check. The perimeter is $2 \cdot 94$ ft$+2 \cdot 50$ ft $=188$ ft$+100$ ft $=288$ ft. The length, 94 ft, is 44 ft more than the width, 50 ft. The answer checks.

State. The length of the basketball court is 94 ft, and the width is 50 ft.

15. ***Familiarize.*** Let $x=$ the number of sheets of nonrecycled paper used and $y=$ the number of recycled sheets.

Translate. We organize the information in a table.

	Nonrecycled sheets	Recycled sheets	Total
Number used	x	y	150
Price	1.9¢	2.4¢	
Total cost	$1.9x$	$2.4y$	\$3.41, or 341¢

We get one equation from the "Number used" row of the table:

$$x+y=150$$

The "Total cost" row yields a second equation. We express all costs in cents:

$$1.9x+2.4y=341$$

After clearing decimals, we have the problem translated to a system of equations.

$$x+y=150, \quad (1)$$
$$19x+24y=3410 \quad (2)$$

Carry out. We use the elimination method to solve the system of equations.

$$-19x-19y=-2850 \quad \text{Multiplying (1) by } -19$$
$$19x+24y=3410$$
$$5y=560$$
$$y=112$$

Substitute 112 for y in (1) and solve for x.

$$x+112=150$$
$$x=38$$

Check. A total of $38+112$, or 150 sheets of paper were used. The total cost was $1.9¢(38)+2.4¢(112)=72.2¢+268.8¢=341¢$, or \$3.41. The answer checks.

State. 38 sheets of nonrecycled paper and 112 sheets of recycled paper were used.

17. ***Familiarize.*** Let $g=$ the number of General Electric bulbs purchased and $s=$ the number of SLi bulbs purchased.

Translate. We organize the information in a table.

	GE bulbs	SLi bulbs	Total
Number purchased	g	s	200
Price	\$7.50	\$5	
Total cost	$7.5g$	$5s$	1150

We get our equation from the "Number purchased" row of the table:

$$g+s=200$$

The "Total cost" row yields a second equation:

$$7.5g+5s=1150$$

After clearing decimals, we have the problem translated to a system of equations:

$$g+s=200, \quad (1)$$
$$75g+50s=11,500 \quad (2)$$

Carry out. We use the elimination method to solve the system of equations.

$$-50g-50s=-10,000 \quad \text{Multiplying (1) by } -50$$
$$75g+50s=11,500$$
$$25g=1500$$
$$g=60$$

Substitute 60 for g in (1) and solve for s.

$$60+s=200$$
$$s=140$$

Check. A total of $60+140$, or 200 bulbs, was purchased. The total cost was $\$7.50(60)+\$5(140)=\$450+\$700=\$1150$. The answer checks.

State. 60 General Electric bulbs and 140 SLi bulbs were purchased.

19. ***Familiarize.*** Let $a=$ the number of Apple cartridges purchased and $h=$ the number of HP cartridges.

Translate. We organize the information in a table.

	Apple	HP	Total
Number purchased	a	h	50
Price	\$30.86	\$43.58	
Total cost	$30.86a$	$43.58h$	1733.80

We get one equation from the "Number purchased" row of the table:

$$a+h=50$$

The "Total cost" row yields a second equation:

$$30.86a+43.58h=1733.80$$

After clearing decimals, we have the problem translated to a system of equations:

$$a+h=50, \quad (1)$$
$$3086a+4358h=173,380 \quad (2)$$

Carry out. We use the elimination method to solve the system of equations.

$$\begin{array}{rl} -3086a-3086h = -154,300 & \text{Multiplying (1) by } -3086 \\ 3086a+4358h = 173,380 & \\ \hline 1272h = 19,080 & \\ h = 15 & \end{array}$$

Substitute 15 for h in (1) and solve for a.

$$a + 15 = 50$$
$$a = 35$$

Check. A total of $35+15$, or 50 cartridges, was sold. The total cost was $\$30.86(35) + \$43.58(15) =$ $\$1080.10 + \$653.70 = \$1733.80$. The answer checks.

State. 35 Apple cartridges and 15 HP cartridges were purchased.

21. ***Familiarize***. Let $k =$ the number of pounds of Kenyan French Roast coffee and $s =$ the number of pounds of Sumatran coffee to be used in the mixture. The value of the mixture will be $\$8.40(20)$, or $168.

Translate. We organize the information in a table.

	Kenyan	Sumatran	Mixture
Number of pounds	k	s	20
Price per pound	$9	$8	$8.40
Value of coffee	$9k$	$8s$	168

The "Number of pounds" row of the table gives us one equation:

$$k + s = 20$$

The "Value of coffee" row yields a second equation:

$$9k + 8s = 168$$

We have translated to a system of equations:

$$k + s = 20, \quad (1)$$
$$9k + 8s = 168 \quad (2)$$

Carry out. We use the elimination method to solve the system of equations.

$$\begin{array}{rl} -8k - 8s = -160 & \text{Multiplying (1) by } -8 \\ 9k + 8s = 168 & \\ \hline k = 8 & \end{array}$$

Substitute 8 for k in (1) and solve for s.

$$8 + s = 20$$
$$s = 12$$

Check. The total mixture contains 8 lb + 12 lb, or 20 lb. Its value is $\$9 \cdot 8 + \$8 \cdot 12 = \$72 + \$96 = \$168$. The answer checks.

State. 8 lb of Kenyan French Roast coffee and 12 lb of Sumatran coffee should be used.

23. Observe that the average of 40% and 10% is 25%: $\dfrac{40\% + 10\%}{2} = \dfrac{50\%}{2} = 25\%$. Thus, the caterer should use equal parts of the 40% and 10% mixtures. Since a 20-lb mixture is desired, the caterer should use 10 lb each of the 40% and the 10% mixture.

25. ***Familiarize***. Let $x =$ the number of pounds of Deep Thought Granola and $y =$ the number of pounds of Oat Dream Granola to be used in the mixture. The amount of nuts and dried fruit in the mixture is 19%(20 lb), or 0.19(20 lb) = 3.8 lb.

Translate. We organize the information in a table.

	Deep Thought	Oat Dream	Mixture
Number of pounds	x	y	20
Percent of nuts and dried fruit	25%	10%	19%
Amount of nuts and dried fruit	$0.25x$	$0.1y$	3.8 lb

We get one equation from the "Number of pounds" row of the table:

$$x + y = 20$$

The last row of the table yields a second equation:

$$0.25x + 0.1y = 3.8$$

After clearing decimals, we have the problem translated to a system of equations:

$$x + y = 20, \quad (1)$$
$$25x + 10y = 380 \quad (2)$$

Carry out. We use the elimination method to solve the system of equations.

$$\begin{array}{rl} -10x - 10y = -200 & \text{Multiplying (1) by } -10 \\ 25x + 10y = 380 & \\ \hline 15x = 180 & \\ x = 12 & \end{array}$$

Substitute 12 for x in (1) and solve for y.

$$12 + y = 20$$
$$y = 8$$

Check. The amount of the mixture is 12 lb + 8 lb, or 20 lb. The amount of nuts and dried fruit in the mixture is 0.25(12 lb) + 0.1(8 lb) = 3 lb + 0.8 lb = 3.8 lb. The answer checks.

State. 12 lb of Deep Thought Granola and 8 lb of Oat Dream Granola should be mixed.

27. ***Familiarize***. Let $x =$ the amount of the 6% loan and $y =$ the amount of the 9% loan. Recall that the formula for simple interest is

$$\text{Interest} = \text{Principal} \cdot \text{Rate} \cdot \text{Time}.$$

Translate. We organize the information in a table.

	6% loan	9% loan	Total
Principal	x	y	\$12,000
Interest Rate	6%	9%	
Time	1 yr	1 yr	
Interest	$0.06x$	$0.09y$	\$855

The "Principal" row of the table gives us one equation:

$$x + y = 12,000$$

The last row of the table yields another equation:

$$0.06x + 0.09y = 855$$

After clearing decimals, we have the problem translated to a system of equations:

$$x + y = 12,000 \quad (1)$$
$$6x + 9y = 85,500 \quad (2)$$

Carry out. We use the elimination method to solve the system of equations.

$$-6x - 6y = -72,000 \quad \text{Multiplying (1) by } -6$$
$$6x + 9y = 85,500$$
$$3y = 13,500$$
$$y = 4500$$

Substitute 4500 for y in (1) and solve for x.

$$x + 4500 = 12,000$$
$$x = 7500$$

Check. The loans total \$7500 + \$4500, or \$12,000. The total interest is 0.06(\$7500) + 0.09(\$4500) = \$450 + \$405 = \$855. The answer checks.

State. The 6% loan was for \$7500, and the 9% loan was for \$4500.

29. *Familiarize.* Let x = the number of liters of Arctic Antifreeze and y = the number of liters of Frost-No-More in the mixture. The amount of alcohol in the mixture is 0.15(20 L) = 3 L.

Translate. We organize the information in a table.

	18% solution	10% solution	Mixture
Number of liters	x	y	20
Percent of alcohol	18%	10%	15%
Amount of alcohol	$0.18x$	$0.1y$	3

We get one equation from the "Number of liters" row of the table:

$$x + y = 20$$

The last row of the table yields a second equation:

$$0.18x + 0.1y = 3$$

After clearing decimals we have the problem translated to a system of equations:

$$x + y = 20, \quad (1)$$
$$18x + 10y = 300 \quad (2)$$

Carry out. We use the elimination method to solve the system of equations.

$$-10x - 10y = -200 \quad \text{Multiplying (1) by } -10$$
$$18x + 10y = 300$$
$$8x = 100$$
$$x = 12.5$$

Substitute 12.5 for x in (1) and solve for y.

$$12.5 + y = 20$$
$$y = 7.5$$

Check. The total amount of the mixture is 12.5 L + 7.5 L or 20 L. The amount of alcohol in the mixture is 0.18(12.5 L) + 0.1(7.5 L) = 2.25 L + 0.75 L = 3 L. The answer checks.

State. 12.5 L of Arctic Antifreeze and 7.5 L of Frost-No-More should be used.

31. *Familiarize.* Let x = the number of gallons of 87-octane gas and y = the number of gallons of 93-octane gas in the mixture. The amount of octane in the mixture can be expressed as 91(12), or 1092.

Translate. We organize the information in a table.

	87-octane	93-octane	Mixture
Number of gallons	x	y	12
Octane rating	87	93	91
Total octane	$87x$	$93y$	1092

We get one equation from the "Number of gallons" row of the table:

$$x + y = 12$$

The last row of the table yields a second equation:

$$87x + 93y = 1092$$

We have a system of equations:

$$x + y = 12, \quad (1)$$
$$87x + 93y = 1092 \quad (2)$$

Carry out. We use the elimination method to solve the system of equations.

$$-87x - 87y = -1044 \quad \text{Multiplying (1) by } -87$$
$$87x + 93y = 1092$$
$$6y = 48$$
$$y = 8$$

Substitute 8 for y in (1) and solve for x.

$$x + 8 = 12$$
$$x = 4$$

Check. The total amount of the mixture is 4 gal + 8 gal, or 12 gal. The amount of octane can be expressed as 87(4) + 93(8) = 348 + 744 = 1092. The answer checks.

State. 4 gal of 87-octane gas and 8 gal of 93-octane gas should be used.

33. ***Familiarize.*** From the bar graph we see that whole milk is 4% milk fat, milk for cream cheese is 8% milk fat, and cream is 30% milk fat. Let x = the number of pounds of whole milk and y = the number of pounds of cream to be used. The mixture contains 8%(200 lb), or 0.08(200 lb) = 16 lb of milk fat.

Translate. We organize the information in a table.

	Whole milk	Cream	Mixture
Number of pounds	x	y	200
Percent of milk fat	4%	30%	8%
Amount of milk fat	$0.04x$	$0.3y$	16 lb

We get one equation from the " Number of pounds" row of the table:

$$x + y = 200$$

The last row of the table yields a second equation:

$$0.04x + 0.3y = 16$$

After clearing decimals, we have the problem translated to a system of equations:

$$x + y = 200, \quad (1)$$
$$4x + 30y = 1600 \quad (2)$$

Carry out. We use the elimination method to solve the system of equations.

$$-4x - 4y = -800 \quad \text{Multiplying (1) by } -4$$
$$4x + 30y = 1600$$
$$26y = 800$$
$$y = \frac{400}{13}, \text{ or } 30\frac{10}{13}$$

Substitute $\frac{400}{13}$ for y in (1) and solve for x.

$$x + \frac{400}{13} = 200$$
$$x = \frac{2200}{13}, \text{ or } 169\frac{3}{13}$$

Check. The total amount of the mixture is $\frac{2200}{13}$ lb $+ \frac{400}{13}$ lb $= \frac{2600}{13}$ lb = 200 lb. The amount of milk fat in the mixture is $0.04\left(\frac{2200}{13}\text{ lb}\right) + 0.3\left(\frac{400}{13}\text{ lb}\right) = \frac{88}{13}\text{ lb} + \frac{120}{13}\text{ lb} = \frac{208}{13}\text{ lb} = 16$ lb.

The answer checks.

State. $169\frac{3}{13}$ lb of whole milk and $30\frac{10}{13}$ lb of cream should be mixed.

35. ***Familiarize.*** We first make a drawing.

Slow train
d kilometers 75 km/h $(t + 2)$ hr

Fast train
d kilometers 125 km/h t hr

From the drawing we see that the distances are the same. Now complete the chart.

	Distance (d)	Rate (r)	Time (t)	
Slow train	d	75	$t + 2$	$\rightarrow d = 75(t+2)$
Fast train	d	125	t	$\rightarrow d = 125t$

Translate. Using $d = rt$ in each row of the table, we get a system of equations:

$$d = 75(t + 2),$$
$$d = 125t$$

Carry out. We solve the system of equations.

$$125t = 75(t + 2) \quad \text{Using substitution}$$
$$125t = 75t + 150$$
$$50t = 150$$
$$t = 3$$

Then $d = 125t = 125 \cdot 3 = 375$

Check. At 125 km/h, in 3 hr the fast train will travel $125 \cdot 3 = 375$ km. At 75 km/h, in 3 + 2, or 5 hr the slow train will travel $75 \cdot 5 = 375$ km. The numbers check.

State. The trains will meet 375 km from the station.

37. ***Familiarize.*** We first make a drawing. Let d = the distance and r = the speed of the boat in still water. Then when the boat travels downstream its speed is $r + 6$, and its speed upstream is $r - 6$. From the drawing we see that the distances are the same.

Downstream, 6 mph current
d mi, $r + 6$, 3 hr

Upstream, 6 mph current
d mi, $r - 6$, 5 hr

Organize the information in a table.

	Distance	Rate	Time
Down-stream	d	$r + 6$	3
Up-stream	d	$r - 6$	5

Translate. Using $d = rt$ in each row of the table, we get a system of equations:

$$d = 3(r + 6), \qquad d = 3r + 18,$$
$$\text{or}$$
$$d = 5(r - 6) \qquad d = 5r - 30$$

Carry out. Solve the system of equations.

$$3r + 18 = 5r - 30 \quad \text{Using substitution}$$
$$18 = 2r - 30$$
$$48 = 2r$$
$$24 = r$$

Check. When $r = 24$, then $r + 6 = 24 + 6 = 30$, and the distance traveled in 3 hr is $3 \cdot 30 = 90$ mi. Also, $r - 6 = 24 - 6 = 18$, and the distance traveled in 5 hr is $18 \cdot 5 = 90$ mi. The answer checks.

State. The speed of the boat in still water is 24 mph.

39. ***Familiarize.*** We make a drawing. Note that the plane's speed traveling toward London is $360 + 50$, or 410 mph, and the speed traveling toward New York City is $360 - 50$, or 310 mph. Also, when the plane is d mi from New York City, it is $3458 - d$ mi from London.

New York City — London
310 mph — t hours — t hours — 410 mph
3458 mi
d — 3458 mi $- d$

Organize the information in a table.

	Distance	Rate	Time
Toward NYC	d	310	t
Toward London	$3458 - d$	410	t

Translate. Using $d = rt$ in each row of the table, we get a system of equations:

$$d = 310t, \quad (1)$$
$$3458 - d = 410t \quad (2)$$

Carry out. We solve the system of equations.

$$3458 - 310t = 410t \quad \text{Using substitution}$$
$$3458 = 720t$$
$$4.8028 \approx t$$

Substitute 4.8028 for t in (1).

$$d \approx 310(4.8028) \approx 1489$$

Check. If the plane is 1489 mi from New York City, it can return to New York City, flying at 310 mph, in $1489/310 \approx 4.8$ hr. If the plane is $3458 - 1489$, or 1969 mi from London, it can fly to London, traveling at 410 mph, in $1969/410 \approx 4.8$ hr. Since the times are the same, the answer checks.

State. The point of no return is about 1489 mi from New York City.

41. ***Familiarize.*** Let l = the length, in feet, and w = the width, in feet. Recall that the formula for the perimeter P of a rectangle with length l and width w is $P = 2l + 2w$.

Translate.

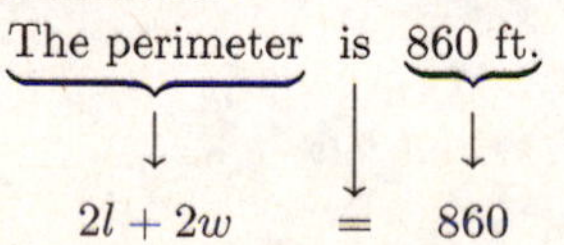

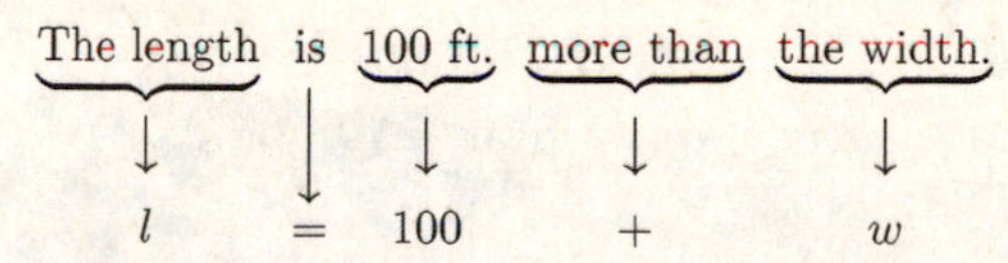

We have translated to a system of equations:

$$2l + 2w = 860, \quad (1)$$
$$l = 100 + w \quad (2)$$

Carry out. We use the substitution method to solve the system of equations.

Substitute $100 + w$ for l in (1) and solve for w.

$$2(100 + w) + 2w = 860$$
$$200 + 2w + 2w = 860$$
$$200 + 4w = 860$$
$$4w = 660$$
$$w = 165$$

Now substitute 165 for w in (2).

$$l = 100 + 165 = 265$$

Check. The perimeter is $2 \cdot 265 \text{ ft} + 2 \cdot 165 \text{ ft} = 530 \text{ ft} + 330 \text{ ft} = 860$ ft. The length, 265 ft, is 100 ft more than the width, 165 ft. The answer checks.

State. The length is 265 ft, and the width is 165 ft.

43. ***Familiarize.*** Let d = the number of properties DeBartolo owned before the merger and s = the number of properties Simon owned.

Translate.

The total number of properties is 183.

$$d + s = 183$$

Simon owned twice as many properties as DeBartolo.

$$s = 2d$$

We have a system of equations.

$$d + s = 183, \quad (1)$$
$$s = 2d \quad (2)$$

Carry out. We use the substitution method to solve the system of equations.

Substitute $2d$ for s in (1) and solve for d.

$$d + 2d = 183$$
$$3d = 183$$
$$d = 61$$

Now substitute 61 for d in (2).

$$s = 2 \cdot 61 = 122$$

Check. The total number of properties is $61 + 122$, or 183. The number of Simon properties, 122, is twice 61, the number of DeBartolo properties. The answer checks.

State. Before the merger DeBartolo owned 61 properties and Simon owned 122 properties.

45. ***Familiarize.*** Let x = the number of 30-sec commercials and y = the number of 60-sec commercials. Then the 30-sec commercials play for $30x$ sec and the 60-sec commercials play for $60y$ sec. Note that $10 \text{ min} = 10 \times 60 \text{ sec} = 600$ sec.

Translate.

The total number of commercials is 12.

$x + y = 12$

The total playing time for the commercials is 600 sec.

$30x + 60y = 600$

We have a system of equations.

$$x + y = 12, \quad (1)$$
$$30x + 60y = 600 \quad (2)$$

Carry out. We use elimination to solve the system of equations.

$$\begin{aligned} -30x - 30y &= -360 \quad \text{Multiplying (1) by } -30 \\ 30x + 60y &= 600 \\ \hline 30y &= 240 \\ y &= 8 \end{aligned}$$

Now substitute 8 for y in (1) and solve for x.

$$x + 8 = 12$$
$$x = 4$$

Check. There is a total of $4+8$, or 12 commercials. They play for $30 \cdot 4 + 60 \cdot 8 = 120 + 480 = 600$ sec. The answer checks.

State. Roscoe played 4 30-sec commercials and 8 60-sec commercials.

47. ***Familiarize.*** The change from the \$9.25 purchase is \$20 − \$9.25, or \$10.75. Let x = the number of quarters and y = the number of fifty-cent pieces. The total value of the quarters, in dollars, is $0.25x$ and the total value of the fifty-cent pieces, in dollars, is $0.50y$.

Translate.

The total number of coins is 30.

$$x + y = 30$$

The total value of the coins is \$10.75.

$$0.25x + 0.50y = 10.75$$

After clearing decimals we have the following system of equations:

$$x + y = 30, \quad (1)$$
$$25x + 50y = 1075 \quad (2)$$

Carry out. We use the elimination method to solve the system of equations.

$$\begin{aligned} -25x - 25y &= -750 \quad \text{Multiplying (1) by } -25 \\ 25x + 50y &= 1075 \\ \hline 25y &= 325 \\ y &= 13 \end{aligned}$$

Substitute 13 for y in (1) and solve for x.

$$x + 13 = 30$$
$$x = 17$$

Check. The total number of coins is $17 + 13$, or 30. The total value of the coins is $\$0.25(17) + \$0.50(13) = \$4.25 + \$6.50 = \$10.75$. The answer checks.

State. There were 17 quarters and 13 fifty-cent pieces.

49. *Writing Exercise*

51.
$$\begin{aligned} 2x - 3y + 12 &= 2 \cdot 5 - 3 \cdot 2 + 12 \\ &= 10 - 6 + 12 \\ &= 4 + 12 \\ &= 16 \end{aligned}$$

53.
$$\begin{aligned} 5a - 7b + 3c &= 5(-2) - 7(3) + 3 \cdot 1 \\ &= -10 - 21 + 3 \\ &= -31 + 3 \\ &= -28 \end{aligned}$$

55.
$$\begin{aligned} 4 - 2y + 3z &= 4 - 2 \cdot \frac{1}{3} + 3 \cdot \frac{1}{4} \\ &= 4 - \frac{2}{3} + \frac{3}{4} \\ &= \frac{48}{12} - \frac{8}{12} + \frac{9}{12} \\ &= \frac{40}{12} + \frac{9}{12} \\ &= \frac{49}{12} \end{aligned}$$

57. *Writing Exercise*

59. ***Familiarize.*** Let x = the number of reams of 0% post-consumer fiber paper purchased and y = the number of reams of 30% post-consumer fiber paper.

Translate. We organize the information in a table.

	0% post-consumer	30% post-consumer	Total
Reams purchased	x	y	60
Percent of post-consumer fiber	0%	30%	20%
Total post-consumer fiber	$0 \cdot x$, or 0	$0.3y$	$0.2(60)$, or 12

We get one equation from the "Reams purchased" row of the table:

$$x + y = 60$$

The last row of the table yields a second equation:

$$0x + 0.3y = 12, \text{ or } 0.3y = 12$$

After clearing the decimal we have the problem translated to a system of equations.

$$x + y = 60, \quad (1)$$
$$3y = 120 \quad (2)$$

Carry out. First we solve (2) for y.

$$3y = 120$$
$$y = 40$$

Now substitute 40 for y in (1) and solve for x.

$$x + 40 = 60$$
$$x = 20$$

Check. The total purchase is $20 + 40$, or 60 reams. The post-consumer fiber can be expressed as $0 \cdot 20 + 0.3(40) = 12$. The answer checks.

State. 20 reams of 0% post-consumer fiber paper and 40 reams of 30% post-consumer fiber paper would have to be purchased.

61. ***Familiarize***. Let $x =$ the amount of the original solution that remains after some of the original solution is drained and replaced with pure antifreeze. Let $y =$ the amount of the original solution that is drained and replaced with pure antifreeze.

Translate. We organize the information in a table. Keep in mind that the table contains information regarding the solution *after* some of the original solution is drained and replaced with pure antifreeze.

	Original Solution	Pure Anti-freeze	New Mixture
Amount of solution	x	y	6.3 L
Percent of antifreeze	30%	100%	50%
Amount of antifreeze in solution	$0.3x$	$1 \cdot y$, or y	0.5(6.3), or 3.15

The "Amount of solution" row gives us one equation:

$$x + y = 6.3 \quad (1)$$

The last row gives us a second equation:

$$0.3x + y = 3.15 \quad (2)$$

After clearing the decimals we have the following system of equations:

$$10x + 10y = 63, \quad (3)$$
$$30x + 100y = 315 \quad (4)$$

Carry out. We use the elimination method.

$$-30x - 30y = -189 \quad \text{Multiplying (3) by } -3$$
$$30x + 100y = 315$$
$$70y = 126$$
$$y = 1.8$$

Although the problem only asks for the amount of pure antifreeze added, we will also find x in order to check.

$$x + 1.8 = 6.3 \quad \text{Substituting 1.8 for } y \text{ in (1)}$$
$$x = 4.5$$

Check. Total amount of new mixture:

$$4.5 + 1.8 = 6.3 \text{ L}$$

Amount of antifreeze in new mixture:

$$0.3(4.5) + 1(1.8) = 1.35 + 1.8 = 3.15 \text{ L}$$

The numbers check.

State. Michelle should drain 1.8 L of the original solution and replace it with pure antifreeze.

63. ***Familiarize***. Let $x =$ the number of members who ordered one book and $y =$ the number of members who ordered two books. Note that the y members ordered a total of $2y$ books.

Translate.

The number of books sold was 880.

$$x + 2y = 880$$

Total sales were \$9840.

$$12x + 20y = 9840$$

We have a system of equations.

$$x + 2y = 880, \quad (1)$$
$$12x + 20y = 9840 \quad (2)$$

Carry out. We use the elimination method.

$$-10x - 20y = -8800 \quad \text{Multiplying (1) by } -10$$
$$12x + 20y = 9840$$
$$2x = 1040$$
$$x = 520$$

Substitute 520 for x in (1) and solve for y.

$$520 + 2y = 880$$
$$2y = 360$$
$$y = 180$$

Check. Total number of books sold: $520 + 2 \cdot 180 = 520 + 360 = 880$

Total sales: $\$12 \cdot 520 + \$20 \cdot 180 = \$6240 + \$3600 = \$9840$

The answer checks.

State. 180 members ordered two books

65. ***Familiarize***. Let $x =$ the number of gallons of pure brown and $y =$ the number of gallons of neutral stain that should be added to the original 0.5 gal. Note that a total of 1 gal of stain needs to be added to bring the amount of stain up to 1.5 gal. The original 0.5 gal of stain contains 20%(0.5 gal), or 0.2(0.5 gal) $= 0.1$ gal of brown stain. The final solution contains 60%(1.5 gal), or 0.6(1.5 gal) $= 0.9$ gal of brown stain. This is composed of the original 0.1 gal and the x gal that are added.

Translate.

The amount of stain added was 1 gal.

$$x + y = 1$$

The amount of brown stain in the final solution is 0.9 gal.

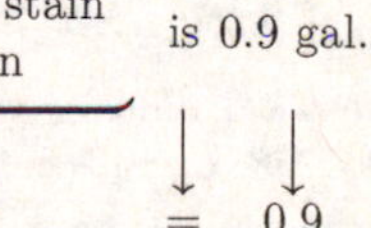

$$0.1 + x = 0.9$$

We have a system of equations.

$$x + y = 1, \quad (1)$$
$$0.1 + x = 0.9 \quad (2)$$

Carry out. First we solve (2) for x.

$$0.1 + x = 0.9$$
$$x = 0.8$$

Then substitute 0.8 for x in (1) and solve for y.

$$0.8 + y = 1$$
$$y = 0.2$$

Check. Total amount of stain: $0.5 + 0.8 + 0.2 = 1.5$ gal

Total amount of brown stain: $0.1 + 0.8 = 0.9$ gal

Total amount of neutral stain: $0.8(0.5)+0.2 = 0.4 + 0.2 = 0.6$ gal $= 0.4(1.5$ gal$)$

The answer checks.

State. 0.8 gal of pure brown and 0.2 gal of neutral stain should be added.

67. ***Familiarize***. Let x and y represent the number of city miles and highway miles that were driven, respectively. Then in city driving, $\frac{x}{18}$ gallons of gasoline are used; in highway driving, $\frac{y}{24}$ gallons are used.

Translate. We organize the information in a table.

Type of driving	City	Highway	Total
Number of miles	x	y	465
Gallons of gasoline used	$\frac{x}{18}$	$\frac{y}{24}$	23

The first row of the table gives us one equation:

$$x + y = 465$$

The second row gives us another equation:

$$\frac{x}{18} + \frac{y}{24} = 23$$

After clearing fractions, we have the following system of equations:

$$x + y = 465, \quad (1)$$
$$24x + 18y = 9936 \quad (2)$$

Solve. We solve the system of equations using the elimination method.

$$\begin{aligned} -18x - 18y &= -8370 \quad \text{Multiplying (1) by } -18 \\ 24x + 18y &= 9936 \\ \hline 6x \qquad &= 1566 \\ x &= 261 \end{aligned}$$

Now substitute 261 for x in Equation (1) and solve for y.

$$261 + y = 465$$
$$y = 204$$

Check. The total mileage is $261 + 204$, or 465. In 216 city miles, 261/18, or 14.5 gal of gasoline are used; in 204 highway miles, 204/24, or 8.5 gal are used. Then a total of $14.5 + 8.5$ or 23 gal of gasoline are used. The answer checks.

State. 261 miles were driven in the city, and 204 miles were driven on the highway.

69. The 1.5 gal mixture contains $0.1 + x$ gal of pure brown stain. (See Exercise 65.). Thus, the function $P(x) = \frac{0.1 + x}{1.5}$ gives the percentage of brown in the mixture as a decimal quantity. Using the Intersect feature, we confirm that when $x = 0.8$, then $P(x) = 0.6$ or 60%.

Exercise Set 8.4

1. The equation is equivalent to one in the form $Ax + By + Cz = D$, so the statement is true.

3. False; see Example 5.

5. True; see Example 6.

7. Substitute $(2, -1, -2)$ into the three equations, using alphabetical order.

$$\begin{array}{c|c} \multicolumn{2}{c}{x + y - 2z = 5} \\ \hline 2 + (-1) - 2(-2) & 5 \\ 2 - 1 + 4 & \end{array}$$
$$5 \stackrel{?}{=} 5 \quad \text{TRUE}$$

$$\begin{array}{c|c} \multicolumn{2}{c}{2x - y - z = 7} \\ \hline 2 \cdot 2 - (-1) - (-2) & 7 \\ 4 + 1 + 2 & \end{array}$$
$$7 \stackrel{?}{=} 7 \quad \text{TRUE}$$

$$\begin{array}{c|c} \multicolumn{2}{c}{-x - 2y + 3z = 6} \\ \hline -2 - 2(-1) + 3(-2) & 6 \\ -2 + 2 - 6 & \end{array}$$
$$-6 \stackrel{?}{=} 6 \quad \text{FALSE}$$

The triple $(2, -1, -2)$ does not make the third equation true, so it is not a solution of the system.

9.
$$\begin{aligned} 2x - y + z &= 10, \quad (1) \\ 4x + 2y - 3z &= 10, \quad (2) \\ x - 3y + 2z &= 8 \quad (3) \end{aligned}$$

1., 2. The equations are already in standard form with no fractions or decimals.

3. Use Equations (1) and (2) to eliminate y:

$$\begin{aligned} 4x - 2y + 2z &= 20 \quad \text{Multiplying (1) by 2} \\ 4x + 2y - 3z &= 10 \quad (2) \\ \hline 8x \qquad - z &= 30 \quad (4) \end{aligned}$$

4. Use a different pair of equations and eliminate y:

$$\begin{aligned} -6x + 3y - 3z &= -30 \quad \text{Multiplying (1) by } -3 \\ x - 3y + 2z &= 8 \quad (3) \\ \hline -5x \qquad - z &= -22 \quad (5) \end{aligned}$$

5. Now solve the system of Equations (4) and (5).

$$8x - z = 30 \quad (4)$$
$$-5x - z = -22 \quad (5)$$

$$8x - z = 30 \quad (4)$$
$$5x + z = 22 \quad \text{Multiplying (5) by } -1$$
$$13x \quad = 52$$
$$x = 4$$

$$8 \cdot 4 - z = 30 \quad \text{Substituting in (4)}$$
$$32 - z = 30$$
$$-z = -2$$
$$z = 2$$

6. Substitute in one of the original equations to find y.

$$2 \cdot 4 - y + 2 = 10 \quad \text{Substituting in (1)}$$
$$10 - y = 10$$
$$-y = 0$$
$$y = 0$$

We obtain $(4, 0, 2)$. This checks, so it is the solution.

11. $x - y + z = 6$, (1)
$2x + 3y + 2z = 2$, (2)
$3x + 5y + 4z = 4$ (3)

1., 2. The equations are already in standard form with no fractions or decimals.

3., 4. We eliminate y from two different pairs of equations.

$$3x - 3y + 3z = 18 \quad \text{Multiplying (1) by 3}$$
$$2x + 3y + 2z = 2 \quad (2)$$
$$5x \quad + 5z = 20 \quad (4)$$

$$5x - 5y + 5z = 30 \quad \text{Multiplying (1) by 5}$$
$$3x + 5y + 4z = 4 \quad (3)$$
$$8x \quad + 9z = 34 \quad (5)$$

5. Now solve the system of Equations (4) and (5).

$$5x + 5z = 20 \quad (4)$$
$$8x + 9z = 34 \quad (5)$$

$$45x + 45z = 180 \quad \text{Multiplying (4) by 9}$$
$$-40x - 45z = -170 \quad \text{Multiplying (5) by } -5$$
$$5x \quad = 10$$
$$x = 2$$

$$5 \cdot 2 + 5z = 20 \quad \text{Substituting in (4)}$$
$$10 + 5z = 20$$
$$5z = 10$$
$$z = 2$$

6. Substitute in one of the original equations to find y.

$$2 - y + 2 = 6 \quad \text{Substituting in (1)}$$
$$4 - y = 6$$
$$-y = 2$$
$$y = -2$$

We obtain $(2, -2, 2)$. This checks, so it is the solution.

13. $6x - 4y + 5z = 31$, (1)
$5x + 2y + 2z = 13$, (2)
$x + y + z = 2$ (3)

1., 2. The equations are already in standard form with no fractions or decimals.

3., 4. We eliminate y from two different pairs of equations.

$$6x - 4y + 5z = 31 \quad (1)$$
$$4x + 4y + 4z = 8 \quad \text{Multiplying (3) by 4}$$
$$10x \quad + 9z = 39 \quad (4)$$

$$5x + 2y + 2z = 13 \quad (2)$$
$$-2x - 2y - 2z = -4 \quad \text{Multiplying (3) by } -2$$
$$3x \quad = 9$$
$$x = 3$$

5. When we used Equations (2) and (3) to eliminate y, we also eliminated z and found that $x = 3$. Substitute 3 for x in Equation (4) to find z.

$$10 \cdot 3 + 9z = 39 \quad \text{Substituting in (4)}$$
$$30 + 9z = 39$$
$$9z = 9$$
$$z = 1$$

6. Substitute in one of the original equations to find y.

$$3 + y + 1 = 2 \quad \text{Substituting in (3)}$$
$$y + 4 = 2$$
$$y = -2$$

We obtain $(3, -2, 1)$. This checks, so it is the solution.

15. $x + y + z = 0$, (1)
$2x + 3y + 2z = -3$, (2)
$-x + 2y - z = 1$ (3)

1., 2. The equations are already in standard form with no fractions or decimals.

3., 4. We eliminate x from two different pairs of equations.

$$-2x - 2y - 2z = 0 \quad \text{Multiplying (1) by } -2$$
$$2x + 3y + 2z = -3 \quad (2)$$
$$y \quad = -3$$

We eliminated not only x but also z and found that $y = -3$.

5., 6. Substitute -3 for y in two of the original equations to produce a system of two equations in two variables. Then solve this system.

$x - 3 + z = 0$ Substituting in (1)

$-x + 2(-3) - z = 1$ Substituting in (3)

Simplifying we have

$$\begin{aligned} x + z &= 3 \\ -x - z &= 7 \\ \hline 0 &= 10 \end{aligned}$$

We get a false equation, so there is no solution.

17. $2x + y - 3z = -4$, (1)

$4x - 2y + z = 9$, (2)

$3x + 5y - 2z = 5$ (3)

1., 2. The equations are already in standard form with no fractions or decimals.

3., 4. We eliminate z from two different pairs of equations.

$$\begin{array}{rll} 2x + y - 3z = & -4 & (1) \\ 12x - 6y + 3z = & 27 & \text{Multiplying (2) by 3} \\ \hline 14x - 5y \quad = & 23 & (4) \end{array}$$

$$\begin{array}{rll} 8x - 4y + 2z = & 18 & \text{Multiplying (2) by 2} \\ 3x + 5y - 2z = & 5 & (3) \\ \hline 11x + y \quad = & 23 & (5) \end{array}$$

5. Now solve the system of Equations (4) and (5).

$14x - 5y = 23$ (4)

$11x + y = 23$ (5)

$$\begin{array}{rll} 14x - 5y = & 23 & (4) \\ 55x + 5y = & 115 & \text{Multiplying (5) by 5} \\ \hline 69x \quad = & 138 & \\ x = & 2 & \end{array}$$

$11 \cdot 2 + y = 23$ Substituting in (5)

$22 + y = 23$

$y = 1$

6. Substitute in one of the original equations to find z.

$4 \cdot 2 - 2 \cdot 1 + z = 9$ Substituting in (2)

$6 + z = 9$

$z = 3$

We obtain $(2, 1, 3)$. This checks, so it is the solution.

19. $2x + y + 2z = 11$, (1)

$3x + 2y + 2z = 8$, (2)

$x + 4y + 3z = 0$ (3)

1., 2. The equations are already in standard form with no fractions or decimals.

3., 4. We eliminate x from two different pairs of equations.

$$\begin{array}{rl} 2x + y + 2z = 11 & (1) \\ -2x - 8y - 6z = 0 & \text{Multiplying (3) by } -2 \\ \hline -7y - 4z = 11 & (4) \end{array}$$

$$\begin{array}{rl} 3x + 2y + 2z = 8 & (2) \\ -3x - 12y - 9z = 0 & \text{Multiplying (3) by } -3 \\ \hline -10y - 7z = 8 & (5) \end{array}$$

5. Now solve the system of Equations (4) and (5).

$-7y - 4z = 11$ (4)

$-10y - 7z = 8$ (5)

$$\begin{array}{rll} -49y - 28z = & 77 & \text{Multiplying (4) by 7} \\ 40y + 28z = & -32 & \text{Multiplying (5) by } -4 \\ \hline -9y \quad = & 45 & \\ y = & -5 & \end{array}$$

$-7(-5) - 4z = 11$ Substituting in (4)

$35 - 4z = 11$

$-4z = -24$

$z = 6$

6. Substitute in one of the original equations to find x.

$x + 4(-5) + 3 \cdot 6 = 0$ Substituting in (3)

$x - 2 = 0$

$x = 2$

We obtain $(2, -5, 6)$. This checks, so it is the solution.

21. $-2x + 8y + 2z = 4$, (1)

$x + 6y + 3z = 4$, (2)

$3x - 2y + z = 0$ (3)

1., 2. The equations are already in standard form with no fractions or decimals.

3., 4. We eliminate z from two different pairs of equations.

$$\begin{array}{rl} -2x + 8y + 2z = 4 & (1) \\ -6x + 4y - 2z = 0 & \text{Multiplying (3) by } -2 \\ \hline -8x + 12y \quad = 4 & (4) \end{array}$$

$$\begin{array}{rl} x + 6y + 3z = 4 & (2) \\ -9x + 6y - 3z = 0 & \text{Multiplying (3) by } -3 \\ \hline -8x + 12y \quad = 4 & (5) \end{array}$$

5. Now solve the system of Equations (4) and (5).

$-8x + 12y = 4$ (4)

$-8x + 12y = 4$ (5)

$$\begin{array}{rll} -8x + 12y = & 4 & (4) \\ 8x - 12y = & -4 & \text{Multiplying (5) by } -1 \\ \hline 0 = & 0 & (6) \end{array}$$

Equation (6) indicates that Equations (1), (2), and (3) are dependent. (Note that if Equation (1) is subtracted from Equation (2), the result is Equation (3).) We could also have concluded that the equations are dependent by observing that Equations (4) and (5) are identical.

23. $4x - y - z = 4,$ (1)

$2x + y + z = -1,$ (2)

$6x - 3y - 2z = 3$ (3)

1., 2. The equations are already in standard form with no fractions or decimals.

3. Add Equations (1) and (2) to eliminate y.

$4x - y - z = 4$ (1)

$2x + y + z = -1$ (2)

$6x \quad = 3$ (4) Adding

4. At this point we can either continue by eliminating y from a second pair of equations or we can solve (4) for x and substitute that value in a different pair of the original equations to obtain a system of two equations in two variables. We take the second option.

$6x = 3$ (4)

$x = \frac{1}{2}$

Substitute $\frac{1}{2}$ for x in (1):

$4\left(\frac{1}{2}\right) - y - z = 4$

$2 - y - z = 4$

$-y - z = 2$ (5)

Substitute $\frac{1}{2}$ for x in (3):

$6\left(\frac{1}{2}\right) - 3y - 2z = 3$

$3 - 3y - 2z = 3$

$-3y - 2z = 0$ (6)

5. Now solve the system of Equations (5) and (6).

$2y + 2z = -4$ Multiplying (5) by -2

$-3y - 2z = 0$ (6)

$-y \quad = -4$

$y = 4$

6. Substitute to find z.

$-4 - z = 2$ Substituting 4 for y in (5)

$-z = 6$

$z = -6$

We obtain $\left(\frac{1}{2}, 4, -6\right)$. This checks, so it is the solution.

25. $r + \frac{3}{2}s + 6t = 2,$

$2r - 3s + 3t = 0.5,$

$r + s + t = 1$

1. All equations are already in standard form.

2. Multiply the first equation by 2 to clear the fraction. Also, multiply the second equation by 10 to clear the decimal.

$2r + 3s + 12t = 4,$ (1)

$20r - 30s + 30t = 5,$ (2)

$r + s + t = 1$ (3)

3., 4. We eliminate s from two different pairs of equations.

$20r + 30s + 120t = 40$ Multiplying (1) by 10

$20r - 30s + 30t = 5$ (2)

$40r \quad + 150t = 45$ (4) Adding

$20r - 30s + 30t = 5$ (2)

$30r + 30s + 30t = 30$ Multiplying (3) by 30

$50r \quad + 60t = 35$ (5) Adding

5. Solve the system of Equations (4) and (5).

$40r + 150t = 45$ (4)

$50r + 60t = 35$ (5)

$200r + 750t = 225$ Multiplying (4) by 5

$-200r - 240t = -140$ Multiplying (5) by -4

$510t = 85$

$t = \frac{85}{510}$

$t = \frac{1}{6}$

$40r + 150\left(\frac{1}{6}\right) = 45$ Substituting $\frac{1}{6}$ for t in (4)

$40r + 25 = 45$

$40r = 20$

$r = \frac{1}{2}$

6. Substitute in one of the original equations to find s.

$\frac{1}{2} + s + \frac{1}{6} = 1$ Substituting $\frac{1}{2}$ for r and $\frac{1}{6}$ for t in (3)

$s + \frac{2}{3} = 1$

$s = \frac{1}{3}$

We obtain $\left(\frac{1}{2}, \frac{1}{3}, \frac{1}{6}\right)$. This checks, so it is the solution.

27. $4a + 9b \quad = 8,$ (1)

$8a \quad + 6c = -1,$ (2)

$6b + 6c = -1$ (3)

1., 2. The equations are already in standard form with no fractions or decimals.

3., 4. Note that there is no c in Equation (1). We will use Equations (2) and (3) to obtain another equation with no c-term.

$8a \quad + 6c = -1$ (2)

$-6b - 6c = 1$ Multiplying (3) by -1

$8a - 6b \quad = 0$ (4) Adding

5. Now solve the system of Equations (1) and (4).

$$\begin{aligned} -8a - 18b &= -16 \quad \text{Multiplying (1) by } -2 \\ 8a - 6b &= 0 \\ \hline -24b &= -16 \\ b &= \frac{2}{3} \end{aligned}$$

$$8a - 6\left(\frac{2}{3}\right) = 0 \quad \text{Substituting } \frac{2}{3} \text{ for } b \text{ in (4)}$$
$$8a - 4 = 0$$
$$8a = 4$$
$$a = \frac{1}{2}$$

6. Substitute in Equation (2) or (3) to find c.

$$8\left(\frac{1}{2}\right) + 6c = -1 \quad \text{Substituting } \frac{1}{2} \text{ for } a \text{ in (2)}$$
$$4 + 6c = -1$$
$$6c = -5$$
$$c = -\frac{5}{6}$$

We obtain $\left(\frac{1}{2}, \frac{2}{3}, -\frac{5}{6}\right)$. This checks, so it is the solution.

29.
$$\begin{aligned} x + y + z &= 57, \quad (1) \\ -2x + y &= 3, \quad (2) \\ x - z &= 6 \quad (3) \end{aligned}$$

1., 2. The equations are already in standard form with no fractions or decimals.

3., 4. Note that there is no z in Equation (2). We will use Equations (1) and (3) to obtain another equation with no z-term.

$$\begin{aligned} x + y + z &= 57 \quad (1) \\ x - z &= 6 \quad (3) \\ \hline 2x + y &= 63 \quad (4) \end{aligned}$$

5. Now solve the system of Equations (2) and (4).

$$\begin{aligned} -2x + y &= 3 \quad (2) \\ 2x + y &= 63 \quad (4) \\ \hline 2y &= 66 \\ y &= 33 \end{aligned}$$

$$2x + 33 = 63 \quad \text{Substituting 33 for y in (4)}$$
$$2x = 30$$
$$x = 15$$

6. Substitute in Equation (1) or (3) to find z.

$$15 - z = 6 \quad \text{Substituting 15 for } x \text{ in (3)}$$
$$9 = z$$

We obtain $(15, 33, 9)$. This checks, so it is the solution.

31.
$$\begin{aligned} a - 3c &= 6, \quad (1) \\ b + 2c &= 2, \quad (2) \\ 7a - 3b - 5c &= 14 \quad (3) \end{aligned}$$

1., 2. The equations are already in standard form with no fractions or decimals.

3., 4. Note that there is no b in Equation (1). We will use Equations (2) and (3) to obtain another equation with no b-term.

$$\begin{aligned} 3b + 6c &= 6 \quad \text{Multiplying (2) by 3} \\ 7a - 3b - 5c &= 14 \quad (3) \\ \hline 7a + c &= 20 \quad (4) \end{aligned}$$

5. Now solve the system of Equations (1) and (4).

$$\begin{aligned} a - 3c &= 6 \quad (1) \\ 7a + c &= 20 \quad (4) \end{aligned}$$

$$\begin{aligned} a - 3c &= 6 \quad (1) \\ 21a + 3c &= 60 \quad \text{Multiplying (4) by 3} \\ \hline 22a &= 66 \\ a &= 3 \end{aligned}$$

$$3 - 3c = 6 \quad \text{Substituting in (1)}$$
$$-3c = 3$$
$$c = -1$$

6. Substitute in Equation (2) or (3) to find b.

$$b + 2(-1) = 2 \quad \text{Substituting in (2)}$$
$$b - 2 = 2$$
$$b = 4$$

We obtain $(3, 4, -1)$. This checks, so it is the solution.

33.
$$\begin{aligned} x + y + z &= 83, \quad (1) \\ y &= 2x + 3, \quad (2) \\ z &= 40 + x \quad (3) \end{aligned}$$

Observe, from Equations (2) and (3), that we can substitute $2+3x$ for y and $40+x$ for z in Equation (1) and solve for x.

$$\begin{aligned} x + y + x &= 83 \\ x + (2x + 3) + (40 + x) &= 83 \\ 4x + 43 &= 83 \\ 4x &= 40 \\ x &= 10 \end{aligned}$$

Now substitute 10 for x in Equation (2).

$$y = 2x + 3 = 2 \cdot 10 + 3 = 20 + 3 = 23$$

Finally, substitute 10 for x in Equation (3).

$$z = 40 + x = 40 + 10 = 50.$$

We obtain $(10, 23, 50)$. This checks, so it is the solution.

35.
$$\begin{aligned} x + z &= 0, \quad (1) \\ x + y + 2z &= 3, \quad (2) \\ y + z &= 2 \quad (3) \end{aligned}$$

1., 2. The equations are already in standard form with no fractions or decimals.

3., 4. Note that there is no y in Equation (1). We use Equations (2) and (3) to obtain another equation with no y-term.

$$\begin{array}{rll} x + y + 2z = & 3 & (2) \\ -y - z = & -2 & \text{Multiplying (3) by } -1 \\ \hline x + z = & 1 & (4) \quad \text{Adding} \end{array}$$

5. Now solve the system of Equations (1) and (4).

$$\begin{array}{rll} x + z = & 0 & (1) \\ -x - z = & -1 & \text{Multiplying (4) by } -1 \\ \hline 0 = & -1 & \text{Adding} \end{array}$$

We get a false equation, or contradiction. There is no solution.

37.

$$\begin{aligned} x + y + z &= 1, \quad (1) \\ -x + 2y + z &= 2, \quad (2) \\ 2x - y &= -1 \quad (3) \end{aligned}$$

1., 2. The equations are already in standard form with no fractions or decimals.

3. Note that there is no z in Equation (3). We will use Equations (1) and (2) to eliminate z:

$$\begin{array}{rll} x + y + z = & 1 & (1) \\ x - 2y - z = & -2 & \text{Multiplying (2) by } -1 \\ \hline 2x - y = & -1 & \text{Adding} \end{array}$$

Equations (3) and (4) are identical, so Equations (1), (2), and (3) are dependent. (We have seen that if Equation (2) is multiplied by -1 and added to Equation (1), the result is Equation (3).)

39. *Writing Exercise*

41. Let x represent the larger number and y represent the smaller number. Then we have $x = 2y$.

43. Let x, $x + 1$, and $x + 2$ represent the numbers. Then we have $x + (x + 1) + (x + 2) = 45$.

45. Let x and y represent the first two numbers and let z represent the third number. Then we have $x + y = 5z$.

47. *Writing Exercise*

49.

$$\frac{x+2}{3} - \frac{y+4}{2} + \frac{z+1}{6} = 0,$$

$$\frac{x-4}{3} + \frac{y+1}{4} - \frac{z-2}{2} = -1,$$

$$\frac{x+1}{2} + \frac{y}{2} + \frac{z-1}{4} = \frac{3}{4}$$

1., 2. We clear fractions and write each equation in standard form.

To clear fractions, we multiply both sides of each equation by the LCM of its denominators. The LCM's are 6, 12, and 4, respectively.

$$\begin{aligned} 6\left(\frac{x+2}{3} - \frac{y+4}{2} + \frac{z+1}{6}\right) &= 6 \cdot 0 \\ 2(x+2) - 3(y+4) + (z+1) &= 0 \\ 2x + 4 - 3y - 12 + z + 1 &= 0 \\ 2x - 3y + z &= 7 \end{aligned}$$

$$\begin{aligned} 12\left(\frac{x-4}{3} + \frac{y+1}{4} - \frac{z-2}{2}\right) &= 12 \cdot (-1) \\ 4(x-4) + 3(y+1) - 6(z-2) &= -12 \\ 4x - 16 + 3y + 3 - 6z + 12 &= -12 \\ 4x + 3y - 6z &= -11 \end{aligned}$$

$$\begin{aligned} 4\left(\frac{x+1}{2} + \frac{y}{2} + \frac{z-1}{4}\right) &= 4 \cdot \frac{3}{4} \\ 2(x+1) + 2(y) + (z-1) &= 3 \\ 2x + 2 + 2y + z - 1 &= 3 \\ 2x + 2y + z &= 2 \end{aligned}$$

The resulting system is

$$\begin{aligned} 2x - 3y + z &= 7, \quad (1) \\ 4x + 3y - 6z &= -11, \quad (2) \\ 2x + 2y + z &= 2 \quad (3) \end{aligned}$$

3., 4. We eliminate z from two different pairs of equations.

$$\begin{array}{rll} 12x - 18y + 6z = & 42 & \text{Multiplying (1) by 6} \\ 4x + 3y - 6z = & -11 & (2) \\ \hline 16x - 15y = & 31 & (4) \quad \text{Adding} \end{array}$$

$$\begin{array}{rll} 2x - 3y + z = & 7 & (1) \\ -2x - 2y - z = & -2 & \text{Multiplying (3) by } -1 \\ \hline -5y = & 5 & (5) \quad \text{Adding} \end{array}$$

5. Solve (5) for y:

$$\begin{aligned} -5y &= 5 \\ y &= -1 \end{aligned}$$

Substitute -1 for y in (4):

$$\begin{aligned} 16x - 15(-1) &= 31 \\ 16x + 15 &= 31 \\ 16x &= 16 \\ x &= 1 \end{aligned}$$

6. Substitute 1 for x and -1 for y in (1):

$$\begin{aligned} 2 \cdot 1 - 3(-1) + z &= 7 \\ 5 + z &= 7 \\ z &= 2 \end{aligned}$$

We obtain $(1, -1, 2)$. This checks, so it is the solution.

51.

$$\begin{aligned} w + x - y + z &= 0, \quad (1) \\ w - 2x - 2y - z &= -5, \quad (2) \\ w - 3x - y + z &= 4, \quad (3) \\ 2w - x - y + 3z &= 7 \quad (4) \end{aligned}$$

The equations are already in standard form with no fractions or decimals.

Start by eliminating z from three different pairs of equations.

$$\begin{array}{rl} w + x - y + z = 0 & (1) \\ w - 2x - 2y - z = -5 & (2) \\ \hline 2w - x - 3y = -5 & (5) \text{ Adding} \end{array}$$

$$\begin{array}{rl} w - 2x - 2y - z = -5 & (2) \\ w - 3x - y + z = 4 & (3) \\ \hline 2w - 5x - 3y = -1 & (6) \text{ Adding} \end{array}$$

$$\begin{array}{rl} 3w - 6x - 6y - 3z = -15 & \text{Multiplying (2) by 3} \\ 2w - x - y + 3z = 7 & (4) \\ \hline 5w - 7x - 7y = -8 & (7) \text{ Adding} \end{array}$$

Now solve the system of equations (5), (6), and (7).

$2w - x - 3y = -5,$ (5)

$2w - 5x - 3y = -1,$ (6)

$5w - 7x - 7y = -8.$ (7)

$$\begin{array}{rl} 2w - x - 3y = -5 & (5) \\ -2w + 5x + 3y = 1 & \text{Multiplying (6) by } -1 \\ \hline 4x = -4 & \\ x = -1 & \end{array}$$

Substituting -1 for x in (5) and (7) and simplifying, we have

$2w - 3y = -6,$ (8)

$5w - 7y = -15.$ (9)

Now solve the system of Equations (8) and (9).

$$\begin{array}{rl} 10w - 15y = -30 & \text{Multiplying (8) by 5} \\ -10w + 14y = 30 & \text{Multiplying (9) by } -2 \\ \hline -y = 0 & \\ y = 0 & \end{array}$$

Substitute 0 for y in Equation (8) or (9) and solve for w.

$2w - 3 \cdot 0 = -6$ Substituting in (8)

$2w = -6$

$w = -3$

Substitute in one of the original equations to find z.

$-3 - 1 - 0 + z = 0$ Substituting in (1)

$-4 + z = 0$

$z = 4$

We obtain $(-3, -1, 0, 4)$. This checks, so it is the solution.

53. $\frac{2}{x} + \frac{2}{y} - \frac{3}{z} = 3,$

$\frac{1}{x} - \frac{2}{y} - \frac{3}{z} = 9,$

$\frac{7}{x} - \frac{2}{y} + \frac{9}{z} = -39$

Let u represent $\frac{1}{x}$, v represent $\frac{1}{y}$, and w represent $\frac{1}{z}$. Substituting, we have

$2u + 2v - 3w = 3,$ (1)

$u - 2v - 3w = 9,$ (2)

$7u - 2v + 9w = -39$ (3)

1., 2. The equations in u, v, and w are in standard form with no fractions or decimals.

3., 4. We eliminate v from two different pairs of equations.

$$\begin{array}{rl} 2u + 2v - 3w = 3 & (1) \\ u - 2v - 3w = 9 & (2) \\ \hline 3u - 6w = 12 & (4) \text{ Adding} \end{array}$$

$$\begin{array}{rl} 2u + 2v - 3w = 3 & (1) \\ 7u - 2v + 9w = -39 & (3) \\ \hline 9u + 6w = -36 & (5) \text{ Adding} \end{array}$$

5. Now solve the system of Equations (4) and (5).

$$\begin{array}{rl} 3u - 6w = 12, & (4) \\ 9u + 6w = -36 & (5) \\ \hline 12u = -24 & \\ u = -2 & \end{array}$$

$3(-2) - 6w = 12$ Substituting in (4)

$-6 - 6w = 12$

$-6w = 18$

$w = -3$

6. Substitute in Equation (1), (2), or (3) to find v.

$2(-2) + 2v - 3(-3) = 3$ Substituting in (1)

$2v + 5 = 3$

$2v = -2$

$v = -1$

Solve for x, y, and z. We substitute -2 for u, -1 for v, and -3 for w.

$$\begin{array}{lll} u = \frac{1}{x} & v = \frac{1}{y} & w = \frac{1}{z} \\ -2 = \frac{1}{x} & -1 = \frac{1}{y} & -3 = \frac{1}{z} \\ x = \frac{1}{2} & y = -1 & z = -\frac{1}{3} \end{array}$$

We obtain $\left(-\frac{1}{2}, -1, -\frac{1}{3}\right)$. This checks, so it is the solution.

55. $5x - 6y + kz = -5,$ (1)

$x + 3y - 2z = 2,$ (2)

$2x - y + 4z = -1$ (3)

Eliminate y from two different pairs of equations.

$$\begin{array}{rl} 5x - 6y + kz = -5 & (1) \\ 2x + 6y - 4z = 4 & \text{Multiplying (2) by 2} \\ \hline 7x + (k-4)z = -1 & (4) \end{array}$$

$$\begin{array}{rl} x + 3y - 2z = 2 & (2) \\ 6x - 3y + 12z = -3 & \text{Multiplying (3) by 3} \\ \hline 7x + 10z = -1 & (5) \end{array}$$

Solve the system of Equations (4) and (5).

$7x + (k-4)z = -1$ (4)

$7x + 10z = -1$ (5)

$$\begin{aligned} -7x - \quad (k-4)z &= \quad 1 \quad \text{Multiplying (4) by } -1 \\ 7x + \quad\quad 10z &= -1 \quad (5) \\ \hline (-k+14)z &= \quad 0 \quad (6) \end{aligned}$$

The system is dependent for the value of k that makes Equation (6) true. This occurs when $-k+14$ is 0. We solve for k:

$$-k + 14 = 0$$
$$14 = k$$

57. $z = b - mx - ny$

Three solutions are $(1, 1, 2)$, $(3, 2, -6)$, and $\left(\frac{3}{2}, 1, 1\right)$. We substitute for x, y, and z and then solve for b, m, and n.

$$2 = b - m - n,$$
$$-6 = b - 3m - 2n,$$
$$1 = b - \frac{3}{2}m - n$$

1., 2. Write the equations in standard form. Also, clear the fraction in the last equation.

$$\begin{aligned} b - \ m - \ n &= \ \ 2, \quad (1) \\ b - 3m - 2n &= -6, \quad (2) \\ 2b - 3m - 2n &= \ \ 2 \quad (3) \end{aligned}$$

3., 4. Eliminate b from two different pairs of equations.

$$\begin{aligned} b - \ m - \ n &= 2 \quad (1) \\ -b + 3m + 2n &= 6 \quad \text{Multiplying (2) by } -1 \\ \hline 2m + n &= 8 \quad (4) \quad \text{Adding} \end{aligned}$$

$$\begin{aligned} -2b + 2m + 2n &= -4 \quad \text{Multiplying (1) by } -2 \\ 2b - 3m - 2n &= \ \ 2 \quad (3) \\ \hline -m \quad\quad &= -2 \quad (5) \quad \text{Adding} \end{aligned}$$

5. We solve Equation (5) for m:

$$-m = -2$$
$$m = 2$$

Substitute in Equation (4) and solve for n.

$$2 \cdot 2 + n = 8$$
$$4 + n = 8$$
$$n = 4$$

6. Substitute in one of the original equations to find b.

$$b - 2 - 4 = 2 \quad \text{Substituting 2 for } m \text{ and 4 for } n \text{ in (1)}$$
$$b - 6 = 2$$
$$b = 8$$

The solution is $(8, 2, 4)$, so the equation is $z = 8 - 2x - 4y$.

Exercise Set 8.5

1. ***Familiarize***. Let x = the first number, y = the second number, and z = the third number.

Translate.

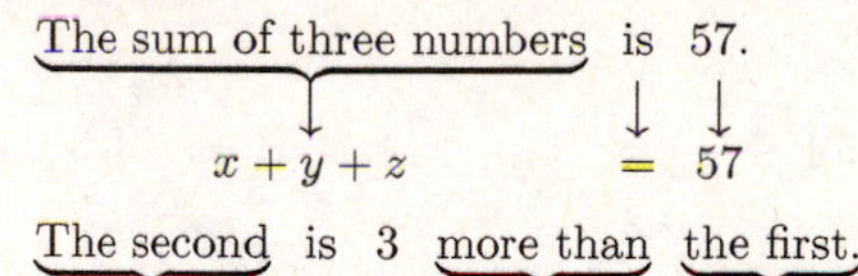

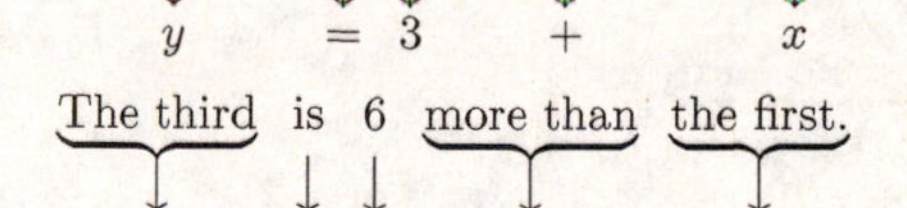

The third is 6 more than the first.

$z = 6 + x$

We now have a system of equations.

$$\begin{aligned} x + y + z &= 57, &\quad \text{or} \quad x + y + z &= 57, \\ y &= 3 + x & -x + y \quad &= 3, \\ z &= 6 + x & -x \quad + z &= 6 \end{aligned}$$

Carry out. Solving the system we get $(16, 19, 22)$.

Check. The sum of the three numbers is $16 + 19 + 22$, or 57. The second number, 19, is three more than the first number, 16. The third number, 22, is 6 more than the first number, 16. The numbers check.

State. The numbers are 16, 19, and 22.

3. ***Familiarize***. Let x = the first number, y = the second number, and z = the third number.

Translate.

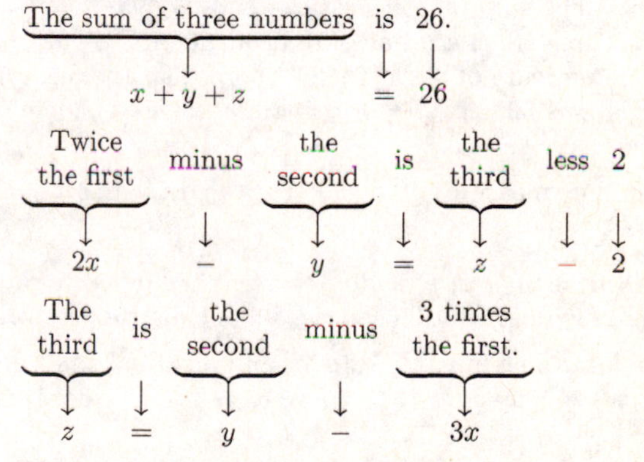

We now have a system of equations.

$$\begin{aligned} x + y + z &= 26, &\quad \text{or} \quad x + y + z &= 26, \\ 2x - y &= z - 2, & 2x - y - z &= -2, \\ z &= y - 3x & 3x - y + z &= 0 \end{aligned}$$

Carry out. Solving the system we get $(8, 21, -3)$.

Check. The sum of the numbers is $8+21-3$, or 26. Twice the first minus the second is $2 \cdot 8 - 21$, or -5, which is 2 less than the third. The second minus three times the first is $21 - 3 \cdot 8$, or -3, which is the third. The numbers check.

State. The numbers are 8, 21, and -3.

5. ***Familiarize***. We first make a drawing.

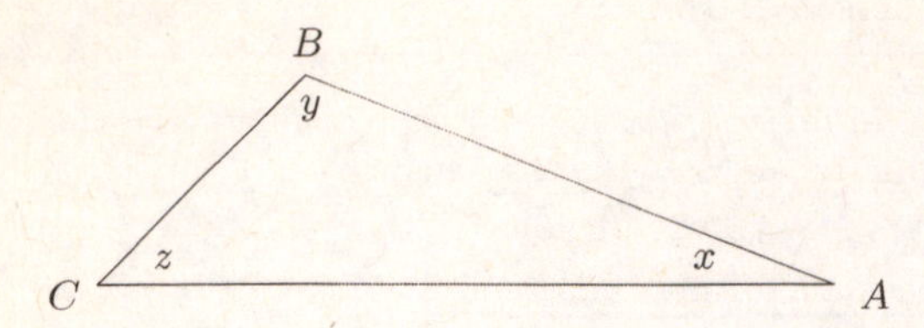

We let x, y, and z represent the measures of angles A, B, and C, respectively. The measures of the angles of a triangle add up to 180°.

Translate.

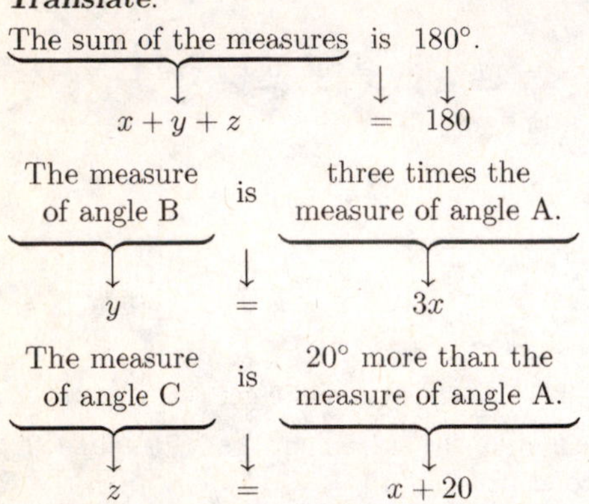

We now have a system of equations.

$$x + y + z = 180,$$
$$y = 3x,$$
$$z = x + 20$$

Carry out. Solving the system we get $(32, 96, 52)$.

Check. The sum of the measures is $32° + 96° + 52°$, or 180°. Three times the measure of angle A is $3 \cdot 32°$, or 96°, the measure of angle B. 20° more than the measure of angle A is $32° + 20°$, or 52°, the measure of angle C. The numbers check.

State. The measures of angles A, B, and C are 32°, 96°, and 52°, respectively.

7. ***Familiarize***. Let x, y, and z represent the monthly rates for an individual adult, a spouse, and a child, respectively.

Translate. The monthly rate for an individual adult and a spouse (a couple) is \$121, so we have

$$x + y = 121.$$

The monthly rate for an individual adult and one child is \$107, so we have

$$x + z = 107.$$

The monthly rate for a couple and one child is \$164, so we have

$$x + y + z = 164.$$

We now have a system of equations.

$$x + y = 121,$$
$$x + z = 107,$$
$$x + y + z = 164$$

Carry out. Solving the system, we get $(64, 57, 43)$.

Check. The monthly rate for a couple is \$64 + \$57, or \$121. For an individual adult and one child the monthly rate is \$64 + \$43, or \$107, and for a couple and one child the rate is \$64 + \$57 + \$43, or \$164. The answer checks.

State. The monthly rate for an individual adult is \$64, for a spouse it is \$57, and for a child it is \$43.

9. ***Familiarize***. Let x, y, and z represent the number of grams of fiber in 1 bran muffin, 1 banana, and a 1-cup serving of Wheaties, respectively.

Translate.

Two bran muffins, 1 banana, and a 1-cup serving of Wheaties contain 9 g of fiber, so we have

$$2x + y + z = 9.$$

One bran muffin, 2 bananas, and a 1-cup serving of Wheaties contain 10.5 g of fiber, so we have

$$x + 2y + z = 10.5$$

Two bran muffins and a 1-cup serving of Wheaties contain 6 g of fiber, so we have

$$2x + z = 6.$$

We now have a system of equations.

$$2x + y + z = 9,$$
$$x + 2y + z = 10.5,$$
$$2x + z = 6$$

Carry out. Solving the system, we get $(1.5, 3, 3)$.

Check. Two bran muffins 1 banana, and a 1-cup serving of Wheaties contain $2(1.5) + 3 + 3$, or 9 g of fiber. One bran muffin, 2 bananas, and a 1-cup serving of Wheaties contain $1.5 + 2 \cdot 3 + 3$, or 10.5 g of fiber. Two bran muffins and a 1-cup serving of Wheaties contain $2(1.5) + 3$, or 6 g of fiber. The answer checks.

State. A bran muffin has 1.5 g of fiber, a banana has 3 g, and a 1-cup serving of Wheaties has 3 g.

11. Observe that the basic model with a sunroof costs \$25,495 and when 4WD is added the price rises to \$27,465. This tells us that the price of 4WD is \$27,465 − \$25,495, or \$1970. Now observe that the basic model with 4WD costs \$26,665 so the basic model costs \$26,665 − \$1970, or \$24,695. Finally, we know that the sunroof costs \$25,495 − \$24,695, or \$800.

13. We know that Elrod, Dot, and Wendy can weld 74 linear feet per hour when working together. We also know that Elrod and Dot together can weld 44 linear feet per hour, which leads to the conclusion that Wendy can weld $74 - 44$, or 30 linear feet per hour alone. We also know that Elrod and Wendy together can weld 50 linear feet per hour. This, along with the earlier conclusion that Wendy can weld 30 linear feet per hour alone, leads to two conclusions: Elrod can weld $50 - 30$, or 20 linear feet per hour alone and Dot can weld $74 - 50$, or 24 linear feet per hour alone.

15. ***Familiarize***. Let x = the number of 12-oz cups, y = the number of 16-oz cups, and z = the number of 20-oz cups that Roz filled. Note that six 144-oz brewers contain $6 \cdot 144$, or 864 oz of coffee. Also, x 12-oz cups contain a total of

$12x$ oz of coffee and bring in $\$1.40x$, y 16-oz cups contain $16y$ oz and bring in $\$1.60y$, and z 20-oz cups contain $20z$ oz and bring in $\$1.70z$.

Translate.

The total number of coffees served was 55.

$$x + y + z = 55$$

The total amount of coffee served was 864 oz.

$$12x + 16y + 20z = 864$$

The total amount collected was \$85.90.

$$1.40x + 1.60y + 1.70z = 85.90$$

Now we have a system of equations.

$$x + y + z = 55,$$
$$12x + 16y + 20z = 864,$$
$$1.40x + 1.60y + 1.70z = 85.90$$

Carry out. Solving the system we get $(17, 25, 13)$.

Check. The total number of coffees served was $17+25+13$, or 55. The total amount of coffee served was $12 \cdot 17 + 16 \cdot 25 + 20 \cdot 13 = 204 + 400 + 260 = 864$ oz. The total amount collected was $\$1.40(17) + \$1.60(25) + \$1.70(13) = \$23.80 + \$40 + \$22.10 = \$85.90$. The numbers check.

State. Roz filled 17 12-oz cups, 25 16-oz cups, and 13 20-oz cups.

17. ***Familiarize.*** Let x, y, and z represent the number of small, medium, and large soft drinks sold, respectively. Then x small drinks brought in $\$1 \cdot x$, or $\$x$; y medium drinks brought in $\$1.15y$; and z large drinks brought in $\$1.30z$.

Translate.

The total number of drinks served was 40.

$$x + y + z = 40$$

The total amount collected was \$45.25.

$$x + 1.15y + 1.30z = 45.25$$

The number of small and large drinks sold was the number of medium drinks sold less 10.

$$x + z = y - 10$$

Now we have a system of equations.

$$x + y + z = 40,$$
$$x + 1.15y + 1.30z = 45.25,$$
$$x + z = y - 10$$

Carry out. Solving the system, we get $(10, 25, 5)$.

Check. The total number of drinks sold was $10 + 25 + 5$, or 40. The total amount they brought in was $\$1(10)+\$1.15(25)+\$1.30(5) = \$10+\$28.75+\$6.50 = \$45.25$. The number of small and large drinks sold, $10 + 5$, or 15, is 10 less than 25, the number of medium drinks sold. The numbers check.

State. 10 small drinks, 25 medium drinks, and 5 large drinks were sold.

19. ***Familiarize.*** Let r = the number of servings of roast beef, p = the number of baked potatoes, and b = the number of servings of broccoli. Then r servings of roast beef contain $300r$ Calories, $20r$ g of protein, and no vitamin C. In p baked potatoes there are $100p$ Calories, $5p$ g of protein, and $20p$ mg of vitamin C. And b servings of broccoli contain $50b$ Calories, $5b$ g of protein, and $100b$ mg of vitamin C. The patient requires 800 Calories, 55 g of protein, and 220 mg of vitamin C.

Translate. Write equations for the total number of calories, the total amount of protein, and the total amount of vitamin C.

$$300r + 100p + 50b = 800 \quad \text{(Calories)}$$
$$20r + 5p + 5b = 55 \quad \text{(protein)}$$
$$20p + 100b = 220 \quad \text{(vitamin C)}$$

We now have a system of equations.

Carry out. Solving the system we get $(2, 1, 2)$.

Check. Two servings of roast beef provide 600 Calories, 40 g of protein, and no vitamin C. One baked potato provides 100 Calories, 5 g of protein, and 20 mg of vitamin C. And 2 servings of broccoli provide 100 Calories, 10 g of protein, and 200 mg of vitamin C. Together, then, they provide 800 Calories, 55 g of protein, and 220 mg of vitamin C. The values check.

State. The dietician should prepare 2 servings of roast beef, 1 baked potato, and 2 servings of broccoli.

21. ***Familiarize.*** Let x, y, and z represent the populations of Asia, Africa, and the rest of the world respectively, in billions, in 2050.

Translate.

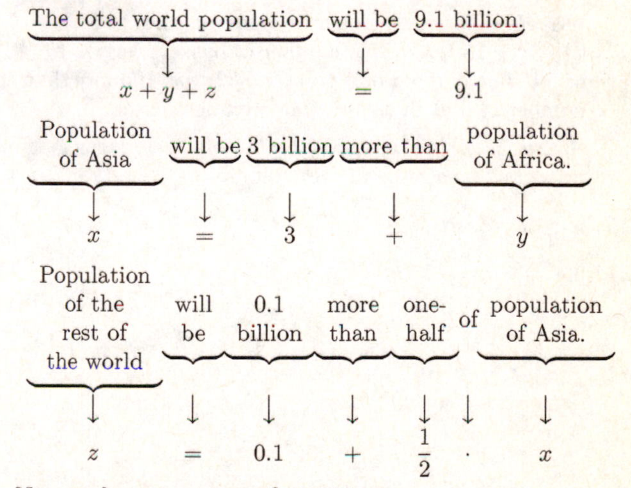

Now we have a system of equations.

$$x + y + z = 9.1,$$
$$x = 3 + y,$$
$$z = 0.1 + \frac{1}{2}x$$

Carry out. Solving the system, we get $(4.8, 1.8, 2.5)$.

Check. The total population will be $4.8 + 1.8 + 2.5$, or 9.1 billion. The population of Asia, 4.8 billion, is 3 billion more than the population of Africa, 1.8 billion. Also, 0.1 billion more than half the population of Asia is $0.1 + \frac{1}{2}(4.8) = 0.1 + 2.4 = 2.5$ billion, the population of the rest of the world. The numbers check.

State. In 2050 the population of Asia will be 4.8 billion, the population of Africa will be 1.8 billion, and the population of the rest of the world will be 2.5 billion.

23. ***Familiarize***. Let x, y, and z represent the number of 2-point field goals, 3-point field goals, and 1-point foul shots made, respectively. The total number of points scored from each of these types of goals is $2x$, $3y$, and z.

Translate.

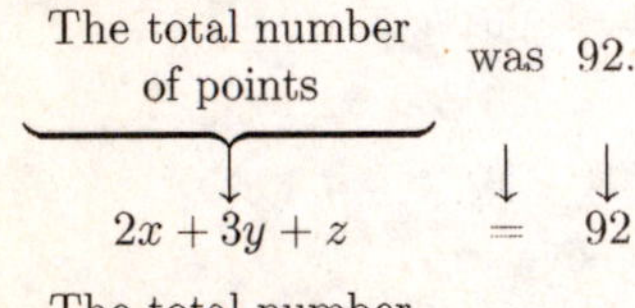

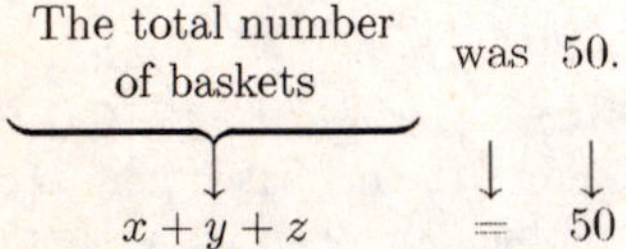

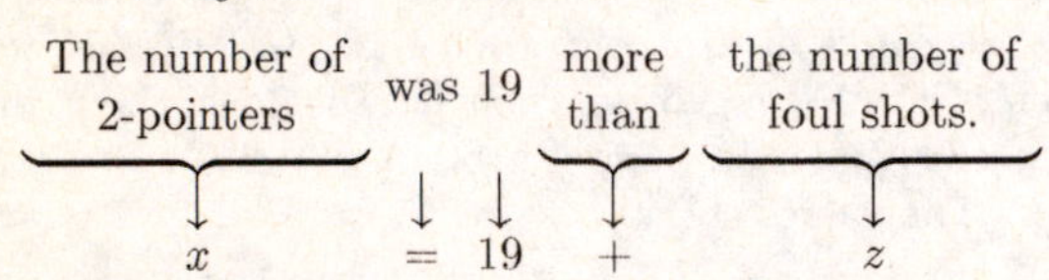

Now we have a system of equations.

$$2x + 3y + z = 92,$$
$$x + y + z = 50,$$
$$x = 19 + z$$

Carry out. Solving the system we get $(32, 5, 13)$.

Check. The total number of points was $2 \cdot 32 + 3 \cdot 5 + 13 = 64 + 15 + 13 = 92$. The number of baskets was $32 + 5 + 13$, or 50. The number of 2-pointers, 32, was 19 more than the number of foul shots, 13. The numbers check.

State. The Knicks made 32 two-point field goals, 5 three-point field goals, and 13 foul shots.

25. *Writing Exercise*

27. $5(-3) + 7 = -15 + 7 = -8$

29. $-6(8) + (-7) = -48 + (-7) = -55$

31. $-7(2x - 3y + 5z) = -7 \cdot 2x - 7(-3y) - 7(5z)$
$= -14x + 21y - 35z$

33.
$$-4(2a + 5b) + 3a + 20b$$
$$= -8a - 20b + 3a + 20b$$
$$= -8a + 3a - 20b + 20b$$
$$= -5a$$

35. *Writing Exercise*

37. ***Familiarize***. Let w, x, y, and z represent the monthly rates for an applicant, a spouse, the first child, and the second child respectively.

Translate.

The rate for the applicant and his or her spouse is \$160/month, so we have

$$w + x = 160.$$

The rate for the applicant, a spouse, and one child is \$203/month, so we have

$$w + x + y = 203.$$

The rate for the applicant, a spouse, and two children is \$243/month, so we have

$$w + x + y + z = 243.$$

The rate for the applicant and one child is \$145/month, so we have

$$w + y = 145.$$

Now we have a system of equations.

$$w + x = 160, \quad (1)$$
$$w + x + y = 203, \quad (2)$$
$$w + x + y + z = 243, \quad (3)$$
$$w + y = 145 \quad (4)$$

Carry out. We solve the system of equations. First substitute 203 for $w + x + y$ in (3) and solve for z.

$$203 + z = 243$$
$$z = 40$$

Next substitute 160 for $w + x$ in (2) and solve for y.

$$160 + y = 203$$
$$y = 43$$

Substitute 43 for y in (4) and solve for w.

$$w + 43 = 145$$
$$w = 102$$

Finally substitute 102 for w in (1) and solve for x.

$$102 + x = 160$$
$$x = 58$$

The solution is $(102, 58, 43, 40)$.

Check. The check is left to the student.

State. The separate monthly rates for an applicant, a spouse, the first child, and the second child are \$102, \$58, \$43, and \$40, respectively.

39. ***Familiarize***. Let w, x, y, and z represent the ages of Tammy, Carmen, Dennis, and Mark respectively.

Translate.

Tammy's age is the sum of the ages of Carmen and Dennis, so we have

$$w = x + y.$$

Carmen's age is 2 more than the sum of the ages of Dennis and Mark, so we have

$$x = 2 + y + z.$$

Dennis's age is four times Mark's age, so we have

$y = 4z.$

The sum of all four ages is 42, so we have

$w + x + y + z = 42.$

Now we have a system of equations.

$w = x + y,$ (1)

$x = 2 + y + z,$ (2)

$y = 4z,$ (3)

$w + x + y + z = 42$ (4)

Carry out. We solve the system of equations. First we will express w, x, and y in terms of z and then solve for z. From (3) we know that $y = 4z$. Substitute $4z$ for y in (2):

$x = 2 + 4z + z = 2 + 5z.$

Substitute $2 + 5z$ for x and $4z$ for y in (1):

$w = 2 + 5z + 4z = 2 + 9z.$

Now substitute $2 + 9z$ for w, $2 + 5z$ for x, and $4z$ for y in (4) and solve for z.

$$2 + 9z + 2 + 5z + 4z + z = 42$$
$$19z + 4 = 42$$
$$19z = 38$$
$$z = 2$$

Then we have:

$w = 2 + 9z = 2 + 9 \cdot 2 = 20,$

$x = 2 + 5z = 2 + 5 \cdot 2 = 12,$ and

$y = 4z = 4 \cdot 2 = 8$

Although we were asked to find only Tammy's age, we found all of the ages so that we can check the result.

Check. The check is left to the student.

State. Tammy is 20 years old.

41. Let T, G, and H represent the number of tickets Tom, Gary, and Hal begin with, respectively. After Hal gives tickets to Tom and Gary, each has the following number of tickets:

Tom: $T + T$, or $2T$,

Gary: $G + G$, or $2G$,

Hal: $H - T - G$.

After Tom gives tickets to Gary and Hal, each has the following number of tickets:

Gary: $2G + 2G$, or $4G$,

Hal: $(H - T - G) + (H - T - G)$, or $2(H - T - G)$,

Tom: $2T - 2G - (H - T - G)$, or $3T - H - G$

After Gary gives tickets to Hal and Tom, each has the following number of tickets:

Hal: $2(H - T - G) + 2(H - T - G)$, or $4(H - T - G)$

Tom: $(3T - H - G) + (3T - H - G)$, or $2(3T - H - G)$,

Gary: $4G - 2(H - T - G) - (3T - H - G)$, or $7G - H - T$.

Since Hal, Tom, and Gary each finish with 40 tickets, we write the following system of equations:

$$4(H - T - G) = 40,$$
$$2(3T - H - G) = 40,$$
$$7G - H - T = 40$$

Solving the system we find that $T = 35$, so Tom started with 35 tickets.

Exercise Set 8.6

1. horizontal; columns; see page 554 in the text.

3. entry; see page 554 in the text.

5. multiple; see page 557 in the text.

7. $9x - 2y = 5,$

$3x - 3y = 11$

Write a matrix using only the constants.

$$\left[\begin{array}{cc|c} 9 & -2 & 5 \\ 3 & -3 & 11 \end{array}\right]$$

Multiply row 2 by 3 to make the first number in row 2 a multiple of 9.

$$\left[\begin{array}{cc|c} 9 & -2 & 5 \\ 9 & -9 & 33 \end{array}\right] \quad \text{New Row 2} = 3(\text{Row 2})$$

Multiply row 1 by -1 and add it to row 2.

$$\left[\begin{array}{cc|c} 9 & -2 & 5 \\ 0 & -7 & 28 \end{array}\right] \quad \text{New Row 2} = -1(\text{Row 1}) + \text{Row 2}$$

Reinserting the variables, we have

$9x - 2y = 5,$ (1)

$-7y = 28.$ (2)

Solve Equation (2) for y.

$$-7y = 28$$
$$y = -4$$

Substitute -4 for y in Equation (1) and solve for x.

$$9x - 2y = 5$$
$$9x - 2(-4) = 5$$
$$9x + 8 = 5$$
$$9x = -3$$
$$x = -\frac{1}{3}$$

The solution is $\left(-\frac{1}{3}, -4\right)$.

9. $x + 4y = 8,$

$3x + 5y = 3$

We first write a matrix using only the constants.

$$\left[\begin{array}{cc|c} 1 & 4 & 8 \\ 3 & 5 & 3 \end{array}\right]$$

Multiply the first row by -3 and add it to the second row.

$$\left[\begin{array}{cc|c} 1 & 4 & 8 \\ 0 & -7 & -21 \end{array}\right] \text{ New Row 2} = -3(\text{Row 1}) + \text{Row 2}$$

Reinserting the variables, we have

$$x + 4y = 8, \quad (1)$$
$$-7y = -21. \quad (2)$$

Solve Equation (2) for y.

$$-7y = -21$$
$$y = 3$$

Substitute 3 for y in Equation (1) and solve for x.

$$x + 4 \cdot 3 = 8$$
$$x + 12 = 8$$
$$x = -4$$

The solution is $(-4, 3)$.

11. $6x - 2y = 4,$

$7x + y = 13$

Write a matrix using only the constants.

$$\left[\begin{array}{cc|c} 6 & -2 & 4 \\ 7 & 1 & 13 \end{array}\right]$$

Multiply the second row by 6 to make the first number in row 2 a multiple of 6.

$$\left[\begin{array}{cc|c} 6 & -2 & 4 \\ 42 & 6 & 78 \end{array}\right] \text{ New Row 2} = 6(\text{Row 2})$$

Now multiply the first row by -7 and add it to the second row.

$$\left[\begin{array}{cc|c} 6 & -2 & 4 \\ 0 & 20 & 50 \end{array}\right] \text{ New Row 2} = -7(\text{Row 1}) + \text{Row 2}$$

Reinserting the variables, we have

$$6x - 2y = 4, \quad (1)$$
$$20y = 50. \quad (2)$$

Solve Equation (2) for y.

$$20y = 50$$
$$y = \frac{5}{2}$$

Substitute $\frac{5}{2}$ for y in Equation (1) and solve for x.

$$6x - 2y = 4$$
$$6x - 2\left(\frac{5}{2}\right) = 4$$
$$6x - 5 = 4$$
$$6x = 9$$
$$x = \frac{3}{2}$$

The solution is $\left(\frac{3}{2}, \frac{5}{2}\right)$.

13. $3x + 2y + 2z = 3,$

$x + 2y - z = 5,$

$2x - 4y + z = 0$

We first write a matrix using only the constants.

$$\left[\begin{array}{ccc|c} 3 & 2 & 2 & 3 \\ 1 & 2 & -1 & 5 \\ 2 & -4 & 1 & 0 \end{array}\right]$$

First interchange rows 1 and 2 so that each number below the first number in the first row is a multiple of that number.

$$\left[\begin{array}{ccc|c} 1 & 2 & -1 & 5 \\ 3 & 2 & 2 & 3 \\ 2 & -4 & 1 & 0 \end{array}\right]$$

Multiply row 1 by -3 and add it to row 2.

Multiply row 1 by -2 and add it to row 3.

$$\left[\begin{array}{ccc|c} 1 & 2 & -1 & 5 \\ 0 & -4 & 5 & -12 \\ 0 & -8 & 3 & -10 \end{array}\right]$$

Multiply row 2 by -2 and add it to row 3.

$$\left[\begin{array}{ccc|c} 1 & 2 & -1 & 5 \\ 0 & -4 & 5 & -12 \\ 0 & 0 & -7 & 14 \end{array}\right]$$

Reinserting the variables, we have

$$x + 2y - z = 5, \quad (1)$$
$$-4y + 5z = -12, \quad (2)$$
$$-7z = 14. \quad (3)$$

Solve (3) for z.

$$-7z = 14$$
$$z = -2$$

Substitute -2 for z in (2) and solve for y.

$$-4y + 5(-2) = -12$$
$$-4y - 10 = -12$$
$$-4y = -2$$
$$y = \frac{1}{2}$$

Substitute $\frac{1}{2}$ for y and -2 for z in (1) and solve for x.

$$x + 2 \cdot \frac{1}{2} - (-2) = 5$$
$$x + 1 + 2 = 5$$
$$x + 3 = 5$$
$$x = 2$$

The solution is $\left(2, \frac{1}{2}, -2\right)$.

15. $p - 2q - 3r = 3,$
$2p - q - 2r = 4,$
$4p + 5q + 6r = 4$

We first write a matrix using only the constants.

$$\left[\begin{array}{ccc|c} 1 & -2 & -3 & 3 \\ 2 & -1 & -2 & 4 \\ 4 & 5 & 6 & 4 \end{array}\right]$$

$$\left[\begin{array}{ccc|c} 1 & -2 & -3 & 3 \\ 0 & 3 & 4 & -2 \\ 0 & 13 & 18 & -8 \end{array}\right]$$ New Row 2 = -2(Row 1) + Row 2; New Row 3 = -4(Row 1) + Row 3

$$\left[\begin{array}{ccc|c} 1 & -2 & -3 & 3 \\ 0 & 3 & 4 & -2 \\ 0 & 39 & 54 & -24 \end{array}\right]$$ New Row 3 = 3(Row 3)

$$\left[\begin{array}{ccc|c} 1 & -2 & -3 & 3 \\ 0 & 3 & 4 & -2 \\ 0 & 0 & 2 & 2 \end{array}\right]$$ New Row 3 = -13(Row 2)+ Row 3

Reinserting the variables, we have

$p - 2q - 3r = 3,$ (1)
$3q + 4r = -2,$ (2)
$2r = 2$ (3)

Solve (3) for r.

$2r = 2$
$r = 1$

Substitute 1 for r in (2) and solve for q.

$3q + 4 \cdot 1 = -2$
$3q + 4 = -2$
$3q = -6$
$q = -2$

Substitute -2 for q and 1 for r in (1) and solve for p.

$p - 2(-2) - 3 \cdot 1 = 3$
$p + 4 - 3 = 3$
$p + 1 = 3$
$p = 2$

The solution is $(2, -2, 1)$.

17. $3p + 2r = 11,$
$q - 7r = 4,$
$p - 6q = 1$

We first write a matrix using only the constants.

$$\left[\begin{array}{ccc|c} 3 & 0 & 2 & 11 \\ 0 & 1 & -7 & 4 \\ 1 & -6 & 0 & 1 \end{array}\right]$$

$$\left[\begin{array}{ccc|c} 1 & -6 & 0 & 1 \\ 0 & 1 & -7 & 4 \\ 3 & 0 & 2 & 11 \end{array}\right]$$ Interchange Row 1 and Row 3

$$\left[\begin{array}{ccc|c} 1 & -6 & 0 & 1 \\ 0 & 1 & -7 & 4 \\ 0 & 18 & 2 & 8 \end{array}\right]$$ New Row 3 = -3(Row 1) + Row 3

$$\left[\begin{array}{ccc|c} 1 & -6 & 0 & 1 \\ 0 & 1 & -7 & 4 \\ 0 & 0 & 128 & -64 \end{array}\right]$$ New Row 3 = -18(Row 2) + Row 3

Reinserting the variables, we have

$p - 6q = 1,$ (1)
$q - 7r = 4,$ (2)
$128r = -64.$ (3)

Solve (3) for r.

$128r = -64$
$r = -\frac{1}{2}$

Substitute $-\frac{1}{2}$ for r in (2) and solve for q.

$q - 7r = 4$
$q - 7\left(-\frac{1}{2}\right) = 4$
$q + \frac{7}{2} = 4$
$q = \frac{1}{2}$

Substitute $\frac{1}{2}$ for q in (1) and solve for p.

$p - 6 \cdot \frac{1}{2} = 1$
$p - 3 = 1$
$p = 4$

The solution is $\left(4, \frac{1}{2}, -\frac{1}{2}\right)$.

19. We will rewrite the equations with the variables in alphabetical order:

$-2w + 2x + 2y - 2z = -10,$
$w + x + y + z = -5,$
$3w + x - y + 4z = -2,$
$w + 3x - 2y + 2z = -6$

Write a matrix using only the constants.

$$\left[\begin{array}{cccc|c} -2 & 2 & 2 & -2 & -10 \\ 1 & 1 & 1 & 1 & -5 \\ 3 & 1 & -1 & 4 & -2 \\ 1 & 3 & -2 & 2 & -6 \end{array}\right]$$

$$\left[\begin{array}{cccc|c} -1 & 1 & 1 & -1 & -5 \\ 1 & 1 & 1 & 1 & -5 \\ 3 & 1 & -1 & 4 & -2 \\ 1 & 3 & -2 & 2 & -6 \end{array}\right]$$ New Row 1 = $\frac{1}{2}$(Row 1)

$$\left[\begin{array}{rrrr|r} -1 & 1 & 1 & -1 & -5 \\ 0 & 2 & 2 & 0 & -10 \\ 0 & 4 & 2 & 1 & -17 \\ 0 & 4 & -1 & 1 & -11 \end{array}\right]$$ New Row 2 = Row 1 + Row 2; New Row 3 = 3(Row 1) +Row 3; New Row 4 = Row 1 + Row 4

$$\left[\begin{array}{rrrr|r} -1 & 1 & 1 & -1 & -5 \\ 0 & 2 & 2 & 0 & -10 \\ 0 & 0 & -2 & 1 & 3 \\ 0 & 0 & -5 & 1 & 9 \end{array}\right]$$ New Row 3 = −2(Row 2) + Row 3; New Row 4 = −2(Row 2) + Row 4

$$\left[\begin{array}{rrrr|r} -1 & 1 & 1 & -1 & -5 \\ 0 & 2 & 2 & 0 & -10 \\ 0 & 0 & -2 & 1 & 3 \\ 0 & 0 & -10 & 2 & 18 \end{array}\right]$$ New Row 4 = 2(Row 4)

$$\left[\begin{array}{rrrr|r} -1 & 1 & 1 & -1 & -5 \\ 0 & 2 & 2 & 0 & -10 \\ 0 & 0 & -2 & 1 & 3 \\ 0 & 0 & 0 & -3 & 3 \end{array}\right]$$ New Row 4 = −5(Row 3) + Row 4

Reinserting the variables, we have

$$\begin{aligned} -w + x + y - z &= -5, && (1) \\ 2x + 2y &= -10, && (2) \\ -2y + z &= 3, && (3) \\ -3z &= 3. && (4) \end{aligned}$$

Solve (4) for z.

$$-3z = 3$$
$$z = -1$$

Substitute -1 for z in (3) and solve for y.

$$-2y + (-1) = 3$$
$$-2y = 4$$
$$y = -2$$

Substitute -2 for y in (2) and solve for x.

$$2x + 2(-2) = -10$$
$$2x - 4 = -10$$
$$2x = -6$$
$$x = -3$$

Substitute -3 for x, -2 for y, and -1 for z in (1) and solve for w.

$$-w + (-3) + (-2) - (-1) = -5$$
$$-w - 3 - 2 + 1 = -5$$
$$-w - 4 = -5$$
$$-w = -1$$
$$w = 1$$

The solution is $(1, -3, -2, -1)$.

21. ***Familiarize.*** Let d = the number of dimes and n = the number of nickels. The value of d dimes is \$0.10d, and the value of n nickels is \$0.05n.

Translate.

Total number of coins is 42.

$$d + n = 42$$

Total value of coins is \$3.

$$0.10d + 0.05n = 3$$

After clearing decimals, we have this system.

$$d + n = 42,$$
$$10d + 5n = 300$$

Carry out. Solve using matrices.

$$\left[\begin{array}{rr|r} 1 & 1 & 42 \\ 10 & 5 & 300 \end{array}\right]$$

$$\left[\begin{array}{rr|r} 1 & 1 & 42 \\ 0 & -5 & -120 \end{array}\right]$$ New Row 2 = −10(Row 1) + Row 2

Reinserting the variables, we have

$$d + n = 42, \quad (1)$$
$$-5n = -120 \quad (2)$$

Solve (2) for n.

$$-5n = -120$$
$$n = 24$$

$d + 24 = 42$ Back-substituting

$$d = 18$$

Check. The sum of the two numbers is 42. The total value is \$0.10(18) + \$0.05(24) = \$1.80 + \$1.20 = \$3. The numbers check.

State. There are 18 dimes and 24 nickels.

23. ***Familiarize.*** We let x represent the number of pounds of the \$4.05 kind and y represent the number of pounds of the \$2.70 kind of granola. We organize the information in a table.

Granola	Number of pounds	Price per pound	Value
\$4.05 kind	x	\$4.05	\$4.05x
\$2.70 kind	y	\$2.70	\$2.70y
Mixture	15	\$3.15	\$3.15 × 15 or \$47.25

Translate.

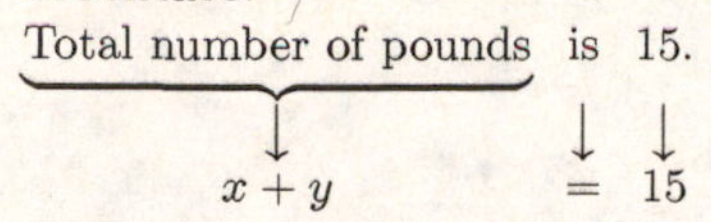
Total number of pounds is 15.

$$x + y = 15$$

Total value of mixture is \$47.25.

$$4.05x + 2.70y = 47.25$$

After clearing decimals, we have this system:

$$x + y = 15,$$
$$405x + 270y = 4725$$

Carry out. Solve using matrices.

$$\left[\begin{array}{cc|c} 1 & 1 & 15 \\ 405 & 270 & 4725 \end{array}\right]$$

$$\left[\begin{array}{cc|c} 1 & 1 & 15 \\ 0 & -135 & -1350 \end{array}\right] \quad \text{New Row 2} = -405(\text{Row 1}) + \text{Row 2}$$

Reinserting the variables, we have

$$x + y = 15, \quad (1)$$
$$-135y = -1350 \quad (2)$$

Solve (2) for y.

$$-135y = -1350$$
$$y = 10$$

Back-substitute 10 for y in (1) and solve for x.

$$x + 10 = 15$$
$$x = 5$$

Check. The sum of the numbers is 15. The total value is \$4.05(5) + \$2.70(10), or \$20.25 + \$27.00, or \$47.25. The numbers check.

State. 5 pounds of the \$4.05 per lb granola and 10 pounds of the \$2.70 per lb granola should be used.

25. ***Familiarize.*** We let x, y, and z represent the amounts invested at 7%, 8%, and 9%, respectively. Recall the formula for simple interest:

Interest = Principal × Rate × Time

Translate. We organize the information in a table.

	First Invest-ment	Second Invest-ment	Third Invest-ment	Total
P	x	y	z	\$2500
R	7%	8%	9%	
T	1 yr	1 yr	1 yr	
I	$0.07x$	$0.08y$	$0.09z$	\$212

The first row gives us one equation:

$$x + y + z = 2500$$

The last row gives a second equation:

$$0.07x + 0.08y + 0.09z = 212$$

Amount invested at 9% is \$1100 more than amount invested at 8%.

$$z = \$1100 + y$$

After clearing decimals, we have this system:

$$x + y + z = 2500,$$
$$7x + 8y + 9z = 21,200,$$
$$-y + z = 1100$$

Carry out. Solve using matrices.

$$\left[\begin{array}{ccc|c} 1 & 1 & 1 & 2500 \\ 7 & 8 & 9 & 21,200 \\ 0 & -1 & 1 & 1100 \end{array}\right]$$

$$\left[\begin{array}{ccc|c} 1 & 1 & 1 & 2500 \\ 0 & 1 & 2 & 3700 \\ 0 & -1 & 1 & 1100 \end{array}\right] \quad \text{New Row 2} = -7(\text{Row 1}) + \text{Row 2}$$

$$\left[\begin{array}{ccc|c} 1 & 1 & 1 & 2500 \\ 0 & 1 & 2 & 3700 \\ 0 & 0 & 3 & 4800 \end{array}\right] \quad \text{New Row 3} = \text{Row 2} + \text{Row 3}$$

Reinserting the variables, we have

$$x + y + z = 2500, \quad (1)$$
$$y + 2z = 3700, \quad (2)$$
$$3z = 4800 \quad (3)$$

Solve (3) for z.

$$3z = 4800$$
$$z = 1600$$

Back-substitute 1600 for z in (2) and solve for y.

$$y + 2 \cdot 1600 = 3700$$
$$y + 3200 = 3700$$
$$y = 500$$

Back-substitute 500 for y and 1600 for z in (1) and solve for x.

$$x + 500 + 1600 = 2500$$
$$x + 2100 = 2500$$
$$x = 400$$

Check. The total investment is \$400 + \$500 + \$1600, or \$2500. The total interest is 0.07(\$400) + 0.08(\$500) + 0.09(\$1600) = \$28 + \$40 + \$144 = \$212. The amount invested at 9%, \$1600, is \$1100 more than the amount invested at 8%, \$500. The numbers check.

State. \$400 is invested at 7%, \$500 is invested at 8%, and \$1600 is invested at 9%.

27. *Writing Exercise*

29. $5(-3) - (-7)4 = -15 - (-28) = -15 + 28 = 13$

31.
$$-2(5 \cdot 3 - 4 \cdot 6) - 3(2 \cdot 7 - 15) + 4(3 \cdot 8 - 5 \cdot 4)$$
$$= -2(15 - 24) - 3(14 - 15) + 4(24 - 20)$$
$$= -2(-9) - 3(-1) + 4(4)$$
$$= 18 + 3 + 16$$
$$= 21 + 16$$
$$= 37$$

33. *Writing Exercise*

35. ***Familiarize.*** Let w, x, y, and z represent the thousand's, hundred's, ten's, and one's digits, respectively.

Translate.

The sum of the digits is 10.

$$w + x + y + z = 10$$

Twice the sum of the thousand's and ten's digits is the sum of the hundred's and one's digits less one.

$$2(w + y) = x + z - 1$$

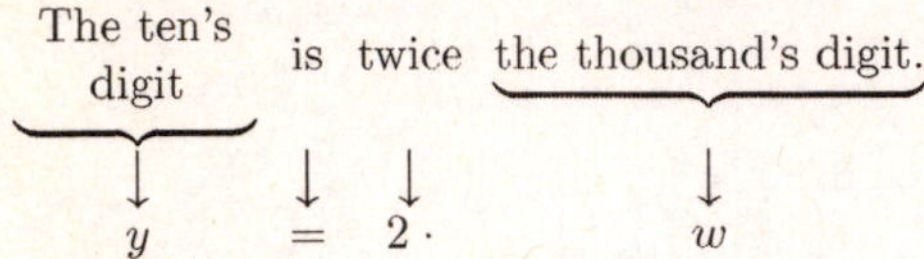

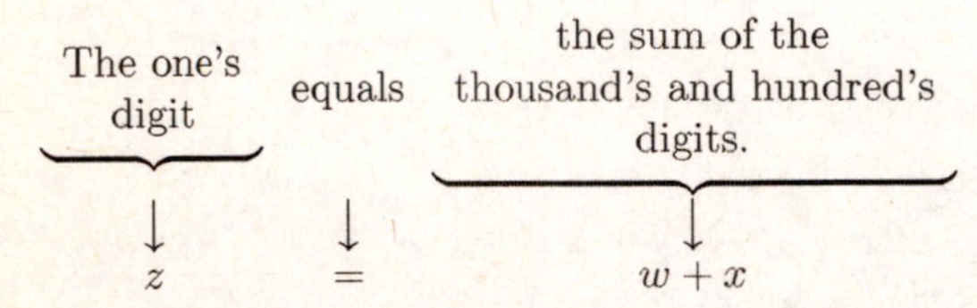

We have a system of equations which can be written as

$$\begin{aligned} w + x + y + z &= 10, \\ 2w - x + 2y - z &= -1, \\ -2w + y &= 0, \\ w + x - z &= 0. \end{aligned}$$

Carry out. We can use matrices to solve the system. We get $(1, 3, 2, 4)$.

Check. The sum of the digits is 10. Twice the sum of 1 and 2 is 6. This is one less than the sum of 3 and 4. The ten's digit, 2, is twice the thousand's digit, 1. The one's digit, 4, equals $1 + 3$. The numbers check.

State. The number is 1324.

Exercise Set 8.7

1. True; see page 559 in the text.

3. False; see page 561 in the text.

5. False; it is the value of the denominator, not the numerator, that indicates whether the equations are dependent. See page 560 in the text.

7. $\begin{vmatrix} 5 & 1 \\ 2 & 4 \end{vmatrix} = 5 \cdot 4 - 2 \cdot 1 = 20 - 2 = 18$

9. $\begin{vmatrix} 6 & -9 \\ 2 & 3 \end{vmatrix} = 6 \cdot 3 - 2(-9) = 18 + 18 = 36$

11. $$\begin{vmatrix} 1 & 4 & 0 \\ 0 & -1 & 2 \\ 3 & -2 & 1 \end{vmatrix}$$

$$\begin{aligned} &= 1\begin{vmatrix} -1 & 2 \\ -2 & 1 \end{vmatrix} - 0\begin{vmatrix} 4 & 0 \\ -2 & 1 \end{vmatrix} + 3\begin{vmatrix} 4 & 0 \\ -1 & 2 \end{vmatrix} \\ &= 1[-1 \cdot 1 - (-2) \cdot 2] - 0 + 3[4 \cdot 2 - (-1) \cdot 0] \\ &= 1 \cdot 3 - 0 + 3 \cdot 8 \\ &= 3 - 0 + 24 \\ &= 27 \end{aligned}$$

13. $$\begin{vmatrix} -1 & -2 & -3 \\ 3 & 4 & 2 \\ 0 & 1 & 2 \end{vmatrix}$$

$$\begin{aligned} &= -1\begin{vmatrix} 4 & 2 \\ 1 & 2 \end{vmatrix} - 3\begin{vmatrix} -2 & -3 \\ 1 & 2 \end{vmatrix} + 0\begin{vmatrix} -2 & -3 \\ 4 & 2 \end{vmatrix} \\ &= -1[4 \cdot 2 - 1 \cdot 2] - 3[-2 \cdot 2 - 1(-3)] + 0 \\ &= -1 \cdot 6 - 3 \cdot (-1) + 0 \\ &= -6 + 3 + 0 \\ &= -3 \end{aligned}$$

15. $$\begin{vmatrix} -4 & -2 & 3 \\ -3 & 1 & 2 \\ 3 & 4 & -2 \end{vmatrix}$$

$$\begin{aligned} &= -4\begin{vmatrix} 1 & 2 \\ 4 & -2 \end{vmatrix} - (-3)\begin{vmatrix} -2 & 3 \\ 4 & -2 \end{vmatrix} + 3\begin{vmatrix} -2 & 3 \\ 1 & 2 \end{vmatrix} \\ &= -4[1(-2) - 4 \cdot 2] + 3[-2(-2) - 4 \cdot 3] + \\ &\quad\; 3(-2 \cdot 2 - 1 \cdot 3) \\ &= -4(-10) + 3(-8) + 3(-7) \\ &= 40 - 24 - 21 = -5 \end{aligned}$$

17. $5x + 8y = 1$,

$3x + 7y = 5$

We compute D, D_x, and D_y.

$$D = \begin{vmatrix} 5 & 8 \\ 3 & 7 \end{vmatrix} = 35 - 24 = 11$$

$$D_x = \begin{vmatrix} 1 & 8 \\ 5 & 7 \end{vmatrix} = 7 - 40 = -33$$

$$D_y = \begin{vmatrix} 5 & 1 \\ 3 & 5 \end{vmatrix} = 25 - 3 = 22$$

Then,

$$x = \frac{D_x}{D} = \frac{-33}{11} = -3$$

and

$$y = \frac{D_y}{D} = \frac{22}{11} = 2.$$

The solution is $(-3, 2)$.

19. $5x - 4y = -3,$

$7x + 2y = 6$

We compute D, D_x, and D_y.

$$D = \begin{vmatrix} 5 & -4 \\ 7 & 2 \end{vmatrix} = 10 - (-28) = 38$$

$$D_x = \begin{vmatrix} -3 & -4 \\ 6 & 2 \end{vmatrix} = -6 - (-24) = 18$$

$$D_y = \begin{vmatrix} 5 & -3 \\ 7 & 6 \end{vmatrix} = 30 - (-21) = 51$$

Then,

$$x = \frac{D_x}{D} = \frac{18}{38} = \frac{9}{19}$$

and

$$y = \frac{D_y}{D} = \frac{51}{38}.$$

The solution is $\left(\frac{9}{19}, \frac{51}{38}\right)$.

21. $3x - y + 2z = 1,$

$x - y + 2z = 3,$

$-2x + 3y + z = 1$

We compute D, D_x, and D_y.

$$D = \begin{vmatrix} 3 & -1 & 2 \\ 1 & -1 & 2 \\ -2 & 3 & 1 \end{vmatrix}$$

$$= 3\begin{vmatrix} -1 & 2 \\ 3 & 1 \end{vmatrix} - 1\begin{vmatrix} -1 & 2 \\ 3 & 1 \end{vmatrix} - 2\begin{vmatrix} -1 & 2 \\ -1 & 2 \end{vmatrix}$$

$$= 3(-7) - 1(-7) - 2(0)$$

$$= -21 + 7 - 0$$

$$= -14$$

$$D_x = \begin{vmatrix} 1 & -1 & 2 \\ 3 & -1 & 2 \\ 1 & 3 & 1 \end{vmatrix}$$

$$= 1\begin{vmatrix} -1 & 2 \\ 3 & 1 \end{vmatrix} - 3\begin{vmatrix} -1 & 2 \\ 3 & 1 \end{vmatrix} + 1\begin{vmatrix} -1 & 2 \\ -1 & 2 \end{vmatrix}$$

$$= 1(-7) - 3(-7) + 1(0)$$

$$= -7 + 21 + 0$$

$$= 14$$

$$D_y = \begin{vmatrix} 3 & 1 & 2 \\ 1 & 3 & 2 \\ -2 & 1 & 1 \end{vmatrix}$$

$$= 3\begin{vmatrix} 3 & 2 \\ 1 & 1 \end{vmatrix} - 1\begin{vmatrix} 1 & 2 \\ 1 & 1 \end{vmatrix} - 2\begin{vmatrix} 1 & 2 \\ 3 & 2 \end{vmatrix}$$

$$= 3 \cdot 1 - 1(-1) - 2(-4)$$

$$= 3 + 1 + 8$$

$$= 12$$

Then,

$$x = \frac{D_x}{D} = \frac{14}{-14} = -1$$

and

$$y = \frac{D_y}{D} = \frac{12}{-14} = -\frac{6}{7}.$$

Substitute in the third equation to find z.

$$-2(-1) + 3\left(-\frac{6}{7}\right) + z = 1$$

$$2 - \frac{18}{7} + z = 1$$

$$-\frac{4}{7} + z = 1$$

$$z = \frac{11}{7}$$

The solution is $\left(-1, -\frac{6}{7}, \frac{11}{7}\right)$.

23. $2x - 3y + 5z = 27,$

$x + 2y - z = -4,$

$5x - y + 4z = 27$

We compute D, D_x, and D_y.

$$D = \begin{vmatrix} 2 & -3 & 5 \\ 1 & 2 & -1 \\ 5 & -1 & 4 \end{vmatrix}$$

$$= 2\begin{vmatrix} 2 & -1 \\ -1 & 4 \end{vmatrix} - 1\begin{vmatrix} -3 & 5 \\ -1 & 4 \end{vmatrix} + 5\begin{vmatrix} -3 & 5 \\ 2 & -1 \end{vmatrix}$$

$$= 2(7) - 1(-7) + 5(-7)$$

$$= 14 + 7 - 35 = -14$$

$$D_x = \begin{vmatrix} 27 & -3 & 5 \\ -4 & 2 & -1 \\ 27 & -1 & 4 \end{vmatrix}$$

$$= 27\begin{vmatrix} 2 & -1 \\ -1 & 4 \end{vmatrix} - (-4)\begin{vmatrix} -3 & 5 \\ -1 & 4 \end{vmatrix} + 27\begin{vmatrix} -3 & 5 \\ 2 & -1 \end{vmatrix}$$

$$= 27(7) + 4(-7) + 27(-7)$$

$$= 189 - 28 - 189$$

$$= -28$$

$$D_y = \begin{vmatrix} 2 & 27 & 5 \\ 1 & -4 & -1 \\ 5 & 27 & 4 \end{vmatrix}$$

$$= 2\begin{vmatrix} -4 & -1 \\ 27 & 4 \end{vmatrix} - 1\begin{vmatrix} 27 & 5 \\ 27 & 4 \end{vmatrix} + 5\begin{vmatrix} 27 & 5 \\ -4 & -1 \end{vmatrix}$$

$$= 2(11) - 1(-27) + 5(-7)$$

$$= 22 + 27 - 35 = 14$$

$$= 14$$

Then,

$$x = \frac{D_x}{D} = \frac{-28}{-14} = 2,$$

and

$$y = \frac{D_y}{D} = \frac{14}{-14} = -1.$$

We substitute in the second equation to find z.

$$2 + 2(-1) - z = -4$$
$$2 - 2 - z = -4$$
$$-z = -4$$
$$z = 4$$

The solution is $(2, -1, 4)$.

25. $r - 2s + 3t = 6,$

$2r - s - t = -3,$

$r + s + t = 6$

We compute D, D_r, and D_s.

$$D = \begin{vmatrix} 1 & -2 & 3 \\ 2 & -1 & -1 \\ 1 & 1 & 1 \end{vmatrix}$$

$$= 1\begin{vmatrix} -1 & -1 \\ 1 & 1 \end{vmatrix} - 2\begin{vmatrix} -2 & 3 \\ 1 & 1 \end{vmatrix} + 1\begin{vmatrix} -2 & 3 \\ -1 & -1 \end{vmatrix}$$

$$= 1(0) - 2(-5) + 1(5)$$
$$= 0 + 10 + 5$$
$$= 15$$

$$D_r = \begin{vmatrix} 6 & -2 & 3 \\ -3 & -1 & -1 \\ 6 & 1 & 1 \end{vmatrix}$$

$$= 6\begin{vmatrix} -1 & -1 \\ 1 & 1 \end{vmatrix} - (-3)\begin{vmatrix} -2 & 3 \\ 1 & 1 \end{vmatrix} + 6\begin{vmatrix} -2 & 3 \\ -1 & -1 \end{vmatrix}$$

$$= 6(0) + 3(-5) + 6(5)$$
$$= 0 - 15 + 30$$
$$= 15$$

$$D_s = \begin{vmatrix} 1 & 6 & 3 \\ 2 & -3 & -1 \\ 1 & 6 & 1 \end{vmatrix}$$

$$= 1\begin{vmatrix} -3 & -1 \\ 6 & 1 \end{vmatrix} - 2\begin{vmatrix} 6 & 3 \\ 6 & 1 \end{vmatrix} + 1\begin{vmatrix} 6 & 3 \\ -3 & -1 \end{vmatrix}$$

$$= 1(3) - 2(-12) + 1(3)$$
$$= 3 + 24 + 3$$
$$= 30$$

Then,

$$r = \frac{D_r}{D} = \frac{15}{15} = 1,$$

and

$$s = \frac{D_s}{D} = \frac{30}{15} = 2.$$

Substitute in the third equation to find t.

$$1 + 2 + t = 6$$
$$3 + t = 6$$
$$t = 3$$

The solution is $(1, 2, 3)$.

27. *Writing Exercise*

29. $0.5x - 2.34 + 2.4x = 7.8x - 9$

$$2.9x - 2.34 = 7.8x - 9$$
$$6.66 = 4.9x$$
$$\frac{6.66}{4.9} = x$$
$$\frac{666}{490} = x$$
$$\frac{333}{245} = x$$

The solution is $\frac{333}{245}$.

31. ***Familiarize.*** We first make a drawing.

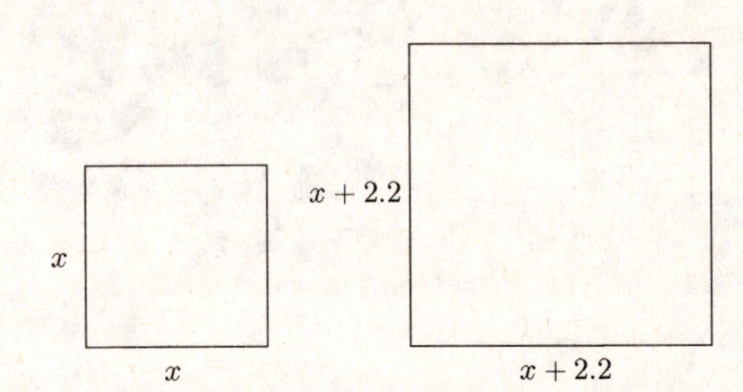

Let x represent the length of a side of the smaller square and $x + 2.2$ the length of a side of the larger square. The perimeter of the smaller square is $4x$. The perimeter of the larger square is $4(x + 2.2)$.

Translate.

The sum of the perimeters is 32.8 ft.

$$4x + 4(x + 2.2) = 32.8$$

Carry out. We solve the equation.

$$4x + 4x + 8.8 = 32.8$$
$$8x = 24$$
$$x = 3$$

Check. If $x = 3$ ft, then $x + 2.2 = 5.2$ ft. The perimeters are $4 \cdot 3$, or 12 ft, and $4(5.2)$, or 20.8 ft. The sum of the two perimeters is $12 + 20.8$, or 32.8 ft. The values check.

State. The wire should be cut into two pieces, one measuring 12 ft and the other 20.8 ft.

33. ***Familiarize.*** Let x represent the number of rolls of insulation required for the Mazzas' attic and let y represent the number of rolls required for the Kranepools' attic.

Translate.

Insulation for Mazzas' attic	is	three and a half	times	insulation for Kranepools' attic.
↓	↓	↓	↓	↓
x	$=$	3.5	$\cdot$	y

Total number of rolls is 36.

$x+y$ $= 36$

We have a system of equations:

$x = 3.5y$, (1)

$x + y = 36$ (2)

Carry out. We use the substitution method to solve the system of equations. First we substitute $3.5y$ for x in Equation (2).

$x + y = 36$ (2)

$3.5y + y = 36$ Substituting

$4.5y = 36$

$y = 8$

Now substitute 8 for y in Equation (1).

$x = 3.5(8) = 28$

Check. The number 28 is three and a half times 8. Also, the total number of rolls is 28 + 8, or 36. The answer checks.

State. The Mazzas' attic requires 28 rolls of insulation, and the Kranepools' attic requires 8 rolls.

35. *Writing Exercise*

37. $\begin{vmatrix} y & -2 \\ 4 & 3 \end{vmatrix} = 44$

$y \cdot 3 - 4(-2) = 44$ Evaluating the determinant

$3y + 8 = 44$

$3y = 36$

$y = 12$

39. $\begin{vmatrix} m+1 & -2 \\ m-2 & 1 \end{vmatrix} = 27$

$(m+1)(1) - (m-2)(-2) = 27$ Evaluating the determinant

$m + 1 + 2m - 4 = 27$

$3m = 30$

$m = 10$

Exercise Set 8.8

1. (b); see page 565 in the text.

3. (e); see page 565 in the text.

5. (h); see page 565 in the text.

7. (g); see page 568 in the text.

9. $C(x) = 45x + 300,000$ $R(x) = 65x$

a) $P(x) = R(x) - C(x)$

$= 65x - (45x + 300,000)$

$= 65x - 45x - 300,000$

$= 20x - 300,000$

b) To find the break-even point we solve the system

$R(x) = 65x,$

$C(x) = 45x + 300,000.$

Since $R(x) = C(x)$ at the break-even point, we can rewrite the system:

$R(x) = 65x,$ (1)

$R(x) = 45x + 300,000$ (2)

We solve using substitution.

$65x = 45x + 300,000$ Substituting $65x$ for $R(x)$ in (2)

$20x = 300,000$

$x = 15,000$

Thus, 15,000 units must be produced and sold in order to break even. The revenue will be $R(65) = 65(15,000)$, or 975,000. The break-even point is (15,000 units, \$975,000).

11. $C(x) = 10x + 120,000$ $R(x) = 60x$

a) $P(x) = R(x) - C(x)$

$= 60x - (10x + 120,000)$

$= 60x - 10x - 120,000$

$= 50x - 120,000$

b) Solve the system

$R(x) = 60x,$

$C(x) = 10x + 120,000.$

Since both $R(x)$ and $C(x)$ are in dollars and they are equal at the break-even point, we can rewrite the system:

$d = 60x,$ (1)

$d = 10x + 120,000$ (2)

We solve using substitution.

$60x = 10x + 120,000$ Substituting $60x$ for d in (2)

$50x = 120,000$

$x = 2400$

Thus, 2400 units must be produced and sold in order to break even. The revenue will be $R(2400) = 60 \cdot 2400 = 144,000$. The break-even point is (2400 units, \$144,000).

13. $C(x) = 40x + 22,500$ $R(x) = 85x$

a) $P(x) = R(x) - C(x)$

$= 85x - (40x + 22,500)$

$= 85x - 40x - 22,500$

$= 45x - 22,500$

b) Solve the system

$R(x) = 85x,$

$C(x) = 40x + 22,500.$

Since both $R(x)$ and $C(x)$ are in dollars and they are equal at the break-even point, we can rewrite the system:

$d = 85x, \quad (1)$

$d = 40x + 22,500 \quad (2)$

We solve using substitution.

$85x = 40x + 22,500$ Substituting $85x$ for d in (2)

$45x = 22,500$

$x = 500$

Thus, 500 units must be produced and sold in order to break even. The revenue will be $R(500) = 85 \cdot 500 = 42,500$. The break-even point is (500 units, \$42,500).

15. $C(x) = 22x + 16,000 \qquad R(x) = 40x$

a) $P(x) = R(x) - C(x)$

$= 40x - (22x + 16,000)$

$= 40x - 22x - 16,000$

$= 18x - 16,000$

b) Solve the system

$R(x) = 40x,$

$C(x) = 22x + 16,000.$

Since both $R(x)$ and $C(x)$ are in dollars and they are equal at the break-even point, we can rewrite the system:

$d = 40x, \quad (1)$

$d = 22x + 16,000 \quad (2)$

We solve using substitution.

$40x = 22x + 16,000$ Substituting $40x$ for d in (2)

$18x = 16,000$

$x \approx 889$ units

Thus, 889 units must be produced and sold in order to break even. The revenue will be $R(889) = 40 \cdot 889 = 35,560$. The break-even point is (889 units, \$35,560).

17. $C(x) = 75x + 100,000 \qquad R(x) = 125x$

a) $P(x) = R(x) - C(x)$

$= 125x - (75x + 100,000)$

$= 125x - 75x - 100,000$

$= 50x - 100,000$

b) Solve the system

$R(x) = 125x,$

$C(x) = 75x + 100,000.$

Since $R(x) = C(x)$ at the break-even point, we can rewrite the system:

$R(x) = 125x, \quad (1)$

$R(x) = 75x + 100,000 \quad (2)$

We solve using substitution.

$125x = 75x + 100,000$ Substituting $125x$ for $R(x)$ in (2)

$50x = 100,000$

$x = 2000$

To break even 2000 units must be produced and sold. The revenue will be $R(2000) = 125 \cdot 2000 = 250,000$. The break-even point is (2000 units, \$250,000).

19. $D(p) = 1000 - 10p,$

$S(p) = 230 + p$

Since both demand and supply are quantities, the system can be rewritten:

$q = 1000 - 10p, \quad (1)$

$q = 230 + p \quad (2)$

Substitute $1000 - 10p$ for q in (2) and solve.

$1000 - 10p = 230 + p$

$770 = 11p$

$70 = p$

The equilibrium price is \$70 per unit. To find the equilibrium quantity we substitute \$70 into either $D(p)$ or $S(p)$.

$D(70) = 1000 - 10 \cdot 70 = 1000 - 700 = 300$

The equilibrium quantity is 300 units.

The equilibrium point is (\$70, 300).

21. $D(p) = 760 - 13p,$

$S(p) = 430 + 2p$

Rewrite the system:

$q = 760 - 13p, \quad (1)$

$q = 430 + 2p \quad (2)$

Substitute $760 - 13p$ for q in (2) and solve.

$760 - 13p = 430 + 2p$

$330 = 15p$

$22 = p$

The equilibrium price is \$22 per unit.

To find the equilibrium quantity we substitute \$22 into either $D(p)$ or $S(p)$.

$S(22) = 430 + 2(22) = 430 + 44 = 474$

The equilibrium quantity is 474 units.

The equilibrium point is (\$22, 474).

23. $D(p) = 7500 - 25p,$

$S(p) = 6000 + 5p$

Rewrite the system:

$q = 7500 - 25p, \quad (1)$

$q = 6000 + 5p \quad (2)$

Substitute $7500 - 25p$ for q in (2) and solve.

$$7500 - 25p = 6000 + 5p$$
$$1500 = 30p$$
$$50 = p$$

The equilibrium price is \$50 per unit.

To find the equilibrium quantity we substitute \$50 into either $D(p)$ or $S(p)$.

$$D(50) = 7500 - 25(50) = 7500 - 1250 = 6250$$

The equilibrium quantity is 6250 units.

The equilibrium point is $(\$50, 6250)$.

25. $D(p) = 1600 - 53p,$

$S(p) = 320 + 75p$

Rewrite the system:

$q = 1600 - 53p,$ (1)

$q = 320 + 75p$ (2)

Substitute $1600 - 53p$ for q in (2) and solve.

$$1600 - 53p = 320 + 75p$$
$$1280 = 128p$$
$$10 = p$$

The equilibrium price is \$10 per unit.

To find the equilibrium quantity we substitute \$10 into either $D(p)$ or $S(p)$.

$$S(10) = 320 + 75(10) = 320 + 750 = 1070$$

The equilibrium quantity is 1070 units.

The equilibrium point is $(\$10, 1070)$.

27. a) $C(x) =$ Fixed costs + Variable costs

$C(x) = 125,300 + 450x,$

where x is the number of computers produced.

b) Each computer sells for \$800. The total revenue is 800 times the number of computers sold. We assume that all computers produced are sold.

$R(x) = 800x$

c) $P(x) = R(x) - C(x)$

$$P(x) = 800x - (125,300 + 450x)$$
$$= 800x - 125,300 - 450x$$
$$= 350x - 125,300$$

d) $P(x) = 350x - 125,300$

$$P(100) = 350(100) - 125,300$$
$$= 35,000 - 125,300$$
$$= -90,300$$

The company will realize a \$90,300 loss when 100 computers are produced and sold.

$$P(400) = 350(400) - 125,300$$
$$= 140,000 - 125,300$$
$$= 14,700$$

The company will realize a profit of \$14,700 from the production and sale of 400 computers.

e) Solve the system

$R(x) = 800x,$

$C(x) = 125,300 + 450x.$

Since both $R(x)$ and $C(x)$ are in dollars and they are equal at the break-even point, we can rewrite the system:

$d = 800x,$ (1)

$d = 125,300 + 450x$ (2)

We solve using substitution.

$800x = 125,300 + 450x$ Substituting $800x$ for d in (2)

$$350x = 125,300$$
$$x = 358$$

The firm will break even if it produces and sells 358 computers and takes in a total of $R(358) = 800 \cdot 358 = \$286,400$ in revenue. Thus, the break-even point is (358 computers, \$286,400).

29. a) $C(x) =$ Fixed costs + Variable costs

$C(x) = 16,404 + 6x,$

where x is the number of caps produced, in dozens.

b) Each dozen caps sell for \$18. The total revenue is 18 times the number of caps sold, in dozens. We assume that all caps produced are sold.

$R(x) = 18x$

c) $P(x) = R(x) - C(x)$

$$P(x) = 18x - (16,404 + 6x)$$
$$= 18x - 16,404 - 6x$$
$$= 12x - 16,404$$

d)
$$P(3000) = 12(3000) - 16,404$$
$$= 36,000 - 16,404$$
$$= 19,596$$

The company will realize a profit of \$19,596 when 3000 dozen caps are produced and sold.

$$P(1000) = 12(1000) - 16,404$$
$$= 12,000 - 16,404$$
$$= -4404$$

The company will realize a \$4404 loss when 1000 dozen caps are produced and sold.

e) Solve the system

$R(x) = 18x,$

$C(x) = 16,404 + 6x.$

Since both $R(x)$ and $C(x)$ are in dollars and they are equal at the break-even point, we can rewrite the system:

$d = 18x,$ (1)

$d = 16,404 + 6x$ (2)

We solve using substitution.

$18x = 16,404 + 6x$ Substituting $18x$ for d in (2)

$12x = 16,404$

$x = 1367$

The firm will break even if it produces and sells 1367 dozen caps and takes in a total of $R(1367) = 18 \cdot 1367 = \$24,606$ in revenue. Thus, the break-even point is (1367 dozen caps, $24,606).

31. *Writing Exercise*

33. $3x - 9 = 27$

$3x = 36$ Adding 9 to both sides

$x = 12$ Dividing both sides by 3

The solution is 12.

35. $4x - 5 = 7x - 13$

$-5 = 3x - 13$ Subtracting $4x$ from both sides

$8 = 3x$ Adding 13 to both sides

$\frac{8}{3} = x$ Dividing both sides by 3

The solution is $\frac{8}{3}$.

37. $7 - 2(x - 8) = 14$

$7 - 2x + 16 = 14$ Removing parentheses

$-2x + 23 = 14$ Collecting like terms

$-2x = -9$ Subtracting 23 from both sides

$x = \frac{9}{2}$ Dividing both sides by -2

The solution is $\frac{9}{2}$.

39. *Writing Exercise*

41. The supply function contains the points ($2, 100) and ($8, 500). We find its equation:

$$m = \frac{500 - 100}{8 - 2} = \frac{400}{6} = \frac{200}{3}$$

$y - y_1 = m(x - x_1)$ Point-slope form

$$y - 100 = \frac{200}{3}(x - 2)$$

$$y - 100 = \frac{200}{3}x - \frac{400}{3}$$

$$y = \frac{200}{3}x - \frac{100}{3}$$

We can equivalently express supply S as a function of price p:

$$S(p) = \frac{200}{3}p - \frac{100}{3}$$

The demand function contains the points ($1, 500) and ($9, 100). We find its equation:

$$m = \frac{100 - 500}{9 - 1} = \frac{-400}{8} = -50$$

$$y - y_1 = m(x - x_1)$$

$$y - 500 = -50(x - 1)$$

$$y - 500 = -50x + 50$$

$$y = -50x + 550$$

We can equivalently express demand D as a function of price p:

$$D(p) = -50p + 550$$

We have a system of equations

$$S(p) = \frac{200}{3}p - \frac{100}{3},$$

$$D(p) = -50p + 550.$$

Rewrite the system:

$$q = \frac{200}{3}p - \frac{100}{3}, \quad (1)$$

$$q = -50p + 550 \quad (2)$$

Substitute $\frac{200}{3}p - \frac{100}{3}$ for q in (2) and solve.

$$\frac{200}{3}p - \frac{100}{3} = -50p + 550$$

$200p - 100 = -150p + 1650$ Multiplying by 3 to clear fractions

$$350p - 100 = 1650$$

$$350p = 1750$$

$$p = 5$$

The equilibrium price is $5 per unit.

To find the equilibrium quantity, we substitute $5 into either $S(p)$ or $D(p)$.

$$D(5) = -50(5) + 550 = -250 + 550 = 300$$

The equilibrium quantity is 300 yo-yo's.

The equilibrium point is ($5, 300 yo-yo's).

43. a) Use a graphing calculator to find the first coordinate of the point of intersection of $y_1 = -14.97x + 987.35$ and $y_2 = 98.55x - 5.13$, to the nearest hundredth. It is 8.74, so the price per unit that should be charged is $8.74.

b) Use a graphing calculator to find the first coordinate of the point of intersection of $y_1 = 87,985 + 5.15x$ and $y_2 = 8.74x$. It is about 24,508.4, so 24,509 units must be sold in order to break even.

Chapter 9

Inequalities and Problem Solving

Exercise Set 9.1

1. If we add 7 to both sides of the inequality $x-7>-2$, we get the inequality $x>5$ so these are equivalent inequalities.

3. If we add $3x$ to both sides of the equation $5x+7=6-3x$, we get the equation $8x+7=6$, so these are equivalent equations.

5. The solution set of $-4t\le 12$ is $\{t|t\ge -3\}$ and the solution set of $t\le -3$ is $\{t|t\le -3\}$. The solution sets are not the same, so the inequalities are not equivalent.

7. The expressions are equivalent by the distributive law.

9. The solution set of $-\frac{1}{2}x<7$ is $\{x|x>-14\}$ and the solution set of $x>14$ is $\{x|x>14\}$. The solution sets are not the same, so the inequalities are not equivalent.

11. $y<6$

Graph: The solutions consist of all real numbers less than 6, so we shade all numbers to the left of 5 and use an open circle at 6 to indicate that it is not a solution.

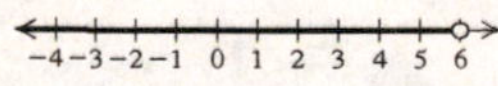

Set builder notation: $\{y|y<6\}$

Interval notation: $(-\infty,6)$

13. $x\ge -4$

Graph: We shade all numbers to the right of -4 and use a solid endpoint at -4 to indicate that it is also a solution.

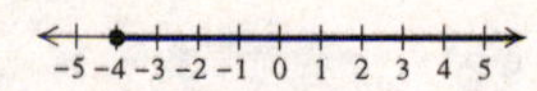

Set builder notation: $\{x|x\ge -4\}$

Interval notation: $[-4,\infty)$

15. $t>-3$

Graph: We shade all numbers to the right of -3 and use an open circle at -3 to indicate that it is not a solution.

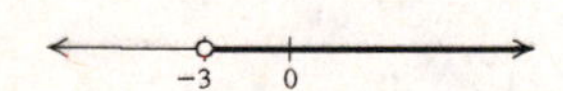

Set builder notation: $\{t|t>-3\}$

Interval notation: $(-3,\infty)$

17. $x\le -7$

Graph: We shade all numbers to the left of -7 and use a solid endpoint at -7 to indicate that it is also a solution.

−7 −6 −5 −4 −3 −2 −1 0 1 2 3

Set builder notation: $\{x|x\le -7\}$

Interval notation: $(-\infty,-7]$

19.
$$\begin{aligned} y-9&>-18 \\ y-9+9&>-18+9 \quad \text{Adding 9} \\ y&>-9 \end{aligned}$$

The solution set is $\{y|y>-9\}$, or $(-9,\infty)$.

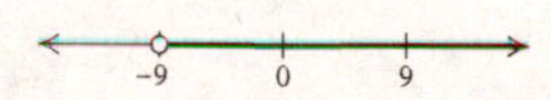

21.
$$\begin{aligned} y-20&\le -6 \\ y-20+20&\le -6+20 \quad \text{Adding 20} \\ y&\le 14 \end{aligned}$$

The solution set is $\{y|y\le 14\}$, or $(-\infty,14]$.

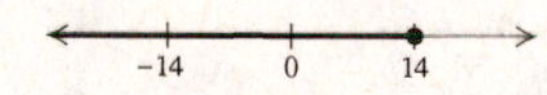

23.
$$\begin{aligned} 9t&<-81 \\ \frac{1}{9}\cdot 9t&<\frac{1}{9}(-81) \quad \text{Multiplying by } \frac{1}{9} \\ t&<-9 \end{aligned}$$

The solution set is $\{t|t<-9\}$, or $(-\infty,-9)$.

−9 0 9

25.
$$\begin{aligned} -9x&\ge -8.1 \\ -\frac{1}{9}(-9x)&\le -\frac{1}{9}(-8.1) \quad \text{Multiplying by } -\frac{1}{9} \text{ and reversing the inequality symbol} \\ x&\le \frac{8.1}{9} \\ x&\le 0.9 \end{aligned}$$

The solution set is $\{x|x\le 0.9\}$, or $(-\infty,0.9]$.

27.
$$\begin{aligned} -\frac{3}{4}x&\ge -\frac{5}{8} \\ -\frac{4}{3}\left(-\frac{3}{4}x\right)&\le -\frac{4}{3}\left(-\frac{5}{8}\right) \quad \text{Multiplying by } -\frac{4}{3} \text{ and reversing the inequality symbol} \\ x&\le \frac{20}{24} \\ x&\le \frac{5}{6} \end{aligned}$$

The solution set is $\left\{x\middle|x\le \frac{5}{6}\right\}$, or $\left(-\infty,\frac{5}{6}\right]$.

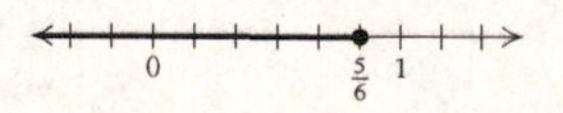

29. $\dfrac{2x+7}{5} < -9$

$5 \cdot \dfrac{2x+7}{5} < 5(-9)$ Multiplying by 5

$2x + 7 < -45$

$2x < -52$ Adding -7

$x < -26$ Dividing by 2

The solution set is $\{x|x < -26\}$, or $(-\infty, -26)$.

−26 0 26

31. $\dfrac{3t-7}{-4} \leq 5$

$-4 \cdot \dfrac{3t-7}{-4} \geq -4 \cdot 5$ Multiplying by -4 and reversing the inequality symbol

$3t - 7 \geq -20$

$3t \geq -13$ Adding 7

$t \geq -\dfrac{13}{3}$ Dividing by 3

The solution set is $\left\{t\middle|t \geq -\dfrac{13}{3}\right\}$, or $\left[-\dfrac{13}{3}, \infty\right)$.

$-\frac{13}{3}$ −4 0 4

33. $f(x) = 2x + 1$, $g(x) = x + 7$

$f(x) \geq g(x)$

$2x + 1 \geq x + 7$

$x + 1 \geq 7$ Adding $-x$

$x \geq 6$ Adding -1

The solution set is $\{x|x \geq 6\}$, or $[6, \infty)$.

−6 0 6

35. $f(x) = 7 - 3x$, $g(x) = 2x - 3$

$f(x) \leq g(x)$

$7 - 3x \leq 2x - 3$

$7 - 5x \leq -3$ Adding $-2x$

$-5x \leq -10$ Adding -7

$x \geq 2$ Multiplying by $-\dfrac{1}{5}$ and reversing the inequality symbol

The solution set is $\{x|x \geq 2\}$, or $[2, \infty)$.

−2 0 2

37. $f(x) = 2x - 7$, $g(x) = 5x - 9$

$f(x) \leq g(x)$

$2x - 7 < 5x - 9$

$-3x - 7 < -9$ Adding $-5x$

$-3x < -2$ Adding 7

$x > \dfrac{2}{3}$ Dividing by -3

The solution set is $\left\{x\middle|x > \dfrac{2}{3}\right\}$, or $\left(\dfrac{2}{3}, \infty\right)$.

$-\frac{1}{3}$ 0 $\frac{1}{3}$ $\frac{2}{3}$ 1 $\frac{4}{3}$

39. $f(x) = \dfrac{3}{8} + 2x$, $g(x) = 3x - \dfrac{1}{8}$

$g(x) \geq f(x)$

$3x - \dfrac{1}{8} \geq \dfrac{3}{8} + 2x$

$x - \dfrac{1}{8} = \dfrac{3}{8}$ Adding $-2x$

$x \geq \dfrac{1}{2}$ Adding $\dfrac{1}{8}$

The solution set is $\left\{x\middle|x \geq \dfrac{1}{2}\right\}$, or $\left[\dfrac{1}{2}, \infty\right)$.

0 $\frac{1}{2}$ 1

41. $4(3y - 2) \geq 9(2y + 5)$

$12y - 8 \geq 18y + 45$

$-6y - 8 \geq 45$

$-6y \geq 53$

$y \leq -\dfrac{53}{6}$

The solution set is $\left\{y\middle|y \leq -\dfrac{53}{6}\right\}$, or $\left(-\infty, -\dfrac{53}{6}\right]$.

43. $5(t - 3) + 4t < 2(7 + 2t)$

$5t - 15 + 4t < 14 + 4t$

$9t - 15 < 14 + 4t$

$5t - 15 < 14$

$5t < 29$

$t < \dfrac{29}{5}$

The solution set is $\left\{t\middle|t < \dfrac{29}{5}\right\}$, or $\left(-\infty, \dfrac{29}{5}\right)$.

45. $5[3m - (m + 4)] > -2(m - 4)$

$5(3m - m - 4) > -2(m - 4)$

$5(2m - 4) > -2(m - 4)$

$10m - 20 > -2m + 8$

$12m - 20 > 8$

$12m > 28$

$m > \dfrac{28}{12}$

$m > \dfrac{7}{3}$

The solution set is $\left\{m\middle|m > \dfrac{7}{3}\right\}$, or $\left(\dfrac{7}{3}, \infty\right)$.

47. $19 - (2x + 3) \leq 2(x + 3) + x$

$$19 - 2x - 3 \leq 2x + 6 + x$$
$$16 - 2x \leq 3x + 6$$
$$16 - 5x \leq 6$$
$$-5x \leq -10$$
$$x \geq 2$$

The solution set is $\{x|x \geq 2\}$, or $[2, \infty)$.

49. $\frac{1}{4}(8y + 4) - 17 < -\frac{1}{2}(4y - 8)$

$$2y + 1 - 17 < -2y + 4$$
$$2y - 16 < -2y + 4$$
$$4y - 16 < 4$$
$$4y < 20$$
$$y < 5$$

The solution set is $\{y|y < 5\}$, or $(-\infty, 5)$.

51. $2[8 - 4(3 - x)] - 2 \geq 8[2(4x - 3) + 7] - 50$

$$2[8 - 12 + 4x] - 2 \geq 8[8x - 6 + 7] - 50$$
$$2[-4 + 4x] - 2 \geq 8[8x + 1] - 50$$
$$-8 + 8x - 2 \geq 64x + 8 - 50$$
$$8x - 10 \geq 64x - 42$$
$$-56x - 10 \geq -42$$
$$-56x \geq -32$$
$$x \leq \frac{32}{56}$$
$$x \leq \frac{4}{7}$$

The solution set is $\left\{x\middle|x \leq \frac{4}{7}\right\}$, or $\left(-\infty, \frac{4}{7}\right]$.

53. ***Familiarize.*** Let m = the number of peak local minutes used. Then the charge for the minutes used is $\$0.022m$ and the total monthly charge is $\$13.55 + \$0.022m$.

Translate. We write an inequality stating that the monthly charge is at least \$39.40.

$$13.55 + 0.022m \geq 39.40$$

Carry out.

$$13.55 + 0.022m \geq 39.40$$
$$0.022m \geq 25.85$$
$$m \geq 1175$$

Check. We can do a partial check by substituting a value for m less than 1175. When $m = 1174$, the monthly charge is $\$13.55 + \$0.022(1174) \approx \$39.38$. This is less than the maximum charge of \$39.40. We cannot check all possible values for m, so we stop here.

State. A customer must speak on the phone for 1175 local peak minutes or more if the maximum charge is to apply.

55. ***Familiarize.*** Let c = the number of checks per month. Then the Anywhere plan will cost $\$0.20c$ per month and the Acu-checking plan will cost $\$2 + \$0.12c$ per month.

Translate. We write an inequality stating that the Acu-checking plan costs less than the Anywhere plan.

$$2 + 0.12c < 0.20c$$

Carry out.

$$2 + 0.12c < 0.20c$$
$$2 < 0.08c$$
$$25 < c$$

Check. We can do a partial check by substituting a value for c less than 25 and a value for c greater than 25. When $c = 24$, the Acu-checking plan costs $\$2 + \$0.12(24)$, or \$4.88, and the Anywhere plan costs \$0.20(24), or \$4.80, so the Anywhere plan is less expensive. When $c = 26$, the Acu-checking plan costs $\$2 + \$0.12(26)$, or \$5.12, and the Anywhere plan costs \$0.20(26), or \$5.20, so Acu-checking is less expensive. We cannot check all possible values for c, so we stop here.

State. The Acu-checking plan costs less for more than 25 checks per month.

57. ***Familiarize.*** We list the given information in a table.

Plan A: Monthly Income	Plan B: Monthly Income
\$400 salary	\$610
8% of sales	5% of sales
Total: 400 + 8% of sales	Total: 610 + 5% of sales

Suppose Toni had gross sales of \$5000 one month. Then under plan A she would earn

$$\$400 + 0.08(\$5000), \text{ or } \$800.$$

Under plan B she would earn

$$\$610 + 0.05(\$5000), \text{ or } \$860.$$

This shows that, for gross sales of \$5000, plan B is better.

If Toni had gross sales of \$10,000 one month, then under plan A she would earn

$$\$400 + 0.08(\$10,000), \text{ or } \$1200.$$

Under plan B she would earn

$$\$610 + 0.05(\$10,000), \text{ or } \$1110.$$

This shows that, for gross sales of \$10,000, plan A is better. To determine all values for which plan A is better we solve an inequality.

Translate.

$$\underbrace{\text{Income from plan A}}_{400 + 0.08s} \quad \underbrace{\text{is greater than}}_{>} \quad \underbrace{\text{income from plan B.}}_{610 + 0.05s}$$

Carry out.

$$400 + 0.08s > 610 + 0.05s$$
$$400 + 0.03s > 610$$
$$0.03s > 210$$
$$s > 7000$$

Check. For $s = \$7000$, the income from plan A is

$$\$400 + 0.08(\$7000), \text{ or } \$960$$

and the income from plan B is

$$\$610 + 0.05(\$7000), \text{ or } \$960.$$

This shows that for sales of $7000 Toni's income is the same from each plan. In the Familiarize step we shows that, for a value less than $7000, plan B is better and, for a value greater than $7000, plan A is better. Since we cannot check all possible values, we stop here.

State. Toni should select plan A for gross sales greater than $7000.

59. ***Familiarize.*** Let p = the number of people in the party. Then plan A will cost $\$30p$ and plan B will cost $\$1300 + \$20(p - 25)$.

Translate. We write an inequality stating that the cost of plan B is less than the cost of plan A.

$$1300 + 20(p - 25) < 30p$$

Carry out.

$$1300 + 20(p - 25) < 30p$$
$$1300 + 20p - 500 < 30p$$
$$800 + 20p < 30p$$
$$800 < 10p$$
$$80 < p$$

Check. We can do a partial check by substituting a value for p less than 80 and a value for p greater than 80. When $p = 79$, plan A costs $\$30 \cdot 79$, or $2370, and plan B costs $\$1300 + \$20(79 - 25)$, or $2380, so plan A costs less. When $p = 81$, plan A costs $\$30 \cdot 81$, or $2430, and plan B costs $\$1300 + \$20(81 - 25)$, or $2420, so plan B costs less. We cannot check all possible values of p, so we stop here.

State. Plan B costs less for parties of more than 80 people.

61. ***Familiarize.*** Let n = the number of people who attend. Then the total receipts are $\$6 \cdot n$, and the amount of receipts over $750 is $\$6 \cdot n - \$750$. The band will receive $750 plus 15% of $\$6 \cdot n - \$750$, or $\$750 + 0.15(\$6 \cdot n - \$750)$.

Translate. We write an inequality stating that the amount the band receives is at least $1200.

$$750 + 0.15(6n - 750) \geq 1200$$

Carry out.

$$750 + 0.15(6n - 750) \geq 1200$$
$$750 + 0.9n - 112.5 \geq 1200$$
$$0.9n + 637.5 \geq 1200$$
$$0.9n \geq 562.5$$
$$n \geq 625$$

Check. When $n = 625$, the band receives $\$75 + 0.15(\$6 \cdot 625 - \$750)$, or $\$750 + 0.15(\$3000)$, or $\$750 + \$450$, or $1200. When $n = 626$, the band receives $\$750 + 0.15(\$6 \cdot 626 - \$750)$, or $\$750 + 0.15(\$3006)$, or $\$750 + \$450.90$, or $1200.90. Since the band receives exactly $1200 when 625 people attend and more than $1200 when 626 people attend, we have performed a partial check. We cannot check all possible solutions, so we stop here.

State. At least 625 people must attend in order for the band to receive at least $1200.

63. a) ***Familiarize.*** Find the values of x for which $R(x) < C(x)$.

Translate.

$$26x < 90,000 + 15x$$

Carry out.

$$11x < 90,000$$
$$x < 8181\frac{9}{11}$$

Check. $R\left(8181\frac{9}{11}\right) = \$212,727.27 = C\left(8181\frac{9}{11}\right)$.

Calculate $R(x)$ and $C(x)$ for some x greater than $8181\frac{9}{11}$ and for some x less than $8181\frac{9}{11}$.

Suppose $x = 8200$:

$$R(x) = 26(8200) = 213,200 \quad \text{and}$$
$$C(x) = 90,000 + 15(8200) = 213,000.$$

In this case $R(x) > C(x)$.

Suppose $x = 8000$:

$$R(x) = 26(8000) = 208,000 \quad \text{and}$$
$$C(x) = 90,000 + 15(8000) = 210,000.$$

In this case $R(x) < C(x)$.

Then for $x < 8181\frac{9}{11}$, $R(x) < C(x)$.

State. We will state the result in terms of integers, since the company cannot sell a fraction of a lamp. For 8181 or fewer lamps the company loses money.

b) Our check in part a) shows that for $x > 8181\frac{9}{11}$, $R(x) > C(x)$ and the company makes a profit. Again, we will state the result in terms of an integer. For more than 8181 lamps the company makes money.

65. *Writing Exercise*

67. $f(x) = \dfrac{3}{x - 2}$

Since $\dfrac{3}{x - 2}$ cannot be computed when $x - 2$ is 0, we solve an equation:

$$x - 2 = 0$$
$$x = 2$$

The domain is $\{x|x \text{ is a real number } and\ x \neq 2\}$.

69. $f(x) = \dfrac{5x}{7 - 2x}$

Since $\dfrac{5x}{7 - 2x}$ cannot be computed when $7 - 2x$ is 0, we solve an equation:

$$7 - 2x = 0$$
$$7 = 2x$$
$$\frac{7}{2} = x$$

The domain is $\left\{x \middle| x \text{ is a real number } and\ x \neq \frac{7}{2}\right\}$.

71. $9x - 2(x - 5) = 9x - 2x + 10 = 7x + 10$

73. *Writing Exercise*

75.
$$\begin{aligned} 3ax + 2x &\geq 5ax - 4 \\ 2x - 2ax &\geq -4 \\ 2x(1 - a) &\geq -4 \\ x(1 - a) &\geq -2 \\ x &\leq -\frac{2}{1-a}, \text{ or } \frac{2}{a-1} \end{aligned}$$

We reversed the inequality symbol when we divided because when $a > 1$, then $1 - a < 0$.

The solution set is $\left\{x \middle| x \leq \frac{2}{a-1}\right\}$.

77.
$$\begin{aligned} a(by - 2) &\geq b(2y + 5) \\ aby - 2a &\geq 2by + 5b \\ aby - 2by &\geq 2a + 5b \\ y(ab - 2b) &\geq 2a + 5b \\ y &\geq \frac{2a + 5b}{ab - 2b}, \text{ or } \frac{2a + 5b}{b(a - 2)} \end{aligned}$$

The inequality symbol remained unchanged when we divided because when $a > 2$ and $b > 0$, then $ab - 2b > 0$.

The solution set is $\left\{y \middle| y \geq \frac{2a + 5b}{b(a - 2)}\right\}$.

79.
$$\begin{aligned} c(2 - 5x) + dx &> m(4 + 2x) \\ 2c - 5cx + dx &> 4m + 2mx \\ -5cx + dx - 2mx &> 4m - 2c \\ x(-5c + d - 2m) &> 4m - 2c \\ x[d - (5c + 2m)] &> 4m - 2c \\ x &> \frac{4m - 2c}{d - (5c + 2m)} \end{aligned}$$

The inequality symbol remained unchanged when we divided because when $5c + 2m < d$, then $d - (5c + 2m) > 0$.

The solution set is $\left\{x \middle| x > \frac{4m - 2c}{d - (5c + 2m)}\right\}$.

81. False. If $a = 2$, $b = 3$, $c = 4$, and $d = 5$, then $2 < 3$ and $4 < 5$ but $2 - 4 = 3 - 5$.

83. *Writing Exercise*

85. $x + 5 \leq 5 + x$

$5 \leq 5$ Subtracting x

We get an inequality that is true for all real numbers x. Thus the solution set is all real numbers.

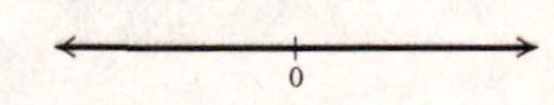

87. $0^2 = 0$, $x^2 > 0$ for $x \neq 0$

The solution is $\{x | x \text{ is a real number and } x \neq 0\}$.

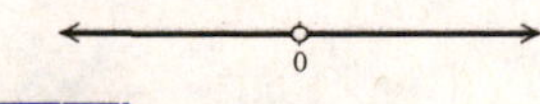

89.

Exercise Set 9.2

1. h

3. f

5. e

7. b

9. c

11. $\{5, 9, 11\} \cap \{9, 11, 18\}$

The numbers 9 and 11 are common to both sets, so the intersection is $\{9, 11\}$.

13. $\{0, 5, 10, 15\} \cup \{5, 15, 20\}$

The numbers in either or both sets are 0, 5, 10, 15, and 20, so the union is $\{0, 5, 10, 15, 20\}$.

15. {a, b, c, d, e, f} ∩ {b, d, f}

The letters b, d, and f are common to both sets, so the intersection is {b, d, f}.

17. {r, s, t} ∪ {r, u, t, s, v}

The letters in either or both sets are r, s, t, u, and v, so the union is {r, s, t, u, v}.

19. $\{3, 6, 9, 12\} \cap \{5, 10, 15\}$

There are no numbers common to both sets, so the solution set has no members. It is $\emptyset$.

21. $\{3, 5, 7\} \cup \emptyset$

The numbers in either or both sets are 3, 5, and 7, so the union is $\{3, 5, 7\}$.

23. $3 < x < 7$

This inequality is an abbreviation for the conjunction $3 < x$ *and* $x < 7$. The graph is the intersection of two separate solution sets: $\{x | 3 < x\} \cap \{x | x < 7\} = \{x | 3 < x < 7\}$.

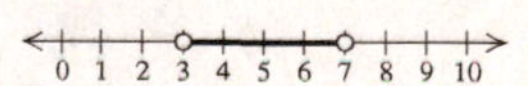

Interval notation: $(3, 7)$

25. $-6 \leq y \leq -2$

This inequality is an abbreviation for the conjunction $-6 \leq y$ *and* $y \leq -2$.

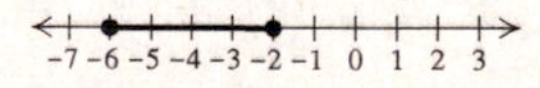

Interval notation: $[-6, -2]$

27. $x < -1$ *or* $x > 4$

The graph of this disjunction is the union of the graphs of the individual solution sets $\{x | x < -1\}$ and $\{x | x > 4\}$.

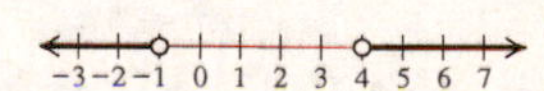

Interval notation: $(-\infty, -2) \cup (3, \infty)$

29. $x \le -2$ *or* $x > 1$

Interval notation: $(-\infty, -2] \cup (1, \infty)$

31. $-4 \le -x < 2$

$4 \ge x > -2$ Multiplying by -1 and reversing the inequality symbols

$-2 < x \le 4$ Rewriting

Interval notation: $(-2, 4]$

33. $x > -2$ *and* $x < 4$

This conjunction can be abbreviated as $-2 < x < 4$.

Interval notation: $(-2, 4)$

35. $5 > a$ *or* $a > 7$

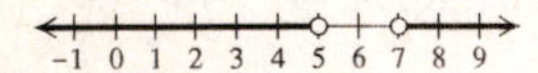

Interval notation: $(-\infty, 5) \cup (7, \infty)$

37. $x \ge 5$ *or* $-x \ge 4$

Multiplying the second inequality by -1 and reversing the inequality symbols, we get $x \ge 5$ *or* $x \le -4$.

Interval notation: $(-\infty - 4] \cup [5, \infty)$

39. $7 > y$ *and* $y \ge -3$

This conjunction can be abbreviated as $-3 \le y < 7$.

Interval notation: $[-3, 7)$

41. $x < 7$ *and* $x \ge 3$

This conjunction can be abbreviated as $3 \le x < 7$.

Interval notation: $[3, 7)$

43. $t < 2$ *or* $t < 5$

Observe that every number that is less than 2 is also less than 5. Then $t < 2$ *or* $t < 5$ is equivalent to $t < 5$ and the graph of this disjunction is the set $\{t | t < 5\}$.

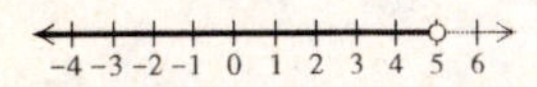

Interval notation: $(-\infty, 5)$

45. $-2 < t + 1 < 8$

$-2 - 1 < t < 8 - 1$

$-3 < t < 7$

The solution set is $\{t | -3 < t < 7\}$, or $(-3, 7)$.

47. $2 < x + 3$ *and* $x + 1 \le 5$

$-1 < x$ *and* $x \le 4$

We can abbreviate the answer as $-1 < x \le 4$. The solution set is $\{x | -1 < x \le 4\}$, or $(-1, 4]$.

49. $-7 \le 2a - 3$ *and* $3a + 1 < 7$

$-4 \le 2a$ *and* $3a < 6$

$-2 \le a$ *and* $a < 2$

We can abbreviate the answer as $-2 \le a < 2$. The solution set is $\{a | -2 \le a < 2\}$, or $[-2, 2)$.

51. $x + 7 \le -2$ *or* $x + 7 \ge -3$

Observe that any real number is either less than or equal to -2 or greater than or equal to -3. Then the solution set is $\{x | x \text{ is a real number}\}$, or $(-\infty, \infty)$.

53. $5 > \dfrac{x-3}{4} > 1$

$20 > x - 3 > 4$ Multiplying by 4

$23 > x > 7$, or

$7 < x < 23$

The solution set is $\{x | 7 < x < 23\}$, or $(7, 23)$.

55. $-7 \le 4x + 5 \le 13$

$-12 \le 4x \le 8$

$-3 \le x \le 2$

The solution set is $\{x | -3 \le x \le 2\}$, or $[-3, 2]$.

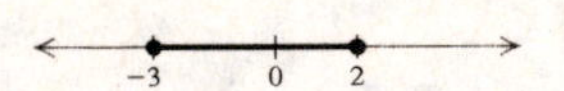

57. $2 \le 3x - 1 \le 8$

$3 \le 3x \le 9$

$1 \le x \le 3$

The solution set is $\{x | 1 \le x \le 3\}$, or $[1, 3]$.

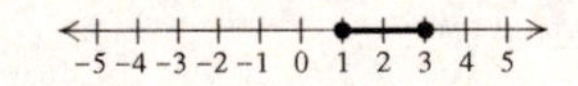

59. $-21 \le -2x - 7 < 0$

$-14 \le -2x < 7$

$7 \ge x > -\dfrac{7}{2}$, or

$-\dfrac{7}{2} < x \le 7$

The solution set is $\left\{x \middle| -\dfrac{7}{2} < x \le 7\right\}$, or $\left(-\dfrac{7}{2}, 7\right]$.

61. $3x - 1 \leq 2$ *or* $3x - 1 \geq 8$

$3x \leq 3$ *or* $3x \geq 9$

$x \leq 1$ *or* $x \geq 3$

The solution set is $\{x|x \leq 1 \text{ or } x \geq 3\}$, or $(-\infty, 1] \cup [3, \infty)$.

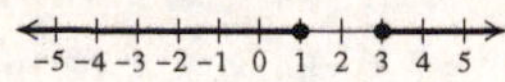

63. $2x - 7 < -3$ *or* $2x - 7 > 5$

$2x < 4$ *or* $2x > 12$

$x < 2$ *or* $x > 6$

The solution set is $\{x|x < 2 \text{ or } x > 6\}$, or $(-\infty, 2) \cup (6, \infty)$.

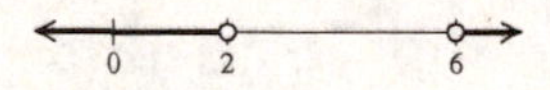

65. $6 > 2a - 1$ *or* $-4 \leq -3a + 2$

$7 > 2a$ *or* $-6 \leq -3a$

$\frac{7}{2} > a$ *or* $2 \geq a$

The solution set is $\left\{a\middle|\frac{7}{2} > a\right\} \cup \{a|2 \geq a\} =$

$\left\{a\middle|\frac{7}{2} > a\right\}$, or $\left\{a\middle|a < \frac{7}{2}\right\}$, or $\left(-\infty, \frac{7}{2}\right)$.

67. $a + 3 < -2$ *and* $3a - 4 < 8$

$a < -5$ *and* $3a < 12$

$a < -5$ *and* $a < 4$

The solution set is $\{a|a < -5\} \cap \{a|a < 4\} =$ $\{a|a < -5\}$, or $(-\infty, -5)$.

69. $3x + 2 < 2$ *or* $4 - 2x < 14$

$3x < 0$ *or* $-2x < 10$

$x < 0$ *or* $x > -5$

The solution set is $\{x|x < 0\} \cup \{x|x > -5\} =$ the set of all real numbers, or $(-\infty, \infty)$.

71. $2t - 7 \leq 5$ *or* $5 - 2t > 3$

$2t \leq 12$ *or* $-2t > -2$

$t \leq 6$ *or* $t < 1$

The solution set is $\{t|t \leq 6\} \cup \{t|t < 1\} = \{t|t \leq 6\}$, or $(-\infty, 6]$.

73. $f(x) = \dfrac{9}{x + 8}$

$f(x)$ cannot be computed when the denominator is 0. Since $x + 8 = 0$ is equivalent to $x = -8$, we have Domain of $f = \{x|x \text{ is a real number } \textit{and } x \neq -8\} = (-\infty, -8) \cup (-8, \infty)$.

75. $f(x) = \sqrt{x - 6}$

The expression $\sqrt{x - 6}$ is not a real number when $x - 6$ is negative. Thus, the domain of f is the set of all x-values for which $x - 6 \geq 0$. Since $x - 6 \geq 0$ is equivalent to $x \geq 6$, we have Domain of $f = [6, \infty)$.

77. $f(x) = \dfrac{x + 3}{2x - 8}$

$f(x)$ cannot be computed when the denominator is 0. Since $2x - 8 = 0$ is equivalent to $x = 4$, we have Domain of $f = \{x|x \text{ is a real number and } x \neq 4\}$, or $(-\infty, 4) \cup (4, \infty)$.

79. $f(x) = \sqrt{2x + 7}$

The expression $\sqrt{2x + 7}$ is not a real number when $2x + 7$ is negative. Thus, the domain of f is the set of all x-values for which $2x + 7 \geq 0$. Since $2x + 7 \geq 0$ is equivalent to $x \geq -\frac{7}{2}$, we have Domain of $f = \left[-\frac{7}{2}, \infty\right)$.

81. $f(x) = \sqrt{8 - 2x}$

The expression $\sqrt{8 - 2x}$ is not a real number when $8 - 2x$ is negative. Thus, the domain of f is the set of all x-values for which $8 - 2x \geq 0$. Since $8 - 2x \geq 0$ is equivalent to $x \leq 4$, we have Domain of $f = (-\infty, 4]$.

83. *Writing Exercise*

85. Graph: $y = 5$

The graph of any constant function $y = c$ is a horizontal line that crosses the vertical axis at $(0, c)$. Thus, the graph of $y = 5$ is a horizontal line that crosses the vertical axis at $(0, 5)$.

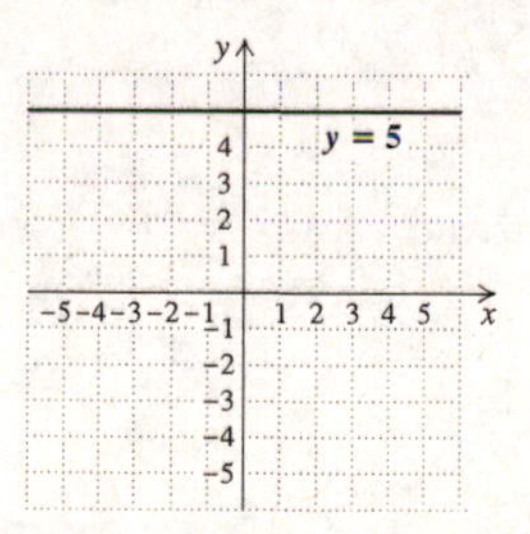

87. Graph $f(x) = |x|$

We make a table of values, plot points, and draw the graph.

x	$f(x)$
-5	5
-2	2
0	0
1	1
4	4

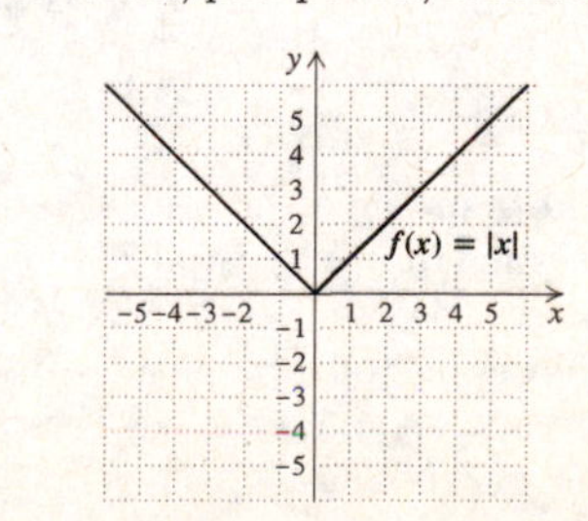

89. Graph both equations.

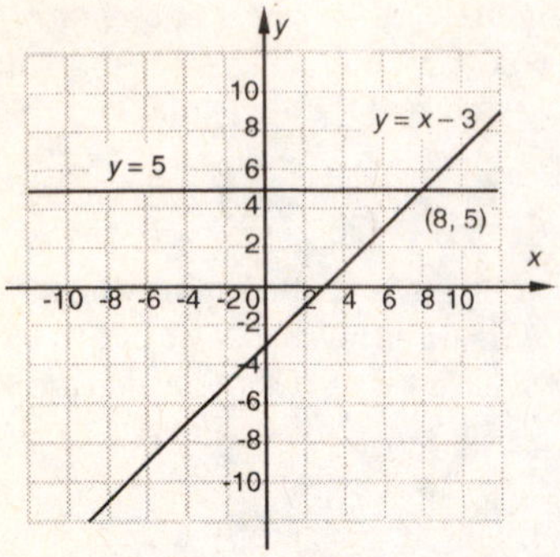

The solution (point of intersection) is apparently $(8, 5)$.

$$\begin{array}{c|c} y = x - 3 & \\ \hline 5 & 8 - 3 \\ \end{array} \qquad \begin{array}{c|c} y = 5 & \\ \hline ? & \\ \end{array}$$

$\overset{?}{5 = 5}$ TRUE $\qquad$ $5 = 5$ TRUE

The solution is $(8, 5)$.

91. *Writing Exercise*

93. From the graph we observe that the values of x for which $2x - 5 > -7$ *and* $2x - 5 < 7$ are $\{x| - 1 < x < 6\}$, or $(-1, 6)$.

95. Solve $19 < P(t) < 30$, or $19 < 0.44t + 10.2 < 30$.

$19 < 0.44t + 10.2 < 30$

$8.8 < 0.44t < 19.8$

$20 < t < 45$

The percentage of childless 40-44-year-old women will be between 19% and 30% from 20 years after 1980 to 45 years after 1980 or from 2000 to 2025.

97. Solve $32 < f(x) < 46$, or $32 < 2(x + 10) < 46$.

$32 < 2(x + 10) < 46$

$32 < 2x + 20 < 46$

$12 < 2x < 26$

$6 < x < 13$

For U.S. dress sizes between 6 and 13, dress sizes in Italy will be between 32 and 46.

99. $10.8 < -0.0433t + 10.49 < 11.5$

$0.31 < -0.0433t < 1.01$

$-7.16 > t > -23.33$

$1988.5 - 7.16 = 1981.34$ and $1988.5 - 23.33 = 1965.17$

Thus, records of 11.5 sec and 10.8 sec were set during 1965 and 1981, respectively. Thus, we have $1965 \le y \le 1981$.

101. Let c = the number of crossings per year. Then at the \$3 per crossing rate, the total cost of c crossings is \$3c. Two six-month passes cost $2 \cdot \$15$, or \$30. The additional \$0.50 per crossing toll brings the total cost of c crossings to $\$30 + \$0.50c$. A one-year pass costs \$150 regardless of the number of crossings.

We write an inequality that states that the cost of c crossings per year using the six-month passes is less than the cost using the \$3 per crossing toll and is less than the cost using the one-year pass. Solve:

$30 + 0.50c < 3c$ *and* $30 + 0.50c < 150$

We get $12 < c$ *and* $c < 240$, or $12 < c < 240$.

For more than 12 crossings but less than 240 crossings per year the six-month passes are the most economical choice.

103. $4m - 8 > 6m + 5$ *or* $5m - 8 < -2$

$-13 > 2m$ *or* $5m < 6$

$-\frac{13}{2} > m$ *or* $m < \frac{6}{5}$

$\left\{m \middle| m < \frac{6}{5}\right\}$, or $\left(-\infty, \frac{6}{5}\right)$

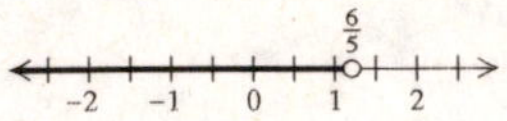

105. $3x < 4 - 5x < 5 + 3x$

$0 < 4 - 8x < 5$

$-4 < -8x < 1$

$\frac{1}{2} > x > -\frac{1}{8}$

The solution set is $\left\{x \middle| -\frac{1}{8} < x < \frac{1}{2}\right\}$, or $\left(-\frac{1}{8}, \frac{1}{2}\right)$.

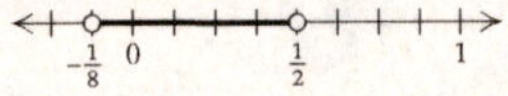

107. Let $a = b = c = 2$. Then $a \le c$ *and* $c \le b$, but $b \not> a$. The given statement is false.

109. If $-a < c$, then $-1(-a) > -1 \cdot c$, or $a > -c$. Then if $a > -c$ and $-c > b$, we have $a > -c > b$, so $a > b$ and the given statement is true.

111. $f(x) = \dfrac{\sqrt{3 - 4x}}{x + 7}$

$3 - 4x \ge 0$ is equivalent to $x \le \frac{3}{4}$ and $x + 7 = 0$ is equivalent to $x = -7$. Then we have Domain of $f = \left\{x \middle| x \le \frac{3}{4} \text{ and } x \ne -7\right\}$, or $(-\infty, -7) \cup \left(-7, \frac{3}{4}\right]$.

113. Observe that the graph of y_1 lies below the graph of y_2 for x in the interval $\left(-\frac{3}{2}, \infty\right)$. Also, the graph of y_3 lies below the graph of y_4 for x in the interval $(-\infty, -7)$. Thus, the solution set is $(-\infty, -7) \cup \left(-\frac{3}{2}, \infty\right)$.

115. 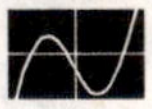

Exercise Set 9.3

1. True; see page 598 in the text.

3. $|0| = 0$, so the statement is false.

5. True; see page 598 in the text.

7. False; see page 603 in the text.

9. $|x| = 7$

$x = -7$ *or* $x = 7$ Using the absolute-value principle

The solution set is $\{-7, 7\}$.

11. $|x| = -6$

The absolute value of a number is always nonnegative. Therefore, the solution set is $\emptyset$.

13. $|p| = 0$

The only number whose absolute value is 0 is 0. The solution set is $\{0\}$.

15. $|t| = 5.5$

$t = -5.5$ *or* $t = 5.5$ Absolute-value principle

The solution set is $\{-5.5, 5.5\}$.

17. $|2x - 3| = 4$

$2x - 3 = -4$ *or* $2x - 3 = 4$ Absolute-value principle

$2x = -1$ *or* $2x = 7$

$x = -\frac{1}{2}$ *or* $x = \frac{7}{2}$

The solution set is $\left\{-\frac{1}{2}, \frac{7}{2}\right\}$.

19. $|3x - 5| = -8$

Absolute value is always nonnegative, so the equation has no solution. The solution set is $\emptyset$.

21. $|x - 2| = 6$

$x - 2 = -6$ *or* $x - 2 = 6$ Absolute-value principle

$x = -4$ *or* $x = 8$

The solution set is $\{-4, 8\}$.

23. $|x - 5| = 3$

$x - 5 = -3$ *or* $x - 5 = 3$

$x = 2$ *or* $x = 8$

The solution set is $\{2, 8\}$.

25. $|x - 7| = 9$

$x - 7 = -9$ *or* $x - 7 = 9$

$x = -2$ *or* $x = 16$

The solution set is $\{-2, 16\}$.

27. $|5x| - 3 = 37$

$|5x| = 40$ Adding 3

$5x = -40$ *or* $5x = 40$

$x = -8$ *or* $x = 8$

The solution set is $\{-8, 8\}$.

29. $7|q| - 2 = 9$

$7|q| = 11$ Adding 2

$|q| = \frac{11}{7}$ Multiplying by $\frac{1}{7}$

$q = -\frac{11}{7}$ *or* $q = \frac{11}{7}$

The solution set is $\left\{-\frac{11}{7}, \frac{11}{7}\right\}$.

31. $\left|\frac{2x - 1}{3}\right| = 5$

$\frac{2x - 1}{3} = -5$ *or* $\frac{2x - 1}{3} = 5$

$2x - 1 = -15$ *or* $2x - 1 = 15$

$2x = -14$ *or* $2x = 16$

$x = -7$ *or* $x = 8$

The solution set is $\{-7, 8\}$.

33. $|m + 5| + 9 = 16$

$|m + 5| = 7$ Adding -9

$m + 5 = -7$ *or* $m + 5 = 7$

$m = -12$ *or* $m = 2$

The solution set is $\{-12, 2\}$.

35. $5 - 2|3x - 4| = -5$

$-2|3x - 4| = -10$

$|3x - 4| = 5$

$3x - 4 = -5$ *or* $3x - 4 = 5$

$3x = -1$ *or* $3x = 9$

$x = -\frac{1}{3}$ *or* $x = 3$

The solution set is $\left\{-\frac{1}{3}, 3\right\}$.

37. $|2x + 6| = 8$

$2x + 6 = -8$ *or* $2x + 6 = 8$

$2x = -14$ *or* $2x = 2$

$x = -7$ *or* $x = 1$

The solution set is $\{-7, 1\}$.

39. $|x| - 3 = 5.7$

$|x| = 8.7$

$x = -8.7$ *or* $x = 8.7$

The solution set is $\{-8.7, 8.7\}$.

41. $\left|\frac{3x-2}{5}\right| = 2$

$\frac{3x-2}{5} = -2$ *or* $\frac{3x-2}{5} = 2$

$3x - 2 = -10$ *or* $3x - 2 = 10$

$3x = -8$ *or* $3x = 12$

$x = -\frac{8}{3}$ *or* $x = 4$

The solution set is $\left\{-\frac{8}{3}, 4\right\}$.

43. $|x+4| = |2x-7|$

$x + 4 = 2x - 7$ *or* $x + 4 = -(2x - 7)$

$4 = x - 7$ *or* $x + 4 = -2x + 7$

$11 = x$ *or* $3x + 4 = 7$

$3x = 3$

$x = 1$

The solution set is $\{1, 11\}$.

45. $|x+4| = |x-3|$

$x + 4 = x - 3$ *or* $x + 4 = -(x - 3)$

$4 = -3$ *or* $x + 4 = -x + 3$

False $\quad 2x = -1$

$x = -\frac{1}{2}$

The solution set is $\left\{-\frac{1}{2}\right\}$.

47. $|3a-1| = |2a+4|$

$3a - 1 = 2a + 4$ *or* $3a - 1 = -(2a + 4)$

$a - 1 = 4$ *or* $3a - 1 = -2a - 4$

$a = 5$ *or* $5a - 1 = -4$

$5a = -3$

$a = -\frac{3}{5}$

The solution set is $\left\{-\frac{3}{5}, 5\right\}$.

49. $|n-3| = |3-n|$

$n - 3 = 3 - n$ *or* $n - 3 = -(3 - n)$

$2n - 3 = 3$ *or* $n - 3 = -3 + n$

$2n = 6$ *or* $-3 = -3$

$n = 3$ True for all real values of n

The solution set is the set of all real numbers.

51. $|7-a| = |a+5|$

$7 - a = a + 5$ *or* $7 - a = -(a + 5)$

$7 = 2a + 5$ *or* $7 - a = -a - 5$

$2 = 2a$ *or* $7 = -5$

$1 = a$ False

The solution set is $\{1\}$.

53. $\left|\frac{1}{2}x - 5\right| = \left|\frac{1}{4}x + 3\right|$

$\frac{1}{2}x - 5 = \frac{1}{4}x + 3$ *or* $\frac{1}{2}x - 5 = -\left(\frac{1}{4}x + 3\right)$

$\frac{1}{4}x - 5 = 3$ *or* $\frac{1}{2}x - 5 = -\frac{1}{4}x - 3$

$\frac{1}{4}x = 8$ *or* $\frac{3}{4}x - 5 = -3$

$x = 32$ *or* $\frac{3}{4}x = 2$

$x = \frac{8}{3}$

The solution set is $\left\{32, \frac{8}{3}\right\}$.

55. $|a| \le 9$

$-9 \le a \le 9$ Part (b)

The solution set is $\{a \mid -9 \le a \le 9\}$, or $[-9, 9]$.

−9 0 9

57. $|x| > 8$

$x < -8$ *or* $8 < x$ Part (c)

The solution set is $\{x \mid x < -8 \text{ or } x > 8\}$, or $(-\infty, -8) \cup (8, \infty)$.

−10 −8 −6 −4 −2 0 2 4 6 8 10

59. $|t| > 0$

$t < 0$ *or* $0 < t$ Part (c)

The solution set is $\{t \mid t < 0 \text{ or } t > 0\}$, or $\{t \mid t \neq 0\}$, or $(-\infty, 0) \cup (0, \infty)$.

−5 −4 −3 −2 −1 0 1 2 3 4 5

61. $|x-1| < 4$

$-4 < x - 1 < 4$ Part (b)

$-3 < x < 5$

The solution set is $\{x \mid -3 < x < 5\}$, or $(-3, 5)$.

−3 0 5

63. $|x+2| \le 6$

$-6 \le x + 2 \le 6$ Part (b)

$-8 \le x \le 4$ Adding -2

The solution set is $\{x \mid -8 \le x \le 4\}$, or $[-8, 4]$.

−10 −8 −6 −4 −2 0 2 4 6 8 10

65. $|x-3| + 2 > 7$

$|x-3| > 5$ Adding -2

$x - 3 < -5$ *or* $5 < x - 3$ Part (c)

$x < -2$ *or* $8 < x$

The solution set is $\{x \mid x < -2 \text{ or } x > 8\}$, or $(-\infty, -2) \cup (8, \infty)$.

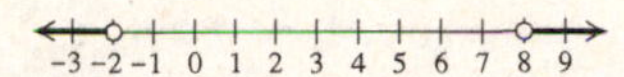

67. $|2y-9| > -5$

Since absolute value is never negative, any value of $2y-9$, and hence any value of y, will satisfy the inequality. The solution set is the set of all real numbers, or $(-\infty, \infty)$.

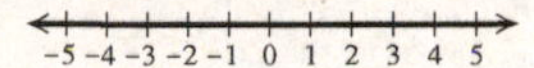

69. $|3a-4|+2 \geq 8$

$$|3a-4| \geq 6 \quad \text{Adding } -2$$
$$3a-4 \leq -6 \quad or \quad 6 \leq 3a-4 \quad \text{Part (c)}$$
$$3a \leq -2 \quad or \quad 10 \leq 3a$$
$$a \leq -\frac{2}{3} \quad or \quad \frac{10}{3} \leq a$$

The solution set is $\left\{a \middle| a \leq -\frac{2}{3} \text{ or } a \geq \frac{10}{3}\right\}$, or $\left(-\infty, -\frac{2}{3}\right] \cup \left[\frac{10}{3}, \infty\right)$.

71. $|y-3| < 12$

$$-12 < y-3 < 12 \quad \text{Part (b)}$$
$$-9 < y < 15 \quad \text{Adding 3}$$

The solution set is $\{y| -9 < y < 15\}$, or $(-9, 15)$.

73. $9-|x+4| \leq 5$

$$-|x+4| \leq -4$$
$$|x+4| \geq 4 \quad \text{Multiplying by } -1$$
$$x+4 \leq -4 \quad or \quad 4 \leq x+4 \quad \text{Part (c)}$$
$$x \leq -8 \quad or \quad 0 \leq x$$

The solution set is $\{x|x \leq -8 \text{ or } x \geq 0\}$, or $(-\infty, -8] \cup [0, \infty)$.

75. $|4-3y| > 8$

$$4-3y < -8 \quad or \quad 8 < 4-3y \quad \text{Part (c)}$$
$$-3y < -12 \quad or \quad 4 < -3y \quad \text{Adding } -4$$
$$y > 4 \quad or \quad -\frac{4}{3} > y \quad \text{Multiplying by } -\frac{1}{3}$$

The solution set is $\left\{y \middle| y < -\frac{4}{3} \text{ or } y > 4\right\}$, or $\left(-\infty, -\frac{4}{3}\right) \cup (4, \infty)$.

77. $|5-4x| < -6$

Absolute value is always nonnegative, so the inequality has no solution. The solution set is $\emptyset$.

79. $\left|\frac{2-5x}{4}\right| \geq \frac{2}{3}$

$$\frac{2-5x}{4} \leq -\frac{2}{3} \quad or \quad \frac{2}{3} \leq \frac{2-5x}{4} \quad \text{Part (c)}$$
$$2-5x \leq -\frac{8}{3} \quad or \quad \frac{8}{3} \leq 2-5x \quad \text{Multiplying by 4}$$
$$-5x \leq -\frac{14}{3} \quad or \quad \frac{2}{3} \leq -5x \quad \text{Adding } -2$$
$$x \geq \frac{14}{15} \quad or \quad -\frac{2}{15} \geq x \quad \text{Multiplying by } -\frac{1}{5}$$

The solution set is $\left\{x \middle| x \leq -\frac{2}{15} \text{ or } x \geq \frac{14}{15}\right\}$, or $\left(-\infty, -\frac{2}{15}\right] \cup \left[\frac{14}{15}, \infty\right)$.

81. $|m+3|+8 \leq 14$

$$|m+3| \leq 6 \quad \text{Adding } -8$$
$$-6 \leq m+3 \leq 6$$
$$-9 \leq m \leq 3$$

The solution set is $\{m| -9 \leq m \leq 3\}$, or $[-9, 3]$.

83. $25-2|a+3| > 19$

$$-2|a+3| > -6$$
$$|a+3| < 3 \quad \text{Multiplying by } -\frac{1}{2}$$
$$-3 < a+3 < 3 \quad \text{Part (b)}$$
$$-6 < a < 0$$

The solution set is $\{a| -6 < a < 0\}$, or $(-6, 0)$.

85. $|2x-3| \leq 4$

$$-4 \leq 2x-3 \leq 4 \quad \text{Part (b)}$$
$$-1 \leq 2x \leq 7 \quad \text{Adding 3}$$
$$-\frac{1}{2} \leq x \leq \frac{7}{2} \quad \text{Multiplying by } \frac{1}{2}$$

The solution set is $\left\{x \middle| -\frac{1}{2} \leq x \leq \frac{7}{2}\right\}$, or $\left[-\frac{1}{2}, \frac{7}{2}\right]$.

87. $5+|3x-4| \geq 16$

$$|3x-4| \geq 11$$
$$3x-4 \leq -11 \quad or \quad 11 \leq 3x-4 \quad \text{Part (c)}$$
$$3x \leq -7 \quad or \quad 15 \leq 3x$$
$$x \leq -\frac{7}{3} \quad or \quad 5 \leq x$$

The solution set is $\left\{x \middle| x \leq -\frac{7}{3} \text{ or } x \geq 5\right\}$, or $\left(-\infty, -\frac{7}{3}\right] \cup [5, \infty)$.

89. $7 + |2x - 1| < 16$

$|2x - 1| < 9$

$-9 < 2x - 1 < 9$ Part (b)

$-8 < 2x < 10$

$-4 < x < 5$

The solution set is $\{x| -4 < x < 5\}$, or $(-4, 5)$.

91. *Writing Exercise*

93. $2x - 3y = 7$, (1)

$3x + 2y = -10$ (2)

We will use the elimination method. First, multiply equation (1) by 2 and equation (2) by 3 and add to eliminate a variable.

$$\begin{aligned} 4x - 6y &= 14 \\ 9x + 6y &= -30 \\ \hline 13x &= -16 \\ x &= -\frac{16}{13} \end{aligned}$$

Now substitute $-\frac{16}{13}$ for x in either of the original equations and solve for y.

$$\begin{aligned} 3x + 2y &= -10 \quad (2) \\ 3\left(-\frac{16}{13}\right) + 2y &= -10 \\ -\frac{48}{13} + 2y &= -10 \\ 2y &= -\frac{130}{13} + \frac{48}{13} \\ 2y &= -\frac{82}{13} \\ y &= -\frac{41}{13} \end{aligned}$$

The solution is $\left(-\frac{16}{13}, -\frac{41}{13}\right)$.

95. $x = -2 + 3y$, (1)

$x - 2y = 2$ (2)

We will use the substitution method. We substitute $-2 + 3y$ for x in equation (2).

$$\begin{aligned} x - 2y &= 2 \quad (2) \\ (-2 + 3y) - 2y &= 2 \quad \text{Substituting} \\ -2 + y &= 2 \\ y &= 4 \end{aligned}$$

Now substitute 4 for y in equation (1) and find x.

$x = -2 + 3 \cdot 4 = -2 + 12 = 10$

The solution is $(10, 4)$.

97. Graph both equations.

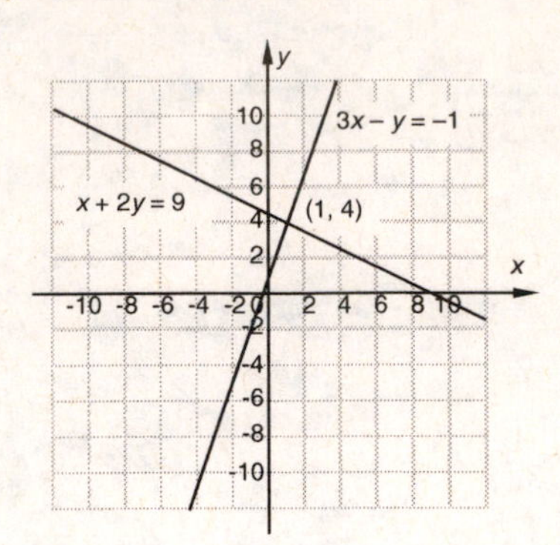

The solution (point of intersection) is apparently $(1, 4)$.

Check:

$x + 2y = 9$		$3x - y = -1$	
$1 + 2 \cdot 4$	9	$3 \cdot 1 - 4$	-1
$1 + 8$		$3 - 4$	
$9 \stackrel{?}{=} 9$ TRUE		$-1 \stackrel{?}{=} -1$ TRUE	

The solution is $(1, 4)$.

99. *Writing Exercise*

101. From the definition of absolute value, $|3t - 5| = 3t - 5$ only when $3t - 5 \geq 0$. Solve $3t - 5 \geq 0$.

$$\begin{aligned} 3t - 5 &\geq 0 \\ 3t &\geq 5 \\ t &\geq \frac{5}{3} \end{aligned}$$

The solution set is $\left\{t \middle| t \geq \frac{5}{3}\right\}$, or $\left[\frac{5}{3}, \infty\right)$.

103. $|x + 2| > x$

The inequality is true for all $x < 0$ (because absolute value must be nonnegative). The solution set in this case is $\{x|x < 0\}$. If $x = 0$, we have $|0 + 2| > 0$, which is true. The solution set in this case is $\{0\}$. If $x > 0$, we have the following:

$$\begin{aligned} x + 2 < -x \quad &or \quad x < x + 2 \\ 2x < -2 \quad &or \quad 0 < 2 \\ x < -1 \quad & \end{aligned}$$

Although $x > 0$ *and* $x < -1$ yields no solution, $x > 0$ and $2 > 0$ (true for all x) yields the solution set $\{x|x > 0\}$ in this case. The solution set for the inequality is $\{x|x < 0\} \cup \{0\} \cup \{x|x > 0\}$, or $\{x|x \text{ is a real number}\}$, or $(-\infty, \infty)$.

105. $|5t - 3| = 2t + 4$

From the definition of absolute value, we know that $2t + 4 \geq 0$, or $t \geq -2$. So we have

$t \geq -2$ *and*

$5t - 3 = -(2t + 4)$ *or* $5t - 3 = 2t + 4$

$5t - 3 = -2t - 4$ *or* $3t = 7$

$7t = -1$ *or* $t = \frac{7}{3}$

$t = -\frac{1}{7}$ *or* $t = \frac{7}{3}$

Since $-\frac{1}{7} \geq -2$ and $\frac{7}{3} \geq -2$, the solution set is $\left\{-\frac{1}{7}, \frac{7}{3}\right\}$.

107. Using part (b), we find that $-3 < x < 3$ is equivalent to $|x| < 3$.

109. $x \leq -6$ *or* $6 \leq x$

$|x| \geq 6$ Using part (c)

111. $x < -8$ *or* $2 < x$

$x + 3 < -5$ *or* $5 < x + 3$ Adding 3

$|x + 3| > 5$ Using part (c)

113. The distance from x to 7 is $|x - 7|$ or $|7 - x|$, so we have $|x - 7| < 2$, or $|7 - x| < 2$.

115. The length of the segment from -1 to 7 is $|-1-7| = |-8| =$ 8 units. The midpoint of the segment is $\frac{-1+7}{2} = \frac{6}{2} = 3$. Thus, the interval extends $8/2$, or 4, units on each side of 3. An inequality for which the closed interval is the solution set is then $|x - 3| \leq 4$.

117. The length of the segment from -7 to -1 is $|-7-(-1)| = |-6| = 6$ units. The midpoint of the segment is $\frac{-7+(-1)}{2} = \frac{-8}{2} = -4$. Thus, the interval extends $6/2$, or 3, units on each side of -4. An inequality for which the open interval is the solution set is $|x - (-4)| < 3$, or $|x + 4| < 3$.

119. $|d - 60 \text{ ft}| \leq 10 \text{ ft}$

$-10 \text{ ft} \leq d - 60 \text{ ft} \leq 10 \text{ ft}$

$50 \text{ ft} \leq d \leq 70 \text{ ft}$

When the bungee jumper is 50 ft above the river, she is $150 - 50$, or 100 ft, from the bridge. When she is 70 ft above the river, she is $150 - 70$, or 80 ft, from the bridge. Thus, at any given time, the bungee jumper is between 80 ft and 100 ft from the bridge.

121. Graph $g(x) = 4$ on the same axes as $f(x) = |2x - 6|$.

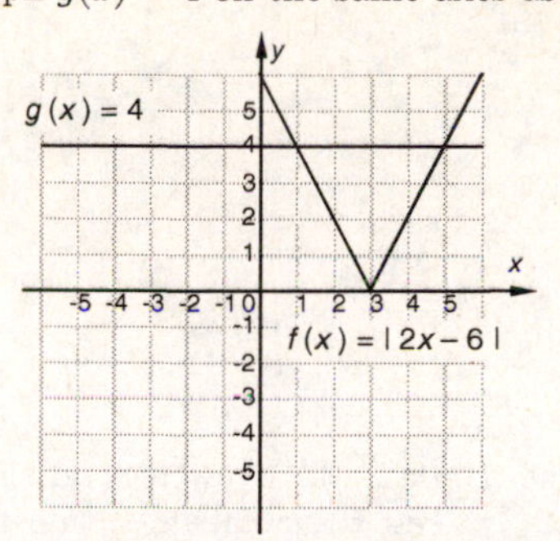

The solution set consists of the x-values for which $(x, f(x))$ is on or below the horizontal line $g(x) = 4$. These x-values comprise the interval $[1, 5]$.

123.

125. *Writing Exercise*

Exercise Set 9.4

1. e; see pages 606 and 614 in the text.

3. d; see pages 611-614 in the text.

5. b; see page 612 in the text.

7. We replace x with -4 and y with 2.

$$2x + 3y < -1$$

$2(-4) + 3 \cdot 2$	-1
$-8 + 6$	
$-2 \stackrel{?}{=} -1$	TRUE

Since $-2 < -1$ is true, $(-4, 2)$ is a solution.

9. We replace x with 8 and y with 14.

$$2y - 3x \geq 9$$

$2 \cdot 14 - 3 \cdot 8$	9
$28 - 24$	
$4 \stackrel{?}{=} 9$	FALSE

Since $4 > 9$ is false, $(8, 14)$ is not a solution.

11. Graph: $y > \frac{1}{2}x$

We first graph the line $y = \frac{1}{2}x$. We draw the line dashed since the inequality symbol is $>$. To determine which half-plane to shade, test a point not on the line. We try $(0, 1)$:

$$y > \frac{1}{2}x$$

1	$\frac{1}{2} \cdot 0$
$1 \stackrel{?}{=} 0$	TRUE

Since $1 > 0$ is true, (0.1) is a solution as are all of the points in the half-plane containing $(0, 1)$. We shade that half-plane and obtain the graph.

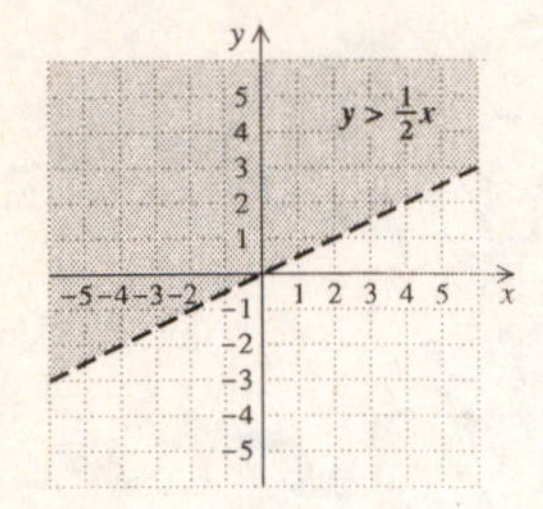

13. Graph: $y \geq x - 3$

First graph the line $y = x - 3$. Draw it solid since the inequality symbol is $\geq$. Test the point $(0, 0)$ to determine if it is a solution.

$$\begin{array}{c|c} \multicolumn{2}{c}{y \geq x - 3} \\ \hline 0 & 0 - 3 \end{array}$$

$0 \stackrel{?}{=} -3$ TRUE

Since $0 \geq -3$ is true, we shade the half-plane that contains $(0, 0)$ and obtain the graph.

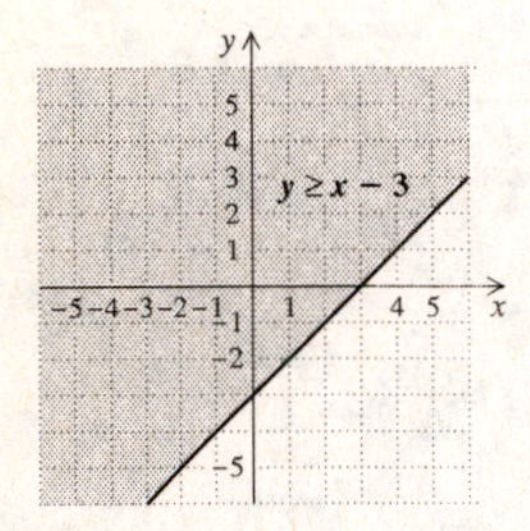

15. Graph: $y \leq x + 5$

First graph the line $y = x + 5$. Draw it solid since the inequality symbol is $\leq$. Test the point $(0, 0)$ to determine if it is a solution.

$$\begin{array}{c|c} \multicolumn{2}{c}{y \leq x + 5} \\ \hline 0 & 0 + 5 \end{array}$$

$0 \stackrel{?}{=} 5$ TRUE

Since $0 \leq 5$ is true, we shade the half-plane that contains $(0, 0)$ and obtain the graph.

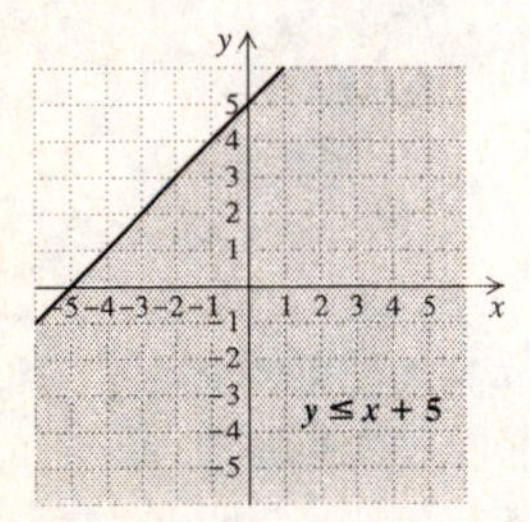

17. Graph: $x - y \leq 4$

First graph the line $x - y = 4$. Draw a solid line since the inequality symbol is $\leq$. Test the point $(0, 0)$ to determine if it is a solution.

$$\begin{array}{c|c} \multicolumn{2}{c}{x - y \leq 4} \\ \hline 0 - 0 & 4 \end{array}$$

$0 \stackrel{?}{=} 4$ TRUE

Since $0 \leq 4$ is true, we shade the half-plane that contains $(0, 0)$ and obtain the graph.

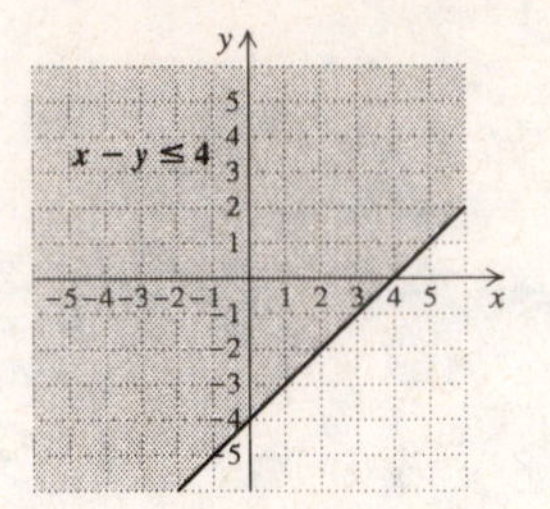

19. Graph: $2x + 3y < 6$

First graph $2x + 3y = 6$. Draw the line dashed since the inequality symbol is $<$. Test the point $(0, 0)$ to determine if it is a solution.

$$\begin{array}{c|c} \multicolumn{2}{c}{2x + 3y < 6} \\ \hline 2 \cdot 0 + 3 \cdot 0 & 6 \end{array}$$

$0 \stackrel{?}{=} 6$ TRUE

Since $0 < 6$ is true, we shade the half-plane containing $(0, 0)$ and obtain the graph.

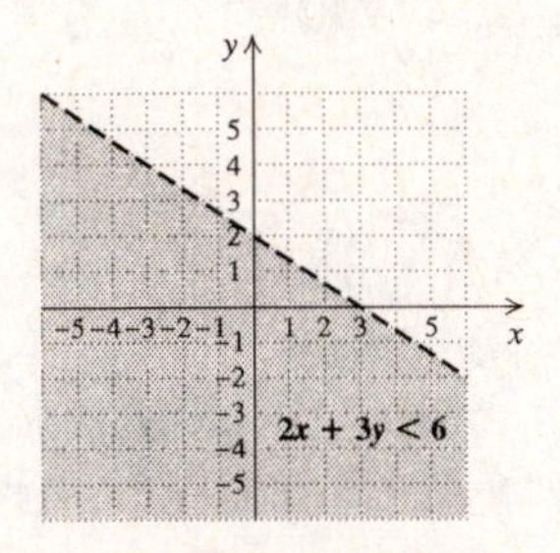

21. Graph: $2x - y \leq 4$

We first graph $2x - y = 4$. Draw the line solid since the inequality symbol is $\leq$. Test the point $(0, 0)$ to determine if it is a solution.

$$\begin{array}{c|c} \multicolumn{2}{c}{2x - y \leq 4} \\ \hline 2 \cdot 0 - 0 & 4 \end{array}$$

$0 \stackrel{?}{=} 4$ TRUE

Since $0 \leq 4$ is true, we shade the half-plane containing $(0, 0)$ and obtain the graph.

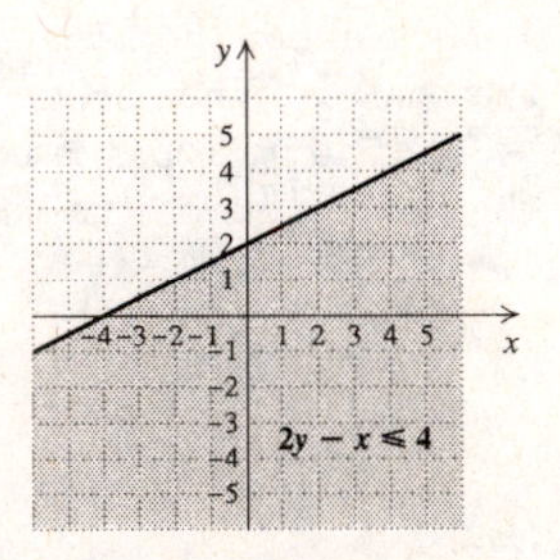

23. Graph: $2x - 2y \geq 8 + 2y$

$2x - 4y \geq 8$

First graph $2x - 4y = 8$. Draw the line solid since the inequality symbol is $\geq$. Test the point $(0, 0)$ to determine if it is a solution.

$$\begin{array}{c|c} \multicolumn{2}{c}{2x - 4y \geq 8} \\ \hline 2 \cdot 0 - 4 \cdot 0 & 8 \end{array}$$

$0 \stackrel{?}{=} 8$ FALSE

Since $0 \geq 8$ is false, we shade the half-plane that does not contain $(0, 0)$ and obtain the graph.

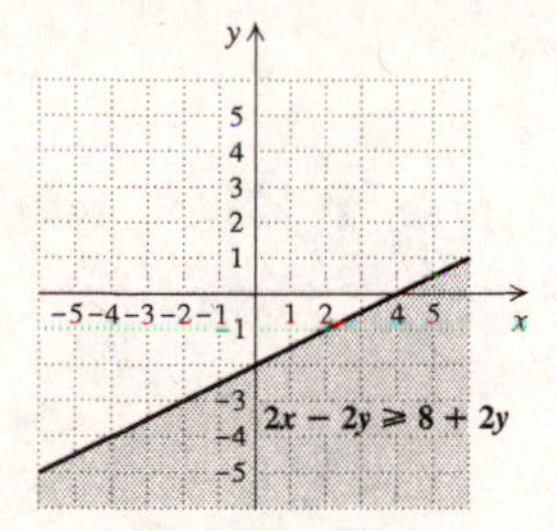

25. Graph: $y \geq 3$

We first graph $y = 3$. Draw the line solid since the inequality symbol is $\geq$. Test the point $(0, 0)$ to determine if it is a solution.

$$\begin{array}{c} y \geq 3 \\ \hline 0 \stackrel{?}{=} 3 \end{array} \quad \text{FALSE}$$

Since $0 \geq 3$ is false, we shade the half-plane that does not contain $(0, 0)$ and obtain the graph.

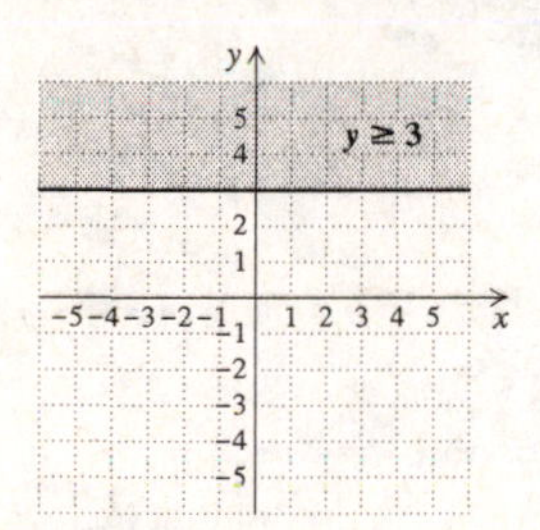

27. Graph: $x \leq 6$

We first graph $x = 6$. We draw the line solid since the inequality symbol is $\leq$. Test the point $(0, 0)$ to determine if it is a solution.

$$\begin{array}{c} x \leq 6 \\ \hline 0 \stackrel{?}{=} 6 \end{array} \quad \text{TRUE}$$

Since $0 \leq 6$ is true, we shade the half-plane containing $(0, 0)$ and obtain the graph.

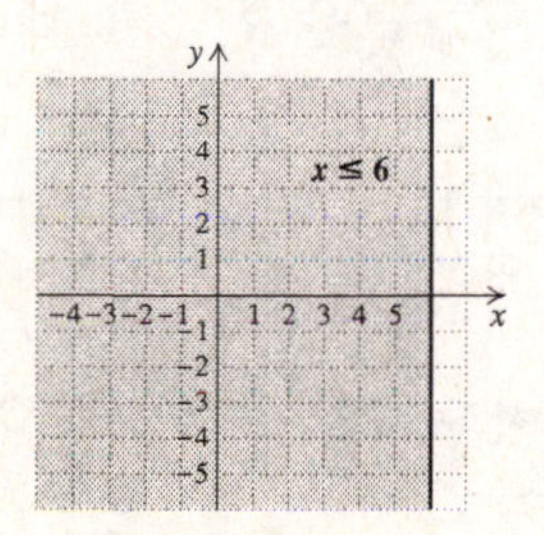

29. Graph: $-2 < y < 7$

This is a system of inequalities:

$-2 < y,$

$y < 7$

The graph of $-2 < y$ is the half-plane above the line $-2 = y$; the graph of $y < 7$ is the half-plane below the line $y = 7$. We shade the intersection of these graphs.

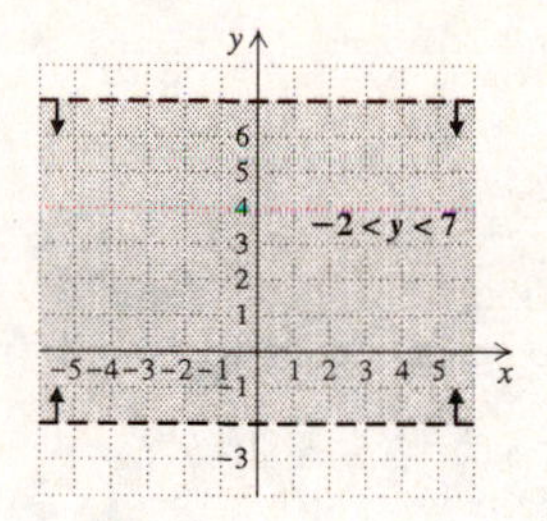

31. Graph: $-4 \leq x \leq 2$

This is a system of inequalities:

$-4 \leq x,$

$x \leq 2$

Graph $-4 \leq x$ and $x \leq 2$. Then shade the intersection of these graphs.

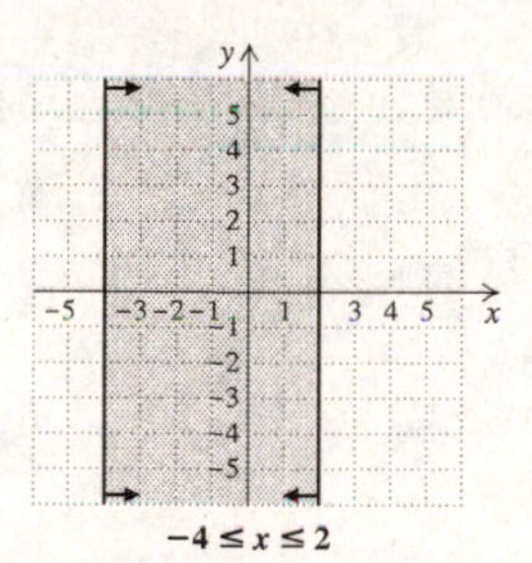

33. Graph: $0 \leq y \leq 3$

This is a system of inequalities:

$0 \leq y,$

$y \leq 3$

Graph $0 \leq y$ and $y \leq 3$.

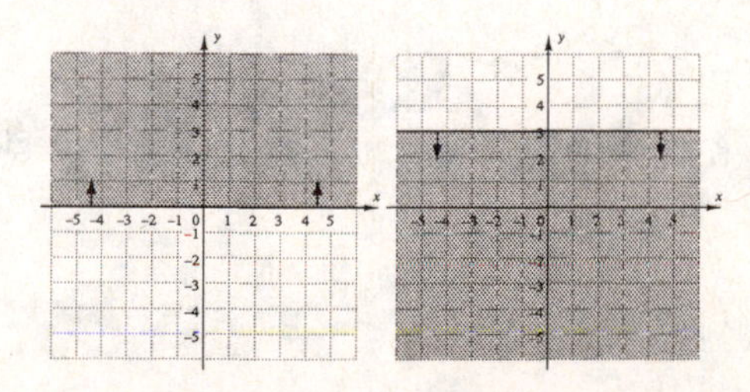

Then we shade the intersection of these graphs.

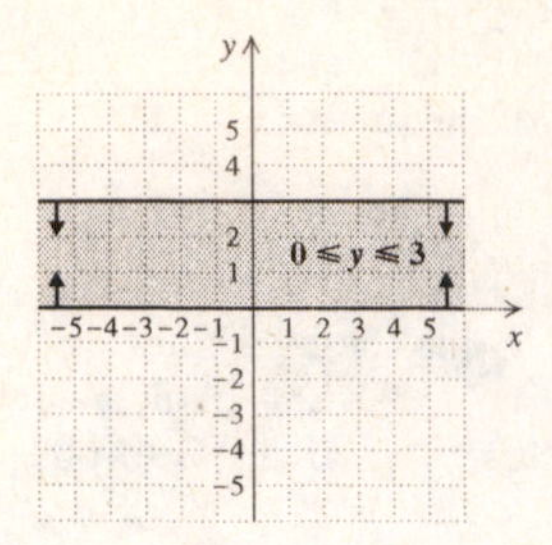

35. Graph: $y < -x$,

$y > x + 2$

We graph the lines $y = -x$ and $y = x + 2$, using dashed lines. We indicate the region for each inequality by the arrows at the ends of the lines. Note where the regions overlap and shade the region of solutions.

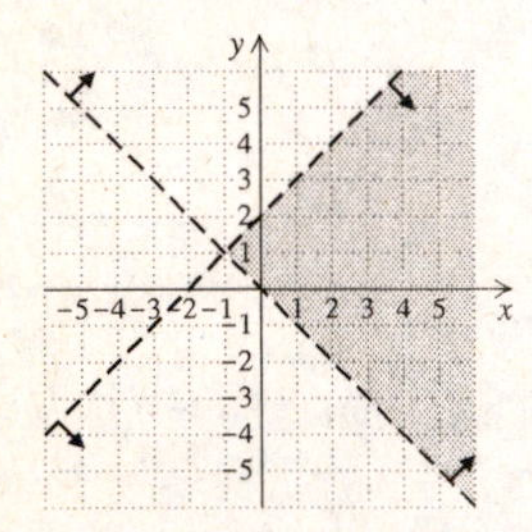

37. Graph: $y \geq x$,

$y \leq 2x - 4$

Graph $y = x$ and $y = 2x - 4$, using solid lines. Indicate the region for each inequality by arrows, and shade the region where they overlap.

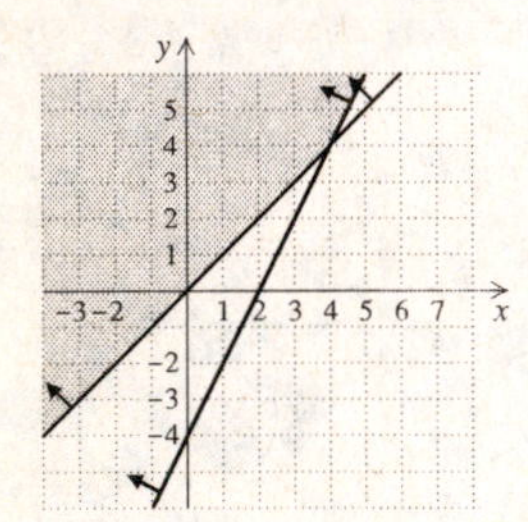

39. Graph: $y \leq -3$,

$x \geq -1$

Graph $y = -3$ and $x = -1$ using solid lines. Indicate the region for each inequality by arrows, and shade the region where they overlap.

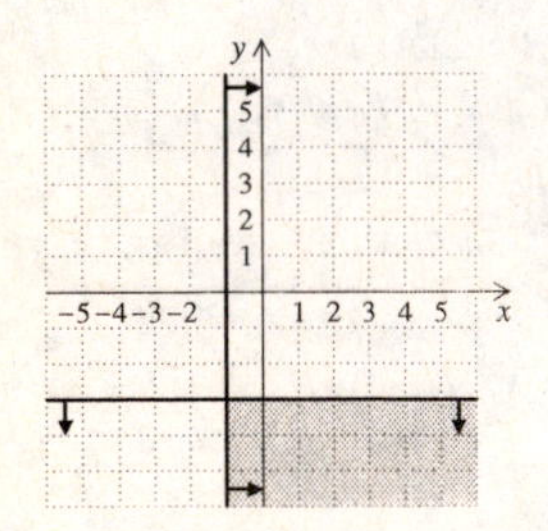

41. Graph: $x > -4$,

$y < -2x + 3$

Graph the lines $x = -4$ and $y = -2x + 3$, using dashed lines. Indicate the region for each inequality by arrows, and shade the region where they overlap.

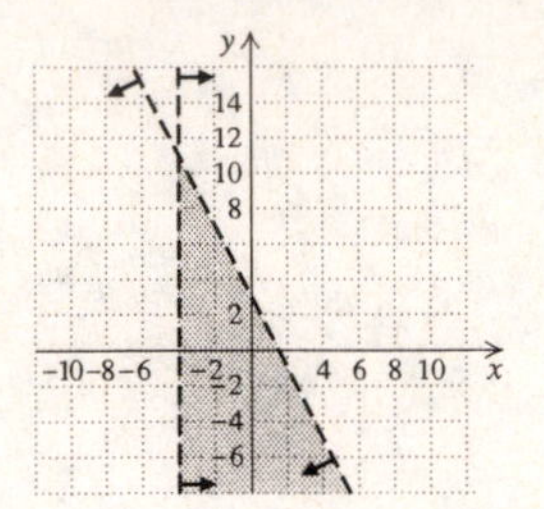

43. Graph: $y \leq 5$,

$y \geq -x + 4$

Graph the lines $y = 5$ and $y = -x+4$, using solid lines. Indicate the region for each inequality by arrows, and shade the region where they overlap.

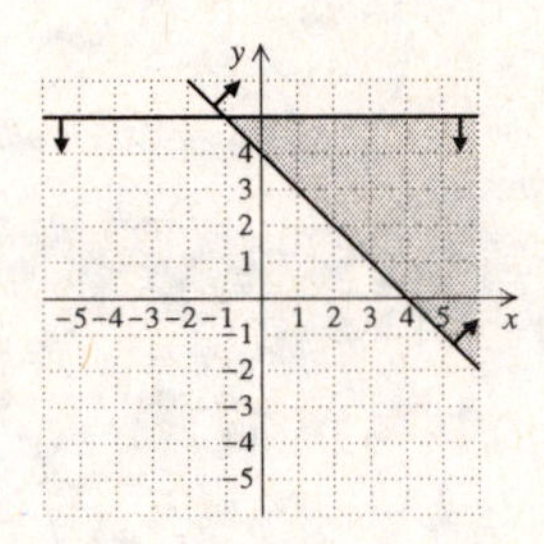

45. Graph: $x + y \leq 6$,

$x - y \leq 4$

Graph the lines $x + y = 6$ and $x - y = 4$, using solid lines. Indicate the region for each inequality by arrows, and shade the region where they overlap.

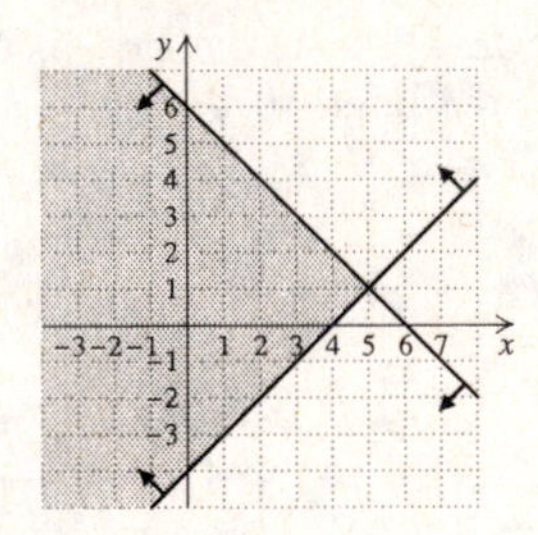

47. Graph: $y + 3x > 0$,

$y + 3x < 2$

Graph the lines $y + 3x = 0$ and $y + 3x = 2$, using dashed lines. Indicate the region for each inequality by arrows, and shade the region where they overlap.

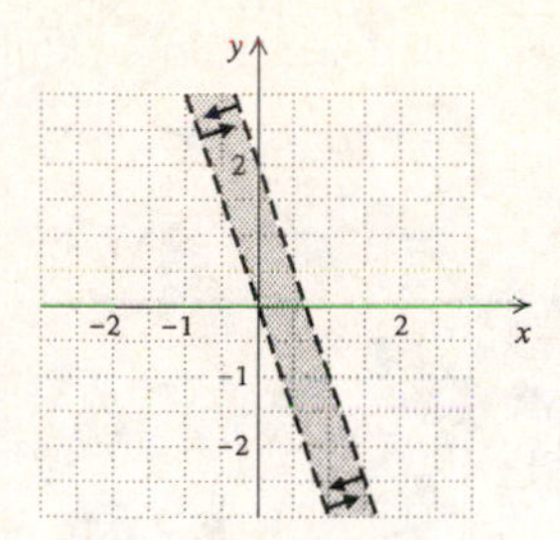

49. Graph: $y \le 2x - 3$, (1)

$y \ge -2x + 1$, (2)

$x \le 5$ (3)

Graph the lines $y = 2x - 3$, $y = -2x + 1$, and $x = 5$ using solid lines. Indicate the region for each inequality by arrows, and shade the region where they overlap.

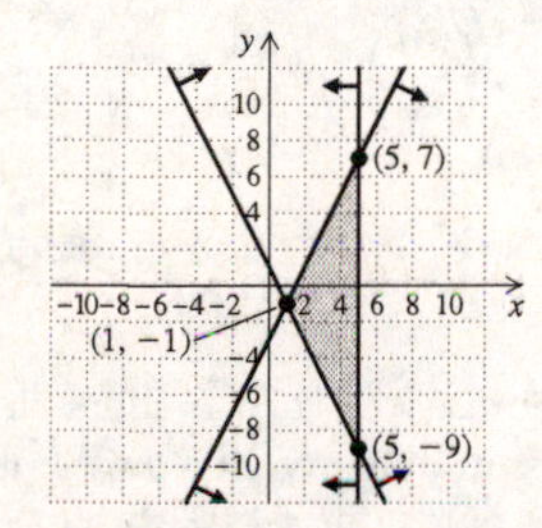

To find the vertex we solve three different systems of related equations.

From (1) and (2) we have $y = 2x - 3$,
$y = -2x + 1$.

Solving, we obtain the vertex $(1, -1)$.

From (1) and (3) we have $y = 2x - 3$,
$x = 5$.

Solving, we obtain the vertex $(5, 7)$.

From (2) and (3) we have $y = -2x + 1$,
$x = 5$.

Solving, we obtain the vertex $(5, -9)$.

51. Graph: $x + 2y \le 12$, (1)

$2x + y \le 12$ (2)

$x \ge 0$, (3)

$y \ge 0$ (4)

Graph the lines $x + 2y = 12$, $2x + y = 12$, $x = 0$, and $y = 0$ using solid lines. Indicate the region for each inequality by arrows, and shade the region where they overlap.

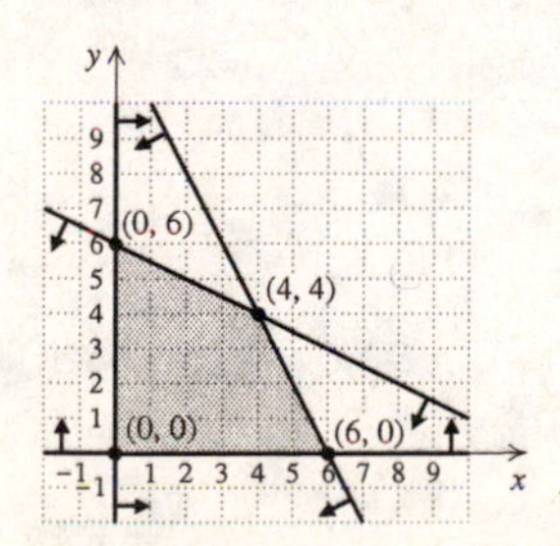

To find the vertices we solve four different systems of equations.

From (1) and (2) we have $x + 2y = 12$,
$2x + y = 12$.

Solving, we obtain the vertex $(4, 4)$.

From (1) and (3) we have $x + 2y = 12$,
$x = 0$.

Solving, we obtain the vertex $(0, 6)$.

From (2) and (4) we have $2x + y = 12$,
$y = 0$.

Solving, we obtain the vertex $(6, 0)$.

From (3) and (4) we have $x = 0$,
$y = 0$.

Solving, we obtain the vertex $(0, 0)$.

53. Graph: $8x + 5y \le 40$, (1)

$x + 2y \le 8$ (2)

$x \ge 0$, (3)

$y \ge 0$ (4)

Graph the lines $8x + 5y = 40$, $x + 2y = 8$, $x = 0$, and $y = 0$ using solid lines. Indicate the region for each inequality by arrows, and shade the region where they overlap.

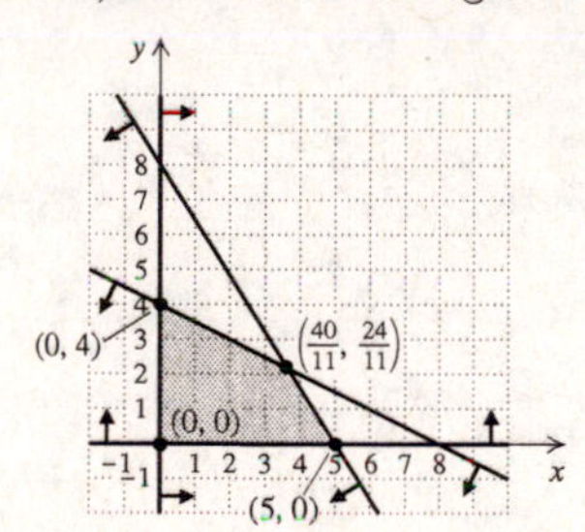

To find the vertices we solve four different systems of equations.

From (1) and (2) we have $8x + 5y = 40$,
$x + 2y = 8$.

Solving, we obtain the vertex $\left(\frac{40}{11}, \frac{24}{11}\right)$.

From (1) and (4) we have $8x + 5y = 40$,
$y = 0$.

Solving, we obtain the vertex $(5, 0)$.

From (2) and (3) we have $x + 2y = 8$,
$x = 0$.

Solving, we obtain the vertex $(0, 4)$.

From (3) and (4) we have $x = 0$,
$y = 0$.

Solving, we obtain the vertex $(0, 0)$.

55. Graph: $y - x \geq 2$, (1)

$y - x \leq 4$, (2)

$2 \leq x \leq 5$ (3)

Think of (3) as two inequalities:

$2 \leq x$, (4)

$x \leq 5$ (5)

Graph the lines $y - x = 2$, $y - x = 4$, $x = 2$, and $x = 5$, using solid lines. Indicate the region for each inequality by arrows, and shade the region where they overlap.

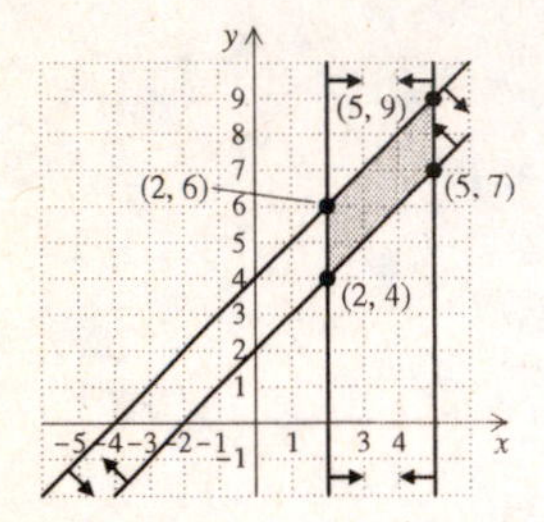

To find the vertices we solve four different systems of equations.

From (1) and (4) we have $y - x = 2$,

$x = 2$.

Solving, we obtain the vertex $(2, 4)$.

From (1) and (5) we have $y - x = 2$,

$x = 5$.

Solving, we obtain the vertex $(5, 7)$.

From (2) and (4) we have $y - x = 4$,

$x = 2$.

Solving, we obtain the vertex $(2, 6)$.

From (2) and (5) we have $y - x = 4$,

$x = 5$.

Solving, we obtain the vertex $(5, 9)$.

57. *Writing Exercise*

59. ***Familiarize***. We let x and y represent the number of pounds of peanuts and fancy nuts in the mixture, respectively. We organize the given information in a table.

Type of nuts	Peanuts	Fancy	Mixture
Amount	x	y	10
Price per pound	\$2.50	\$7	
Value	$2.5x$	$7y$	40

Translate. We get a system of equations from the first and third rows of the table.

$$x + y = 10,$$
$$2.5x + 7y = 40$$

Clearing decimals we have

$$x + y = 10, \quad (1)$$
$$25x + 70y = 400. \quad (2)$$

Carry out. We use the elimination method. Multiply Equation (1) by -25 and add.

$$\begin{aligned} -25x - 25y &= -250 \\ 25x + 70y &= 400 \\ \hline 45y &= 150 \\ y &= \frac{10}{3}, \text{ or } 3\frac{1}{3} \end{aligned}$$

Substitute $\frac{10}{3}$ for y in Equation (1) and solve for x.

$$x + y = 10$$
$$x + \frac{10}{3} = 10$$
$$x = \frac{20}{3}, \text{ or } 6\frac{2}{3}$$

Check. The sum of $6\frac{2}{3}$ and $3\frac{1}{3}$ is 10. The value of the mixture is $2.5\left(\frac{20}{3}\right) + 7\left(\frac{10}{3}\right)$, or $\frac{50}{3} + \frac{70}{3}$, or \$40. These numbers check.

State. $6\frac{2}{3}$ lb of peanuts and $3\frac{1}{3}$ lb of fancy nuts should be used.

61. ***Familiarize***. Let x = the number of card holders tickets that were sold and y = the number of noncard holders tickets. We arrange the information in a table.

	Card holders	Noncard holders	Total
Price	\$2.50	\$4	
Number sold	x	y	203
Money taken in	$2.5x$	$4y$	\$620

Translate. The last two rows of the table give us two equations. The total number of tickets sold was 203, so we have

$$x + y = 203.$$

The total amount of money collected was \$620, so we have

$$2.5x + 4y = 620.$$

We can multiply the second equation on both sides by 10 to clear the decimal. The resulting system is

$$x + y = 203, \quad (1)$$
$$25x + 40y = 6200. \quad (2)$$

Carry out. We use the elimination method. We multiply on both sides of Equation (1) by -25 and then add.

$$\begin{aligned} -25x - 25y &= -5075 \quad \text{Multiplying by } -25 \\ 25x + 40y &= 6200 \\ \hline 15y &= 1125 \\ y &= 75 \end{aligned}$$

We go back to Equation (1) and substitute 75 for y.

$$x + y = 203$$
$$x + 75 = 203$$
$$x = 128$$

Check. The number of tickets sold was $128 + 75$, or 203. The money collected was $\$2.50(128) + \$4(75)$, or $\$320 + \$300$, or \$620. These numbers check.

State. 128 card holders tickets and 75 noncard holders tickets were sold.

63. ***Familiarize***. The formula for the area of a triangle with base b and height h is $A = \frac{1}{2}bh$.

Translate. Substitute 200 for A and 16 for b in the formula.

$$A = \frac{1}{2}bh$$
$$200 = \frac{1}{2} \cdot 16 \cdot h$$

Carry out. We solve the equation.

$$200 = \frac{1}{2} \cdot 16 \cdot h$$
$$200 = 8h \quad \text{Multiplying}$$
$$25 = h \quad \text{Dividing by 8 on both sides}$$

Check. The area of a triangle with base 16 ft and height 25 ft is $\frac{1}{2} \cdot 16 \cdot 25$, or 200 ft^2. The answer checks.

State. The seed can fill a triangle that is 25 ft tall.

65. *Writing Exercise*

67. Graph: $x + y > 8$,
$x + y \leq -2$

Graph the line $x + y = 8$ using a dashed line and graph $x + y = -2$, using a solid line. Indicate the region for each inequality by arrows. The regions do not overlap (the solution set is $\emptyset$), so we do not shade any portion of the graph.

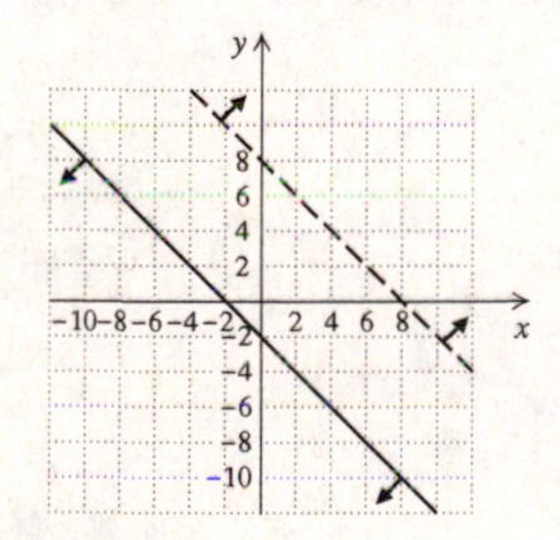

69. Graph: $x - 2y \leq 0$,
$-2x + y \leq 2$,
$x \leq 2$,
$y \leq 2$,
$x + y \leq 4$

Graph the five inequalities above, and shade the region where they overlap.

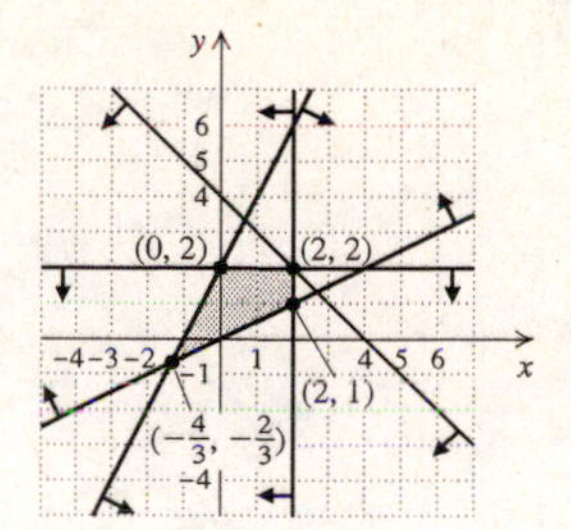

71. Both the width and the height must be positive, so we have

$w > 0$,
$h > 0$.

To be checked as luggage, the sum of the width, height, and length cannot exceed 62 in., so we have

$w + h + 30 \leq 62$, or
$w + h \leq 32$.

The girth is represented by $2w+2h$ and the length is 30 in. In order to meet postal regulations the sum of the girth and the length cannot exceed 130 in., so we have:

$2w + 2h + 30 < 130$, or
$2w + 2h \leq 100$, or
$w + h \leq 50$

Thus, have a system of inequalities:

$w > 0$,
$h > 0$
$w + h < 32$,
$w + h \leq 50$

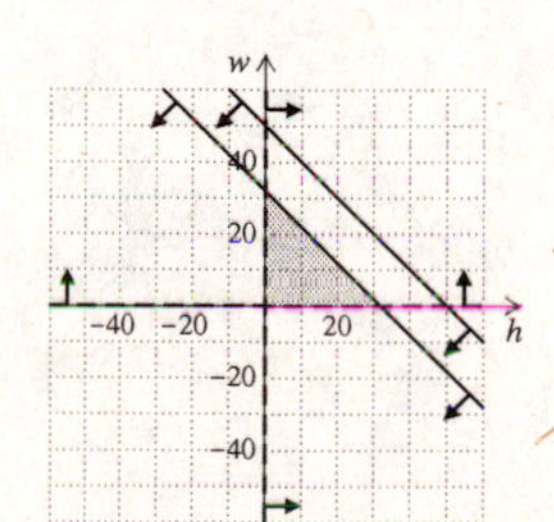

73. Graph: $35c + 75a > 1000$,
$c \geq 0$,
$a \geq 0$

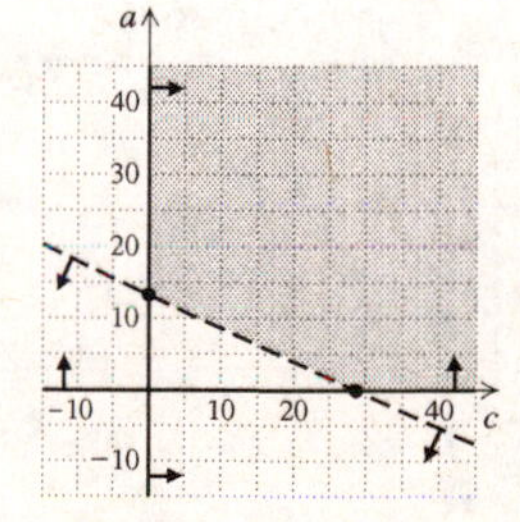

75. a)

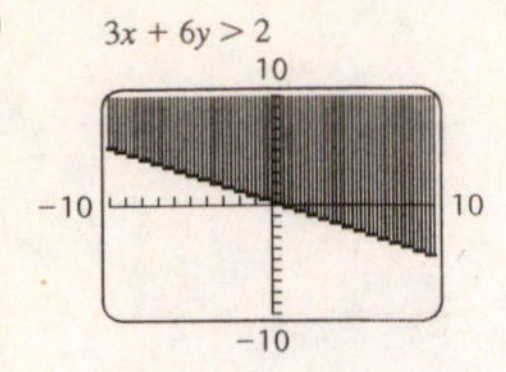

b)

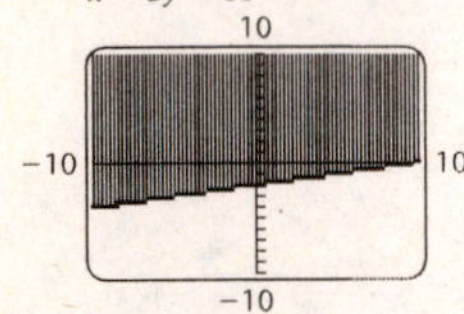

c)

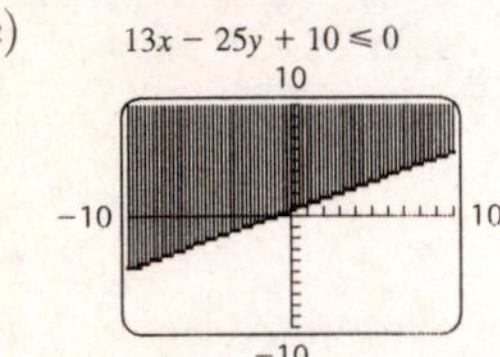

d)

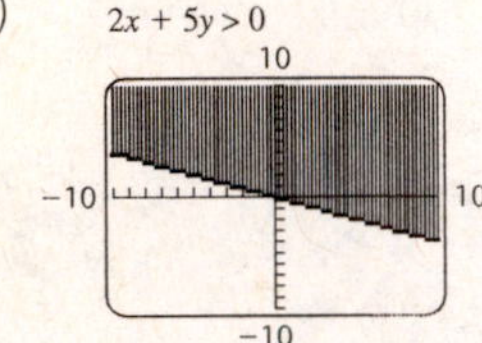

Exercise Set 9.5

1. minimized; see page 619 in the text.

3. constraints; see page 619 in the text.

5. feasible; see page 620 in the text.

7. Find the maximum and minimum values of
$F = 2x + 14y$,

subject to

$$5x + 3y \le 34, \quad (1)$$
$$3x + 5y \le 30, \quad (2)$$
$$x \ge 0, \quad (3)$$
$$y \ge 0. \quad (4)$$

Graph the system of inequalities and find the coordinates of the vertices.

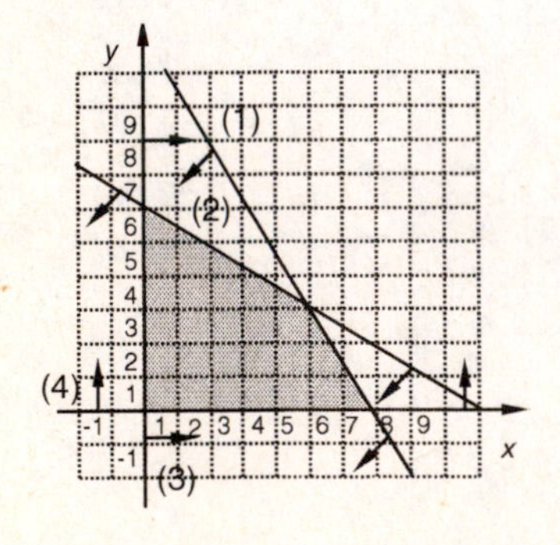

To find one vertex we solve the system

$$x = 0,$$
$$y = 0.$$

This vertex is $(0, 0)$.

To find a second vertex we solve the system

$$5x + 3y = 34,$$
$$y = 0.$$

This vertex is $\left(\frac{34}{5}, 0\right)$.

To find a third vertex we solve the system

$$5x + 3y = 34,$$
$$3x + 5y = 30.$$

This vertex is $(5, 3)$.

To find the fourth vertex we solve the system

$$3x + 5y = 30,$$
$$x = 0.$$

This vertex is $(0, 6)$.

Now find the value of F at each of these points.

Vertex (x, y)	$F = 2x + 14y$	
$(0, 0)$	$2 \cdot 0 + 14 \cdot 0 = 0 + 0 = 0$	⟵Minimum
$\left(\frac{34}{5}, 0\right)$	$2 \cdot \frac{34}{5} + 14 \cdot 0 = \frac{68}{5} + 0 = 13\frac{3}{5}$	
$(5, 3)$	$2 \cdot 5 + 14 \cdot 3 = 10 + 42 = 52$	
$(0, 6)$	$2 \cdot 0 + 14 \cdot 6 = 0 + 84 = 84$	⟵Maximum

The maximum value of F is 84 when $x = 0$ and $y = 6$. The minimum value of F is 0 when $x = 0$ and $y = 0$.

9. Find the maximum and minimum values of
$P = 8x - y + 20$,

subject to

$$6x + 8y \le 48, \quad (1)$$
$$0 \le y \le 4, \quad (2)$$
$$0 \le x \le 7. \quad (3)$$

Think of (2) as $0 \le y$, (4)
$y \le 4$. (5)

Think of (3) as $0 \le x$, (6)
$x \le 7$. (7)

Graph the system of inequalities.

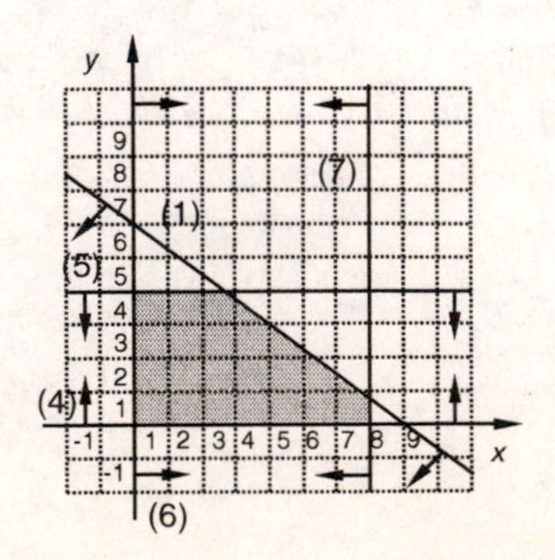

To determine the coordinates of the vertices, we solve the following systems:

$x = 0,$ $\quad$ $x = 7,$ $\quad$ $6x + 8y = 48,$

$y = 0;$ $\quad$ $y = 0;$ $\quad$ $x = 7;$

$6x + 8y = 48,$ $\quad$ $x = 0,$

$y = 4;$ $\quad$ $y = 4$

The vertices are $(0,0)$, $(7,0)$, $\left(7, \frac{3}{4}\right)$, $\left(\frac{8}{3}, 4\right)$, and $(0,4)$, respectively. Compute the value of P at each of these points.

Vertex (x, y)	$P = 8x - y + 20$	
$(0,0)$	$8 \cdot 0 - 0 + 20 =$ $0 - 0 + 20 = 20$	
$(7,0)$	$8 \cdot 7 - 0 + 20 =$ $56 - 0 + 20 =$ 76	← Maximum
$\left(7, \frac{3}{4}\right)$	$8 \cdot 7 - \frac{3}{4} + 20 =$ $56 - \frac{3}{4} + 20 = 75\frac{1}{4}$	
$\left(\frac{8}{3}, 4\right)$	$8 \cdot \frac{8}{3} - 4 + 20 =$ $\frac{64}{3} - 4 + 20 = 37\frac{1}{3}$	
$(0,4)$	$8 \cdot 0 - 4 + 20 =$ $0 - 4 + 20 =$ 16	← Minimum

The maximum is 76 when $x = 7$ and $y = 0$. The minimum is 16 when $x = 0$ and $y = 4$.

11. Find the maximum and minimum values of

$F = 2y - 3x,$

subject to

$y \leq 2x + 1, \quad (1)$

$y \geq -2x + 3, \quad (2)$

$x \leq 3 \quad (3)$

Graph the system of inequalities and find the coordinates of the vertices.

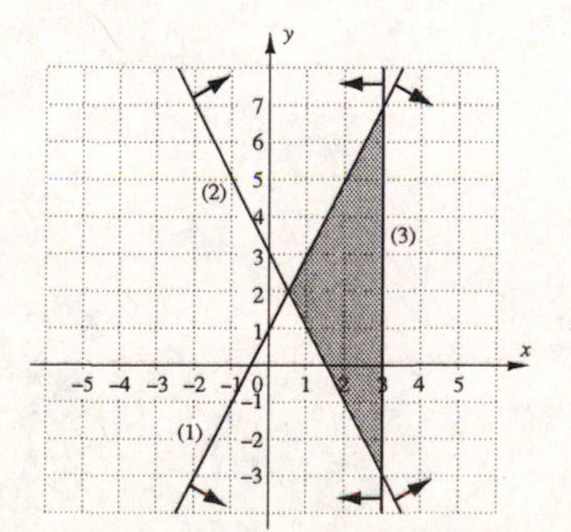

To determine the coordinates of the vertices, we solve the following systems:

$y = 2x + 1,$ $\quad$ $y = 2x + 1,$ $\quad$ $y = -2x + 3,$

$y = -2x + 3;$ $\quad$ $x = 3;$ $\quad$ $x = 3$

The solutions of the systems are $\left(\frac{1}{2}, 2\right)$, $(3,7)$, and $(3,-3)$, respectively. Now find the value of F at each of these points.

Vertex (x, y)	$F = 2y - 3x$	
$\left(\frac{1}{2}, 2\right)$	$2 \cdot 2 - 3 \cdot \frac{1}{2} = \frac{5}{2}$	
$(3,7)$	$2 \cdot 7 - 3 \cdot 3 = 5$	← Maximum
$(3,-3)$	$2(-3) - 3 \cdot 3 = -15$	← Minimum

The maximum value is 5 when $x = 3$ and $y = 7$. The minimum value is -15 when $x = 3$ and $y = -3$.

13. ***Familiarize***. Let $x =$ the number of gallons of gasoline the car uses and $y =$ the number of gallons the moped uses.

Translate. The mileage m is given by

$M = 20x + 100y.$

We wish to maximize M subject to these constraints:

$x + y \leq 12,$

$0 \leq x \leq 10,$

$0 \leq y \leq 3.$

Carry out. We graph the system of inequalities, determine the vertices, and evaluate M at each vertex.

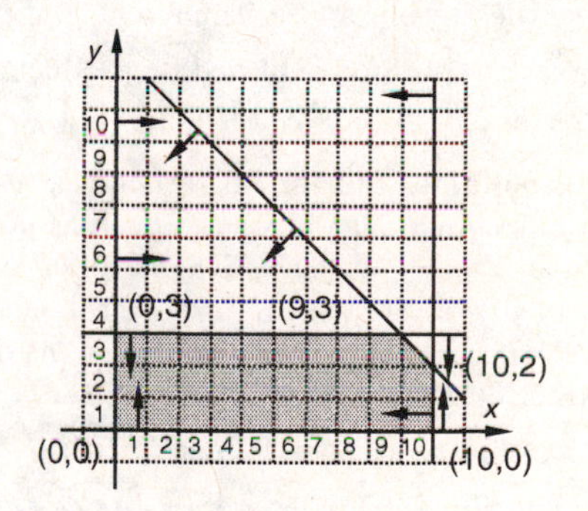

Vertex	$M = 20x + 100y$
$(0,0)$	0
$(0,3)$	300
$(9,3)$	480
$(10,2)$	400
$(10,0)$	200

The largest mileage in the table is 480, obtained when the car uses 9 gallons of gasoline and the moped uses 3 gallons.

Check. Go over the algebra and arithmetic.

State. The maximum number of miles is 480 when the car uses 9 gal of gasoline and the moped uses 3 gal.

15. ***Familiarize.*** Let $x =$ the number of units of lumber and $y =$ the number of units of plywood produced per week.

Translate. The profit P is given by

$P = \$20x + \$30y$.

We wish to maximize P subject to these constraints:

$x + y \leq 400,$

$x \geq 100,$

$y \geq 150.$

Carry out. We graph the system of inequalities, determine the vertices, and evaluate P at each vertex.

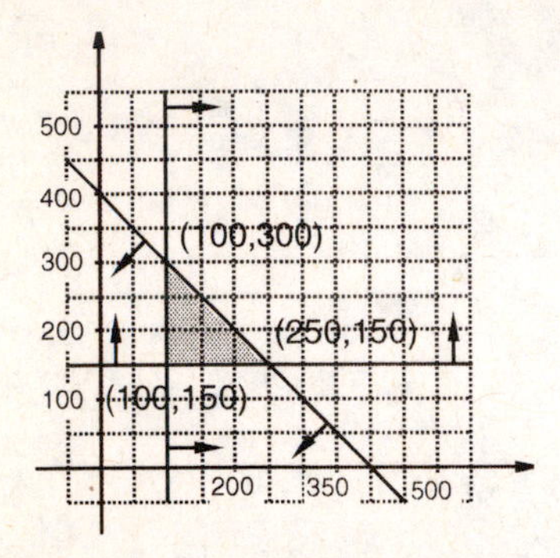

Vertex	$P = \$20x + \$30y$
(100, 150)	$\$20 \cdot 100 + \$30 \cdot 150 = \$6500$
(100, 300)	$\$20 \cdot 100 + \$30 \cdot 300 = \$11,000$
(250, 150)	$\$20 \cdot 250 + \$30 \cdot 150 = \$9500$

The greatest profit in the table is \$11,000, obtained when 100 units of lumber and 300 units of plywood are produced.

Check. Go over the algebra and arithmetic.

State. The maximum profit is achieved by producing 100 units of lumber and 300 units of plywood.

17. In order to earn the most interest Rosa should invest the entire \$40,000. She should also invest as much as possible in the type of investment that has the higher interest rate. Thus, she should invest \$22,000 in corporate bonds and the remaining \$18,000 in municipal bonds. The maximum income is $0.08(\$22,000) + 0.075(\$18,000) = \$3110$.

We can also solve this problem as follows.

Let $x =$ the amount invested in corporate bonds and $y =$ the amount invested in municipal bonds. Find the maximum value of

$I = 0.08x + 0.075y$

subject to

$x + y \leq \$40,000,$

$\$6000 \leq x \leq \$22,000$

$0 \leq y \leq \$30,000.$

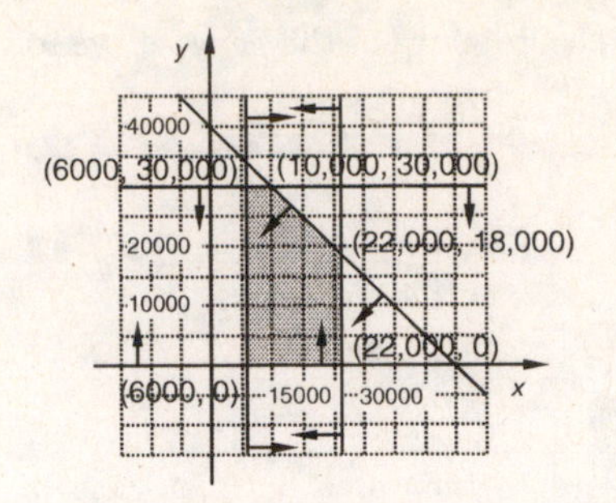

Vertex	$I = 0.08x + 0.075y$
(\$6000, \$0)	\$480
(\$6000, \$30,000)	\$2730
(\$10,000, \$30,000)	\$3050
(\$22,000, \$18,000)	\$3110
(\$22,000, \$0)	\$1760

The maximum income of \$3110 occurs when \$22,000 is invested in corporate bonds and \$18,000 is invested in municipal bonds.

19. ***Familiarize.*** We organize the information in a table. Let $x =$ the number of matching questions and $y =$ the number of essay questions you answer.

Type	Number of points for each	Number answered	Total points
Matching	10	$3 \leq x \leq 12$	$10x$
Essay	25	$4 \leq y \leq 15$	$25y$
Total		$x + y \leq 20$	$10x + 25y$

Since Phil can answer no more than a total of 20 questions, we have the inequality $x + y \leq 20$ in the "Number answered" column. The expression $10x + 25y$ in the "Total points" column gives the total score on the test.

Translate. The score S is given by

$S = 10x + 25y.$

We wish to maximize S subject to these facts (constraints) about x and y.

$3 \leq x \leq 12,$

$4 \leq y \leq 15,$

$x + y \leq 20.$

Carry out. We graph the system of inequalities, determine the vertices, and evaluate S at each vertex.

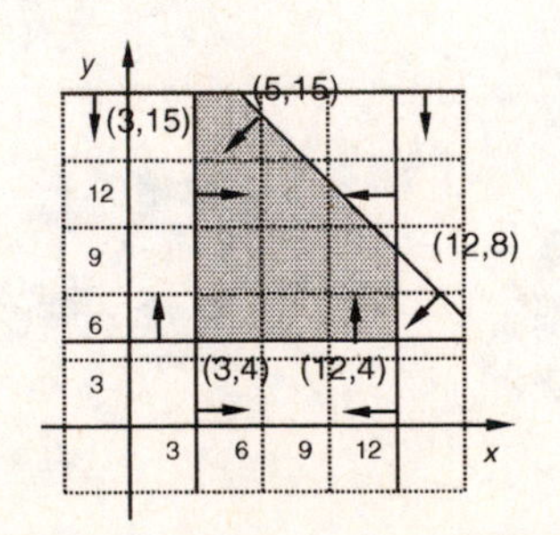

Vertex	$S = 4x + 7y$
$(3,4)$	$10 \cdot 3 + 25 \cdot 4 = 130$
$(3,15)$	$10 \cdot 3 + 25 \cdot 15 = 405$
$(5,15)$	$10 \cdot 5 + 25 \cdot 15 = 425$
$(12,8)$	$10 \cdot 12 + 25 \cdot 8 = 320$
$(12,4)$	$10 \cdot 12 + 25 \cdot 4 = 220$

The greatest score in the table is 425, obtained when 5 matching questions and 15 essay questions are answered correctly.

Check. Go over the algebra and arithmetic.

State. The maximum score is 425 points when 5 matching questions and 15 essay questions are answered correctly.

21. ***Familiarize***. Let x = the Merlot acreage and y = the Cabernet acreage.

Translate. The profit P is given by

$P = \$400x + \$300y$.

We wish to maximize P subject to these constraints:

$x + y \leq 240,$

$2x + y \leq 320,$

$x \geq 0,$

$y \geq 0.$

Carry out. We graph the system of inequalities, determine the vertices, and evaluate P at each vertex.

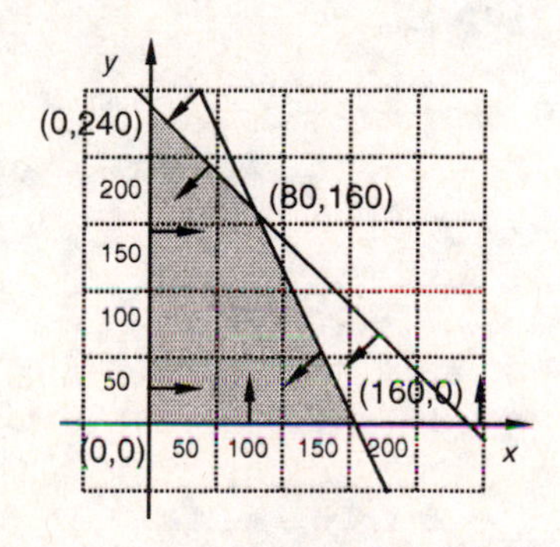

Vertex	$P = \$400x + \$300y$
$(0,0)$	$0
$(0,240)$	$72,000
$(80,160)$	$80,000
$(160,0)$	$64,000

Check. Go over the algebra and arithmetic.

State. The maximum profit occurs by planting 80 acres of Merlot grapes and 160 acres of Cabernet graphs.

23. ***Familiarize***. Let x = the number of knit suits and y = the number of worsted suits made per day.

Translate. The profit P is given by

$P = \$68x + \$62y$.

We wish to maximize P subject to these constraints.

$2x + 4y \leq 20,$

$4x + 2y \leq 16,$

$x \geq 0,$

$y \geq 0.$

Carry out. Graph the system of inequalities, determine the vertices, the evaluate P at each vertex.

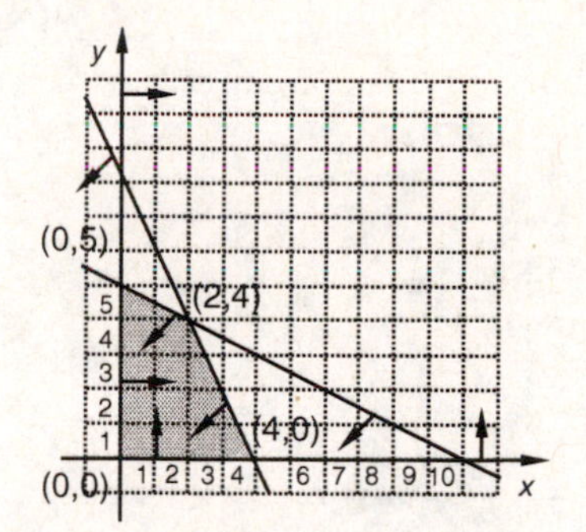

Vertex	$P = \$68x + \$62y$
$(0,0)$	$\$68 \cdot 0 + \$62 \cdot 0 = \$0$
$(0,5)$	$\$68 \cdot 0 + \$62 \cdot 5 = \$310$
$(2,4)$	$\$68 \cdot 2 + \$62 \cdot 4 = \$384$
$(4,0)$	$\$68 \cdot 4 + \$62 \cdot 0 = \$272$

Check. Go over the algebra and arithmetic.

State. The maximum profit per day is $384 when 2 knit suits and 4 worsted suits are made.

25. *Writing Exercise*

27.

$$\begin{aligned} &5x^3 - 4x^2 - 7x + 2 \\ &= 5(-2)^3 - 4(-2)^2 - 7(-2) + 2 \\ &= 5(-8) - 4(4) - 7(-2) + 2 \\ &= -40 - 16 + 14 + 2 \\ &= -40 \end{aligned}$$

29. $3(2x - 5) + 4(x + 5) = 6x - 15 + 4x + 20 = 10x + 5$

31. $6x - 3(x + 2) = 6x - 3x - 6 = 3x - 6$

33. *Writing Exercise*

35. ***Familiarize***. Let x represent the number of T3 planes and y represent the number of S5 planes. Organize the information in a table.

Plane	Number of planes	Passengers		
		First	Tourist	Economy
T3	x	$40x$	$40x$	$120x$
S5	y	$80y$	$30y$	$40y$

Plane	Cost per mile
T3	$30x$
S5	$25y$

Translate. Suppose C is the total cost per mile. Then $C = 30x + 25y$. We wish to minimize C subject to these facts (constraints) about x and y.

$$40x + 80y \geq 2000,$$
$$40x + 30y \geq 1500,$$
$$120x + 40y \geq 2400,$$
$$x \geq 0,$$
$$y \geq 0$$

Carry out. Graph the system of inequalities, determine the vertices, and evaluate C at each vertex.

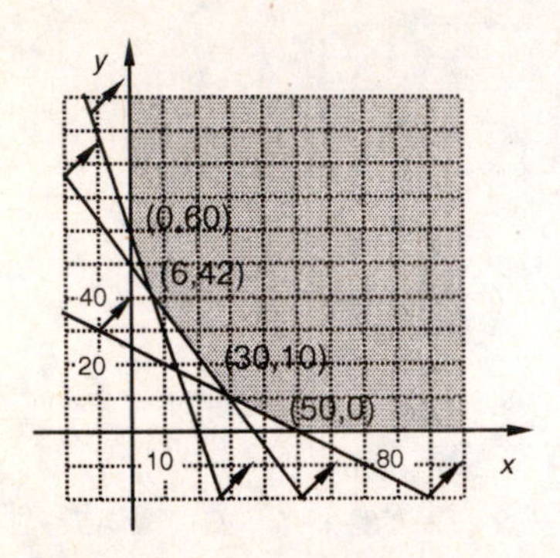

Vertex	$C = 30x + 25y$
$(0, 60)$	$30(0) + 25(60) = 1500$
$(6, 42)$	$30(6) + 25(42) = 1230$
$(30, 10)$	$30(30) + 25(10) = 1150$
$(50, 0)$	$30(50) + 25(0) = 1500$

Check. Go over the algebra and arithmetic.

State. In order to minimize the operating cost, 30 T3 planes and 10 S5 planes should be used.

37. ***Familiarize***. Let x = the number of chairs and y = the number of sofas produced.

Translate. Find the maximum value of

$$I = \$80x + \$1200y$$

subject to

$$20x + 100y \leq 1900,$$
$$x + 50y \leq 500,$$
$$2x + 20y \leq 240,$$
$$x \geq 0,$$
$$y \geq 0.$$

Carry out. Graph the system of inequalities, determine the vertices, and evaluate I at each vertex.

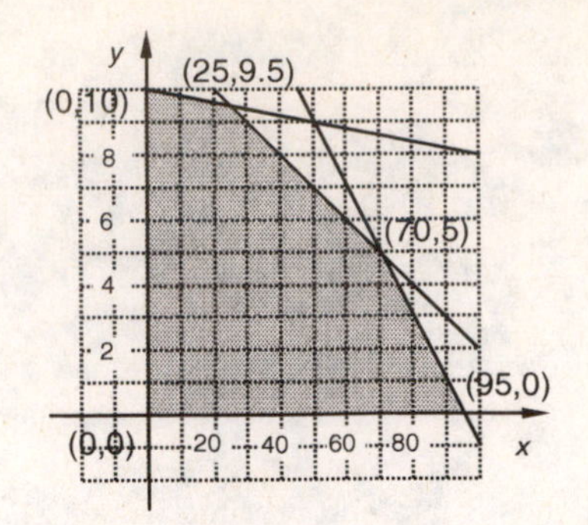

Vertex	$I = \$80x + \$1200y$
$(0, 0)$	$0
$(0, 10)$	$12,000
$(25, 9.5)$	$13,400
$(70, 5)$	$11,600
$(95, 0)$	$7600

Check. Go over the algebra and arithmetic.

State. The maximum income of $13,400 occurs when 25 chairs and 9.5 sofas are made. A more practical answer is that the maximum income of $12,800 is achieved when 25 chairs and 9 sofas are made.

Chapter 10

Exponents and Radicals

Exercise Set 10.1

1. two; see page 634 in the text.

3. positive; see Example 4.

5. irrational; see page 635 in the text.

7. nonnegative; see page 639 in the text.

9. The square roots of 49 are 7 and -7, because $7^2 = 49$ and $(-7)^2 = 49$.

11. The square roots of 144 are 12 and -12, because $12^2 = 144$ and $(-12)^2 = 144$.

13. The square roots of 400 are 20 and -20, because $20^2 = 400$ and $(-20)^2 = 400$.

15. The square roots of 900 are 30 and -30, because $30^2 = 900$ and $(-30)^2 = 900$.

17. $-\sqrt{\dfrac{36}{49}} = -\dfrac{6}{7}$ Since $\sqrt{\dfrac{36}{49}} = \dfrac{6}{7}$, $-\sqrt{\dfrac{36}{49}} = -\dfrac{6}{7}$.

19. $\sqrt{441} = 21$ Remember, $\sqrt{}$ indicates the principle square root.

21. $-\sqrt{\dfrac{16}{81}} = -\dfrac{4}{9}$ Since $\sqrt{\dfrac{16}{81}} = \dfrac{4}{9}$, $-\sqrt{\dfrac{16}{81}} = -\dfrac{4}{9}$.

23. $\sqrt{0.04} = 0.2$

25. $-\sqrt{-0.0025} = -0.05$

27. $5\sqrt{p^2+4}$

The radicand is the expression written under the radical sign, $p^2 + 4$.

Since the index is not written, we know it is 2.

29. $x^2y^2\sqrt{\dfrac{x}{y+4}}$

The radicand is the expression written under the radical sign, $\dfrac{x}{y+4}$.

The index is 3.

31.
$$\begin{aligned} f(t) &= \sqrt{5t-10} \\ f(6) &= \sqrt{5\cdot 6-10} = \sqrt{20} \\ f(2) &= \sqrt{5\cdot 2-10} = \sqrt{0} = 0 \\ f(1) &= \sqrt{5\cdot 1-10} = \sqrt{-5} \end{aligned}$$

Since negative numbers do not have real-number square roots, $f(1)$ does not exist.

$$f(-1) = \sqrt{5(-1)-10} = \sqrt{-15}$$

Since negative numbers do not have real-number square roots, $f(-1)$ does not exist.

33.
$$\begin{aligned} t(x) &= -\sqrt{2x+1} \\ t(4) &= -\sqrt{2\cdot 4+1} = -\sqrt{9} = -3 \\ t(0) &= -\sqrt{2\cdot 0+1} = -\sqrt{1} = -1 \\ t(-1) &= -\sqrt{2(-1)+1} = -\sqrt{-1}; \end{aligned}$$

$t(-1)$ does not exist.

$$t\left(-\frac{1}{2}\right) = -\sqrt{2\left(-\frac{1}{2}\right)+1} = -\sqrt{0} = 0$$

35.
$$\begin{aligned} f(t) &= \sqrt{t^2+1} \\ f(0) &= \sqrt{0^2+1} = \sqrt{1} = 1 \\ f(-1) &= \sqrt{(-1)^2+1} = \sqrt{2} \\ f(-10) &= \sqrt{(-10)^2+1} = \sqrt{101} \end{aligned}$$

37.
$$\begin{aligned} g(x) &= \sqrt{x^3+9} \\ g(-2) &= \sqrt{(-2)^3+9} = \sqrt{1} = 1 \\ g(-3) &= \sqrt{(-3)^3+9} = \sqrt{-18}; \end{aligned}$$

$g(-3)$ does not exist.

$$g(3) = \sqrt{3^3+9} = \sqrt{36} = 6$$

39. $\sqrt{36x^2} = \sqrt{(6x)^2} = |6x| = 6|x|$

Since x might be negative, absolute-value notation is necessary.

41. $\sqrt{(-6b)^2} = |-6b| = |-6|\cdot|b| = 6|b|$

Since b might be negative, absolute-value notation is necessary.

43. $\sqrt{(8-t)^2} = |8-t|$

Since $8-t$ might be negative, absolute-value notation is necessary.

45. $\sqrt{y^2+16y+64} = \sqrt{(y+8)^2} = |y+8|$

Since $y+8$ might be negative, absolute-value notation is necessary.

47. $\sqrt{4x^2+28x+49} = \sqrt{(2x+7)^2} = |2x+7|$

Since $2x+7$ might be negative, absolute-value notation is necessary.

49. $\sqrt[4]{256} = 4$ Since $4^4 = 256$

51. $\sqrt[5]{-1} = -1$ Since $(-1)^5 = -1$

53. $\sqrt[5]{-\dfrac{32}{243}} = -\dfrac{2}{3}$ Since $\left(-\dfrac{2}{3}\right)^5 = -\dfrac{32}{243}$

55. $\sqrt[6]{x^6} = |x|$

The index is even. Use absolute-value notation since x could have a negative value.

57. $\sqrt[4]{(6a)^4} = |6a| = 6|a|$

The index is even. Use absolute-value notation since a could have a negative value.

59. $\sqrt[10]{(-6)^{10}} = |-6| = 6$

61. $\sqrt[414]{(a+b)^{414}} = |a+b|$

The index is even. Use absolute-value notation since $a+b$ could have a negative value.

63. $\sqrt{a^{22}} = |a^{11}|$ Note that $(a^{11})^2 = a^{22}$; a^{11} could have a negative value.

65. $\sqrt{-25}$ is not a real number, so $\sqrt{-25}$ cannot be simplified.

67. $\sqrt{16x^2} = \sqrt{(4x)^2} = 4x$ Assuming x is nonnegative

69. $\sqrt{(3c)^2} = 3c$ Assuming $3c$ is nonnegative

71. $\sqrt{(a+1)^2} = a+1$ Assuming $a+1$ is nonnegative

73. $\sqrt{4x^2+8x+4} = \sqrt{4(x^2+2x+1)} =$
$\sqrt{[2(x+1)]^2} = 2(x+1)$, or $2x+2$

75. $\sqrt{9t^2-12t+4} = \sqrt{(3t-2)^2} = 3t-2$

77. $\sqrt[3]{27} = 3$ $(3^3 = 27)$

79. $\sqrt[4]{16x^4} = \sqrt[4]{(2x)^4} = 2x$

81. $\sqrt[3]{-216} = -6$ $[(-6)^3 = -216]$

83. $-\sqrt[3]{-125y^3} = -(-5y)$ $[(-5y)^3 = -125y^3]$
$= 5y$

85. $\sqrt{t^{18}} = \sqrt{(t^9)^2} = t^9$

87. $\sqrt{(x-2)^8} = \sqrt{[(x-2)^4]^2} = (x-2)^4$

89.
$$\begin{aligned} f(x) &= \sqrt[3]{x+1} \\ f(7) &= \sqrt[3]{7+1} = \sqrt[3]{8} = 2 \\ f(26) &= \sqrt[3]{26+1} = \sqrt[3]{27} = 3 \\ f(-9) &= \sqrt[3]{-9+1} = \sqrt[3]{-8} = -2 \\ f(-65) &= \sqrt[3]{-65+1} = \sqrt[3]{-64} = -4 \end{aligned}$$

91.
$$\begin{aligned} g(t) &= \sqrt[4]{t-3} \\ g(19) &= \sqrt[4]{19-3} = \sqrt[4]{16} = 2 \\ g(-13) &= \sqrt[4]{-13-3} = \sqrt[4]{-16}; \\ &\quad g(-13) \text{ does not exist.} \\ g(1) &= \sqrt[4]{1-3} = \sqrt[4]{-2}; \\ &\quad g(1) \text{ does not exist.} \\ g(84) &= \sqrt[4]{84-3} = \sqrt[4]{81} = 3 \end{aligned}$$

93. $f(x) = \sqrt{x-6}$

Since the index is even, the radicand, $x-6$, must be nonnegative. We solve the inequality:

$$\begin{aligned} x - 6 &\geq 0 \\ x &\geq 6 \end{aligned}$$

Domain of $f = \{x|x \geq 6\}$, or $[6, \infty)$

95. $g(t) = \sqrt[4]{t+8}$

Since the index is even, the radicand, $t+8$, must be nonnegative. We solve the inequality:

$$\begin{aligned} t + 8 &\geq 0 \\ t &\geq -8 \end{aligned}$$

Domain of $g = \{t|t \geq -8\}$, or $[-8, \infty)$

97. $g(x) = \sqrt[4]{2x-10}$

Since the index is even, the radicand, $2x-10$, must be nonnegative. We solve the inequality:

$$\begin{aligned} 2x - 10 &\geq 0 \\ 2x &\geq 10 \\ x &\geq 5 \end{aligned}$$

Domain of $g = \{x|x \geq 5\}$, or $[5, \infty)$

99. $f(t) = \sqrt[5]{8-3t}$

Since the index is odd, the radicand can be any real number.

Domain of $f = \{t|t$ is a real number$\}$, or $(-\infty, \infty)$

101. $h(z) = -\sqrt[6]{5z+2}$

Since the index is even, the radicand, $5z+2$, must be nonnegative. We solve the inequality:

$$\begin{aligned} 5z + 2 &\geq 0 \\ 5z &\geq -2 \\ z &\geq -\frac{2}{5} \end{aligned}$$

Domain of $h = \left\{z \middle| z \geq -\frac{2}{5}\right\}$, or $\left[-\frac{2}{5}, \infty\right)$

103. $f(t) = 7 + \sqrt[8]{t^8}$

Since we can compute $7+\sqrt[8]{t^8}$ for any real number t, the domain is the set of real numbers, or $\{x|x$ is a real number$\}$, or $(-\infty, \infty)$.

105. *Writing Exercise*

107. $(a^3b^2c^5)^3 = a^{3\cdot3}b^{2\cdot3}c^{5\cdot3} = a^9b^6c^{15}$

109. $(2a^{-2}b^3c^{-4})^{-3} = 2^{-3}a^{-2(-3)}b^{3(-3)}c^{-4(-3)} =$
$\frac{1}{2^3}a^6b^{-9}c^{12} = \frac{a^6c^{12}}{8b^9}$

111. $\frac{8x^{-2}y^5}{4x^{-6}z^{-2}} = \frac{8}{4}x^{-2-(-6)}y^5z^2 = 2x^4y^5z^2$

113. *Writing Exercise*

115. $M = -5 + \sqrt{6.7x - 444}$

a) Substitute 300 for x.

$$\begin{aligned} M &= -5 + \sqrt{6.7(300) - 444} \\ &= -5 + \sqrt{1566} \\ &\approx 34.6 \text{ lb} \end{aligned}$$

b) Substitute 100 for x.

$$\begin{aligned} M &= -5 + \sqrt{6.7(100) - 444} \\ &= -5 + \sqrt{226} \\ &\approx 10.0 \text{ lb} \end{aligned}$$

c) Substitute 200 for x.

$$\begin{aligned} M &= -5 + \sqrt{6.7(200) - 444} \\ &= -5 + \sqrt{896} \\ &\approx 24.9 \text{ lb} \end{aligned}$$

d) Substitute 400 for x.

$$\begin{aligned} M &= -5 + \sqrt{6.7(400) - 444} \\ &= -5 + \sqrt{2236} \\ &\approx 42.3 \text{ lb} \end{aligned}$$

117. $f(x) = \sqrt{x+5}$

Since the index is even, the radicand, $x + 5$, must be nonnegative. Solve:

$$\begin{aligned} x + 5 &\geq 0 \\ x &\geq -5 \end{aligned}$$

Domain of $f = \{x|x \geq -5\}$, or $[-5, \infty)$

Make a table of values, keeping in mind that x must be -5 or greater. Plot these points and draw the graph.

x	$f(x)$
-5	0
-4	1
-1	2
1	2.4
3	2.8
4	3

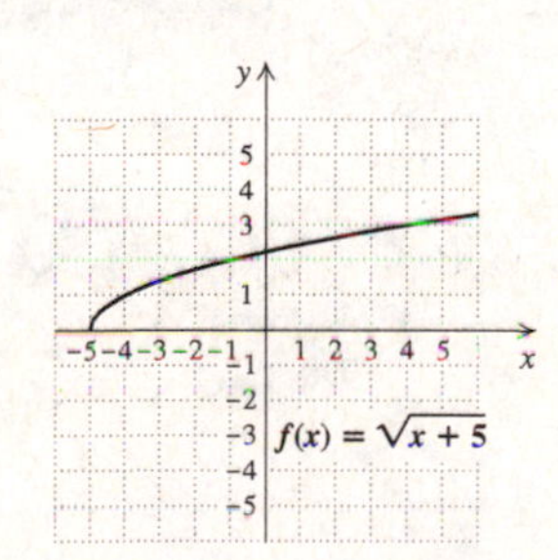

119. $g(x) = \sqrt{x} - 2$

Since the index is even, the radicand, x, must be nonnegative, so we have $x \geq 0$.

Domain of $g = \{x|x \geq 0\}$, or $[0, \infty)$

Make a table of values, keeping in mind that x must be nonnegative. Plot these points and draw the graph.

x	$g(x)$
0	-2
1	-1
4	0
6	0.5
8	0.8

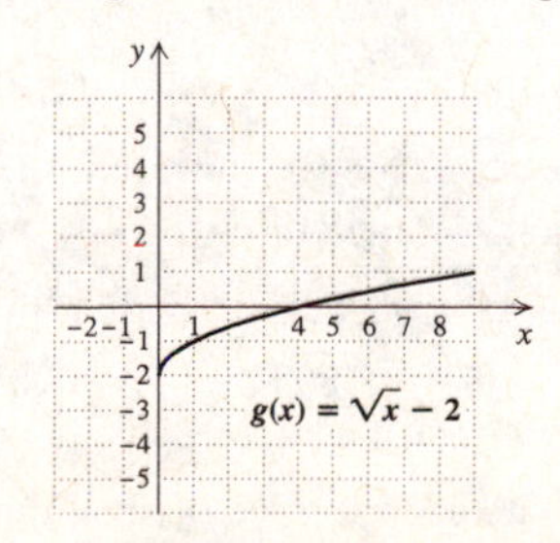

121. $f(x) = \dfrac{\sqrt{x+3}}{\sqrt[4]{2-x}}$

In the numerator we must have $x + 3 \geq 0$, or $x \geq -3$, and in the denominator we must have $2 - x > 0$, or $x < 2$. Thus, we have $x \geq -3$ *and* $x < 2$, so

Domain of $f = \{x| -3 \leq x < 2\}$, or $[-3, 2)$.

123.

Exercise Set 10.2

1. Choice (g) is correct because $a^{m/n} = \sqrt[n]{a^m}$.

3. $x^{-5/2} = \dfrac{1}{x^{5/2}} = \dfrac{1}{(\sqrt{x})^5}$, so choice (e) is correct.

5. $x^{1/5} \cdot x^{2/5} = x^{1/5+2/5} = x^{3/5}$, so choice (a) is correct.

7. Choice (b) is correct because $\sqrt[n]{a^m}$ and $(\sqrt[n]{a})^m$ are equivalent.

9. $x^{1/6} = \sqrt[6]{x}$

11. $(16)^{1/2} = \sqrt{16} = 4$

13. $81^{1/4} = \sqrt[4]{81} = 3$

15. $9^{1/2} = \sqrt{9} = 3$

17. $(xyz)^{1/3} = \sqrt[3]{xyz}$

19. $(a^2b^2)^{1/5} = \sqrt[5]{a^2b^2}$

21. $t^{2/5} = \sqrt[5]{t^2}$

23. $16^{3/4} = \sqrt[4]{16^3} = (\sqrt[4]{16})^3 = 2^3 = 8$

25. $27^{4/3} = \sqrt[3]{27^4} = (\sqrt[3]{27})^4 = 3^4 = 81$

27. $(81x)^{3/4} = \sqrt[4]{(81x)^3} = \sqrt[4]{81^3x^3}$, or $\sqrt[4]{81^3} \cdot \sqrt[4]{x^3} = (\sqrt[4]{81})^3 \cdot \sqrt[4]{x^3} = 3^3\sqrt[4]{x^3} = 27\sqrt[4]{x^3}$

29. $(25x^4)^{3/2} = \sqrt{(25x^4)^3} = \sqrt{25^3 \cdot x^{12}} = \sqrt{25^3} \cdot \sqrt{x^{12}} = (\sqrt{25})^3x^6 = 5^3x^6 = 125x^6$

31. $\sqrt[3]{20} = 20^{1/3}$

33. $\sqrt{17} = 17^{1/2}$

35. $\sqrt{x^3} = x^{3/2}$

37. $\sqrt[5]{m^2} = m^{2/5}$

39. $\sqrt[4]{cd} = (cd)^{1/4}$ Parentheses are required.

41. $\sqrt[5]{xy^2z} = (xy^2z)^{1/5}$

43. $(\sqrt{3mn})^3 = (3mn)^{3/2}$

45. $(\sqrt[7]{8x^2y})^5 = (8x^2y)^{5/7}$

47. $\dfrac{2x}{\sqrt[3]{z^2}} = \dfrac{2x}{z^{2/3}}$

49. $x^{-1/3} = \dfrac{1}{x^{1/3}}$

51. $(2rs)^{-3/4} = \frac{1}{(2rs)^{3/4}}$

53. $\left(\frac{1}{16}\right)^{-3/4} = \left(\frac{16}{1}\right)^{3/4} = (2^4)^{3/4} = 2^{4(3/4)} = 2^3 = 8$

55. $\frac{2c}{a^{-3/5}} = 2a^{3/5}c$

57. $5x^{-2/3}y^{4/5}z = 5 \cdot \frac{1}{x^{2/3}} \cdot y^{4/5} \cdot z = \frac{5y^{4/5}z}{x^{2/3}}$

59. $3^{-5/2}a^3b^{-7/3} = \frac{1}{3^{5/2}} \cdot a^3 \cdot \frac{1}{b^{7/3}} = \frac{a^3}{3^{5/2}b^{7/3}}$

61. $\left(\frac{2ab}{3c}\right)^{-5/6} = \left(\frac{3c}{2ab}\right)^{5/6}$ Finding the reciprocal of the base and changing the sign of the exponent

63. $\frac{6a}{\sqrt[4]{b}} = \frac{6a}{b^{1/4}}$

65. $7^{3/4} \cdot 7^{1/8} = 7^{3/4+1/8} = 7^{6/8+1/8} = 7^{7/8}$

We added exponents after finding a common denominator.

67. $\frac{3^{5/8}}{3^{-1/8}} = 3^{5/8-(-1/8)} = 3^{5/8+1/8} = 3^{6/8} = 3^{3/4}$

We subtracted exponents and simplified.

69. $\frac{5.2^{-1/6}}{5.2^{-2/3}} = 5.2^{-1/6-(-2/3)} = 5.2^{-1/6+2/3} =$

$5.2^{-1/6+4/6} = 5.2^{3/6} = 5.2^{1/2}$

We subtracted exponents after finding a common denominator. Then we simplified.

71. $(10^{3/5})^{2/5} = 10^{3/5 \cdot 2/5} = 10^{6/25}$

We multiplied exponents.

73. $a^{2/3} \cdot a^{5/4} = a^{2/3+5/4} = a^{8/12+15/12} = a^{23/12}$

We added exponents after finding a common denominator.

75. $(64^{3/4})^{4/3} = 64^{\frac{3}{4}\cdot\frac{4}{3}} = 64^1 = 64$

77. $(m^{2/3}n^{-1/4})^{1/2} = m^{2/3 \cdot 1/2}n^{-1/4 \cdot 1/2} = m^{1/3}n^{-1/8} =$

$m^{1/3} \cdot \frac{1}{n^{1/8}} = \frac{m^{1/3}}{n^{1/8}}$

79. $\sqrt[6]{x^4} = x^{4/6}$ Converting to exponential notation

$= x^{2/3}$ Simplifying the exponent

$= \sqrt[3]{x^2}$ Returning to radical notation

81. $\sqrt[4]{a^{12}} = a^{12/4}$ Converting to exponential notation

$= a^3$ Simplifying

83. $\sqrt[5]{a^{10}} = a^{10/5}$ Converting to exponential notation

$= a^2$ Simplifying

85. $(\sqrt[7]{xy})^{14} = (xy)^{14/7}$ Converting to exponential notation

$= (xy)^2$ Simplifying the exponent

$= x^2y^2$ Using the laws of exponents

87. $\sqrt[4]{(7a)^2} = (7a)^{2/4}$ Converting to exponential notation

$= (7a)^{1/2}$ Simplifying the exponent

$= \sqrt{7a}$ Returning to radical notation

89. $(\sqrt[8]{2x})^6 = (2x)^{6/8}$ Converting to exponential notation

$= (2x)^{3/4}$ Simplifying the exponent

$= \sqrt[4]{(2x)^3}$ Returning to radical notation

$= \sqrt[4]{8x^3}$ Using the laws of exponents

91. $\sqrt[3]{\sqrt[6]{a}} = \sqrt[3]{a^{1/6}}$ Converting to

$= (a^{1/6})^{1/3}$ exponential notation

$= a^{1/18}$ Using the laws of exponents

$= \sqrt[18]{a}$ Returning to radical notation

93. $\sqrt[4]{(xy)^{12}} = (xy)^{12/4}$ Converting to exponential notation

$= (xy)^3$ Simplifying the exponent

$= x^3y^3$ Using the laws of exponents

95. $(\sqrt[5]{a^2b^4})^{15} = (a^2b^4)^{15/5}$ Converting to exponential notation

$= (a^2b^4)^3$ Simplifying the exponent

$= a^6b^{12}$ Using the laws of exponents

97. $\sqrt[3]{\sqrt[4]{xy}} = \sqrt[3]{(xy)^{1/4}}$ Converting to

$= [(xy)^{1/4}]^{1/3}$ exponential notation

$= (xy)^{1/12}$ Using a law of exponents

$= \sqrt[12]{xy}$ Returning to radical notation

99. *Writing Exercise*

101. $3x(x^3 - 2x^2) + 4x^2(2x^2 + 5x)$

$= 3x^4 - 6x^3 + 8x^4 + 20x^3$

$= 11x^4 + 14x^3$

103. $(3a - 4b)(5a + 3b)$

$= 3a \cdot 5a + 3a \cdot 3b - 4b \cdot 5a - 4b \cdot 3b$

$= 15a^2 + 9ab - 20ab - 12b^2$

$= 15a^2 - 11ab - 12b^2$

105. ***Familiarize.*** Let p = the selling price of the home.

Translate.

0.5% of the selling price is $467.50

$0.005p = 467.50$

Carry out. We solve the equation.

$0.005p = 467.50$

$p = 93,500$ Dividing by 0.005

Check. 0.5% of $93,500 is 0.005($93,500), or $467.50. The answer checks.

State. The selling price of the home was $93,500.

107. *Writing Exercise*

109. $\sqrt{x\sqrt[3]{x^2}} = \sqrt{x \cdot x^{2/3}} = (x^{5/3})^{1/2} = x^{5/6} = \sqrt[6]{x^5}$

111. $\sqrt[12]{p^2+2pq+q^2} = \sqrt[12]{(p+q)^2} = [(p+q)^2]^{1/12} = (p+q)^{2/12} = (p+q)^{1/6} = \sqrt[6]{p+q}$

113. $2^{7/12} \approx 1.498 \approx 1.5$ so the G that is 7 half steps above middle C has a frequency that is about 1.5 times that of middle C.

115. $P = \dfrac{r^{1.83}}{r^{1.83}+\sigma^{1.83}}$

$= \dfrac{799^{1.83}}{799^{1.83}+749^{1.83}}$

$\approx 53.0\%$ Using a calculator

117. $r(1900) = 10^{-12}2^{-1900/5700} \approx 10^{-12}(0.7937)$, or 7.937×10^{-13}. The ratio is about 7.937×10^{-13} to 1.

119.

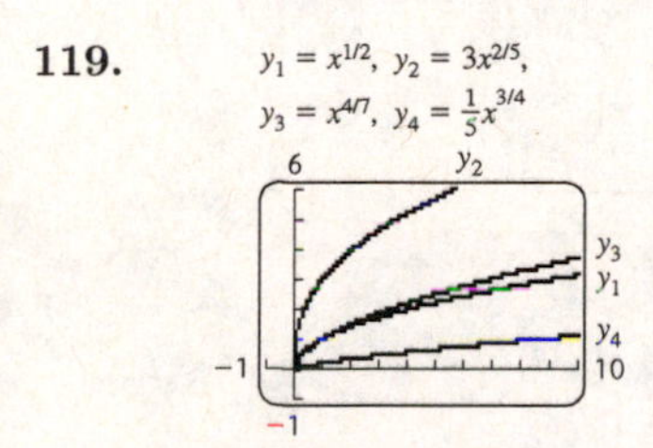

Exercise Set 10.3

1. True; see page 651 in the text.

3. False; see page 652 in the text.

5. True; see page 652 in the text.

7. $\sqrt{5}\sqrt{7} = \sqrt{5 \cdot 7} = \sqrt{35}$

9. $\sqrt[3]{7}\sqrt[3]{2} = \sqrt[3]{7 \cdot 2} = \sqrt[3]{14}$

11. $\sqrt[4]{6}\sqrt[4]{3} = \sqrt[4]{6 \cdot 3} = \sqrt[4]{18}$

13. $\sqrt{2x}\sqrt{13y} = \sqrt{2x \cdot 13y} = \sqrt{26xy}$

15. $\sqrt[5]{8y^3}\sqrt[5]{10y} = \sqrt[5]{8y^3 \cdot 10y} = \sqrt[5]{80y^4}$

17. $\sqrt{y-b}\sqrt{y+b} = \sqrt{(y-b)(y+b)} = \sqrt{y^2-b^2}$

19. $\sqrt[3]{0.7y}\sqrt[3]{0.3y} = \sqrt[3]{0.7y \cdot 0.3y} = \sqrt[3]{0.21y^2}$

21. $\sqrt[5]{x-2}\sqrt[5]{(x-2)^2} = \sqrt[5]{(x-2)(x-2)^2} = \sqrt[5]{(x-2)^3}$

23. $\sqrt{\dfrac{7}{t}}\sqrt{\dfrac{s}{11}} = \sqrt{\dfrac{7}{t} \cdot \dfrac{s}{11}} = \sqrt{\dfrac{7s}{11t}}$

25. $\sqrt[7]{\dfrac{x-3}{4}}\sqrt[7]{\dfrac{5}{x+2}} = \sqrt[7]{\dfrac{x-3}{4} \cdot \dfrac{5}{x+2}} = \sqrt[7]{\dfrac{5x-15}{4x+8}}$

27. $\sqrt{18}$

$= \sqrt{9 \cdot 2}$ 9 is the largest perfect square factor of 18.

$= \sqrt{9} \cdot \sqrt{2}$

$= 3\sqrt{2}$

29. $\sqrt{27}$

$= \sqrt{9 \cdot 3}$ 9 is the largest perfect square factor of 27.

$= \sqrt{9} \cdot \sqrt{3}$

$= 3\sqrt{3}$

31. $\sqrt{8} = \sqrt{4 \cdot 2} = \sqrt{4} \cdot \sqrt{2} = 2\sqrt{2}$

33. $\sqrt{198} = \sqrt{9 \cdot 22} = \sqrt{9} \cdot \sqrt{22} = 3\sqrt{22}$

35. $\sqrt{36a^4b}$

$= \sqrt{36a^4 \cdot b}$ $36a^4$ is a perfect square.

$= \sqrt{36a^4} \cdot \sqrt{b}$ Factoring into two radicals

$= 6a^2\sqrt{b}$ Taking the square root of $36a^4$

37. $\sqrt[3]{8x^3y^2}$

$= \sqrt[3]{8x^3 \cdot y^2}$ $8x^3$ is a perfect cube.

$= \sqrt[3]{8x^3} \cdot \sqrt[3]{y^2}$ Factoring into two radicals

$= 2x\sqrt[3]{y^2}$ Taking the cube root of $8x^3$

39. $\sqrt[3]{-16x^6}$

$= \sqrt[3]{-8x^6 \cdot 2}$ $-8x^6$ is a perfect cube.

$= \sqrt[3]{-8x^6} \cdot \sqrt[3]{2}$

$= -2x^2\sqrt[3]{2}$ Taking the cube root of $-8x^6$

41. $f(x) = \sqrt[3]{125x^5}$

$= \sqrt[3]{125x^3 \cdot x^2}$

$= \sqrt[3]{125x^3} \cdot \sqrt[3]{x^2}$

$= 5x\sqrt[3]{x^2}$

43. $f(x) = \sqrt{49(x-3)^2}$ $49(x-3)^2$ is a perfect square.

$= |7(x-3)|$, or $7|x-3|$

45. $f(x) = \sqrt{5x^2-10x+5}$

$= \sqrt{5(x^2-2x+1)}$

$= \sqrt{5(x-1)^2}$

$= \sqrt{(x-1)^2} \cdot \sqrt{5}$

$= |x-1|\sqrt{5}$

47. $\sqrt{a^6b^7}$

$= \sqrt{a^6 \cdot b^6 \cdot b}$ Identifying the largest even powers of a and b

$= \sqrt{a^6}\sqrt{b^6}\sqrt{b}$ Factoring into several radicals

$= a^3b^3\sqrt{b}$

49. $\sqrt[3]{x^5y^6z^{10}}$

$= \sqrt[3]{x^3 \cdot x^2 \cdot y^6 \cdot z^9 \cdot z}$ Identifying the largest perfect-cube powers of x, y, and z

$= \sqrt[3]{x^3} \cdot \sqrt[3]{y^6} \cdot \sqrt[3]{z^9} \cdot \sqrt[3]{x^2z}$ Factoring into several radicals

$= xy^2z^3\sqrt[3]{x^2z}$

51. $\sqrt[5]{-32a^7b^{11}} = \sqrt[5]{-32 \cdot a^5 \cdot a^2 \cdot b^{10} \cdot b} =$
$\sqrt[5]{-32}\sqrt[5]{a^5}\sqrt[5]{b^{10}}\sqrt[5]{a^2b} = -2ab^2\sqrt[5]{a^2b}$

53. $\sqrt[5]{x^{13}y^8z^{17}} = \sqrt[5]{x^{10} \cdot x^3 \cdot y^5 \cdot y^3 \cdot z^{15} \cdot z^2} =$
$\sqrt[5]{x^{10}}\sqrt[5]{y^5}\sqrt[5]{z^{15}}\sqrt[5]{x^3y^3z^2} = x^2yz^3\sqrt[5]{x^3y^3z^2}$

55. $\sqrt[3]{-80a^{14}} = \sqrt[3]{-8 \cdot 10 \cdot a^{12} \cdot a^2} =$
$\sqrt[3]{-8} \cdot \sqrt[3]{a^{12}} \cdot \sqrt[3]{10a^2} = -2a^4\sqrt[3]{10a^2}$

57. $\sqrt{6}\sqrt{3} = \sqrt{6 \cdot 3} = \sqrt{18} = \sqrt{9 \cdot 2} = 3\sqrt{2}$

59. $\sqrt{15}\sqrt{21} = \sqrt{15 \cdot 21} = \sqrt{315} = \sqrt{9 \cdot 35} = 3\sqrt{35}$

61. $\sqrt[3]{9}\sqrt[3]{3} = \sqrt[3]{9 \cdot 3} = \sqrt[3]{27} = 3$

63. $\sqrt{18a^3}\sqrt{18a^3} = \sqrt{(18a^3)^2} = 18a^3$

65. $\sqrt[3]{5a^2}\sqrt[3]{2a} = \sqrt[3]{5a^2 \cdot 2a} = \sqrt[3]{10a^3} = \sqrt[3]{a^3 \cdot 10} = a\sqrt[3]{10}$

67. $\sqrt{2x^5}\sqrt{10x^2} = \sqrt{20x^7} = \sqrt{4x^6 \cdot 5x} = 2x^3\sqrt{5x}$

69. $\sqrt[3]{s^2t^4}\sqrt[3]{s^4t^6} = \sqrt[3]{s^6t^{10}} = \sqrt[3]{s^6t^9 \cdot t} = s^2t^3\sqrt[3]{t}$

71. $\sqrt[3]{(x+5)^2}\sqrt[3]{(x+5)^4} = \sqrt[3]{(x+5)^6} = (x+5)^2$

73. $\sqrt[4]{20a^3b^7}\sqrt[4]{4a^2b^5} = \sqrt[4]{80a^5b^{12}} = \sqrt[4]{16a^4b^{12} \cdot 5a} =$
$2ab^3\sqrt[4]{5a}$

75. $\sqrt[5]{x^3(y+z)^6}\sqrt[5]{x^3(y+z)^4} = \sqrt[5]{x^6(y+z)^{10}} =$
$\sqrt[5]{x^5(y+z)^{10} \cdot x} = x(y+z)^2\sqrt[5]{x}$

77. *Writing Exercise*

79.
$$\frac{3x}{16y} + \frac{5y}{64x}, \text{ LCD is } 64xy$$
$$= \frac{3x}{16y} \cdot \frac{4x}{4x} + \frac{5y}{64x} \cdot \frac{y}{y}$$
$$= \frac{12x^2}{64xy} + \frac{5y^2}{64xy}$$
$$= \frac{12x^2 + 5y^2}{64xy}$$

81.
$$\frac{4}{x^2-9} - \frac{7}{2x-6}$$
$$= \frac{4}{(x+3)(x-3)} - \frac{7}{2(x-3)}, \text{ LCD is } 2(x+3)(x-3)$$
$$= \frac{4}{(x+3)(x-3)} \cdot \frac{2}{2} - \frac{7}{2(x-3)} \cdot \frac{x+3}{x+3}$$
$$= \frac{8}{2(x+3)(x-3)} - \frac{7(x+3)}{2(x+3)(x-3)}$$
$$= \frac{8 - 7(x+3)}{2(x+3)(x-3)}$$
$$= \frac{8 - 7x - 21}{2(x+3)(x-3)}$$
$$= \frac{-7x - 13}{2(x+3)(x-3)}$$

83. $\frac{9a^4b^7}{3a^2b^5} = \frac{9}{3}a^{4-2}b^{7-5} = 3a^2b^2$

85. *Writing Exercise*

87.
$$R(x) = \frac{1}{2}\sqrt[4]{\frac{x \cdot 3.0 \times 10^6}{\pi^2}}$$
$$R(5 \times 10^4) = \frac{1}{2}\sqrt[4]{\frac{5 \times 10^4 \cdot 3.0 \times 10^6}{\pi^2}}$$
$$= \frac{1}{2}\sqrt[4]{\frac{15 \times 10^{10}}{\pi^2}}$$
$$\approx 175.6 \text{ mi}$$

89. a) $T_w = 33 - \frac{(10.45 + 10\sqrt{8} - 8)(33 - 7)}{22}$
$\approx -3.3°$ C

b) $T_w = 33 - \frac{(10.45 + 10\sqrt{12} - 12)(33 - 0)}{22}$
$\approx -16.6°$ C

c) $T_w = 33 - \frac{(10.45 + 10\sqrt{14} - 14)[33 - (-5)]}{22}$
$\approx -25.5°$ C

d) $T_w = 33 - \frac{(10.45 + 10\sqrt{15} - 15)[33 - (-23)]}{22}$
$\approx -54.0°$ C

91. $(\sqrt[3]{25x^4})^4 = \sqrt[3]{(25x^4)^4} = \sqrt[3]{25^4x^{16}} =$
$\sqrt[3]{25^3 \cdot 25 \cdot x^{15} \cdot x} = \sqrt[3]{25^3}\sqrt[3]{x^{15}}\sqrt[3]{25x} =$
$25x^5\sqrt[3]{25x}$

93. $(\sqrt{a^3b^5})^7 = \sqrt{(a^3b^5)^7} = \sqrt{a^{21}b^{35}} =$
$\sqrt{a^{20} \cdot a \cdot b^{34} \cdot b} = \sqrt{a^{20}}\sqrt{b^{34}}\sqrt{ab} = a^{10}b^{17}\sqrt{ab}$

95.

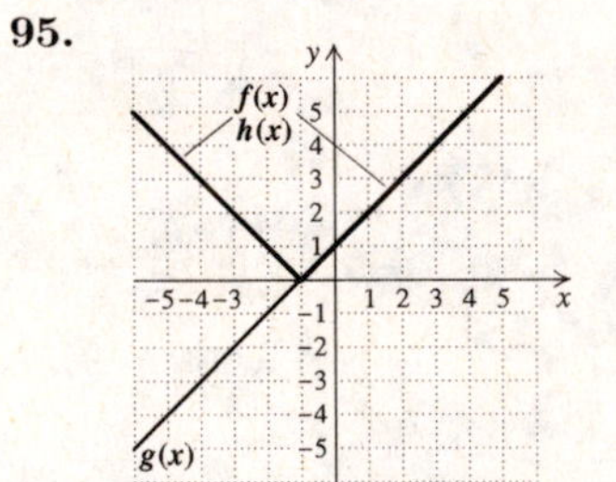

We see that $f(x) = h(x)$ and $f(x) \neq g(x)$.

97. $g(x) = x^2 - 6x + 8$

We must have $x^2 - 6x + 8 \geq 0$, or $(x-2)(x-4) \geq 0$. We graph $y = x^2 - 6x + 8$.

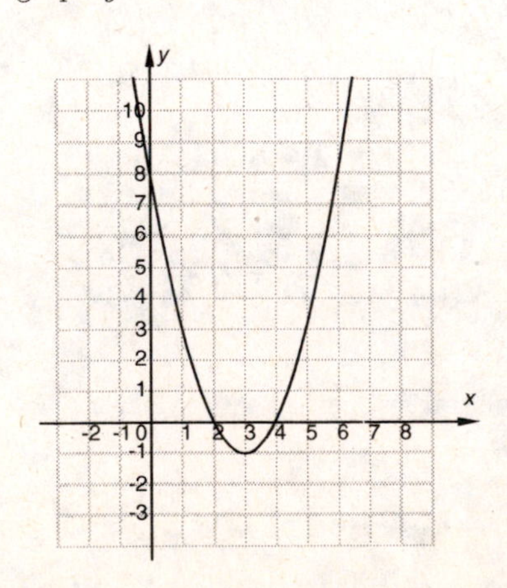

From the graph we see that $y \geq 0$ for $x \leq 2$ or $x \geq 4$, so the domain of g is $\{x|x \leq 2 \text{ or } x \geq 4\}$, or $(-\infty, 2] \cup [4, \infty)$.

99. $\sqrt[5]{4a^{3k+2}}\sqrt[5]{8a^{6-k}} = 2a^4$

$\sqrt[5]{32a^{2k+8}} = 2a^4$

$2\sqrt[5]{a^{2k+8}} = 2a^4$

$\sqrt[5]{a^{2k+8}} = a^4$

$a^{\frac{2k+8}{5}} = a^4$

Since the base is the same, the exponents must be equal. We have:

$\frac{2k+8}{5} = 4$

$2k+8 = 20$

$2k = 12$

$k = 6$

101. *Writing Exercise*

Exercise Set 10.4

1. $\sqrt[3]{\frac{a^2}{b^6}} = \frac{\sqrt[3]{a^2}}{\sqrt[3]{b^6}} = \frac{\sqrt[3]{a^2}}{b^2}$, so choice (e) is correct.

3. $\sqrt[5]{\frac{a^6}{b^4}} = \sqrt[5]{\frac{a^6}{b^4} \cdot \frac{b}{b}} = \sqrt[5]{\frac{a^6 b}{b^4 \cdot b}}$, so choice (f) is correct.

5. $\frac{\sqrt[5]{a^2}}{\sqrt[5]{b^2}} = \frac{\sqrt[5]{a^2}}{\sqrt[5]{b^2}} \cdot \frac{\sqrt[5]{b^3}}{\sqrt[5]{b^3}} = \frac{\sqrt[5]{a^2b^3}}{\sqrt[5]{b^5}}$, so choice (h) is the correct.

7. $\frac{\sqrt[5]{a^2}}{\sqrt[5]{b^3}} = \frac{\sqrt[5]{a^2}}{\sqrt[5]{b^3}} \cdot \frac{\sqrt[5]{b^2}}{\sqrt[5]{b^2}} = \frac{\sqrt[5]{a^2}\sqrt[5]{b^2}}{\sqrt[5]{b^5}}$, so choice (a) is correct.

9. $\sqrt{\frac{36}{25}} = \frac{\sqrt{36}}{\sqrt{25}} = \frac{6}{5}$

11. $\sqrt[3]{\frac{64}{27}} = \frac{\sqrt[3]{64}}{\sqrt[3]{27}} = \frac{4}{3}$

13. $\sqrt{\frac{49}{y^2}} = \frac{\sqrt{49}}{\sqrt{y^2}} = \frac{7}{y}$

15. $\sqrt{\frac{36y^3}{x^4}} = \frac{\sqrt{36y^3}}{\sqrt{x^4}} = \frac{\sqrt{36y^2 \cdot y}}{\sqrt{x^4}} = \frac{\sqrt{36y^2}\sqrt{y}}{\sqrt{x^4}} = \frac{6y\sqrt{y}}{x^2}$

17. $\sqrt[3]{\frac{27a^4}{8b^3}} = \frac{\sqrt[3]{27a^4}}{\sqrt[3]{8b^3}} = \frac{\sqrt[3]{27a^3 \cdot a}}{\sqrt[3]{8b^3}} = \frac{\sqrt[3]{27a^3}\sqrt[3]{a}}{\sqrt[3]{8b^3}} = \frac{3a\sqrt[3]{a}}{2b}$

19. $\sqrt[4]{\frac{16a^4}{b^4c^8}} = \frac{\sqrt[4]{16a^4}}{\sqrt[4]{b^4c^8}} = \frac{2a}{bc^2}$

21. $\sqrt[4]{\frac{a^5b^8}{c^{10}}} = \frac{\sqrt[4]{a^5b^8}}{\sqrt[4]{c^{10}}} = \frac{\sqrt[4]{a^4b^8 \cdot a}}{\sqrt[4]{c^8 \cdot c^2}} = \frac{\sqrt[4]{a^4b^8}\sqrt[4]{a}}{\sqrt[4]{c^8}\sqrt[4]{c^2}} = \frac{ab^2\sqrt[4]{a}}{c^2\sqrt[4]{c^2}}$, or $\frac{ab^2}{c^2}\sqrt[4]{\frac{a}{c^2}}$

23. $\sqrt[5]{\frac{32x^6}{y^{11}}} = \frac{\sqrt[5]{32x^6}}{\sqrt[5]{y^{11}}} = \frac{\sqrt[5]{32x^5 \cdot x}}{\sqrt[5]{y^{10} \cdot y}} = \frac{\sqrt[5]{32x^5} \cdot \sqrt[5]{x}}{\sqrt[5]{y^{10}}\sqrt[5]{y}} = \frac{2x\sqrt[5]{x}}{y^2\sqrt[5]{y}}$, or $\frac{2x}{y^2}\sqrt[5]{\frac{x}{y}}$

25. $\sqrt[6]{\frac{x^6y^8}{z^{15}}} = \frac{\sqrt[6]{x^6y^8}}{\sqrt[6]{z^{15}}} = \frac{\sqrt[6]{x^6y^6 \cdot y^2}}{\sqrt[6]{z^{12} \cdot z^3}} = \frac{\sqrt[6]{x^6y^6}\sqrt[6]{y^2}}{\sqrt[6]{z^{12}}\sqrt[6]{z^3}} = \frac{xy\sqrt[6]{y^2}}{z^2\sqrt[6]{z^3}}$, or $\frac{xy}{z^2}\sqrt[6]{\frac{y^2}{z^3}}$

27. $\frac{\sqrt{35x}}{\sqrt{7x}} = \sqrt{\frac{35x}{7x}} = \sqrt{5}$

29. $\frac{\sqrt[3]{270}}{\sqrt[3]{10}} = \sqrt[3]{\frac{270}{10}} = \sqrt[3]{27} = 3$

31. $\frac{\sqrt{40xy^3}}{\sqrt{8x}} = \sqrt{\frac{40xy^3}{8x}} = \sqrt{5y^3} = \sqrt{y^2 \cdot 5y} = \sqrt{y^2}\sqrt{5y} = y\sqrt{5y}$

33. $\frac{\sqrt[3]{96a^4b^2}}{\sqrt[3]{12a^2b}} = \sqrt[3]{\frac{96a^4b^2}{12a^2b}} = \sqrt[3]{8a^2b} = \sqrt[3]{8}\sqrt[3]{a^2b} = 2\sqrt[3]{a^2b}$

35. $\frac{\sqrt{100ab}}{5\sqrt{2}} = \frac{1}{5}\frac{\sqrt{100ab}}{\sqrt{2}} = \frac{1}{5}\sqrt{\frac{100ab}{2}} = \frac{1}{5}\sqrt{50ab} = \frac{1}{5}\sqrt{25 \cdot 2ab} = \frac{1}{5} \cdot 5\sqrt{2ab} = \sqrt{2ab}$

37. $\frac{\sqrt[4]{48x^9y^{13}}}{\sqrt[4]{3xy^{-2}}} = \sqrt[4]{\frac{48x^9y^{13}}{3xy^{-2}}} = \sqrt[4]{16x^8y^{15}} = \sqrt[4]{16x^8y^{12}}\sqrt[4]{y^3} = 2x^2y^3\sqrt[4]{y^3}$

39. $\frac{\sqrt[3]{x^3-y^3}}{\sqrt[3]{x-y}} = \sqrt[3]{\frac{x^3-y^3}{x-y}} = \sqrt[3]{\frac{(x-y)(x^2+xy+y^2)}{x-y}} = \sqrt[3]{\frac{\cancel{(x-y)}(x^2+xy+y^2)}{\cancel{x-y}}} = \sqrt[3]{x^2+xy+y^2}$

41. $\sqrt{\frac{3}{2}} = \sqrt{\frac{3}{2} \cdot \frac{2}{2}} = \sqrt{\frac{6}{4}} = \frac{\sqrt{6}}{\sqrt{4}} = \frac{\sqrt{6}}{2}$

43. $\frac{6\sqrt{5}}{5\sqrt{3}} = \frac{6\sqrt{5}}{5\sqrt{3}} \cdot \frac{\sqrt{3}}{\sqrt{3}} = \frac{6\sqrt{15}}{5 \cdot 3} = \frac{2\sqrt{15}}{5}$

45. $\sqrt[3]{\frac{16}{9}} = \sqrt[3]{\frac{16}{9} \cdot \frac{3}{3}} = \sqrt[3]{\frac{48}{27}} = \frac{\sqrt[3]{8 \cdot 6}}{\sqrt[3]{27}} = \frac{2\sqrt[3]{6}}{3}$

47. $\frac{\sqrt[3]{3a}}{\sqrt[3]{5c}} = \frac{\sqrt[3]{3a}}{\sqrt[3]{5c}} \cdot \frac{\sqrt[3]{5^2c^2}}{\sqrt[3]{5^2c^2}} = \frac{\sqrt[3]{75ac^2}}{\sqrt[3]{5^3c^3}} = \frac{\sqrt[3]{75ac^2}}{5c}$

49. $\frac{\sqrt[3]{5y^4}}{\sqrt[3]{6x^4}} = \frac{\sqrt[3]{5y^4}}{\sqrt[3]{6x^4}} \cdot \frac{\sqrt[3]{36x^2}}{\sqrt[3]{36x^2}} = \frac{\sqrt[3]{y^3 \cdot 180x^2y}}{\sqrt[3]{216x^6}} = \frac{y\sqrt[3]{180x^2y}}{6x^2}$

51. $\sqrt[3]{\frac{2}{x^2y}} = \sqrt[3]{\frac{2}{x^2y} \cdot \frac{xy^2}{xy^2}} = \sqrt[3]{\frac{2xy^2}{x^3y^3}} = \frac{\sqrt[3]{2xy^2}}{\sqrt[3]{x^3y^3}} = \frac{\sqrt[3]{2xy^2}}{xy}$

53. $\sqrt{\frac{7a}{18}} = \sqrt{\frac{7a}{18} \cdot \frac{2}{2}} = \sqrt{\frac{14a}{36}} = \frac{\sqrt{14a}}{\sqrt{36}} = \frac{\sqrt{14a}}{6}$

55. $\sqrt{\frac{9}{20x^2y}} = \sqrt{\frac{9}{20x^2y} \cdot \frac{5y}{5y}} = \sqrt{\frac{9 \cdot 5y}{100x^2y^2}} = \frac{\sqrt{9 \cdot 5y}}{\sqrt{100x^2y^2}} = \frac{3\sqrt{5y}}{10xy}$

57. $\sqrt{\frac{10ab^2}{72a^3b}} = \sqrt{\frac{5b}{36a^2}} = \frac{\sqrt{5b}}{6a}$

59. $\frac{\sqrt{5}}{\sqrt{7x}} = \frac{\sqrt{5}}{\sqrt{7x}} \cdot \frac{\sqrt{5}}{\sqrt{5}} = \frac{\sqrt{25}}{\sqrt{35x}} = \frac{5}{\sqrt{35x}}$

61. $\sqrt{\frac{14}{21}} = \sqrt{\frac{2}{3}} = \sqrt{\frac{2}{3} \cdot \frac{2}{2}} = \sqrt{\frac{4}{6}} = \frac{\sqrt{4}}{\sqrt{6}} = \frac{2}{\sqrt{6}}$

63. $\frac{4\sqrt{13}}{3\sqrt{7}} = \frac{4\sqrt{13}}{3\sqrt{7}} \cdot \frac{\sqrt{13}}{\sqrt{13}} = \frac{4\sqrt{169}}{3\sqrt{91}} = \frac{4 \cdot 13}{3\sqrt{91}} = \frac{52}{3\sqrt{91}}$

65. $\frac{\sqrt[3]{7}}{\sqrt[3]{2}} = \frac{\sqrt[3]{7}}{\sqrt[3]{2}} \cdot \frac{\sqrt[3]{7^2}}{\sqrt[3]{7^2}} = \frac{\sqrt[3]{7^3}}{\sqrt[3]{98}} = \frac{7}{\sqrt[3]{98}}$

67. $\sqrt{\frac{7x}{3y}} = \sqrt{\frac{7x}{3y} \cdot \frac{7x}{7x}} = \frac{\sqrt{(7x)^2}}{\sqrt{21xy}} = \frac{7x}{\sqrt{21xy}}$

69. $\sqrt[3]{\frac{2a^5}{5b}} = \sqrt[3]{\frac{2a^5}{5b} \cdot \frac{4a}{4a}} = \sqrt[3]{\frac{8a^6}{20ab}} = \frac{2a^2}{\sqrt[3]{20ab}}$

71. $\sqrt{\frac{x^3y}{2}} = \sqrt{\frac{x^3y}{2} \cdot \frac{xy}{xy}} = \sqrt{\frac{x^4y^2}{2xy}} = \frac{\sqrt{x^4y^2}}{\sqrt{2xy}} = \frac{x^2y}{\sqrt{2xy}}$

73. *Writing Exercise*

75. $\frac{3}{x-5} \cdot \frac{x-1}{x+5} = \frac{3(x-1)}{(x-5)(x+5)}$

77. $\frac{a^2 - 8a + 7}{a^2 - 49} = \frac{(a-1)(a-7)}{(a+7)(a-7)}$

$= \frac{(a-1)\cancel{(a-7)}}{(a+7)\cancel{(a-7)}}$

$= \frac{a-1}{a+7}$

79. $(5a^3b^4)^3 = 5^3(a^3)^3(b^4)^3 = 125a^{3\cdot3}b^{4\cdot3} = 125a^9b^{12}$

81. *Writing Exercise*

83. a) $T = 2\pi\sqrt{\frac{65}{980}} \approx 1.62$ sec

b) $T = 2\pi\sqrt{\frac{98}{980}} \approx 1.99$ sec

c) $T = 2\pi\sqrt{\frac{120}{980}} \approx 2.20$ sec

85. $\frac{(\sqrt[3]{81mn^2})^2}{(\sqrt[3]{mn})^2} = \frac{\sqrt[3]{(81mn^2)^2}}{\sqrt[3]{(mn)^2}}$

$= \frac{\sqrt[3]{6561m^2n^4}}{\sqrt[3]{m^2n^2}}$

$= \sqrt[3]{\frac{6561m^2n^4}{m^2n^2}}$

$= \sqrt[3]{6561n^2}$

$= \sqrt[3]{729 \cdot 9n^2}$

$= \sqrt[3]{729}\sqrt[3]{9n^2}$

$= 9\sqrt[3]{9n^2}$

87. $\sqrt{a^2-3} - \frac{a^2}{\sqrt{a^2-3}}$

$= \sqrt{a^2-3} - \frac{a^2}{\sqrt{a^2-3}} \cdot \frac{\sqrt{a^2-3}}{\sqrt{a^2-3}}$

$= \sqrt{a^2-3} - \frac{a^2\sqrt{a^2-3}}{a^2-3}$

$= \sqrt{a^2-3} \cdot \frac{a^2-3}{a^2-3} - \frac{a^2\sqrt{a^2-3}}{a^2-3}$

$= \frac{a^2\sqrt{a^2-3} - 3\sqrt{a^2-3} - a^2\sqrt{a^2-3}}{a^2-3}$

$= \frac{-3\sqrt{a^2-3}}{a^2-3}$, or $\frac{-3}{\sqrt{a^2-3}}$

89. Step 1: $\sqrt[n]{x} = x^{1/n}$, by definition;

Step 2: $\left(\frac{x}{y}\right)^n = \frac{x^n}{y^n}$, raising a quotient to a power;

Step 3: $x^{1/n} = \sqrt[n]{x}$, by definition

91. $f(x) = \sqrt{18x^3}$, $g(x) = \sqrt{2x}$

$(f/g)(x) = \frac{f(x)}{g(x)} = \frac{\sqrt{18x^3}}{\sqrt{2x}} = \sqrt{\frac{18x^3}{2x}} = \sqrt{9x^2} = 3x$

$\sqrt{2x}$ is defined for $2x \geq 0$, or $x \geq 0$. To avoid division by 0, we must exclude 0 from the domain. Thus, the domain of $f/g = \{x|x$ is a real number and $x > 0\}$, or $(0, \infty)$.

93. $f(x) = \sqrt{x^2-9}$, $g(x) = \sqrt{x-3}$

$(f/g)(x) = \frac{f(x)}{g(x)} = \frac{\sqrt{x^2-9}}{\sqrt{x-3}} = \sqrt{\frac{x^2-9}{x-3}} =$

$\sqrt{\frac{(x+3)(x-3)}{x-3}} = \sqrt{x+3}$

$\sqrt{x-3}$ is defined for $x - 3 \geq 0$, or $x \geq 3$. To avoid division by 0 we must exclude 3 from the domain. Thus, the domain of $f/g = \{x|x$ is a real number and $x > 3\}$, or $(3, \infty)$.

Exercise Set 10.5

1. To add radical expressions, the indices and the radicands must be the same.

3. To find a product by adding exponents, the bases must be the same.

5. To rationalize the <u>numerator</u> of $\dfrac{\sqrt{a+0.1}-\sqrt{a}}{0.1}$, we multiply by a form of 1, using the <u>conjugate</u> of $\sqrt{a+0.1}-\sqrt{a}$, or $\sqrt{a+0.1}+\sqrt{a}$, to write 1.

7. $2\sqrt{5}+7\sqrt{5}=(2+7)\sqrt{5}=9\sqrt{5}$

9. $7\sqrt[3]{4}-5\sqrt[3]{4}=(7-5)\sqrt[3]{4}=2\sqrt[3]{4}$

11. $\sqrt[3]{y}+9\sqrt[3]{y}=(1+9)\sqrt[3]{y}=10\sqrt[3]{y}$

13. $8\sqrt{2}-6\sqrt{2}+5\sqrt{2}=(8-6+5)\sqrt{2}=7\sqrt{2}$

15. $9\sqrt[3]{7}-\sqrt{3}+4\sqrt[3]{7}+2\sqrt{3}=$
$(9+4)\sqrt[3]{7}+(-1+2)\sqrt{3}=13\sqrt[3]{7}+\sqrt{3}$

17.
$4\sqrt{27}-3\sqrt{3}$
$=4\sqrt{9\cdot 3}-3\sqrt{3}$ Factoring the
$=4\sqrt{9}\cdot\sqrt{3}-3\sqrt{3}$ first radical
$=4\cdot 3\sqrt{3}-3\sqrt{3}$ Taking the square root of 9
$=12\sqrt{3}-3\sqrt{3}$
$=9\sqrt{3}$ Combining like radicals

19.
$3\sqrt{45}+7\sqrt{20}$
$=3\sqrt{9\cdot 5}+7\sqrt{4\cdot 5}$ Factoring the
$=3\sqrt{9}\cdot\sqrt{5}+7\sqrt{4}\cdot\sqrt{5}$ radicals
$=3\cdot 3\sqrt{5}+7\cdot 2\sqrt{5}$ Taking the square roots
$=9\sqrt{5}+14\sqrt{5}$
$=23\sqrt{5}$ Combining like radicals

21. $3\sqrt[3]{16}+\sqrt[3]{54}=3\sqrt[3]{8\cdot 2}+\sqrt[3]{27\cdot 2}=$
$3\sqrt[3]{8}\cdot\sqrt[3]{2}+\sqrt[3]{27}\cdot\sqrt[3]{2}=3\cdot 2\sqrt[3]{2}+3\sqrt[3]{2}=$
$6\sqrt[3]{2}+3\sqrt[3]{2}=9\sqrt[3]{2}$

23. $\sqrt{5a}+2\sqrt{45a^3}=\sqrt{5a}+2\sqrt{9a^2\cdot 5a}=$
$\sqrt{5a}+2\sqrt{9a^2}\cdot\sqrt{5a}=\sqrt{5a}+2\cdot 3a\sqrt{5a}=$
$\sqrt{5a}+6a\sqrt{5a}=(1+6a)\sqrt{5a}$

25. $\sqrt[3]{6x^4}+\sqrt[3]{48x}=\sqrt[3]{x^3\cdot 6x}+\sqrt[3]{8\cdot 6x}=$
$\sqrt[3]{x^3}\cdot\sqrt[3]{6x}+\sqrt[3]{8}\cdot\sqrt[3]{6x}=x\sqrt[3]{6x}+2\sqrt[3]{6x}=$
$(x+2)\sqrt[3]{6x}$

27. $\sqrt{4a-4}+\sqrt{a-1}=\sqrt{4(a-4)}+\sqrt{a-1}=$
$\sqrt{4}\sqrt{a-1}+\sqrt{a-1}=2\sqrt{a-1}+\sqrt{a-1}=3\sqrt{a-1}$

29. $\sqrt{x^3-x^2}+\sqrt{9x-9}=\sqrt{x^2(x-1)}+\sqrt{9(x-1)}=$
$\sqrt{x^2}\cdot\sqrt{x-1}+\sqrt{9}\cdot\sqrt{x-1}=$
$x\sqrt{x-1}+3\sqrt{x-1}=(x+3)\sqrt{x-1}$

31. $\sqrt{3}(4+\sqrt{3})=\sqrt{3}\cdot 4+\sqrt{3}\cdot\sqrt{3}=4\sqrt{3}+3$

33. $3\sqrt{5}(\sqrt{5}-\sqrt{2})=3\sqrt{5}\cdot\sqrt{5}-3\sqrt{5}\cdot\sqrt{2}=$
$3\cdot 5-3\sqrt{10}=15-3\sqrt{10}$

35. $\sqrt{2}(3\sqrt{10}-2\sqrt{2})=\sqrt{2}\cdot 3\sqrt{10}-\sqrt{2}\cdot 2\sqrt{2}=$
$3\sqrt{20}-2\cdot 2=3\sqrt{4\cdot 5}-4=6\sqrt{5}-4$

37. $\sqrt[3]{3}(\sqrt[3]{9}-4\sqrt[3]{21})=\sqrt[3]{3}\cdot\sqrt[3]{9}-\sqrt[3]{3}\cdot 4\sqrt[3]{21}=$
$\sqrt[3]{27}-4\sqrt[3]{63}=3-4\sqrt[3]{63}$

39. $\sqrt[3]{a}(\sqrt[3]{a^2}+\sqrt[3]{24a^2})=\sqrt[3]{a}\cdot\sqrt[3]{a^2}+\sqrt[3]{a}\sqrt[3]{24a^2}=$
$\sqrt[3]{a^3}+\sqrt[3]{24a^3}=\sqrt[3]{a^3}+\sqrt[3]{8a^3\cdot 3}=$
$a+2a\sqrt[3]{3}$

41. $(2+\sqrt{6})(5-\sqrt{6})=2\cdot 5-2\sqrt{6}+5\sqrt{6}-\sqrt{6}\cdot\sqrt{6}=$
$10+3\sqrt{6}-6=4+3\sqrt{6}$

43. $(\sqrt{2}+\sqrt{7})(\sqrt{3}-\sqrt{7})=\sqrt{2}\cdot\sqrt{3}-\sqrt{2}\cdot\sqrt{7}+\sqrt{7}\cdot\sqrt{3}-\sqrt{7}\cdot\sqrt{7}=$
$\sqrt{6}-\sqrt{14}+\sqrt{21}-7$

45. $(3-\sqrt{5})(3+\sqrt{5})=3^2-(\sqrt{5})^2=9-5=4$

47. $(\sqrt{6}+\sqrt{8})(\sqrt{6}-\sqrt{8})=(\sqrt{6})^2-(\sqrt{8})^2=6-8=-2$

49. $(3\sqrt{7}+2\sqrt{5})(2\sqrt{7}-4\sqrt{5})=$
$3\sqrt{7}\cdot 2\sqrt{7}-3\sqrt{7}\cdot 4\sqrt{5}+2\sqrt{5}\cdot 2\sqrt{7}-2\sqrt{5}\cdot 4\sqrt{5}=$
$6\cdot 7-12\sqrt{35}+4\sqrt{35}-8\cdot 5=42-8\sqrt{35}-40=$
$2-8\sqrt{35}$

51. $(2+\sqrt{3})^2=2^2+2\cdot 2\cdot\sqrt{3}+(\sqrt{3})^2=4+4\sqrt{3}+3=7+4\sqrt{3}$

53. $(\sqrt{3}-\sqrt{2})^2=(\sqrt{3})^2-2\cdot\sqrt{3}\cdot\sqrt{2}+(\sqrt{2})^2=$
$3-2\sqrt{6}+2=5-2\sqrt{6}$

55. $(\sqrt{2t}+\sqrt{5})^2=(\sqrt{2t})^2+2\cdot\sqrt{2t}\cdot\sqrt{5}+(\sqrt{5})^2=$
$2t+2\sqrt{10t}+5$

57. $(3-\sqrt{x+5})^2=3^2-2\cdot 3\cdot\sqrt{x+5}+(\sqrt{x+5})^2=$
$9-6\sqrt{x+5}+x+5=14-6\sqrt{x+5}+x$

59. $(2\sqrt[4]{7}-\sqrt[4]{6})(3\sqrt[4]{9}+2\sqrt[4]{5})=$
$2\sqrt[4]{7}\cdot 3\sqrt[4]{9}+2\sqrt[4]{7}\cdot 2\sqrt[4]{5}-\sqrt[4]{6}\cdot 3\sqrt[4]{9}-\sqrt[4]{6}\cdot 2\sqrt[4]{5}=$
$6\sqrt[4]{63}+4\sqrt[4]{35}-3\sqrt[4]{54}-2\sqrt[4]{30}$

61. $\dfrac{5}{4-\sqrt{3}}=\dfrac{5}{4-\sqrt{3}}\cdot\dfrac{4+\sqrt{3}}{4+\sqrt{3}}=\dfrac{5(4+\sqrt{3})}{(4-\sqrt{3})(4+\sqrt{3})}=$
$\dfrac{20+5\sqrt{3}}{4^2-(\sqrt{3})^2}=\dfrac{20+5\sqrt{3}}{16-3}=\dfrac{20+5\sqrt{3}}{13}$

63. $\dfrac{2+\sqrt{5}}{6+\sqrt{3}}=\dfrac{2+\sqrt{5}}{6+\sqrt{3}}\cdot\dfrac{6-\sqrt{3}}{6-\sqrt{3}}=$
$\dfrac{(2+\sqrt{5})(6-\sqrt{3})}{(6+\sqrt{3})(6-\sqrt{3})}=\dfrac{12-2\sqrt{3}+6\sqrt{5}-\sqrt{15}}{36-3}=$
$\dfrac{12-2\sqrt{3}+6\sqrt{5}-\sqrt{15}}{33}$

65. $\dfrac{\sqrt{a}}{\sqrt{a}+\sqrt{b}}=\dfrac{\sqrt{a}}{\sqrt{a}+\sqrt{b}}\cdot\dfrac{\sqrt{a}-\sqrt{b}}{\sqrt{a}-\sqrt{b}}=$
$\dfrac{\sqrt{a}(\sqrt{a}-\sqrt{b})}{(\sqrt{a}+\sqrt{b})(\sqrt{a}-\sqrt{b})}=\dfrac{a-\sqrt{ab}}{a-b}$

67. $\dfrac{\sqrt{7}-\sqrt{3}}{\sqrt{3}-\sqrt{7}}=\dfrac{-1(\sqrt{3}-\sqrt{7})}{\sqrt{3}-\sqrt{7}}=-1\cdot\dfrac{\sqrt{3}-\sqrt{7}}{\sqrt{3}-\sqrt{7}}=$
$-1\cdot 1=-1$

69. $\dfrac{3\sqrt{2}-\sqrt{7}}{4\sqrt{2}+2\sqrt{5}} = \dfrac{3\sqrt{2}-\sqrt{7}}{4\sqrt{2}+2\sqrt{5}} \cdot \dfrac{4\sqrt{2}-2\sqrt{5}}{4\sqrt{2}-2\sqrt{5}} =$

$\dfrac{(3\sqrt{2}-\sqrt{7})(4\sqrt{2}-2\sqrt{5})}{(4\sqrt{2}+2\sqrt{5})(4\sqrt{2}-2\sqrt{5})} =$

$\dfrac{12\cdot 2-6\sqrt{10}-4\sqrt{14}+2\sqrt{35}}{16\cdot 2-4\cdot 5} =$

$\dfrac{24-6\sqrt{10}-4\sqrt{14}+2\sqrt{35}}{32-20} = \dfrac{24-6\sqrt{10}-4\sqrt{14}+2\sqrt{35}}{12} =$

$\dfrac{2(12-3\sqrt{10}-2\sqrt{14}+\sqrt{35})}{2\cdot 6} = \dfrac{12-3\sqrt{10}-2\sqrt{14}+\sqrt{35}}{6}$

71. $\dfrac{\sqrt{7}+2}{5} = \dfrac{\sqrt{7}+2}{5} \cdot \dfrac{\sqrt{7}-2}{\sqrt{7}-2} =$

$\dfrac{(\sqrt{7}+2)(\sqrt{7}-2)}{5(\sqrt{7}-2)} = \dfrac{(\sqrt{7})^2-2^2}{5\sqrt{7}-10} =$

$\dfrac{7-4}{5\sqrt{7}-10} = \dfrac{3}{5\sqrt{7}-10}$

73. $\dfrac{\sqrt{6}-2}{\sqrt{3}+7} = \dfrac{\sqrt{6}-2}{\sqrt{3}+7} \cdot \dfrac{\sqrt{6}+2}{\sqrt{6}+2} =$

$\dfrac{(\sqrt{6}-2)(\sqrt{6}+2)}{(\sqrt{3}+7)(\sqrt{6}+2)} = \dfrac{6-4}{\sqrt{18}+2\sqrt{3}+7\sqrt{6}+14} =$

$\dfrac{2}{3\sqrt{2}+2\sqrt{3}+7\sqrt{6}+14}$

75. $\dfrac{\sqrt{x}-\sqrt{y}}{\sqrt{x}+\sqrt{y}} = \dfrac{\sqrt{x}-\sqrt{y}}{\sqrt{x}+\sqrt{y}} \cdot \dfrac{\sqrt{x}+\sqrt{y}}{\sqrt{x}+\sqrt{y}} =$

$\dfrac{(\sqrt{x}-\sqrt{y})(\sqrt{x}+\sqrt{y})}{(\sqrt{x}+\sqrt{y})(\sqrt{x}+\sqrt{y})} = \dfrac{x-y}{x+2\sqrt{xy}+y}$

77. $\dfrac{\sqrt{a+h}-\sqrt{a}}{h} = \dfrac{\sqrt{a+h}-\sqrt{a}}{h} \cdot \dfrac{\sqrt{a+h}+\sqrt{a}}{\sqrt{a+h}+\sqrt{a}} =$

$\dfrac{(\sqrt{a+h}-\sqrt{a})(\sqrt{a+h}+\sqrt{a})}{h(\sqrt{a+h}+\sqrt{a})} = \dfrac{a+h-a}{h(\sqrt{a+h}+\sqrt{a})} =$

$\dfrac{h}{h(\sqrt{a+h}+\sqrt{a})} = \dfrac{1}{\sqrt{a+h}+\sqrt{a}}$

79. $\sqrt{a}\sqrt[4]{a^3}$

$= a^{1/2}\cdot a^{3/4}$ Converting to exponential notation

$= a^{5/4}$ Adding exponents

$= a^{1+1/4}$ Writing 5/4 as a mixed number

$= a\cdot a^{1/4}$ Factoring

$= a\sqrt[4]{a}$ Returning to radical notation

81. $\sqrt[5]{b^2}\sqrt{b^3}$

$= b^{2/5}\cdot b^{3/2}$ Converting to exponential notation

$= b^{19/10}$ Adding exponents

$= b^{1+9/10}$ Writing 19/10 as a mixed number

$= b\cdot b^{9/10}$ Factoring

$= b\sqrt[10]{b^9}$ Returning to radical notation

83. $\sqrt{xy^3}\sqrt[3]{x^2y} = (xy^3)^{1/2}(x^2y)^{1/3}$

$= (xy^3)^{3/6}(x^2y)^{2/6}$

$= [(xy^3)^3(x^2y)^2]^{1/6}$

$= \sqrt[6]{x^3y^9\cdot x^4y^2}$

$= \sqrt[6]{x^7y^{11}}$

$= \sqrt[6]{x^6y^6\cdot xy^5}$

$= xy\sqrt[6]{xy^5}$

85. $\sqrt[4]{9ab^3}\sqrt{3a^4b} = (9ab^3)^{1/4}(3a^4b)^{1/2}$

$= (9ab^3)^{1/4}(3a^4b)^{2/4}$

$= [(9ab^3)(3a^4b)^2]^{1/4}$

$= \sqrt[4]{9ab^3\cdot 9a^8b^2}$

$= \sqrt[4]{81a^9b^5}$

$= \sqrt[4]{81a^8b^4\cdot ab}$

$= 3a^2b\sqrt[4]{ab}$

87. $\sqrt{a^4b^3c^4}\sqrt[3]{ab^2c} = (a^4b^3c^4)^{1/2}(ab^2c)^{1/3}$

$= (a^4b^3c^4)^{3/6}(ab^2c)^{2/6}$

$= [(a^4b^3c^4)^3(ab^2c)^2]^{1/6}$

$= \sqrt[6]{a^{12}b^9c^{12}\cdot a^2b^4c^2}$

$= \sqrt[6]{a^{14}b^{13}c^{14}}$

$= \sqrt[6]{a^{12}b^{12}c^{12}\cdot a^2bc^2}$

$= a^2b^2c^2\sqrt[6]{a^2bc^2}$

89. $\dfrac{\sqrt[3]{a^2}}{\sqrt[4]{a}}$

$= \dfrac{a^{2/3}}{a^{1/4}}$ Converting to exponential notation

$= a^{2/3-1/4}$ Subtracting exponents

$= a^{5/12}$ Converting back

$= \sqrt[12]{a^5}$ to radical notation

91. $\dfrac{\sqrt[4]{x^2y^3}}{\sqrt[3]{xy}}$

$= \dfrac{(x^2y^3)^{1/4}}{(xy)^{1/3}}$ Converting to exponential notation

$= \dfrac{x^{2/4}y^{3/4}}{x^{1/3}y^{1/3}}$ Using the power and product rules

$= x^{2/4-1/3}y^{3/4-1/3}$ Subtracting exponents

$= x^{2/12}y^{5/12}$

$= (x^2y^5)^{1/2}$ Converting back to

$= \sqrt[12]{x^2y^5}$ radical notation

93. $\dfrac{\sqrt{ab^3}}{\sqrt[5]{a^2b^3}}$

$= \dfrac{(ab^3)^{1/2}}{(a^2b^3)^{1/5}}$ Converting to exponential notation

$= \dfrac{a^{1/2}b^{3/2}}{a^{2/5}b^{3/5}}$

$= a^{1/10}b^{9/10}$ Subtracting exponents

$= (ab^9)^{1/10}$ Converting back to

$= \sqrt[10]{ab^9}$ radical notation

95. $\dfrac{\sqrt[4]{(3x-1)^3}}{\sqrt[5]{(3x-1)^3}}$

$= \dfrac{(3x-1)^{3/4}}{(3x-1)^{3/5}}$ Converting to exponential notation

$= (3x-1)^{3/4-3/5}$ Subtracting exponents

$= (3x-1)^{3/20}$ Converting back

$= \sqrt[20]{(3x-1)^3}$ to radical notation

97. $\dfrac{\sqrt[3]{(2x+1)^2}}{\sqrt[5]{(2x+1)^2}}$

$= \dfrac{(2x+1)^{2/3}}{(2x+1)^{2/5}}$ Converting to exponential notation

$= (2x+1)^{2/3-2/5}$ Subtracting exponents

$= (2x+1)^{4/15}$ Converting back to

$= \sqrt[15]{(2x+1)^4}$ radical notation

99. $\sqrt[3]{x^2y}(\sqrt{xy} - \sqrt[5]{xy^3})$

$= (x^2y)^{1/3}[(xy)^{1/2} - (xy^3)^{1/5}]$

$= x^{2/3}y^{1/3}(x^{1/2}y^{1/2} - x^{1/5}y^{3/5})$

$= x^{2/3}y^{1/3}x^{1/2}y^{1/2} - x^{2/3}y^{1/3}x^{1/5}y^{3/5}$

$= x^{2/3+1/2}y^{1/3+1/2} - x^{2/3+1/5}y^{1/3+3/5}$

$= x^{7/6}y^{5/6} - x^{13/15}y^{14/15}$

$= x^{1\frac{1}{6}}y^{\frac{5}{6}} - x^{13/15}y^{14/15}$

Writing a mixed numeral

$= x \cdot x^{1/6}y^{5/6} - x^{13/15}y^{14/15}$

$= x(xy^5)^{1/6} - (x^{13}y^{14})^{1/15}$

$= x\sqrt[6]{xy^5} - \sqrt[15]{x^{13}y^{14}}$

101. $(m + \sqrt[3]{n^2})(2m + \sqrt[4]{n})$

$= (m + n^{2/3})(2m + n^{1/4})$ Converting to exponential notation

$= 2m^2 + mn^{1/4} + 2mn^{2/3} + n^{2/3}n^{1/4}$ Using FOIL

$= 2m^2 + mn^{1/4} + 2mn^{2/3} + n^{2/3+1/4}$ Adding exponents

$= 2m^2 + mn^{1/4} + 2mn^{2/3} + n^{11/12}$

$= 2m^2 + m\sqrt[4]{n} + 2m\sqrt[3]{n^2} + \sqrt[12]{n^{11}}$ Converting back to radical notation

103. $f(x) = \sqrt[4]{x}$, $g(x) = \sqrt[4]{2x} - \sqrt[4]{x^{11}}$

$(f \cdot g)(x) = \sqrt[4]{x}(\sqrt[4]{2x} - \sqrt[4]{x^{11}})$

$= \sqrt[4]{2x^2} - \sqrt[4]{x^{12}}$

$= \sqrt[4]{2x^2} - x^3$

105. $f(x) = x + \sqrt{7}$, $g(x) = x - \sqrt{7}$

$(f \cdot g)(x) = (x + \sqrt{7})(x - \sqrt{7})$

$= x^2 - (\sqrt{7})^2$

$= x^2 - 7$

107. $f(x) = x^2$

$f(5+\sqrt{2}) = (5+\sqrt{2})^2 = 25 + 10\sqrt{2} + (\sqrt{2})^2 =$
$25 + 10\sqrt{2} + 2 = 27 + 10\sqrt{2}$

109. $f(x) = x^2$

$f(\sqrt{3}-\sqrt{5}) = (\sqrt{3}-\sqrt{5})^2 =$
$(\sqrt{3})^2 - 2\cdot\sqrt{3}\cdot\sqrt{5} + (\sqrt{5})^2 =$
$3 - 2\sqrt{15} + 5 = 8 - 2\sqrt{15}$

111. *Writing Exercise*

113. $\dfrac{12x}{x-4} - \dfrac{3x^2}{x+4} = \dfrac{384}{x^2-16}$

$\dfrac{12x}{x-4} - \dfrac{3x^2}{x+4} = \dfrac{384}{(x+4)(x-4)}$, LCM is $(x+4)(x-4)$.

Note that $x \neq -4$ and $x \neq 4$.

$(x+4)(x-4)\left[\dfrac{12x}{x-4} - \dfrac{3x^2}{x+4}\right] = (x+4)(x-4)\cdot\dfrac{384}{(x+4)(x-4)}$

$12x(x+4) - 3x^2(x-4) = 384$

$12x^2 + 48x - 3x^3 + 12x^2 = 384$

$-3x^3 + 24x^2 + 48x - 384 = 0$

$-3(x^3 - 8x^2 - 16x + 128) = 0$

$-3[x^2(x-8) - 16(x-8)] = 0$

$-3(x-8)(x^2-16) = 0$

$-3(x-8)(x+4)(x-4) = 0$

$x - 8 = 0$ *or* $x + 4 = 0$ *or* $x - 4 = 0$

$x = 8$ *or* $x = -4$ *or* $x = 4$

Check: For 8:

$$\begin{array}{c|c} \dfrac{12x}{x-4} - \dfrac{3x^2}{x+4} & \dfrac{384}{x^2-16} \\ \hline \dfrac{12\cdot 8}{8-4} - \dfrac{3\cdot 8^2}{8+4} & \dfrac{384}{8^2-16} \\ \dfrac{96}{4} - \dfrac{192}{12} & \dfrac{384}{48} \\ 24 - 16 & 8 \end{array}$$

$8 \stackrel{?}{=} 8$ TRUE

8 is a solution.

For -4:

$$\frac{12x}{x-4} - \frac{3x^2}{x+4} = \frac{384}{x^2-16}$$

$$\frac{12(-4)}{-4-4} - \frac{3(-4)^2}{-4+4} \;\Big|\; \frac{384}{(-4)^2-16}$$

$$\frac{-48}{-8} - \frac{48}{0} \stackrel{?}{=} \frac{384}{16-16} \quad \text{UNDEFINED}$$

-4 is not a solution.

For 4:

$$\frac{12x}{x-4} - \frac{3x^2}{x+4} = \frac{384}{x^2-16}$$

$$\frac{12\cdot 4}{4-4} - \frac{3\cdot 4^2}{4+4} \;\Big|\; \frac{384}{4^2-16}$$

$$\frac{48}{0} - \frac{48}{8} \stackrel{?}{=} \frac{384}{16-16} \quad \text{UNDEFINED}$$

4 is not a solution.

The checks confirm that -4 and 4 are not solutions. The solution is 8.

115.

$$5x^2 - 6x + 1 = 0$$
$$(5x-1)(x-1) = 0$$
$$5x - 1 = 0 \quad or \quad x - 1 = 0$$
$$5x = 1 \quad or \quad x = 1$$
$$x = \frac{1}{5} \quad or \quad x = 1$$

The solutions are $\frac{1}{5}$ and 1.

117. ***Familiarize***. Let x = the number.

Translate.

A number plus its square is 20.

$$x + x^2 = 20$$

Carry out. We solve the equation.

$$x + x^2 = 20$$
$$x^2 + x - 20 = 0$$
$$(x+5)(x-4) = 0$$
$$x + 5 = 0 \quad or \quad x - 4 = 0$$
$$x = -5 \quad or \quad x = 4$$

Check. $-5+(-5)^2 = 20$ and $4+4^2 = 20$, so both numbers check.

State. The number is -5 or 4.

119. *Writing Exercise*

121.

$$\begin{aligned} f(x) &= \sqrt{20x^2+4x^3} - 3x\sqrt{45+9x} + \sqrt{5x^2+x^3} \\ &= \sqrt{4x^2(5+x)} - 3x\sqrt{9(5+x)} + \sqrt{x^2(5+x)} \\ &= \sqrt{4x^2}\sqrt{5+x} - 3x\sqrt{9}\sqrt{5+x} + \sqrt{x^2}\sqrt{5+x} \\ &= 2x\sqrt{5+x} - 3x\cdot 3\sqrt{5+x} + x\sqrt{5+x} \\ &= 2x\sqrt{5+x} - 9x\sqrt{5+x} + x\sqrt{5+x} \\ &= -6x\sqrt{5+x} \end{aligned}$$

123.

$$\begin{aligned} f(x) &= \sqrt[4]{x^5-x^4} + 3\sqrt[4]{x^9-x^8} \\ &= \sqrt[4]{x^4(x-1)} + 3\sqrt[4]{x^8(x-1)} \\ &= \sqrt[4]{x^4}\cdot\sqrt[4]{x-1} + 3\sqrt[4]{x^8}\sqrt[4]{x-1} \\ &= x\sqrt[4]{x-1} + 3x^2\sqrt[4]{x-1} \\ &= (x+3x^2)\sqrt[4]{x-1} \end{aligned}$$

125.

$$\frac{1}{2}\sqrt{36a^5bc^4} - \frac{1}{2}\sqrt[3]{64a^4bc^6} + \frac{1}{6}\sqrt{144a^3bc^6} =$$

$$\frac{1}{2}\sqrt{36a^4c^4\cdot ab} - \frac{1}{2}\sqrt[3]{64a^3c^6\cdot ab} + \frac{1}{6}\sqrt{144a^2c^6\cdot ab} =$$

$$\frac{1}{2}(6a^2c^2)\sqrt{ab} - \frac{1}{2}(4ac^2)\sqrt[3]{ab} + \frac{1}{6}(12ac^3)\sqrt{ab} =$$

$$3a^2c^2\sqrt{ab} - 2ac^2\sqrt[3]{ab} + 2ac^3\sqrt{ab}$$

$$(3a^2c^2 + 2ac^3)\sqrt{ab} - 2ac^2\sqrt[3]{ab}, \text{ or}$$

$$ac^2[(3a+2c)\sqrt{ab} - 2\sqrt[3]{ab}]$$

127.

$$\begin{aligned} &\sqrt{27a^5(b+1)}\sqrt[3]{81a(b+1)^4} \\ &= [27a^5(b+1)]^{1/2}[81a(b+1)^4]^{1/3} \\ &= [27a^5(b+1)]^{3/6}[81a(b+1)^4]^{2/6} \\ &= \{[3^3a^5(b+1)]^3[3^4a(b+1)^4]^2\}^{1/6} \\ &= \sqrt[6]{3^9a^{15}(b+1)^3\cdot 3^8a^2(b+1)^8} \\ &= \sqrt[6]{3^{17}a^{17}(b+1)^{11}} \\ &= \sqrt[6]{3^{12}a^{12}(b+1)^6\cdot 3^5a^5(b+1)^5} \\ &= 3^2a^2(b+1)\sqrt[6]{3^5a^5(b+1)^5}, \text{ or} \\ &\quad 9a^2(b+1)\sqrt[6]{243a^5(b+1)^5} \end{aligned}$$

129.

$$\frac{\frac{1}{\sqrt{w}} - \sqrt{w}}{\frac{\sqrt{w}+1}{\sqrt{w}}} = \frac{\frac{1}{\sqrt{w}} - \sqrt{w}}{\frac{\sqrt{w}+1}{\sqrt{w}}}\cdot\frac{\sqrt{w}}{\sqrt{w}} = \frac{1-w}{\sqrt{w}+1} =$$

$$\frac{1-w}{\sqrt{w}+1}\cdot\frac{\sqrt{w}-1}{\sqrt{w}-1} = \frac{\sqrt{w}-1-w\sqrt{w}+w}{w-1} =$$

$$\frac{(w-1)-\sqrt{w}(w-1)}{w-1} = \frac{(w-1)(1-\sqrt{w})}{w-1} =$$

$$1-\sqrt{w}$$

131. $x - 5 = (\sqrt{x})^2 - (\sqrt{5})^2 = (\sqrt{x}+\sqrt{5})(\sqrt{x}-\sqrt{5})$

133. $x - a = (\sqrt{x})^2 - (\sqrt{a})^2 = (\sqrt{x}+\sqrt{a})(\sqrt{x}-\sqrt{a})$

135.

$$(\sqrt{x+2} - \sqrt{x-2})^2 =$$

$$x + 2 - 2\sqrt{(x+2)(x-2)} + x - 2 =$$

$$x + 2 - 2\sqrt{x^2-4} + x - 2 = 2x - 2\sqrt{x^2-4s}$$

Exercise Set 10.6

1. False; if $x^2 = 25$, then $x = 5$ or $x = -5$.

3. True by the principle of powers

5. If we add 8 to both sides of $\sqrt{x} - 8 = 7$, we get $\sqrt{x} = 15$, so the statement is true.

7.
$$\begin{aligned} \sqrt{5x-2} &= 7 \\ (\sqrt{5x-2})^2 &= 7^2 \quad \text{Principle of powers (squaring)} \\ 5x-2 &= 49 \\ 5x &= 51 \\ x &= \frac{51}{5} \end{aligned}$$

Check:
$$\begin{array}{c|c} \sqrt{5x-2} & = 7 \\ \hline \sqrt{5\cdot\frac{51}{5}-2} & 7 \\ \sqrt{49} & \\ 7 & \stackrel{?}{=} 7 \quad \text{TRUE} \end{array}$$

The solution is $\frac{51}{5}$.

9.
$$\begin{aligned} \sqrt{3x}+1 &= 6 \\ \sqrt{3x} &= 5 \quad \text{Adding to isolate the radical} \\ (\sqrt{3x})^2 &= 5^2 \quad \text{Principle of powers (squaring)} \\ 3x &= 25 \\ x &= \frac{25}{3} \end{aligned}$$

Check:
$$\begin{array}{c|c} \sqrt{3x}+1 & = 6 \\ \hline \sqrt{3\cdot\frac{25}{3}}+1 & 6 \\ 5+1 & \\ 6 & \stackrel{?}{=} 6 \quad \text{TRUE} \end{array}$$

The solution is $\frac{25}{3}$.

11.
$$\begin{aligned} \sqrt{y+1}-5 &= 8 \\ \sqrt{y+1} &= 13 \quad \text{Adding to isolate the radical} \\ (\sqrt{y+1})^2 &= 13^2 \quad \text{Principle of powers (squaring)} \\ y+1 &= 169 \\ y &= 168 \end{aligned}$$

Check:
$$\begin{array}{c|c} \sqrt{y+1}-5 & = 8 \\ \hline \sqrt{168+1}-5 & 8 \\ 13-5 & \\ 8 & \stackrel{?}{=} 8 \quad \text{TRUE} \end{array}$$

The solution is 168.

13.
$$\begin{aligned} \sqrt{x-7}+3 &= 10 \\ \sqrt{x-7} &= 7 \quad \text{Adding to isolate the radical} \\ (\sqrt{x-7})^2 &= 7^2 \quad \text{Principle of powers (squaring)} \\ x-7 &= 49 \\ x &= 56 \end{aligned}$$

Check:
$$\begin{array}{c|c} \sqrt{x-7}+3 & = 10 \\ \hline \sqrt{56-7}+3 & 10 \\ \sqrt{49}+3 & \\ 7+3 & \\ 10 & \stackrel{?}{=} 10 \quad \text{TRUE} \end{array}$$

The solution is 56.

15.
$$\begin{aligned} \sqrt[3]{x+5} &= 2 \\ (\sqrt[3]{x+5})^3 &= 2^3 \\ x+5 &= 8 \\ x &= 3 \end{aligned}$$

Check:
$$\begin{array}{c|c} \sqrt[3]{x+5} & = 2 \\ \hline \sqrt[3]{3+5} & 2 \\ \sqrt[3]{8} & \\ 2 & \stackrel{?}{=} 2 \quad \text{TRUE} \end{array}$$

The solution is 3.

17.
$$\begin{aligned} \sqrt[4]{y-1} &= 3 \\ (\sqrt[4]{y-1})^4 &= 3^4 \\ y-1 &= 81 \\ y &= 82 \end{aligned}$$

Check:
$$\begin{array}{c|c} \sqrt[4]{y-1} & = 3 \\ \hline \sqrt[4]{82-1} & 3 \\ \sqrt[4]{81} & \\ 3 & \stackrel{?}{=} 3 \quad \text{TRUE} \end{array}$$

The solution is 82.

19.
$$\begin{aligned} 3\sqrt{x} &= x \\ (3\sqrt{x})^2 &= x^2 \\ 9x &= x^2 \\ 0 &= x^2-9x \\ 0 &= x(x-9) \end{aligned}$$
$x=0$ *or* $x=9$

Check:

For 0:
$$\begin{array}{c|c} 3\sqrt{x} & = x \\ \hline 3\sqrt{0} & 0 \\ 3\cdot 0 & \\ 0 & \stackrel{?}{=} 0 \quad \text{TRUE} \end{array}$$

For 9:
$$\begin{array}{c|c} 3\sqrt{x} & = x \\ \hline 3\sqrt{9} & 9 \\ 3\cdot 3 & \\ 9 & \stackrel{?}{=} 9 \quad \text{TRUE} \end{array}$$

The solutions are 0 and 9.

21.
$$\begin{aligned} 2y^{1/2} - 7 &= 9 \\ 2\sqrt{y} - 7 &= 9 \\ 2\sqrt{y} &= 16 \\ \sqrt{y} &= 8 \\ (\sqrt{y})^2 &= 8^2 \\ y &= 64 \end{aligned}$$

Check:
$$\begin{array}{r|l} \multicolumn{2}{c}{2y^{1/2} - 7 = 9} \\ \hline 2 \cdot 64^{1/2} - 7 & 9 \\ 2 \cdot 8 - 7 & \\ \multicolumn{2}{c}{9 \stackrel{?}{=} 9 \quad \text{TRUE}} \end{array}$$

The solution is 64.

23.
$$\begin{aligned} \sqrt[3]{x} &= -3 \\ (\sqrt[3]{x})^3 &= (-3)^3 \\ x &= -27 \end{aligned}$$

Check:
$$\begin{array}{r|l} \multicolumn{2}{c}{\sqrt[3]{x} = -3} \\ \hline \sqrt{-27} & -3 \\ \multicolumn{2}{c}{-3 \stackrel{?}{=} -3 \quad \text{TRUE}} \end{array}$$

The solution is -27.

25.
$$\begin{aligned} t^{1/3} - 2 &= 3 \\ t^{1/3} &= 5 \\ (t^{1/3})^3 &= 5^3 \quad \text{Principle of powers} \\ t &= 125 \end{aligned}$$

Check:
$$\begin{array}{r|l} \multicolumn{2}{c}{t^{1/3} - 2 = 3} \\ \hline 125^{1/3} - 2 & 3 \\ 5 - 2 & \\ \multicolumn{2}{c}{3 \stackrel{?}{=} 3 \quad \text{TRUE}} \end{array}$$

The solution is 125.

27.
$$\begin{aligned} (y-3)^{1/2} &= -2 \\ \sqrt{y-3} &= -2 \end{aligned}$$

This equation has no solution, since the principal square root is never negative.

29.
$$\begin{aligned} \sqrt[4]{3x+1} - 4 &= -1 \\ \sqrt[4]{3x+1} &= 3 \\ (\sqrt[4]{3x+1})^4 &= 3^4 \\ 3x + 1 &= 81 \\ 3x &= 80 \\ x &= \frac{80}{3} \end{aligned}$$

Check:
$$\begin{array}{r|l} \multicolumn{2}{c}{\sqrt[4]{3x+1} - 4 = -1} \\ \hline \sqrt[4]{3 \cdot \frac{80}{3} + 1} - 4 & -1 \\ \sqrt[4]{81} - 4 & \\ 3 - 4 & \\ \multicolumn{2}{c}{-1 \stackrel{?}{=} -1 \quad \text{TRUE}} \end{array}$$

The solution is $\frac{80}{3}$.

31.
$$\begin{aligned} (x+7)^{1/3} &= 4 \\ [(x+7)^{1/3}]^3 &= 4^3 \\ x + 7 &= 64 \\ x &= 57 \end{aligned}$$

Check:
$$\begin{array}{r|l} \multicolumn{2}{c}{(x+7)^{1/3} = 4} \\ \hline (57+7)^{1/3} & 4 \\ 64^{1/3} & \\ \multicolumn{2}{c}{4 \stackrel{?}{=} 4 \quad \text{TRUE}} \end{array}$$

The solution is 57.

33.
$$\begin{aligned} \sqrt[3]{3y+6} + 2 &= 3 \\ \sqrt[3]{3y+6} &= 1 \\ (\sqrt[3]{3y+6})^3 &= 1^3 \\ 3y + 6 &= 1 \\ 3y &= -5 \\ y &= -\frac{5}{3} \end{aligned}$$

Check:
$$\begin{array}{r|l} \multicolumn{2}{c}{\sqrt[3]{3y+6} + 2 = 3} \\ \hline \sqrt[3]{3\left(-\frac{5}{3}\right) + 6} + 2 & 3 \\ \sqrt[3]{1} + 2 & \\ 1 + 2 & \\ \multicolumn{2}{c}{3 \stackrel{?}{=} 3 \quad \text{TRUE}} \end{array}$$

The solution is $-\frac{5}{3}$.

35.
$$\begin{aligned} \sqrt{3t+4} &= \sqrt{4t+3} \\ (\sqrt{3t+4})^2 &= (\sqrt{4t+3})^2 \\ 3t + 4 &= 4t + 3 \\ 4 &= t + 3 \\ 1 &= t \end{aligned}$$

Check:
$$\begin{array}{r|l} \multicolumn{2}{c}{\sqrt{3t+4} = \sqrt{4t+3}} \\ \hline \sqrt{3 \cdot 1 + 4} & \sqrt{4 \cdot 1 + 3} \\ \multicolumn{2}{c}{\sqrt{7} \stackrel{?}{=} \sqrt{7} \quad \text{TRUE}} \end{array}$$

The solution is 1.

37.
$$3(4-t)^{1/4} = 6^{1/4}$$
$$[3(4-t)^{1/4}]^4 = (6^{1/4})^4$$
$$81(4-t) = 6$$
$$324 - 81t = 6$$
$$-81t = -318$$
$$t = \frac{106}{27}$$

The number $\frac{106}{27}$ checks and is the solution.

39. $3 + \sqrt{5-x} = x$
$$\sqrt{5-x} = x - 3$$
$$(\sqrt{5-x})^2 = (x-3)^2$$
$$5 - x = x^2 - 6x + 9$$
$$0 = x^2 - 5x + 4$$
$$0 = (x-1)(x-4)$$
$$x - 1 = 0 \ or \ x - 4 = 0$$
$$x = 1 \ or \ \ x = 4$$

Check:

For 1: $3 + \sqrt{5-x} = x$

$3+\sqrt{5-1}$	1
$3+\sqrt{4}$	
$3+2$	

$5 \stackrel{?}{=} 1$ FALSE

For 4: $3 + \sqrt{5-x} = x$

$3+\sqrt{5-4}$	4
$3+\sqrt{1}$	
$3+1$	

$4 \stackrel{?}{=} 4$ TRUE

Since 4 checks but 1 does not, the solution is 4.

41. $\sqrt{4x-3} = 2 + \sqrt{2x-5}$ One radical is already isolated.

$(\sqrt{4x-3})^2 = (2+\sqrt{2x-5})^2$ Squaring both sides
$$4x - 3 = 4 + 4\sqrt{2x-5} + 2x - 5$$
$$2x - 2 = 4\sqrt{2x-5}$$
$$x - 1 = 2\sqrt{2x-5}$$
$$x^2 - 2x + 1 = 8x - 20$$
$$x^2 - 10x + 21 = 0$$
$$(x-7)(x-3) = 0$$
$$x - 7 = 0 \ or \ x - 3 = 0$$
$$x = 7 \ or \ \ x = 3$$

Both numbers check. The solutions are 7 and 3.

43. $\sqrt{20-x} + 8 = \sqrt{9-x} + 11$

$\sqrt{20-x} = \sqrt{9-x} + 3$ Isolating one radical

$(\sqrt{20-x})^2 = (\sqrt{9-x}+3)^2$ Squaring both sides
$$20 - x = 9 - x + 6\sqrt{9-x} + 9$$

$2 = 6\sqrt{9-x}$ Isolating the remaining radical

$1 = 3\sqrt{9-x}$ Multiplying by $\frac{1}{2}$

$1^2 = (3\sqrt{9-x})^2$ Squaring both sides
$$1 = 9(9-x)$$
$$1 = 81 - 9x$$
$$-80 = -9x$$
$$\frac{80}{9} = x$$

The number $\frac{80}{9}$ checks and is the solution.

45. $\sqrt{x+2} + \sqrt{3x+4} = 2$

$\sqrt{x+2} = 2 - \sqrt{3x+4}$ Isolating one radical
$$(\sqrt{x+2})^2 = (2-\sqrt{3x+4})^2$$
$$x + 2 = 4 - 4\sqrt{3x+4} + 3x + 4$$

$-2x - 6 = -4\sqrt{3x+4}$ Isolating the remaining radical

$x + 3 = 2\sqrt{3x+4}$ Multiplying by $-\frac{1}{2}$
$$(x+3)^2 = (2\sqrt{3x+4})^2$$
$$x^2 + 6x + 9 = 4(3x+4)$$
$$x^2 + 6x + 9 = 12x + 16$$
$$x^2 - 6x - 7 = 0$$
$$(x-7)(x+1) = 0$$
$$x - 7 = 0 \ \text{or} \ x + 1 = 0$$
$$x = 7 \ \text{or} \ x = -1$$

Check:

For 7:

$\sqrt{x+2} + \sqrt{3x+4} = 2$

$\sqrt{7+2}+\sqrt{3\cdot 7+4}$	2
$\sqrt{9}+\sqrt{25}$	

$8 \stackrel{?}{=} 2$ FALSE

For -1:

$\sqrt{x+2} + \sqrt{3x+4} = 2$

$\sqrt{-1+2}+\sqrt{3\cdot(-1)+4}$	2
$\sqrt{1}+\sqrt{1}$	

$2 \stackrel{?}{=} 2$ TRUE

Since -1 checks but 7 does not, the solution is -1.

47. We must have $f(x) = 2$, or $\sqrt{x} + \sqrt{x-9} = 1$.

$$\sqrt{x} + \sqrt{x-9} = 1$$
$$\sqrt{x-9} = 1 - \sqrt{x} \quad \text{Isolating one radical term}$$
$$(\sqrt{x-9})^2 = (1-\sqrt{x})^2$$
$$x - 9 = 1 - 2\sqrt{x} + x$$
$$-10 = -2\sqrt{x} \quad \text{Isolating the remaining radical term}$$
$$5 = \sqrt{x}$$
$$25 = x$$

This value does not check. There is no solution, so there is no value of x for which $f(x) = 1$.

49.
$$\sqrt{t-2} - \sqrt{4t+1} = -3$$
$$\sqrt{t-2} = \sqrt{4t+1} - 3$$
$$(\sqrt{t-2})^2 = (\sqrt{4t+1} - 3)^2$$
$$t - 2 = 4t + 1 - 6\sqrt{4t+1} + 9$$
$$-3t - 12 = -6\sqrt{4t+1}$$
$$t + 4 = 2\sqrt{4t+1}$$
$$(t+4)^2 = (2\sqrt{4t+1})^2$$
$$t^2 + 8t + 16 = 4(4t+1)$$
$$t^2 + 8t + 16 = 16t + 4$$
$$t^2 - 8t + 12 = 0$$
$$(t-2)(t-6) = 0$$
$$t - 2 = 0 \quad or \quad t - 6 = 0$$
$$t = 2 \quad or \quad t = 6$$

Both numbers check, so we have $f(t) = -3$ when $t = 2$ and when $t = 6$.

51. We must have $\sqrt{2x-3} = \sqrt{x+7} - 2$.

$$\sqrt{2x-3} = \sqrt{x+7} - 2$$
$$(\sqrt{2x-3})^2 = (\sqrt{x+7} - 2)^2$$
$$2x - 3 = x + 7 - 4\sqrt{x+7} + 4$$
$$x - 14 = -4\sqrt{x+7}$$
$$(x-14)^2 = (-4\sqrt{x+7})^2$$
$$x^2 - 28x + 196 = 16(x+7)$$
$$x^2 - 28x + 196 = 16x + 112$$
$$x^2 - 44x + 84 = 0$$
$$(x-2)(x-42) = 0$$
$$x = 2 \quad or \quad x = 42$$

Since 2 checks but 42 does not, we have $f(x) = g(x)$ when $x = 2$.

53. We must have $4 - \sqrt{t-3} = (t+5)^{1/2}$.

$$4 - \sqrt{t-3} = (t+5)^{1/2}$$
$$(4 - \sqrt{t-3})^2 = [(t+5)^{1/2}]^2$$
$$16 - 8\sqrt{t-3} + t - 3 = t + 5$$
$$-8\sqrt{t-3} = -8$$
$$\sqrt{t-3} = 1$$
$$(\sqrt{t-3})^2 = 1^2$$
$$t - 3 = 1$$
$$t = 4$$

The number 4 checks, so we have $f(t) = g(t)$ when $t = 4$.

55. *Writing Exercise*

57. ***Familiarize***. Let h = the height of the triangle, in inches. Then $h + 2$ = the base. Recall that the formula for the area of a triangle with base b and height h is $A = \frac{1}{2}bh$.

Translate. Substitute in the formula.

$$31\frac{1}{2} = \frac{1}{2}(h+2)(h)$$

Carry out. We solve the equation.

$$31\frac{1}{2} = \frac{1}{2}(h+2)(h)$$
$$\frac{63}{2} = \frac{1}{2}(h+2)(h)$$
$$63 = (h+2)(h) \quad \text{Multiplying by 2}$$
$$63 = h^2 + 2h$$
$$0 = h^2 + 2h - 63$$
$$0 = (h+9)(h-7)$$
$$h + 9 = 0 \quad or \quad h - 7 = 0$$
$$h = -9 \quad or \quad h = 7$$

Check. Since the height of the triangle cannot be negative we check only 7. If the height is 7 in., then the base is $7 + 2$, or 9 in., and the area is $\frac{1}{2} \cdot 9 \cdot 7 = \frac{63}{2} = 31\frac{1}{2}$ in^2. The answer checks.

State. The height of the triangle is 7 in., and the base is 9 in.

59. Graph $f(x) = \frac{2}{3}x - 5$.

The y-intercept is $(0, -5)$. The slope is $\frac{2}{3}$. We plot $(0, -5)$. From that point we go up 2 units and right 3 units and plot a second point at $(3, -3)$. Then we draw the graph.

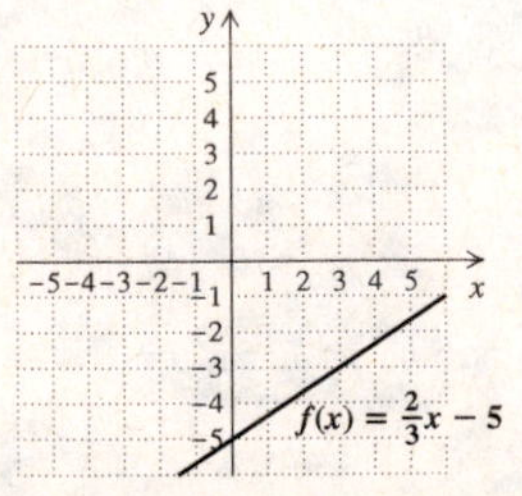

To check that the line is drawn correctly, we calculate the coordinates of another point on the line. For $x = 6$, we have

$$f(6) = \frac{2}{3} \cdot 6 - 5 = 4 - 5 = -1.$$

Since the point $(6, -1)$ appears to lie on the line, we have a check.

61. Graph $F(x) < -2x + 4$

First graph $F(x) = -2x + 4$ using a dashed line since the inequality symbol is $<$. Because the inequality is of the form $y < mx + b$, we shade below the line.

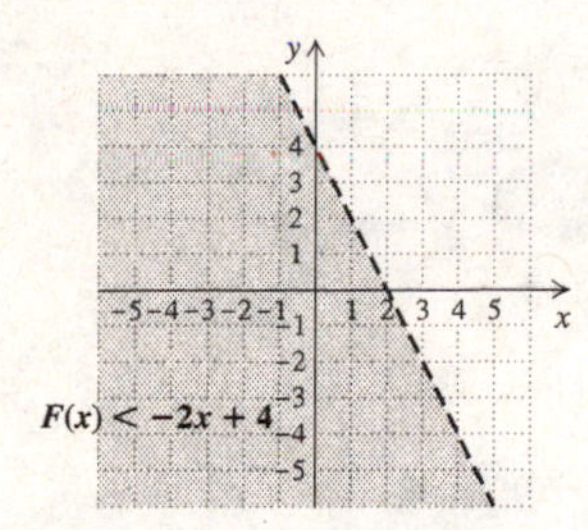

63. *Writing Exercise*

65. Substitute 1880 for $S(t)$ and solve for t.

$$1880 = 1087.7\sqrt{\frac{9t + 2617}{2457}}$$

$$1.7284 \approx \sqrt{\frac{9t + 2617}{2457}} \quad \text{Dividing by 1087.7}$$

$$(1.7284)^2 \approx \left(\sqrt{\frac{9t + 2617}{2457}}\right)^2$$

$$2.9874 \approx \frac{9t + 2617}{2457}$$

$$7340.0418 \approx 9t + 2617$$

$$4723.0418 \approx 9t$$

$$524.7824 \approx t$$

The temperature is about 524.8°C.

67.

$$S = 1087.7\sqrt{\frac{9t + 2617}{2457}}$$

$$\frac{S}{1087.7} = \sqrt{\frac{9t + 2617}{2457}}$$

$$\left(\frac{S}{1087.7}\right)^2 = \left(\sqrt{\frac{9t + 2617}{2457}}\right)^2$$

$$\frac{S^2}{1087.7^2} = \frac{9t + 2617}{2457}$$

$$\frac{2457S^2}{1087.7^2} = 9t + 2617$$

$$\frac{2457S^2}{1087.7^2} - 2617 = 9t$$

$$\frac{1}{9}\left(\frac{2457S^2}{1087.7^2} - 2617\right) = t$$

69. $d(n) = 0.75\sqrt{2.8n}$

Substitute 84 for $d(n)$ and solve for n.

$$84 = 0.75\sqrt{2.8n}$$

$$112 = \sqrt{2.8n}$$

$$(112)^2 = (\sqrt{2.8n})^2$$

$$12,544 = 2.8n$$

$$4480 = n$$

About 4480 rpm will produce peak performance.

71.

$$v = \sqrt{2gr}\sqrt{\frac{h}{r + h}}$$

$$v^2 = 2gr \cdot \frac{h}{r + h} \quad \text{Squaring both sides}$$

$$v^2(r + h) = 2grh \quad \text{Multiplying by } r+h$$

$$v^2r + v^2h = 2grh$$

$$v^2h = 2grh - v^2r$$

$$v^2h = r(2gh - v^2)$$

$$\frac{v^2h}{2gh - v^2} = r$$

73. $D(h) = 1.2\sqrt{h}$

$$10.2 = 1.2\sqrt{h}$$

$$8.5 = \sqrt{h}$$

$$(8.5)^2 = (\sqrt{h})^2$$

$$72.25 = h$$

The sailor must climb 72.25 ft above sea level.

75.

$$\frac{x + \sqrt{x + 1}}{x - \sqrt{x + 1}} = \frac{5}{11}$$

$$11(x + \sqrt{x + 1}) = 5(x - \sqrt{x + 1})$$

$$11x + 11\sqrt{x + 1} = 5x - 5\sqrt{x + 1}$$

$$16\sqrt{x + 1} = -6x$$

$$8\sqrt{x + 1} = -3x$$

$$(8\sqrt{x + 1})^2 = (-3x)^2$$

$$64(x + 1) = 9x^2$$

$$64x + 64 = 9x^2$$

$$0 = 9x^2 - 64x - 64$$

$$0 = (9x + 8)(x - 8)$$

$$9x + 8 = 0 \quad \text{or} \quad x - 8 = 0$$

$$9x = -8 \quad \text{or} \quad x = 8$$

$$x = -\frac{8}{9} \quad \text{or} \quad x = 8$$

Since $-\frac{8}{9}$ checks but 8 does not, the solution is $-\frac{8}{9}$.

77.

$$(z^2 + 17)^{3/4} = 27$$

$$[(z^2 + 17)^{3/4}]^{4/3} = (3^3)^{4/3}$$

$$z^2 + 17 = 3^4$$

$$z^2 + 17 = 81$$

$$z^2 - 64 = 0$$

$$(z + 8)(z - 8) = 0$$

$z = -8$ *or* $z = 8$

Both -8 and 8 check. They are the solutions.

79.
$$\begin{aligned}
\sqrt{8-b} &= b\sqrt{8-b}\\
(\sqrt{8-b})^2 &= (b\sqrt{8-b})^2\\
(8-b) &= b^2(8-b)\\
0 &= b^2(8-b)-(8-b)\\
0 &= (8-b)(b^2-1)\\
0 &= (8-b)(b+1)(b-1)
\end{aligned}$$
$$8-b=0 \text{ or } b+1=0 \text{ or } b-1=0$$
$$8=b \text{ or } \quad b=-1 \text{ or } \quad b=1$$

Since the numbers 8 and 1 check but -1 does not, 8 and 1 are the solutions.

81. We find the values of x for which $g(x) = 0$.
$$6x^{1/2}+6x^{-1/2}-37=0$$
$$6\sqrt{x}+\frac{6}{\sqrt{x}}=37$$
$$\left(6\sqrt{x}+\frac{6}{\sqrt{x}}\right)^2=37^2$$
$$36x+72+\frac{36}{x}=1369$$
$$36x^2+72x+36=1369x \quad \text{Multiplying by } x$$
$$36x^2-1297x+36=0$$
$$(36x-1)(x-36)=0$$
$$36x-1=0 \quad \textit{or} \quad x-36=0$$
$$36x=1 \quad \textit{or} \quad x=36$$
$$x=\frac{1}{36} \quad \textit{or} \quad x=36$$

Both numbers check. The x-intercepts are $\left(\frac{1}{36},0\right)$ and $(36,0)$.

83. 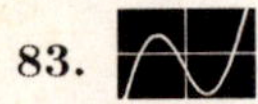

85.

Exercise Set 10.7

1. right; hypotenuse; see page 681 in the text.

3. square roots; see page 681 in the text.

5. 30°-60°-90°; leg; see page 684 in the text.

7. $a = 5, \quad b = 3$

Find c.
$$c^2 = a^2+b^2 \quad \text{Pythagorean theorem}$$
$$c^2 = 5^2+3^2 \quad \text{Substituting}$$
$$c^2 = 25+9$$
$$c^2 = 34$$
$$c = \sqrt{34} \quad \text{Exact answer}$$
$$c \approx 5.831 \quad \text{Approximation}$$

9. $a = 9, \quad b = 9$

Observe that the legs have the same length, so this is an isosceles right triangle. Then we know that the length of the hypotenuse is the length of a leg times $\sqrt{2}$, or $9\sqrt{2}$, or approximately 12.728.

11. $b = 12, \quad c = 13$

Find a.
$$a^2+b^2=c^2 \quad \text{Pythagorean theorem}$$
$$a^2+12^2=13^2 \quad \text{Substituting}$$
$$a^2+144=169$$
$$a^2=25$$
$$a=5$$

13.
$$a^2+b^2=c^2 \quad \text{Pythagorean theorem}$$
$$(4\sqrt{3})^2+b^2=8^2$$
$$16\cdot 3+b^2=64$$
$$48+b^2=64$$
$$b^2=16$$
$$b=4$$

The other leg is 4 m long.

15.
$$a^2+b^2=c^2 \quad \text{Pythagorean theorem}$$
$$1^2+b^2=(\sqrt{20})^2 \quad \text{Substituting}$$
$$1+b^2=20$$
$$b^2=19$$
$$b=\sqrt{19}$$
$$b\approx 4.359$$

The length of the other leg is $\sqrt{19}$ in., or about 4.359 in.

17. Observe that the length of the hypotenuse, $\sqrt{2}$, is $\sqrt{2}$ times the length of the given leg, 1. Thus, we have an isosceles right triangle and the length of the other leg is also 1.

19. From the drawing in the text we see that we have a right triangle with legs of 150 ft and 200 ft. Let $d =$ the length of the diagonal, in feet. We use the Pythagorean theorem to find d.
$$150^2+200^2=d^2$$
$$22,500+40,000=d^2$$
$$62,500=d^2$$
$$250=d$$

Clare travels 250 ft across the parking lot.

21. We make a drawing and let $d =$ the distance from home plate to second base.

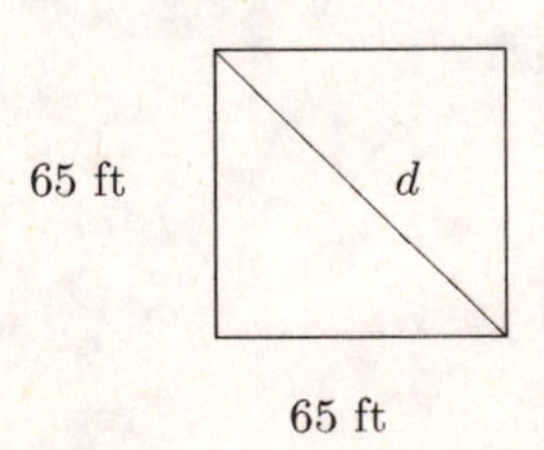

Note that we have an isosceles right triangle. Then the length of the hypotenuse is the length of a leg times $\sqrt{2}$, or $65\sqrt{2}$ ft. This is about 91.924 ft.

(We could also have used the Pythagorean theorem, solving $65^2 + 65^2 = d^2$.)

23. We make a drawing similar to the one in the text.

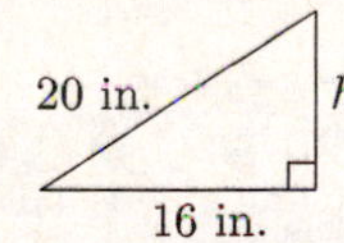

We use the Pythagorean theorem to find h.

$$16^2 + h^2 = 20^2$$
$$256 + h^2 = 400$$
$$h^2 = 144$$
$$h = 12$$

The height of the screen is 12 in.

25. First we will find the diagonal distance, d, in feet, across the room. We make a drawing.

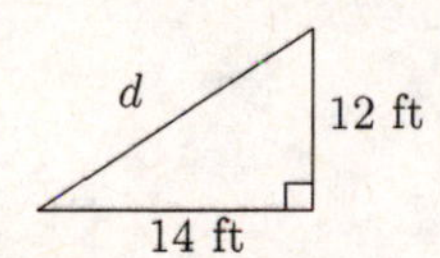

Now we use the Pythagorean theorem.

$$12^2 + 14^2 = d^2$$
$$144 + 196 = d^2$$
$$340 = d^2$$
$$\sqrt{340} = d$$
$$18.439 \approx d$$

Recall that 4 ft of slack is required on each end. Thus, $\sqrt{340}+2\cdot 4$, or $(\sqrt{340}+8)$ ft, of wire should be purchased. This is about 26.439 ft.

27. The diagonal is the hypotenuse of a right triangle with legs of 70 paces and 40 paces. First we use the Pythagorean theorem to find the length d of the diagonal, in paces.

$$70^2 + 40^2 = d^2$$
$$4900 + 1600 = d^2$$
$$6500 = d^2$$
$$\sqrt{6500} = d$$
$$80.623 \approx d$$

If Marissa walks along two sides of the quad she takes 70+40, or 110 paces. Then by using the diagonal she saves $(110 - \sqrt{6500})$ paces. This is approximately $110 - 80.623$, or 29.377 paces.

29. Since one acute angle is 45°, this is an isosceles right triangle with one leg = 5. Then the other leg = 5 also. And the hypotenuse is the length of the a leg times $\sqrt{2}$, or $5\sqrt{2}$.

Exact answer: Leg = 5, hypotenuse = $5\sqrt{2}$

Approximation: hypotenuse ≈ 7.071

31. This is a 30°-60°-90° right triangle with hypotenuse 14. We find the legs:

$$2a = 14, \text{ so } a = 7 \text{ and } a\sqrt{3} = 7\sqrt{3}$$

Exact answer: shorter leg = 7; longer leg = $7\sqrt{3}$

Approximation: longer leg ≈ 12.124

33. This is a 30°-60°-90° right triangle with one leg = 15. We substitute to find the length of the other leg, a, and the hypotenuse, c.

$$b = a\sqrt{3}$$
$$15 = a\sqrt{3}$$
$$\frac{15}{\sqrt{3}} = a$$
$$\frac{15\sqrt{3}}{3} = a \quad \text{Rationalizing the denominator}$$
$$5\sqrt{3} = a \quad \text{Simplifying}$$
$$c = 2a$$
$$c = 2\cdot 5\sqrt{3}$$
$$c = 10\sqrt{3}$$

Exact answer: $a = 5\sqrt{3},\ c = 10\sqrt{3}$

Approximations: $a \approx 8.660,\ c \approx 17.321$

35. This is an isosceles right triangle with hypotenuse 13. The two legs have the same length, a.

$$a\sqrt{2} = 13$$
$$a = \frac{13}{\sqrt{2}} = \frac{13\sqrt{2}}{2}$$

Exact answer: $\dfrac{13\sqrt{2}}{2}$

Approximation: 9.192

37. This is a 30°-60°-90° triangle with the shorter leg = 14. We find the longer leg and the hypotenuse.

$$a\sqrt{3} = 14\sqrt{3}, \text{ and } 2a = 2\cdot 14 = 28.$$

Exact answer: longer leg = $14\sqrt{3}$, hypotenuse = 28

Approximation: longer leg ≈ 24.249

39.

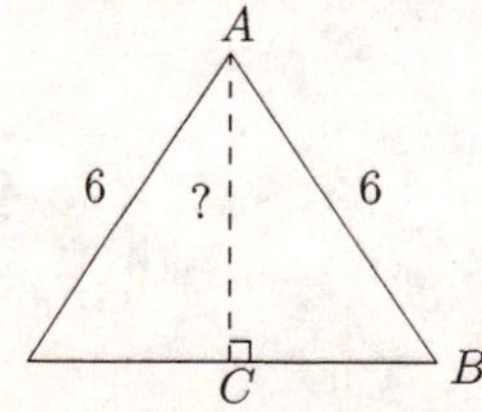

This is an equilateral triangle, so all the angles are 60°. The altitude bisects one angle and one side. Then triangle ABC is a 30-60-90 right triangle with the shorter leg of length 6/2, or 3, and hypotenuse of length 6. Then the length of the other leg is the length of the shorter leg times $\sqrt{3}$:

Exact answer: $3\sqrt{3}$

Approximation: 5.196

41.

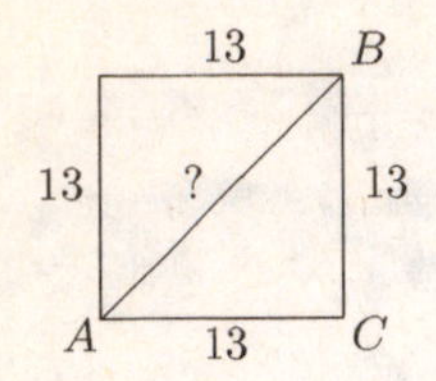

Triangle ABC is an isosceles right triangle with legs of length 13. Then the length of the hypotenuse is the length of a leg times $\sqrt{2}$.

Exact answer: $13\sqrt{2}$

Approximation: 18.385

43.

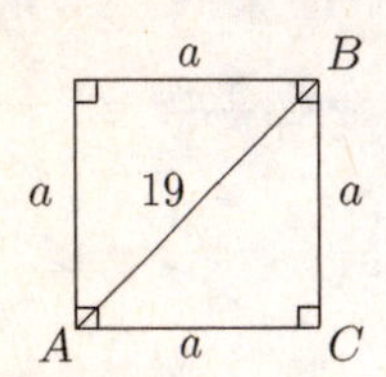

Triangle ABC is an isosceles right triangle with hypotenuse 19. Then the length of a side of the square a is the length of the legs of the triangle. We have:

$$a\sqrt{2} = 19$$
$$a = \frac{19}{\sqrt{2}}, \text{ or}$$
$$a = \frac{19\sqrt{2}}{2}$$

Exact answer: $a = \dfrac{19\sqrt{2}}{2}$

Approximation: $a \approx 13.435$

45. We will express all distances in feet. Recall that 1 mi = 5280 ft.

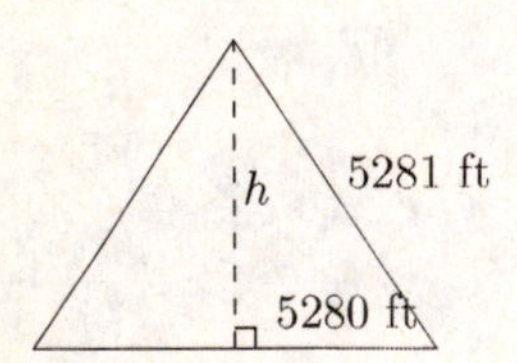

We use the Pythagorean theorem to find h.

$$h^2 + (5280)^2 = (5281)^2$$
$$h^2 + 27,878,400 = 27,888,961$$
$$h^2 = 10,561$$
$$h = \sqrt{10,561}$$
$$h \approx 102.767$$

The height of the bulge is $\sqrt{10,561}$ ft, or about 102.767 ft.

47.

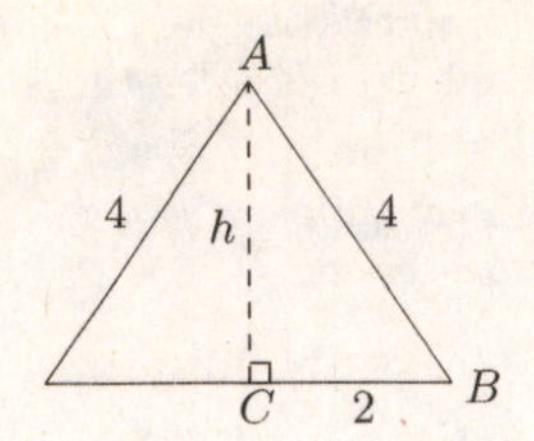

The entrance is an equilateral triangle, so all the angles are 60°. The altitude bisects one angle and one side. Then triangle ABC is a 30-60-90 right triangle with the shorter leg of length 4/2, or 2, and hypotenuse of length 4. Then the height of the tent is the length of the shorter leg times $\sqrt{3}$.

Exact answer: $h = 2\sqrt{3}$ ft

Approximation: $h \approx 3.464$ ft

49.

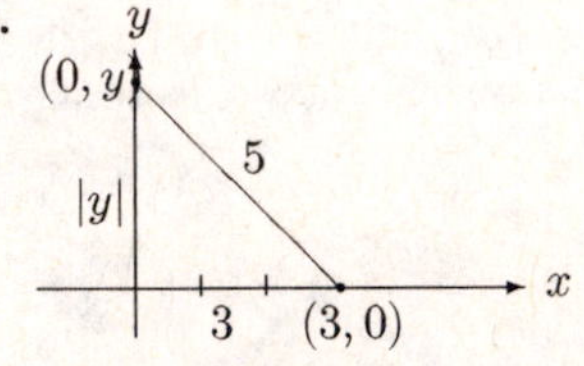

$$|y|^2 + 3^2 = 5^2$$
$$y^2 + 9 = 25$$
$$y^2 = 16$$
$$y = \pm 4$$

The points are $(0, -4)$ and $(0, 4)$.

51. *Writing Exercise*

53. $47(-1)^{19} = 47(-1) = -47$

55. $x^3 - 9x = x \cdot x^2 - 9 \cdot x = x(x^2 - 9) = x(x+3)(x-3)$

57. $|3x - 5| = 7$

$$3x - 5 = 7 \quad or \quad 3x - 5 = -7$$
$$3x = 12 \quad or \quad 3x = -2$$
$$x = 4 \quad or \quad x = -\frac{2}{3}$$

The solution set is $\left\{4, -\frac{2}{3}\right\}$.

59. *Writing Exercise*

61. The length of a side of the hexagon is 72/6, or 12 cm. Then the shaded region is a triangle with base 12 cm. To find the height of the triangle, note that it is the longer leg of a 30°-60°-90°right triangle. Thus its length is the length of the length of the shorter leg times $\sqrt{3}$. The length of the shorter leg is half the length of the base, $\frac{1}{2} \cdot 12$ cm, or 6 cm, so the length of the longer leg is $6\sqrt{3}$ cm. Now we find the area of the triangle.

$$\begin{aligned} A &= \frac{1}{2}bh \\ &= \frac{1}{2}(12\text{ cm})(6\sqrt{3}\text{ cm}) \\ &= 36\sqrt{3}\text{ cm}^2 \\ &\approx 62.354\text{ cm}^2 \end{aligned}$$

63.

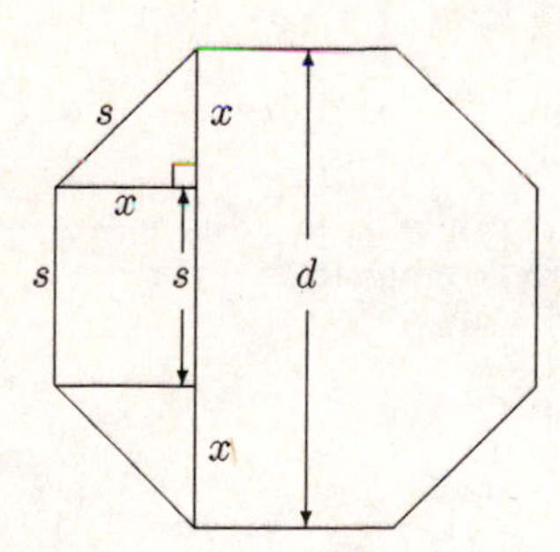

$d = s + 2x$

Use the Pythagoran theorem to find x.

$$\begin{aligned} x^2 + x^2 &= s^2 \\ 2x^2 &= s^2 \\ x^2 &= \frac{s^2}{2} \\ x &= \frac{s}{\sqrt{2}} = \frac{s}{\sqrt{2}} \cdot \frac{\sqrt{2}}{2} = \frac{s\sqrt{2}}{2} \end{aligned}$$

Then $d = s + 2x = s + 2\left(\frac{s\sqrt{2}}{2}\right) = s + s\sqrt{2}$.

65.

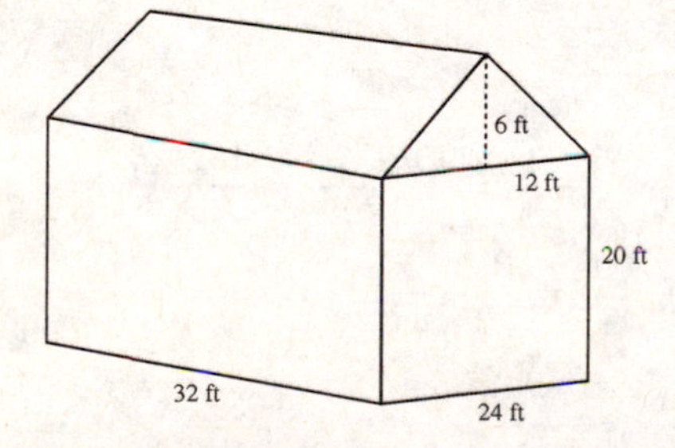

The area to be painted consists of two 20 ft by 24 ft rectangles, two 20 ft by 32 ft rectangles, and two triangles with height 6 ft and base 24 ft. The area of the two 20 ft by 24 ft rectangles is $2 \cdot 20\text{ ft} \cdot 24\text{ ft} = 960\text{ ft}^2$. The area of the two 20 ft by 32 ft rectangles is $2 \cdot 20\text{ ft} \cdot 32\text{ ft} = 1280\text{ ft}^2$. The area of the two triangles is $2 \cdot \frac{1}{2} \cdot 24\text{ ft} \cdot 6\text{ ft} = 144\text{ ft}^2$. Thus, the total area to be painted is $960\text{ ft}^2 + 1280\text{ ft}^2 + 144\text{ ft}^2 = 2384\text{ ft}^2$.

One gallon of paint covers a minimum of 450 ft^2, so we divide to determine how many gallons of paint are required: $\frac{2384}{450} \approx 5.3$. Thus, 5 gallons of paint should be bought to paint the house. This answer assumes that the total area of the doors and windows is 134 ft^2 or more. $(5 \cdot 450 = 2250$ and $2384 = 2250 + 134)$

67. First we find the radius of a circle with an area of 6160 ft^2. This is the length of the hose.

$$\begin{aligned} A &= \pi r^2 \\ 6160 &= \pi r^2 \\ \frac{6160}{\pi} &= r^2 \\ \sqrt{\frac{6160}{\pi}} &= r \\ 44.28 &\approx r \end{aligned}$$

Now we make a drawing of the room.

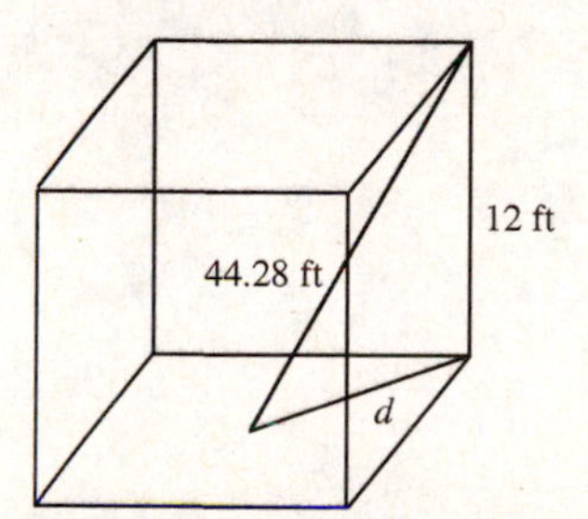

We use the Pythagorean theorem to find d.

$$\begin{aligned} d^2 + 12^2 &= 44.28^2 \\ d^2 + 144 &= 1960.7184 \\ d^2 &= 1816.7184 \\ d &\approx 42.623 \end{aligned}$$

Now we make a drawing of the floor of the room.

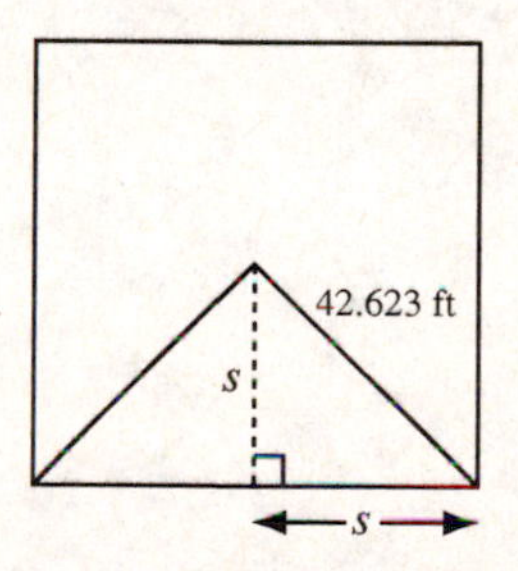

We have an isosceles right triangle with hypotenuse 42.623 ft. We find the length of a side s.

$$\begin{aligned} s\sqrt{2} &= 42.623 \\ s &= \frac{42.623}{\sqrt{2}} \approx 30.14\text{ ft} \end{aligned}$$

Then the length of a side of the room is $2s = 2(30.14\text{ ft}) = 60.28$ ft; so the dimensions of the largest square room that meets the given conditions is 60.28 ft by 60.28 ft.

Exercise Set 10.8

1. False; see page 689 in the text.

3. True; see page 690 in the text.

5. True; see page 690 in the text.

7. False; see Exercises 59-64.

9. $\sqrt{-36} = \sqrt{-1 \cdot 36} = \sqrt{-1} \cdot \sqrt{36} = i \cdot 6 = 6i$

11. $\sqrt{-13} = \sqrt{-1 \cdot 13} = \sqrt{-1} \cdot \sqrt{13} = i\sqrt{13}$, or $\sqrt{13}i$

13. $\sqrt{-18} = \sqrt{-1} \cdot \sqrt{9} \cdot \sqrt{2} = i \cdot 3 \cdot \sqrt{2} = 3i\sqrt{2}$, or $3\sqrt{2}i$

15. $\sqrt{-3} = \sqrt{-1 \cdot 3} = \sqrt{-1} \cdot \sqrt{3} = i\sqrt{3}$, or $\sqrt{3}i$

17. $\sqrt{-81} = \sqrt{-1 \cdot 81} = \sqrt{-1} \cdot \sqrt{81} = i \cdot 9 = 9i$

19. $\sqrt{-300} = \sqrt{-1} \cdot \sqrt{100} \cdot \sqrt{3} = i \cdot 10 \cdot \sqrt{3} = 10i\sqrt{3}$, or $10\sqrt{3}i$

21. $6 - \sqrt{-84} = 6 - \sqrt{-1 \cdot 4 \cdot 21} = 6 - i \cdot 2\sqrt{21} = 6 - 2i\sqrt{21}$, or $6 - 2\sqrt{21}i$

23. $-\sqrt{-76} + \sqrt{-125} = -\sqrt{-1 \cdot 4 \cdot 19} + \sqrt{-1 \cdot 25 \cdot 5} =$
$-i \cdot 2\sqrt{19} + i \cdot 5\sqrt{5} = -2i\sqrt{19} + 5i\sqrt{5} = (-2\sqrt{19} + 5\sqrt{5})i$

25. $\sqrt{-18} - \sqrt{-100} = \sqrt{-1 \cdot 9 \cdot 2} - \sqrt{-1 \cdot 100} =$
$i \cdot 3\sqrt{2} - i \cdot 10 = 3i\sqrt{2} - 10i = (3\sqrt{2} - 10)i$

27. $(6 + 7i) + (5 + 3i)$
$= (6 + 5) + (7 + 3)i$ Combining the real and the imaginary parts
$= 11 + 10i$

29. $(9 + 8i) - (5 + 3i) = (9 - 5) + (8 - 3)i$
$= 4 + 5i$

31. $(7 - 4i) - (5 - 3i) = (7 - 5) + [-4 - (-3)]i = 2 - i$

33. $(-5 - i) - (7 + 4i) = (-5 - 7) + (-1 - 4)i = -12 - 5i$

35. $7i \cdot 6i = 42 \cdot i^2 = 42(-1) = -42$

37. $(-4i)(-6i) = 24 \cdot i^2 = 24(-1) = -24$

39. $\sqrt{-36}\sqrt{-9} = \sqrt{-1} \cdot \sqrt{36} \cdot \sqrt{-1} \cdot \sqrt{9} = i \cdot 6 \cdot i \cdot 3 =$
$i^2 \cdot 18 = -1 \cdot 18 = -18$

41. $\sqrt{-5}\sqrt{-2} = \sqrt{-1} \cdot \sqrt{5} \cdot \sqrt{-1} \cdot \sqrt{2} =$
$i \cdot \sqrt{5} \cdot i \cdot \sqrt{2} = i^2 \cdot \sqrt{10} = -1 \cdot \sqrt{10} = -\sqrt{10}$

43. $\sqrt{-6}\sqrt{-21} = \sqrt{-1} \cdot \sqrt{6} \cdot \sqrt{-1} \cdot \sqrt{21} =$
$i \cdot \sqrt{6} \cdot i \cdot \sqrt{21} = i^2\sqrt{126} = -1 \cdot \sqrt{9 \cdot 14} = -3\sqrt{14}$

45. $5i(2 + 6i) = 5i \cdot 2 + 5i \cdot 6i = 10i + 30i^2 =$
$10i - 30 = -30 + 10i$

47. $-7i(3 - 4i) = -7i \cdot 3 - 7i(-4i) = -21i + 28i^2 =$
$-21i - 28 = -28 - 21i$

49. $(1 + i)(3 + 2i) = 3 + 2i + 3i + 2i^2 =$
$3 + 2i + 3i - 2 = 1 + 5i$

51. $(6 - 5i)(3 + 4i) = 18 + 24i - 15i - 20i^2 =$
$18 + 24i - 15i + 20 = 38 + 9i$

53. $(7 - 2i)(2 - 6i) = 14 - 42i - 4i + 12i^2 =$
$14 - 42i - 4i - 12 = 2 - 46i$

55. $(-2 + 3i)(-2 + 5i) = 4 - 10i - 6i + 15i^2 =$
$4 - 10i - 6i - 15 = -11 - 16i$

57. $(-5 - 4i)(3 + 7i) = -15 - 35i - 12i - 28i^2 =$
$-15 - 35i - 12i + 28 = 13 - 47i$

59. $(4 - 2i)^2 = 4^2 - 2 \cdot 4 \cdot 2i + (2i)^2 = 16 - 16i + 4i^2 =$
$16 - 16i - 4 = 12 - 16i$

61. $(2 + 3i)^2 = 2^2 + 2 \cdot 2 \cdot 3i + (3i)^2 = 4 + 12i + 9i^2 =$
$4 + 12i - 9 = -5 + 12i$

63. $(-2 + 3i)^2 = (-2)^2 + 2(-2)(3i) + (3i)^2 =$
$4 - 12i + 9i^2 = 4 - 12i - 9 = -5 - 12i$

65. $\frac{7}{4 + i}$
$= \frac{7}{4 + i} \cdot \frac{4 - i}{4 - i}$ Multiplying by 1, using the conjugate
$= \frac{28 - 7i}{16 - i^2}$
$= \frac{28 - 7i}{16 - (-1)}$ $i^2 = -1$
$= \frac{28 - 7i}{17}$
$= \frac{28}{17} - \frac{7}{17}i$

67. $\frac{2}{3 - 2i} = \frac{2}{3 - 2i} \cdot \frac{3 + 2i}{3 + 2i}$ Multiplying by 1, using the conjugate
$= \frac{6 + 4i}{9 - 4i^2}$
$= \frac{6 + 4i}{9 - 4(-1)}$
$= \frac{6 + 4i}{13}$
$= \frac{6}{13} + \frac{4}{13}i$

69. $\frac{2i}{5 + 3i} = \frac{2i}{5 + 3i} \cdot \frac{5 - 3i}{5 - 3i} = \frac{10i - 6i^2}{25 - 9i^2} = \frac{10i + 6}{25 + 9} =$
$\frac{10i + 6}{34} = \frac{6}{34} + \frac{10}{34}i = \frac{3}{17} + \frac{5}{17}i$

71. $\frac{5}{6i} = \frac{5}{6i} \cdot \frac{i}{i} = \frac{5i}{6i^2} = \frac{5i}{-6} = -\frac{5}{6}i$

73. $\frac{5 - 3i}{4i} = \frac{5 - 3i}{4i} \cdot \frac{i}{i} = \frac{5i - 3i^2}{4i^2} = \frac{5i + 3}{-4} =$
$-\frac{3}{4} - \frac{5}{4}i$

75. $\frac{7i + 14}{7i} = \frac{7i}{7i} + \frac{14}{7i} = 1 + \frac{2}{i} = 1 + \frac{2}{i} \cdot \frac{i}{i} =$
$1 + \frac{2i}{i^2} = 1 + \frac{2i}{-1} = 1 - 2i$

77. $\frac{4 + 5i}{3 - 7i} = \frac{4 + 5i}{3 - 7i} \cdot \frac{3 + 7i}{3 + 7i} = \frac{12 + 28i + 15i + 35i^2}{9 - 49i^2} =$
$\frac{12 + 28i + 15i - 35}{9 + 49} = \frac{-23 + 43i}{58} = -\frac{23}{58} + \frac{43}{58}i$

79. $\frac{2 + 3i}{2 + 5i} = \frac{2 + 3i}{2 + 5i} \cdot \frac{2 - 5i}{2 - 5i} = \frac{4 - 10i + 6i - 15i^2}{4 - 25i^2} =$
$\frac{4 - 10i + 6i + 15}{4 + 25} = \frac{19 - 4i}{29} = \frac{19}{29} - \frac{4}{29}i$

81. $\frac{3-2i}{4+3i} = \frac{3-2i}{4+3i} \cdot \frac{4-3i}{4-3i} = \frac{12-9i-8i+6i^2}{16-9i^2} =$
$\frac{12-9i-8i-6}{16+9} = \frac{6-17i}{25} = \frac{6}{25} - \frac{17}{25}i$

83. $i^7 = i^6 \cdot i = (i^2)^3 \cdot i = (-1)^3 \cdot i = -1 \cdot i = -i$

85. $i^{24} = (i^2)^{12} = (-1)^{12} = 1$

87. $i^{42} = (i^2)^{21} = (-1)^{21} = -1$

89. $i^9 = (i^2)^4 \cdot i = (-1)^4 \cdot i = 1 \cdot i = i$

91. $(-i)^6 = (-1 \cdot i)^6 = (-1)^6 \cdot i^6 = 1 \cdot i^6 = (i^2)^3 = (-1)^3 = -1$

93. $(5i)^3 = 5^3 \cdot i^3 = 125 \cdot i^2 \cdot i = 125(-1)(i) = -125i$

95. $i^2 + i^4 = -1 + (i^2)^2 = -1 + (-1)^2 = -1 + 1 = 0$

97. *Writing Exercise*

99. Graph $f(x) = 3x - 5$.

The y-intercept is $(0, -5)$ and the slope is 3, or $\frac{3}{1}$. First plot the y-intercept and, from there, go up 3 units and right 1 unit to $(1, -2)$. Plot this point and draw the graph.

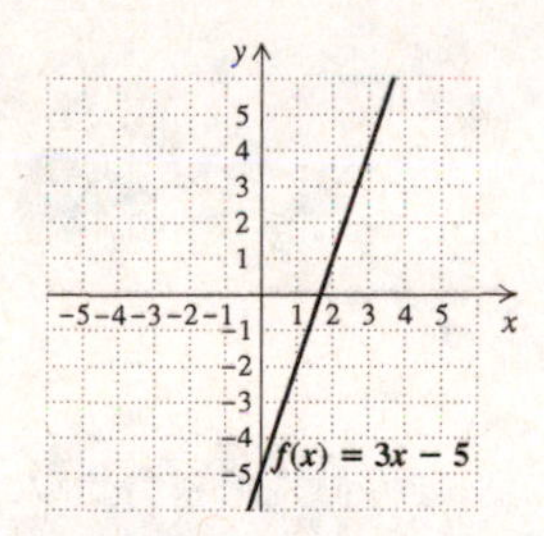

To check that the line is drawn correctly, we find the coordinates of another point on the line. When $x = 2$, we have:

$f(2) = 3 \cdot 2 - 5 = 6 - 5 = 1.$

Since the point $(2, 1)$ appears to be on the line, we have a check.

101. Graph $F(x) = x^2$.

We select numbers for x and find the corresponding values of y. Then we plot these ordered pairs and draw the graph.

x	$F(x)$
-2	4
-1	1
0	0
1	1
2	4

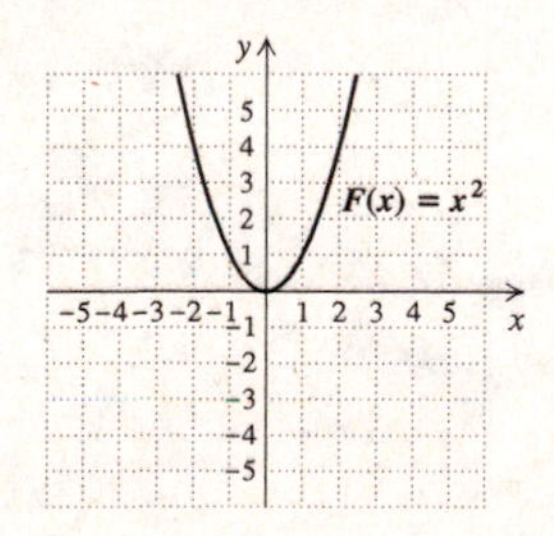

103. $28 = 3x^2 - 17x$

$0 = 3x^2 - 17x - 28$

$0 = (3x + 4)(x - 7)$

$3x + 4 = 0 \quad or \quad x - 7 = 0$

$3x = -4 \quad or \quad x = 7$

$x = -\frac{4}{3} \quad or \quad x = 7$

Both values check. The solutions are $-\frac{4}{3}$ and 7.

105. *Writing Exercise*

107. $g(3i) = \frac{(3i)^4 - (3i)^2}{3i - 1} = \frac{81i^4 - 9i^2}{-1 + 3i} = \frac{81 + 9}{-1 + 3i} =$
$\frac{90}{-1+3i} = \frac{90}{-1+3i} \cdot \frac{-1-3i}{-1-3i} = \frac{90(-1-3i)}{1-9i^2} =$
$\frac{90(-1-3i)}{1+9} = \frac{90(-1-3i)}{10} = \frac{9 \cdot \cancel{10}(-1-3i)}{\cancel{10}} =$
$9(-1-3i) = -9 - 27i$

109. First we simplify $g(z)$.

$g(z) = \frac{z^4 - z^2}{z - 1} = \frac{z^2(z^2 - 1)}{z - 1} = \frac{z^2(z+1)(z-1)}{z-1} =$
$\frac{z^2(z+1)\cancel{(z-1)}}{\cancel{z-1}} = z^2(z+1)$

Now we substitute.

$g(5i - 1) = (5i - 1)^2(5i - 1 + 1) =$
$(25i^2 - 10i + 1)(5i) =$
$(-25 - 10i + 1)(5i) = (-24 - 10i)(5i) =$
$-120i - 50i^2 = 50 - 120i$

111. $\frac{1}{\frac{1-i}{10} - \left(\frac{1-i}{10}\right)^2} = \frac{1}{\frac{1-i}{10} - \left(\frac{-2i}{100}\right)} =$
$\frac{1}{\frac{1-i}{10} + \frac{i}{50}} = \frac{1}{\frac{1-i}{10} + \frac{i}{50}} \cdot \frac{50}{50} = \frac{50}{5 - 5i + i} =$
$\frac{50}{5-4i} = \frac{50}{5-4i} \cdot \frac{5+4i}{5+4i} = \frac{250+200i}{41} = \frac{250}{41} + \frac{200}{41}i$

113. $(1-i)^3(1+i)^3 =$
$(1-i)(1+i) \cdot (1-i)(1+i) \cdot (1-i)(1+i) =$
$(1-i^2)(1-i^2)(1-i^2) = (1+1)(1+1)(1+1) =$
$2 \cdot 2 \cdot 2 = 8$

115. $\frac{6}{1 + \frac{3}{i}} = \frac{6}{\frac{i+3}{i}} = \frac{6i}{i+3} = \frac{6i}{i+3} \cdot \frac{-i+3}{-i+3} =$
$\frac{-6i^2 + 18i}{-i^2 + 9} = \frac{6 + 18i}{10} = \frac{6}{10} + \frac{18}{10}i = \frac{3}{5} + \frac{9}{5}i$

117. $\frac{i - i^{38}}{1+i} = \frac{i - (i^2)^{19}}{1+i} = \frac{i - (-1)^{19}}{1+i} = \frac{i - (-1)}{1+i} =$
$\frac{i+1}{1+i} = 1$

Chapter 11

Quadratic Functions and Equations

Exercise Set 11.1

1. $\sqrt{k}$; $-\sqrt{k}$

3. $t+3$; $t+3$

5. 25; 5

7. $4x^2 = 20$

$x^2 = 5$ Multiplying by $\frac{1}{4}$

$x = \sqrt{5}$, or $x = -\sqrt{5}$ Using the principle of square roots

The solutions are $\sqrt{5}$ and $-\sqrt{5}$, or $\pm\sqrt{5}$.

9. $9x^2 + 16 = 0$

$x^2 = -\frac{16}{9}$ Isolating x^2

$x = \sqrt{-\frac{16}{9}}$ *or* $x = -\sqrt{-\frac{16}{9}}$

$x = \sqrt{\frac{16}{9}}\sqrt{-1}$ *or* $x = -\sqrt{\frac{16}{9}}\sqrt{-1}$

$x = \frac{4}{3}i$ *or* $x = -\frac{4}{3}i$

The solutions are $\frac{4}{3}i$ and $-\frac{4}{3}i$, or $\pm\frac{4}{3}i$.

11. $5t^2 - 7 = 0$

$t^2 = \frac{7}{5}$

$t = \sqrt{\frac{7}{5}}$ *or* $t = -\sqrt{\frac{7}{5}}$ Principle of square roots

$t = \sqrt{\frac{7}{5}\cdot\frac{5}{5}}$ *or* $t = -\sqrt{\frac{7}{5}\cdot\frac{5}{5}}$ Rationalizing denominators

$t = \frac{\sqrt{35}}{5}$ *or* $t = -\frac{\sqrt{35}}{5}$

The solutions are $\sqrt{\frac{7}{5}}$ and $-\sqrt{\frac{7}{5}}$. This can also be written as $\pm\sqrt{\frac{7}{5}}$ or, if we rationalize the denominator, $\pm\frac{\sqrt{35}}{5}$.

13. $(x-1)^2 = 49$

$x - 1 = 7$ *or* $x - 1 = -7$ Principle of square roots

$x = 8$ *or* $x = -6$

The solutions are 8 and -6.

15. $(a-13)^2 = 18$

$a - 13 = \sqrt{18}$ *or* $a - 13 = -\sqrt{18}$ Principle of square roots

$a - 13 = 3\sqrt{2}$ *or* $a - 13 = -3\sqrt{2}$

$a = 13 + 3\sqrt{2}$ *or* $a = 13 - 3\sqrt{2}$

The solutions are $13 + 3\sqrt{2}$ and $13 - 3\sqrt{2}$, or $13 \pm 3\sqrt{2}$.

17. $(x+1)^2 = -9$

$x + 1 = \sqrt{-9}$ *or* $x + 1 = -\sqrt{-9}$

$x + 1 = 3i$ *or* $x + 1 = -3i$

$x = -1 + 3i$ *or* $x = -1 - 3i$

The solutions are $-1 + 3i$ and $-1 - 3i$, or $-1 \pm 3i$.

19. $\left(y + \frac{3}{4}\right)^2 = \frac{17}{16}$

$y + \frac{3}{4} = \pm\frac{\sqrt{17}}{4}$

$y = -\frac{3}{4} \pm \frac{\sqrt{17}}{4}$, or $\frac{-3 \pm \sqrt{17}}{4}$

The solutions are $-\frac{3}{4} \pm \frac{\sqrt{17}}{4}$, or $\frac{-3 \pm \sqrt{17}}{4}$.

21. $x^2 - 10x + 25 = 64$

$(x-5)^2 = 64$

$x - 5 = \pm 8$

$x = 5 \pm 8$

$x = 13$ *or* $x = -3$

The solutions are 13 and -13.

23. $f(x) = 16$

$(x-5)^2 = 16$ Substituting

$x - 5 = 4$ *or* $x - 5 = -4$

$x = 9$ *or* $x = 1$

The solutions are 9 and 1.

25. $F(t) = 13$

$(t+4)^2 = 13$ Substituting

$t + 4 = \sqrt{13}$ *or* $t + 4 = -\sqrt{13}$

$t = -4 + \sqrt{13}$ *or* $t = -4 - \sqrt{13}$

The solutions are $-4 + \sqrt{13}$ and $-4 - \sqrt{13}$, or $-4 \pm \sqrt{13}$.

27. $g(x) = x^2 + 14x + 49$

Observe first that $g(0) = 49$. Also observe that when $x = -14$, then $x^2 + 14x = (-14)^2 - (14)(14) = (14)^2 - (14)^2 = 0$, so $g(-14) = 49$ as well. Thus, we have $x = 0$ or $x = 14$.

We can also do this problem as follows.

$$g(x) = 49$$
$$x^2 + 14x + 49 = 49 \quad \text{Substituting}$$
$$(x+7)^2 = 49$$
$$x+7 = 7 \quad or \quad x+7 = -7$$
$$x = 0 \quad or \quad x = -14$$

The solutions are 0 and -14.

29. $x^2 + 16x$

We take half the coefficient of x and square it: Half of 16 is 8, and $8^2 = 64$. We add 64.

$x^2 + 16x + 64 = (x+8)^2$

31. $t^2 - 10t$

We take half the coefficient of t and square it:

Half of -10 is -5, and $(-5)^2 = 25$. We add 25.

$t^2 - 10t + 25 = (t-5)^2$

33. $x^2 + 3x$

We take half the coefficient of x and square it:

$\frac{1}{2}(3) = \frac{3}{2}$ and $\left(\frac{3}{2}\right)^2 = \frac{9}{4}$. We add $\frac{9}{4}$.

$x^2 + 3x + \frac{9}{4} = \left(x + \frac{3}{2}\right)^2$

35. $t^2 - 9t$

We take half the coefficient of t and square it:

$\frac{1}{2}(-9) = -\frac{9}{2}$, and $\left(-\frac{9}{2}\right)^2 = \frac{81}{4}$. We add $\frac{81}{4}$.

$t^2 - 9t + \frac{81}{4} = \left(t - \frac{9}{2}\right)^2$

37. $x^2 + \frac{2}{5}x$

$\frac{1}{2} \cdot \frac{2}{5} = \frac{1}{5}$, and $\left(\frac{1}{5}\right)^2 = \frac{1}{25}$. We add $\frac{1}{25}$.

$x^2 + \frac{2}{5}x + \frac{1}{25} = \left(x + \frac{1}{5}\right)^2$

39. $t^2 - \frac{5}{6}t$

$\frac{1}{2}\left(-\frac{5}{6}\right) = -\frac{5}{12}$, and $\left(-\frac{5}{12}\right)^2 = \frac{25}{144}$. We add $\frac{25}{144}$.

$t^2 - \frac{5}{6}t + \frac{25}{144} = \left(t - \frac{5}{12}\right)^2$

41.

$$x^2 + 6x = 7$$
$$x^2 + 6x + 9 = 7 + 9 \quad \text{Adding 9 to both sides to complete the square}$$
$$(x+3)^2 = 16 \quad \text{Factoring}$$
$$x + 3 = \pm 4 \quad \text{Principle of square roots}$$
$$x = -3 \pm 4$$
$$x = -3 + 4 \quad or \quad x = -3 - 4$$
$$x = 1 \quad or \quad x = -7$$

The solutions are 1 and -7.

43.

$$t^2 - 10t = -24$$
$$t^2 - 10t + 25 = -24 + 25 \quad \text{Adding 25 to both sides to complete the square}$$
$$(t-5)^2 = 1 \quad \text{Factoring}$$
$$t - 5 = \pm 1 \quad \text{Principle of square roots}$$
$$t = 5 \pm 1$$
$$t = 5 + 1 \quad or \quad x = 5 - 1$$
$$t = 6 \quad or \quad x = 4$$

The solutions are 6 and 4.

45.

$$x^2 + 10x + 9 = 0$$
$$x^2 + 10x = -9 \quad \text{Adding } -9 \text{ to both sides}$$
$$x^2 + 10x + 25 = -9 + 25 \quad \text{Completing the square}$$
$$(x+5)^2 = 16$$
$$x + 5 = \pm 4$$
$$x = -5 \pm 4$$
$$x = -5 + 4 \quad or \quad x = -5 - 4$$
$$x = -1 \quad or \quad x = -9$$

The solutions are -1 and -9.

47.

$$t^2 + 8t - 3 = 0$$
$$t^2 + 8t = 3$$
$$t^2 + 8t + 16 = 3 + 16$$
$$(t+4)^2 = 19$$
$$t + 4 = \pm\sqrt{19}$$
$$t = -4 \pm \sqrt{19}$$

The solutions are $-4 \pm \sqrt{19}$.

49. The value of $f(x)$ must be 0 at any x-intercepts.

$$f(x) = 0$$
$$x^2 + 6x + 7 = 0$$
$$x^2 + 6x = -7$$
$$x^2 + 6x + 9 + -7 + 9$$
$$(x+3)^2 = 2$$
$$x + 3 = \pm\sqrt{2}$$
$$x = -3 \pm \sqrt{2}$$

The x-intercepts are $(-3 - \sqrt{2}, 0)$ and $(-3 + \sqrt{2}, 0)$.

51. The value of $g(x)$ must be 0 at any x-intercepts.

$$g(x) = 0$$
$$x^2 + 12x + 25 = 0$$
$$x^2 + 12x = -25$$
$$x^2 + 12x + 36 = -25 + 36$$
$$(x+6)^2 = 11$$
$$x + 6 = \pm\sqrt{11}$$
$$x = -6 \pm \sqrt{11}$$

The x-intercepts are $(-6 - \sqrt{11}, 0)$ and $(-6 + \sqrt{11}, 0)$.

53. The value of $f(x)$ must be 0 at any x-intercepts.

$$\begin{aligned} f(x) &= 0 \\ x^2 - 10x - 22 &= 0 \\ x^2 - 10x &= 22 \\ x^2 - 10x + 25 &= 22 + 25 \\ (x-5)^2 &= 47 \\ x - 5 &= \pm\sqrt{47} \\ x &= 5 \pm \sqrt{47} \end{aligned}$$

The x-intercepts are $(5 - \sqrt{47}, 0)$ and $(5 + \sqrt{47}, 0)$.

55.

$$\begin{aligned} 9x^2 + 18x &= -8 \\ x^2 + 2x &= -\frac{8}{9} \quad \text{Dividing both sides by 9} \\ x^2 + 2x + 1 &= -\frac{8}{9} + 1 \\ (x+1)^2 &= \frac{1}{9} \\ x + 1 &= \pm\frac{1}{3} \\ x &= -1 \pm \frac{1}{3} \end{aligned}$$

$$x = -1 - \frac{1}{3} \quad \text{or} \quad x = -1 + \frac{1}{3}$$

$$x = -\frac{4}{3} \quad \text{or} \quad x = -\frac{2}{3}$$

The solutions are $-\frac{4}{3}$ and $-\frac{2}{3}$.

57.

$$\begin{aligned} 3x^2 - 5x - 2 &= 0 \\ 3x^2 - 5x &= 2 \\ x^2 - \frac{5}{3}x &= \frac{2}{3} \quad \text{Dividing both sides by 3} \\ x^2 - \frac{5}{3}x + \frac{25}{36} &= \frac{2}{3} + \frac{25}{36} \\ \left(x - \frac{5}{6}\right)^2 &= \frac{49}{36} \\ x - \frac{5}{6} &= \pm\frac{7}{6} \\ x &= \frac{5}{6} \pm \frac{7}{6} \end{aligned}$$

$$x = \frac{5}{6} - \frac{7}{6} \quad \text{or} \quad x = \frac{5}{6} + \frac{7}{6}$$

$$x = -\frac{1}{3} \quad \text{or} \quad x = 2$$

The solutions are $-\frac{1}{3}$ and 2.

59.

$$\begin{aligned} 5x^2 + 4x - 3 &= 0 \\ 5x^2 + 4x &= 3 \\ x^2 + \frac{4}{5}x &= \frac{3}{5} \quad \text{Dividing both sides by 5} \\ x^2 + \frac{4}{5}x + \frac{4}{25} &= \frac{3}{5} + \frac{4}{25} \\ \left(x + \frac{2}{5}\right)^2 &= \frac{19}{25} \\ x + \frac{2}{5} &= \pm\frac{\sqrt{19}}{5} \\ x &= -\frac{2}{5} \pm \frac{\sqrt{19}}{5}, \text{ or } \frac{-2 \pm \sqrt{19}}{5} \end{aligned}$$

The solutions are $-\frac{2}{5} \pm \frac{\sqrt{19}}{5}$, or $\frac{-2 \pm \sqrt{19}}{5}$.

61. The value of $f(x)$ must be 0 at any x-intercepts.

$$\begin{aligned} f(x) &= 0 \\ 4x^2 + 2x - 3 &= 0 \\ 4x^2 + 2x &= 3 \\ x^2 + \frac{1}{2}x &= \frac{3}{4} \quad \text{Dividing both sides by 4} \\ x^2 + \frac{1}{2}x + \frac{1}{16} &= \frac{3}{4} + \frac{1}{16} \\ \left(x + \frac{1}{4}\right)^2 &= \frac{13}{16} \\ x + \frac{1}{4} &= \pm\frac{\sqrt{13}}{4} \\ x &= -\frac{1}{4} \pm \frac{\sqrt{13}}{4}, \text{ or } \frac{-1 \pm \sqrt{13}}{4} \end{aligned}$$

The x-intercepts are $\left(-\frac{1}{4} - \frac{\sqrt{13}}{4}, 0\right)$ and $\left(-\frac{1}{4} + \frac{\sqrt{13}}{4}, 0\right)$, or $\left(\frac{-1-\sqrt{13}}{4}, 0\right)$ and $\left(\frac{-1+\sqrt{13}}{4}, 0\right)$.

63. The value of $g(x)$ must be 0 at any x-intercepts.

$$\begin{aligned} g(x) &= 0 \\ 2x^2 - 3x - 1 &= 0 \\ 2x^2 - 3x &= 1 \\ x^2 - \frac{3}{2}x &= \frac{1}{2} \quad \text{Dividing both sides by 2} \\ x^2 - \frac{3}{2}x + \frac{9}{16} &= \frac{1}{2} + \frac{9}{16} \\ \left(x - \frac{3}{4}\right)^2 &= \frac{17}{16} \\ x - \frac{3}{4} &= \pm\frac{\sqrt{17}}{4} \\ x &= \frac{3}{4} \pm \frac{\sqrt{17}}{4}, \text{ or } \frac{3 \pm \sqrt{17}}{4} \end{aligned}$$

The x-intercepts are $\left(\frac{3}{4} - \frac{\sqrt{17}}{4}, 0\right)$ and $\left(\frac{3}{4} + \frac{\sqrt{17}}{4}, 0\right)$, or $\left(\frac{3-\sqrt{17}}{4}, 0\right)$ and $\left(\frac{3+\sqrt{17}}{4}, 0\right)$.

65. ***Familiarize***. We are already familiar with the compound-interest formula.

Translate. We substitute into the formula.

$$A = P(1+r)^t$$
$$2420 = 2000(1+r)^2$$

Carry out. We solve for r.

$$2420 = 2000(1+r)^2$$
$$\frac{2420}{2000} = (1+r)^2$$
$$\frac{121}{100} = (1+r)^2$$
$$\pm\sqrt{\frac{121}{100}} = 1+r$$
$$\pm\frac{11}{10} = 1+r$$
$$-\frac{10}{10} + \frac{11}{10} = r$$
$$\frac{1}{10} = r \text{ or } -\frac{21}{10} = r$$

Check. Since the interest rate cannot be negative, we need only check $\frac{1}{10}$, or 10%. If \$2000 were invested at 10% interest, compounded annually, then in 2 years it would grow to $\$2000(1.1)^2$, or \$2420. The number 10% checks.

State. The interest rate is 10%.

67. ***Familiarize***. We are already familiar with the compound-interest formula.

Translate. We substitute into the formula.

$$A = P(1+r)^t$$
$$1805 = 1280(1+r)^2$$

Carry out. We solve for r.

$$1805 = 1280(1+r)^2$$
$$\frac{1805}{1280} = (1+r)^2$$
$$\frac{361}{256} = (1+r)^2$$
$$\pm\frac{19}{16} = 1+r$$
$$-\frac{16}{16} \pm \frac{19}{16} = r$$
$$\frac{3}{16} = r \text{ or } -\frac{35}{16} = r$$

Check. Since the interest rate cannot be negative, we need only check $\frac{3}{16}$ or 18.75%. If \$1280 were invested at 18.75% interest, compounded annually, then in 2 years it would grow to $\$1280(1.1875)^2$, or \$1805. The number 18.75% checks.

State. The interest rate is 18.75%.

69. ***Familiarize***. We are already familiar with the compound-interest formula.

Translate. We substitute into the formula.

$$A = P(1+r)^t$$
$$6760 = 6250(1+r)^2$$

Carry out. We solve for r.

$$\frac{6760}{6250} = (1+r)^2$$
$$\frac{676}{625} = (1+r)^2$$
$$\pm\frac{26}{25} = 1+r$$
$$-\frac{25}{25} \pm \frac{26}{25} = r$$
$$\frac{1}{25} = r \text{ or } -\frac{51}{25} = r$$

Check. Since the interest rate cannot be negative, we need only check $\frac{1}{25}$, or 4%. If \$6250 were invested at 4% interest, compounded annually, then in 2 years it would grow to $\$6250(1.04)^2$, or \$6760. The number 4% checks.

State. The interest rate is 4%.

71. ***Familiarize***. We will use the formula $s = 16t^2$.

Translate. We substitute into the formula.

$$s = 16t^2$$
$$1053 = 16t^2$$

Carry out. We solve for t.

$$1053 = 16t^2$$
$$\frac{1053}{16} = t^2$$
$$\sqrt{\frac{1053}{16}} = t \quad \text{Principle of square roots; rejecting the negative square root}$$
$$8.1 \approx t$$

Check. Since $16(8.1)^2 = 1049.76 \approx 1053$, our answer checks.

State. It would take an object about 8.1 sec to fall freely from the Royal Gorge bridge.

73. ***Familiarize***. We will use the formula $s = 16t^2$.

Translate. We substitute into the formula.

$$s = 16t^2$$
$$1454 = 16t^2$$

Carry out. We solve for t.

$$1454 = 16t^2$$
$$\frac{1454}{16} = t^2$$
$$\sqrt{\frac{1454}{16}} = t \quad \text{Principle of square roots; rejecting the negative square root}$$
$$9.5 \approx t$$

Check. Since $16(9.5)^2 = 1444 \approx 1454$, our answer checks.

State. It would take an object about 9.5 sec to fall freely from the top of the Sears Tower.

75. *Writing Exercise*

77. $at^2 - bt = 3 \cdot 4^2 - 5 \cdot 4$
$= 3 \cdot 16 - 5 \cdot 4$
$= 48 - 20$
$= 28$

79. $\sqrt[3]{270} = \sqrt[3]{27 \cdot 10} = \sqrt[3]{27}\sqrt[3]{10} = 3\sqrt[3]{10}$

81. $f(x) = \sqrt{3x-5}$
$f(10) = \sqrt{3 \cdot 10 - 5} = \sqrt{30-5} = \sqrt{25} = 5$

83. *Writing Exercise*

85. In order for $x^2 + bx + 81$ to be a square, the following must be true:

$$\left(\frac{b}{2}\right)^2 = 81$$
$$\frac{b^2}{4} = 81$$
$$b^2 = 324$$
$$b = 18 \text{ or } b = -18$$

87. We see that x is a factor of each term, so x is also a factor of $f(x)$. We have $f(x) = x(2x^4 - 9x^3 - 66x^2 + 45x + 280)$. Since $x^2 - 5$ is a factor of $f(x)$ it is also a factor of $2x^4 - 9x^3 - 66x^2 + 45x + 280$. We divide to find another factor.

$$\begin{array}{r|l} & 2x^2 - 9x - 56 \\ x^2-5 & 2x^4 - 9x^3 - 66x^2 + 45x + 280 \\ & \underline{2x^4 \qquad\quad - 10x^2} \\ & -9x^3 - 56x^2 + 45x \\ & \underline{-9x^3 \qquad\quad + 45x} \\ & -56x^2 \qquad + 280 \\ & \underline{-56x^2 \qquad + 280} \\ & 0 \end{array}$$

Then we have $f(x) = x(x^2-5)(2x^2-9x-56)$, or $f(x) = x(x^2-5)(2x+7)(x-8)$. Now we find the values of a for which $f(a) = 0$.

$$f(a) = 0$$
$$a(a^2-5)(2a+7)(a-8) = 0$$
$$a=0 \text{ or } a^2-5=0 \text{ or } 2a+7=0 \text{ or } a-8=0$$
$$a=0 \text{ or } a^2=5 \text{ or } 2a=-7 \text{ or } a=8$$
$$a=0 \text{ or } a=\pm\sqrt{5} \text{ or } a=-\frac{7}{2} \text{ or } a=8$$

The solutions are 0, $\sqrt{5}$, $-\sqrt{5}$, $-\frac{7}{2}$, and 8.

89. ***Familiarize***. It is helpful to list information in a chart and make a drawing. Let r represent the speed of the fishing boat. Then $r - 7$ represents the speed of the barge.

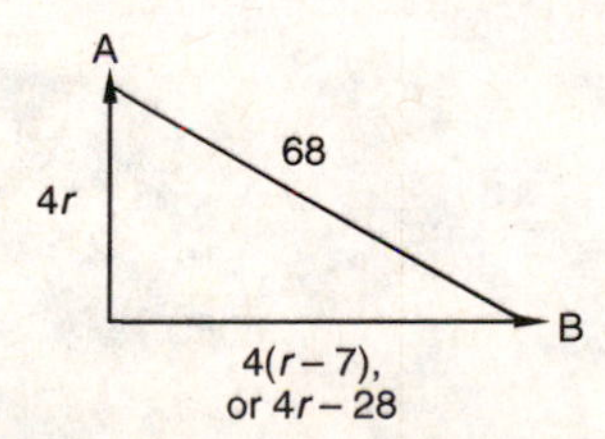

Boat	r	t	d
Fishing	r	4	$4r$
Barge	$r-7$	4	$4(r-7)$

Translate. We use the Pythagorean equation:

$$a^2 + b^2 = c^2$$
$$(4r-28)^2 + (4r)^2 = 68^2$$

Carry out.

$$(4r-28)^2 + (4r)^2 = 68^2$$
$$16r^2 - 224r + 784 + 16r^2 = 4624$$
$$32r^2 - 224r - 3840 = 0$$
$$r^2 - 7r - 120 = 0$$
$$(r+8)(r-15) = 0$$
$$r+8=0 \text{ or } r-15=0$$
$$r=-8 \text{ or } r=15$$

Check. We check only 15 since the speeds of the boats cannot be negative. If the speed of the fishing boat is 15 km/h, then the speed of the barge is $15 - 7$, or 8 km/h, and the distances they travel are $4 \cdot 15$ (or 60) and $4 \cdot 8$ (or 32).

$60^2 + 32^2 = 3600 + 1024 = 4624 = 68^2$

The values check.

State. The speed of the fishing boat is 15 km/h, and the speed of the barge is 8 km/h.

91. 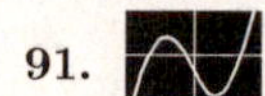

93. *Writing Exercise*

Exercise Set 11.2

1. True; see page 717 in the text.

3. False; see page 717 in the text.

5. False; the quadratic formula yields at most two solutions.

7. $x^2 + 7x - 3 = 0$

$a = 1, b = 7, c = -3$

$$x = \frac{-b \pm \sqrt{b^2-4ac}}{2a}$$
$$x = \frac{-7 \pm \sqrt{7^2 - 4 \cdot 1 \cdot (-3)}}{2 \cdot 1} = \frac{-7 \pm \sqrt{49+12}}{2}$$
$$x = \frac{-7 \pm \sqrt{61}}{2} = -\frac{7}{2} \pm \frac{\sqrt{61}}{2}$$

The solutions are $-\frac{7}{2} - \frac{\sqrt{61}}{2}$ and $-\frac{7}{2} + \frac{\sqrt{61}}{2}$.

9.
$$3p^2 = 18p - 6$$
$$3p^2 - 18p + 6 = 0$$
$$p^2 - 6p + 2 = 0 \quad \text{Dividing by 3}$$
$$a = 1, b = -6, c = 2$$

$$p = \frac{-b \pm \sqrt{b^2 - 4ac}}{2a}$$

$$p = \frac{-(-6) \pm \sqrt{(-6)^2 - 4 \cdot 1 \cdot 2}}{2 \cdot 1} = \frac{6 \pm \sqrt{36 - 8}}{2}$$

$$p = \frac{6 \pm \sqrt{28}}{2} = \frac{6 \pm 2\sqrt{7}}{2}$$

$$p = \frac{6}{2} \pm \frac{2\sqrt{7}}{2} = 3 \pm \sqrt{7}$$

The solutions are $3 + \sqrt{7}$ and $3 - \sqrt{7}$.

11. $x^2 - x + 1 = 0$

$a = 1,\ b = -1,\ c = 1$

$$x = \frac{-b \pm \sqrt{b^2 - 4ac}}{2a}$$

$$x = \frac{-(-1) \pm \sqrt{(-1)^2 - 4 \cdot 1 \cdot 1}}{2 \cdot 1} = \frac{1 \pm \sqrt{1 - 4}}{2}$$

$$x = \frac{1 \pm \sqrt{-3}}{2} = \frac{1 \pm i\sqrt{3}}{2} = \frac{1}{2} \pm \frac{\sqrt{3}}{2}i$$

The solutions are $\frac{1}{2} + \frac{\sqrt{3}}{2}i$ and $\frac{1}{2} - \frac{\sqrt{3}}{2}i$.

13. $x^2 + 13 = 4x$

$x^2 - 4x + 13 = 0$

$a = 1,\ b = -4,\ c = 13$

$$x = \frac{-b \pm \sqrt{b^2 - 4ac}}{2a}$$

$$x = \frac{-(-4) \pm \sqrt{(-4)^2 - 4 \cdot 1 \cdot 13}}{2 \cdot 1} = \frac{4 \pm \sqrt{16 - 52}}{2}$$

$$x = \frac{4 \pm \sqrt{-36}}{2} = \frac{4 \pm 6i}{2}$$

$$x = \frac{4}{2} \pm \frac{6}{2}i = 2 \pm 3i$$

The solutions are $2 + 3i$ and $2 - 3i$.

15. $h^2 + 4 = 6h$

$h^2 - 6h + 4 = 0$

$a = 1,\ b = -6,\ c = 4$

$$x = \frac{-(-6) \pm \sqrt{(-6)^2 - 4 \cdot 1 \cdot 4}}{2 \cdot 1} = \frac{6 \pm \sqrt{36 - 16}}{2}$$

$$x = \frac{6 \pm \sqrt{20}}{2} = \frac{6 \pm \sqrt{4 \cdot 5}}{2} = \frac{6 \pm 2\sqrt{5}}{2} = \frac{6}{2} \pm \frac{2\sqrt{5}}{2}$$

$$x = 3 \pm \sqrt{5}$$

The solutions are $3 + \sqrt{5}$ and $3 - \sqrt{5}$.

17. $\frac{1}{x^2} - 3 = \frac{8}{x}$, LCD is x^2

$$x^2\left(\frac{1}{x^2} - 3\right) = x^2 \cdot \frac{8}{x}$$

$$x^2 \cdot \frac{1}{x^2} - x^2 \cdot 3 = 8x$$

$$1 - 3x^2 = 8x$$

$$0 = 3x^2 + 8x - 1$$

$a = 3,\ b = 8,\ c = -1$

$$x = \frac{-8 \pm \sqrt{8^2 - 4 \cdot 3 \cdot (-1)}}{2 \cdot 3} = \frac{-8 \pm \sqrt{64 + 12}}{6}$$

$$x = \frac{-8 \pm \sqrt{76}}{6} = \frac{-8 \pm \sqrt{4 \cdot 19}}{6} = \frac{-8 \pm 2\sqrt{19}}{6}$$

$$x = \frac{-4 \pm \sqrt{19}}{3} = -\frac{4}{3} \pm \frac{\sqrt{19}}{3}$$

The solutions are $-\frac{4}{3} - \frac{\sqrt{19}}{3}$ and $-\frac{4}{3} + \frac{\sqrt{19}}{3}$.

19. $3x + x(x - 2) = 4$

$3x + x^2 - 2x = 4$

$x^2 + x = 4$

$x^2 + x - 4 = 0$

$a = 1,\ b = 1,\ c = -4$

$$x = \frac{-1 \pm \sqrt{1^2 - 4 \cdot 1 \cdot (-4)}}{2 \cdot 1} = \frac{-1 \pm \sqrt{1 + 16}}{2}$$

$$x = \frac{-1 \pm \sqrt{17}}{2} = -\frac{1}{2} \pm \frac{\sqrt{17}}{2}$$

The solutions are $-\frac{1}{2} - \frac{\sqrt{17}}{2}$ and $-\frac{1}{2} + \frac{\sqrt{17}}{2}$.

21. $12x^2 + 9t = 1$

$12t^2 + 9t - 1 = 0$

$a = 12,\ b = 9,\ c = -1$

$$t = \frac{-9 \pm \sqrt{9^2 - 4 \cdot 12 \cdot (-1)}}{2 \cdot 12} = \frac{-9 \pm \sqrt{81 + 48}}{24}$$

$$t = \frac{-9 \pm \sqrt{129}}{24} = -\frac{9}{24} \pm \frac{\sqrt{129}}{24} = -\frac{3}{8} \pm \frac{\sqrt{129}}{24}$$

The solutions are $-\frac{3}{8} - \frac{\sqrt{129}}{24}$ and $-\frac{3}{8} + \frac{\sqrt{129}}{24}$.

23. $25x^2 - 20x + 4 = 0$

$(5x - 2)(5x - 2) = 0$

$5x - 2 = 0$ *or* $5x - 2 = 0$

$5x = 2$ *or* $5x = 2$

$x = \frac{2}{5}$ *or* $x = \frac{2}{5}$

The solution is $\frac{2}{5}$.

25. $7x(x + 2) + 5 = 3x(x + 1)$

$7x^2 + 14x + 5 = 3x^2 + 3x$

$4x^2 + 11x + 5 = 0$

$a = 4,\ b = 11,\ c = 5$

$$x = \frac{-11 \pm \sqrt{11^2 - 4 \cdot 4 \cdot 5}}{2 \cdot 4} = \frac{-11 \pm \sqrt{121 - 80}}{8}$$

$$x = \frac{-11 \pm \sqrt{41}}{8} = -\frac{11}{8} \pm \frac{\sqrt{41}}{8}$$

The solutions are $-\frac{11}{8} - \frac{\sqrt{41}}{8}$ and $-\frac{11}{8} + \frac{\sqrt{41}}{8}$.

27. $14(x-4)-(x+2)=(x+2)(x-4)$

$$14x-56-x-2=x^2-2x-8 \quad \text{Removing parentheses}$$
$$13x-58=x^2-2x-8$$
$$0=x^2-15x+50$$
$$0=(x-10)(x-5)$$
$$x-10=0 \quad or \quad x-5=0$$
$$x=10 \quad or \quad x=5$$

The solutions are 10 and 5.

29. $5x^2=13x+17$

$$5x^2-13x-17=0$$
$$a=5,\ b=-13,\ c=-17$$
$$x=\frac{-(-13)\pm\sqrt{(-13)^2-4(5)(-17)}}{2\cdot 5}$$
$$x=\frac{13\pm\sqrt{169+340}}{10}=\frac{13\pm\sqrt{509}}{10}$$
$$x=\frac{13}{10}\pm\frac{\sqrt{509}}{10}$$

The solutions are $\frac{13}{10}-\frac{\sqrt{509}}{10}$ and $\frac{13}{10}+\frac{\sqrt{509}}{10}$.

31. $x^2+9=4x$

$$x^2-4x+9=0$$
$$a=1,\ b=-4,\ c=9$$
$$x=\frac{-(-4)\pm\sqrt{(-4)^2-4\cdot1\cdot9}}{2\cdot1}=\frac{4\pm\sqrt{16-36}}{2}$$
$$x=\frac{4\pm\sqrt{-20}}{2}=\frac{4\pm\sqrt{-4\cdot5}}{2}$$
$$x=\frac{4\pm2i\sqrt{5}}{2}=\frac{4}{2}\pm\frac{2\sqrt{5}}{2}i=2\pm i\sqrt{5}$$

The solutions are $2+\sqrt{5}i$ and $2-\sqrt{5}i$.

33. $x^3-8=0$

$$x^3-2^3=0$$
$$(x-2)(x^2+2x+4)=0$$
$$x-2=0 \quad or \quad x^2+2x+4=0$$
$$x=2 \quad or \quad x=\frac{-2\pm\sqrt{2^2-4\cdot1\cdot4}}{2\cdot1}$$
$$x=2 \quad or \quad x=\frac{-2\pm\sqrt{-12}}{2}=\frac{-2\pm2i\sqrt{3}}{2}$$
$$x=2 \quad or \quad x=-\frac{2}{2}\pm\frac{2\sqrt{3}}{2}i$$
$$x=2 \quad or \quad x=-1\pm\sqrt{3}i$$

The solutions are 2, $-1+\sqrt{3}i$, and $-1-\sqrt{3}i$.

35. $f(x)=0$

$$3x^2-5x+2=0$$
$$(3x-2)(x-1)=0$$
$$3x-2=0 \quad or \quad x-1=0$$
$$3x=2 \quad or \quad x=1$$
$$x=\frac{2}{3} \quad or \quad x=1$$

$f(x)=0$ for $x=\frac{2}{3}$ and $x=1$.

37. $f(x)=1$

$$\frac{7}{x}+\frac{7}{x+4}=1 \quad \text{Substituting}$$
$$x(x+4)\left(\frac{7}{x}+\frac{7}{x+4}\right)=x(x+4)\cdot1$$

Multiplying by the LCD

$$7(x+4)+7x=x^2+4x$$
$$7x+28+7x=x^2+4x$$
$$14x+28=x^2+4x$$
$$0=x^2-10x-28$$
$$a=1,\ b=-10,\ c=-28$$
$$x=\frac{-(-10)\pm\sqrt{(-10)^2-4\cdot1\cdot(-28)}}{2\cdot1}$$
$$x=\frac{10\pm\sqrt{100+112}}{2}=\frac{10\pm\sqrt{212}}{2}$$
$$x=\frac{10\pm\sqrt{4\cdot53}}{2}=\frac{10\pm2\sqrt{53}}{2}$$
$$x=5\pm\sqrt{53}$$

$f(x)=1$ for $x=5+\sqrt{53}$ and $x=5-\sqrt{53}$.

39. $F(x)=G(x)$

$$\frac{x+3}{x}=\frac{x-4}{3} \quad \text{Substituting}$$
$$3x\left(\frac{x+3}{x}\right)=3x\left(\frac{x-4}{3}\right) \quad \text{Multiplying by the LCD}$$
$$3x+9=x^2-4x$$
$$0=x^2-7x-9$$
$$a=1,\ b=-7,\ c=-9$$
$$x=\frac{-(-7)\pm\sqrt{(-7)^2-4\cdot1\cdot(-9)}}{2\cdot1}$$
$$x=\frac{7\pm\sqrt{49+36}}{2}=\frac{7\pm\sqrt{85}}{2}$$
$$x=\frac{7}{2}\pm\frac{\sqrt{85}}{2}$$

$F(x)=G(x)$ for $x=\frac{7}{2}-\frac{\sqrt{85}}{2}$ and $x=\frac{7}{2}+\frac{\sqrt{85}}{2}$.

41. $f(x)=g(x)$

$$\frac{15-2x}{6}=\frac{3}{x}, \quad \text{LCD is } 6x$$
$$6x\cdot\frac{15-2x}{6}=6x\cdot\frac{3}{x}$$
$$x(15-2x)=6\cdot3$$
$$15x-2x^2=18$$
$$0=2x^2-15x+18$$
$$0=(2x-3)(x-6)$$
$$2x-3=0 \quad or \quad x-6=0$$
$$2x=3 \quad or \quad x=6$$
$$x=\frac{3}{2} \quad or \quad x=6$$

$f(x)=g(x)$ for $x=\frac{3}{2}$ and $x=6$.

43. $x^2 + 4x - 7 = 0$

$a = 1,\ b = 4,\ c = -7$

$$x = \frac{-4 \pm \sqrt{4^2 - 4 \cdot 1 \cdot (-7)}}{2 \cdot 1} = \frac{-4 \pm \sqrt{16 + 28}}{2}$$

$$x = \frac{-4 \pm \sqrt{44}}{2}$$

Using a calculator we find that $\frac{-4 + \sqrt{44}}{2} \approx$ 1.31662479 and $\frac{-4 - \sqrt{44}}{2} \approx -5.31662479$.

The solutions are approximately 1.31662479 and −5.31662479.

45. $x^2 - 6x + 4 = 0$

$a = 1,\ b = -6,\ c = 4$

$$x = \frac{-(-6) \pm \sqrt{(-6)^2 - 4 \cdot 1 \cdot 4}}{2 \cdot 1} = \frac{6 \pm \sqrt{36 - 16}}{2}$$

$$x = \frac{6 \pm \sqrt{20}}{2}$$

Using a calculator we find that $\frac{6 + \sqrt{20}}{2} \approx$ 5.236067978 and $\frac{6 - \sqrt{20}}{2} \approx 0.7639320225$.

The solutions are approximately 5.236067978 and 0.7639320225.

47. $2x^2 - 3x - 7 = 0$

$a = 2,\ b = -3,\ c = -7$

$$x = \frac{-(-3) \pm \sqrt{(-3)^2 - 4 \cdot 2 \cdot (-7)}}{2 \cdot 2}$$

$$x = \frac{3 \pm \sqrt{9 + 56}}{4} = \frac{3 \pm \sqrt{65}}{4}$$

Using a calculator we find that $\frac{3 + \sqrt{65}}{4} \approx$ 2.765564437 and $\frac{3 - \sqrt{65}}{4} \approx -1.265564437$.

The solutions are approximately 2.765564437 and −1.265564437.

49. *Writing Exercise*

51. ***Familiarize***. Let x = the number of pounds of Kenyan coffee and y = the number of pounds of Kona coffee in the mixture. We organize the information in a table.

Type of Coffee	Kenyan	Kona	Mixture
Price per pound	\$6.75	\$11.25	\$8.55
Number of pounds	x	y	50
Total cost	\$6.75x	\$11.25y	\$8.55 × 50, or \$427.50

Translate. From the last two rows of the table we get a system of equations.

$$x + y = 50,$$
$$6.75x + 11.25y = 427.50$$

Solve. Solving the system of equations, we get (30, 20).

Check. The total number of pounds in the mixture is 30 + 20, or 50. The total cost of the mixture is \$6.75(30) + \$11.25(20) = \$427.50. The values check.

State. The mixture should consist of 30 lb of Kenyan coffee and 20 lb of Kona coffee.

53. $\sqrt{27a^2b^5} \cdot \sqrt{6a^3b} = \sqrt{27a^2b^5 \cdot 6a^3b} = \sqrt{162a^5b^6} = \sqrt{81a^4b^6 \cdot 2a} = \sqrt{81a^4b^6}\sqrt{2a} = 9a^2b^3\sqrt{2a}$

55.
$$\frac{\frac{3}{x-1}}{\frac{1}{x+1} + \frac{2}{x-1}}$$
$$= \frac{\frac{3}{x-1}}{\frac{1}{x+1} + \frac{2}{x-1}} \cdot \frac{(x-1)(x+1)}{(x-1)(x+1)}$$
$$= \frac{3(x+1)}{x - 1 + 2(x+1)}$$
$$= \frac{3x+3}{x - 1 + 2x + 2}$$
$$= \frac{3x+3}{3x+1},\ \text{or}\ \frac{3(x+1)}{3x+1}$$

57. *Writing Exercise*

59. $f(x) = \frac{x^2}{x-2} + 1$

To find the x-coordinates of the x-intercepts of the graph of f, we solve $f(x) = 0$.

$$\frac{x^2}{x-2} + 1 = 0$$
$$x^2 + x - 2 = 0 \quad \text{Multiplying by } x - 2$$
$$(x+2)(x-1) = 0$$
$$x = -2 \ \text{ or } \ x = 1$$

The x-intercepts are $(-2, 0)$ and $(1, 0)$.

61.
$$f(x) = g(x)$$
$$\frac{x^2}{x-2} + 1 = \frac{4x-2}{x-2} + \frac{x+4}{2} \quad \text{Substituting}$$
$$2(x-2)\left(\frac{x^2}{x-2} + 1\right) = 2(x-2)\left(\frac{4x-2}{x-2} + \frac{x+4}{2}\right) \quad \text{Multiplying by the LCD}$$
$$2x^2 + 2(x-2) = 2(4x-2) + (x-2)(x+4)$$
$$2x^2 + 2x - 4 = 8x - 4 + x^2 + 2x - 8$$
$$2x^2 + 2x - 4 = x^2 + 10x - 12$$
$$x^2 - 8x + 8 = 0$$

$a = 1,\ b = -8,\ c = 8$

$$x = \frac{-(-8) \pm \sqrt{(-8)^2 - 4\cdot 1\cdot 8}}{2\cdot 1} = \frac{8 \pm \sqrt{64-32}}{2}$$

$$x = \frac{8 \pm \sqrt{32}}{2} = \frac{8 \pm \sqrt{16\cdot 2}}{2} = \frac{8 \pm 4\sqrt{2}}{2}$$

$$x = \frac{8}{2} \pm \frac{4\sqrt{2}}{2} = 4 \pm 2\sqrt{2}$$

The solutions are $4 + 2\sqrt{2}$ and $4 - 2\sqrt{2}$.

63. $z^2 + 0.84z - 0.4 = 0$

$a = 1$, $b = 0.84$, $c = -0.4$

$$z = \frac{-0.84 \pm \sqrt{(0.84)^2 - 4\cdot 1\cdot (-0.4)}}{2\cdot 1}$$

$$z = \frac{-0.84 \pm \sqrt{2.3056}}{2}$$

$$z = \frac{-0.84 + \sqrt{2.3056}}{2} \approx 0.3392101158$$

$$z = \frac{-0.84 - \sqrt{2.3056}}{2} \approx -1.179210116$$

The solutions are approximately 0.3392101158 and -1.179210116.

65. $\sqrt{2}x^2 + 5x + \sqrt{2} = 0$

$$x = \frac{-5 \pm \sqrt{5^2 - 4\cdot\sqrt{2}\cdot\sqrt{2}}}{2\sqrt{2}} = \frac{-5 \pm \sqrt{17}}{2\sqrt{2}}, \text{ or}$$

$$x = \frac{-5 \pm \sqrt{17}}{2\sqrt{2}}\cdot\frac{\sqrt{2}}{\sqrt{2}} = \frac{-5\sqrt{2} \pm \sqrt{34}}{4}$$

The solutions are $\dfrac{-5\sqrt{2} \pm \sqrt{34}}{4}$.

67.

$$kx^2 + 3x - k = 0$$

$$k(-2)^2 + 3(-2) - k = 0 \quad \text{Substituting } -2 \text{ for } x$$

$$4k - 6 - k = 0$$

$$3k = 6$$

$$k = 2$$

$$2x^2 + 3x - 2 = 0 \quad \text{Substituting 2 for } k$$

$$(2x - 1)(x + 2) = 0$$

$$2x - 1 = 0 \quad or \quad x + 2 = 0$$

$$x = \frac{1}{2} \quad or \quad x = -2$$

The other solution is $\dfrac{1}{2}$.

69. 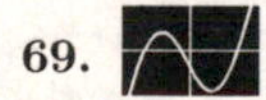

Exercise Set 11.3

1. ***Familiarize***. We first make a drawing, labeling it with the known and unknown information. We can also organize the information in a table. We let r represent the speed and t the time for the first part of the trip.

r mph t hr $r-10$ mph $4-t$ hr
120 mi   100 mi

Trip	Distance	Speed	Time
1st part	120	r	t
2nd part	100	$r-10$	$4-t$

Translate. Using $r = \dfrac{d}{t}$, we get two equations from the table, $r = \dfrac{120}{t}$ and $r - 10 = \dfrac{100}{4-t}$.

Carry out. We substitute $\dfrac{120}{t}$ for r in the second equation and solve for t.

$$\frac{120}{t} - 10 = \frac{100}{4-t}, \quad \text{LCD is } t(4-t)$$

$$t(4-t)\left(\frac{120}{t} - 10\right) = t(4-t)\cdot\frac{100}{4-t}$$

$$120(4-t) - 10t(4-t) = 100t$$

$$480 - 120t - 40t + 10t^2 = 100t$$

$$10t^2 - 260t + 480 = 0 \quad \text{Standard form}$$

$$t^2 - 26t + 48 = 0 \quad \text{Multiplying by } \frac{1}{10}$$

$$(t-2)(t-24) = 0$$

$$t = 2 \quad or \quad t = 24$$

Check. Since the time cannot be negative (If $t = 24$, $4 - t = -20$.), we check only 2 hr. If $t = 2$, then $4 - t = 2$. The speed of the first part is $\dfrac{120}{2}$, or 60 mph. The speed of the second part is $\dfrac{100}{2}$, or 50 mph. The speed of the second part is 10 mph slower than the first part. The value checks.

State. The speed of the first part was 60 mph, and the speed of the second part was 50 mph.

3. ***Familiarize***. We first make a drawing. We also organize the information in a table. We let r = the speed and t = the time of the slower trip.

200 mi r mph t hr
200 mi $r+10$ mph $t-1$ hr

Trip	Distance	Speed	Time
Slower	200	r	t
Faster	200	$r+10$	$t-1$

Translate. Using $t = d/r$, we get two equations from the table:

$$t = \frac{200}{r} \quad \text{and} \quad t - 1 = \frac{200}{r+10}$$

Carry out. We substitute $\dfrac{200}{r}$ for t in the second equation and solve for r.

$$\frac{200}{r} - 1 = \frac{200}{r+10}, \text{ LCD is } r(r+10)$$

$$r(r+10)\left(\frac{200}{r} - 1\right) = r(r+10) \cdot \frac{200}{r+10}$$

$$200(r+10) - r(r+10) = 200r$$

$$200r + 2000 - r^2 - 10r = 200r$$

$$0 = r^2 + 10r - 2000$$

$$0 = (r+50)(r-40)$$

$r = -50$ *or* $r = 40$

Check. Since negative speed has no meaning in this problem, we check only 40. If $r = 40$, then the time for the slower trip is $\frac{200}{40}$, or 5 hours. If $r = 40$, then $r + 10 = 50$ and the time for the faster trip is $\frac{200}{50}$, or 4 hours. This is 1 hour less time than the slower trip took, so we have an answer to the problem.

State. The speed is 40 mph.

5. ***Familiarize***. We make a drawing and then organize the information in a table. We let r = the speed and t = the time of the Cessna.

600 mi r mph t hr

1000 mi $r + 50$ mph $t + 1$ hr

Plane	Distance	Speed	Time
Cessna	600	r	t
Beechcraft	1000	$r + 50$	$t + 1$

Translate. Using $t = d/r$, we get two equations from the table:

$$t = \frac{600}{r} \quad \text{and} \quad t + 1 = \frac{1000}{r+50}$$

Carry out. We substitute $\frac{600}{r}$ for t in the second equation and solve for r.

$$\frac{600}{r} + 1 = \frac{1000}{r+50}, \text{ LCD is } r(r+50)$$

$$r(r+50)\left(\frac{600}{r} + 1\right) = r(r+50) \cdot \frac{1000}{r+50}$$

$$600(r+50) + r(r+50) = 1000r$$

$$600r + 30,000 + r^2 + 50r = 1000r$$

$$r^2 - 350r + 30,000 = 0$$

$$(r-150)(r-200) = 0$$

$r = 150$ *or* $r = 200$

Check. If $r = 150$, then the Cessna's time is $\frac{600}{150}$, or 4 hr and the Beechcraft's time is $\frac{1000}{150+50}$, or $\frac{1000}{200}$, or 5 hr. If $r = 200$, then the Cessna's time is $\frac{600}{200}$, or 3 hr and the Beechcraft's time is $\frac{1000}{200+50}$, or $\frac{1000}{250}$, or 4 hr. Since the Beechcraft's time is 1 hr longer in each case, both values check. There are two solutions.

State. The speed of the Cessna is 150 mph and the speed of the Beechcraft is 200 mph; or the speed of the Cessna is 200 mph and the speed of the Beechcraft is 250 mph.

7. ***Familiarize***. We make a drawing and then organize the information in a table. We let r represent the speed and t the time of the trip to Hillsboro.

Hillsboro

40 mi r mph t hr

40 mi $r - 6$ mph $14 - t$ hr

Trip	Distance	Speed	Time
To Hillsboro	40	r	t
Return	40	$r - 6$	$14 - t$

Translate. Using $t = \frac{d}{r}$, we get two equations from the table,

$$t = \frac{40}{r} \quad \text{and} \quad 14 - t = \frac{40}{r-6}.$$

Carry out. We substitute $\frac{40}{r}$ for t in the second equation and solve for r.

$$14 - \frac{40}{r} = \frac{40}{r-6}, \text{ LCD is } r(r-6)$$

$$r(r-6)\left(14 - \frac{40}{r}\right) = r(r-6) \cdot \frac{40}{r-6}$$

$$14r(r-6) - 40(r-6) = 40r$$

$$14r^2 - 84r - 40r + 240 = 40r$$

$$14r^2 - 164r + 240 = 0$$

$$7r^2 - 82r + 120 = 0$$

$$(7r - 12)(r - 10) = 0$$

$r = \frac{12}{7}$ *or* $r = 10$

Check. Since negative speed has no meaning in this problem (If $r = \frac{12}{7}$, then $r - 6 = -\frac{30}{7}$.), we check only 10 mph. If $r = 10$, then the time of the trip to Hillsboro is $\frac{40}{10}$, or 4 hr. The speed of the return trip is $10 - 6$, or 4 mph, and the time is $\frac{40}{4}$, or 10 hr. The total time for the round trip is 4 hr + 10 hr, or 14 hr. The value checks.

State. Naoki's speed on the trip to Hillsboro was 10 mph and it was 4 mph on the return trip.

9. ***Familiarize***. We make a drawing and organize the information in a table. Let r represent the speed of the boat in still water, and let t represent the time of the trip upriver.

60 mi $r - 3$ mph t hr   Upriver

Downriver   60 mi $r + 3$ mph $9 - t$ hr

Trip	Distance	Speed	Time
Upriver	60	$r-3$	t
Downriver	60	$r+3$	$9-t$

Translate. Using $t = \frac{d}{r}$, we get two equations from the table,

$$t = \frac{60}{r-3} \text{ and } 9 - t = \frac{60}{r+3}.$$

Carry out. We substitute $\frac{60}{r-3}$ for t in the second equation and solve for r.

$$9 - \frac{60}{r-3} = \frac{60}{r+3}$$
$$(r-3)(r+3)\left(9 - \frac{60}{r-3}\right) = (r-3)(r+3)\cdot\frac{60}{r+3}$$
$$9(r-3)(r+3) - 60(r+3) = 60(r-3)$$
$$9r^2 - 81 - 60r - 180 = 60r - 180$$
$$9r^2 - 120r - 81 = 0$$
$$3r^2 - 40r - 27 = 0 \quad \text{Dividing by 3}$$

We use the quadratic formula.

$$r = \frac{-(-40) \pm \sqrt{(-40)^2 - 4\cdot 3\cdot(-27)}}{2\cdot 3}$$
$$r = \frac{40 \pm \sqrt{1924}}{6}$$

$r \approx 14$ *or* $r \approx -0.6$

Check. Since negative speed has no meaning in this problem, we check only 14 mph. If $r \approx 14$, then the speed upriver is about $14-3$, or 11 mph, and the time is about $\frac{60}{11}$, or 5.5 hr. The speed downriver is about $14+3$, or 17 mph, and the time is about $\frac{60}{17}$, or 3.5 hr. The total time of the round trip is $5.5+3.5$, or 9 hr. The value checks.

State. The speed of the boat in still water is about 14 mph.

11. ***Familiarize.*** Let x represent the time it takes the spring to fill the pool. Then $x-6$ represents the time it takes the well to fill the pool. It takes them 4 hr to fill the pool working together, so they can fill $\frac{1}{4}$ of the pool in 1 hr. The spring will fill $\frac{1}{x}$ of the pool in 1 hr, and the well will fill $\frac{1}{x-6}$ of the pool in 1 hr.

Translate. We have an equation.

$$\frac{1}{x} + \frac{1}{x-6} = \frac{1}{4}$$

Carry out. We solve the equation.

We multiply by the LCD, $4x(x-6)$.

$$4x(x-6)\left(\frac{1}{x} + \frac{1}{x-6}\right) = 4x(x-6)\cdot\frac{1}{4}$$
$$4(x-6) + 4x = x(x-6)$$
$$4x - 24 + 4x = x^2 - 6x$$
$$0 = x^2 - 14x + 24$$
$$0 = (x-2)(x-12)$$

$x = 2$ *or* $x = 12$

Check. Since negative time has no meaning in this problem, 2 is not a solution $(2-6=-4)$. We check only 12 hr. This is the time it would take the spring working alone. Then the well would take $12-6$, or 6 hr working alone. The well would fill $4\left(\frac{1}{6}\right)$, or $\frac{2}{3}$, of the pool in 4 hr, and the spring would fill $4\left(\frac{1}{12}\right)$, or $\frac{1}{3}$, of the pool in 4 hr. Thus in 4 hr they would fill $\frac{2}{3} + \frac{1}{3}$ of the pool. This is all of it, so the numbers check.

State. It takes the spring, working alone, 12 hr to fill the pool.

13. We make a drawing and then organize the information in a table. We let r represent Ellen's speed in still water. Then $r-2$ is the speed upstream and $r+2$ is the speed downstream. Using $t = \frac{d}{r}$, we let $\frac{1}{r-2}$ represent the time upstream and $\frac{1}{r+2}$ represent the time downstream.

1 mi $r-2$ mph → Upstream

Downstream ← 1 mi $r+2$ mph

Trip	Distance	Speed	Time
Upstream	1	$r-2$	$\frac{1}{r-2}$
Downstream	1	$r+2$	$\frac{1}{r+2}$

Translate. The time for the round trip is 1 hour. We now have an equation.

$$\frac{1}{r-2} + \frac{1}{r+2} = 1$$

Carry out. We solve the equation. We multiply by the LCD, $(r-2)(r+2)$.

$$(r-2)(r+2)\left(\frac{1}{r-2} + \frac{1}{r+2}\right) = (r-2)(r+2)\cdot 1$$
$$(r+2) + (r-2) = (r-2)(r+2)$$
$$2r = r^2 - 4$$
$$0 = r^2 - 2r - 4$$

$a = 1,\ b = -2,\ c = -4$

$$r = \frac{-(-2) \pm \sqrt{(-2)^2 - 4\cdot 1(-4)}}{2\cdot 1}$$
$$r = \frac{2 \pm \sqrt{4+16}}{2} = \frac{2 \pm \sqrt{20}}{2}$$
$$r = \frac{2 \pm 2\sqrt{5}}{2} = 1 \pm \sqrt{5}$$

$1 + \sqrt{5} \approx 1 + 2.236 \approx 3.24$

$1 - \sqrt{5} \approx 1 - 2.236 \approx -1.24$

Check. Since negative speed has no meaning in this problem, we check only 3.24 mph. If $r \approx 3.24$, then $r-2 \approx 1.24$ and $r+2 \approx 5.24$. The time it takes to travel upstream is approximately $\frac{1}{1.24}$, or 0.806 hr, and the time it takes to travel downstream is approximately $\frac{1}{5.24}$, or 0.191 hr. The total time is 0.997 which is approximately 1 hour. The value checks.

State. Ellen's speed in still water is approximately 3.24 mph.

15. $A = 4\pi r^2$

$\frac{A}{4\pi} = r^2$ Dividing by 4π

$\frac{1}{2}\sqrt{\frac{A}{\pi}} = r$ Taking the positive square root

17. $A = 2\pi r^2 + 2\pi rh$

$0 = 2\pi r^2 + 2\pi rh - A$ Standard form

$a = 2\pi,\ b = 2\pi h,\ c = -A$

$r = \frac{-2\pi h \pm \sqrt{(2\pi h)^2 - 4 \cdot 2\pi \cdot (-A)}}{2 \cdot 2\pi}$ Using the quadratic formula

$r = \frac{-2\pi h \pm \sqrt{4\pi^2 h^2 + 8\pi A}}{4\pi}$

$r = \frac{-2\pi h \pm 2\sqrt{\pi^2 h^2 + 2\pi A}}{4\pi}$

$r = \frac{-\pi h \pm \sqrt{\pi^2 h^2 + 2\pi A}}{2\pi}$

Since taking the negative square root would result in a negative answer, we take the positive one.

$r = \frac{-\pi h + \sqrt{\pi^2 h^2 + 2\pi A}}{2\pi}$

19. $N = \frac{kQ_1Q_2}{s^2}$

$Ns^2 = kQ_1Q_2$ Multiplying by s^2

$s^2 = \frac{kQ_1Q_2}{N}$ Dividing by N

$s = \sqrt{\frac{kQ_1Q_2}{N}}$ Taking the positive square root

21. $T = 2\pi\sqrt{\frac{l}{g}}$

$\frac{T}{2\pi} = \sqrt{\frac{l}{g}}$ Multiplying by $\frac{1}{2\pi}$

$\frac{T^2}{4\pi^2} = \frac{l}{g}$ Squaring

$gT^2 = 4\pi^2 l$ Multiplying by $4\pi^2 g$

$g = \frac{4\pi^2 l}{T^2}$ Multiplying by $\frac{1}{T^2}$

23. $a^2 + b^2 + c^2 = d^2$

$c^2 = d^2 - a^2 - b^2$ Subtracting a^2 and b^2

$c = \sqrt{d^2 - a^2 - b^2}$ Taking the positive square root

25. $s = v_0 t + \frac{gt^2}{2}$

$0 = \frac{gt^2}{2} + v_0 t - s$ Standard form

$a = \frac{g}{2},\ b = v_0,\ c = -s$

$t = \frac{-v_0 \pm \sqrt{v_0^2 - 4\left(\frac{g}{2}\right)(-s)}}{2\left(\frac{g}{2}\right)}$

$t = \frac{-v_0 \pm \sqrt{v_0^2 + 2gs}}{g}$

Since taking the negative square root would result in a negative answer, we take the positive one.

$t = \frac{-v_0 + \sqrt{v_0^2 + 2gs}}{g}$

27. $N = \frac{1}{2}(n^2 - n)$

$N = \frac{1}{2}n^2 - \frac{1}{2}n$

$0 = \frac{1}{2}n^2 - \frac{1}{2}n - N$

$a = \frac{1}{2},\ b = -\frac{1}{2},\ c = -N$

$n = \frac{-\left(-\frac{1}{2}\right) \pm \sqrt{\left(-\frac{1}{2}\right)^2 - 4 \cdot \frac{1}{2} \cdot (-N)}}{2\left(\frac{1}{2}\right)}$

$n = \frac{1}{2} \pm \sqrt{\frac{1}{4} + 2N}$

$n = \frac{1}{2} \pm \sqrt{\frac{1 + 8N}{4}}$

$n = \frac{1}{2} \pm \frac{1}{2}\sqrt{1 + 8N}$

Since taking the negative square root would result in a negative answer, we take the positive one.

$n = \frac{1}{2} + \frac{1}{2}\sqrt{1 + 8N}$, or $\frac{1 + \sqrt{1 + 8N}}{2}$

29. $V = 3.5\sqrt{h}$

$V^2 = 12.25h$ Squaring

$\frac{V^2}{12.25} = h$

31. $at^2 + bt + c = 0$

The quadratic formula gives the result.

$t = \frac{-b \pm \sqrt{b^2 - 4ac}}{2a}$

33. a) ***Familiarize and Translate***. From Example 4, we know

$$t = \frac{-v_0 + \sqrt{{v_0}^2 + 19.6s}}{9.8}.$$

Carry out. Substituting 500 for s and 0 for v_0, we have

$$t = \frac{0 + \sqrt{0^2 + 19.6(500)}}{9.8}$$

$$t \approx 10.1$$

Check. Substitute 10.1 for t and 0 for v_0 in the original formula. (See Example 4.)

$$s = 4.9t^2 + v_0t = 4.9(10.1)^2 + 0 \cdot (10.1)^2 \approx 500$$

The answer checks.

State. It takes the bolt about 10.1 sec to reach the ground.

b) *Familiarize and Translate.* From Example 4, we know

$$t = \frac{-v_0 + \sqrt{v_0^2 + 19.6s}}{9.8}.$$

Carry out. Substitute 500 for s and 30 for v_0.

$$t = \frac{-30 + \sqrt{30^2 + 19.6(500)}}{9.8}$$

$$t \approx 7.49$$

Check. Substitute 30 for v_0 and 7.49 for t in the original formula. (See Example 4.)

$$s = 4.9t^2 + v_0t = 4.9(7.49)^2 + (30)(7.49) \approx 500$$

The answer checks.

State. It takes the ball about 7.49 sec to reach the ground.

c) *Familiarize and Translate.* We will use the formula in Example 4, $s = 4.9t^2 + v_0t$.

Carry out. Substitute 5 for t and 30 for v_0.

$$s = 4.9(5)^2 + 30(5) = 272.5$$

Check. We can substitute 30 for v_0 and 272.5 for s in the form of the formula we used in part (b).

$$t = \frac{-v_0 + \sqrt{v_0^2 + 19.6s}}{9.8} = \frac{-30 + \sqrt{(30)^2 + 19.6(272.5)}}{9.8} = 5$$

The answer checks.

State. The object will fall 272.5 m.

35. *Familiarize.* We will use the formula $4.9t^2 = s$.

Translate. Substitute 40 for s.

$$4.9t^2 = 40$$

Carry out. We solve the equation.

$$4.9t^2 = 40$$

$$t^2 = \frac{40}{4.9}$$

$$t = \sqrt{\frac{40}{4.9}}$$

$$t \approx 2.9$$

Check. Substitute 2.9 for t in the formula.

$$s = 4.9(2.9)^2 = 41.209 \approx 40$$

The answer checks.

State. Jesse will fall for about 2.9 sec before the cord begins to stretch.

37. *Familiarize and Translate.* From Example 3, we know

$$T = \frac{\sqrt{3V}}{12}.$$

Carry out. Substituting 45 for V, we have

$$T = \frac{\sqrt{3 \cdot 45}}{12}$$

$$T \approx 0.968$$

Check. Substitute 0.968 for T in the original formula. (See Example 3.)

$$48T^2 = V$$

$$48(0.968)^2 = V$$

$$45 \approx V$$

The answer checks.

State. Steve Francis' hang time is about 0.968 sec.

39. *Familiarize and Translate.* We will use the formula in Example 4, $s = 4.9t^2 + v_0t$.

Carry out. Solve the formula for v_0.

$$s - 4.9t^2 = v_0t$$

$$\frac{s - 4.9t^2}{t} = v_0$$

Now substitute 51.6 for s and 3 for t.

$$\frac{51.6 - 4.9(3)^2}{3} = v_0$$

$$2.5 = v_0$$

Check. Substitute 3 for t and 2.5 for v_0 in the original formula.

$$s = 4.9(3)^2 + 2.5(3) = 51.6$$

The solution checks.

State. The initial velocity is 2.5 m/sec.

41. *Familiarize and Translate.* From Exercise 32 we know that

$$r = -1 + \frac{-P_2 + \sqrt{P_2^2 + 4P_1A}}{2P_1},$$

where A is the total amount in the account after two years, P_1 is the amount of the original deposit, P_2 is deposited at the beginning of the second year, and r is the annual interest rate.

Carry out. Substitute 3000 for P_1, 1700 for P_2, and 5253.70 for A.

$$r = -1 + \frac{-1700 + \sqrt{(1700)^2 + 4(3000)(5253.70)}}{2(3000)}$$

Using a calculator, we have $r = 0.07$.

Check. Substitute in the original formula in Exercise 32.

$$P_1(1 + r)^2 + P_2(1 + r) = A$$

$$3000(1.07)^2 + 1700(1.07) = A$$

$$5253.70 = A$$

The answer checks.

State. The annual interest rate is 0.07, or 7%.

43. *Writing Exercise*

45. $b^2 - 4ac = 6^2 - 4 \cdot 5 \cdot 7$
$= 36 - 4 \cdot 5 \cdot 7$
$= 36 - 140$
$= -104$

47. $\dfrac{x^2 + xy}{2x} = \dfrac{x(x+y)}{2x}$
$= \dfrac{x(x+y)}{2 \cdot x}$
$= \dfrac{\cancel{x}(x+y)}{2 \cdot \cancel{x}}$
$= \dfrac{x+y}{2}$

49. $\dfrac{3+\sqrt{45}}{6} = \dfrac{3+\sqrt{9 \cdot 5}}{6} = \dfrac{3+3\sqrt{5}}{6} = \dfrac{\cancel{3}(1+\sqrt{5})}{\cancel{3} \cdot 2} = \dfrac{1+\sqrt{5}}{2}$

51. *Writing Exercise*

53.
$$A = 6.5 - \frac{20.4t}{t^2+36}$$
$$(t^2+36)A = (t^2+36)\left(6.5 - \frac{20.4t}{t^2+36}\right)$$
$$At^2 + 36A = (t^2+36)(6.5) - (t^2+36)\left(\frac{20.4t}{t^2+36}\right)$$
$$At^2 + 36A = 6.5t^2 + 234 - 20.4t$$
$$At^2 - 6.5t^2 + 20.4 + 36A - 234 = 0$$
$$(A-6.5)t^2 + 20.4t + (36A - 234) = 0$$
$a = A - 6.5$, $b = 20.4$, $c = 36A - 234$
$$t = \frac{-20.4 \pm \sqrt{(20.4)^2 - 4(A-6.5)(36A-234)}}{2(A-6.5)}$$
$$t = \frac{-20.4 \pm \sqrt{416.16 - 144A^2 + 1872A - 6084}}{2(A-6.5)}$$
$$t = \frac{-20.4 \pm \sqrt{-144A^2 + 1872A - 5667.84}}{2(A-6.5)}$$
$$t = \frac{-20.4 \pm \sqrt{144(-A^2 + 13A - 39.36)}}{2(A-6.5)}$$
$$t = \frac{-20.4 \pm 12\sqrt{-A^2 + 13A - 39.36}}{2(A-6.5)}$$
$$t = \frac{2(-10.2 \pm 6\sqrt{-A^2 + 13A - 39.36})}{2(A-6.5)}$$
$$t = \frac{-10.2 \pm 6\sqrt{-A^2 + 13A - 39.36}}{A-6.5}$$

55. ***Familiarize.*** Let a = the number. Then $a - 1$ is 1 less than a and the reciprocal of that number is $\dfrac{1}{a-1}$. Also, 1 more than the number is $a + 1$.

Translate.

The reciprocal of 1 less than a number	is	1 more than the number.
↓	↓	↓
$\dfrac{1}{(a-1)}$	$=$	$a+1$

Carry out. We solve the equation.
$$\frac{1}{a-1} = a+1, \text{ LCD is } a-1$$
$$(a-1) \cdot \frac{1}{a-1} = (a-1)(a+1)$$
$$1 = a^2 - 1$$
$$2 = a^2$$
$$\pm\sqrt{2} = a$$

Check. $\dfrac{1}{\sqrt{2}-1} \approx 2.4142 \approx \sqrt{2}+1$ and $\dfrac{1}{-\sqrt{2}-1} \approx -0.4142 \approx -\sqrt{2}+1$. The answers check.

State. The numbers are $\sqrt{2}$ and $-\sqrt{2}$, or $\pm\sqrt{2}$.

57.
$$\frac{w}{l} = \frac{l}{w+l}$$
$$l(w+l) \cdot \frac{w}{l} = l(w+l) \cdot \frac{l}{w+l}$$
$$w(w+l) = l^2$$
$$w^2 + lw = l^2$$
$$0 = l^2 - lw - w^2$$
Use the quadratic formula with $a = 1$, $b = -w$, and $c = -w^2$.
$$l = \frac{-(-w) \pm \sqrt{(-w)^2 - 4 \cdot 1(-w^2)}}{2 \cdot 1}$$
$$l = \frac{w \pm \sqrt{w^2 + 4w^2}}{2} = \frac{w \pm \sqrt{5w^2}}{2}$$
$$l = \frac{w \pm w\sqrt{5}}{2}$$
Since $\dfrac{w - w\sqrt{5}}{2}$ is negative we use the positive square root:
$l = \dfrac{w + w\sqrt{5}}{2}$ Then $L(A) = \sqrt{\dfrac{A}{2}}$.

59. $mn^4 - r^2pm^3 - r^2n^2 + p = 0$
Let $u = n^2$. Substitute and rearrange.
$mu^2 - r^2u - r^2pm^3 + p = 0$
$a = m$, $b = -r^2$, $c = -r^2pm^3 + p$
$$u = \frac{-(-r^2) \pm \sqrt{(-r^2)^2 - 4 \cdot m(-r^2pm^3 + p)}}{2 \cdot m}$$
$$u = \frac{r^2 \pm \sqrt{r^4 + 4m^4r^2p - 4mp}}{2m}$$
$$n^2 = \frac{r^2 \pm \sqrt{r^4 + 4m^4r^2p - 4mp}}{2m}$$
$$n = \pm\sqrt{\frac{r^2 \pm \sqrt{r^4 + 4m^4r^2p - 4mp}}{2m}}$$

61. Let s represent a length of a side of the cube, let S represent the surface area of the cube, and let A represent the surface area of the sphere. Then the diameter of the sphere is s, so the radius r is $s/2$. From Exercise 15, we know, $A = 4\pi r^2$, so when $r = s/2$ we have $A = 4\pi\left(\dfrac{s}{2}\right)^2 = 4\pi \cdot \dfrac{s^2}{4} = \pi s^2$. From the formula for the surface area of a cube (See Exercise 16.) we know that $S = 6s^2$, so $\dfrac{S}{6} = s^2$ and then $A = \pi \cdot \dfrac{S}{6}$, or $A(S) = \dfrac{\pi S}{6}$.

Exercise Set 11.4

1. discriminant; see page 730 in the text.

3. two; see page 731 in the text.

5. rational; see page 731 in the text.

7. $x^2 - 7x + 5 = 0$

$a = 1,\ b = -7,\ c = 5$

We substitute and compute the discriminant.

$$\begin{aligned} b^2 - 4ac &= (-7)^2 - 4 \cdot 1 \cdot 5 \\ &= 49 - 20 \\ &= 29 \end{aligned}$$

Since the discriminant is a positive number that is not a perfect square, there are two irrational solutions.

9. $x^2 + 3 = 0$

$a = 1,\ b = 0,\ c = 3$

We substitute and compute the discriminant.

$$\begin{aligned} b^2 - 4ac &= 0^2 - 4 \cdot 1 \cdot 3 \\ &= -12 \end{aligned}$$

Since the discriminant is negative, there are two imaginary-number solutions.

11. $x^2 - 5 = 0$

$a = 1,\ b = 0,\ c = -5$

We substitute and compute the discriminant.

$$\begin{aligned} b^2 - 4ac &= 0^2 - 4 \cdot 1 \cdot (-5) \\ &= 20 \end{aligned}$$

Since the discriminant is a positive number that is not a perfect square, there are two irrational solutions.

13. $4x^2 + 8x - 5 = 0$

$a = 4,\ b = 8,\ c = -5$

We substitute and compute the discriminant.

$$\begin{aligned} b^2 - 4ac &= 8^2 - 4 \cdot 4 \cdot (-5) \\ &= 64 + 80 \\ &= 144 \end{aligned}$$

Since the discriminant is a positive number and a perfect square, there are two rational solutions.

15. $x^2 + 4x + 6 = 0$

$a = 1,\ b = 4,\ c = 6$

We substitute and compute the discriminant.

$$\begin{aligned} b^2 - 4ac &= 4^2 - 4 \cdot 1 \cdot 6 \\ &= 16 - 24 \\ &= -8 \end{aligned}$$

Since the discriminant is negative, there are two imaginary-number solutions.

17. $9t^2 - 48t + 64 = 0$

$a = 9,\ b = -48,\ c = 64$

We substitute and compute the discriminant.

$$\begin{aligned} b^2 - 4ac &= (-48)^2 - 4 \cdot 9 \cdot 64 \\ &= 2304 - 2304 \\ &= 0 \end{aligned}$$

Since the discriminant is 0, there is just one solution and it is a rational number.

19. $10x^2 - x - 2 = 0$

$a = 10,\ b = -1,\ c = -2$

We substitute and compute the discriminant.

$$\begin{aligned} b^2 - 4ac &= (-1)^2 - 4 \cdot 10 \cdot (-2) \\ &= 1 + 80 = 81 \end{aligned}$$

Since the discriminant is a positive number and a perfect square, there are two rational solutions.

21. $9t^2 - 3t = 0$

Observe that we can factor $9t^2 - 3t$. This tells us that there are two rational solutions. We could also do this problem as follows.

$a = 9,\ b = -3,\ c = 0$

We substitute and compute the discriminant.

$$\begin{aligned} b^2 - 4ac &= (-3)^2 - 4 \cdot 9 \cdot 0 \\ &= 9 - 0 \\ &= 9 \end{aligned}$$

Since the discriminant is a positive number and a perfect square, there are two rational solutions.

23. $x^2 + 4x = 8$

$x^2 + 4x - 8 = 0$ Standard form

$a = 1,\ b = 4,\ c = -8$

We substitute and compute the discriminant.

$$\begin{aligned} b^2 - 4ac &= 4^2 - 4 \cdot 1 \cdot (-8) \\ &= 16 + 32 = 48 \end{aligned}$$

Since the discriminant is a positive number that is not a perfect square, there are two irrational solutions.

25. $2a^2 - 3a = -5$

$2a^2 - 3a + 5 = 0$ Standard form

$a = 2,\ b = -3,\ c = 5$

We substitute and compute the discriminant.

$$\begin{aligned} b^2 - 4ac &= (-3)^2 - 4 \cdot 2 \cdot 5 \\ &= 9 - 40 \\ &= -31 \end{aligned}$$

Since the discriminant is negative, there are two imaginary-number solutions.

27. $y^2 + \frac{9}{4} = 4y$

$y^2 - 4y + \frac{9}{4} = 0$ Standard form

$a = 1,\ b = -4,\ c = \frac{9}{4}$

We substitute and compute the discriminant.

$b^2 - 4ac = (-4)^2 - 4 \cdot 1 \cdot \frac{9}{4}$

$= 16 - 9$

$= 7$

The discriminant is a positive number that is not a perfect square. There are two irrational solutions.

29. The solutions are -7 and 3.

$x = -7$ *or* $x = 3$

$x + 7 = 0$ *or* $x - 3 = 0$

$(x + 7)(x - 3) = 0$ Principle of zero products

$x^2 + 4x - 21 = 0$ FOIL

31. The only solution is 3. It must be a repeated solution.

$x = 3$ *or* $x = 3$

$x - 3 = 0$ *or* $x - 3 = 0$

$(x - 3)(x - 3) = 0$ Principle of zero products

$x^2 - 6x + 9 = 0$ FOIL

33. The solutions are -1 and 3.

$x = -1$ *or* $x = -3$

$x + 1 = 0$ *or* $x + 3 = 0$

$(x + 1)(x + 3) = 0$

$x^2 + 4x + 3 = 0$

35. The solutions are 5 and $\frac{3}{4}$.

$x = 5$ *or* $x = \frac{3}{4}$

$x - 5 = 0$ *or* $x - \frac{3}{4} = 0$

$(x - 5)\left(x - \frac{3}{4}\right) = 0$

$x^2 - \frac{3}{4}x - 5x + \frac{15}{4} = 0$

$x^2 - \frac{23}{4}x + \frac{15}{4} = 0$

$4x^2 - 23x + 15 = 0$ Multiplying by 4

37. The solutions are $-\frac{1}{4}$ and $-\frac{1}{2}$.

$x = -\frac{1}{4}$ *or* $x = -\frac{1}{2}$

$x + \frac{1}{4} = 0$ *or* $x + \frac{1}{2} = 0$

$\left(x + \frac{1}{4}\right)\left(x + \frac{1}{2}\right) = 0$

$x^2 + \frac{1}{2}x + \frac{1}{4}x + \frac{1}{8} = 0$

$x^2 + \frac{3}{4}x + \frac{1}{8} = 0$

$8x^2 + 6x + 1 = 0$ Multiplying by 8

39. The solutions are 2, 4 and -0.4.

$x = 2.4$ *or* $x = -0.4$

$x - 2.4 = 0$ *or* $x + 0.4 = 0$

$(x - 2.4)(x + 0.4) = 0$

$x^2 + 0.4x - 2.4x - 0.96 = 0$

$x^2 - 2x - 0.96 = 0$

41. The solutions are $-\sqrt{3}$ and $\sqrt{3}$.

$x = -\sqrt{3}$ *or* $x = \sqrt{3}$

$x + \sqrt{3} = 0$ *or* $x - \sqrt{3} = 0$

$(x + \sqrt{3})(x - \sqrt{3}) = 0$

$x^2 - 3 = 0$

43. The solutions are $2\sqrt{5}$ and $-2\sqrt{5}$.

$x = 2\sqrt{5}$ *or* $x = -2\sqrt{5}$

$x - 2\sqrt{5} = 0$ *or* $x + 2\sqrt{5} = 0$

$(x - 2\sqrt{5})(x + 2\sqrt{5}) = 0$

$x^2 - (2\sqrt{5})^2 = 0$

$x^2 - 4 \cdot 5 = 0$

$x^2 - 20 = 0$

45. The solutions are $4i$ and $-4i$.

$x = 4i$ *or* $x = -4i$

$x - 4i = 0$ *or* $x + 4i = 0$

$(x - 4i)(x + 4i) = 0$

$x^2 - (4i)^2 = 0$

$x^2 + 16 = 0$

47. The solutions are $2 - 7i$ and $2 + 7i$.

$x = 2 - 7i$ *or* $x = 2 + 7i$

$x - 2 + 7i = 0$ *or* $x - 2 - 7i = 0$

$(x - 2) + 7i = 0$ *or* $(x - 2) - 7i = 0$

$[(x - 2) + 7i][(x - 2) - 7i] = 0$

$(x - 2)^2 - (7i)^2 = 0$

$x^2 - 4x + 4 - 49i^2 = 0$

$x^2 - 4x + 4 + 49 = 0$

$x^2 - 4x + 53 = 0$

49. The solutions are $3 - \sqrt{14}$ and $3 + \sqrt{14}$.

$x=3-\sqrt{14}$ *or* $x=3+\sqrt{14}$

$x-3+\sqrt{14}=0$ *or* $x-3-\sqrt{14}=0$

$(x-3)+\sqrt{14}=0$ *or* $(x-3)-\sqrt{14}=0$

$[(x - 3) + \sqrt{14}][(x - 3) - \sqrt{14}] = 0$

$(x - 3)^2 - (\sqrt{14})^2 = 0$

$x^2 - 6x + 9 - 14 = 0$

$x^2 - 6x - 5 = 0$

51. The solutions are $1-\frac{\sqrt{21}}{3}$ and $1+\frac{\sqrt{21}}{3}$.

$$x=1-\frac{\sqrt{21}}{3} \quad or \quad x=1+\frac{\sqrt{21}}{3}$$
$$x-1+\frac{\sqrt{21}}{3}=0 \quad or \quad x-1-\frac{\sqrt{21}}{3}=0$$
$$(x-1)+\frac{\sqrt{21}}{3}=0 \quad or \quad (x-1)-\frac{\sqrt{21}}{3}=0$$
$$\left[(x-1)+\frac{\sqrt{21}}{3}\right]\left[(x-1)-\frac{\sqrt{21}}{3}\right]=0$$
$$(x-1)^2-\left(\frac{\sqrt{21}}{3}\right)^2=0$$
$$x^2-2x+1-\frac{21}{9}=0$$
$$x^2-2x+1-\frac{7}{3}=0$$
$$x^2-2x-\frac{4}{3}=0$$
$$3x^2-6x-4=0 \quad \text{Multiplying by 3}$$

53. The solutions are -2, 1, and 5.

$$x=-2 \quad or \quad x=1 \quad or \quad x=5$$
$$x+2=0 \quad or \quad x-1=0 \quad or \quad x-5=0$$
$$(x+2)(x-1)(x-5)=0$$
$$(x^2+x-2)(x-5)=0$$
$$x^3+x^2-2x-5x^2-5x+10=0$$
$$x^3-4x^2-7x+10=0$$

55. The solutions are -1, 0, and 3.

$$x=-1 \quad or \quad x=0 \quad or \quad x=3$$
$$x+1=0 \quad or \quad x=0 \quad or \quad x-3=0$$
$$(x+1)(x)(x-3)=0$$
$$(x^2+x)(x-3)=0$$
$$x^3-3x^2+x^2-3x=0$$
$$x^3-2x^2-3x=0$$

57. *Writing Exercise*

59. $(3a^2)^4=3^4(a^2)^4=81a^{2\cdot 4}=81a^8$

61. $f(x)=x^2-7x-8$

We find the values of x for which $f(x)=0$.

$$x^2-7x-8=0$$
$$(x-8)(x+1)=0$$
$$x-8=0 \quad or \quad x+1=0$$
$$x=8 \quad or \quad x=-1$$

The x-intercepts are $(8,0)$ and $(-1,0)$.

63. ***Familiarize***. Let x and y represent the number of 30-sec and 60-sec commercials, respectively. Then the amount of time for the 30-sec commercials was $30x$ sec, or $\frac{30x}{60}=\frac{x}{2}$ min. The amount of time for the 60-sec commercials was $60x$ sec, or $\frac{60x}{60}=x$ min.

Translate. Rewording, we write two equations. We will express time in minutes.

Total number of commercials is 12.

$$x+y=12$$

Time for 30-sec commercials is total commercial time less 6 min.

$$\frac{x}{2}=\frac{x}{2}+x-6$$

Carry out. Solving the system of equations we get $(6,6)$.

Check. If there are six 30-sec and six 60-sec commercials, the total number of commercials is 12. The amount of time for six 30-sec commercials is 180 sec, or 3 min, and for six 60-sec commercials is 360 sec, or 6 min. The total commercial time is 9 min, and the amount of time for 30-sec commercials is 6 min less than this. The numbers check.

State. There were six 30-sec commercials.

65. *Writing Exercise*

67. The graph includes the points $(-3,0)$, $(0,-3)$, and $(1,0)$. Substituting in $y=ax^2+bx+c$, we have three equations.

$$0=9a-3b+c,$$
$$-3=c,$$
$$0=a+b+c$$

The solution of this system of equations is $a=1$, $b=2$, $c=-3$.

69. a) $kx^2-2x+k=0$; one solution is -3

We first find k by substituting -3 for x.

$$k(-3)^2-2(-3)+k=0$$
$$9k+6+k=0$$
$$10k=-6$$
$$k=-\frac{6}{10}$$
$$k=-\frac{3}{5}$$

b) Now substitute $-\frac{3}{5}$ for k in the original equation.

$$-\frac{3}{5}x^2-2x+\left(-\frac{3}{5}\right)=0$$
$$3x^2+10x+3=0 \quad \text{Multiplying by } -5$$
$$(3x+1)(x+3)=0$$
$$x=-\frac{1}{3} \text{ or } x=-3$$

The other solution is $-\frac{1}{3}$.

71. a) $x^2 - (6+3i)x + k = 0$; one solution is 3.

We first find k by substituting 3 for x.

$$3^2 - (6+3i)3 + k = 0$$
$$9 - 18 - 9i + k = 0$$
$$-9 - 9i + k = 0$$
$$k = 9 + 9i$$

b) Now we substitute $9+9i$ for k in the original equation.

$$x^2 - (6+3i)x + (9+9i) = 0$$
$$x^2 - (6+3i)x + 3(3+3i) = 0$$
$$[x - (3+3i)][x-3] = 0$$
$$x = 3+3i \text{ or } x = 3$$

The other solution is $3+3i$.

73. The solutions of $ax^2+bx+c=0$ are $x = \dfrac{-b \pm \sqrt{b^2-4ac}}{2a}$. When there is just one solution, $b^2 - 4ac = 0$, so $x = \dfrac{-b \pm 0}{2a} = -\dfrac{b}{2a}$.

75. We substitute $(-3,0)$, $\left(\frac{1}{2},0\right)$, and $(0,-12)$ in $f(x) = ax^2+bx+c$ and get three equations.

$$0 = 9a - 3b + c,$$
$$0 = \frac{1}{4}a + \frac{1}{2}b + c,$$
$$-12 = c$$

The solution of this system of equations is $a = 8$, $b = 20$, $c = -12$.

77. If $1-\sqrt{5}$ and $3+2i$ are two solutions, then $1+\sqrt{5}$ and $3-2i$ are also solutions. The equation of lowest degree that has these solutions is found as follows.

$$[x-(1-\sqrt{5})][x-(1+\sqrt{5})][x-(3+2i)][x-(3-2i)] = 0$$
$$(x^2-2x-4)(x^2-6x+13) = 0$$
$$x^4 - 8x^3 + 21x^2 - 2x - 52 = 0$$

79. *Writing Exercise*

Exercise Set 11.5

1. $x^6 = (x^3)^2$, so (f) is an appropriate choice.

3. $x^8 = (x^4)^2$, so (h) is an appropriate choice.

5. $x^{4/3} = (x^{2/3})^2$, so (g) is an appropriate choice.

7. $x^{-4/3} = (x^{-2/3})^2$, so (e) is an appropriate choice.

9. $x^4 - 5x^2 + 4 = 0$

Let $u = x^2$ and $u^2 = x^4$.

$u^2 - 5u + 4 = 0$ Substituting u for x^2

$(u-1)(u-4) = 0$

$u - 1 = 0$ *or* $u - 4 = 0$

$u = 1$ *or* $u = 4$

Now replace u with x^2 and solve these equations.

$x^2 = 1$ *or* $x^2 = 4$

$x = \pm 1$ *or* $x = \pm 2$

The numbers 1, -1, 2, and -2 check. They are the solutions.

11. $x^4 - 9x^2 + 20 = 0$

Let $u = x^2$ and $u^2 = x^4$.

$u^2 - 9u + 20 = 0$ Substituting

$(u-4)(u-5) = 0$

$u = 4$ *or* $u = 5$

Now replace u with x^2 and solve these equations:

$x^2 = 4$ *or* $x^2 = 5$

$x = \pm 2$ *or* $x = \pm\sqrt{5}$

The numbers 2, -2, $\sqrt{5}$, and $-\sqrt{5}$ check. They are the solutions.

13. $4t^4 - 19t^2 + 12 = 0$

Let $u = t^2$ and $u^2 = t^4$.

$4u^2 - 19u + 12 = 0$ Substituting

$(4u-3)(u-4) = 0$

$4u - 3 = 0$ *or* $u - 4 = 0$

$u = \frac{3}{4}$ *or* $u = 4$

Now replace u with t^2 and solve these equations:

$t^2 = \frac{3}{4}$ *or* $t^2 = 4$

$t = \pm\frac{\sqrt{3}}{2}$ *or* $t = \pm 2$

The numbers $\frac{\sqrt{3}}{2}$, $-\frac{\sqrt{3}}{2}$, 2, and -2 check. They are the solutions.

15. $r - 2\sqrt{r} - 6 = 0$

Let $u = \sqrt{r}$ and $u^2 = r$.

$$u^2 - 2u - 6 = 0$$
$$u = \frac{-(-2) \pm \sqrt{(-2)^2 - 4\cdot 1\cdot(-6)}}{2\cdot 1}$$
$$u = \frac{2 \pm \sqrt{28}}{2} = \frac{2+2\sqrt{7}}{2}$$
$$u = 1 \pm \sqrt{7}$$

Replace u with $\sqrt{r}$ and solve these equations:

$\sqrt{r} = 1+\sqrt{7}$ *or* $\sqrt{r} = 1-\sqrt{7}$

$(\sqrt{r})^2 = (1+\sqrt{7})^2$ No solution: $1-\sqrt{7}$ is negative

$r = 1 + 2\sqrt{7} + 7$

$r = 8 + 2\sqrt{7}$

The number $8+2\sqrt{7}$ checks. It is the solution.

17. $(x^2-7)^2 - 3(x^2-7) + 2 = 0$

Let $u = x^2 - 7$ and $u^2 = (x^2-7)^2$.

$u^2 - 3u + 2 = 0$ Substituting

$(u-1)(u-2) = 0$

$u = 1 \quad or \quad u = 2$

$x^2 - 7 = 1 \quad or \quad x^2 - 7 = 2$ Replacing u with $x^2 - 7$

$x^2 = 8 \quad or \quad x^2 = 9$

$x = \pm\sqrt{8} \quad or \quad x = \pm 3$

$x = \pm 2\sqrt{2} \quad or \quad x = \pm 3$

The numbers $2\sqrt{2}$, $-2\sqrt{3}$, 3, and -3 check. They are the solutions.

19. $(1+\sqrt{x})^2 + 5(1+\sqrt{x}) + 6 = 0$

Let $u = 1 + \sqrt{x}$ and $u^2 = (1+\sqrt{x})^2$.

$u^2 + 5u + 6 = 0$ Substituting

$(u+3)(u+2) = 0$

$u = -3 \quad or \quad u = -2$

$1 + \sqrt{x} = -3 \quad or \quad 1 + \sqrt{x} = -2$ Replacing u with $1 + \sqrt{x}$

$\sqrt{x} = -4 \quad or \quad \sqrt{x} = -3$

Since the principal square root cannot be negative, this equation has no solution.

21. $x^{-2} - x^{-1} - 6 = 0$

Let $u = x^{-1}$ and $u^2 = x^{-2}$.

$u^2 - u - 6 = 0$ Substituting

$(u-3)(u+2) = 0$

$u = 3$ or $u = -2$

Now we replace u with x^{-1} and solve these equations:

$x^{-1} = 3 \quad or \quad x^{-1} = -2$

$\frac{1}{x} = 3 \quad or \quad \frac{1}{x} = -2$

$\frac{1}{3} = x \quad or \quad -\frac{1}{2} = x$

Both $\frac{1}{3}$ and $-\frac{1}{2}$ check. They are the solutions.

23. $4x^{-2} + x^{-1} - 5 = 0$

Let $u = x^{-1}$ and $u^2 = x^{-2}$.

$4u^2 + u - 5 = 0$ Substituting

$(4u+5)(u-1) = 0$

$u = -\frac{5}{4}$ or $u = 1$

Now we replace u with x^{-1} and solve these equations:

$x^{-1} = -\frac{5}{4} \quad or \quad x^{-1} = 1$

$\frac{1}{x} = -\frac{5}{4} \quad or \quad \frac{1}{x} = 1$

$4 = -5x \quad or \quad 1 = x$

$-\frac{4}{5} = x \quad or \quad 1 = x$

The numbers $-\frac{4}{5}$ and 1 check. They are the solutions.

25. $t^{2/3} + t^{1/3} - 6 = 0$

Let $u = t^{1/3}$ and $u^2 = t^{2/3}$.

$u^2 + u - 6 = 0$ Substituting

$(u+3)(u-2) = 0$

$u = -3$ or $u = 2$

Now we replace u with $t^{1/3}$ and solve these equations:

$t^{1/3} = -3 \quad or \quad t^{1/3} = 2$

$t = (-3)^3 \quad or \quad t = 2^3$ Raising to the third power

$t = -27 \quad or \quad t = 8$

Both -27 and 8 check. They are the solutions.

27. $y^{1/3} - y^{1/6} - 6 = 0$

Let $u = y^{1/6}$ and $u^2 = y^{2/3}$.

$u^2 - u - 6 = 0$ Substituting

$(u-3)(u+2) = 0$

$u = 3$ or $u = -2$

Now we replace u with $y^{1/6}$ and solve these equations:

$y^{1/6} = 3 \quad or \quad y^{1/6} = -2$

$\sqrt[6]{y} = 3 \quad or \quad \sqrt[6]{y} = -2$

$y = 3^6$ This equation has no

$y = 729$ solution since principal sixth roots are never negative.

The number 729 checks and is the solution.

29. $t^{1/3} + 2t^{1/6} = 3$

$t^{1/3} + 2t^{1/6} - 3 = 0$

Let $u = t^{1/6}$ and $u^2 = t^{2/6} = t^{1/3}$.

$u^2 + 2u - 3 = 0$ Substituting

$(u+3)(u-1) = 0$

$u = -3 \quad or \quad u = 1$

$t^{1/6} = -3 \quad or \quad t^{1/6} = 1$ Substituting $t^{1/6}$ for u

No solution $\quad t = 1$

The number 1 checks and is the solution.

31. $(3-\sqrt{x})^2 - 10(3-\sqrt{x}) + 23 = 0$

Let $u = 3 - \sqrt{x}$ and $u^2 = (3-\sqrt{x})^2$.

$u^2 - 10u + 23 = 0$ Substituting

$$u = \frac{-(-10) \pm \sqrt{(-10)^2 - 4 \cdot 1 \cdot 23}}{2 \cdot 1}$$

$$u = \frac{10 \pm \sqrt{8}}{2} = \frac{2 \cdot 5 \pm 2\sqrt{2}}{2}$$

$$u = 5 \pm \sqrt{2}$$

$u = 5 + \sqrt{2}$ or $u = 5 - \sqrt{2}$

Now we replace u with $3 - \sqrt{x}$ and solve these equations:

$3 - \sqrt{x} = 5 + \sqrt{2} \quad or \quad 3 - \sqrt{x} = 5 - \sqrt{2}$

$-\sqrt{x} = 2 + \sqrt{2} \quad or \quad -\sqrt{x} = 2 - \sqrt{2}$

$\sqrt{x} = -2 - \sqrt{2} \quad or \quad \sqrt{x} = -2 + \sqrt{2}$

Since both $-2 - \sqrt{2}$ and $-2 + \sqrt{2}$ are negative and principal square roots are never negative, the equation has no solution.

33. $16\left(\frac{x-1}{x-8}\right)^2+8\left(\frac{x-1}{x-8}\right)+1=0$

Let $u=\frac{x-1}{x-8}$ and $u^2=\left(\frac{x-1}{x-8}\right)^2$.

$16u^2+8u+1=0$ Substituting

$(4u+1)(4u+1)=0$

$u=-\frac{1}{4}$

Now we replace u with $\frac{x-1}{x-8}$ and solve this equation:

$\frac{x-1}{x-8}=-\frac{1}{4}$

$4x-4=-x+8$ Multiplying by $4(x-8)$

$5x=12$

$x=\frac{12}{5}$

The number $\frac{12}{5}$ checks and is the solution.

35. The x-intercepts occur where $f(x)=0$. Thus, we must have $5x+13\sqrt{x}-6=0$.

Let $u=\sqrt{x}$ and $u^2=x$.

$5u^2+13u-6=0$ Substituting

$(5u-2)(u+3)=0$

$u=\frac{2}{5}$ *or* $u=-3$

Now replace u with $\sqrt{x}$ and solve these equations:

$\sqrt{x}=\frac{2}{5}$ *or* $\sqrt{x}=-3$

$x=\frac{4}{25}$ No solution

The number $\frac{4}{25}$ checks. Thus, the x-intercept is $\left(\frac{4}{25},0\right)$.

37. The x-intercepts occur where $f(x)=0$. Thus, we must have $(x^2-3x)^2-10(x^2-3x)+24=0$.

Let $u=x^2-3x$ and $u^2=(x^2-3x)^2$.

$u^2-10u+24=0$ Substituting

$(u-6)(u-4)=0$

$u=6$ *or* $u=4$

Now replace u with x^2-3x and solve these equations:

$x^2-3x=6$ *or* $x^2-3x=4$

$x^2-3x-6=0$ *or* $x^2-3x-4=0$

$x=\frac{-(-3)\pm\sqrt{(-3)^2-4(1)(-6)}}{2\cdot 1}$ *or* $(x-4)(x+1)=0$

$x=\frac{3}{2}\pm\frac{\sqrt{33}}{2}$ *or* $x=4$ *or* $x=-1$

All four numbers check. Thus, the x-intercepts are $\left(\frac{3}{2}+\frac{\sqrt{33}}{2},0\right)$, $\left(\frac{3}{2}-\frac{\sqrt{33}}{2},0\right)$, $(4,0)$, and $(-1,0)$.

39. The x-intercepts occur where $f(x)=0$. Thus, we must have $x^{2/5}+x^{1/5}-6=0$.

Let $u=x^{1/5}$ and $u^2=x^{2/5}$.

$u^2+u-6=0$ Substituting

$(u+3)(u-2)=0$

$u=-3$ *or* $u=2$

$x^{1/5}=-3$ *or* $x^{1/5}=2$ Replacing u with $x^{1/5}$

$x=-243$ *or* $x=32$ Raising to the fifth power

Both -243 and 32 check. Thus, the x-intercepts are $(-243,0)$ and $(32,0)$.

41. $f(x)=\left(\frac{x^2+2}{x}\right)^4+7\left(\frac{x^2+2}{x}\right)^2+5$

Observe that, for all real numbers x, each term is positive. Thus, there are no real-number values of x for which $f(x)=0$ and hence no x-intercepts.

43. *Writing Exercise*

45. Graph $f(x)=\frac{3}{2}x$.

We find some ordered pairs, plot points, and draw the graph.

x	y
-4	-6
-2	-3
0	0
2	3
4	6

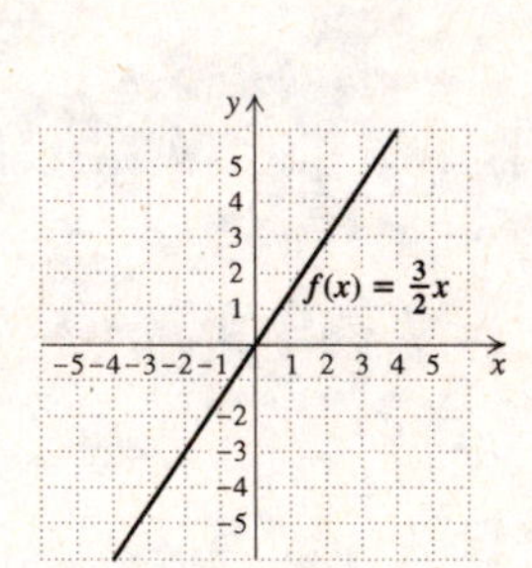

47. Graph $g(x)=\frac{2}{x}$.

We find some ordered pairs, plot points, and draw the graph. Note that we cannot use 0 as a first coordinate since division by 0 is undefined.

x	y
-4	$-\frac{1}{2}$
-2	-1
$-\frac{1}{2}$	-4
$\frac{1}{2}$	4
2	1
4	$\frac{1}{2}$

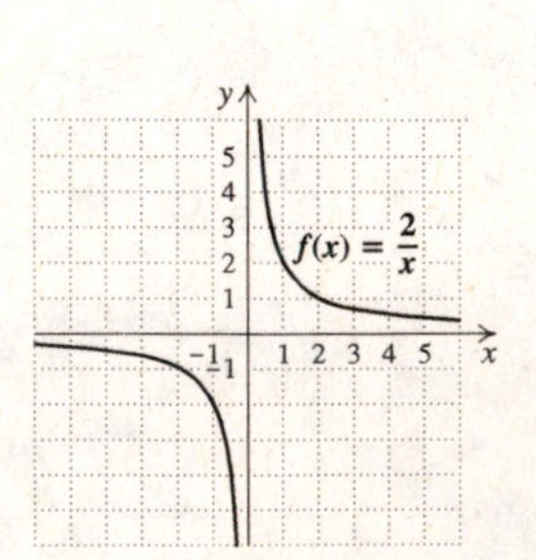

49. ***Familiarize.*** Let a = the number of pounds of Hiker's Mix in the mixture and b = the number of pounds of Trail Snax. We organize the information in a table.

Solution	Hiker's Mix	Trail Snax	Mixture
Number of pounds	a	b	12
Percent of peanuts	18%	45%	36%
Pounds of peanuts	$0.18a$	$0.45b$	$0.36(12)$, or 4.32 lb

From the first row of the table we get one equation:

$$a + b = 12$$

We get a second equation from the last row of the table:

$$0.18a + 0.45b = 4.32$$

After clearing decimals, we have the following system of equations:

$$a + b = 12, \quad (1)$$
$$18a + 45b = 432 \quad (2)$$

Carry out. We use the elimination method. First we multiply equation (1) by -18 and then add.

$$\begin{array}{r} -18a - 18b = -216 \\ 18a + 45b = 432 \\ \hline 27b = 216 \\ b = 8 \end{array}$$

Now we substitute 8 for b in one of the original equations and solve for a.

$$a + b = 12 \quad (1)$$
$$a + 8 = 12$$
$$a = 4$$

Check. If 4 lb of Hiker's Mix and 8 lb of Trail Snax are used, the mixture has 4 + 8, or 12 lb. The amount of peanuts in 4 lb of Hiker's Mix is 0.18(4), or 0.72 lb. The amount of peanuts in 8 lb of Trail Snax is 0.45(8), or 3.6 lb. Then the amount of peanuts in the mixture is 0.72 + 3.6, or 4.32 lb. The answer checks.

State. The mixture should contain 4 lb of Hiker's Mix and 8 lb of Trail Snax.

51. *Writing Exercise*

53. $5x^4 - 7x^2 + 1 = 0$

Let $u = x^2$ and $u^2 = x^4$.

$5u^2 - 7u + 1 = 0$ Substituting

$$u = \frac{-(-7) \pm \sqrt{(-7)^2 - 4 \cdot 5 \cdot 1}}{2 \cdot 5}$$

$$u = \frac{7 \pm \sqrt{29}}{10}$$

$$x^2 = \frac{7 \pm \sqrt{29}}{10} \quad \text{Replacing } u \text{ with } x^2$$

$$x = \pm\sqrt{\frac{7 \pm \sqrt{29}}{10}}$$

All four numbers check and are the solutions.

55. $(x^2 - 4x - 2)^2 - 13(x^2 - 4x - 2) + 30 = 0$

Let $u = x^2 - 4x - 2$ and $u^2 = (x^2 - 4x - 2)^2$.

$u^2 - 13u + 30 = 0$ Substituting

$(u - 3)(u - 10) = 0$

$u = 3$ *or* $u = 10$

$x^2 - 4x - 2 = 3$ *or* $x^2 - 4x - 2 = 10$

Replacing u with $x^2 - 4x - 2$

$x^2 - 4x - 5 = 0$ *or* $x^2 - 4x - 12 = 0$

$(x - 5)(x + 1) = 0$ *or* $(x - 6)(x + 2) = 0$

$x = 5$ *or* $x = -1$ *or* $x = 6$ *or* $x = -2$

All four numbers check and are the solutions.

57. $\dfrac{x}{x-1} - 6\sqrt{\dfrac{x}{x-1}} - 40 = 0$

Let $u = \sqrt{\dfrac{x}{x-1}}$ and $u^2 = \dfrac{x}{x-1}$.

$u^2 - 6u - 40 = 0$ Substituting

$(u - 10)(u + 4) = 0$

$u = 10$ *or* $u = -4$

$\sqrt{\dfrac{x}{x-1}} = 10$ *or* $\sqrt{\dfrac{x}{x-1}} = -4$

$\dfrac{x}{x-1} = 100$ *or* No solution

$x = 100x - 100$ Multiplying by $(x - 1)$

$100 = 99x$

$\dfrac{100}{99} = x$

The number $\dfrac{100}{99}$ checks. It is the solution.

59. $a^5(a^2 - 25) + 13a^3(25 - a^2) + 36a(a^2 - 25) = 0$

$a^5(a^2 - 25) - 13a^3(a^2 - 25) + 36a(a^2 - 25) = 0$

$a(a^2 - 25)(a^4 - 13a^2 + 36) = 0$

$a(a^2 - 25)(a^2 - 4)(a^2 - 9) = 0$

$a=0$ *or* $a^2 - 25=0$ *or* $a^2 - 4=0$ *or* $a^2 - 9 = 0$

$a=0$ *or* $a^2=25$ *or* $a^2=4$ *or* $a^2 = 9$

$a=0$ *or* $a=\pm 5$ *or* $a=\pm 2$ *or* $a = \pm 3$

All seven numbers check. The solutions are 0, 5, -5, 2, -2, 3, and -3.

61. $x^6 - 28x^3 + 27 = 0$

Let $u = x^3$.

$u^2 - 28u + 27 = 0$

$(u - 27)(u - 1) = 0$

$u = 27$ *or* $u = 1$

$x^3 = 27$ *or* $x^3 = 1$

$x^3 - 27 = 0$ *or* $x^3 - 1 = 0$

First we solve $x^3 - 27 = 0$.

$x^3 - 27 = 0$

$(x - 3)(x^2 + 3x + 9) = 0$

$$x-3=0 \quad or \quad x^2+3x+9=0$$
$$x=3 \quad or \quad x=\frac{-3\pm\sqrt{3^2-4\cdot 1\cdot 9}}{2\cdot 1}$$
$$x=3 \quad or \quad x=\frac{-3\pm\sqrt{-27}}{2}$$
$$x=3 \quad or \quad x=-\frac{3}{2}\pm\frac{3\sqrt{3}}{2}i$$

Next we solve $x^3-1=0$.

$$x^3-1=0$$
$$(x-1)(x^2+x+1)=0$$
$$x-1=0 \quad or \quad x^2+x+1=0$$
$$x=1 \quad or \quad x=\frac{-1\pm\sqrt{1^2-4\cdot 1\cdot 1}}{2\cdot 1}$$
$$x=1 \quad or \quad x=\frac{-1\pm\sqrt{-3}}{2}$$
$$x=1 \quad or \quad x=-\frac{1}{2}\pm\frac{\sqrt{3}}{2}i$$

All six numbers check.

63. 

Exercise Set 11.6

1. The graph of $f(x)=2(x-1)^2+3$ has vertex $(1,3)$ and opens up. Choice (h) is correct.

3. The graph of $f(x)=2(x+1)^2+3$ has vertex $(-1,3)$ and opens up. Choice (f) is correct.

5. The graph of $f(x)=-2(x+1)^2+3$ has vertex $(-1,3)$ and opens down. Choice (b) is correct.

7. The graph of $f(x)=2(x+1)^2-3$ has vertex $(-1,-3)$ and opens up. Choice (e) is correct.

9. $f(x)=x^2$

See Example 1 in the text.

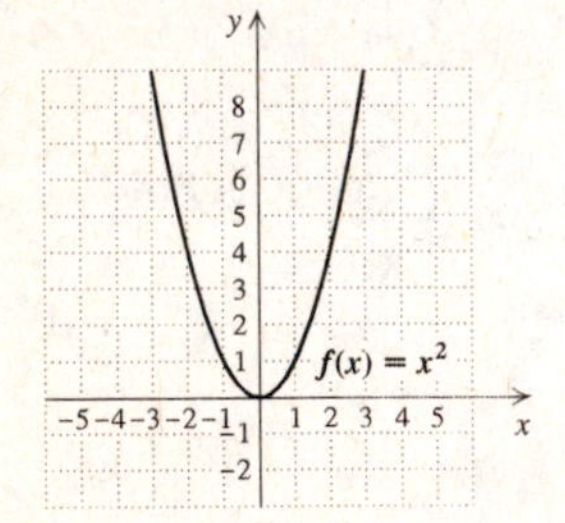

11. $f(x)=-2x^2$

We choose some numbers for x and compute $f(x)$ for each one. Then we plot the ordered pairs $(x, f(x))$ and connect them with a smooth curve.

x	$f(x)=-4x^2$
0	0
1	-2
2	-8
-1	-2
-2	-8

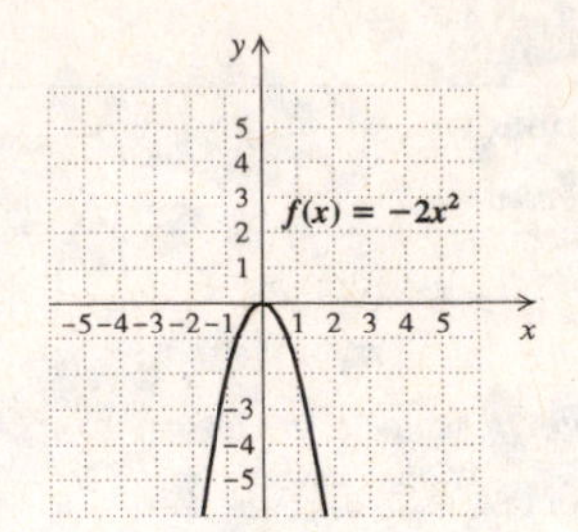

13. $g(x)=\frac{1}{3}x^2$

x	$g(x)=\frac{1}{3}x^2$
0	0
1	$\frac{1}{3}$
2	$\frac{4}{3}$
3	3
-1	$\frac{1}{3}$
-2	$\frac{4}{3}$
-3	3

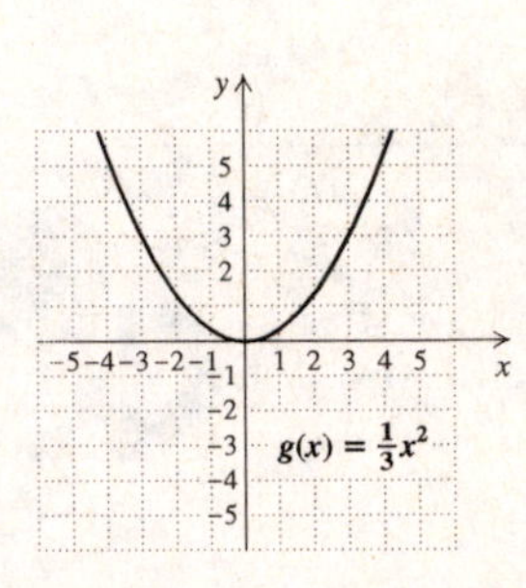

15. $h(x)=-\frac{1}{3}x^2$

Observe that the graph of $h(x)=-\frac{1}{3}x^2$ is the reflection of the graph of $g(x)=\frac{1}{3}x^2$ across the x-axis. We graphed $g(x)$ in Exercise 13, so we can use it to graph $h(x)$. If we did not make this observation we could find some ordered pairs, plot points, and connect them with a smooth curve.

x	$h(x)=-\frac{1}{3}x^2$
0	0
1	$-\frac{1}{3}$
2	$-\frac{4}{3}$
3	-3
-1	$-\frac{1}{3}$
-2	$-\frac{4}{3}$
-3	-3

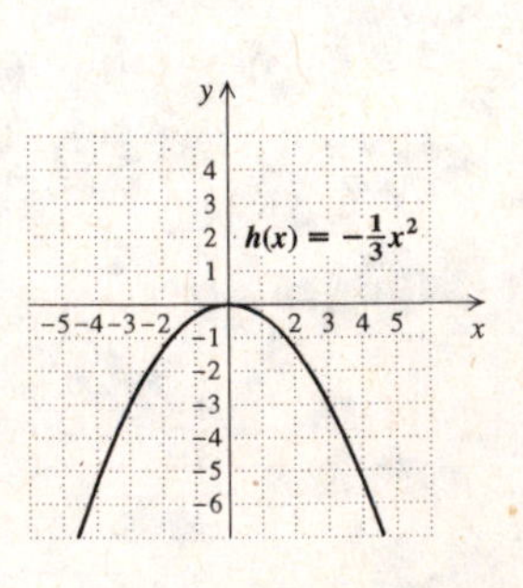

17. $f(x) = \frac{5}{2}x^2$

x	$f(x) = \frac{5}{2}x^2$
0	0
1	$\frac{5}{2}$
2	10
−1	$\frac{5}{2}$
−2	10

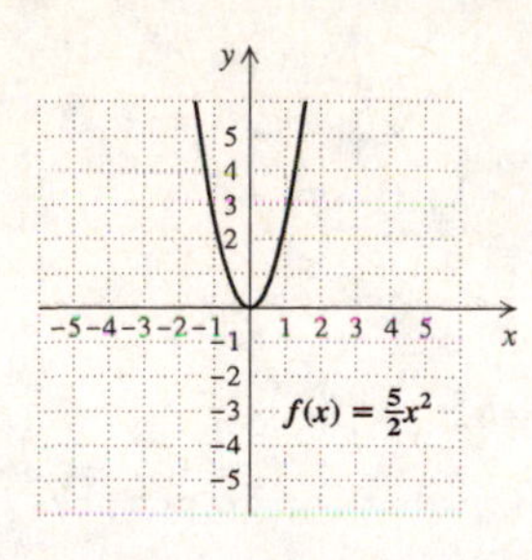

19. $g(x) = (x+1)^2 = [x-(-1)]^2$

We know that the graph of $g(x) = (x+1)^2$ looks like the graph of $f(x) = x^2$ (see Exercise 9) but moved to the left 1 unit.

Vertex: $(-1, 0)$, axis of symmetry: $x = -1$

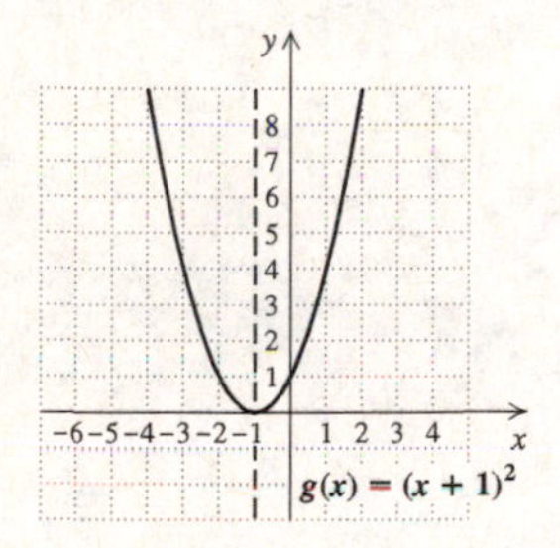

21. $f(x) = (x-2)^2$

The graph of $f(x) = (x-2)^2$ looks like the graph of $f(x) = x^2$ (see Exercise 9) but moved to the right 2 units.

Vertex: $(2, 0)$, axis of symmetry: $x = 2$

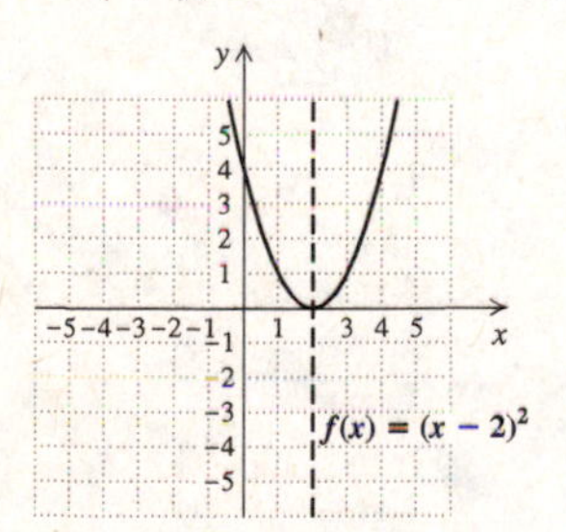

23. $h(x) = (x-3)^2$

The graph of $h(x) = (x-3)^2$ looks like the graph of $f(x) = x^2$ (see Exercise 9) but moved to the right 3 units.

Vertex: $(3, 0)$, axis of symmetry: $x = 3$

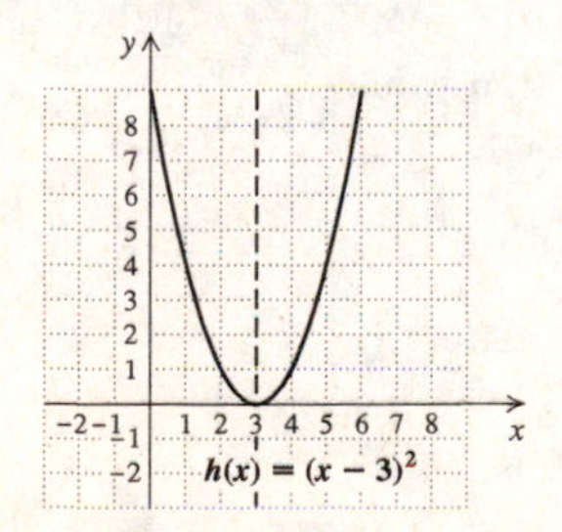

25. $g(x) = -(x+1)^2$

The graph of $g(x) = -(x+1)^2$ looks like the graph of $f(x) = x^2$ (see Exercise 9) but moved to the left 1 unit. It will also open downward because of the negative coefficient, -1.

Vertex: $(-1, 0)$, axis of symmetry: $x = -1$

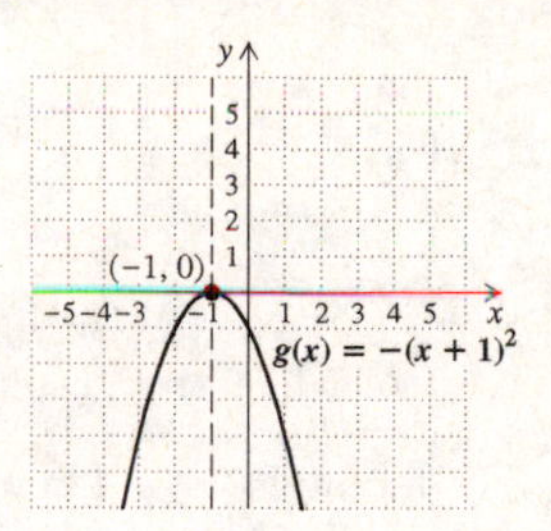

27. $f(x) = -(x-2)^2$

The graph of $f(x) = -(x-2)^2$ looks like the graph of $f(x) = x^2$ (see Exercise 9) but moved to the right 2 units. It will also open downward because of the negative coefficient, -1.

Vertex: $(2, 0)$, axis of symmetry: $x = 2$

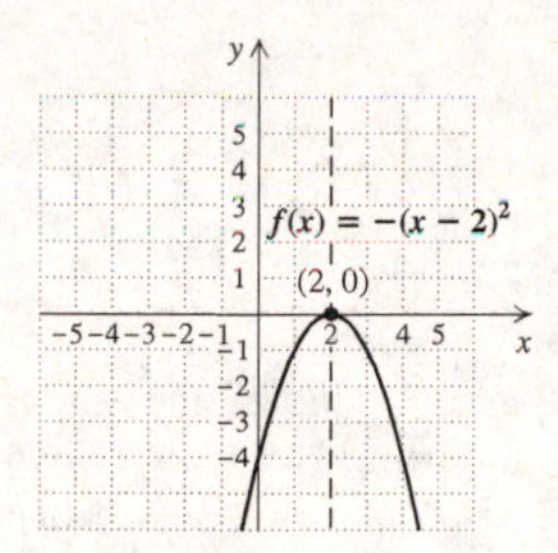

29. $f(x) = 2(x+1)^2$

The graph of $f(x) = 2(x+1)^2$ looks like the graph of $h(x) = 2x^2$ (see graph following Example 1) but moved to the left 1 unit.

Vertex: $(-1, 0)$, axis of symmetry: $x = -1$

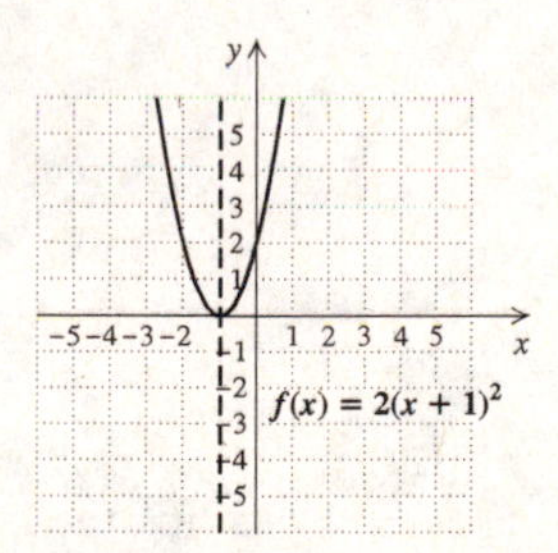

31. $h(x) = -\frac{1}{2}(x-4)^2$

The graph of $h(x) = -\frac{1}{2}(x-4)^2$ looks like the graph of $g(x) = \frac{1}{2}x^2$ (see graph following Example 1) but moved to the right 4 units. It will also open downward because of the negative coefficient, $-\frac{1}{2}$.

Vertex: $(4, 0)$, axis of symmetry: $x = 4$

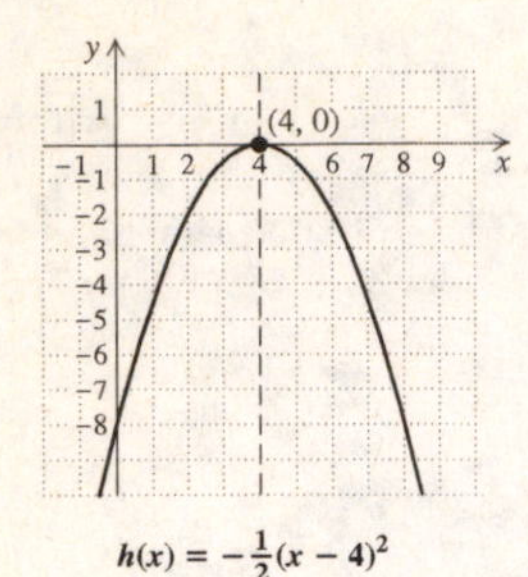

33. $f(x) = \dfrac{1}{2}(x-1)^2$

The graph of $f(x) = \dfrac{1}{2}(x-1)^2$ looks like the graph of $g(x) = \dfrac{1}{2}x^2$ (see graph following Example 1) but moved to the right 1 unit.

Vertex: $(1, 0)$, axis of symmetry: $x = 1$

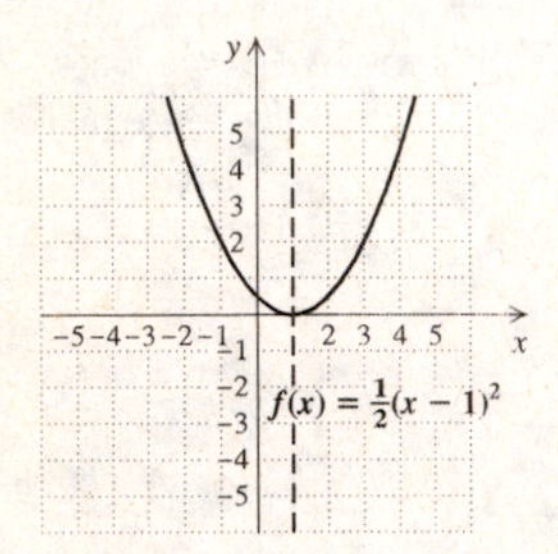

35. $f(x) = -2(x+5)^2 = -2[x-(-5)]^2$

The graph of $f(x) = -2(x+5)^2$ looks like the graph of $h(x) = 2x^2$ (see the graph following Example 1) but moved to the left 5 units. It will also open downward because of the negative coefficient, -2.

Vertex: $(-5, 0)$, axis of symmetry: $x = -5$

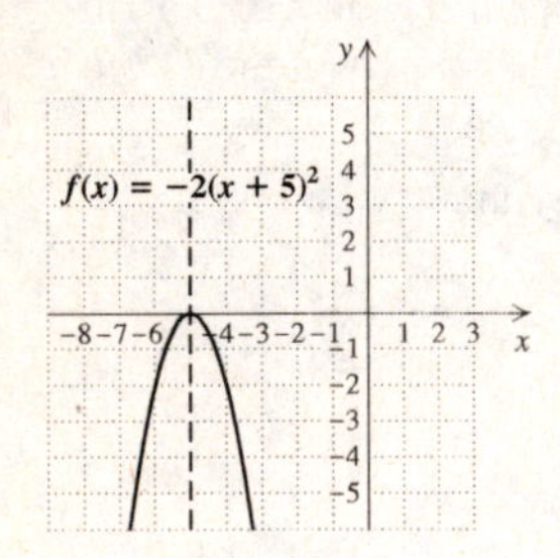

37. $h(x) = -3\left(x - \dfrac{1}{2}\right)^2$

The graph of $h(x) = -3\left(x - \dfrac{1}{2}\right)^2$ looks like the graph of $f(x) = -3x^2$ (see Exercise 12) but moved to the right $\dfrac{1}{2}$ unit.

Vertex: $\left(\dfrac{1}{2}, 0\right)$, axis of symmetry: $x = \dfrac{1}{2}$

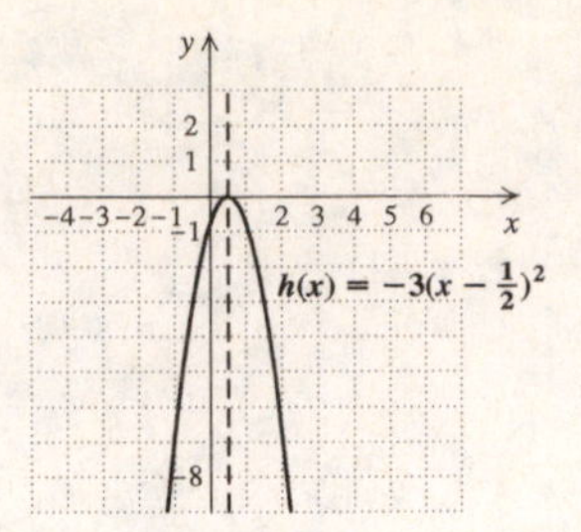

39. $f(x) = (x-5)^2 + 2$

We know that the graph looks like the graph of $f(x) = x^2$ (see Example 1) but moved to the right 5 units and up 2 units. The vertex is $(5, 2)$, and the axis of symmetry is $x = 5$. Since the coefficient of $(x-5)^2$ is positive $(1 > 0)$, there is a minimum function value, 2.

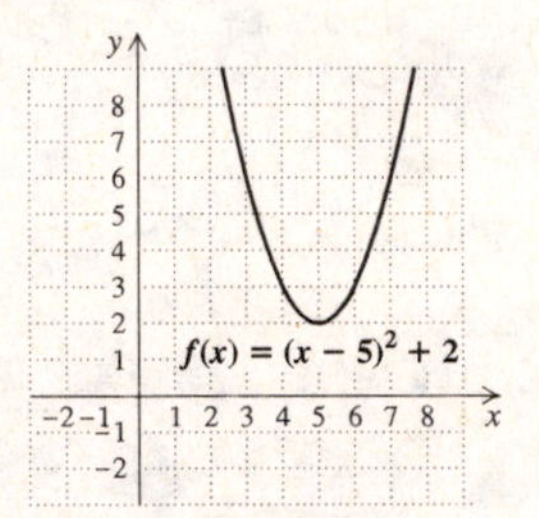

41. $f(x) = (x+1)^2 - 3$

We know that the graph looks like the graph of $f(x) = x^2$ (see Example 1) but moved to the left 1 unit and down 3 units. The vertex is $(-1, -3)$, and the axis of symmetry is $x = -1$. Since the coefficient of $(x+1)^2$ is positive $(1 > 0)$, there is a minimum function value, -3.

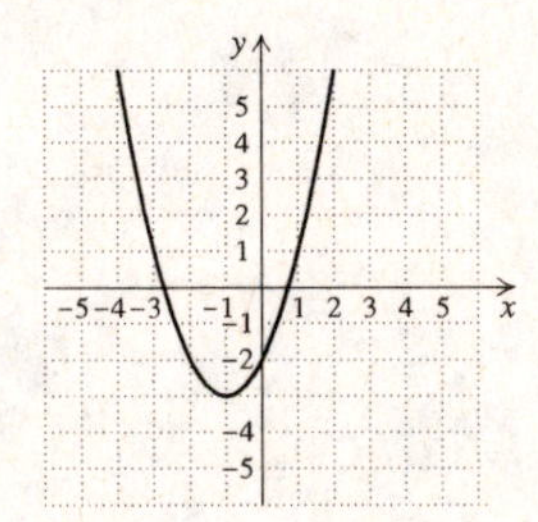

$f(x) = (x+1)^2 - 3$

43. $g(x) = (x+4)^2 + 1$

We know that the graph looks like the graph of $f(x) = x^2$ (see Example 1) but moved to the left 4 units and up 1 unit. The vertex is $(-4, 1)$, and the axis of symmetry is $x = -4$. Since the coefficient of $(x+4)^2$ is positive $(1 > 0)$, there is a minimum function value, 1.

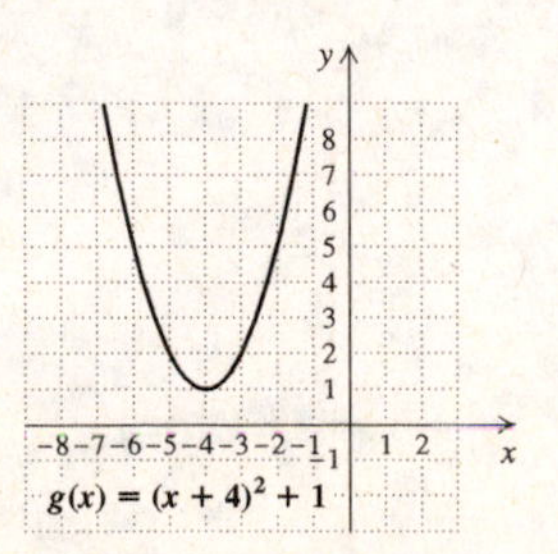

45. $h(x) = -2(x-1)^2 - 3$

We know that the graph looks like the graph of $h(x) = 2x^2$ (see graph following Example 1) but moved to the right 1 unit and down 3 units and turned upside down. The vertex is $(1, -3)$, and the axis of symmetry is $x = 1$. The maximum function value is -3.

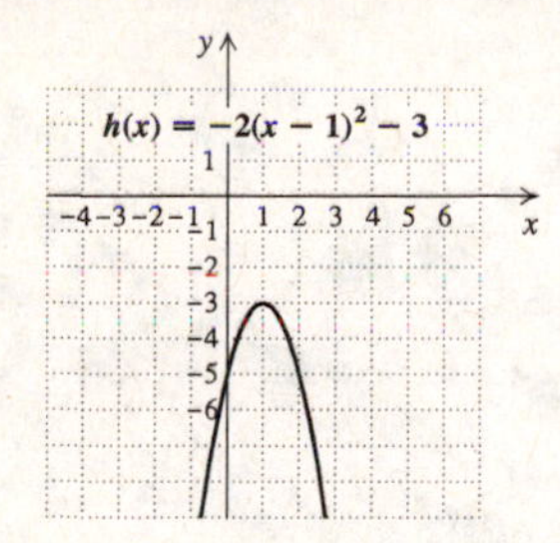

47. $f(x) = 2(x+4)^2 + 1$

We know that the graph looks like the graph of $f(x) = 2x^2$ (see graph following Example 1) but moved to the left 4 units and up 1 unit. The vertex is $(-4, 1)$, the axis of symmetry is $x = -4$, and the minimum function value is 1.

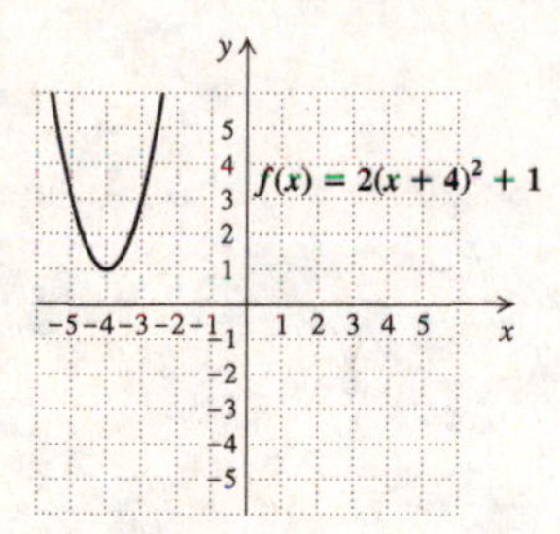

49. $g(x) = -\frac{3}{2}(x-1)^2 + 4$

We know that the graph looks like the graph of $f(x) = \frac{3}{2}x^2$ (see Exercise 18) but moved to the right 1 unit and up 4 units and turned upside down. The vertex is $(1, 4)$, the axis of symmetry is $x = 1$, and the maximum function value is 4.

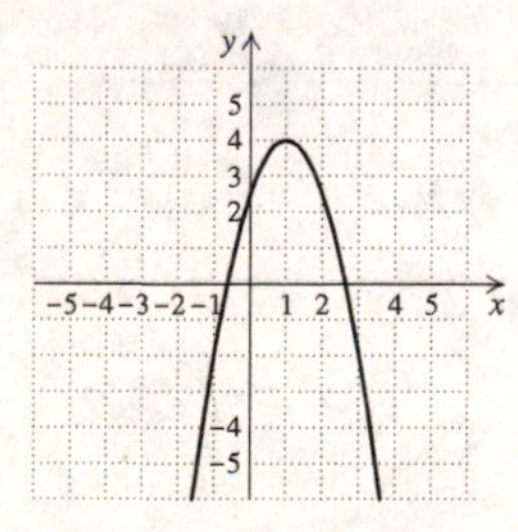

51. $f(x) = 6(x-8)^2 + 7$

This function is of the form $f(x) = a(x-h)^2 + k$ with $a = 6$, $h = 8$, and $k = 7$. The vertex is (h, k), or $(8, 7)$. The axis of symmetry is $x = h$, or $x = 8$. Since $a > 0$, then k, or 7, is the minimum function value.

53. $h(x) = -\frac{2}{7}(x+6)^2 + 11$

This function is of the form $f(x) = a(x-h)^2 + k$ with $a = -\frac{2}{7}$, $h = -6$, and $k = 11$. The vertex is (h, k), or $(-6, 11)$. The axis of symmetry is $x = h$, or $x = -6$. Since $a < 0$, then k, or 11, is the maximum function value.

55. $f(x) = 7\left(x + \frac{1}{4}\right)^2 - 13$

This function is of the form $f(x) = a(x-h)^2+k$ with $a = 7$, $h = -\frac{1}{4}$, and $k = -13$. The vertex is (h, k), or $\left(-\frac{1}{4}, -13\right)$. The axis of symmetry is $x = h$, or $x = -\frac{1}{4}$. Since $a > 0$, then k, or -13, is the minimum function value.

57. $f(x) = \sqrt{2}(x + 4.58)^2 + 65\pi$

This function is of the form $f(x) = a(x-h)^2 + k$ with $a = \sqrt{2}$, $h = -4.58$, and $k = 65\pi$. The vertex is (h, k), or $(-4.58, 65\pi)$. The axis of symmetry is $x = h$, or $x = -4.58$. Since $a > 0$, then k, or 65π, is the minimum function value.

59. *Writing Exercise*

61. Graph $2x - 7y = 28$.

Find the x-intercept.

$$2x - 7\cdot 0 = 28$$
$$2x = 28$$
$$x = 14$$

The x-intercept is $(14, 0)$.

Find the y-intercept.

$$2\cdot 0 - 7y = 28$$
$$-7y = 28$$
$$y = -4$$

The y-intercept is $(0, -4)$.

Plot the intercepts and draw a line through them. A third point can be plotted as a check.

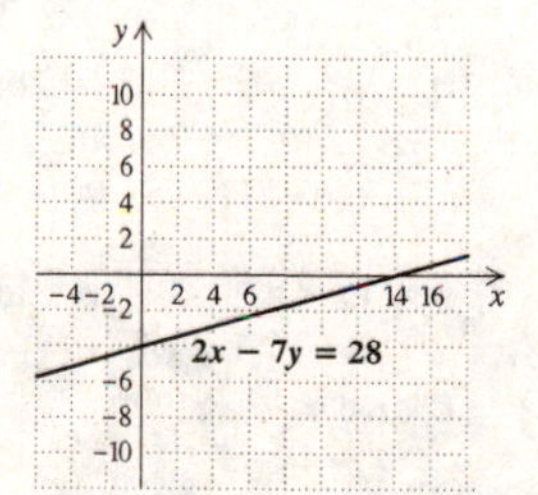

63. $3x + 4y = -19,$ (1)

$7x - 6y = -29$ (2)

Multiply Equation (1) by 3 and multiply Equation (2) by 2. Then add the equations to eliminate the y-term.

$$\begin{aligned} 9x + 12y &= -57 \\ 14x - 12y &= -58 \\ \hline 23x \quad &= -115 \\ x &= -5 \end{aligned}$$

Now substitute -5 for x in one of the original equations and solve for y. We use Equation (1).

$$\begin{aligned} 3(-5) + 4y &= -19 \\ -15 + 4y &= -19 \\ 4y &= -4 \\ y &= -1 \end{aligned}$$

The pair $(-5, -1)$ checks and it is the solution.

65. $x^2 + 5x$

We take half the coefficient of x and square it.

$$\frac{1}{2} \cdot 5 = \frac{5}{2}, \left(\frac{5}{2}\right)^2 = \frac{25}{4}$$

Then we have $x^2 + 5x + \frac{25}{4} = \left(x + \frac{5}{2}\right)^2$.

67. *Writing Exercise*

69. The equation will be of the form $f(x) = \frac{3}{5}(x - h)^2 + k$ with $h = 4$ and $k = 1$:

$$f(x) = \frac{3}{5}(x - 4)^2 + 1$$

71. The equation will be of the form $f(x) = \frac{3}{5}(x - h)^2 + k$ with $h = 3$ and $k = -1$:

$$f(x) = \frac{3}{5}(x - 3)^2 + (-1), \text{ or}$$

$$f(x) = \frac{3}{5}(x - 3)^2 - 1$$

73. The equation will be of the form $f(x) = \frac{3}{5}(x - h)^2 + k$ with $h = -2$ and $k = -5$:

$$f(x) = \frac{3}{5}[x - (-2)]^2 + (-5), \text{ or}$$

$$f(x) = \frac{3}{5}(x + 2)^2 - 5$$

75. Since there is a minimum at $(2, 0)$, the parabola will have the same shape as $f(x) = 2x^2$. It will be of the form $f(x) = 2(x-h)^2+k$ with $h = 2$ and $k = 0$: $f(x) = 2(x-2)^2$

77. Since there is a maximum at $(0, 3)$, the parabola will have the same shape as $g(x) = -2x^2$. It will be of the form $g(x) = -2(x - h)^2 + k$ with $h = 0$ and $k = 3$: $g(x) = -2(x - 0)^2 + 3$, or $g(x) = -2x^2 + 3$

79. The maximum value of $g(x)$ is 1 and occurs at the point $(5, 1)$, so for $F(x)$ we have $h = 5$ and $k = 1$. $F(x)$ has the same shape as $f(x)$ and has a minimum, so $a = 3$. Thus, $F(x) = 3(x - 5)^2 + 1$.

81. The graph of $y = f(x - 1)$ looks like the graph of $y = f(x)$ moved 1 unit to the right.

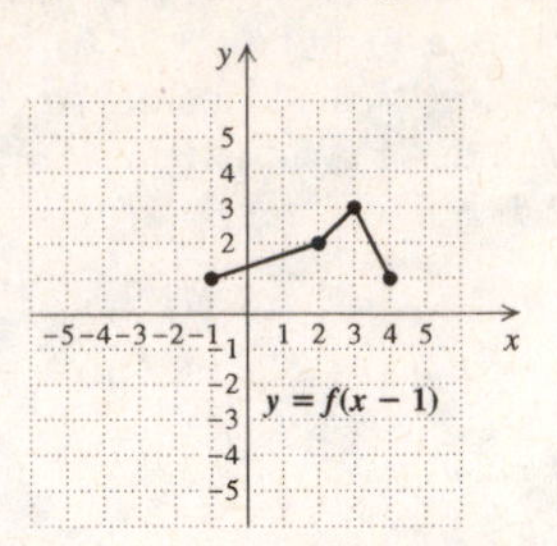

83. The graph of $y = f(x) + 2$ looks like the graph of $y = f(x)$ moved up 2 units.

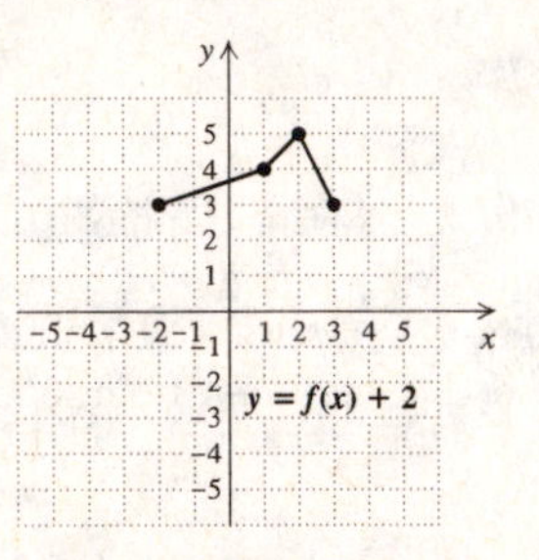

85. The graph of $y = f(x + 3) - 2$ looks like the graph of $y = f(x)$ moved 3 units to the left and also moved down 2 units.

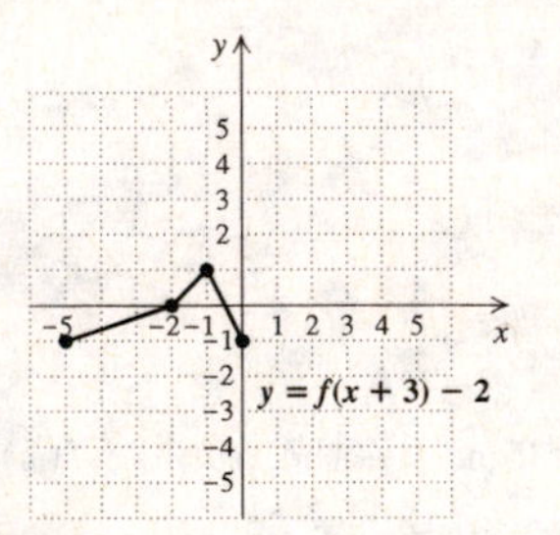

87.

89. *Writing Exercise*

Exercise Set 11.7

1. 9

3. 18

5. 3

7. $\frac{5}{2}$; -4

9.
$$\begin{aligned} f(x) &= x^2 + 4x + 5 \\ &= (x^2 + 4x + 4 - 4) + 5 \quad \text{Adding } 4 - 4 \\ &= (x^2 + 4x + 4) - 4 + 5 \quad \text{Regrouping} \\ &= (x + 2)^2 + 1 \end{aligned}$$

The vertex is $(-2, 1)$, the axis of symmetry is $x = -2$, and the graph opens upward since the coefficient 1 is positive. We plot a few points as a check and draw the curve.

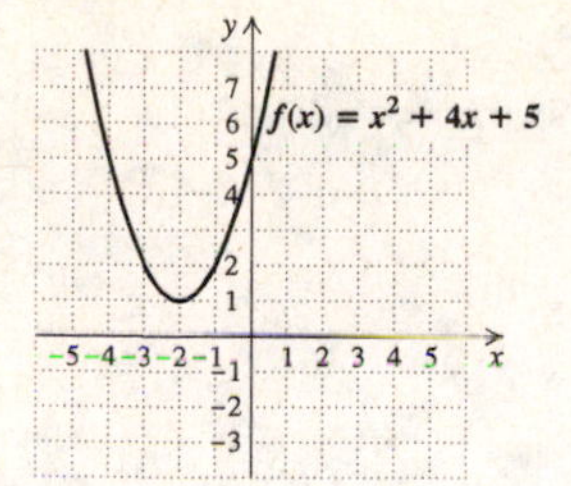

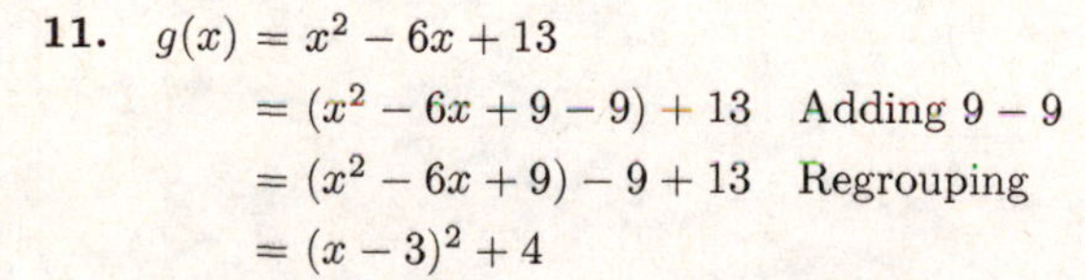

11. $g(x) = x^2 - 6x + 13$

$= (x^2 - 6x + 9 - 9) + 13$ Adding $9 - 9$

$= (x^2 - 6x + 9) - 9 + 13$ Regrouping

$= (x - 3)^2 + 4$

The vertex is $(3, 4)$, the axis of symmetry is $x = 3$, and the graph opens upward since the coefficient 1 is positive. We plot a few points as a check and draw the curve.

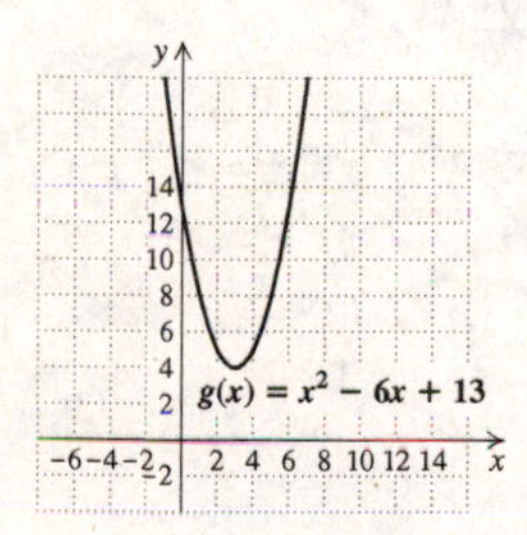

13. $f(x) = x^2 + 8x + 20$

$= (x^2 + 8x + 16 - 16) + 20$ Adding $16 - 16$

$= (x^2 + 8x + 16) - 16 + 20$ Regrouping

$= (x + 4)^2 + 4$

The vertex is $(-4, 4)$, the axis of symmetry is $x = -4$, and the graph opens upward since the coefficient 1 is positive.

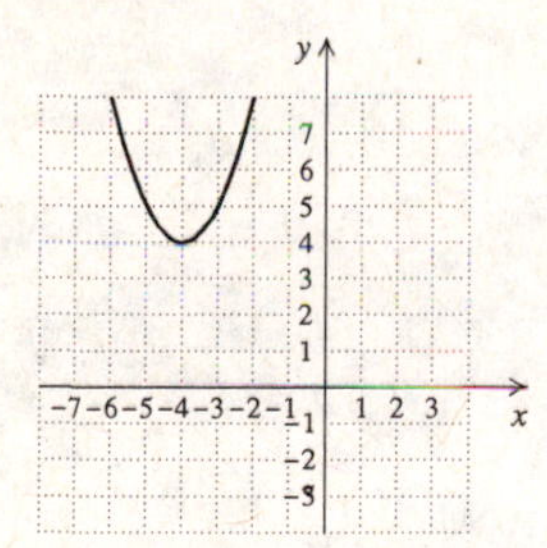

15. $h(x) = 2x^2 - 16x + 25$

$= 2(x^2 - 8x) + 25$ Factoring 2 from the first two terms

$= 2(x^2 - 8x + 16 - 16) + 25$ Adding 16–16 inside the parentheses

$= 2(x^2 - 8x + 16) + 2(-16) + 25$ Distributing to obtain a trinomial square

$= 2(x - 4)^2 - 7$

The vertex is $(4, -7)$, the axis of symmetry is $x = 4$, and the graph opens upward since the coefficient 2 is positive.

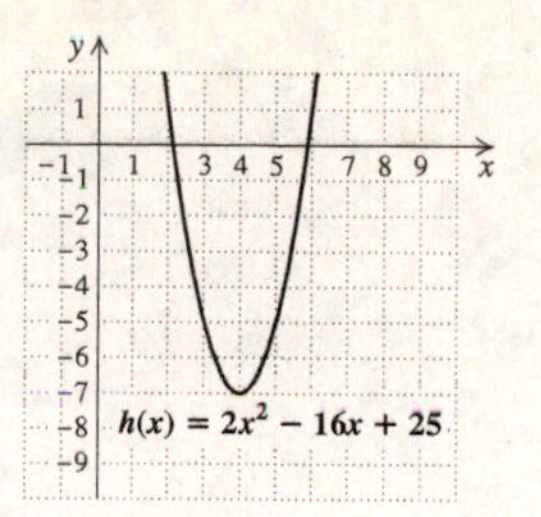

17. $f(x) = -x^2 + 2x + 5$

$= -(x^2 - 2x) + 5$ Factoring -1 from the first two terms

$= -(x^2 - 2x + 1 - 1) + 5$ Adding $1 - 1$ inside the parentheses

$= -(x^2 - 2x + 1) - (-1) + 5$

$= -(x - 1)^2 + 6$

The vertex is $(1, 6)$, the axis of symmetry is $x = 1$, and the graph opens downward since the coefficient -1 is negative.

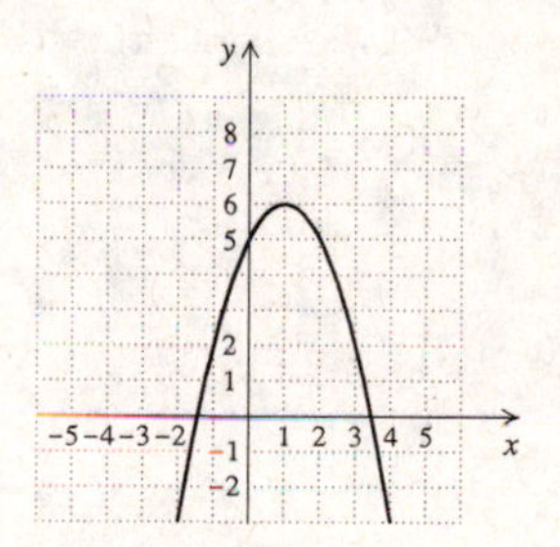

19. $g(x) = x^2 + 3x - 10$

$= \left(x^2 + 3x + \frac{9}{4} - \frac{9}{4}\right) - 10$

$= \left(x^2 + 3x + \frac{9}{4}\right) - \frac{9}{4} - 10$

$= \left(x + \frac{3}{2}\right)^2 - \frac{49}{4}$

The vertex is $\left(-\frac{3}{2}, -\frac{49}{4}\right)$, the axis of symmetry is $x = -\frac{3}{2}$, and the graph opens upward since the coefficient 1 is positive.

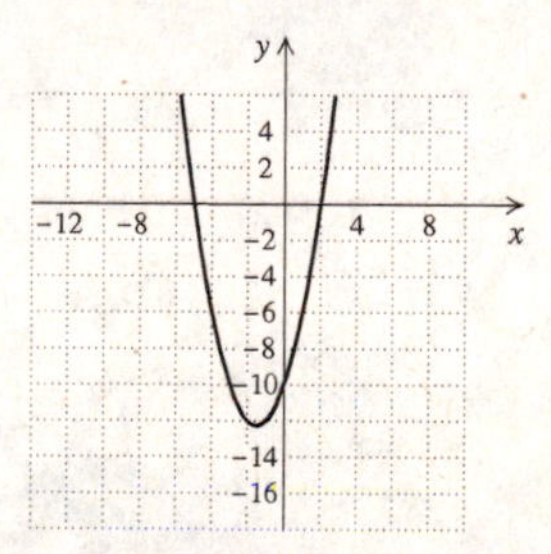

21. $f(x) = 3x^2 - 24x + 50$

$= 3(x^2 - 8x) + 50$ Factoring

$= 3(x^2 - 8x + 16 - 16) + 50$ Adding $16 - 16$ inside the parentheses

$= 3(x^2 - 8x + 16) - 3 \cdot 16 + 50$

$= 3(x - 4)^2 + 2$

The vertex is $(4, 2)$, the axis of symmetry is $x = 4$, and the graph opens upward since the coefficient 3 is positive.

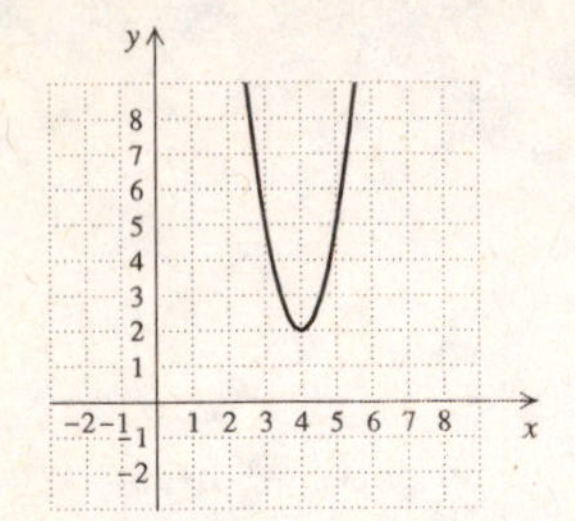

23. 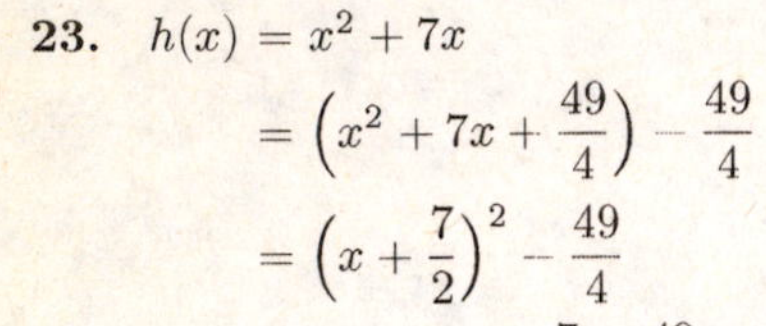

$$\begin{aligned} h(x) &= x^2 + 7x \\ &= \left(x^2 + 7x + \frac{49}{4}\right) - \frac{49}{4} \\ &= \left(x + \frac{7}{2}\right)^2 - \frac{49}{4} \end{aligned}$$

The vertex is $\left(-\frac{7}{2}, -\frac{49}{4}\right)$, the axis of symmetry is $x = -\frac{7}{2}$, and the graph opens upward since the coefficient 1 is positive.

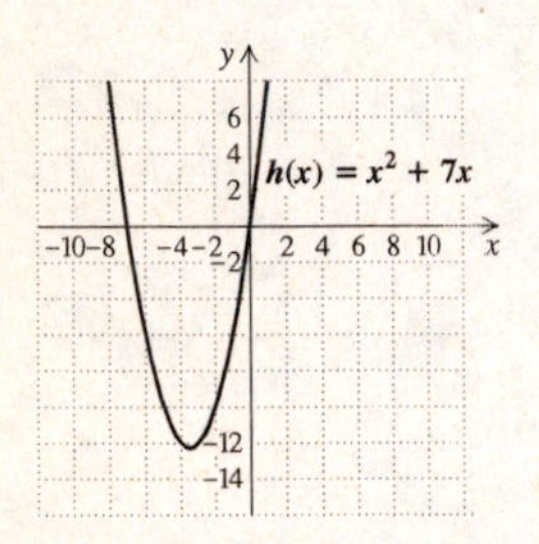

25.

$$\begin{aligned} f(x) &= -2x^2 - 4x - 6 \\ &= -2(x^2 + 2x) - 6 \quad \text{Factoring} \\ &= -2(x^2 + 2x + 1 - 1) - 6 \\ &\qquad \text{Adding } 1 - 1 \text{ inside the parentheses} \\ &= -2(x^2 + 2x + 1) - 2(-1) - 6 \\ &= -2(x + 1)^2 - 4 \end{aligned}$$

The vertex is $(-1, -4)$, the axis of symmetry is $x = -1$, and the graph opens downward since the coefficient -2 is negative.

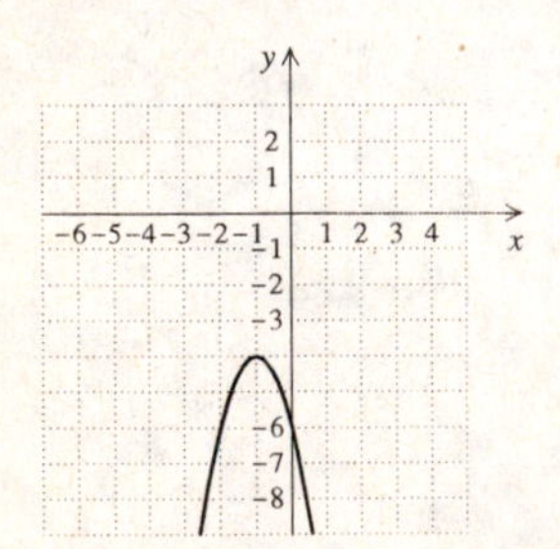

27.

$$\begin{aligned} g(x) &= 2x^2 - 8x + 3 \\ &= 2(x^2 - 4x) + 3 \quad \text{Factoring} \\ &= 2(x^2 - 4x + 4 - 4) + 3 \\ &\qquad \text{Adding } 4 - 4 \text{ inside the parentheses} \\ &= 2(x^2 - 4x + 4) + 2(-4) + 3 \\ &= 2(x - 2)^2 - 5 \end{aligned}$$

The vertex is $(2, -5)$, the axis of symmetry is $x = 2$, and the graph opens upward since the coefficient 2 is positive.

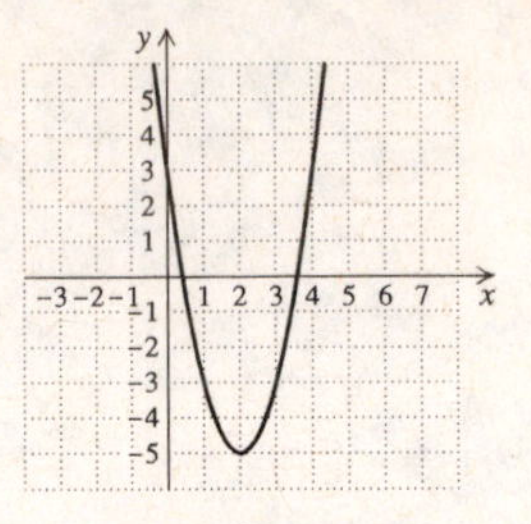

$g(x) = 2x^2 - 8x + 3$

29.

$$\begin{aligned} f(x) &= -3x^2 + 5x - 2 \\ &= -3\left(x^2 - \frac{5}{3}x\right) - 2 \quad \text{Factoring} \\ &= -3\left(x^2 - \frac{5}{3}x + \frac{25}{36} - \frac{25}{36}\right) - 2 \\ &\qquad \text{Adding } \frac{25}{36} - \frac{25}{36} \text{ inside the parentheses} \\ &= -3\left(x^2 - \frac{5}{3}x + \frac{25}{36}\right) - 3\left(-\frac{25}{36}\right) - 2 \\ &= -3\left(x - \frac{5}{6}\right)^2 + \frac{1}{12} \end{aligned}$$

The vertex is $\left(\frac{5}{6}, \frac{1}{12}\right)$, the axis of symmetry is $x = \frac{5}{6}$, and the graph opens downward since the coefficient -3 is negative.

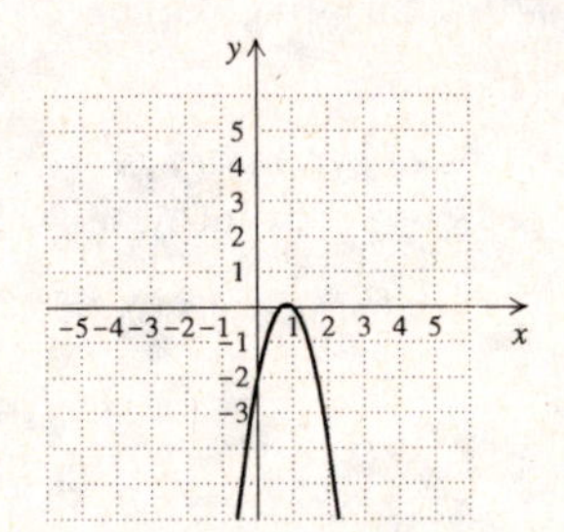

31.

$$\begin{aligned} h(x) &= \frac{1}{2}x^2 + 4x + \frac{19}{3} \\ &= \frac{1}{2}(x^2 + 8x) + \frac{19}{3} \quad \text{Factoring} \\ &= \frac{1}{2}(x^2 + 8x + 16 - 16) + \frac{19}{3} \\ &\qquad \text{Adding } 16 - 16 \text{ inside the parentheses} \\ &= \frac{1}{2}(x^2 + 8x + 16) + \frac{1}{2}(-16) + \frac{19}{3} \\ &= \frac{1}{2}(x + 4)^2 - \frac{5}{3} \end{aligned}$$

The vertex is $\left(-4, -\frac{5}{3}\right)$, the axis of symmetry is $x = -4$, and the graph opens upward since the coefficient $\frac{1}{2}$ is positive.

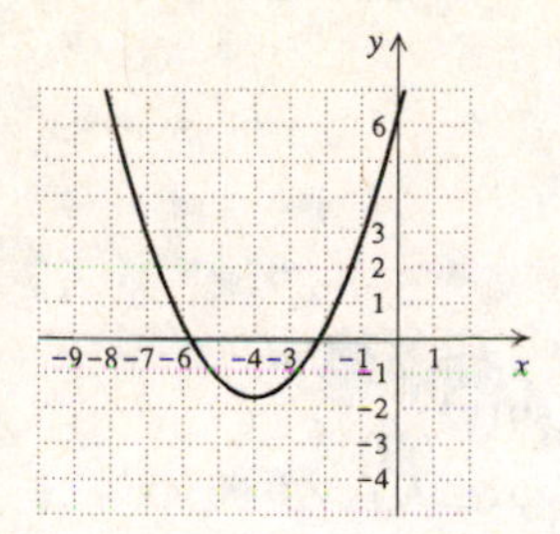

33. $f(x) = x^2 - 6x + 3$

To find the x-intercepts, solve the equation $0 = x^2 - 6x + 3$. Use the quadratic formula.

$$x = \frac{-(-6) \pm \sqrt{(-6)^2 - 4 \cdot 1 \cdot 3}}{2 \cdot 1}$$

$$x = \frac{6 \pm \sqrt{24}}{2} = \frac{6 \pm 2\sqrt{6}}{2} = 3 \pm \sqrt{6}$$

The x-intercepts are $(3 - \sqrt{6}, 0)$ and $(3 + \sqrt{6}, 0)$.

The y-intercept is $(0, f(0))$, or $(0, 3)$.

35. $g(x) = -x^2 + 2x + 3$

To find the x-intercepts, solve the equation $0 = -x^2 + 2x + 3$. We factor.

$0 = -x^2 + 2x + 3$

$0 = x^2 - 2x - 3$ Multiplying by -1

$0 = (x - 3)(x + 1)$

$x = 3$ *or* $x = -1$

The x-intercepts are $(-1, 0)$ and $(3, 0)$.

The y-intercept is $(0, g(0))$, or $(0, 3)$.

37. $f(x) = x^2 - 9x$

To find the x-intercepts, solve the equation $0 = x^2 - 9x$. We factor.

$0 = x^2 - 9x$

$0 = x(x - 9)$

$x = 0$ *or* $x = 9$

The x-intercepts are $(0, 0)$ and $(9, 0)$.

Since $(0, 0)$ is an x-intercept, we observe that $(0, 0)$ is also the y-intercept.

39. $h(x) = -x^2 + 4x - 4$

To find the x-intercepts, solve the equation $0 = -x^2 + 4x - 4$. We factor.

$0 = -x^2 + 4x - 4$

$0 = x^2 - 4x + 4$ Multiplying by -1

$0 = (x - 2)(x - 2)$

$x = 2$ or $x = 2$

The x-intercept is $(2, 0)$.

The y-intercept is $(0, h(0))$, or $(0, -4)$.

41. $f(x) = 2x^2 - 4x + 6$

To find the x-intercepts, solve the equation $0 = 2x^2 - 4x + 6$. We use the quadratic formula.

$$x = \frac{-(-4) \pm \sqrt{(-4)^2 - 4 \cdot 2 \cdot 6}}{2 \cdot 2}$$

$$x = \frac{4 \pm \sqrt{-32}}{4} = \frac{4 \pm 4i\sqrt{2}}{2} = 2 \pm 2i\sqrt{2}$$

There are no real-number solutions, so there is no x-intercept.

The y-intercept is $(0, f(0))$, or $(0, 6)$.

43. *Writing Exercise*

45. $5x - 3y = 16$, (1)

$4x + 2y = 4$ (2)

Multiply equation (1) by 2 and equation (2) by 3 and add.

$$\begin{array}{r} 10x - 6y = 32 \\ 12x + 6y = 12 \\ \hline 22x \qquad = 44 \\ x = 2 \end{array}$$

Substitute 2 for x in one of the original equations and solve for y.

$4x + 2y = 4$ (1)

$4 \cdot 2 + 2y = 4$

$8 + 2y = 4$

$2y = -4$

$y = -2$

The solution is $(2, -2)$.

47. $4a - 5b + c = 3$, (1)

$3a - 4b + 2c = 3$, (2)

$a + b - 7c = -2$ (3)

First multiply equation (1) by -2 and add it to equation (2).

$$\begin{array}{r} -8a + 10b - 2c = -6 \\ 3a - 4b + 2c = 3 \\ \hline -5a + 6b \qquad = -3 \quad (4) \end{array}$$

Next multiply equation (1) by 7 and add it to equation (3).

$$\begin{array}{r} 28a - 35b + 7c = 21 \\ a + b - 7c = -2 \\ \hline 29a - 34b \qquad = 19 \quad (5) \end{array}$$

Now we solve the system of equations (4) and (5). Multiply equation (4) by 29 and equation (5) by 5 and add.

$$\begin{array}{r} -145a + 174b = -87 \\ 145a - 170b = 95 \\ \hline 4b = 8 \\ b = 2 \end{array}$$

Substitute 2 for b in equation (4) and solve for a.

$$-5a + 6 \cdot 2 = -3$$
$$-5a + 12 = -3$$
$$-5a = -15$$
$$a = 3$$

Now substitute 3 for a and 2 for b in equation (1) and solve for c.

$$4 \cdot 3 - 5 \cdot 2 + c = 3$$
$$12 - 10 + c = 3$$
$$2 + c = 3$$
$$c = 1$$

The solution is $(3, 2, 1)$.

49.
$$\sqrt{4x-4} = \sqrt{x+4} + 1$$
$$4x - 4 = x + 4 + 2\sqrt{x+4} + 1 \quad \text{Squaring both sides}$$
$$3x - 9 = 2\sqrt{x+4}$$
$$9x^2 - 54x + 81 = 4(x+4) \quad \text{Squaring both sides again}$$
$$9x^2 - 54x + 81 = 4x + 16$$
$$9x^2 - 58x + 65 = 0$$
$$(9x - 13)(x - 5) = 0$$
$$x = \frac{13}{9} \quad \text{or} \quad x = 5$$

Check: For $x = \frac{13}{9}$:

$\sqrt{4x-4}$	$= \sqrt{x+4} + 1$
$\sqrt{4\left(\frac{13}{9}\right) - 4}$	$\sqrt{\frac{13}{9} + 4} + 1$
$\sqrt{\frac{16}{9}}$	$\sqrt{\frac{49}{9}} + 1$
$\frac{4}{3}$	$\frac{7}{3} + 1$

$$\frac{4}{3} \stackrel{?}{=} \frac{10}{3} \quad \text{FALSE}$$

For $x = 5$:

$\sqrt{4x-4}$	$= \sqrt{x+4} + 1$
$\sqrt{4 \cdot 5 - 4}$	$\sqrt{5+4} + 1$
$\sqrt{16}$	$\sqrt{9} + 1$
4	$3 + 1$

$$4 \stackrel{?}{=} 4 \quad \text{TRUE}$$

5 checks, but $\frac{13}{9}$ does not. The solution is 5.

51. *Writing Exercise*

53. a)
$$f(x) = 2.31x^2 - 3.135x - 5.89$$
$$= 2.31(x^2 - 1.357142857x) - 5.89$$
$$= 2.31(x^2 - 1.357142857x + 0.460459183 - 0.460459183) - 5.89$$
$$= 2.31(x^2 - 1.357142857x + 0.460459183) + 2.31(-0.460459183) - 5.89$$
$$= 2.31(x - 0.678571428)^2 - 6.953660714$$

Since the coefficient 2.31 is positive, the function has a minimum value. It is -6.953660714.

b) To find the x-intercepts, solve $0 = 2.31x^2 - 3.135x - 5.89$.

$$x = \frac{-(-3.135) \pm \sqrt{(-3.135)^2 - 4(2.31)(-5.89)}}{2(2.31)}$$
$$x \approx \frac{3.135 \pm 8.015723611}{4.62}$$
$$x \approx -1.056433682 \quad \text{or} \quad x \approx 2.413576539$$

The x-intercepts are $(-1.056433682, 0)$ and $(2.413576539, 0)$.

The y-intercept is $(0, f(0))$, or $(0, -5.89)$.

55. $f(x) = x^2 - x - 6$

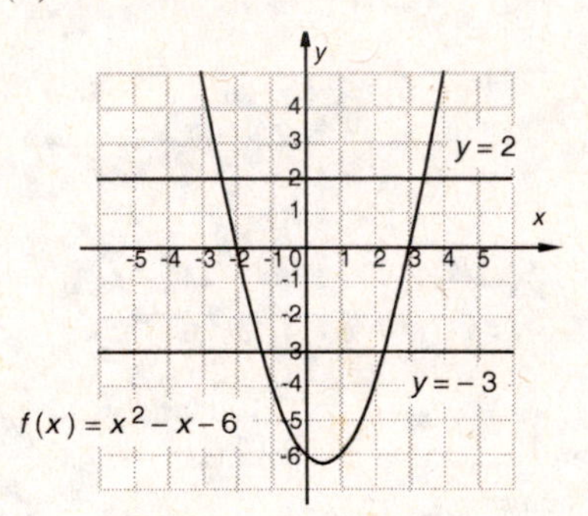

a) The solutions of $x^2 - x - 6 = 2$ are the first coordinates of the points of intersection of the graphs of $f(x) = x^2 - x - 6$ and $y = 2$. From the graph we see that the solutions are approximately -2.4 and 3.4.

b) The solutions of $x^2 - x - 6 = -3$ are the first coordinates of the points of intersection of the graphs of $f(x) = x^2 - x - 6$ and $y = -3$. From the graph we see that the solutions are approximately -1.3 and 2.3.

57.
$$f(x) = mx^2 - nx + p$$
$$= m\left(x^2 - \frac{n}{m}x\right) + p$$
$$= m\left(x^2 - \frac{n}{m}x + \frac{n^2}{4m^2} - \frac{n^2}{4m^2}\right) + p$$
$$= m\left(x - \frac{n}{2m}\right)^2 - \frac{n^2}{4m} + p$$
$$= m\left(x - \frac{n}{2m}\right)^2 + \frac{-n^2 + 4mp}{4m}, \text{ or}$$
$$m\left(x - \frac{n}{2m}\right)^2 + \frac{4mp - n^2}{4m}$$

59. The horizontal distance from $(-1,0)$ to $(3,-5)$ is $|3-(-1)|$, or 4, so by symmetry the other x-intercept is $(3+4,0)$, or $(7,0)$. Substituting the three ordered pairs $(-1,0)$, $(3,-5)$, and $(7,0)$ in the equation $f(x) = ax^2 + bx + c$ yields a system of equations:

$$0 = a - b + c,$$
$$-5 = 9a + 3b + c,$$
$$0 = 49a + 7b + c$$

The solution of this system of equations is

$\left(\frac{5}{16}, -\frac{15}{8}, -\frac{35}{16}\right)$, so $f(x) = \frac{5}{16}x^2 - \frac{15}{8}x - \frac{35}{16}$.

If we complete the square we find that this function can also be expressed as $f(x) = \frac{5}{16}(x-3)^2 - 5$.

61. $f(x) = |x^2 - 1|$

We plot some points and draw the curve. Note that it will lie entirely on or above the x-axis since absolute value is never negative.

x	$f(x)$
-3	8
-2	3
-1	0
0	1
1	0
2	3
3	8

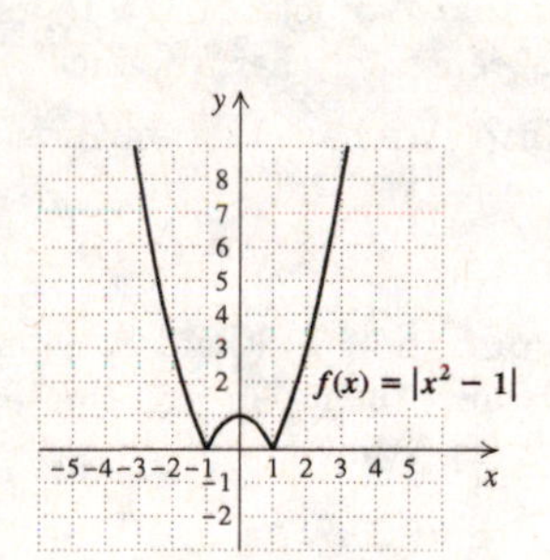

63. $f(x) = |2(x-3)^2 - 5|$

We plot some points and draw the curve. Note that it will lie entirely on or above the x-axis since absolute value is never negative.

x	$f(x)$
-1	27
0	13
1	3
2	3
3	5
4	3
5	3
6	13

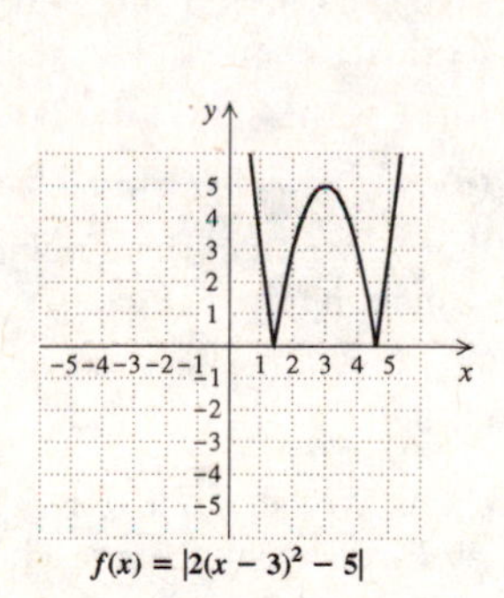

Exercise Set 11.8

1. (e)

3. (c)

5. (d)

7. ***Familiarize and Translate.*** We are given the function $N(x) = -0.4x^2 + 9x + 11$.

Carry out. To find the value of x for which $N(x)$ is a maximum, we first find $-\frac{b}{2a}$:

$$-\frac{b}{2a} = -\frac{9}{2(-0.4)} = 11.25$$

Now we find the maximum value of the function $N(11.25)$:

$N(11.25) = -0.4(11.25)^2 + 9(11.25) + 11 = 61.625$

Check. We can go over the calculations again. We could also solve the problem again by completing the square. The answer checks.

State. Daily ticket sales will peak 11 days after the concert is announced. About 62 tickets will be sold that day.

9. ***Familiarize and Translate.*** We want to find the value of x for which $C(x) = 0.1x^2 - 0.7x + 2.425$ is a minimum.

Carry out. We complete the square.

$$C(x) = 0.1(x^2 - 7x + 12.25) + 2.425 - 1.225$$
$$C(x) = 0.1(x - 3.5)^2 + 1.2$$

The minimum function value of 1.2 occurs when $x = 3.5$.

Check. Check a function value for x less than 3.5 and for x greater than 3.5.

$$C(3) = 0.1(3)^2 - 0.7(3) + 2.425 = 1.225$$
$$C(4) = 0.1(4)^2 - 0.7(4) + 2.425 = 1.225$$

Since 1.2 is less than these numbers, it looks as though we have a minimum.

State. The minimum average cost is \$1.2 hundred, or \$120. To achieve the minimum cost, 3.5 hundred, or 350 Dobros should be built.

11. ***Familiarize.*** We make a drawing and label it.

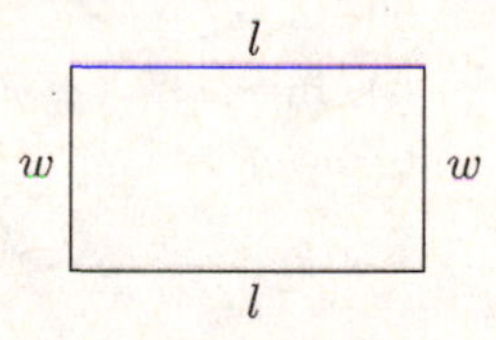

Perimeter: $2l + 2w = 128$ in.

Area: $A = l \cdot w$

Translate. We have a system of equations.

$$2l + 2w = 128,$$
$$A = lw$$

Carry out. Solving the first equation for l, we get $l = 64 - w$. Substituting for l in the second equation we get a quadratic function A:

$$A = (64 - w)w$$
$$A = -w^2 + 64w$$

Completing the square, we get

$$A = -(w - 32)^2 + 1024.$$

The maximum function value is 1024. It occurs when w is 32. When $w = 32$, $l = 64 - 32$, or 32.

Check. We check a function value for w less than 32 and for w greater than 32.

$$A(31) = -31^2 + 64 \cdot 31 = 1023$$
$$A(33) = -33^2 + 64 \cdot 33 = 1023$$

Since 1024 is greater than these numbers, it looks as though we have a maximum.

State. The maximum area occurs when the dimensions are 32 in. by 32 in.

13. ***Familiarize***. We make a drawing and label it.

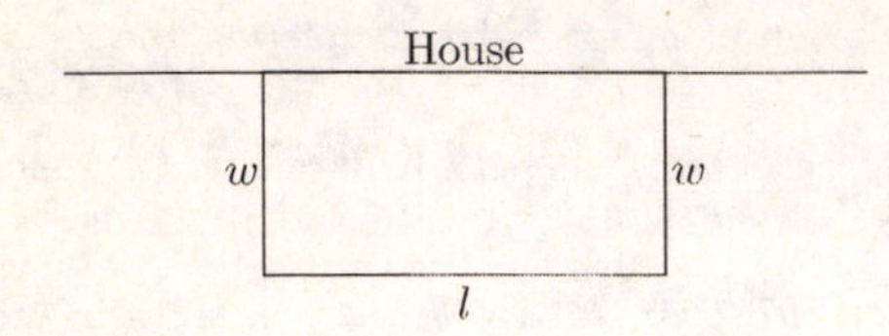

Translate. We have two equations.

$$l + 2w = 60,$$
$$A = lw$$

Carry out. Solve the first equation for l.

$$l = 60 - 2w$$

Substitute for l in the second equation.

$$A = (60 - 2w)w$$
$$A = -2w^2 + 60w$$

Completing the square, we get

$$A = -2(w - 15)^2 + 450.$$

The maximum function value of 450 occurs when $w = 15$. When $w = 15$, $l = 60 - 2 \cdot 15 = 30$.

Check. Check a function value for w less than 15 and for w greater than 15.

$$A(14) = -2 \cdot 14^2 + 60 \cdot 14 = 448$$
$$A(16) = -2 \cdot 16^2 + 60 \cdot 16 = 448$$

Since 450 is greater than these numbers, it looks as though we have a maximum.

State. The maximum area of 450 ft^2 will occur when the dimensions are 15 ft by 30 ft.

15. ***Familiarize***. Let x represent the height of the file and y represent the width. We make a drawing.

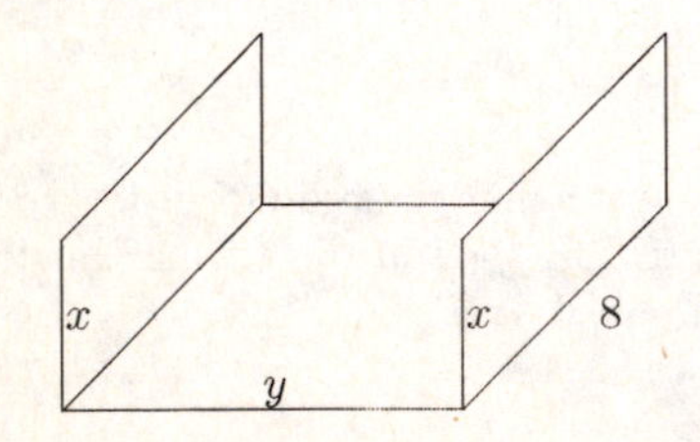

Translate. We have two equations.

$$2x + y = 14$$
$$V = 8xy$$

Carry out. Solve the first equation for y.

$$y = 14 - 2x$$

Substitute for y in the second equation.

$$V = 8x(14 - 2x)$$
$$V = -16x^2 + 112x$$

Completing the square, we get

$$V = -16\left(x - \frac{7}{2}\right)^2 + 196.$$

The maximum function value of 196 occurs when $x = \frac{7}{2}$. When $x = \frac{7}{2}$, $y = 14 - 2 \cdot \frac{7}{2} = 7$.

Check. Check a function value for x less than $\frac{7}{2}$ and for x greater than $\frac{7}{2}$.

$$V(3) = -16 \cdot 3^2 + 112 \cdot 3 = 192$$
$$V(4) = -16 \cdot 4^2 + 112 \cdot 4 = 192$$

Since 196 is greater than these numbers, it looks as though we have a maximum.

State. The file should be $\frac{7}{2}$ in., or 3.5 in., tall.

17. ***Familiarize***. We let x and y represent the numbers, and we let P represent their product.

Translate. We have two equations.

$$x + y = 18,$$
$$P = xy$$

Carry out. Solving the first equation for y, we get $y = 18 - x$. Substituting for y in the second equation we get a quadratic function P:

$$P = x(18 - x)$$
$$P = -x^2 + 18x$$

Completing the square, we get

$$P = -(x - 9)^2 + 81.$$

The maximum function value is 81. It occurs when $x = 9$. When $x = 9$, $y = 18 - 9$, or 9.

Check. We can check a function value for x less than 9 and for x greater than 9.

$$P(10) = -10^2 + 18 \cdot 10 = 80$$
$$P(8) = -8^2 + 18 \cdot 8 = 80$$

Since 81 is greater than these numbers, it looks as though we have a maximum.

State. The maximum product of 81 occurs for the numbers 9 and 9.

19. ***Familiarize***. We let x and y represent the two numbers, and we let P represent their product.

Translate. We have two equations.

$$x - y = 8,$$
$$P = xy$$

Carry out. Solve the first equation for x.

$$x = 8 + y$$

Substitute for x in the second equation.

$$P = (8 + y)y$$
$$P = y^2 + 8y$$

Completing the square, we get

$$P = (y + 4)^2 - 16.$$

The minimum function value is -16. It occurs when $y = -4$. When $y = -4$, $x = 8 + (-4)$, or 4.

Check. Check a function value for y less than -4 and for y greater than -4.

$$P(-5) = (-5)^2 + 8(-5) = -15$$
$$P(-3) = (-3)^2 + 8(-3) = -15$$

Since -16 is less than these numbers, it looks as though we have a minimum.

State. The minimum product of -16 occurs for the numbers 4 and -4.

21. From the results of Exercises 17 and 18, we might observe that the numbers are -5 and -5 and that the maximum product is 25. We could also solve this problem as follows.

Familiarize. We let x and y represent the two numbers, and we let P represent their product.

Translate. We have two equations.

$$x + y = -10,$$
$$P = xy$$

Carry out. Solve the first equation for y.

$$y = -10 - x$$

Substitute for y in the second equation.

$$P = x(-10 - x)$$
$$P = -x^2 - 10x$$

Completing the square, we get

$$P = -(x + 5)^2 + 25$$

The maximum function value is 25. It occurs when $x = -5$. When $x = -5$, $y = -10 - (-5)$, or -5.

Check. Check a function value for x less than -5 and for x greater than -5.

$$P(-6) = -(-6)^2 - 10(-6) = 24$$
$$P(-4) = -(-4)^2 - 10(-4) = 24$$

Since 25 is greater than these numbers, it looks as though we have a maximum.

State. The maximum product of 25 occurs for the numbers -5 and -5.

23. The data points rise. The graph does not appear to represent a quadratic function in which the data points would rise and then fall or vice versa. Thus a linear function $f(x) = mx + b$ might be used to model the data.

25. The data points rise and then fall. The graph appears to represent a quadratic function that opens downward. Thus a quadratic function $f(x) = ax^2 + bx + c$, $a < 0$, might be used to model the data.

27. The data points fall and then rise. The graph appears to represent a quadratic function that opens upward. Thus a quadratic function $f(x) = ax^2 + bx + c$, $a > 0$, might be used to model the data.

29. The data points rise and then fall. The graph appears to represent a quadratic function that opens downward. Thus a quadratic function $f(x) = ax^2 + bx + c$, $a < 0$, might be used to model the data.

31. The data points appear to represent the right half of a quadratic function that opens upward. Thus a quadratic function $f(x) = ax^2 + bx + c$, $a > 0$, might be used to model the data.

33. The data points rise. The graph does not appear to represent a quadratic function in which the data points would rise and then fall or vice versa. Thus a linear function $f(x) = mx + b$ might be used to model the data.

35. We look for a function of the form $f(x) = ax^2 + bx + c$. Substituting the data points, we get

$$4 = a(1)^2 + b(1) + c,$$
$$-2 = a(-1)^2 + b(-1) + c,$$
$$13 = a(2)^2 + b(2) + c,$$

or

$$4 = a + b + c,$$
$$-2 = a - b + c,$$
$$13 = 4a + 2b + c.$$

Solving this system, we get

$$a = 2,\ b = 3,\ \text{and}\ c = -1.$$

Therefore the function we are looking for is

$$f(x) = 2x^2 + 3x - 1.$$

37. We look for a function of the form $f(x) = ax^2 + bx + c$. Substituting the data points, we get

$$0 = a(2)^2 + b(2) + c,$$
$$3 = a(4)^2 + b(4) + c,$$
$$-5 = a(12)^2 + b(12) + c,$$

or

$$0 = 4a + 2b + c,$$
$$3 = 16a + 4b + c,$$
$$-5 = 144a + 12b + c.$$

Solving this system, we get

$$a = -\frac{1}{4},\ b = 3,\ c = -5.$$

Therefore the function we are looking for is

$$f(x) = -\frac{1}{4}x^2 + 3x - 5.$$

39. a) ***Familiarize.*** We look for a function of the form $A(s) = as^2 + bs + c$, where $A(s)$ represents the number of nighttime accidents (for every 200 million km) and s represents the travel speed (in km/h).

Translate. We substitute the given values of s and $A(s)$.

$$400 = a(60)^2 + b(60) + c,$$
$$250 = a(80)^2 + b(80) + c,$$
$$250 = a(100)^2 + b(100) + c,$$

or

$$400 = 3600a + 60b + c,$$
$$250 = 6400a + 80b + c,$$
$$250 = 10,000a + 100b + c.$$

Carry out. Solving the system of equations, we get

$a = \frac{3}{16}$, $b = -\frac{135}{4}$, $c = 1750$.

Check. Recheck the calculations.

State. The function

$A(s) = \frac{3}{16}s^2 - \frac{135}{4}s + 1750$ fits the data.

b) Find $A(50)$.

$$A(50) = \frac{3}{16}(50)^2 - \frac{135}{4}(50) + 1750 = 531.25$$

About 531 accidents occur at 50 km/h.

41. ***Familiarize.*** Think of a coordinate system placed on the drawing in the text with the origin at the point where the arrow is released. Then three points on the arrow's parabolic path are $(0,0)$, $(63,27)$, and $(126,0)$. We look for a function of the form $h(d) = ad^2 + bd + c$, where $h(d)$ represents the arrow's height and d represents the distance the arrow has traveled horizontally.

Translate. We substitute the values given above for d and $h(d)$.

$$0 = a \cdot 0^2 + b \cdot 0 + c,$$
$$27 = a \cdot 63^2 + b \cdot 63 + c,$$
$$0 = a \cdot 126^2 + b \cdot 126 + c$$

or

$$0 = c,$$
$$27 = 3969a + 63b + c,$$
$$0 = 15,876a + 126b + c$$

Carry out. Solving the system of equations, we get

$a \approx -0.0068$, $b \approx 0.8571$, and $c = 0$.

Check. Recheck the calculations.

State. The function $h(d) = -0.0068d^2 + 0.8571d$ expresses the arrow's height as a function of the distance it has traveled horizontally.

43. *Writing Exercise*

45.
$$\frac{x}{x^2+17x+72} - \frac{8}{x^2+15x+56}$$
$$= \frac{x}{(x+8)(x+9)} - \frac{8}{(x+8)(x+7)}$$
$$= \frac{x}{(x+8)(x+9)} \cdot \frac{x+7}{x+7} - \frac{8}{(x+8)(x+7)} \cdot \frac{x+9}{x+9}$$
$$= \frac{x(x+7) - 8(x+9)}{(x+8)(x+9)(x+7)}$$
$$= \frac{x^2+7x-8x-72}{(x+8)(x+9)(x+7)}$$
$$= \frac{x^2-x-72}{(x+8)(x+9)(x+7)} = \frac{(x-9)\cancel{(x+8)}}{\cancel{(x+8)}(x+9)(x+7)}$$
$$= \frac{x-9}{(x+9)(x+7)}$$

47.
$$\frac{t^2-4}{t^2-7t-8} \cdot \frac{t^2-64}{t^2-5t+6} = \frac{(t^2-4)(t^2-64)}{(t^2-7t-8)(t^2-5t+6)}$$
$$= \frac{(t+2)(t-2)(t+8)(t-8)}{(t+1)(t-8)(t-2)(t-3)}$$
$$= \frac{(t+2)\cancel{(t-2)}(t+8)\cancel{(t-8)}}{(t+1)\cancel{(t-8)}\cancel{(t-2)}(t-3)}$$
$$= \frac{(t+2)(t+8)}{(t+1)(t-3)}$$

49.
$$5x - 9 < 31$$
$$5x < 40$$
$$x < 8$$

The solutions set is $\{x|x < 8\}$, or $(-\infty, 8)$.

51. *Writing Exercise*

53. ***Familiarize.*** Position the bridge on a coordinate system as shown with the vertex of the parabola at $(0,30)$.

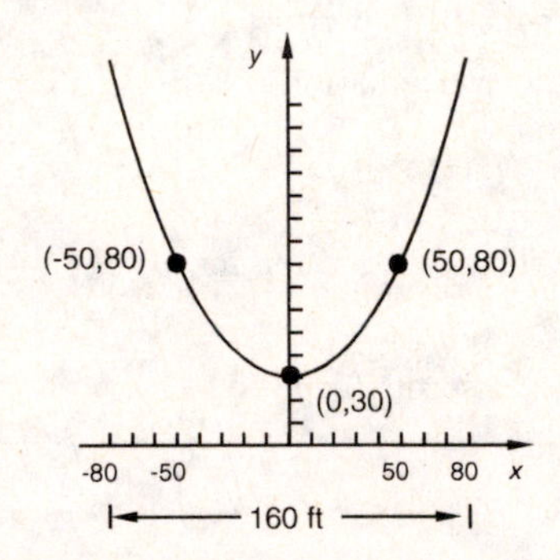

We find a function of the form $y = ax^2+bx+c$ which represents the parabola containing the points $(0,30)$, $(-50,80)$, and $(50,80)$.

Translate. Substitute for x and y.

$$30 = a \cdot 0^2 + b \cdot 0 + c,$$
$$80 = a(-50)^2 + b(-50) + c,$$
$$80 = a(50)^2 + b(50) + c,$$

or

$$30 = c,$$
$$80 = 2500a - 50b + c,$$
$$80 = 2500a + 50b + c.$$

Carry out. Solving the system of equations, we get $a = 0.02$, $b = 0$, $c = 30$.

The function $y = 0.02x^2 + 30$ represents the parabola.

Because the cable supports are 160 ft apart, the tallest supports are positioned 160/2, or 80 ft, to the left and right of the midpoint. This means that the longest vertical cables occur at $x = -80$ and $x = 80$. For $x = \pm 80$,

$$y = 0.02(\pm 80)^2 + 30$$
$$= 128 + 30$$
$$= 158 \text{ ft}$$

Check. We go over the calculations.

State. The longest vertical cables are 158 ft long.

55. ***Familiarize.*** Let x represent the number of 25¢ increases in the admission price. Then $10 + 0.25x$ represents the admission price, and $80 - x$ represents the corresponding average attendance. Let R represent the total revenue.

Translate. Since the total revenue is the product of the cover charge and the number attending a show, we have the following function for the amount of money the owner makes.

$R(x) = (10 + 0.25x)(80 - x)$, or

$R(x) = -0.25x^2 + 10x + 800$

Carry out. Completing the square, we get

$R(x) = -0.25(x - 20)^2 + 900$

The maximum function value of 900 occurs when $x = 20$. The owner should charge $\$10 + \$0.25(20)$, or \$15.

Check. We check a function value for x less than 20 and for x greater than 20.

$R(19) = -0.25(19)^2 + 10 \cdot 19 + 800 = 899.75$

$R(21) = -0.25(21)^2 + 10 \cdot 21 + 800 = 899.75$

Since 900 is greater than these numbers, it looks as though we have a maximum.

State. The owner should charge \$15.

57. ***Familiarize.*** We add labels to the drawing in the text.

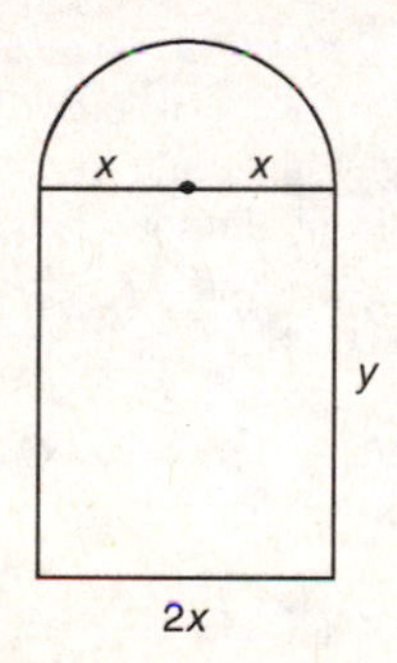

The perimeter of the semicircular portion of the window is $\frac{1}{2} \cdot 2\pi x$, or πx. The perimeter of the rectangular portion is $y + 2x + y$, or $2x + 2y$. The area of the semicircular portion of the window is $\frac{1}{2} \cdot \pi x^2$, or $\frac{\pi}{2}x^2$. The area of the rectangular portion is $2xy$.

Translate. We have two equations, one giving the perimeter of the window and the other giving the area.

$\pi x + 2x + 2y = 24,$

$A = \frac{\pi}{2}x^2 + 2xy$

Carry out. Solve the first equation for y.

$$\pi x + 2x + 2y = 24$$
$$2y = 24 - \pi x - 2x$$
$$y = 12 - \frac{\pi x}{2} - x$$

Substitute for y in the second equation.

$$A = \frac{\pi}{2}x^2 + 2x\left(12 - \frac{\pi x}{2} - x\right)$$
$$A = \frac{\pi}{2}x^2 + 24x - \pi x^2 - 2x^2$$
$$A = -2x^2 - \frac{\pi}{2}x^2 + 24x$$
$$A = -\left(2x + \frac{\pi}{2}\right)x^2 + 24x$$

Completing the square, we get

$$A = -\left(2 + \frac{\pi}{2}\right)\left(x^2 + \frac{24}{-\left(2 + \frac{\pi}{2}\right)}x\right)$$
$$A = -\left(2 + \frac{\pi}{2}\right)\left(x^2 - \frac{48}{4 + \pi}x\right)$$
$$A = -\left(2 + \frac{\pi}{2}\right)\left(x - \frac{24}{4 + \pi}\right)^2 + \left(\frac{24}{4 + \pi}\right)^2$$

The maximum function value occurs when $x = \frac{24}{4 + \pi}$. When $x = \frac{24}{4 + \pi}$,

$$y = 12 - \frac{\pi}{2}\left(\frac{24}{4 + \pi}\right) - \frac{24}{4 + \pi} =$$
$$\frac{48 + 12\pi}{4 + \pi} - \frac{12\pi}{4 + \pi} - \frac{24}{4 + \pi} = \frac{24}{4 + \pi}.$$

Check. Recheck the calculations.

State. The radius of the circular portion of the window and the height of the rectangular portion should each be $\frac{24}{4 + \pi}$ ft.

59. a) Enter the data and use the quadratic regression operation on a graphing calculator. We get $c(x) = 261.875x^2 - 882.5642857x + 2134.571429$, where x is the number of years after 1992.

b) In 2008, $x = 2008 - 1992 = 16$.

$c(16) \approx 55,053$ cars (rounding down)

Exercise Set 11.9

1. The solutions of $(x - 3)(x + 2) = 0$ are 3 and -2 and for a test value in $[-2, 3]$, say 0, $(x - 3)(x + 2)$ is negative so the statement is true. (Note that the endpoints must be included in the solution set because the inequality symbol is $\leq$.)

3. The solutions of $(x - 1)(x - 6) = 0$ are 1 and 6. For a value of x less than 1, say 0, $(x - 1)(x - 6)$ is positive; for a value of x greater than 6, say 7, $(x - 1)(x - 6)$ is also positive. Thus the statement is true. (Note that the endpoints of the intervals are not included because the inequality symbol is $>$.)

5. Since $x - 5 = 0$ when $x = 5$ and $x + 4 = 0$ when $x = -4$, the statement is true.

7. The only critical point is 5 and for a value of x in $[5, \infty)$, say 6, $\frac{3}{x - 5}$ is positive, so the statement is false.

9. $(x+4)(x-3) < 0$

The solutions of $(x+4)(x-3) = 0$ are -4 and 3. They are not solutions of the inequality, but they divide the real-number line in a natural way. The product $(x+4)(x-3)$ is positive or negative, for values other than -4 and 3, depending on the signs of the factors $x+4$ and $x-3$.

$x+4 > 0$ when $x > -4$ and $x+4 < 0$ when $x < -4$.

$x-3 > 0$ when $x > 3$ and $x-3 < 0$ when $x < 3$.

We make a diagram.

Sign of $x+4$	$-$	$+$	$+$
Sign of $x-3$	$-$	$-$	$+$
Sign of product	$+$	$-$	$+$
		-4	3

For the product $(x+4)(x-3)$ to be negative, one factor must be positive and the other negative. We see from the diagram that numbers satisfying $-4 < x < 3$ are solutions. The solution set of the inequality is $(-4, 3)$ or $\{x|-4 < x < 3\}$.

11. $(x+7)(x-2) \geq 0$

The solutions of $(x+7)(x-2) = 0$ are -7 and 2. They divide the number line into three intervals as shown:

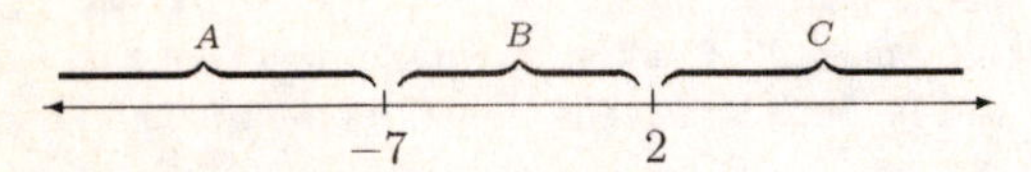

We try test numbers in each interval.

A: Test -8, $f(-8) = (-8+7)(-8-2) = 10$

B: Test 0, $f(0) = (0+7)(0-2) = -14$

C: Test 3, $f(3) = (3+7)(3-2) = 10$

Since $f(-8)$ and $f(3)$ are positive, the function value will be positive for all numbers in the intervals containing -8 and 3. The inequality symbol is $\leq$, so we need to include the endpoints. The solution set is $(-\infty, -7] \cup [2, \infty)$, or $\{x|x \leq -7 \text{ or } x \geq 2\}$.

13. $x^2 - x - 2 > 0$

$(x+1)(x-2) > 0$ Factoring

The solutions of $(x+1)(x-2) = 0$ are -1 and 2. They divide the number line into three intervals as shown:

A B C

-1 2

We try test numbers in each interval.

A: Test -2, $f(-2) = (-2+1)(-2-2) = 4$

B: Test 0, $f(0) = (0+1)(0-2) = -2$

C: Test 3, $f(3) = (3+1)(3-2) = 4$

Since $f(-2)$ and $f(3)$ are positive, the function value will be positive for all numbers in the intervals containing -2 and 3. The solution set is $(-\infty, -1) \cup (2, \infty)$, or $\{x|x < -1 \text{ or } x > 2\}$.

15. $x^2 + 4x + 4 < 0$

$(x+2)^2 < 0$

Observe that $(x+2)^2 \geq 0$ for all values of x. Thus, the solution set is $\emptyset$.

17.

$$x^2 - 4x < 12$$
$$x^2 - 4x - 12 < 0$$
$$(x-6)(x+2) < 0$$

The solutions of $(x-6)(x+2) = 0$ are 6 and -2. They are not solutions of the inequality, but they divide the real-number line in a natural way. The product $(x-6)(x+2)$ is positive or negative, for values other than 6 and -2, depending on the signs of the factors $x-6$ and $x+2$.

$x-6 > 0$ when $x > 6$ and $x-6 < 0$ when $x < 6$.

$x+2 > 0$ when $x > -2$ and $x+2 < 0$ when $x < -2$.

We make a diagram.

Sign of $x-6$	$-$	$-$	$+$
Sign of $x+2$	$-$	$+$	$+$
Sign of product	$+$	$-$	$+$
		-2	6

For the product $(x-6)(x+2)$ to be negative, one factor must be positive and the other negative. The only situation in the diagram for which this happens is when $-2 < x < 6$. The solution set of the inequality is $(-2, 6)$, or $\{x|-2 < x < 6\}$.

19. $3x(x+2)(x-2) < 0$

The solutions of $3x(x+2)(x-2) = 0$ are 0, -2, and 2. They divide the real-number line into four intervals as shown:

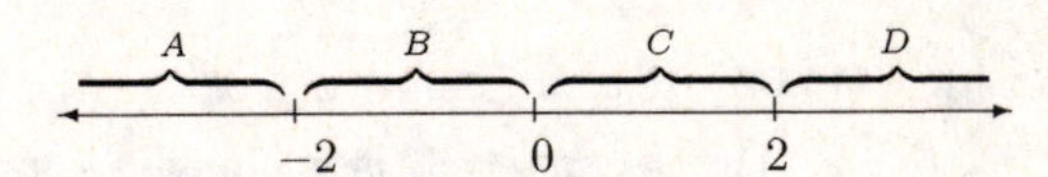

We try test numbers in each interval.

A: Test -3, $f(-3) = 3(-3)(-3+2)(-3-2) = -45$

B: Test -1, $f(-1) = 3(-1)(-1+2)(-1-2) = 9$

C: Test 1, $f(1) = 3(1)(1+2)(1-2) = -9$

D: Test 3, $f(3) = 3(3)(3+2)(3-2) = 45$

Since $f(-3)$ and $f(1)$ are negative, the function value will be negative for all numbers in the intervals containing -3 and 1. The solution set is $(-\infty, -2) \cup (0, 2)$, or $\{x|x < -2 \text{ or } 0 < x < 2\}$.

21. $(x-1)(x+2)(x-4) \geq 0$

The solutions of $(x-1)(x+2)(x-4) = 0$ are 1, -2, and 4. They divide the real-number line in a natural way. The product $(x-1)(x+2)(x-4)$ is positive or negative depending on the signs of $x-1$, $x+2$, and $x-4$.

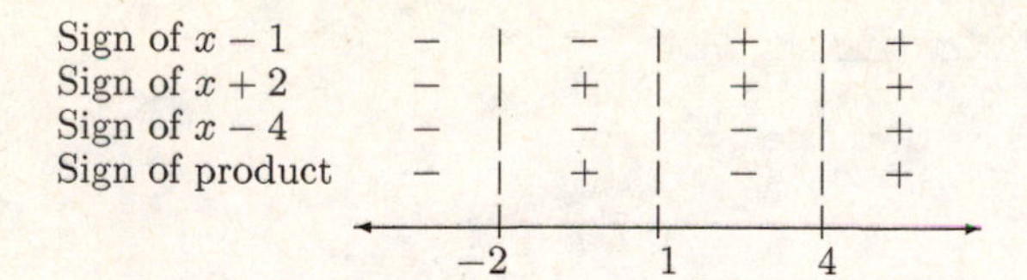

A product of three numbers is positive when all three factors are positive or when two are negative and one is positive. Since the $\geq$ symbol allows for equality, the endpoints -2, 1, and 4 are solutions. From the chart we see that the solution set is $[-2,1]\cup[4,\infty)$, or $\{x|-2\leq x\leq 1 \text{ or } x\geq 4\}$.

23.
$$f(x)\leq 3$$
$$x^2-1\leq 3$$
$$x^2-4\leq 0$$
$$(x-2)(x+2)\leq 0$$

The solutions of $(x-2)(x+2)=0$ are 2 and -2. They divide the real-number line as shown below.

Sign of $x-2$	−	−	+
Sign of $x+2$	−	+	+
Sign of product	+	−	+

-2 2

Because the inequality symbol is $\leq$, we must include the endpoints in the solution set. From the chart we see that the solution set is $[-2,2]$, or $\{x|-2\leq x\leq 2\}$.

25.
$$g(x)>0$$
$$(x-2)(x-3)(x+1)>0$$

The solutions of $(x-2)(x-3)(x+1)=0$ are 2, 3, and -1. They divide the real-number line into four intervals as shown below.

A B C D

-1 2 3

We try test numbers in each interval.

A: Test -2, $f(-2)=(-2-2)(-2-3)(-2+1)=-20$

B: Test 0, $f(0)=(0-2)(0-3)(0+1)=6$

C: Test $\frac{5}{2}$, $f\left(\frac{5}{2}\right)=\left(\frac{5}{2}-2\right)\left(\frac{5}{2}-3\right)\left(\frac{5}{2}+1\right)=-\frac{7}{8}$

D: Test 4, $f(4)=(4-2)(4-3)(4+1)=10$

The function value will be positive for all numbers in intervals B and D. The solution set is $(-1,2)\cup(3,\infty)$, or $\{x|-1<x<2 \text{ or } x>3\}$.

27.
$$F(x)\leq 0$$
$$x^3-7x^2+10x\leq 0$$
$$x(x^2-7x+10)\leq 0$$
$$x(x-2)(x-5)\leq 0$$

The solutions of $x(x-2)(x-5)=0$ are 0, 2, and 5. They divide the real-number line as shown below.

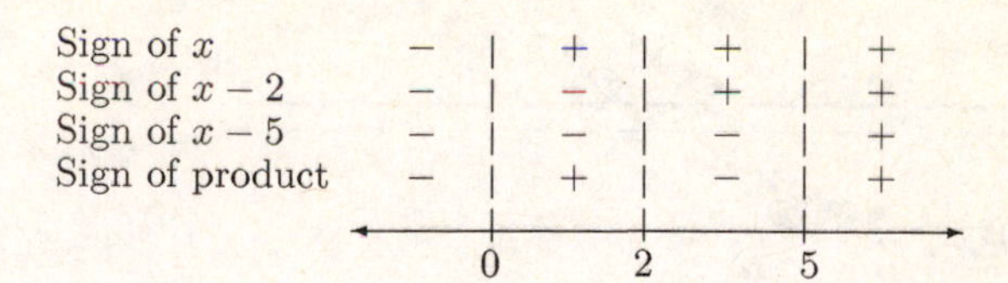

Because the inequality symbol is $\leq$ we must include the endpoints in the solution set. From the chart we see that the solution set is $(-\infty,0]\cup[2,5]$ or $\{x|x\leq 0 \text{ or } 2\leq x\leq 5\}$.

29. $\frac{1}{x+5}<0$

We write the related equation by changing the $<$ symbol to $=$:

$$\frac{1}{x+5}=0$$

We solve the related equation.

$$(x+5)\cdot\frac{1}{x+5}=(x+5)\cdot 0$$
$$1=0$$

The related equation has no solution.

Next we find the values that make the denominator 0 by setting the denominate equal to 0 and solving:

$$x+5=0$$
$$x=-5$$

We use -5 to divide the number line into two intervals as shown:

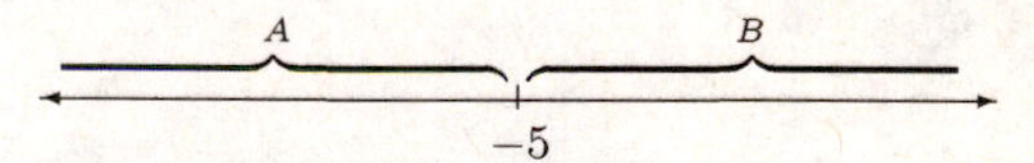

We try test numbers in each interval.

A: Test -6, $\frac{1}{-6+5}=\frac{1}{-1}=-1<0$

The number -6 is a solution of the inequality, so the interval A is part of the solution set.

B: Test 0, $\frac{1}{0+5}=\frac{1}{5}\not<0$

The number 0 is not a solution of the inequality, so the interval B is not part of the solution set. The solution set is $(-\infty,-5)$, or $\{x|x<-5\}$.

31. $\frac{x+1}{x-3}\geq 0$

Solve the related equation.

$$\frac{x+1}{x-3}=0$$
$$x+1=0$$
$$x=-1$$

Find the values that make the denominator 0.

$$x-3=0$$
$$x=3$$

Use the numbers -1 and 3 to divide the number line into intervals as shown:

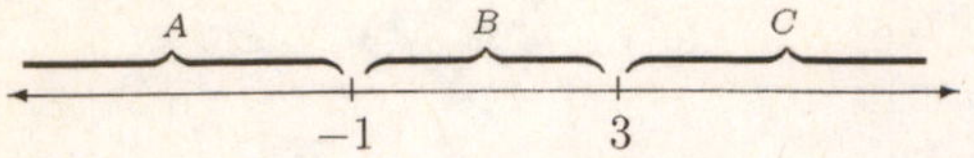

Try test numbers in each interval.

A: Test -2, $\dfrac{-2+1}{-2-3}=\dfrac{-1}{-5}=\dfrac{1}{5}>0$

The number -2 is a solution of the inequality, so the interval A is part of the solution set.

B: Test 0, $\dfrac{0+1}{0-3}=\dfrac{1}{-3}=-\dfrac{1}{3}\not>0$

The number 0 is not a solution of the inequality, so the interval B is not part of the solution set.

C: Test 4, $\dfrac{4+1}{4-3}=\dfrac{5}{1}=5>0$

The number 4 is a solution of the inequality, so the interval C is part of the solution set.

The solution set includes intervals A and C. The number -1 is also included since the inequality symbol is $\geq$ and -1 is the solution of the related equation. The number 3 is not included since $\dfrac{x+1}{x-3}$ is undefined for $x=3$. The solution set is $(-\infty,-1]\cup(3,\infty)$, or $\{x|x\leq -1 \text{ or } x>3\}$.

33. $\dfrac{x+1}{x+6}\geq 1$

Solve the related equation.

$$\frac{x+1}{x+6}=1$$
$$x+1=x+6$$
$$1=6$$

The related equation has no solution.

Find the values that make the denominator 0.

$$x+6=0$$
$$x=-6$$

Use the number -6 to divide the number line into two intervals.

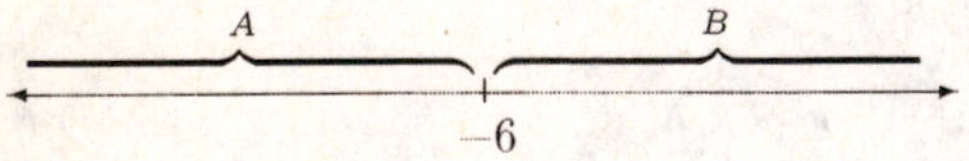

Try test numbers in each interval.

A: Test -7, $\dfrac{-7+1}{-7+6}=\dfrac{-6}{-1}=6\geq 1$.

The number -7 is a solution of the inequality, so the interval A is part of the solution set.

B: Test 0, $\dfrac{0+1}{0+6}=\dfrac{1}{6}\not\geq 1$

The number 0 is not a solution of the inequality, so the interval B is not part of the solution set. The number -6 is not included in the solution set since $\dfrac{x+1}{x+6}$ is undefined for $x=-6$.

The solution set is $(-\infty,-6)$, or $\{x|x<-6\}$.

35. $\dfrac{(x-2)(x+1)}{x-5}\leq 0$

Solve the related equation.

$$\frac{(x-2)(x+1)}{x-5}=0$$
$$(x-2)(x+1)=0$$
$$x=2 \text{ or } x=-1$$

Find the values that make the denominator 0.

$$x-5=0$$
$$x=5$$

Use the numbers 2, -1, and 5 to divide the number line into intervals as shown:

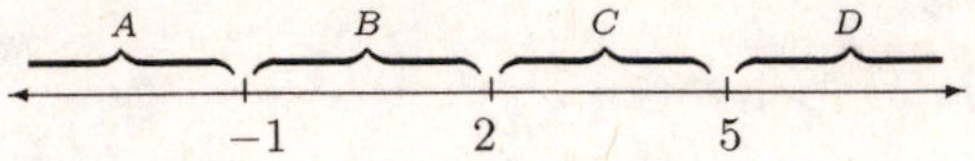

Try test numbers in each interval.

A: Test -2, $\dfrac{(-2-2)(-2+1)}{-2-5}=\dfrac{-4(-1)}{-7}=$

$-\dfrac{4}{7}\leq 0$

Interval A is part of the solution set.

B: Test 0, $\dfrac{(0-2)(0+1)}{0-5}=\dfrac{-2\cdot 1}{-5}=\dfrac{2}{5}\not\leq 0$

Interval B is not part of the solution set.

C: Test 3, $\dfrac{(3-2)(3+1)}{3-5}=\dfrac{1\cdot 4}{-2}=-2\leq 0$

Interval C is part of the solution set.

D: Test 6, $\dfrac{(6-2)(6+1)}{6-5}=\dfrac{4\cdot 7}{1}=28\not\leq 0$

Interval D is not part of the solution set.

The solution set includes intervals A and C. The numbers -1 and 2 are also included since the inequality symbol is $\leq$ and -1 and 2 are the solutions of the related equation. The number 5 is not included since $\dfrac{(x-2)(x+1)}{x-5}$ is undefined for $x=5$.

The solution set is $(-\infty,-1]\cup[2,5)$, or $\{x|x\leq -1 \text{ or } 2\leq x<5\}$.

37. $\dfrac{x}{x+3}\geq 0$

Solve the related equation.

$$\frac{x}{x+3}=0$$
$$x=0$$

Find the values that make the denominator 0.

$$x+3=0$$
$$x=-3$$

Use the numbers 0 and -3 to divide the number line into intervals as shown.

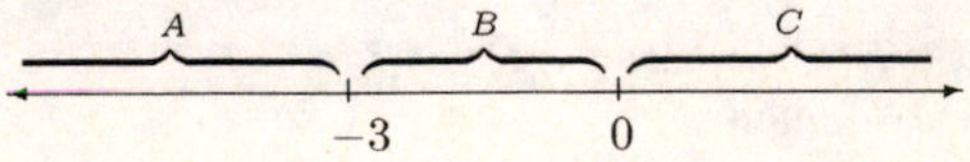

Try test numbers in each interval.

A: Test -4, $\dfrac{-4}{-4+3} = \dfrac{-4}{-1} = 4 \geq 0$

Interval A is part of the solution set.

B: Test -1, $\dfrac{-1}{-1+3} = \dfrac{-1}{2} = -\dfrac{1}{2} \ngeq 0$

Interval B is not part of the solution set.

C: Test 1, $\dfrac{1}{1+3} = \dfrac{1}{4} \geq 0$

The interval C is part of the solution set.

The solution set includes intervals A and C. The number 0 is also included since the inequality symbol is $\geq$ and 0 is the solution of the related equation. The number -3 is not included since $\dfrac{x}{x+3}$ is undefined for $x = -3$. The solution set is $(-\infty, -3) \cup [0, \infty)$, or $\{x|x < -3 \text{ or } x \geq 0\}$.

39. $\dfrac{x-5}{x} < 1$

Solve the related equation.

$$\frac{x-5}{x} = 1$$
$$x - 5 = x$$
$$-5 = 0$$

The related equation has no solution.

Find the values that make the denominator 0.

$$x = 0$$

Use the number 0 to divide the number line into two intervals as shown.

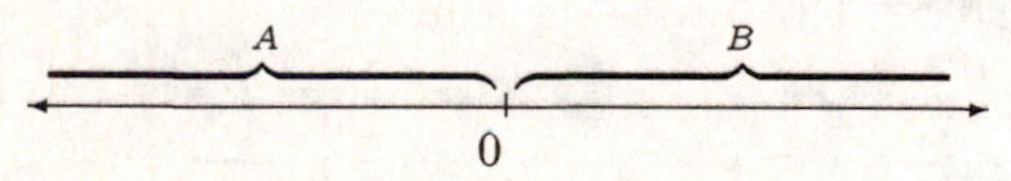

Try test numbers in each interval.

A: Test -1, $\dfrac{-1-5}{-1} = \dfrac{-6}{-1} = 6 \nless 1$

Interval A is not part of the solution set.

B: Test 1, $\dfrac{1-5}{1} = \dfrac{-4}{1} = -4 < 1$

Interval B is part of the solution set.

The solution set is $(0, \infty)$ or $\{x|x > 0\}$.

41. $\dfrac{x-1}{(x-3)(x+4)} \leq 0$

Solve the related equation.

$$\frac{x-1}{(x-3)(x+4)} = 0$$
$$x - 1 = 0$$
$$x = 1$$

Find the values that make the denominator 0.

$$(x-3)(x+4) = 0$$
$$x = 3 \text{ or } x = -4$$

Use the numbers 1, 3, and -4 to divide the number line into intervals as shown:

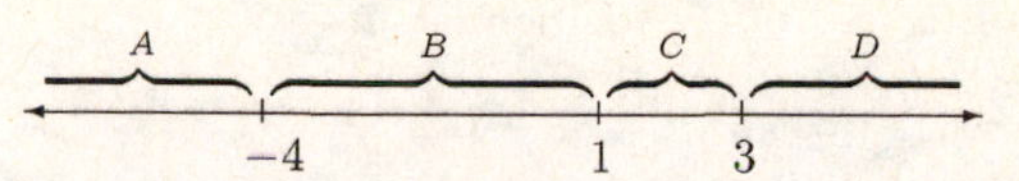

Try test numbers in each interval.

A: Test -5, $\dfrac{-5-1}{(-5-3)(-5+4)} = \dfrac{-6}{-8(-1)} = -\dfrac{3}{4} < 0$

Interval A is part of the solution set.

B: Test 0, $\dfrac{0-1}{(0-3)(0+4)} = \dfrac{-1}{-3 \cdot 4} = \dfrac{1}{12} \nless 0$

Interval B is not part of the solution set.

C: Test 2, $\dfrac{2-1}{(2-3)(2+4)} = \dfrac{1}{-1 \cdot 6} = -\dfrac{1}{6} < 0$

Interval C is part of the solution set.

D: Test 4, $\dfrac{4-1}{(4-3)(4+4)} = \dfrac{3}{1 \cdot 8} = \dfrac{3}{8} \nless 0$

Interval D is not part of the solution set.

The solution set includes intervals A and C. The number 1 is also included since the inequality symbol is $\leq$ and 1 is the solution of the related equation. The numbers -4 and 3 are not included since $\dfrac{x-1}{(x-3)(x+4)}$ is undefined for $x = -4$ and for $x = 3$.

The solution set is $(-\infty, -4) \cup [1, 3)$, or $\{x|x < -4 \text{ or } 1 \leq x < 3\}$.

43. $f(x) \geq 0$

$$\frac{5-2x}{4x+3} \geq 0$$

Solve the related equation.

$$\frac{5-2x}{4x+3} = 0$$
$$5 - 2x = 0$$
$$5 = 2x$$
$$\frac{5}{2} = x$$

Find the values that make the denominator 0.

$$4x + 3 = 0$$
$$4x = -3$$
$$x = -\frac{3}{4}$$

Use the numbers $\dfrac{5}{2}$ and $-\dfrac{3}{4}$ to divide the number line as shown:

A B C

$-\dfrac{3}{4}$ $\dfrac{5}{2}$

Try test numbers in each interval.

A: Test -1, $\dfrac{5-2(-1)}{4(-1)+3} = -7 \ngeq 0$

Interval A is not part of the solution set.

B: Test 0, $\dfrac{5-2 \cdot 0}{4 \cdot 0 + 3} = \dfrac{5}{3} \geq 0$

Interval B is part of the solution set.

C: Test 3, $\dfrac{5-2\cdot 3}{4\cdot 3+3} = -\dfrac{1}{15} \not\geq 0$

Interval C is not part of the solution set.

The solution set includes interval B. The number $\dfrac{5}{2}$ is also included since the inequality symbol is $\geq$ and $\dfrac{5}{2}$ is the solution of the related equation. The number $-\dfrac{3}{4}$ is not included since $\dfrac{5-2x}{4x+3}$ is undefined for $x = -\dfrac{3}{4}$. The solution set is $\left(-\dfrac{3}{4}, \dfrac{5}{2}\right]$, or $\left\{x \middle| -\dfrac{3}{4} < x \leq \dfrac{5}{2}\right\}$.

45. $G(x) \leq 1$

$\dfrac{1}{x-2} \leq 1$

Solve the related equation.

$\dfrac{1}{x-2} = 1$

$1 = x - 2$

$3 = x$

Find the values of x that make the denominator 0.

$x - 2 = 0$

$x = 2$

Use the numbers 2 and 3 to divide the number line as shown.

A B C

2 3

Try a test number in each interval.

A: Test 0, $\dfrac{1}{0-2} = -\dfrac{1}{2} \leq 1$

Interval A is part of the solution set.

B: Test $\dfrac{5}{2}$, $\dfrac{1}{\frac{5}{2}-2} = \dfrac{1}{\frac{1}{2}} = 2 \not\leq 1$

Interval B is not part of the solution set.

C: Test 4, $\dfrac{1}{4-2} = \dfrac{1}{2} \leq 1$

Interval C is part of the solution set.

The solution set includes intervals A and B. The number 3 is also included since the inequality symbol is $\leq$ and 3 is the solution of the related equation. The number 2 is not included since $\dfrac{1}{x-2}$ is undefined for $x = 2$. The solution set is $(-\infty, 2) \cup [3, \infty)$, or $\{x | x < 2 \text{ or } x \geq 3\}$.

47. *Writing Exercise*

49. $(2a^3b^2c^4)^3 = 2^3(a^3)^3(b^2)^3(c^4)^3 = 8a^{3\cdot 3}b^{2\cdot 3}c^{4\cdot 3} = 8a^9b^6c^{12}$

51. $2^{-5} = \dfrac{1}{2^5} = \dfrac{1}{32}$

53. $f(x) = 3x^2$

$f(a+1) = 3(a+1)^2 = 3(a^2+2a+1) = 3a^2+6a+3$

55. *Writing Exercise*

57. $x^2 + 2x < 5$

$x^2 + 2x - 5 < 0$

Using the quadratic formula, we find that the solutions of the related equation are $x = -1 \pm \sqrt{6}$. These numbers divide the real-number line into three intervals as shown:

A B C

$-1-\sqrt{6}$ $-1+\sqrt{6}$

We try test numbers in each interval.

A: Test -4, $f(-4) = (-4)^2 + 2(-4) - 5 = 3$

B: Test 0, $f(0) = 0^2 + 2\cdot 0 - 5 = -5$

C: Test 2, $f(2) = 2^2 + 2\cdot 2 - 5 = 3$

The function value will be negative for all numbers in interval B. The solution set is $(-1-\sqrt{6}, -1+\sqrt{6})$, or $\{x | -1-\sqrt{6} < x < -1+\sqrt{6}\}$.

59. $x^4 + 3x^2 \leq 0$

$x^2(x^2+3) \leq 0$

$x^2 = 0$ for $x = 0$, $x^2 > 0$ for $x \neq 0$, $x^2 + 3 > 0$ for all x

The solution set is $\{0\}$.

61. a) $-3x^2 + 630x - 6000 > 0$

$x^2 - 210x + 2000 < 0$ Multiplying by $-\dfrac{1}{3}$

$(x-200)(x-10) < 0$

The solutions of $f(x) = (x-200)(x-10) = 0$ are 200 and 10. They divide the number line as shown:

A B C

10 200

A: Test 0, $f(0) = 0^2 - 210\cdot 0 + 2000 = 2000$

B: Test 20, $f(20) = 20^2 - 210\cdot 20 + 2000 = -1800$

C: Test 300, $f(300) = 300^2 - 210\cdot 300 + 2000 = 29{,}000$

The company makes a profit for values of x such that $10 < x < 200$, or for values of x in the interval $(10, 200)$.

b) See part (a). Keep in mind that x must be nonnegative since negative numbers have no meaning in this application.

The company loses money for values of x such that $0 \leq x < 10$ or $x > 200$, or for values of x in the interval $[0, 10) \cup (200, \infty)$.

63. We find values of n such that $N \geq 66$ *and* $N \leq 300$.

For $N \geq 66$:

$\dfrac{n(n-1)}{2} \geq 66$

$n(n-1) \geq 132$

$n^2 - n - 132 \geq 0$

$(n-12)(n+11) \geq 0$

The solutions of $f(n) = (n-12)(n+11) = 0$ are 12 and -11. They divide the number line as shown:

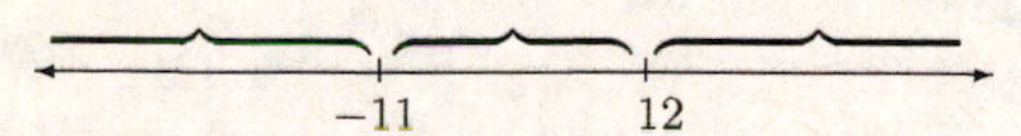

However, only positive values of n have meaning in this exercise so we need only consider the intervals shown below:

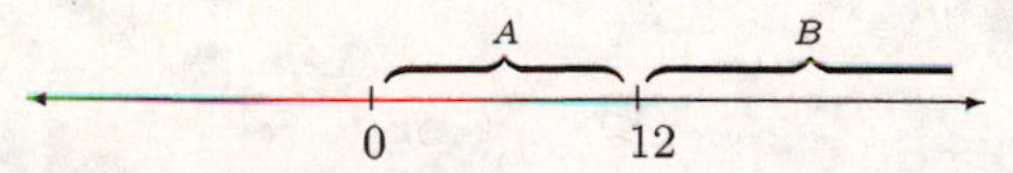

A: Test 1, $f(1) = 1^2 - 1 - 132 = -132$

B: Test 20, $f(20) = 20^2 - 20 - 132 = 248$

Thus, $N \geq 66$ for $\{n|n \geq 12\}$.

For $N \leq 300$:

$$\frac{n(n-1)}{2} \leq 300$$
$$n(n-1) \leq 600$$
$$n^2 - n - 600 \leq 0$$
$$(n-25)(n+24) \leq 0$$

The solutions of $f(n) = (n-25)(n+24) = 0$ are 25 and -24. They divide the number line as shown:

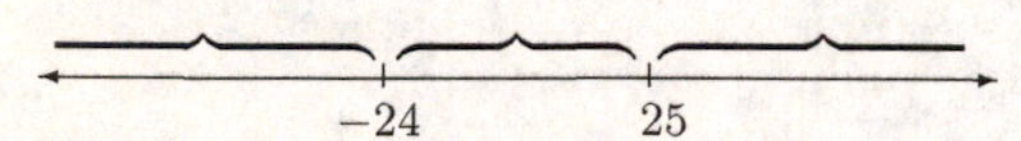

However, only positive values of n have meaning in this exercise so we need only consider the intervals shown below:

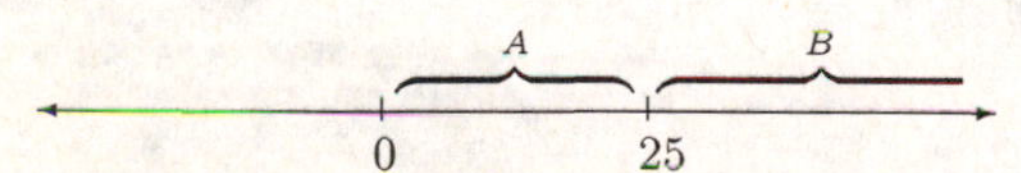

A: Test 1, $f(1) = 1^2 - 1 - 600 = -600$

B: Test 30, $f(30) = 30^2 - 30 - 600 = 270$

Thus, $N \leq 300$ (and $n > 0$) for $\{n|0 < n \leq 25\}$.

Then $66 \leq N \leq 300$ for $\{n|n$ is an integer and $12 \leq n \leq 25\}$.

65. From the graph we determine the following:

The solutions of $f(x) = 0$ are -2, 1, and 3.

The solution of $f(x) < 0$ is$(-\infty, -2) \cup (1, 3)$, or $\{x|x < -2 \text{ or } 1 < x < 3\}$.

The solution of $f(x) > 0$ is$(-2, 1) \cup (3, \infty)$, or $\{x| -2 < x < 1 \text{ or } x > 3\}$.

67. From the graph we determine the following:

$f(x)$ has no zeros.

The solutions $f(x) < 0$ are $(-\infty, 0)$, or $\{x|x < 0\}$.

The solutions of $f(x) > 0$ are $(0, \infty)$, or $\{x|x > 0\}$.

69. From the graph we determine the following:

The solutions of $f(x) = 0$ are -1 and 0.

The solutions of $f(x) < 0$ are $(-\infty, -3) \cup (-1, 0)$, or $\{x|x < -3 \text{ or } -1 < x < 0\}$.

The solutions of $f(x) > 0$ are $(-3, -1) \cup (0, 2) \cup (2, \infty)$, or $\{x| -3 < x < -1 \text{ or } 0 < x < 2 \text{ or } x > 2\}$.

71.

Chapter 12

Exponential and Logarithmic Functions

Exercise Set 12.1

1. True; see page 783 in the text.

3. $(g \circ f)(x) = g(f(x)) = x^2 + 3 \neq (x+3)^2$, so the statement is false.

5. False; see page 786 in the text.

7. True; see page 787 in the text.

9. $$\begin{aligned}(f \circ g)(1) &= f(g(1)) = f(2 \cdot 1 - 3)\\ &= f(-1) = (-1)^2 + 1\\ &= 1 + 1 = 2\end{aligned}$$
$$\begin{aligned}(g \circ f)(1) &= g(f(1)) = g(1^2 + 1)\\ &= g(2) = 2 \cdot 2 - 3 = 1\end{aligned}$$
$$\begin{aligned}(f \circ g)(x) &= f(g(x)) = f(2x - 3)\\ &= (2x-3)^2 + 1\\ &= 4x^2 - 12x + 9 + 1\\ &= 4x^2 - 12x + 10\end{aligned}$$
$$\begin{aligned}(g \circ f)(x) &= g(f(x)) = g(x^2 + 1)\\ &= 2(x^2 + 1) - 3\\ &= 2x^2 + 2 - 3\\ &= 2x^2 - 1\end{aligned}$$

11. $$\begin{aligned}(f \circ g)(1) &= f(g(1)) = f(2 \cdot 1^2 - 7)\\ &= f(-5) = -5 - 3 = -8\end{aligned}$$
$$\begin{aligned}(g \circ f)(1) &= g(f(1)) = g(1 - 3)\\ &= g(-2) = 2(-2)^2 - 7\\ &= 8 - 7 = 1\end{aligned}$$
$$\begin{aligned}(f \circ g)(x) &= f(g(x)) = f(2x^2 - 7)\\ &= 2x^2 - 7 - 3 = 2x^2 - 10\end{aligned}$$
$$\begin{aligned}(g \circ f)(x) &= g(f(x)) = g(x - 3)\\ &= 2(x-3)^2 - 7\\ &= 2(x^2 - 6x + 9) - 7\\ &= 2x^2 - 12x + 18 - 7\\ &= 2x^2 - 12x + 11\end{aligned}$$

13. $$\begin{aligned}(f \circ g)(1) &= f(g(1)) = f\left(\frac{1}{1^2}\right)\\ &= f(1) = 1 + 7 = 8\end{aligned}$$
$$\begin{aligned}(g \circ f)(1) &= g(f(1)) = g(1 + 7)\\ &= g(8) = \frac{1}{8^2} = \frac{1}{64}\end{aligned}$$
$$\begin{aligned}(f \circ g)(x) &= f(g(x))\\ &= f\left(\frac{1}{x^2}\right) = \frac{1}{x^2} + 7\end{aligned}$$
$$\begin{aligned}(g \circ f)(x) &= g(f(x))\\ &= g(x + 7) = \frac{1}{(x+7)^2}\end{aligned}$$

15. $$\begin{aligned}(f \circ g)(1) &= f(g(1)) = f(1 + 3)\\ &= f(4) = \sqrt{4} = 2\end{aligned}$$
$$\begin{aligned}(g \circ f)(1) &= g(f(1)) = g(\sqrt{1})\\ &= g(1) = 1 + 3 = 4\end{aligned}$$
$$(f \circ g)(x) = f(g(x)) = f(x + 3) = \sqrt{x + 3}$$
$$(g \circ f)(x) = g(f(x)) = g(\sqrt{x}) = \sqrt{x} + 3$$

17. $$\begin{aligned}(f \circ g)(1) &= f(g(1)) = f\left(\frac{1}{1}\right)\\ &= f(1) = \sqrt{4 \cdot 1} = \sqrt{4} = 2\end{aligned}$$
$$\begin{aligned}(g \circ f)(1) &= g(f(1)) = g(\sqrt{4 \cdot 1})\\ &= g(\sqrt{4}) = g(2) = \frac{1}{2}\end{aligned}$$
$$\begin{aligned}(f \circ g)(x) &= f(g(x)) = f\left(\frac{1}{x}\right)\\ &= \sqrt{4 \cdot \frac{1}{x}} = \sqrt{\frac{4}{x}}\end{aligned}$$
$$(g \circ f)(x) = g(f(x)) = g(\sqrt{4x}) = \frac{1}{\sqrt{4x}}$$

19. $$\begin{aligned}(f \circ g)(1) &= f(g(1)) = f(\sqrt{1 - 1})\\ &= f(\sqrt{0}) = f(0) = 0^2 + 4 = 4\end{aligned}$$
$$\begin{aligned}(g \circ f)(1) &= g(f(1)) = g(1^2 + 4)\\ &= g(5) = \sqrt{5 - 1} = \sqrt{4} = 2\end{aligned}$$
$$\begin{aligned}(f \circ g)(x) &= f(g(x)) = f(\sqrt{x - 1})\\ &= (\sqrt{x-1})^2 + 4 = x - 1 + 4 = x + 3\end{aligned}$$
$$\begin{aligned}(g \circ f)(x) &= g(f(x)) = g(x^2 + 4)\\ &= \sqrt{x^2 + 4 - 1} = \sqrt{x^2 + 3}\end{aligned}$$

21. $h(x) = (7 + 5x)^2$

This is $7 + 5x$ raised to the second power, so the two most obvious functions are $f(x) = x^2$ and $g(x) = 7 + 5x$.

23. $h(x) = \sqrt{2x + 7}$

We have $2x + 7$ and take the square root of their expression, so the two most obvious functions are $f(x) = \sqrt{x}$ and $g(x) = 2x + 7$.

25. $h(x) = \dfrac{2}{x - 3}$

This is 2 divided by $x - 3$, so two functions that can be used are $f(x) = \dfrac{2}{x}$ and $g(x) = x - 3$.

27. The graph of $f(x) = x - 5$ is shown below.

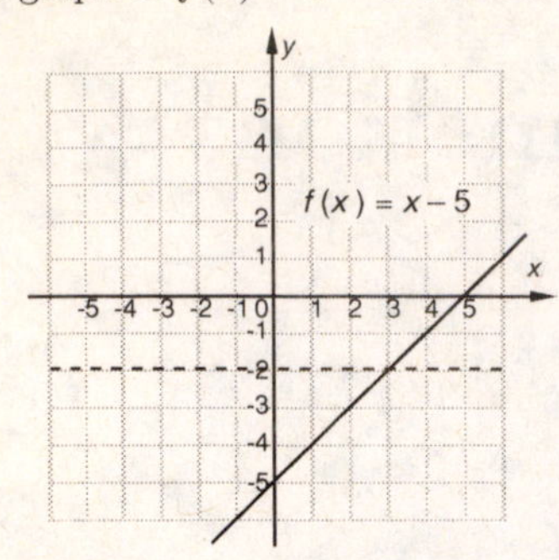

Since there is no horizontal line that crosses the graph more than once, the function is one-to-one.

29. $f(x) = x^2 + 1$

Observe that the graph of this function is a parabola that opens up. Thus, there are many horizontal lines that cross the graph more than once, so the function is not one-to-one. We can also draw the graph as shown below.

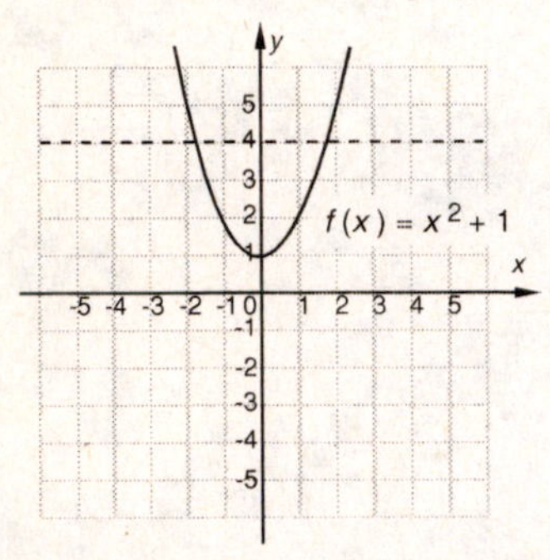

There are many horizontal lines that cross the graph more than once. In particular, the line $y = 4$ crosses the graph more than once. The function is not one-to-one.

31. Since there is no horizontal line that crosses the graph more than once, the function is one-to-one.

33. There are many horizontal lines that cross the graph more than once, so the function is not one-to-one.

35. a) The function $f(x) = x + 4$ is a linear function that is not constant, so it passes the horizontal-line test. Thus, f is one-to-one.

b) Replace $f(x)$ by y: $y = x + 4$

Interchange x and y: $x = y + 4$

Solve for y: $x - 4 = y$

Replace y by $f^{-1}(x)$: $f^{-1}(x) = x - 4$

37. a) The function $f(x) = 2x$ is a linear function that is not constant, so it passes the horizontal-line test. Thus, f is one-to-one.

b) Replace $f(x)$ by y: $y = 2x$

Interchange x and y: $x = 2y$

Solve for y: $\dfrac{x}{2} = y$

Replace y by $f^{-1}(x)$: $f^{-1}(x) = \dfrac{x}{2}$

39. a) The function $g(x) = 3x - 1$ is a linear function that is not constant, so it passes the horizontal-line test. Thus, g is one-to-one.

b) Replace $g(x)$ by y: $y = 3x - 1$

Interchange variables: $x = 3y - 1$

Solve for y: $x + 1 = 3y$

$$\frac{x+1}{3} = y$$

Replace y by $g^{-1}(x)$: $g^{-1}(x) = \dfrac{x+1}{3}$

41. a) The function $f(x) = \dfrac{1}{2}x + 1$ is a linear function that is not constant, so it passes the horizontal-line test. Thus, f is one-to-one.

b) Replace $f(x)$ by y: $y = \dfrac{1}{2}x + 1$

Interchange variables: $x = \dfrac{1}{2}y + 1$

Solve for y:
$$x = \frac{1}{2}y + 1$$
$$x - 1 = \frac{1}{2}y$$
$$2x - 2 = y$$

Replace y by $f^{-1}(x)$: $f^{-1}(x) = 2x - 2$

43. a) The graph of $g(x) = x^2 + 5$ is shown below. There are many horizontal lines that cross the graph more than once. For example, the line $y = 8$ crosses the graph more than once. The function is not one-to-one.

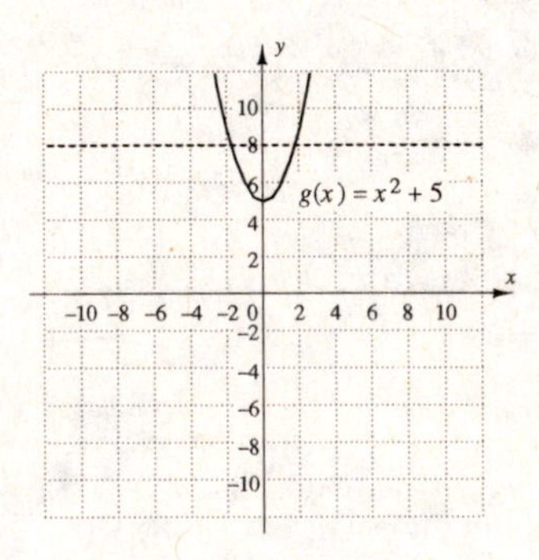

45. a) The function $h(x) = -2x + 4$ is a linear function that is not constant, so it passes the horizontal-line test. Thus, h is one-to-one.

b) Replace $h(x)$ by y: $y = -2x + 4$

Interchange variables: $x = -2y + 4$

Solve for y:
$$x = -2y + 4$$
$$x - 4 = -2y$$
$$\frac{x-4}{-2} = y$$

Replace y by $h^{-1}(x)$: $h^{-1}(x) = \dfrac{x-4}{-2}$

47. a) The graph of $f(x) = \dfrac{1}{x}$ is shown below. It passes the horizontal-line test, so the function is one-to-one.

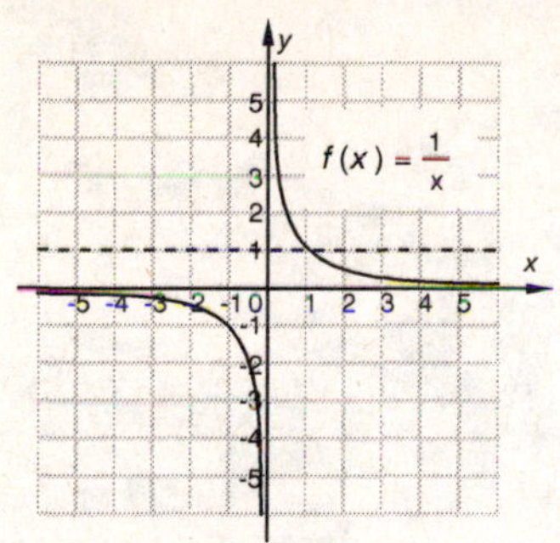

b) Replace $f(x)$ by y: $y = \dfrac{1}{x}$

Interchange x and y: $x = \dfrac{1}{y}$

Solve for y: $xy = 1$

$$y = \frac{1}{x}$$

Replace y by $f^{-1}(x)$: $f^{-1}(x) = \dfrac{1}{x}$

49. a) The graph of $G(x) = 4$ is shown below. The horizontal line $y = 4$ crosses the graph more than once, so the function is not one-to-one.

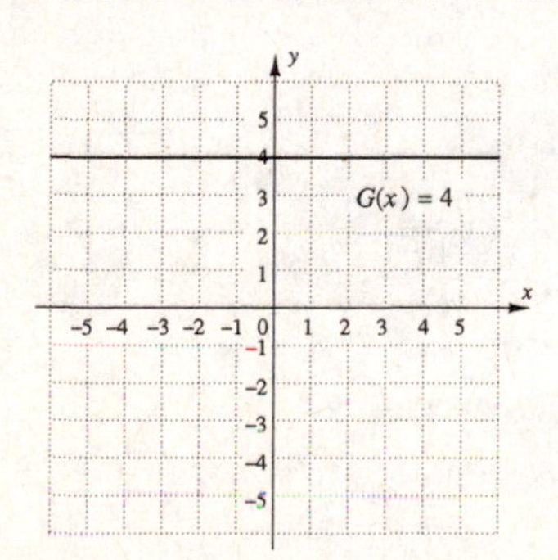

51. a) The function $f(x) = \dfrac{2x+1}{3} = \dfrac{2}{3}x + \dfrac{1}{3}$ is a linear function that is not constant, so it passes the horizontal-line test. Thus, f is one-to-one.

b) Replace $f(x)$ by y: $y = \dfrac{2x+1}{3}$

Interchange x and y: $x = \dfrac{2y+1}{3}$

Solve for y: $3x = 2y + 1$

$$3x - 1 = 2y$$

$$\frac{3x-1}{2} = y$$

Replace y by $f^{-1}(x)$: $f^{-1}(x) = \dfrac{3x-1}{2}$

53. a) The graph of $f(x) = x^3 - 5$ is shown below. It passes the horizontal-line test, so the function is one-to-one.

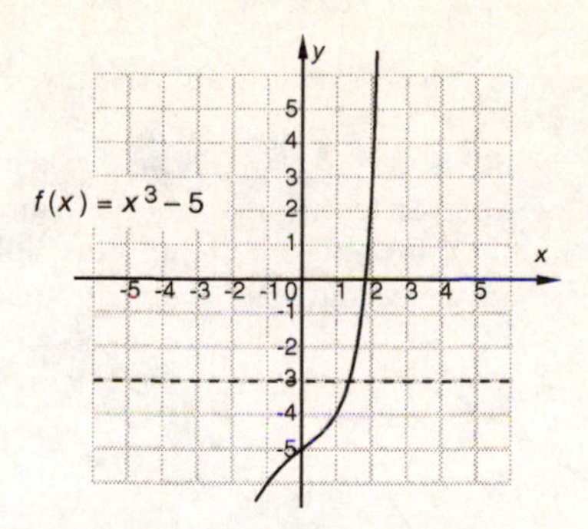

b) Replace $f(x)$ by y: $y = x^3 - 5$

Interchange x and y: $x = y^3 - 5$

Solve for y: $x + 5 = y^3$

$$\sqrt[3]{x+5} = y$$

Replace y by $f^{-1}(x)$: $f^{-1}(x) = \sqrt[3]{x+5}$

55. a) The graph of $g(x) = (x-2)^3$ is shown below. It passes the horizontal-line test, so the function is one-to-one.

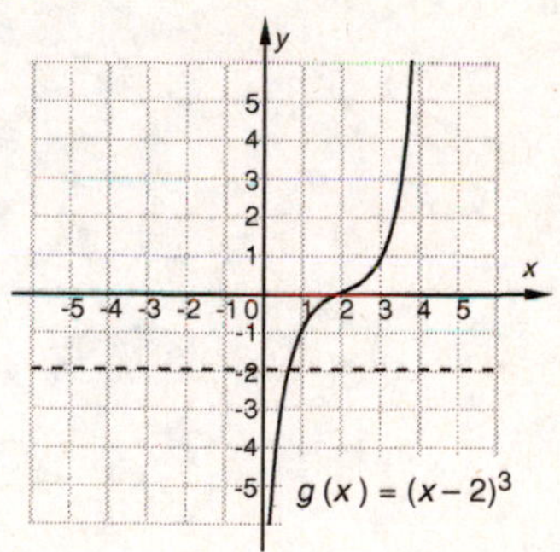

b) Replace $g(x)$ by y: $y = (x-2)^3$

Interchange x and y: $x = (y-2)^3$

Solve for y: $\sqrt[3]{x} = y - 2$

$$\sqrt[3]{x} + 2 = y$$

Replace y by $g^{-1}(x)$: $g^{-1}(x) = \sqrt[3]{x} + 2$

57. a) The graph of $f(x) = \sqrt{x}$ is shown below. It passes the horizontal-line test, so the function is one-to-one.

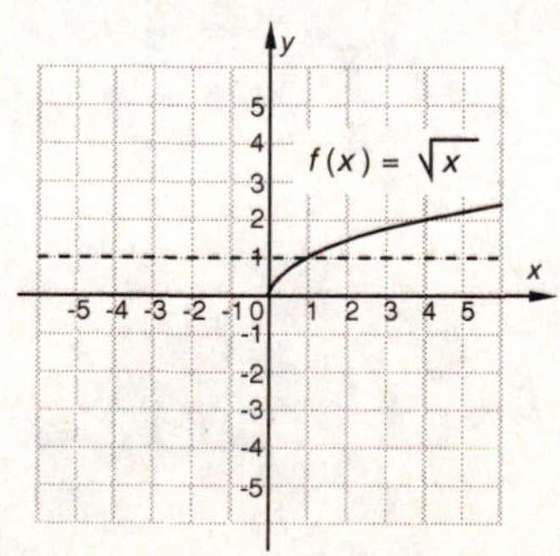

b) Replace $f(x)$ by y: $y = \sqrt{x}$ (Note that $f(x) \geq 0$.)

Interchange x and y: $x = \sqrt{y}$

Solve for y: $x^2 = y$

Replace y by $f^{-1}(x)$: $f^{-1}(x) = x^2$, $x \geq 0$

59. First graph $f(x) = \frac{2}{3}x + 4$. Then graph the inverse function by reflecting the graph of $f(x) = \frac{2}{3}x + 4$ across the line $y = x$. The graph of the inverse function can also be found by first finding a formula for the inverse, substituting to find function values, and then plotting points.

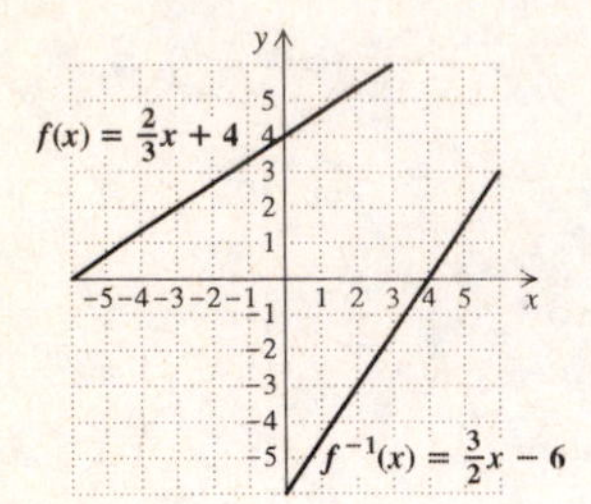

61. Follow the procedure described in Exercise 59 to graph the function and its inverse.

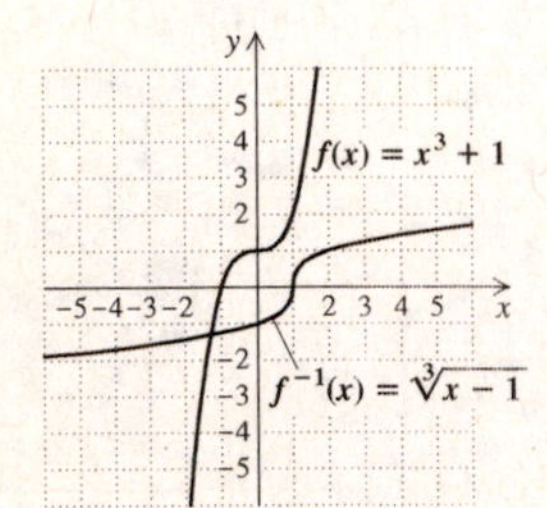

63. Use the procedure described in Exercise 59 to graph the function and its inverse.

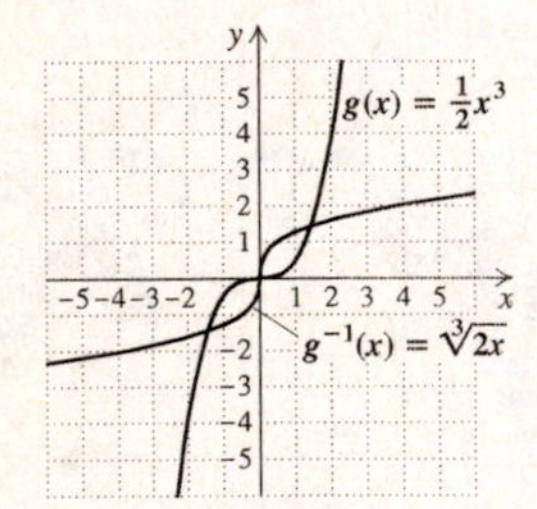

65. Use the procedure described in Exercise 59 to graph the function and its inverse.

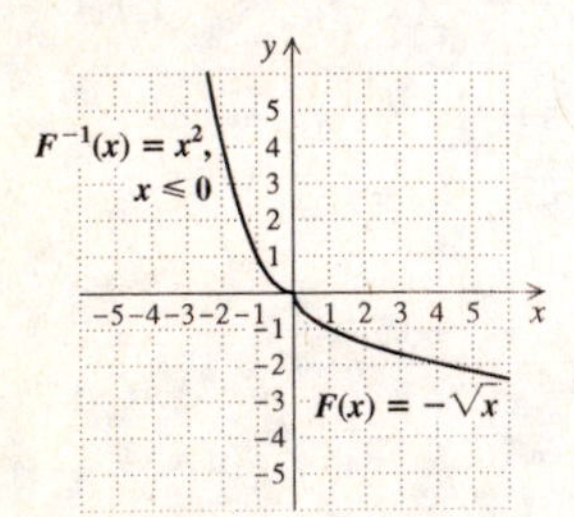

67. Use the procedure described in Exercise 59 to graph the function and its inverse.

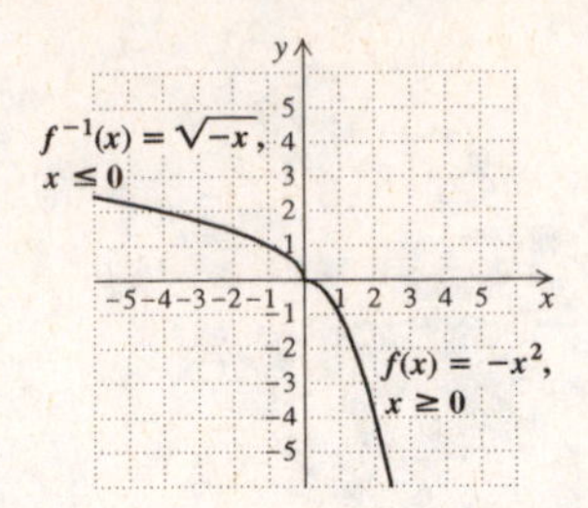

69. We check to see that $(f^{-1} \circ f)(x) = x$ and $(f \circ f^{-1})(x) = x$.

$(f^{-1} \circ f)(x) = f^{-1}(f(x)) = f^{-1}(\sqrt[3]{x - 4}) =$
$(\sqrt[3]{x - 4})^3 + 4 = x - 4 + 4 = x$

$(f \circ f^{-1})(x) = f(f^{-1}(x)) = f(x^3 + 4) = \sqrt[3]{x^3 + 4 - 4} =$
$\sqrt[3]{x^3} = x$

71. We check to see that $f^{-1} \circ f(x) = x$ and $f \circ f^{-1}(x) = x$.

$$f^{-1} \circ f(x) = f^{-1}(f(x)) = f^{-1}\left(\frac{1 - x}{x}\right) = \frac{1}{\frac{1 - x}{x} + 1} =$$

$$\frac{1}{\frac{1 - x}{x} + 1} \cdot \frac{x}{x} = \frac{x}{1 - x + x} = \frac{x}{1} = x$$

$$f \circ f^{-1}(x) = f(f^{-1}(x)) = f\left(\frac{1}{x + 1}\right) = \frac{1 - \frac{1}{x + 1}}{\frac{1}{x + 1}} =$$

$$\frac{1 - \frac{1}{x + 1}}{\frac{1}{x + 1}} \cdot \frac{x + 1}{x + 1} = \frac{x + 1 - 1}{1} = \frac{x}{1} = x$$

73. a) $f(8) = 2(8 + 12) = 2 \cdot 20 = 40$

Size 40 in Italy corresponds to size 8 in the U.S.

$f(10) = 2(10 + 12) = 2 \cdot 22 = 44$

Size 44 in Italy corresponds to size 10 in the U.S.

$f(14) = 2(14 + 12) = 2 \cdot 26 = 52$

Size 52 in Italy corresponds to size 14 in the U.S.

$f(18) = 2(18 + 12) = 2 \cdot 30 = 60$

Size 60 in Italy corresponds to size 18 in the U.S.

b) The function $f(x) = 2(x + 12)$ is a linear function that is not constant, so it passes the horizontal line test and has an inverse that is a function.

Replace $f(x)$ by y: $y = 2(x + 12)$

Interchange x and y: $x = 2(y + 12)$

Solve for y:

$$x = 2(y + 12)$$
$$x = 2y + 24$$
$$x - 24 = 2y$$
$$\frac{x - 24}{2} = y$$

Replace y by $f^{-1}(x)$: $f^{-1}(x) = \frac{x - 24}{2}$, or $\frac{x}{2} - 12$

c) $f^{-1}(40) = \frac{40-24}{2} = \frac{16}{2} = 8$

Size 8 in the U.S. corresponds to size 40 in Italy.

$f^{-1}(44) = \frac{44-24}{2} = \frac{20}{2} = 10$

Size 10 in the U.S. corresponds to size 44 in Italy.

$f^{-1}(52) = \frac{52-24}{2} = \frac{28}{2} = 14$

Size 14 in the U.S. corresponds to size 52 in Italy.

$f^{-1}(60) = \frac{60-24}{2} = \frac{36}{2} = 18$

Size 18 in the U.S. corresponds to size 60 in Italy.

75. *Writing Exercise*

77.
$$\begin{aligned}(a^5b^4)^2(a^3b^5) &= (a^5)^2(b^4)^2(a^3b^5)\\ &= a^{5\cdot 2}b^{4\cdot 2}a^3b^5\\ &= a^{10}b^8a^3b^5\\ &= a^{10+3}b^{8+5}\\ &= a^{13}b^{13}\end{aligned}$$

79. $27^{4/3} = (3^3)^{4/3} = 3^{3\cdot\frac{4}{3}} = 3^4 = 81$

81.
$$\begin{aligned}x &= \frac{2}{3}y - 7\\ x+7 &= \frac{2}{3}y\\ \frac{3}{2}(x+7) &= y\end{aligned}$$

83. *Writing Exercise*

85. Reflect the graph of f across the line $y = x$.

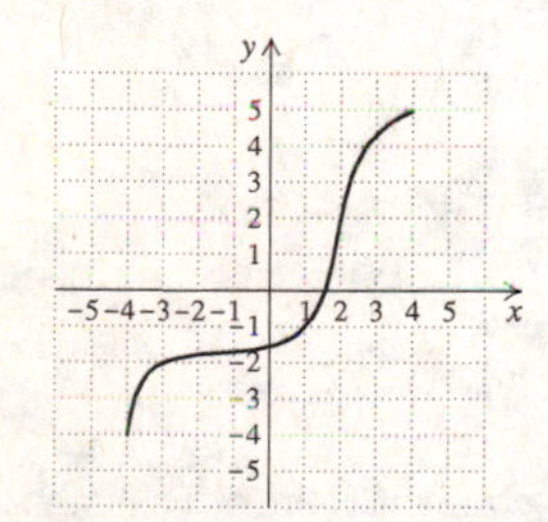

87. From Exercise 73(b), we know that a function that converts dress sizes in Italy to those in the United States is $g(x) = \frac{x-24}{2}$. From Exercise 74, we know that a function that converts dress sizes in the United States to those in France is $f(x) = x + 32$. Then a function that converts dress sizes in Italy to those in France is

$$\begin{aligned}h(x) &= (f\circ g)(x)\\ h(x) &= f\left(\frac{x-24}{2}\right)\\ h(x) &= \frac{x-24}{2} + 32\\ h(x) &= \frac{x}{2} - 12 + 32\\ h(x) &= \frac{x}{2} + 20.\end{aligned}$$

89. *Writing Exercise*

91. Suppose that $h(x) = (f\circ g)(x)$. First note that for $I(x) = x$, $(f\circ I)(x) = f(I(x))$ for any function f.

i)
$$\begin{aligned}((g^{-1}\circ f^{-1})\circ h)(x) &= ((g^{-1}\circ f^{-1})\circ(f\circ g))(x)\\ &= ((g^{-1}\circ(f^{-1}\circ f))\circ g)(x)\\ &= ((g^{-1}\circ I)\circ g)(x)\\ &= (g^{-1}\circ g)(x) = x\end{aligned}$$

ii)
$$\begin{aligned}(h\circ(g^{-1}\circ f^{-1}))(x) &= ((f\circ g)\circ(g^{-1}\circ f^{-1}))(x)\\ &= ((f\circ(g\circ g^{-1}))\circ f^{-1})(x)\\ &= ((f\circ I)\circ f^{-1})(x)\\ &= (f\circ f^{-1})(x) = x\end{aligned}$$

Therefore, $(g^{-1}\circ f^{-1})(x) = h^{-1}(x)$.

93. $(f\circ g)(x) = x$ and $(g\circ f)(x) = x$, so the functions are inverses.

95. $(f\circ g)(x) \neq x$, so the functions are not inverses. (It is also true that $(g\circ f)(x) \neq x$.)

97. (1) C; (2) A; (3) B; (4) D

99. *Writing Exercise*

Exercise Set 12.2

1. True; see page 797 in the text.

3. True; the graph of $y = f(x-3)$ is a translation of the graph of $y = f(x)$ 3 units to the right.

5. False; the graph of $y = 3^x$ crosses the y-axis at $(0, 1)$.

7. Graph: $y = f(x) = 3^x$

We compute some function values, thinking of y as $f(x)$, and keep the results in a table.

$f(0) = 3^0 = 1$

$f(1) = 3^1 = 3$

$f(2) = 3^2 = 9$

$f(-1) = 3^{-1} = \frac{1}{3^1} = \frac{1}{3}$

$f(-2) = 3^{-2} = \frac{1}{3^2} = \frac{1}{9}$

x	y, or $f(x)$
0	1
1	3
2	9
−1	$\frac{1}{3}$
−2	$\frac{1}{9}$

Next we plot these points and connect them with a smooth curve.

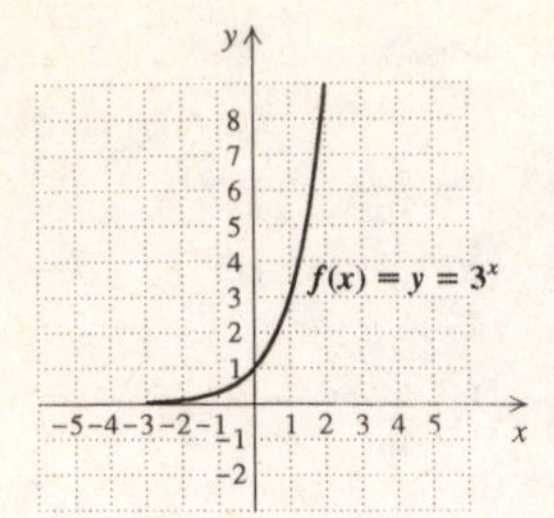

9. Graph: $y = 6^x$

We compute some function values, thinking of y as $f(x)$, and keep the results in a table.

$f(0) = 6^0 = 1$

$f(1) = 6^1 = 6$

$f(2) = 6^2 = 36$

$f(-1) = 6^{-1} = \frac{1}{6^1} = \frac{1}{6}$

$f(-2) = 6^{-2} = \frac{1}{6^2} = \frac{1}{36}$

x	y, or $f(x)$
0	1
1	6
2	36
-1	$\frac{1}{6}$
-2	$\frac{1}{36}$

Next we plot these points and connect them with a smooth curve.

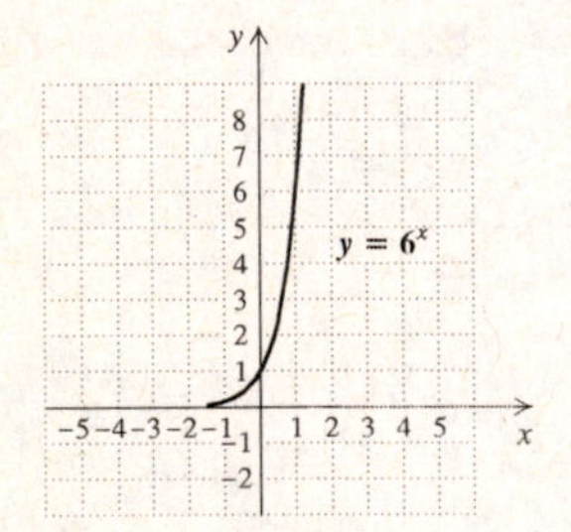

11. Graph: $y = 2^x + 1$

We compute some function values, thinking of y as $f(x)$, and keep the results in a table.

$f(-4) = 2^{-4} + 1 = \frac{1}{2^4} + 1 = \frac{1}{16} + 1 = 1\frac{1}{16}$

$f(-2) = 2^{-2} + 1 = \frac{1}{2^2} + 1 = \frac{1}{4} + 1 = 1\frac{1}{4}$

$f(0) = 2^0 + 1 = 1 + 1 = 2$

$f(1) = 2^1 + 1 = 2 + 1 = 3$

$f(2) = 2^2 + 1 = 4 + 1 = 5$

x	y, or $f(x)$
-4	$1\frac{1}{16}$
-2	$1\frac{1}{4}$
0	2
1	3
2	5

Next we plot these points and connect them with a smooth curve.

13. Graph: $y = 3^x - 2$

We compute some function values, thinking of y as $f(x)$, and keep the results in a table.

$f(-3) = 3^{-3} - 2 = \frac{1}{3^3} - 2 = \frac{1}{27} - 2 = -\frac{53}{27}$

$f(-1) = 3^{-1} - 2 = \frac{1}{3} - 2 = -\frac{5}{3}$

$f(0) = 3^0 - 2 = 1 - 2 = -1$

$f(1) = 3^1 - 2 = 3 - 2 = 1$

$f(2) = 3^2 - 2 = 9 - 2 = 7$

x	y, or $f(x)$
-3	$-\frac{53}{27}$
-1	$-\frac{5}{3}$
0	-1
1	1
2	7

Next we plot these points and connect them with a smooth curve.

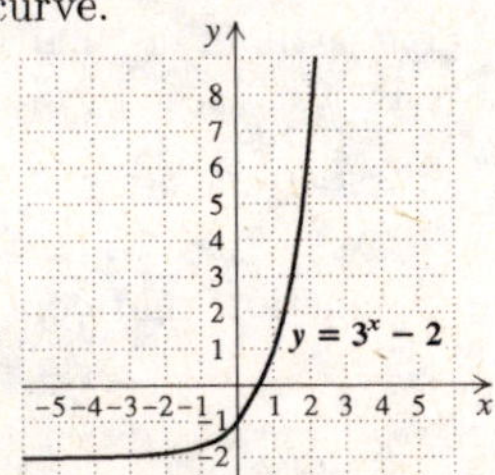

15. Graph: $y = 2^x - 5$

We construct a table of values, thinking of y as $f(x)$. Then we plot the points and connect them with a smooth curve.

$f(0) = 2^0 - 5 = 1 - 5 = -4$

$f(1) = 2^1 - 5 = 2 - 5 = -3$

$f(2) = 2^2 - 5 = 4 - 5 = -1$

$f(3) = 2^3 - 5 = 8 - 5 = 3$

$f(-1) = 2^{-1} - 5 = \frac{1}{2} - 5 = -\frac{9}{2}$

$f(-2) = 2^{-2} - 5 = \frac{1}{4} - 5 = -\frac{19}{4}$

$f(-4) = 2^{-4} - 5 = \frac{1}{16} - 5 = -\frac{79}{16}$

x	y, or $f(x)$
0	-4
1	-3
2	-1
3	3
-1	$-\frac{9}{2}$
-2	$-\frac{19}{4}$
-4	$-\frac{79}{16}$

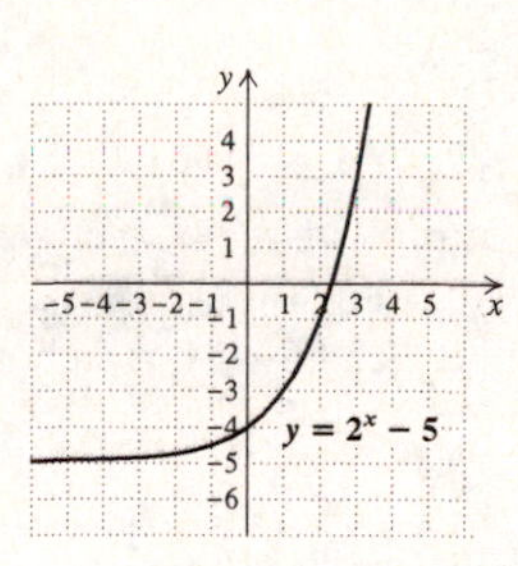

17. Graph: $y = 2^{x-2}$

We construct a table of values, thinking of y as $f(x)$. Then we plot the points and connect them with a smooth curve.

$f(0) = 2^{0-2} = 2^{-2} = \frac{1}{4}$

$f(-1) = 2^{-1-2} = 2^{-3} = \frac{1}{2^3} = \frac{1}{8}$

$f(-2) = 2^{-2-2} = 2^{-4} = \frac{1}{2^4} = \frac{1}{16}$

$f(1) = 2^{1-2} = 2^{-1} = \frac{1}{2}$

$f(2) = 2^{2-2} = 2^0 = 1$

$f(3) = 2^{3-2} = 2^1 = 2$

$f(4) = 2^{4-2} = 2^2 = 4$

x	y, or $f(x)$
0	$\frac{1}{4}$
-1	$\frac{1}{8}$
-2	$\frac{1}{16}$
1	$\frac{1}{2}$
2	1
3	2
4	4

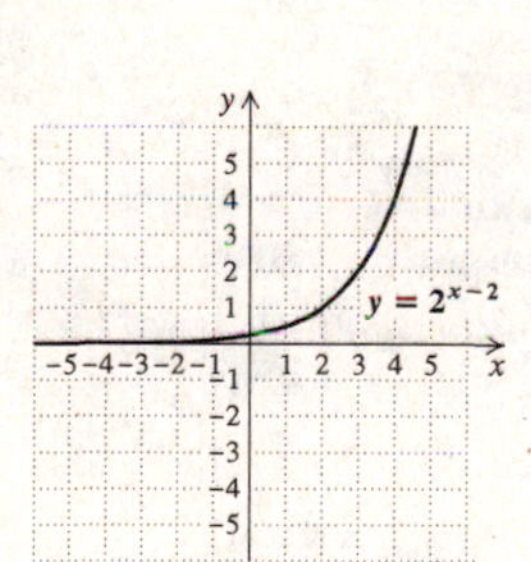

19. Graph: $y = 2^{x+1}$

We construct a table of values, thinking of y as $f(x)$. Then we plot the points and connect them with a smooth curve.

$f(-3) = 2^{-3+1} = 2^{-2} = \frac{1}{4}$

$f(-1) = 2^{-1+1} = 2^0 = 1$

$f(0) = 2^{0+1} = 2^1 = 2$

$f(1) = 2^{1+1} = 2^2 = 4$

x	y, or $f(x)$
-3	$\frac{1}{4}$
-1	1
0	2
1	4

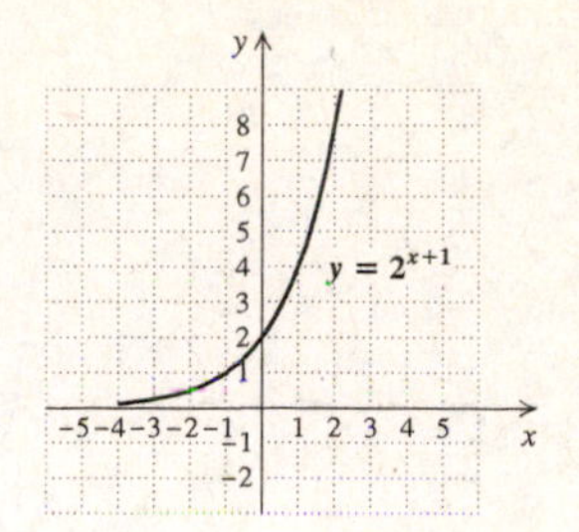

21. Graph: $y = \left(\frac{1}{4}\right)^x$

We construct a table of values, thinking of y as $f(x)$. Then we plot the points and connect them with a smooth curve.

$f(0) = \left(\frac{1}{4}\right)^0 = 1$

$f(1) = \left(\frac{1}{4}\right)^1 = \frac{1}{4}$

$f(2) = \left(\frac{1}{4}\right)^2 = \frac{1}{16}$

$f(-1) = \left(\frac{1}{4}\right)^{-1} = \frac{1}{\frac{1}{4}} = 4$

$f(-2) = \left(\frac{1}{4}\right)^{-2} = \frac{1}{\frac{1}{16}} = 16$

x	y, or $f(x)$
0	1
1	$\frac{1}{4}$
2	$\frac{1}{16}$
-1	4
-2	16

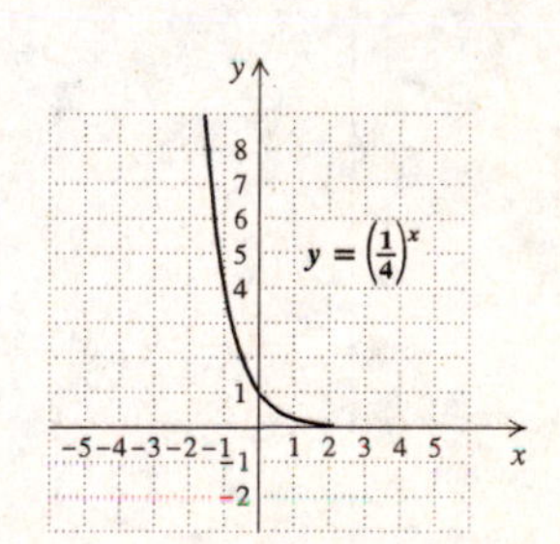

23. Graph: $y = \left(\frac{1}{3}\right)^x$

We construct a table of values, thinking of y as $f(x)$. Then we plot the points and connect them with a smooth curve.

$f(0) = \left(\frac{1}{3}\right)^0 = 1$

$f(1) = \left(\frac{1}{3}\right)^1 = \frac{1}{3}$

$f(2) = \left(\frac{1}{3}\right)^2 = \frac{1}{9}$

$f(3) = \left(\frac{1}{3}\right)^3 = \frac{1}{27}$

$f(-1) = \left(\frac{1}{3}\right)^{-1} = \frac{1}{\left(\frac{1}{3}\right)^1} = \frac{1}{\frac{1}{3}} = 3$

$f(-2) = \left(\frac{1}{3}\right)^{-2} = \frac{1}{\left(\frac{1}{3}\right)^2} = \frac{1}{\frac{1}{9}} = 9$

$f(-3) = \left(\frac{1}{3}\right)^{-3} = \frac{1}{\left(\frac{1}{3}\right)^3} = \frac{1}{\frac{1}{27}} = 27$

x	y, or $f(x)$
0	1
1	$\frac{1}{3}$
2	$\frac{1}{9}$
3	$\frac{1}{27}$
-1	3
-2	9
-3	27

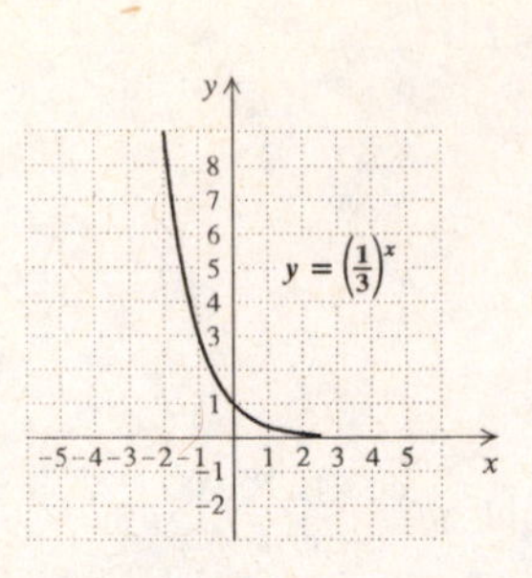

25. Graph: $y = 2^{x+1} - 3$

We construct a table of values, thinking of y as $f(x)$. Then we plot the points and connect them with a smooth curve.

$$f(0) = 2^{0+1} - 3 = 2 - 3 = -1$$
$$f(1) = 2^{1+1} - 3 = 4 - 3 = 1$$
$$f(2) = 2^{2+1} - 3 = 8 - 3 = 5$$
$$f(-1) = 2^{-1+1} - 3 = 1 - 3 = -2$$
$$f(-2) = 2^{-2+1} - 3 = \frac{1}{2} - 3 = -\frac{5}{2}$$
$$f(-3) = 2^{-3+1} - 3 = \frac{1}{4} - 3 = -\frac{11}{4}$$

x	y, or $f(x)$
0	-1
1	1
2	5
-1	-2
-2	$-\frac{5}{2}$
-3	$-\frac{11}{4}$

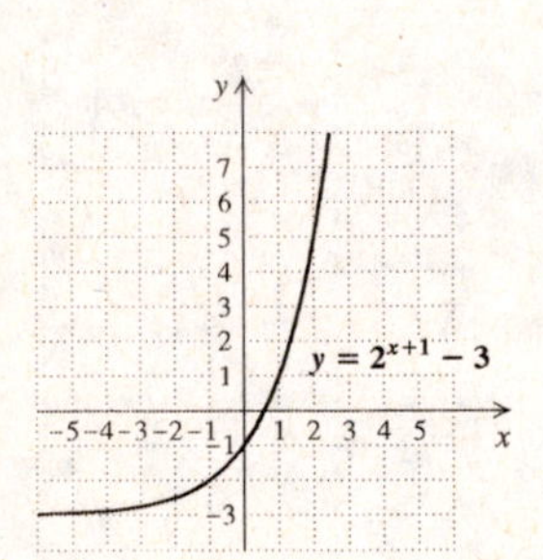

27. Graph: $x = 6^y$

We can find ordered pairs by choosing values for y and then computing values for x.

For $y = 0$, $x = 6^0 = 1$.

For $y = 1$, $x = 6^1 = 6$.

For $y = -1$, $x = 6^{-1} = \frac{1}{6^1} = \frac{1}{6}$.

For $y = -2$, $x = 6^{-2} = \frac{1}{6^2} = \frac{1}{36}$.

x	y
1	0
6	1
$\frac{1}{6}$	-1
$\frac{1}{36}$	-2

(1) Choose values for y.

(2) Compute values for x.

We plot the points and connect them with a smooth curve.

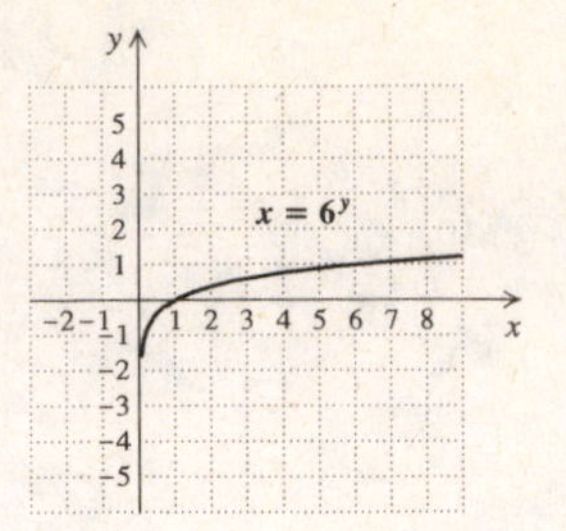

29. Graph: $x = 3^{-y} = \left(\frac{1}{3}\right)^y$

We can find ordered pairs by choosing values for y and then computing values for x. Then we plot these points and connect them with a smooth curve.

For $y = 0$, $x = \left(\frac{1}{3}\right)^0 = 1$.

For $y = 1$, $x = \left(\frac{1}{3}\right)^1 = \frac{1}{3}$.

For $y = 2$, $x = \left(\frac{1}{3}\right)^2 = \frac{1}{9}$.

For $y = -1$, $x = \left(\frac{1}{3}\right)^{-1} = \frac{1}{\frac{1}{3}} = 3$.

For $y = -2$, $x = \left(\frac{1}{3}\right)^{-2} = \frac{1}{\frac{1}{9}} = 9$.

x	y
1	0
$\frac{1}{3}$	1
$\frac{1}{9}$	2
3	-1
9	-2

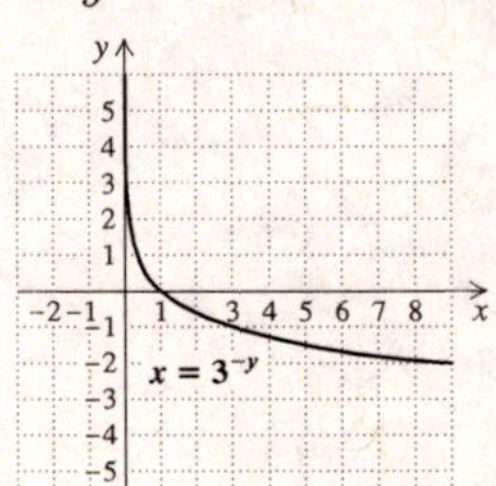

31. Graph: $x = 4^y$

We can find ordered pairs by choosing values for y and then computing values for x. Then we plot these points and connect them with a smooth curve.

For $y = 0$, $x = 4^0 = 1$.

For $y = 1$, $x = 4^1 = 4$.

For $y = 2$, $x = 4^2 = 16$.

For $y = -1$, $x = 4^{-1} = \frac{1}{4}$.

For $y = -2$, $x = 4^{-2} = \frac{1}{16}$.

x	y
1	0
4	1
16	2
$\frac{1}{4}$	-1
$\frac{1}{16}$	-2

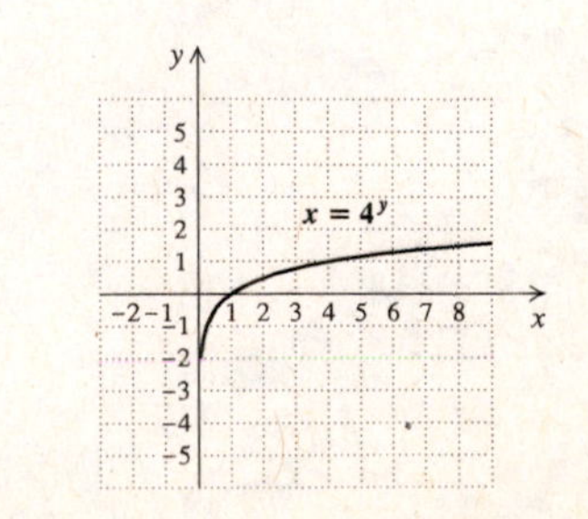

33. Graph: $x = \left(\frac{4}{3}\right)^y$

We can find ordered pairs by choosing values for y and then computing values for x. Then we plot these points and connect them with a smooth curve.

For $y = 0$, $x = \left(\frac{4}{3}\right)^0 = 1$.

For $y = 1$, $x = \left(\frac{4}{3}\right)^1 = \frac{4}{3}$.

For $y = 2$, $x = \left(\frac{4}{3}\right)^2 = \frac{16}{9}$.

For $y = 3$, $x = \left(\frac{4}{3}\right)^3 = \frac{64}{27}$.

For $y = -1$, $x = \left(\frac{4}{3}\right)^{-1} = \frac{3}{4}$.

For $y = -2$, $x = \left(\frac{4}{3}\right)^{-2} = \left(\frac{3}{4}\right)^2 = \frac{9}{16}$.

For $y = -3$, $x = \left(\frac{4}{3}\right)^{-3} = \left(\frac{3}{4}\right)^3 = \frac{27}{64}$.

x	y
1	0
$\frac{4}{3}$	1
$\frac{16}{9}$	2
$\frac{64}{27}$	3
$\frac{3}{4}$	−1
$\frac{9}{16}$	−2
$\frac{27}{64}$	−3

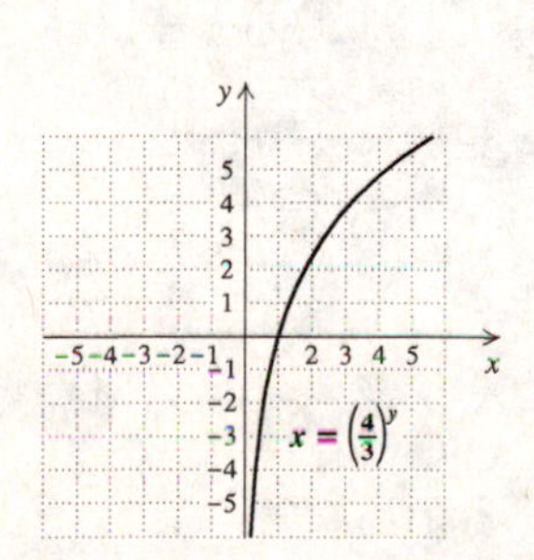

35. Graph $y = 3^x$ (see Exercise 8) and $x = 3^y$ (see Exercise 28) using the same set of axes.

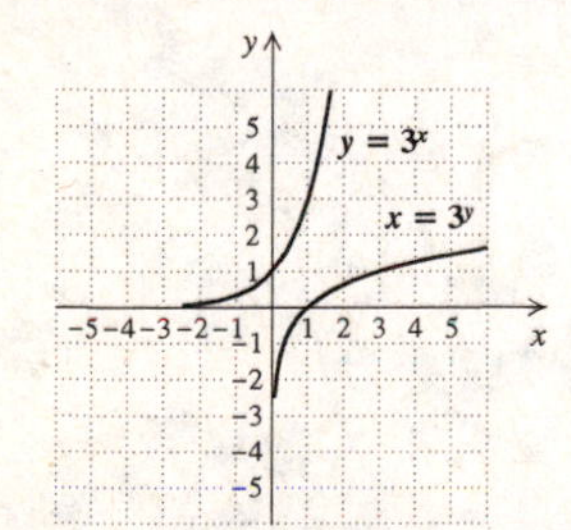

37. Graph $y = \left(\frac{1}{2}\right)^x$ (see Exercise 24) and $x = \left(\frac{1}{2}\right)^y$ (see Exercise 30) using the same set of axes.

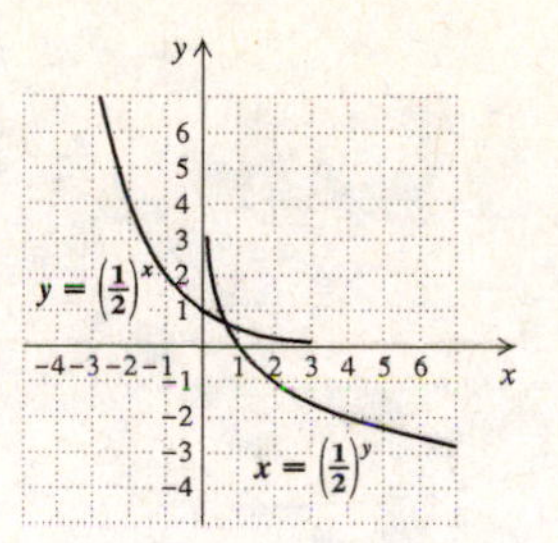

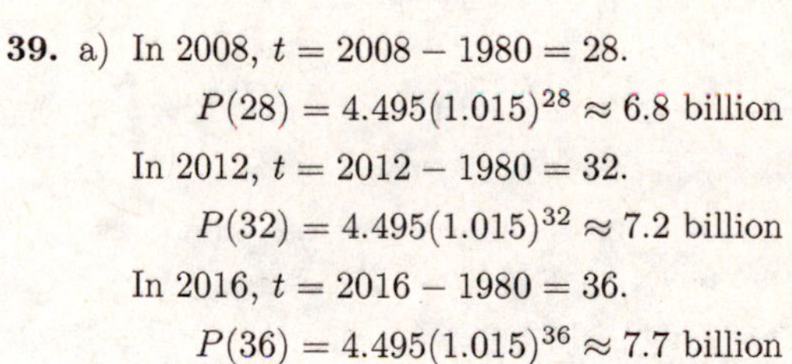

39. a) In 2008, $t = 2008 - 1980 = 28$.

$P(28) = 4.495(1.015)^{28} \approx 6.8$ billion

In 2012, $t = 2012 - 1980 = 32$.

$P(32) = 4.495(1.015)^{32} \approx 7.2$ billion

In 2016, $t = 2016 - 1980 = 36$.

$P(36) = 4.495(1.015)^{36} \approx 7.7$ billion

b) Use the function values computed in part (a) and others, if desired, and draw the graph.

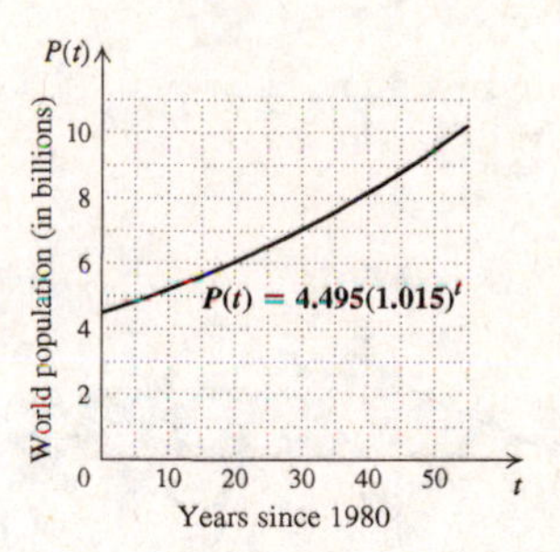

41. a) $P(1) = 21.4(0.914)^1 \approx 19.6\%$

$P(3) = 21.4(0.914)^3 \approx 16.3\%$

1 yr = 12 months; $P(12) = 21.4(0.914)^{12} \approx 7.3\%$

b)

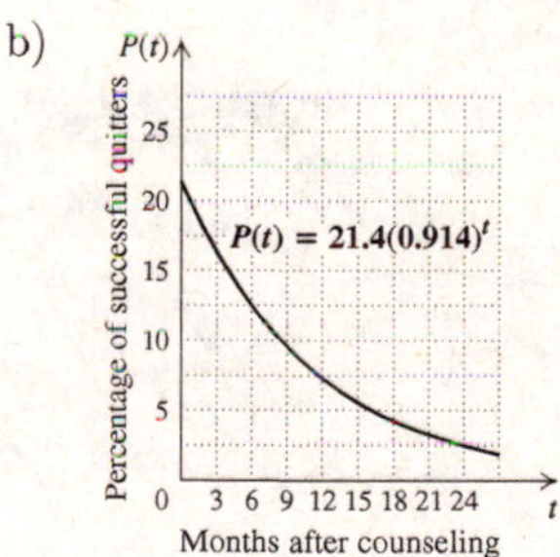

43. a) In 1930, $t = 1930 - 1900 = 30$.

$$P(t) = 150(0.960)^t$$
$$P(30) = 150(0.960)^{30}$$
$$\approx 44.079$$

In 1930, about 44.079 thousand, or 44,079, humpback whales were alive.

In 1960, $t = 1960 - 1900 = 60$.

$$P(t) = 150(0.960)^t$$
$$P(60) = 150(0.960)^{60}$$
$$\approx 12.953$$

In 1960, about 12.953 thousand, or 12,953, humpback whales were alive.

b) Plot the points found in part (a), (30, 44,079) and (60, 12,953) and additional points as needed and graph the function.

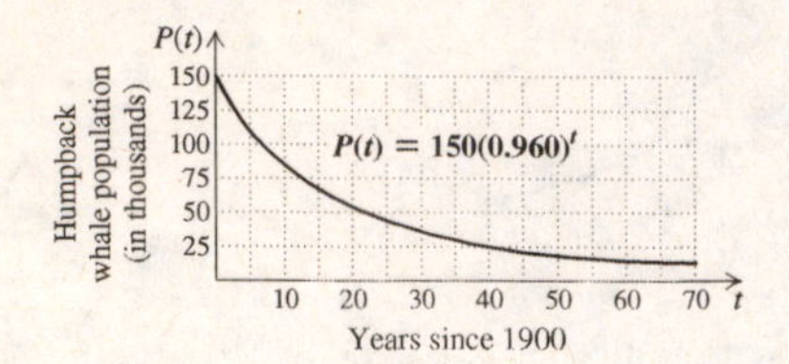

45. a) In 1992, $t = 1992 - 1982 = 10$.

$P(10) = 5.5(1.047)^{10} \approx 8.706$

In 1992, about 8.706 thousand, or 8706, humpback whales were alive.

In 2004, $t = 2004 - 1982 = 22$.

$P(22) = 5.5(1.047)^{22} \approx 15.107$

In 2004, about 15.107 thousand, or 15,107, humpback whales were alive.

b) Use the function values computed in part (a) and others, if desired, and draw the graph.

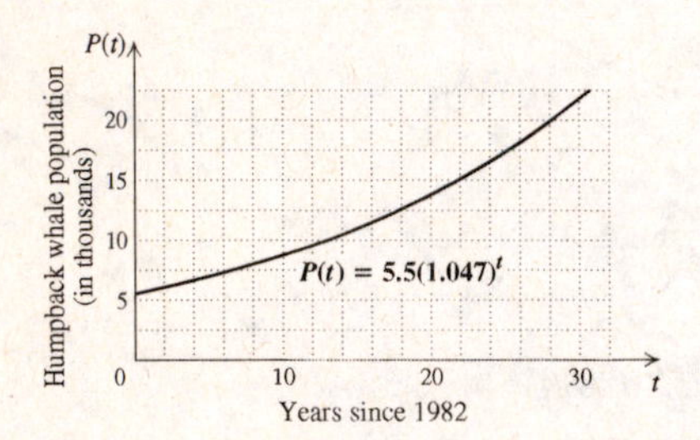

47. a) $A(5) = 10 \cdot 34^5 = 454,354,240 \text{ cm}^2$

$A(7) = 10 \cdot 34^7 = 525,233,501,400 \text{ cm}^2$

b) Use the function values computed in part (a) and others, if desired, and draw the graph.

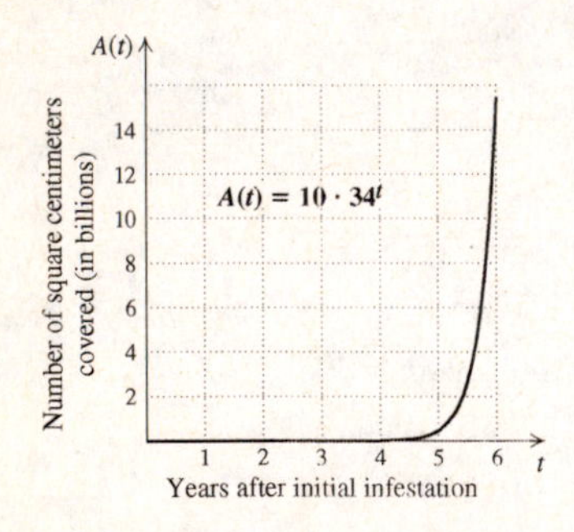

49. *Writing Exercise*

51. $5^{-2} = \frac{1}{5^2} = \frac{1}{25}$

53. $1000^{2/3} = (10^3)^{2/3} = 10^{3 \cdot \frac{2}{3}} = 10^2 = 100$

55. $\frac{10a^8b^7}{2a^2b^4} = \frac{10}{2}a^{8-2}b^{7-4} = 5a^6b^3$

57. *Writing Exercise*

59. Since the bases are the same, the one with the larger exponent is the larger number. Thus $\pi^{2.4}$ is larger.

61. Graph: $f(x) = 3.8^x$

Use a calculator with a power key to construct a table of values. (We will round values of $f(x)$ to the nearest hundredth.) Then plot these points and connect them with a smooth curve.

x	y
0	1
1	3.8
2	14.44
3	54.872
−1	0.26
−2	0.7

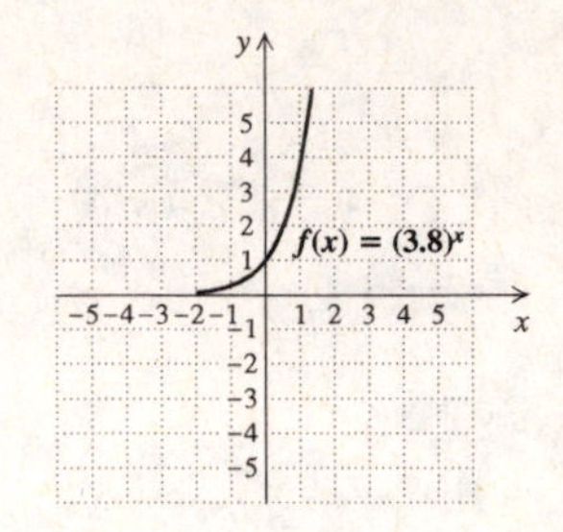

63. Graph: $y = 2^x + 2^{-x}$

Construct a table of values, thinking of y as $f(x)$. Then plot these points and connect them with a curve.

$f(0) = 2^0 + 2^{-0} = 1 + 1 = 2$

$f(1) = 2^1 + 2^{-1} = 2 + \frac{1}{2} = 2\frac{1}{2}$

$f(2) = 2^2 + 2^{-2} = 4 + \frac{1}{4} = 4\frac{1}{4}$

$f(3) = 2^3 + 2^{-3} = 8 + \frac{1}{8} = 8\frac{1}{8}$

$f(-1) = 2^{-1} + 2^{-(-1)} = \frac{1}{2} + 2 = 2\frac{1}{2}$

$f(-2) = 2^{-2} + 2^{-(-2)} = \frac{1}{4} + 4 = 4\frac{1}{4}$

$f(-3) = 2^{-3} + 2^{-(-3)} = \frac{1}{8} + 8 = 8\frac{1}{8}$

x	y, or $f(x)$
0	2
1	$2\frac{1}{2}$
2	$4\frac{1}{4}$
3	$8\frac{1}{8}$
−1	$2\frac{1}{2}$
−2	$4\frac{1}{4}$
−3	$8\frac{1}{8}$

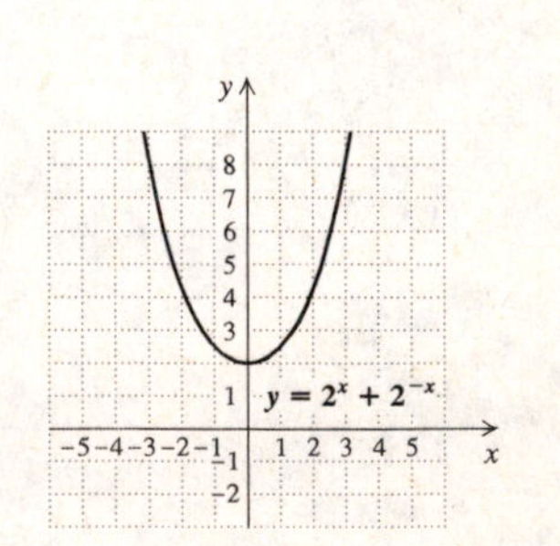

65. Graph: $y = |2^x - 2|$

We construct a table of values, thinking of y as $f(x)$. Then plot these points and connect them with a curve.

$$f(0) = |2^0 - 2| = |1 - 2| = |-1| = 1$$
$$f(1) = |2^1 - 2| = |2 - 2| = |0| = 0$$
$$f(2) = |2^2 - 2| = |4 - 2| = |2| = 2$$
$$f(3) = |2^3 - 2| = |8 - 2| = |6| = 6$$
$$f(-1) = |2^{-1} - 2| = \left|\frac{1}{2} - 2\right| = \left|-\frac{3}{2}\right| = \frac{3}{2}$$
$$f(-3) = |2^{-3} - 2| = \left|\frac{1}{8} - 2\right| = \left|-\frac{15}{8}\right| = \frac{15}{8}$$
$$f(-5) = |2^{-5} - 2| = \left|\frac{1}{32} - 2\right| = \left|-\frac{63}{32}\right| = \frac{63}{32}$$

x	y, or $f(x)$
0	1
1	0
2	2
3	6
-1	$\frac{3}{2}$
-3	$\frac{15}{8}$
-5	$\frac{63}{32}$

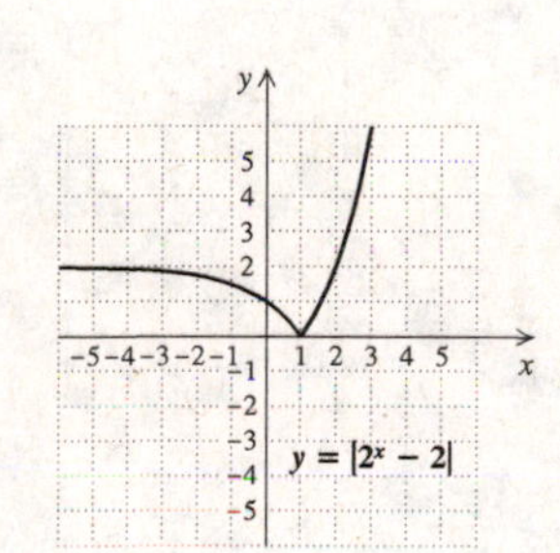

67. Graph: $y = |2x^2 - 1|$

We construct a table of values, thinking of y as $f(x)$. Then we plot these points and connect them with a curve.

$$f(0) = |2^{0^2} - 1| = |1 - 1| = 0$$
$$f(1) = |2^{1^2} - 1| = |2 - 1| = 1$$
$$f(2) = |2^{2^2} - 1| = |16 - 1| = 15$$
$$f(-1) = |2^{(-1)^2} - 1| = |2 - 1| = 1$$
$$f(-2) = |2^{(-2)^2} - 1| = |16 - 1| = 15$$

x	y, or $f(x)$
0	0
1	1
2	15
-1	1
-2	15

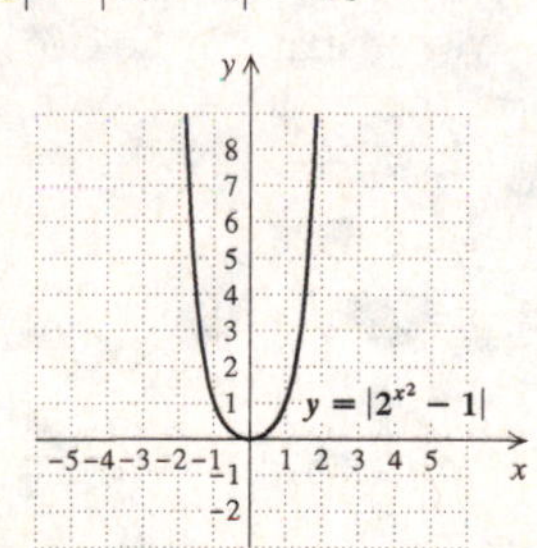

69. $y = 3^{-(x-1)}$

x	y
0	3
1	1
2	$\frac{1}{3}$
3	$\frac{1}{9}$
-1	9

$x = 3^{-(y-1)}$

x	y
3	0
1	1
$\frac{1}{3}$	2
$\frac{1}{9}$	3
9	-1

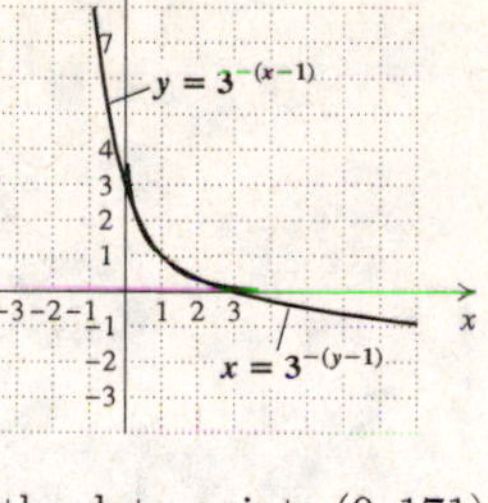

71. Enter the data points $(0, 171)$, $(2, 1099)$, and $(4, 2697)$ and then use the ExpReg option from the STAT CALC menu of a graphing calculator to find an exponential function that models the data:

$A(t) = 200.7624553(1.992834389)^t$, where $A(t)$ is total sales, in millions of dollars, t years after 1997.

In 2008, $t = 2008 - 1997 = 11$.

$A(11) \approx \$395,244.4657$ million, or \$395,244,465,700

73. *Writing Exercise*

75.

Exercise Set 12.3

1. $5^2 = 25$, so choice (g) is correct.

3. $5^1 = 5$, so choice (a) is correct.

5. The exponent to which we raise 5 to get 5^x is x, so choice (b) is correct.

7. $8 = 2^x$ is equivalent to $\log_2 8 = x$, so choice (e) is correct.

9. $\log_{10} 1000$ is the exponent to which we raise 10 to get 1000. Since $10^3 = 1000$, $\log_{10} 1000 = 3$.

11. $\log_2 16$ is the exponent to which we raise 2 to get 16. Since $2^4 = 16$, $\log_2 16 = 4$.

13. $\log_3 81$ is the exponent to which we raise 3 to get 81. Since $3^4 = 81$, $\log_3 81 = 4$.

15. $\log_4 \frac{1}{16}$ is the exponent to which we raise 4 to get $\frac{1}{16}$. Since $4^{-2} = \frac{1}{16}$, $\log_4 \frac{1}{16} = -2$.

17. Since $7^{-1} = \frac{1}{7}$, $\log_7 \frac{1}{7} = -1$.

19. Since $5^4 = 625$, $\log_5 625 = 4$.

21. Since $8^1 = 8$, $\log_8 8 = 1$.

23. Since $8^0 = 1$, $\log_8 1 = 0$.

25. $\log_9 9^5$ is the exponent to which we raise 9 to get 9^5. Clearly, this power is 5, so $\log_9 9^5 = 5$.

27. Since $10^{-2} = \frac{1}{100} = 0.01$, $\log_{10} 0.01 = -2$.

29. Since $9^{1/2} = 3$, $\log_9 3 = \frac{1}{2}$.

31. Since $9 = 3^2$ and $(3^2)^{3/2} = 3^3 = 27$, $\log_9 27 = \frac{3}{2}$.

33. Since $1000 = 10^3$ and $(10^3)^{2/3} = 10^2 = 100$, $\log_{1000} 100 = \frac{2}{3}$.

35. Since $\log_5 7$ is the power to which we raise 5 to get 7, then 5 raised to this power is 7. That is, $5^{\log_5 7} = 7$.

37. Graph: $y = \log_{10} x$

The equation $y = \log_{10} x$ is equivalent to $10^y = x$. We can find ordered pairs by choosing values for y and computing the corresponding x-values.

For $y = 0$, $x = 10^0 = 1$.

For $y = 1$, $x = 10^1 = 10$.

For $y = 2$, $x = 10^2 = 100$.

For $y = -1$, $x = 10^{-1} = \frac{1}{10}$.

For $y = -2$, $x = 10^{-2} = \frac{1}{100}$.

x, or 10^y	y
1	0
10	1
100	2
$\frac{1}{10}$	-1
$\frac{1}{100}$	-2

(1) Select y.

(2) Compute x.

We plot the set of ordered pairs and connect the points with a smooth curve.

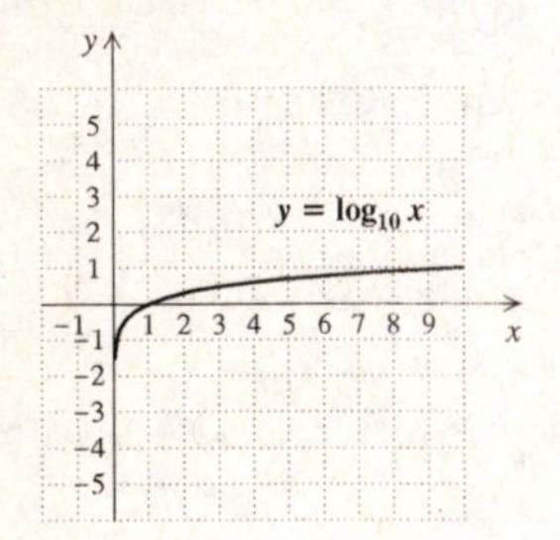

39. Graph: $y = \log_3 x$

The equation $y = \log_3 x$ is equivalent to $3^y = x$. We can find ordered pairs by choosing values for y and computing the corresponding x-values.

For $y = 0$, $x = 3^0 = 1$.

For $y = 1$, $x = 3^1 = 3$.

For $y = 2$, $x = 3^2 = 9$.

For $y = -1$, $x = 3^{-1} = \frac{1}{3}$.

For $y = -2$, $x = 3^{-2} = \frac{1}{9}$.

x, or 3^y	y
1	0
3	1
9	2
$\frac{1}{3}$	-1
$\frac{1}{9}$	-2

We plot the set of ordered pairs and connect the points with a smooth curve.

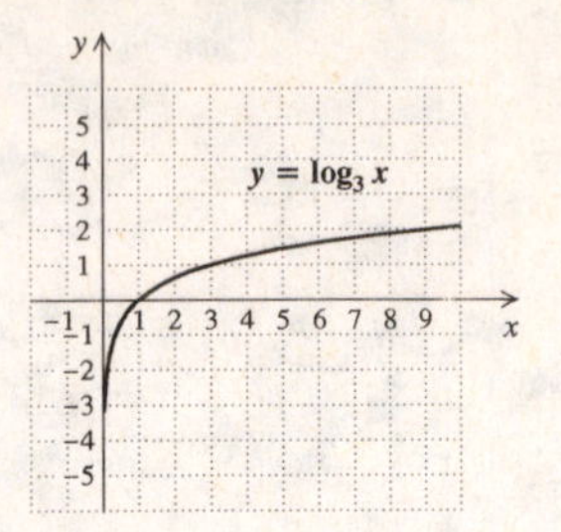

41. Graph: $f(x) = \log_6 x$

Think of $f(x)$ as y. Then $y = \log_6 x$ is equivalent to $6^y = x$. We find ordered pairs by choosing values for y and computing the corresponding x-values. Then we plot the points and connect them with a smooth curve.

For $y = 0$, $x = 6^0 = 1$.

For $y = 1$, $x = 6^1 = 6$.

For $y = 2$, $x = 6^2 = 36$.

For $y = -1$, $x = 6^{-1} = \frac{1}{6}$.

For $y = -2$, $x = 6^{-2} = \frac{1}{36}$.

x, or 6^y	y
1	0
6	1
36	2
$\frac{1}{6}$	-1
$\frac{1}{36}$	-2

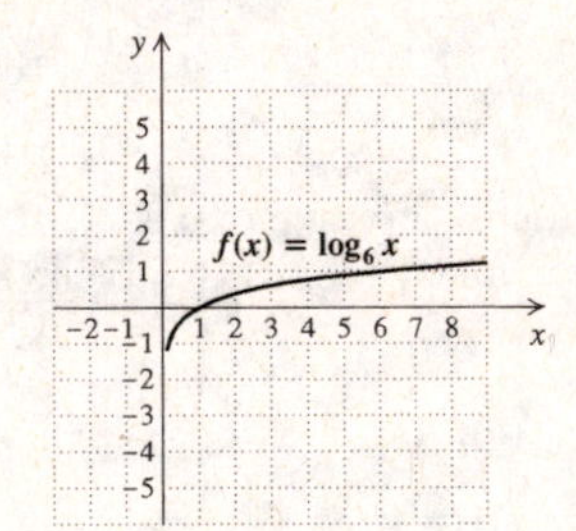

43. Graph: $f(x) = \log_{2.5} x$

Think of $f(x)$ as y. Then $y = \log_{2.5} x$ is equivalent to $2.5^y = x$. We construct a table of values, plot these points and connect them with a smooth curve.

For $y = 0, x = 2.5^0 = 1$.

For $y = 1, x = 2.5^1 = 2.5$.

For $y = 2, x = 2.5^2 = 6.25$.

For $y = 3, x = 2.5^3 = 15.625$.

For $y = -1, x = 2.5^{-1} = 0.4$.

For $y = -2, x = 2.5^{-2} = 0.16$.

x, or 2.5^y	y
1	0
2.5	1
6.25	2
15.625	3
0.4	-1
0.16	-2

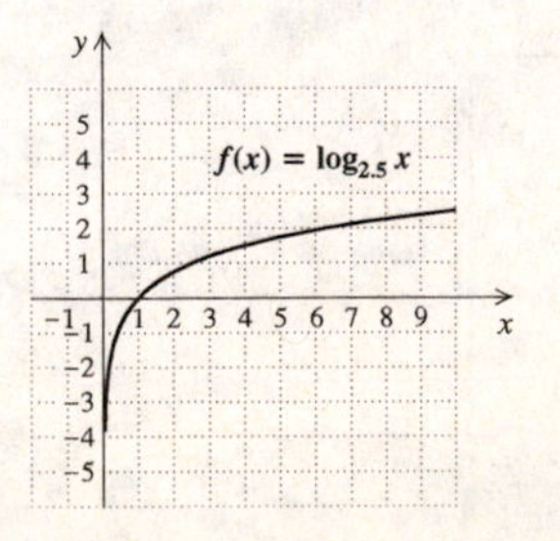

45. Graph $f(x) = 3^x$ (see Exercise Set 9.2, Exercise 7) and $f^{-1}(x) = \log_3 x$ (see Exercise 39 above) on the same set of axes.

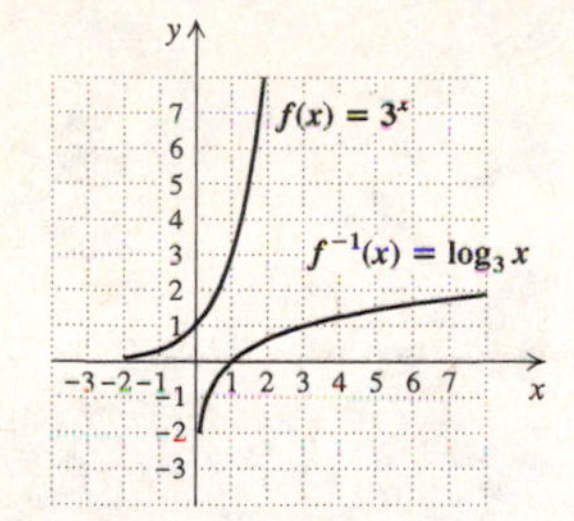

47. $t = \log_5 9 \Rightarrow 5^t = 9$

The base remains the same. The logarithm is the exponent.

49. $\log_5 25 = 2 \Rightarrow 5^2 = 25$

The logarithm is the exponent. The base remains the same.

51. $\log_{10} 0.1 = -1$ is equivalent to $10^{-1} = 0.1$.

53. $\log_{10} 7 = 0.845$ is equivalent to $10^{0.845} = 7$.

55. $\log_c m = 8$ is equivalent to $c^8 = m$.

57. $\log_t Q = r$ is equivalent to $t^r = Q$.

59. $\log_e 0.25 = -1.3863$ is equivalent to $e^{-1.3863} = 0.25$.

61. $\log_r T = -x$ is equivalent to $r^{-x} = T$.

63. $10^2 = 100 \Rightarrow 2 = \log_{10} 100$

The exponent is the logarithm. The base remains the same.

65. $4^{-5} = \frac{1}{1024} \Rightarrow -5 = \log_4 \frac{1}{1024}$

The exponent is the logarithm. The base remains the same.

67. $16^{3/4} = 8$ is equivalent to $\frac{3}{4} = \log_{16} 8$.

69. $10^{0.4771} = 3$ is equivalent to $0.4771 = \log_{10} 3$.

71. $z^m = 6$ is equivalent to $m = \log_z 6$.

73. $p^m = V$ is equivalent to $m = \log_p V$.

75. $e^3 = 20.0855$ is equivalent to $3 = \log_e 20.0855$.

77. $e^{-4} = 0.0183$ is equivalent to $-4 = \log_e 0.0183$.

79. $\log_3 x = 2$

$3^2 = x$ Converting to an exponential equation

$9 = x$ Computing 3^2

81. $\log_5 125 = x$

$5^x = 125$ Converting to an exponential equation

$5^x = 5^3$

$x = 3$ The exponents must be the same.

83. $\log_2 16 = x$

$2^x = 16$ Converting to an exponential equation

$2^x = 2^4$

$x = 4$ The exponents must be the same.

85. $\log_x 7 = 1$

$x^1 = 7$ Converting to an exponential equation

$x = 7$ Siimplifying x^1

87. $\log_3 x = -2$

$3^{-2} = x$ Converting to an exponential equation

$\frac{1}{9} = x$ Simplifying

89. $\log_{32} x = \frac{2}{5}$

$32^{2/5} = x$ Converting to an exponential equation

$(2^5)^{2/5} = x$

$4 = x$

91. *Writing Exercise*

93. $\frac{x^{12}}{x^4} = x^{12-4} = x^8$

95. $(a^4b^6)(a^3b^2) = a^{4+3}b^{6+2} = a^7b^8$

97. $\dfrac{\frac{3}{x} - \frac{2}{xy}}{\frac{2}{x^2} + \frac{1}{xy}}$

The LCD of all the denominators is x^2y. We multiply numerator and denominator by the LCD.

$$\frac{\frac{3}{x} - \frac{2}{xy}}{\frac{2}{x^2} + \frac{1}{xy}} \cdot \frac{x^2y}{x^2y} = \frac{\left(\frac{3}{x} - \frac{2}{xy}\right)x^2y}{\left(\frac{2}{x^2} + \frac{1}{xy}\right)x^2y}$$

$$= \frac{\frac{3}{x} \cdot x^2y - \frac{2}{xy} \cdot x^2y}{\frac{2}{x^2} \cdot x^2y + \frac{1}{xy} \cdot x^2y}$$

$$= \frac{3xy - 2x}{2y + x}, \text{ or}$$

$$\frac{x(3y-2)}{2y+x}$$

99. *Writing Exercise*

101. Graph: $y = \left(\frac{3}{2}\right)^x$

x	y, or $\left(\frac{3}{2}\right)^x$
0	1
1	$\frac{3}{2}$
2	$\frac{9}{4}$
3	$\frac{27}{8}$
−1	$\frac{2}{3}$
−2	$\frac{4}{9}$

Graph: $y = \log_{3/2} x$, or

$x = \left(\frac{3}{2}\right)^y$

x, or $\left(\frac{3}{2}\right)^y$	y
1	0
$\frac{3}{2}$	1
$\frac{9}{4}$	2
$\frac{27}{8}$	3
$\frac{2}{3}$	−1
$\frac{4}{9}$	−2

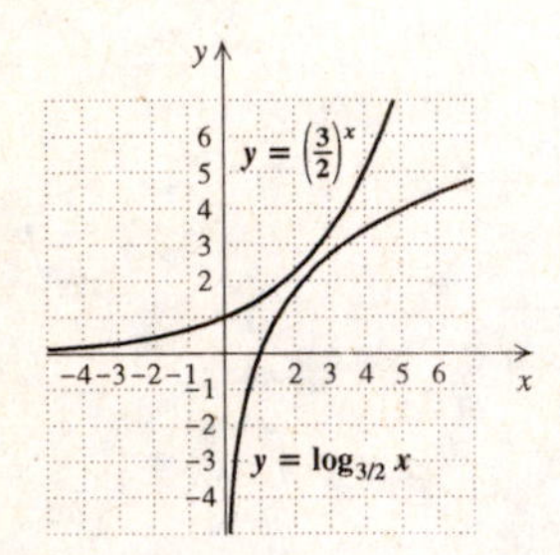

103. Graph: $y = \log_3 |x+1|$

x	y
0	0
2	1
8	2
−2	0
−4	1
−9	2

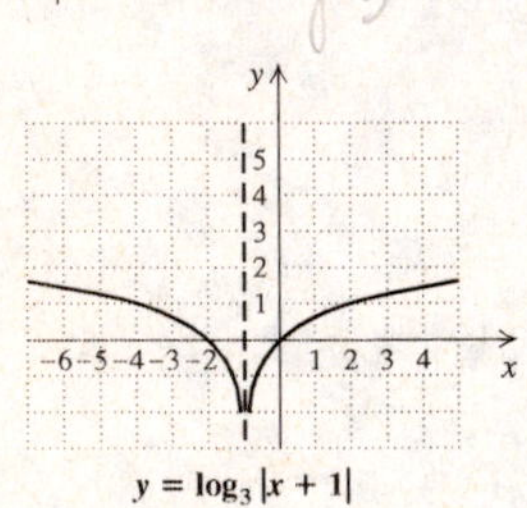

105.
$$\begin{aligned} \log_4(3x-2) &= 2 \\ 4^2 &= 3x-2 \\ 16 &= 3x-2 \\ 18 &= 3x \\ 6 &= x \end{aligned}$$

107.
$$\begin{aligned} \log_{10}(x^2+21x) &= 2 \\ 10^2 &= x^2+21x \\ 0 &= x^2+21x-100 \\ 0 &= (x+25)(x-4) \end{aligned}$$

$x = -25$ *or* $x = 4$

109. Let $\log_{1/5} 25 = x$. Then

$$\begin{aligned} \left(\frac{1}{5}\right)^x &= 25 \\ (5^{-1})^x &= 25 \\ 5^{-x} &= 5^2 \\ -x &= 2 \\ x &= -2. \end{aligned}$$

Thus, $\log_{1/5} 25 = -2$.

111.
$$\begin{aligned} &\log_{10}(\log_4(\log_3 81)) \\ &= \log_{10}(\log_4 4) \qquad (\log_3 81 = 4) \\ &= \log_{10} 1 \qquad (\log_4 4 = 1) \\ &= 0 \end{aligned}$$

113. Let $b = 0$, $x = 1$, and $y = 2$. Then $0^1 = 0^2$, but $1 \neq 2$. Let $b = 1$, $x = 1$, and $y = 2$. Then $1^1 = 1^2$, but $1 \neq 2$.

Exercise Set 12.4

1. Use the product rule for logarithms.

$\log_7 20 = \log_7(5 \cdot 4) = \log_7 5 + \log_7 4$; choice (e) is correct.

3. Use the quotient rule for logarithms.

$\log_7 \frac{5}{4} = \log_7 5 - \log_7 4$; choice (a) is correct.

5. The exponent to which we raise 7 to get 1 is 0, so choice (c) is correct.

7. $\log_3(81 \cdot 27) = \log_3 81 + \log_3 27$ Using the product rule

9. $\log_4(64 \cdot 16) = \log_4 64 + \log_4 16$ Using the product rule

11. $\log_c rst$

$= \log_c r + \log_c s + \log_c t$ Using the product rule

13. $\log_a 5 + \log_a 14 = \log_a(5 \cdot 14)$ Using the product rule

The result can also be expressed as $\log_a 70$.

15. $\log_c t + \log_c y = \log_c(t \cdot y)$ Using the product rule

17. $\log_a r^8 = 8 \log_a r$ Using the power rule

19. $\log_c y^6 = 6 \log_c y$ Using the power rule

21. $\log_b C^{-3} = -3 \log_b C$ Using the power rule

23. $\log_2 \frac{25}{13} = \log_2 25 - \log_2 13$ Using the quotient rule

25. $\log_b \frac{m}{n} = \log_b m - \log_b n$ Using the quotient rule

27. $\log_a 17 - \log_a 6$

$= \log_a \frac{17}{6}$

29. $\log_b 36 - \log_b 4$

$= \log_b \frac{36}{4}$, Using the quotient rule

or $\log_b 9$

31. $\log_a 7 - \log_z 18 = \log_a \frac{7}{18}$ Using the quotient rule

33. $\log_a (xyz)$

$= \log_a x + \log_a y + \log_a z$ Using the product rule

35. $\log_a (x^3 z^4)$

$= \log_a x^3 + \log_a z^4$ Using the product rule

$= 3\log_a x + 4\log_a z$ Using the power rule

37. $\log_a (x^2 y^{-2} z)$

$= \log_a x^2 + \log_a y^{-2} + \log_a z$ Using the product rule

$= 2\log_a x - 2\log_a y + \log_a z$ Using the power rule

39. $\log_a \frac{x^4}{y^3 z}$

$= \log_a x^4 - \log_a y^3 z$ Using the quotient rule

$= \log_a x^4 - (\log_a y^3 + \log_a z)$ Using the product rule

$= \log_a x^4 - \log_a y^3 - \log_a z$ Removing parentheses

$= 4\log_a x - 3\log_a y - \log_a z$ Using the power rule

41. $\log_b \frac{xy^2}{wz^3}$

$= \log_b xy^2 - \log_b wz^3$ Using the quotient rule

$= \log_b x + \log_b y^2 - (\log_b w + \log_b z^3)$

Using the product rule

$= \log_b x + \log_b y^2 - \log_b w - \log_b z^3$

Removing parentheses

$= \log_b x + 2\log_b y - \log_b w - 3\log_b z$

Using the power rule

43. $\log_a \sqrt{\frac{x^7}{y^5 z^8}}$

$= \log_a \left(\frac{x^7}{y^5 z^8}\right)^{1/2}$

$= \frac{1}{2}\log_a \frac{x^7}{y^5 z^8}$ Using the power rule

$= \frac{1}{2}(\log_a x^7 - \log_a y^5 z^8)$ Using the quotient rule

$= \frac{1}{2}\left[\log_a x^7 - (\log_a y^5 + \log_a z^8)\right]$

Using the product rule

$= \frac{1}{2}(\log_a x^7 - \log_a y^5 - \log_a z^8)$

Removing parentheses

$= \frac{1}{2}(7\log_a x - 5\log_a y - 8\log_a z)$

Using the power rule

45. $\log_a \sqrt[3]{\frac{x^6 y^3}{a^2 z^7}}$

$= \log_a \left(\frac{x^6 y^3}{a^2 z^7}\right)^{1/3}$

$= \frac{1}{3}\log_a \frac{x^6 y^3}{a^2 z^7}$ Using the power rule

$= \frac{1}{3}(\log_a x^6 y^3 - \log_a a^2 z^7)$ Using the quotient rule

$= \frac{1}{3}[\log_a x^6 + \log_a y^3 - (\log_a a^2 + \log_a z^7)]$

Using the product rule

$= \frac{1}{3}(\log_a x^6 + \log_a y^3 - \log_a a^2 - \log_a z^7)$

Removing parentheses

$= \frac{1}{3}(\log_a x^6 + \log_a y^3 - 2 - \log_a z^7)$

2 is the number to which we raise a to get a^2.

$= \frac{1}{3}(6\log_a x + 3\log_a y - 2 - 7\log_a z)$

Using the power rule

47. $8\log_a x + 3\log_a z$

$= \log_a x^8 + \log_a z^3$ Using the power rule

$= \log_a (x^8 z^3)$ Using the product rule

49. $\log_a x^2 - 2\log_a \sqrt{x}$

$= \log_a x^2 - \log_a (\sqrt{x})^2$ Using the power rule

$= \log_a x^2 - \log_a x$ $(\sqrt{x})^2 = x$

$= \log_a \frac{x^2}{x}$ Using the quotient rule

$= \log_a x$ Simplifying

51. $\frac{1}{2}\log_a x + 5\log_a y - 2\log_a x$

$= \log_a x^{1/2} + \log_a y^5 - \log_a x^2$ Using the power rule

$= \log_a x^{1/2}y^5 - \log_a x^2$ Using the product rule

$= \log_a \frac{x^{1/2}y^5}{x^2}$ Using the quotient rule

The result can also be expressed as $\log_a \frac{\sqrt{x}y^5}{x^2}$ or as $\log_a \frac{y^5}{x^{3/2}}$.

53. $\log_a (x^2 - 4) - \log_a (x + 2)$

$= \log_a \frac{x^2 - 4}{x + 2}$ Using the quotient rule

$= \log_a \frac{(x+2)(x-2)}{x+2}$

$= \log_a \frac{\cancel{(x+2)}(x-2)}{\cancel{x+2}}$ Simplifying

$= \log_a (x - 2)$

55. $\log_b 15 = \log_b (3 \cdot 5)$

$= \log_b 3 + \log_b 5$ Using the product rule

$= 0.792 + 1.161$

$= 1.953$

57. $\log_b \frac{3}{5} = \log_b 3 - \log_b 5$ Using the quotient rule
$= 0.792 - 1.161$
$= -0.369$

59. $\log_b \frac{1}{5} = \log_b 1 - \log_b 5$ Using the quotient rule
$= 0 - 1.161$ $(\log_b 1 = 0)$
$= -1.161$

61. $\log_b \sqrt{b^3} = \log_b b^{3/2} = \frac{3}{2}$ 3/2 is the number to which we raise b to get $b^{3/2}$.

63. $\log_b 8$

Since 8 cannot be expressed using the numbers 1, 3, and 5, we cannot find $\log_b 8$ using the given information.

65. $\log_t t^7 = 7$ 7 is the exponent to which we raise t to get t^7.

67. $\log_e e^m = m$ m is the exponent to which we raise e to get e^m.

69. *Writing Exercise*

71. Graph $f(x) = \sqrt{x} - 3$.

We construct a table of values, plot points, and connect them with a smooth curve. Note that we must choose nonnegative values of x in order for $\sqrt{x}$ to be a real number.

x	$f(x)$
0	−3
1	−2
4	−1
9	0

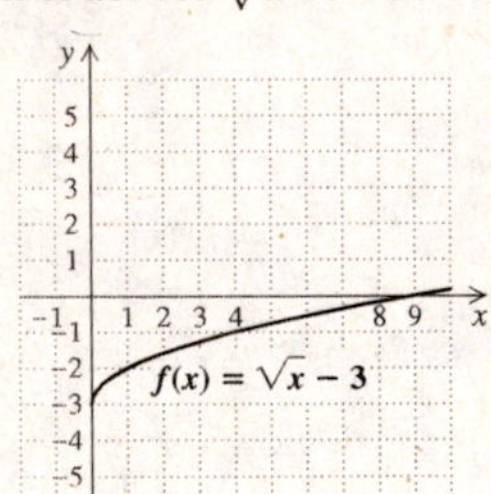

73. Graph $g(x) = \sqrt[3]{x} + 1$.

We construct a table of values, plot points, and connect them with a smooth curve.

x	$g(x)$
−8	−1
−1	0
0	1
1	2
8	3

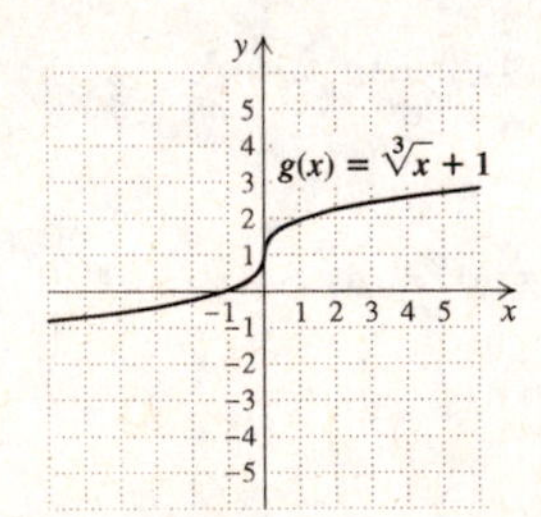

75. $(a^3b^2)^5(a^2b^7) = (a^{3\cdot5}b^{2\cdot5})(a^2b^7) = a^{15}b^{10}a^2b^7 = a^{15+2}b^{10+7} = a^{17}b^{17}$

77. *Writing Exercise*

79. $\log_a (x^8 - y^8) - \log_a (x^2 + y^2)$
$= \log_a \frac{x^8 - y^8}{x^2 + y^2}$
$= \log_a \frac{(x^4 + y^4)(x^2 + y^2)(x + y)(x - y)}{x^2 + y^2}$
$= \log_a [(x^4 + y^4)(x^2 - y^2)]$ Simplifying
$= \log_a (x^6 - x^4y^2 + x^2y^4 - y^6)$

81. $\log_a \sqrt{1 - s^2}$
$= \log_a (1 - s^2)^{1/2}$
$= \frac{1}{2} \log_a (1 - s^2)$
$= \frac{1}{2} \log_a [(1 - s)(1 + s)]$
$= \frac{1}{2} \log_a (1 - s) + \frac{1}{2} \log_a (1 + s)$

83. $\log_a \frac{\sqrt[3]{x^2z}}{\sqrt[3]{y^2z^{-2}}}$
$= \log_a \left(\frac{x^2z^3}{y^2}\right)^{1/3}$
$= \frac{1}{3}(\log_a x^2z^3 - \log_a y^2)$
$= \frac{1}{3}(2 \log_a x + 3 \log_a z - 2 \log_a y)$
$= \frac{1}{3}[2 \cdot 2 + 3 \cdot 4 - 2 \cdot 3]$
$= \frac{1}{3}(10)$
$= \frac{10}{3}$

85. $\log_a x = 2$, so $a^2 = x$.

Let $\log_{1/a} x = n$ and solve for n.

$\log_{1/a} a^2 = n$ Substituting a^2 for x
$\left(\frac{1}{a}\right)^n = a^2$
$(a^{-1})^n = a^2$
$a^{-n} = a^2$
$-n = 2$
$n = -2$

Thus, $\log_{1/a} x = -2$ when $\log_a x = 2$.

87. True; $\log_a(Q + Q^2) = \log_a[Q(1 + Q)] = \log_a Q + \log_a(1 + Q) = \log_a Q + \log_a(Q + 1)$.

Exercise Set 12.5

1. True; see page 817 in the text.

3. True; see page 819 in the text.

5. Using the change-of-base formula with $a = e$, $b = 2$, and $M = 9$, we see that the statement is true.

7. True; see Example 7.

9. 0.7782

11. 1.8621

13. Since $10^3 = 1000$, $\log 1000 = 3$.

15. -0.2782

17. 1.7986

19. 199.5262

21. 1.4894

23. 0.0011

25. 1.6094

27. 4.0431

29. -5.0832

31. 96.7583

33. 15.0293

35. 0.0305

37. 109.9472

39. We will use common logarithms for the conversion. Let $a = 10$, $b = 6$, and $M = 92$ and substitute in the change-of-base formula.

$$\log_b M = \frac{\log_a M}{\log_a b}$$
$$\log_6 92 = \frac{\log_{10} 92}{\log_{10} 6} \approx \frac{1.963787827}{0.7781512504} \approx 2.5237$$

41. We will use common logarithms for the conversion. Let $a = 10$, $b = 2$, and $M = 100$ and substitute in the change-of-base formula.

$$\log_2 100 = \frac{\log_{10} 100}{\log_{10} 2} \approx \frac{2}{0.3010} \approx 6.6439$$

43. We will use natural logarithms for the conversion. Let $a = e$, $b = 7$, and $M = 65$ and substitute in the change-of-base formula.

$$\log_7 65 = \frac{\ln 65}{\ln 7} \approx \frac{4.1744}{1.9459} \approx 2.1452$$

45. We will use natural logarithms for the conversion. Let $a = e$, $b = 0.5$, and $M = 5$ and substitute in the change-of-base formula.

$$\log_{0.5} 5 = \frac{\ln 5}{\ln 0.5} \approx \frac{1.6094}{-0.6931} \approx -2.3219$$

47. We will use common logarithms for the conversion. Let $a = 10$, $b = 2$, and $M = 0.2$ and substitute in the change-of-base formula.

$$\log_2 0.2 = \frac{\log_{10} 0.2}{\log_{10} 2} \approx \frac{-0.6990}{0.3010} \approx -2.3219$$

49. We will use natural logarithms for the conversion. Let $a = e$, $b = \pi$, and $M = 58$ and substitute in the change-of-base formula.

$$\log_\pi 58 = \frac{\ln 58}{\ln \pi} \approx \frac{4.0604}{1.1447} \approx 3.5471$$

51. Graph: $f(x) = e^x$

We find some function values with a calculator. We use these values to plot points and draw the graph.

x	e^x
0	1
1	2.7
2	7.4
3	20.1
-1	0.4
-2	0.1

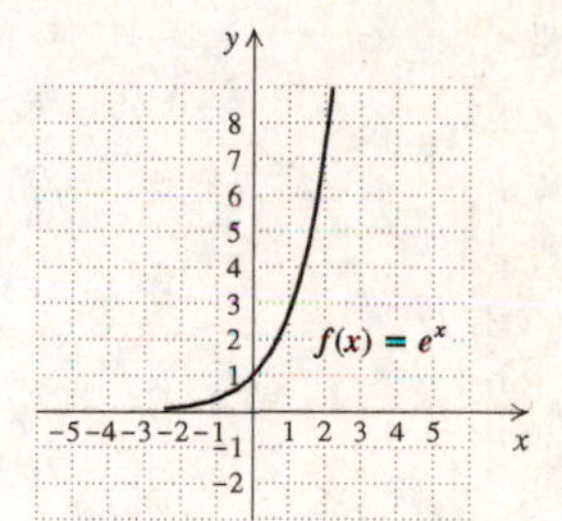

The domain is the set of real numbers and the range is $(0, \infty)$.

53. Graph: $f(x) = e^x + 3$

We find some function values, plot points, and draw the graph.

x	$e^x + 3$
0	4
1	5.72
2	10.39
-1	3.37
-2	3.14

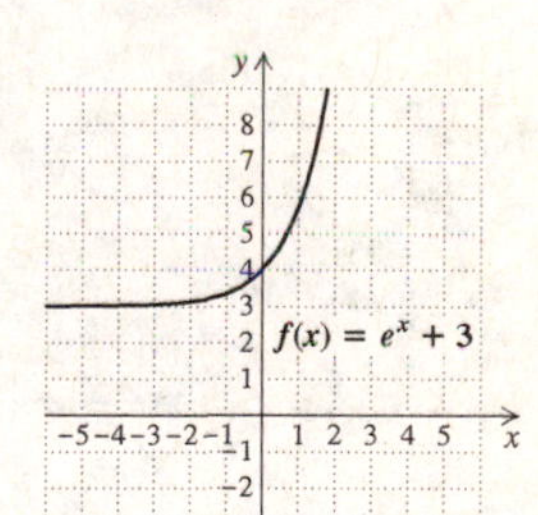

The domain is the set of real numbers and the range is $(3, \infty)$.

55. Graph: $f(x) = e^x - 2$

We find some function values, plot points, and draw the graph.

x	$e^x - 2$
0	-1
1	0.72
2	5.4
-1	-1.6
-2	-1.9

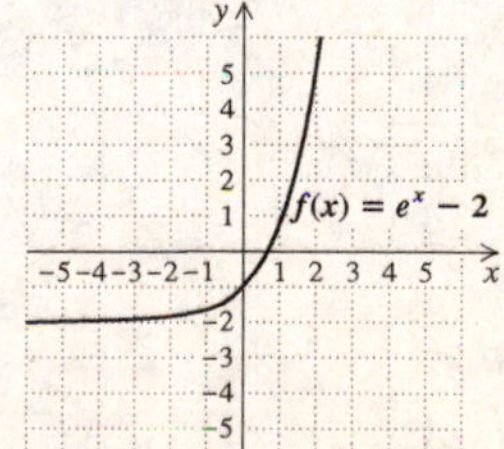

The domain is the set of real numbers and the range is $(-2, \infty)$.

57. Graph: $f(x) = 0.5e^x$

We find some function values, plot points, and draw the graph.

x	$0.5e^x$
0	0.5
1	1.36
2	3.69
-1	0.18
-2	0.07

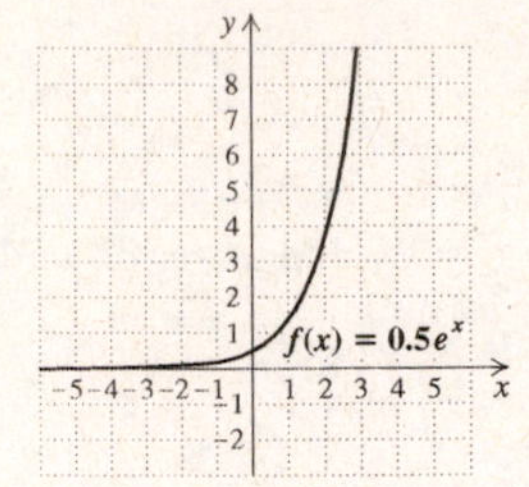

The domain is the set of real numbers and the range is $(0, \infty)$.

59. Graph: $f(x) = 0.5e^{2x}$

We find some function values, plot points, and draw the graph.

x	$0.5e^{2x}$
0	0.5
1	3.69
2	27.30
-1	0.07
-2	0.01

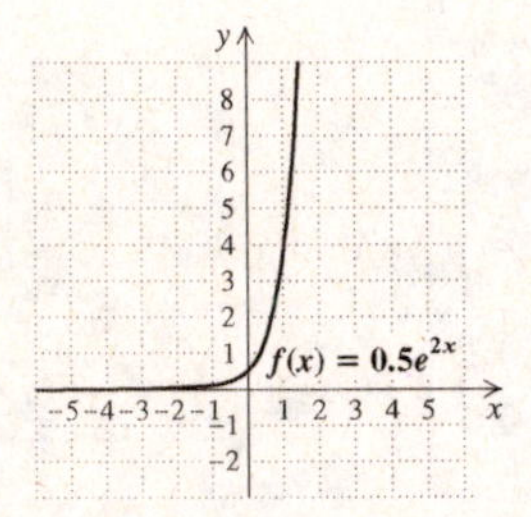

The domain is the set of real numbers and the range is $(0, \infty)$.

61. Graph: $f(x) = e^{x-3}$

We find some function values, plot points, and draw the graph.

x	e^{x-3}
0	0.05
2	0.37
3	1
4	2.72
-2	0.01

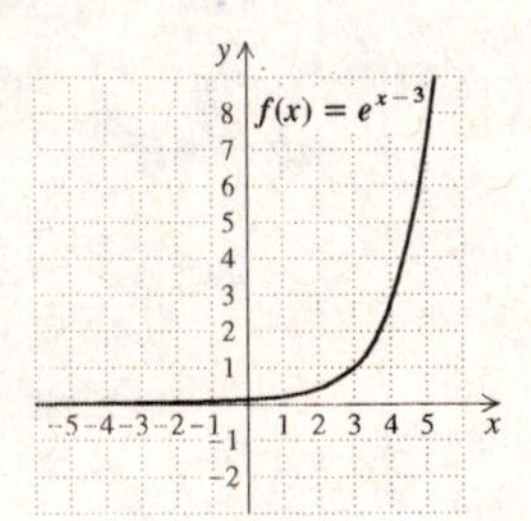

The domain is the set of real numbers and the range is $(0, \infty)$.

63. Graph: $f(x) = e^{x+2}$

We find some function values, plot points, and draw the graph.

x	e^{x+2}
0	7.39
-1	2.72
-2	1
-3	0.37
-4	0.14

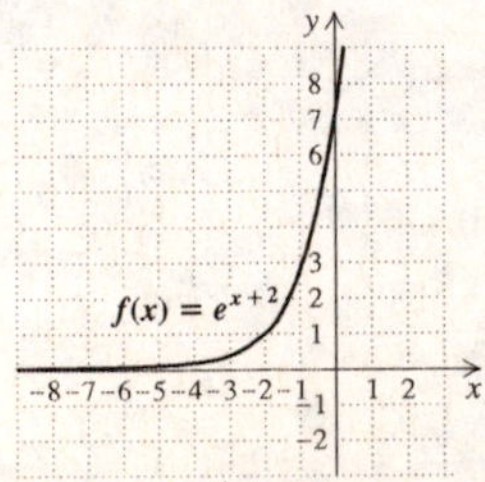

The domain is the set of real numbers and the range is $(0, \infty)$.

65. Graph: $f(x) = -e^x$

We find some function values, plot points, and draw the graph.

x	$-e^x$
0	-1
1	-2.72
2	-7.39
-1	-0.37
-3	-0.05

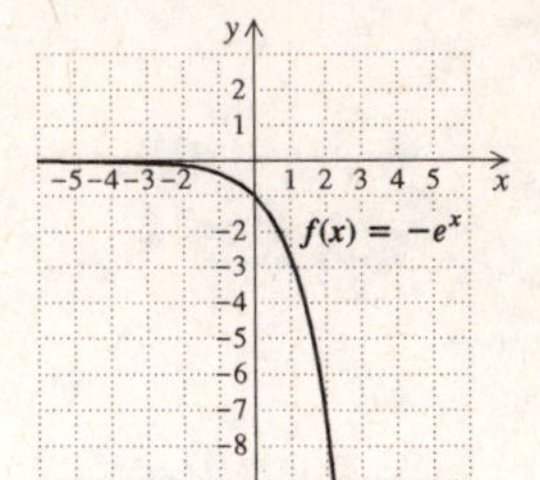

The domain is the set of real numbers and the range is $(-\infty, 0)$.

67. Graph: $g(x) = \ln x + 1$

We find some function values, plot points, and draw the graph.

x	$\ln x + 1$
0.5	0.31
1	1
3	2.10
5	2.61
7	2.95

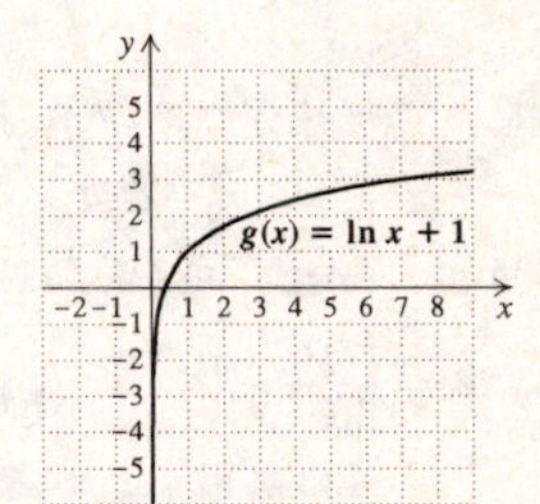

The domain is $(0, \infty)$ and the range is the set of real numbers.

69. Graph: $g(x) = \ln x - 2$

x	$\ln x - 2$
1	-2
2	-1.31
3	-0.90
4	-0.61
5	-0.39

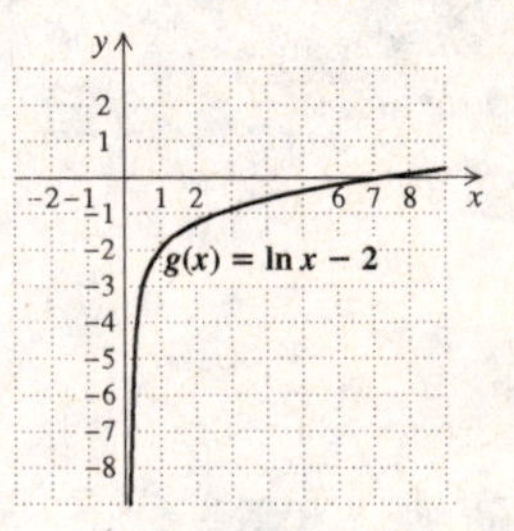

The domain is $(0, \infty)$ and the range is the set of real numbers.

71. Graph: $f(x) = 2 \ln x$

x	$2 \ln x$
0.5	-1.4
1	0
2	1.4
3	2.2
4	2.8
5	3.2
6	3.6

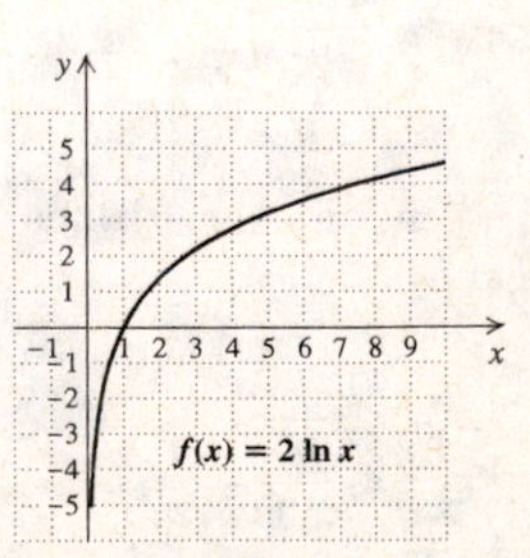

The domain is $(0, \infty)$ and the range is the set of real numbers.

73. Graph: $g(x) = -2 \ln x$

x	$-2 \ln x$
0.5	1.4
1	0
2	-1.4
3	-2.2
4	-2.8
5	-3.2
6	-3.6

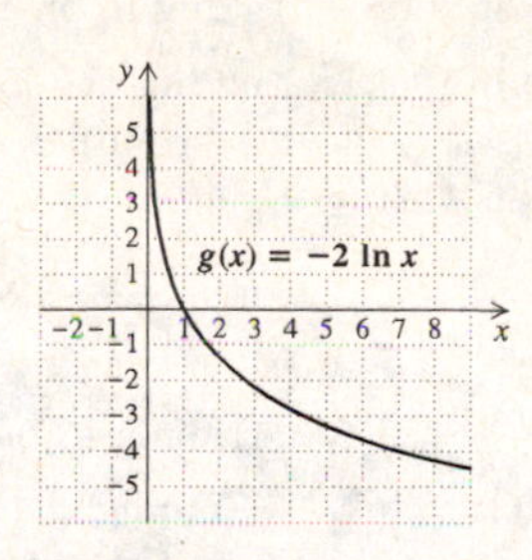

The domain is $(0, \infty)$ and the range is the set of real numbers.

75. Graph: $f(x) = \ln(x + 2)$

We find some function values, plot points, and draw the graph.

x	$\ln(x + 2)$
0	0.69
1	1.10
3	1.61
5	1.95
-1	0
-2	Undefined

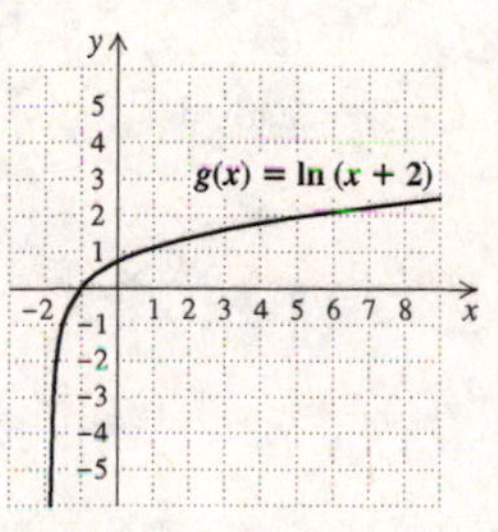

The domain is $(-2, \infty)$ and the range is the set of real numbers.

77. Graph: $g(x) = \ln(x - 1)$

We find some function values, plot points, and draw the graph.

x	$\ln(x - 1)$
1.1	-2.30
2	0
3	0.69
4	1.10
6	1.61

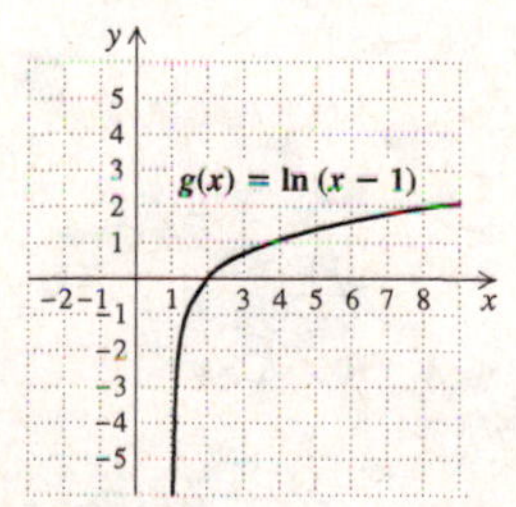

The domain is $(1, \infty)$ and the range is the set of real numbers.

79. *Writing Exercise*

81.
$$\begin{aligned} 4x^2 - 25 &= 0 \\ (2x + 5)(2x - 5) &= 0 \\ 2x + 5 = 0 \quad &\text{or} \quad 2x - 5 = 0 \\ 2x = -5 \quad &\text{or} \quad 2x = 5 \\ x = -\frac{5}{2} \quad &\text{or} \quad x = \frac{5}{2} \end{aligned}$$

The solutions are $-\frac{5}{2}$ and $\frac{5}{2}$.

83.
$$\begin{aligned} 17x - 15 &= 0 \\ 17x &= 15 \\ x &= \frac{15}{17} \end{aligned}$$

The solution is $\frac{15}{17}$.

85. $x^{1/2} - 6x^{1/4} + 8 = 0$

Let $u = x^{1/4}$.

$$\begin{aligned} u^2 - 6u + 8 &= 0 \quad \text{Substituting} \\ (u - 4)(u - 2) &= 0 \\ u = 4 \quad &\text{or} \quad u = 2 \\ x^{1/4} = 4 \quad &\text{or} \quad x^{1/4} = 2 \\ x = 256 \quad &\text{or} \quad x = 16 \quad \text{Raising both sides to the fourth power} \end{aligned}$$

Both numbers check. The solutions are 256 and 16.

87. *Writing Exercise*

89. We use the change-of-base formula.

$$\begin{aligned} \log_6 81 &= \frac{\log 81}{\log 6} \\ &= \frac{\log 3^4}{\log(2 \cdot 3)} \\ &= \frac{4 \log 3}{\log 2 + \log 3} \\ &\approx \frac{4(0.477)}{0.301 + 0.477} \\ &\approx 2.452 \end{aligned}$$

91. We use the change-of-base formula.

$$\begin{aligned} \log_{12} 36 &= \frac{\log 36}{\log 12} \\ &= \frac{\log(2 \cdot 3)^2}{\log(2^2 \cdot 3)} \\ &= \frac{2 \log(2 \cdot 3)}{\log 2^2 + \log 3} \\ &= \frac{2(\log 2 + \log 3)}{2 \log 2 + \log 3} \\ &\approx \frac{2(0.301 + 0.477)}{2(0.301) + 0.477} \\ &\approx 1.442 \end{aligned}$$

93. Use the change-of-base formula with $a = e$ and $b = 10$. We obtain

$$\log M = \frac{\ln M}{\ln 10}.$$

95.
$$\begin{aligned} \log(492x) &= 5.728 \\ 10^{5.728} &= 492x \\ \frac{10^{5.728}}{492} &= x \\ 1086.5129 &\approx x \end{aligned}$$

97.
$$\begin{aligned} \log 692 + \log x &= \log 3450 \\ \log x &= \log 3450 - \log 692 \\ \log x &= \log \frac{3450}{692} \\ x &= \frac{3450}{692} \\ x &\approx 4.9855 \end{aligned}$$

99. (a) Domain: $\{x|x > 0\}$, or $(0, \infty)$;
range: $\{y|y < 0.5135\}$, or $(-\infty, 0.5135)$;

(b) $[-1, 5, -10, 5]$;

(c)

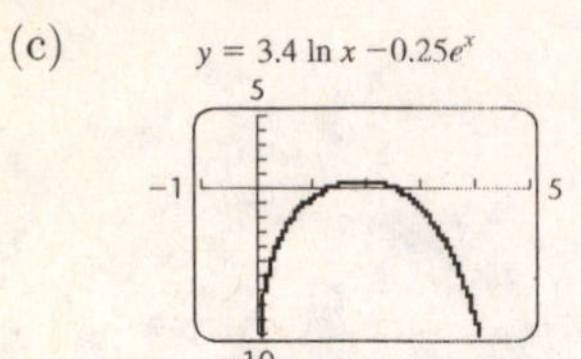

101. (a) Domain $\{x|x > 0\}$, or $(0, \infty)$;
range: $\{y|y > -0.2453\}$, or $(-0.2453, \infty)$

(b) $[-1, 5, -1, 10]$;

(c)

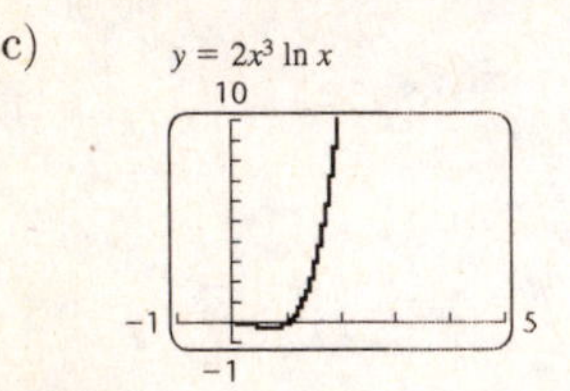

103.

Exercise Set 12.6

1. If we take the common logarithm on both sides, we see that choice (e) is correct.

3. $\ln x = 3$ means that 3 is the exponent to which we raise e to get x, so choice (f) is correct.

5. By the product rule for logarithms, $\log_5 x + \log_5(x - 2) = \log_5[x(x - 2)] = \log_5(x^2 - 2x)$, so choice (b) is correct.

7. By the quotient rule for logarithms, $\ln x - \ln(x - 2) = \ln \dfrac{x}{x - 2}$, so choice (g) is correct.

9.
$$\begin{aligned} 2^x &= 19 \\ \log 2^x &= \log 19 \\ x \log 2 &= \log 19 \\ x &= \frac{\log 19}{\log 2} \\ x &\approx 4.248 \end{aligned}$$

The solution is $\dfrac{\log 19}{\log 2}$, or approximately 4.248.

11.
$$\begin{aligned} 8^{x-1} &= 17 \\ \log 8^{x-1} &= \log 17 \\ (x - 1) \log 8 &= \log 17 \\ x - 1 &= \frac{\log 17}{\log 8} \\ x &= \frac{\log 17}{\log 8} + 1 \\ x &\approx 2.362 \end{aligned}$$

The solution is $\dfrac{\log 17}{\log 8} + 1$, or approximately 2.362.

13.
$$\begin{aligned} e^t &= 1000 \\ \ln e^t &= \ln 1000 && \text{Taking ln on both sides} \\ t &= \ln 1000 && \text{Finding the logarithm of the base to a power} \\ t &\approx 6.908 && \text{Using a calculator} \end{aligned}$$

15.
$$\begin{aligned} e^{0.03t} + 2 &= 7 \\ e^{0.03t} &= 5 \\ \ln e^{0.03t} &= \ln 5 && \text{Taking ln on both sides} \\ 0.03t &= \ln 5 && \text{Finding the logarithm of the base to a power} \\ t &= \frac{\ln 5}{0.03} \\ t &\approx 53.648 \end{aligned}$$

17.
$$\begin{aligned} 5 &= 3^{x+1} \\ \log 5 &= \log 3^{x+1} \\ \log 5 &= (x + 1) \log 3 \\ \log 5 &= x \log 3 + \log 3 \\ \log 5 - \log 3 &= x \log 3 \\ \frac{\log 5 - \log 3}{\log 3} &= x, \text{ or} \\ \frac{\log 5}{\log 3} - 1 &= x \\ 0.465 &\approx x \end{aligned}$$

19. $2^{x+3} = 16$

Observe that $16 = 2^4$. Equating exponents, we have $x + 3 = 4$, or $x = 1$.

21.
$$\begin{aligned} 4.9^x - 87 &= 0 \\ 4.9^x &= 87 \\ \log 4.9^x &= \log 87 \\ x \log 4.9 &= \log 87 \\ x &= \frac{\log 87}{\log 4.9} \\ x &\approx 2.810 \end{aligned}$$

23.
$$\begin{aligned} 19 &= 2e^{4x} \\ \frac{19}{2} &= e^{4x} \\ \ln\left(\frac{19}{2}\right) &= \ln e^{4x} \\ \ln\left(\frac{19}{2}\right) &= 4x \\ \frac{\ln\left(\frac{19}{2}\right)}{4} &= x \\ 0.563 &\approx x \end{aligned}$$

25. $7 + 3e^{5x} = 13$
$$3e^{5x} = 6$$
$$e^{5x} = 2$$
$$\ln e^{5x} = \ln 2$$
$$5x = \ln 2$$
$$x = \frac{\ln 2}{5}$$
$$x \approx 0.139$$

27. $\log_3 x = 4$

$x = 3^4$ Writing an equivalent exponential equation

$x = 81$

29. $\log_2 x = -3$

$x = 2^{-3}$ Writing an equivalent exponential equation

$x = \frac{1}{2^3}$, or $\frac{1}{8}$

31. $\ln x = 5$

$x = e^5$ Writing an equivalent exponential equation

$x \approx 148.413$

33. $\log_8 x = \frac{1}{3}$
$$x = 8^{1/3} = \sqrt[3]{8}$$
$$x = 2$$

35. $\ln 4x = 3$
$$4x = e^3$$
$$x = \frac{e^3}{4} \approx 5.021$$

37. $\log x = 2.5$ The base is 10.
$$x = 10^{2.5}$$
$$x \approx 316.228$$

39. $\ln(2x + 1) = 4$
$$2x + 1 = e^4$$
$$2x = e^4 - 1$$
$$x = \frac{e^4 - 1}{2} \approx 26.799$$

41. $\ln x = 1$
$$x = e \approx 2.718$$

43. $5 \ln x = -15$
$$\ln x = -3$$
$$x = e^{-3} \approx 0.050$$

45. $\log_2(8 - 6x) = 5$
$$8 - 6x = 2^5$$
$$8 - 6x = 32$$
$$-6x = 24$$
$$x = -4$$

The answer checks. The solution is -4.

47. $\log(x - 9) + \log x = 1$ The base is 10.

$\log_{10}[(x - 9)(x)] = 1$ Using the product rule
$$x(x - 9) = 10^1$$
$$x^2 - 9x = 10$$
$$x^2 - 9x - 10 = 0$$
$$(x + 1)(x - 10) = 0$$
$$x = -1 \text{ or } x = 10$$

Check: For -1:

$\log(x - 9) + \log x = 1$
$\log(-1 + 9) + \log(-1) \stackrel{?}{=} 1$ FALSE

For 10:

$\log(x - 9) + \log x$	1
$\log(10 - 9) + \log(10)$	1
$\log 1 + \log 10$	
$0 + 1$	

$1 \stackrel{?}{=} 1$ TRUE

The number -1 does not check, because negative numbers do not have logarithms. The solution is 10.

49. $\log x - \log(x + 3) = 1$ The base is 10.

$\log_{10} \frac{x}{x + 3} = 1$ Using the quotient rule
$$\frac{x}{x + 3} = 10^1$$
$$x = 10(x + 3)$$
$$x = 10x + 30$$
$$-9x = 30$$
$$x = -\frac{10}{3}$$

The number $-\frac{10}{3}$ does not check. The equation has no solution.

51.
$$\log_4(x + 3) = 2 + \log_4(x - 5)$$
$$\log_4(x + 3) - \log_4(x - 5) = 2$$

$\log_4 \frac{x + 3}{x - 5} = 2$ Using the quotient rule
$$\frac{x + 3}{x - 5} = 4^2$$
$$\frac{x + 3}{x - 5} = 16$$
$$x + 3 = 16(x - 5)$$
$$x + 3 = 16x - 80$$
$$83 = 15x$$
$$\frac{83}{15} = x$$

The number $\frac{83}{15}$ checks. It is the solution.

53. $\log_7(x+1) + \log_7(x+2) = \log_7 6$

$\log_7[(x+1)(x+2)] = \log_7 6$ Using the product rule

$\log_7(x^2+3x+2) = \log_7 6$

$x^2+3x+2 = 6$ Using the property of logarithmic equality

$x^2+3x-4 = 0$

$(x+4)(x-1) = 0$

$x = -4$ *or* $x = 1$

The number 1 checks, but -4 does not. The solution is 1.

55. $\log_5(x+4) + \log_5(x-4) = \log_5 20$

$\log_5[(x+4)(x-4)] = \log_5 20$ Using the product rule

$\log_5(x^2-16) = \log_5 20$

$x^2-16 = 20$ Using the property of logarithmic equality

$x^2 = 36$

$x = \pm 6$

The number 6 checks, but -6 does not. The solution is 6.

57. $\ln(x+5) + \ln(x+1) = \ln 12$

$\ln[(x+5)(x+1)] = \ln 12$

$\ln(x^2+6x+5) = \ln 12$

$x^2+6x+5 = 12$

$x^2+6x-7 = 0$

$(x+7)(x-1) = 0$

$x = -7$ *or* $x = 1$

The number -7 does not check, but 1 does. The solution is 1.

59. $\log_2(x+3) + \log_2(x-3) = 4$

$\log_2[(x+3)(x-3)] = 4$

$(x+3)(x-3) = 2^4$

$x^2-9 = 16$

$x^2 = 25$

$x = \pm 5$

The number 5 checks, but -5 does not. The solution is 5.

61. $\log_{12}(x+5) - \log_{12}(x-4) = \log_{12} 3$

$\log_{12}\frac{x+5}{x-4} = \log_{12} 3$

$\frac{x+5}{x-4} = 3$ Using the property of logarithmic equality

$x+5 = 3(x-4)$

$x+5 = 3x-12$

$17 = 2x$

$\frac{17}{2} = x$

The number $\frac{17}{2}$ checks and is the solution.

63. $\log_2(x-2) + \log_2 x = 3$

$\log_2[(x-2)(x)] = 3$

$x(x-2) = 2^3$

$x^2-2x = 8$

$x^2-2x-8 = 0$

$(x-4)(x+2) = 0$

$x = 4$ *or* $x = -2$

The number 4 checks, but -2 does not. The solution is 4.

65. *Writing Exercise*

67. $y = kx$

$7.2 = k(0.8)$ Substituting

$9 = k$ Variation constant

$y = 9x$ Equation of variation

69.

$$T = 2\pi\sqrt{\frac{L}{32}}$$

$$\frac{T}{2\pi} = \sqrt{\frac{L}{32}}$$

$$\left(\frac{T}{2\pi}\right)^2 = \left(\sqrt{\frac{L}{32}}\right)^2$$

$$\frac{T^2}{4\pi^2} = \frac{L}{32}$$

$$32 \cdot \frac{T^2}{4\pi^2} = L$$

$$\frac{8T^2}{\pi^2} = L$$

71. ***Familiarize.*** Let t = the time, in hours, it takes Joni and Miles to key in the score, working together. Then in t hours Joni does $\frac{t}{2}$ of the job, Miles does $\frac{t}{3}$, and together they do 1 entire job.

Translate.

$$\frac{t}{2} + \frac{t}{3} = 1$$

Carry out. We solve the equation. First we multiply by the LCD, 6.

$$6\left(\frac{t}{2} + \frac{t}{3}\right) = 6 \cdot 1$$

$$6 \cdot \frac{t}{2} + 6 \cdot \frac{t}{3} = 6$$

$$3t + 2t = 6$$

$$5t = 6$$

$$t = \frac{6}{5}$$

Check. In $\frac{6}{5}$ hr Joni does $\frac{6/5}{2}$, or $\frac{3}{5}$ of the job, and Miles does $\frac{6/5}{3}$, or $\frac{2}{5}$ of the job. Together they do $\frac{3}{5} + \frac{2}{5}$ or 1 entire job. The answer checks.

State. It takes Joni and Miles $\frac{6}{5}$ hr, or $1\frac{1}{5}$ hr, to do the job, working together.

73. *Writing Exercise*

75.
$$27^x = 81^{2x-3}$$
$$(3^3)^x = (3^4)^{2x-3}$$
$$3^{3x} = 3^{8x-12}$$
$$3x = 8x - 12$$
$$12 = 5x$$
$$\frac{12}{5} = x$$

The solution is $\frac{12}{5}$.

77.
$$\log_x(\log_3 27) = 3$$
$$\log_3 27 = x^3$$
$$3 = x^3 \quad (\log_3 27 = 3)$$
$$\sqrt[3]{3} = x$$

The solution is $\sqrt[3]{3}$.

79.
$$x \cdot \log\frac{1}{8} = \log 8$$
$$x \cdot \log 8^{-1} = \log 8$$
$$x(-\log 8) = \log 8 \quad \text{Using the power rule}$$
$$x = -1$$

The solution is -1.

81.
$$2^{x^2+4x} = \frac{1}{8}$$
$$2^{x^2+4x} = \frac{1}{2^3}$$
$$2^{x^2+4x} = 2^{-3}$$
$$x^2 + 4x = -3$$
$$x^2 + 4x + 3 = 0$$
$$(x+3)(x+1) = 0$$
$$x = -3 \text{ or } x = -1$$

The solutions are -3 and -1.

83.
$$\log_5 |x| = 4$$
$$|x| = 5^4$$
$$|x| = 625$$
$$x = 625 \text{ or } x = -625$$

The solutions are 625 and -625.

85.
$$\log\sqrt{2x} = \sqrt{\log 2x}$$
$$\log(2x)^{1/2} = \sqrt{\log 2x}$$
$$\frac{1}{2}\log 2x = \sqrt{\log 2x}$$
$$\frac{1}{4}(\log 2x)^2 = \log 2x \quad \text{Squaring both sides}$$
$$\frac{1}{4}(\log 2x)^2 - \log 2x = 0$$

Let $u = \log 2x$.

$$\frac{1}{4}u^2 - u = 0$$
$$u\left(\frac{1}{4}u - 1\right) = 0$$
$$u = 0 \quad \text{or} \quad \frac{1}{4}u - 1 = 0$$
$$u = 0 \quad \text{or} \quad \frac{1}{4}u = 1$$
$$u = 0 \quad \text{or} \quad u = 4$$
$$\log 2x = 0 \quad \text{or} \quad \log 2x = 4 \quad \text{Replacing } u \text{ with } \log 2x$$
$$2x = 10^0 \quad \text{or} \quad 2x = 10^4$$
$$2x = 1 \quad \text{or} \quad 2x = 10{,}000$$
$$x = \frac{1}{2} \quad \text{or} \quad x = 5000$$

Both numbers check. The solutions are $\frac{1}{2}$ and 5000.

87.
$$3^{x^2} \cdot 3^{4x} = \frac{1}{27}$$
$$3^{x^2+4x} = 3^{-3}$$
$$x^2 + 4x = -3 \quad \text{The exponents must be equal.}$$
$$x^2 + 4x + 3 = 0$$
$$(x+1)(x+3) = 0$$
$$x = -1 \text{ or } x = -3$$

Both numbers check. The solutions are -1 and -3.

89.
$$\log x^{\log x} = 25$$
$$\log x(\log x) = 25 \quad \text{Using the power rule}$$
$$(\log x)^2 = 25$$
$$\log x = \pm 5$$
$$x = 10^5 \quad \text{or} \quad x = 10^{-5}$$
$$x = 100{,}000 \quad \text{or} \quad x = \frac{1}{100{,}000}$$

Both numbers check. The solutions are 100,000 and $\frac{1}{100{,}000}$.

91.
$$(81^{x-2})(27^{x+1}) = 9^{2x-3}$$
$$[(3^4)^{x-2}][(3^3)^{x+1}] = (3^2)^{2x-3}$$
$$(3^{4x-8})(3^{3x+3}) = 3^{4x-6}$$
$$3^{7x-5} = 3^{4x-6}$$
$$7x - 5 = 4x - 6$$
$$3x = -1$$
$$x = -\frac{1}{3}$$

The solution is $-\frac{1}{3}$.

93.
$$2^y = 16^{x-3} \quad \text{and} \quad 3^{y+2} = 27^x$$
$$2^y = (2^4)^{x-3} \quad \text{and} \quad 3^{y+2} = (3^3)^x$$
$$y = 4x - 12 \quad \text{and} \quad y + 2 = 3x$$
$$12 = 4x - y \quad \text{and} \quad 2 = 3x - y$$

Solving this system of equations we get $x = 10$ and $y = 28$. Then $x + y = 10 + 28 = 38$.

95. Find the first coordinate of the point of intersection of $y_1 = \ln x$ and $y_2 = \log x$. The value of x for which the natural logarithm of x is the same as the common logarithm of x is 1.

Exercise Set 12.7

1. a) Replace $S(t)$ with 2800 and solve for t.

$$S(t) = 200 \cdot 2^t$$
$$2800 = 200 \cdot 2^t$$
$$14 = 2^t$$
$$\ln 14 = \ln 2^t$$
$$\ln 14 = t \ln 2$$
$$\frac{\ln 14}{\ln 2} = t$$
$$4 \approx t$$

Sales of DVD players first reached $2800 million about 4 yr after 1997, or in 2001.

b) $S(0) = 200 \cdot 2^0 = 200 \cdot 1 = 200$, so to find the doubling time we replace $S(t)$ with 400 and solve for t.

$$400 = 200 \cdot 2^t$$
$$2 = 2^t$$
$$1 = t \quad \text{The exponents must be the same.}$$

The doubling time is about 1 year.

3. a) Find $N(21)$.

$$N(x) = 1337(0.9)^x$$
$$N(21) = 1337(0.9)^{21}$$
$$N(21) \approx 146.293$$

We estimate that there are about 146.293 thousand, or 146,293,000 21-year-old skateboarders.

b) $6300 = 6.3$ thousand; substitute 6.3 for $N(x)$ and solve for x.

$$6.3 = 1337(0.9)^x$$
$$0.0047 \approx 0.9^x$$
$$\log 0.0047 \approx \log 0.9^x$$
$$\log 0.0047 \approx x \log 0.9$$
$$\frac{\log 0.0047}{\log 0.9} \approx x$$
$$51 \approx x$$

At about age 51 there are only 6300 skateboarders.

5. a) Replace $A(t)$ with 35,000 and solve for t.

$$A(t) = 29,000(1.03)^t$$
$$35,000 = 29,000(1.03)^t$$
$$1.207 \approx (1.03)^t$$
$$\log 1.207 \approx \log(1.03)^t$$
$$\log 1.207 \approx t \log 1.03$$
$$\frac{\log 1.207}{\log 1.03} \approx t$$
$$6.4 \approx t$$

The amount due will reach $35,000 after about 6.4 years.

b) Replace $A(t)$ with 2(29,000), or 58,000, and solve for t.

$$58,000 = 29,000(1.03)^t$$
$$2 = (1.03)^t$$
$$\log 2 = \log(1.03)^t$$
$$\log 2 = t \log 1.03$$
$$\frac{\log 2}{\log 1.03} = t$$
$$23.4 \approx t$$

The doubling time is about 23.4 years.

7. a) Substitute 50 for $P(t)$ and solve for t.

$$P(t) = 63.03(0.95)^t$$
$$50 = 63.03(0.95)^t$$
$$0.7933 \approx 0.95^t$$
$$\ln 0.7933 \approx \ln 0.95^t$$
$$\ln 0.7933 \approx t \ln 0.95$$
$$\frac{\ln 0.7933}{\ln 0.95} \approx t$$
$$4.5 \approx t$$

The percentage of phones that are land lines will be 50% about 4.5 yr after 2000, so the percentage will drop below 50% in about 2005.

b) Substitute 25 for $P(t)$ and solve for t.

$$25 = 63.03(0.95)^t$$
$$0.3966 \approx 0.95^t$$
$$\ln 0.3966 \approx \ln 0.95^t$$
$$\ln 0.3966 \approx t \ln 0.95$$
$$\frac{\ln 0.3966}{\ln 0.95} \approx t$$
$$18 \approx t$$

25% of phones will be land lines about 18 yr after 2000, so the percentage will drop below 25% in about 2018.

9. a) $P(t)$ is given in thousands, so we substitute 30 for $P(t)$ and solve for t.

$$P(t) = 5.5(1.047)^t$$
$$30 = 5.5(1.047)^t$$
$$5.455 \approx 1.047^t$$
$$\log 5.455 \approx \log 1.047^t$$
$$\log 5.455 \approx t \log 1.047$$
$$\frac{\log 5.455}{\log 1.047} \approx t$$
$$37 \approx t$$

The humpback whale population will reach 30,000 about 37 yr after 1982, or in 2019.

b) $P(0) = 5.5(1.047)^0 = 5.5(1) = 5.5$ and $2(5.5) = 11$, so we substitute 11 for $P(t)$ and solve for t.

$$11 = 5.5(1.047)^t$$
$$2 = 1.047^t$$
$$\log 2 = \log 1.047^t$$
$$\log 2 = t \log 1.047$$
$$\frac{\log 2}{\log 1.047} = t$$
$$15.1 \approx t$$

The doubling time is about 15.1 yr.

11. $\text{pH} = -\log[H^+]$

$= -\log[1.3 \times 10^{-5}]$

$\approx -(-4.886057)$ Using a calculator

≈ 4.9

The pH of fresh-brewed coffee is about 4.9.

13. $\text{pH} = -\log[H^+]$

$7.0 = -\log[H^+]$

$-7.0 = \log[H^+]$

$10^{-7.0} = [H^+]$ Converting to an exponential equation

The hydrogen ion concentration is 10^{-7} moles per liter.

15. $L = 10 \cdot \log \frac{I}{I_0}$

$= 10 \cdot \log \frac{3.2 \times 10^{-6}}{10^{-12}}$

$= 10 \cdot \log(3.2 \times 10^6)$

$\approx 10(6.5)$

≈ 65

The intensity of sound in normal conversation is about 65 decibels.

17. $L = 10 \cdot \log \frac{I}{I_0}$

$105 = 10 \cdot \log \frac{I}{10^{-12}}$

$10.5 = \log \frac{I}{10^{-12}}$

$10.5 = \log I - \log 10^{-12}$ Using the quotient rule

$10.5 = \log I - (-12)$ $(\log 10^a = a)$

$10.5 = \log I + 12$

$-1.5 = \log I$

$10^{-1.5} = I$ Converting to an exponential equation

$3.2 \times 10^{-2} \approx I$

The intensity of the sound is $10^{-1.5}$ W/m^2, or about 3.2×10^{-2} W/m^2.

19. a) Substitute 0.025 for k:

$P(t) = P_0\, e^{0.025t}$

b) To find the balance after one year, replace P_0 with 5000 and t with 1. We find $P(1)$:

$P(1) = 5000\, e^{0.025(1)} = 5000\, e^{0.025} \approx \$5126.58$

To find the balance after 2 years, replace P_0 with 5000 and t with 2. We find $P(2)$:

$P(2) = 5000\, e^{0.025(2)} = 5000\, e^{0.05} \approx \$5256.36$

c) To find the doubling time, replace P_0 with 5000 and $P(t)$ with 10,000 and solve for t.

$10,000 = 5000\, e^{0.025t}$

$2 = e^{0.025t}$

$\ln 2 = \ln e^{0.025t}$ Taking the natural logarithm on both sides

$\ln 2 = 0.025t$ Finding the logarithm of the base to a power

$\frac{\ln 2}{0.025} = t$

$27.7 \approx t$

The investment will double in about 27.7 years.

21. a) $P(t) = 292.80e^{0.009t}$, where $P(t)$ is in millions and t is the number of years after 2004.

b) In 2005, $t = 2005 - 2004 = 1$. Find $P(1)$.

$P(1) = 292.80e^{0.009(1)} = 292.80e^{0.009} \approx 295.45$

The U.S. population will be about 295.45 million in 2005.

c) Substitute 325 for $P(t)$ and solve for t.

$325 = 292.80e^{0.009t}$

$1.1100 \approx e^{0.009t}$

$\ln 1.1100 \approx \ln e^{0.009t}$

$\ln 1.1100 \approx 0.009t$

$\frac{\ln 1.1100}{0.009} \approx t$

$12 \approx t$

The U.S. population will reach 325 million about 12 yr after 2004, or in 2016.

23. The exponential growth function is $S(t) = S_0e^{0.103t}$. We replace $S(t)$ with $2S_0$ and solve for t.

$2S_0 = S_0e^{0.103t}$

$2 = e^{0.103t}$

$\ln 2 = \ln e^{0.103t}$

$\ln 2 = 0.103t$

$\frac{\ln 2}{0.103} = t$

$6.7 \approx t$

The doubling time for iPod sales is about 6.7 months.

25. $Y(x) = 67.17 \ln \frac{x}{4.5}$

a) $Y(7) = 67.17 \ln \frac{7}{4.5} \approx 30$

The world population will reach 7 billion about 30 yr after 1980, or in 2010.

b) $Y(8) = 67.17 \ln \frac{8}{4.5} \approx 39$

The world population will reach 8 billion about 39 yr after 1980, or in 2019.

c) Plot the points found in parts (a) and (b) and others as necessary and draw the graph.

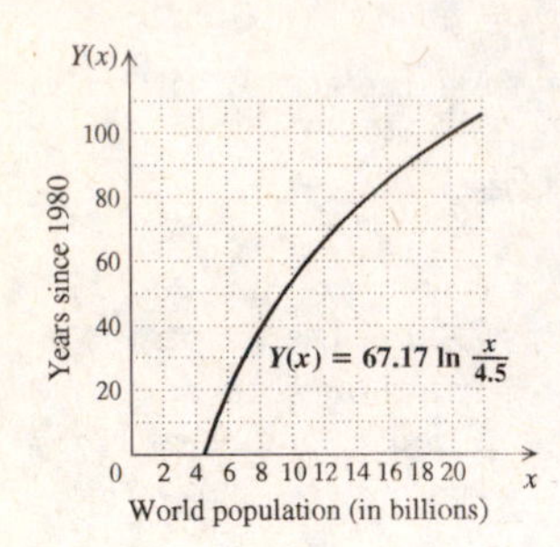

27. a) $S(0) = 68 - 20\log(0+1) = 68 - 20 \log 1 = 68 - 20(0) = 68\%$

b) $S(4) = 68 - 20\log(4+1) = 68 - 20 \log 5 \approx 68 - 20(0.69897) \approx 54\%$

$S(24) = 68 - 20\log(24+1) = 68 - 20 \log 25 \approx 68 - 20(1.39794) \approx 40\%$

c) Using the values we computed in parts (a) and (b) and any others we wish to calculate, we sketch the graph:

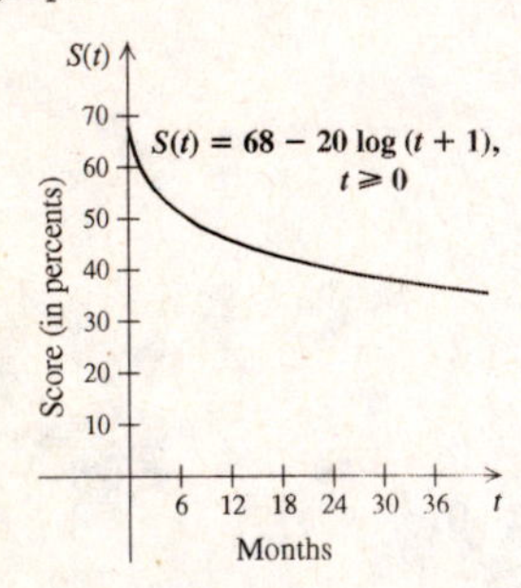

d)
$$\begin{aligned} 50 &= 68 - 20 \log(t+1) \\ -18 &= -20 \log(t+1) \\ 0.9 &= \log(t+1) \\ 10^{0.9} &= t+1 \\ 7.9 &\approx t+1 \\ 6.9 &\approx t \end{aligned}$$

After about 6.9 months, the average score was 50.

29. a) We use the growth equation $N(t) = N_0 e^{kt}$, where t is the number of years since 2000. In 2000, at $t = 0$, 17 people were infected. We substitute 17 for N_0:

$$N(t) = 17e^{kt}.$$

To find the exponential growth rate k, observe that 1 year later 29 people were infected.

$$\begin{aligned} N(1) &= 17e^{k\cdot 1} && \text{Substituting 1 for } t \\ 29 &= 17e^{k} && \text{Substituting 29 for } N(1) \\ 1.706 &\approx e^{k} \\ \ln 1.706 &\approx \ln e^{k} \\ \ln 1.706 &\approx k \\ 0.534 &\approx k \end{aligned}$$

The exponential function is $N(t) = 17e^{0.534t}$, where t is the number of years since 2000.

b) In 2006, $t = 2006 - 2000$, or 6. Find $N(6)$.

$$\begin{aligned} N(6) &= 17e^{0.534(6)} \\ &= 17e^{3.204} \\ &\approx 418.7 \end{aligned}$$

Approximately 419 people will be infected in 2006.

31. a) Let $P(t)$ represent farmland, in millions of acres, and let t represent the number of years after 1990.

$$\begin{aligned} P(t) &= 987e^{-kt} \\ P(12) &= 987e^{-k\cdot 12} \\ 941 &= 987e^{-12k} \\ 0.9534 &\approx e^{-12k} \\ \ln 0.9534 &\approx \ln e^{-12k} \\ \ln 0.9534 &\approx -12k \\ \frac{\ln 0.9534}{-12} &\approx k \\ 0.004 &\approx k \end{aligned}$$

The exponential function is $P(t) = 987e^{-0.004t}$, where $P(t)$ and t are as described above.

b) In 2008, $t = 2008 - 1990 = 18$.

$$\begin{aligned} P(18) &= 987e^{-0.004(18)} \\ &= 987e^{-0.072} \\ &\approx 918 \end{aligned}$$

In 2008 there will be about 918 million acres of U.S. farmland.

c) Substitute 800 for $P(t)$ and solve for t.

$$\begin{aligned} 800 &= 987e^{-0.004t} \\ 0.8105 &\approx e^{-0.004t} \\ \ln 0.8105 &\approx \ln e^{-0.004t} \\ \ln 0.8105 &\approx -0.004t \\ \frac{\ln 0.8105}{-0.004} &\approx t \\ 53 &\approx t \end{aligned}$$

There will be 800 million acres of U.S. farmland about 53 yr after 1990, or in 2043.

33. We will use the function derived in Example 7:

$$P(t) = P_0 e^{-0.00012t}$$

If the scrolls had lost 22.3% of their carbon-14 from an initial amount P_0, then $77.7\%(P_0)$ is the amount present. To find the age t of the scrolls, we substitute $77.7\%(P_0)$, or $0.777P_0$, for $P(t)$ in the function above and solve for t.

$$0.777P_0 = P_0e^{-0.00012t}$$
$$0.777 = e^{-0.00012t}$$
$$\ln 0.777 = \ln\ e^{-0.00012t}$$
$$-0.2523 \approx -0.00012t$$
$$t \approx \frac{-0.2523}{-0.00012} \approx 2103$$

The scrolls are about 2103 years old.

35. The function $P(t) = P_0\, e^{-kt}$, $k > 0$, can be used to model decay. For iodine-131, $k = 9.6\%$, or 0.096. To find the half-life we substitute 0.096 for k and $\frac{1}{2}\ P_0$ for $P(t)$, and solve for t.

$$\frac{1}{2}\ P_0 = P_0\, e^{-0.096t},\ \text{or}\ \frac{1}{2} = e^{-0.096t}$$
$$\ln\frac{1}{2} = \ln\ e^{-0.096t} = -0.096t$$
$$t = \frac{\ln 0.5}{-0.096} \approx \frac{-0.6931}{-0.096} \approx 7.2 \text{ days}$$

37. The function $P(t) = P_0e^{-kt}$, $k > 0$, can be used to model decay. We substitute $\frac{1}{2}P_0$ for $P(t)$ and 1 for t and solve for the decay rate k.

$$\frac{1}{2}P_0 = P_0e^{-k\cdot 1}$$
$$\frac{1}{2} = e^{-k}$$
$$\ln\frac{1}{2} = \ln e^{-k}$$
$$-0.693 \approx -k$$
$$0.693 \approx k$$

The decay rate is 0.693, or 69.3% per year.

39. a) We start with the exponential growth equation

$V(t) = V_0\, e^{kt}$, where t is the number of years after 1991.

Substituting 451,000 for V_0, we have

$V(t) = 451,000\, e^{kt}$.

To find the exponential growth rate k, observe that the card sold for \$1.1 million, or \$1,100,000 in 2000, or 9 years after 1991. We substitute and solve for k.

$$V(9) = 451,000\, e^{k\cdot 9}$$
$$1,100,000 = 451,000\, e^{9k}$$
$$2.4390 \approx e^{9k}$$
$$\ln 2.4390 \approx \ln e^{9k}$$
$$\ln 2.4390 \approx 9k$$
$$\frac{\ln 2.4390}{9} \approx k$$
$$0.099 \approx k$$

Thus, the exponential growth function is $V(t) = 451,000e^{0.099t}$, where t is the number of years after 1991.

b) In 2006, $t = 2006 - 1991 = 15$

$V(15) = 451,000e^{0.099(15)} \approx 1,991,149$

The card's value in 2006 will be about \$1.99 million

c) Substitute 2(\$451,000), or \$902,000 for $V(t)$ and solve for t.

$$902,000 = 451,000\, e^{0.099t}$$
$$2 = e^{0.099t}$$
$$\ln\ 2 = \ln\ e^{0.099t}$$
$$\ln\ 2 = 0.099t$$
$$\frac{\ln\ 2}{0.099} = t$$
$$7.0 \approx t$$

The doubling time is about 7.0 years.

d) Substitute \$3,000,000 for $V(t)$ and solve for t.

$$3,000,000 = 451,000\, e^{0.099t}$$
$$6.6519 \approx e^{0.099t}$$
$$\ln 6.6519 \approx \ln\ e^{0.099t}$$
$$\ln 6.6519 \approx 0.099t$$
$$\frac{\ln 6.6519}{0.099} \approx t$$
$$19 \approx t$$

The value of the card will first exceed \$3,000,000 about 19 years after 1991, or in 2010.

41. *Writing Exercise*

43. Graph $y = x^2 - 8x$.

First we find the vertex.

$$-\frac{b}{2a} = -\frac{-8}{2\cdot 1} = 4$$

When $x = 4$, $y = 4^2 - 8\cdot 4 = 16 - 32 = -16$.

The vertex is $(4, -16)$ and the axis of symmetry is $x = 4$. We plot a few points on either side of the vertex and graph the parabola.

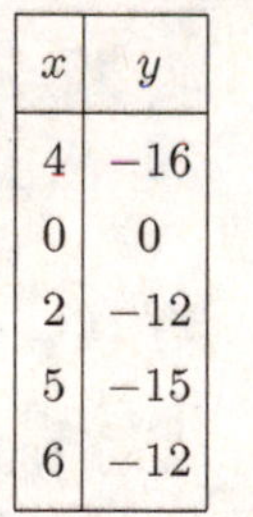

x	y
4	−16
0	0
2	−12
5	−15
6	−12

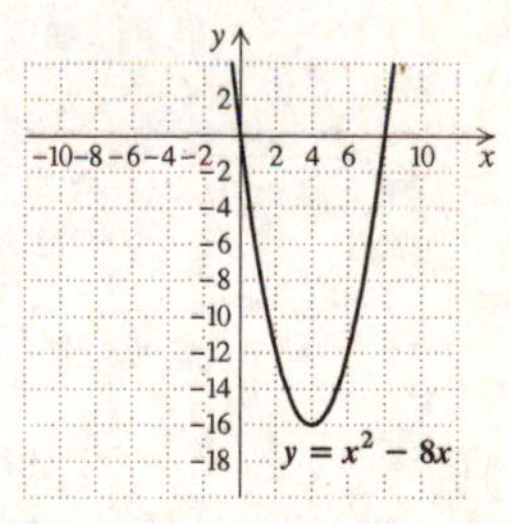

45. Graph $f(x) = 3x^2 - 5x - 1$

First we find the vertex.

$$-\frac{b}{2a} = -\frac{-5}{2\cdot 3} = \frac{5}{6}$$

$$f\left(\frac{5}{6}\right) = 3\left(\frac{5}{6}\right)^2 - 5\cdot\frac{5}{6} - 1 = -\frac{37}{12}$$

The vertex is $\left(\frac{5}{6}, -\frac{37}{12}\right)$ and the axis of symmetry is $x = \frac{5}{6}$. We plot a few points on either side of the vertex and graph the parabola.

x	$f(x)$
$\frac{5}{6}$	$-\frac{37}{12}$
0	-1
-1	7
2	1
3	11

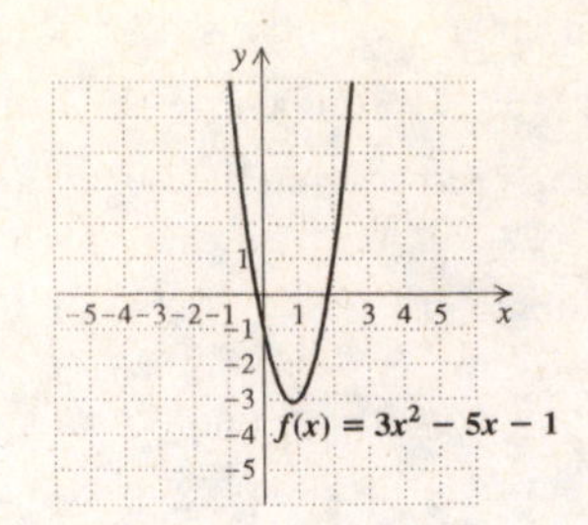

47.

$$x^2 - 8x = 7$$

$$x^2 - 8x + 16 = 7 + 16 \quad \text{Adding } \left[\frac{1}{2}(-8)\right]^2$$

$$(x-4)^2 = 23$$

$$x - 4 = \pm\sqrt{23}$$

$$x = 4 \pm \sqrt{23}$$

The solutions are $4 \pm \sqrt{23}$.

49. *Writing Exercise*

51. We will use the exponential growth function $V(t) = V_0 e^{kt}$, where t is the number of years after 2004 and $V(t)$ is in millions of dollars. Substitute 24 for $V(t)$, 0.04 for k, and 6 for t and solve for V_0.

$$24 = V_0 e^{0.04(6)}$$

$$24 = V_0 e^{0.24}$$

$$\frac{24}{e^{0.24}} = V_0$$

$$18.9 \approx V_0$$

About \$18.9 million would need to be invested.

53. From Exercise 7 we know that the percentage of U.S. phone lines that are land lines is given by $P(t) = 63.03(0.95)^t$, where t is the number of years after 2000. Then the percentage of U.S. phone lines that are cellular t years after 2000 is given by the function $P(t) = 100 - 63.03(0.95)^t$.

55. First we find k. When $t = 24,360$, $P(t) = 0.5P_0$.

$$0.5P_0 = P_0 e^{-k \cdot 24,360}$$

$$0.5 = e^{-24,360k}$$

$$\ln 0.5 = \ln e^{-24,360k}$$

$$\ln 0.5 = -24,360k$$

$$\frac{\ln 0.5}{-24,360} = k$$

$$0.0000285 \approx k$$

Now we have a function for the decay of plutonium-239.

$$P(t) = P_0 e^{-0.0000285t}$$

If a fuel rod has lost 90% of its plutonium, then 10% of the initial amount is still present. We substitute and solve for t.

$$0.1P_0 = P_0 e^{-0.0000285t}$$

$$0.1 = e^{-0.0000285t}$$

$$\ln 0.1 = \ln e^{-0.0000285t}$$

$$\ln 0.1 = -0.0000285t$$

$$\frac{\ln 0.1}{-0.0000285} = t$$

$$80,792 \approx t$$

It will take about 80,792 yr for the fuel rod of plutonium -239 to lose 90% of its radioactivity.

57. Consider an exponential growth function $P(t) = P_0 e^{kt}$. Suppose that at time T, $P(T) = 2P_0$.

Solve for T:

$$2P_0 = P_0 e^{kT}$$

$$2 = e^{kT}$$

$$\ln 2 = \ln e^{kT}$$

$$\ln 2 = kT$$

$$\frac{\ln 2}{k} = T$$

59. *Writing Exercise*

Chapter 13

Conic Sections

Exercise Set 13.1

1. $(x-2)^2+(y+5)^2=9$, or $(x-2)^2+[y-(-5)]^2=3^2$, is the equation of a circle with center $(2,-5)$ and radius 3, so choice (f) is correct.

3. $(x-5)^2+(y+2)^2=9$, or $(x-5)^2+[y-(-2)]^2=3^2$, is the equation of a circle with center $(5,-2)$ and radius 3, so choice (g) is correct.

5. $y=(x-2)^2-5$ is the equation of a parabola with vertex $(2,-5)$ that opens upward, so choice (c) is correct.

7. $x=(y-2)^2-5$ is the equation of a parabola with vertex $(-5,2)$ that opens to the right, so choice (d) is correct.

9. $y=-x^2$

This is equivalent to $y=-(x-0)^2+0$. The vertex is $(0,0)$.

We choose some x-values on both sides of the vertex and compute the corresponding values of y. The graph opens down, because the coefficient of x^2, -1, is negative.

x	y
0	0
1	-1
2	-4
-1	-1
-2	-4

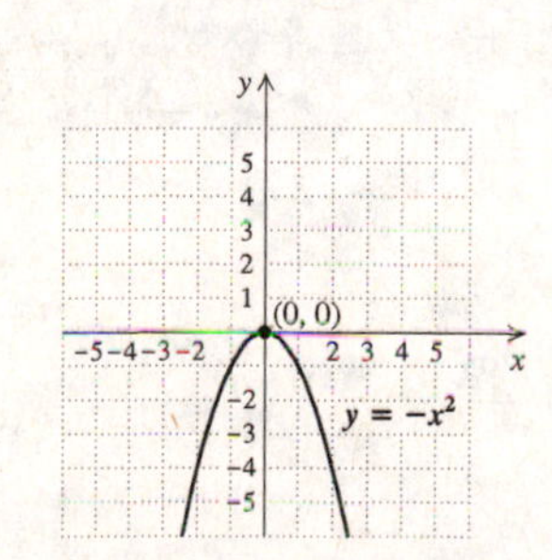

11. $y=-x^2+4x-5$

We can find the vertex by computing the first coordinate, $x=-b/2a$, and then substituting to find the second coordinate:

$$x=-\frac{b}{2a}=-\frac{4}{2(-1)}=2$$

$$y=-x^2+4x-5=-(2)^2+4(2)-5=-1$$

The vertex is $(2,-1)$.

We choose some x-values and compute the corresponding values for y. The graph opens downward because the coefficient of x^2, -1, is negative.

x	y
2	-1
3	-2
4	-5
1	-2
0	-5

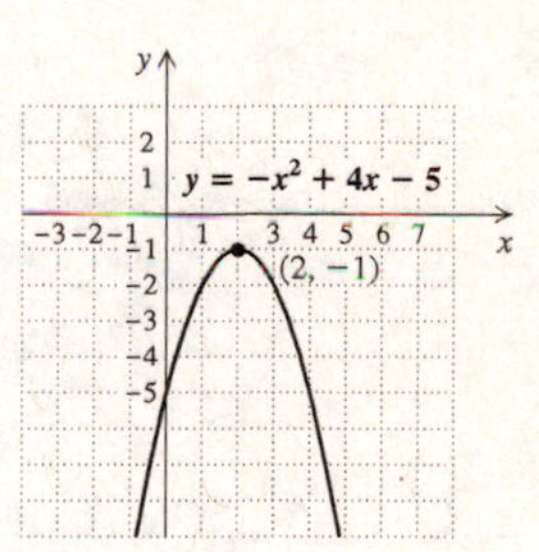

13. $x=y^2-4y+2$

We find the vertex by completing the square.

$$x=(y^2-4y+4)+2-4$$

$$x=(y-2)^2-2$$

The vertex is $(-2,2)$.

To find ordered pairs, we choose values for y and compute the corresponding values of x. The graph opens to the right, because the coefficient of y^2, 1, is positive.

x	y
7	-1
2	0
-1	1
-2	2
-1	3

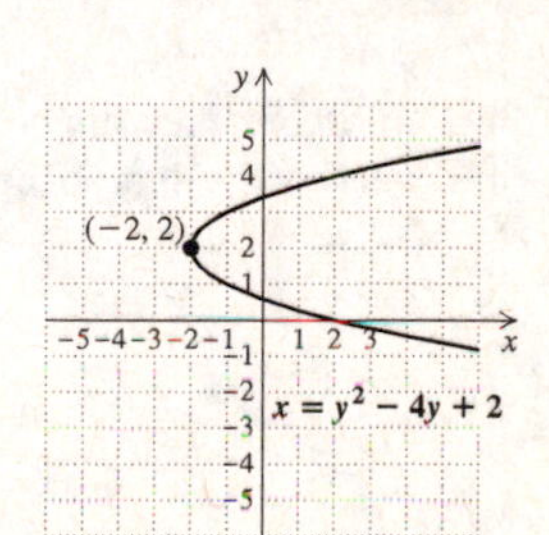

15. $x=y^2+3$

$$x=(y-0)^2+3$$

The vertex is $(3,0)$.

To find the ordered pairs, we choose y-values and compute the corresponding values for x. The graph opens to the right, because the coefficient of y^2, 1, is positive.

x	y
3	0
4	1
7	2
4	-1
7	-2

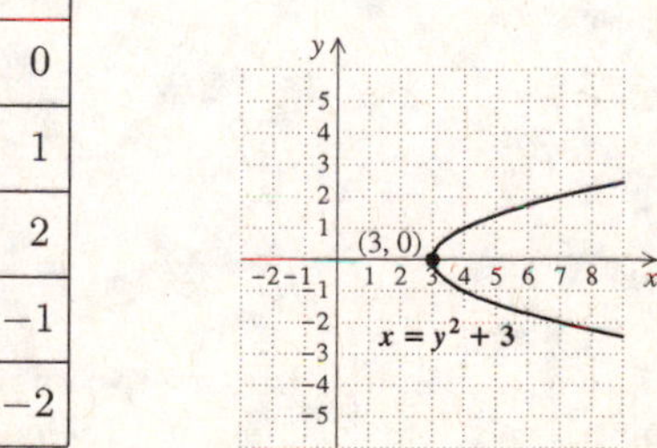

17. $x = -\frac{1}{2}y^2$

$x = -\frac{1}{2}(y-0)^2 + 0$

The vertex is $(0,0)$.

We choose y-values and compute the corresponding values for x. The graph opens to the left, because the coefficient of y^2, $-\frac{1}{2}$, is negative.

x	y
0	0
−2	2
−8	4
−2	−2
−8	−4

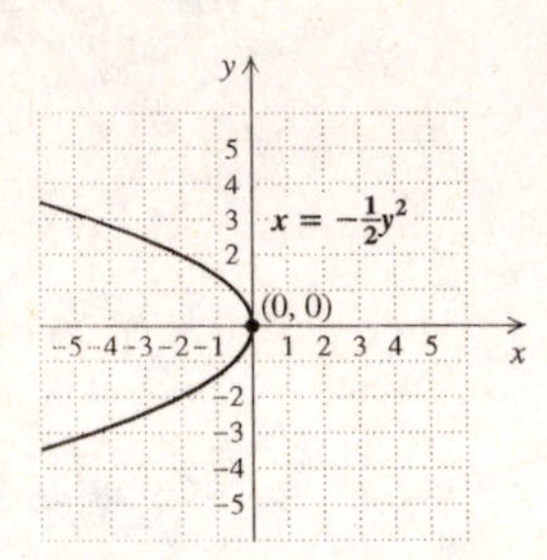

19. $x = -y^2 - 4y$

We find the vertex by computing the second coordinate, $y = -b/2a$, and then substituting to find the first coordinate:

$$y = -\frac{b}{2a} = -\frac{-4}{2(-1)} = -2$$

$$x = -y^2 - 4y = (-2)^2 - 4(-2) = 4$$

The vertex is $(4,-2)$.

We choose y-values and compute the corresponding values for x. The graph opens to the left, because the coefficient of y^2, -1, is negative.

x	y
4	−2
−5	1
0	0
3	−1
3	−3

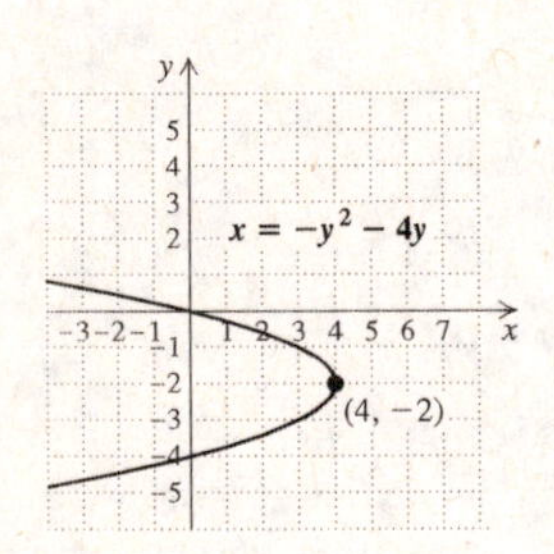

21. $x = 4 - y - y^2$

We find the vertex by completing the square.

$$x = -(y^2 + y) + 4$$

$$x = -\left(y^2 + y + \frac{1}{4}\right) + 4 + \frac{1}{4}$$

$$x = -\left(y + \frac{1}{2}\right)^2 + \frac{17}{4}$$

The vertex is $\left(\frac{17}{4}, -\frac{1}{2}\right)$.

We choose y-values and compute the corresponding values for x. The graph opens to the left, because the coefficient of y^2, -1, is negative.

x	y
$\frac{17}{4}$	$-\frac{1}{2}$
4	0
2	1
−2	2
4	−1
2	−2
−2	−3

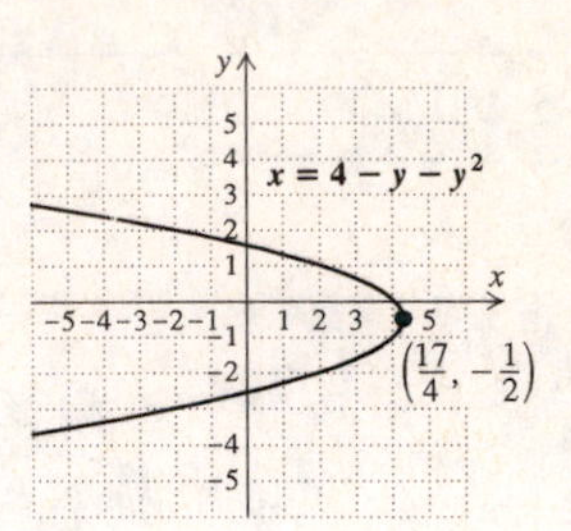

23. $y = x^2 - 2x + 1$

$y = (x-1)^2 + 0$

The vertex is $(1,0)$.

We choose x-values and compute the corresponding values for y. The graph opens upward, because the coefficient of x^2, 1, is positive.

x	y
1	0
0	1
−1	4
2	1
3	4

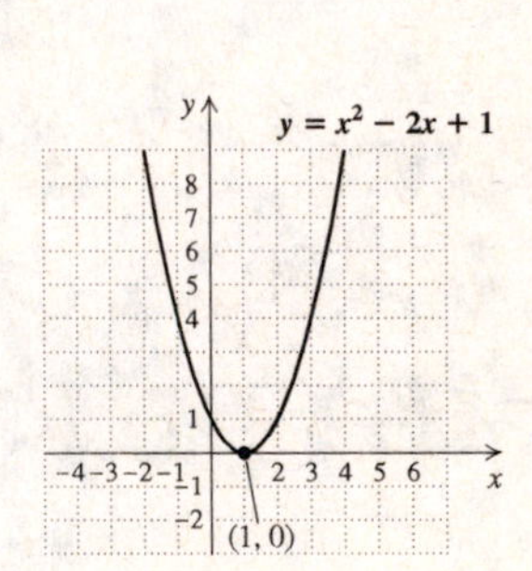

25. $x = -y^2 + 2y - 1$

We find the vertex by computing the second coordinate, $y = -b/2a$, and then substituting to find the first coordinate.

$$y = -\frac{b}{2a} = -\frac{2}{2(-1)} = 1$$

$$x = -y^2 + 2y - 1 = -(1)^2 + 2(1) - 1 = 0$$

The vertex is $(0,1)$.

We choose y-values and compute the corresponding values for x. The graph opens to the left, because the coefficient of y^2, -1, is negative.

x	y
−4	3
−1	2
−1	0
−4	−1
−4	3

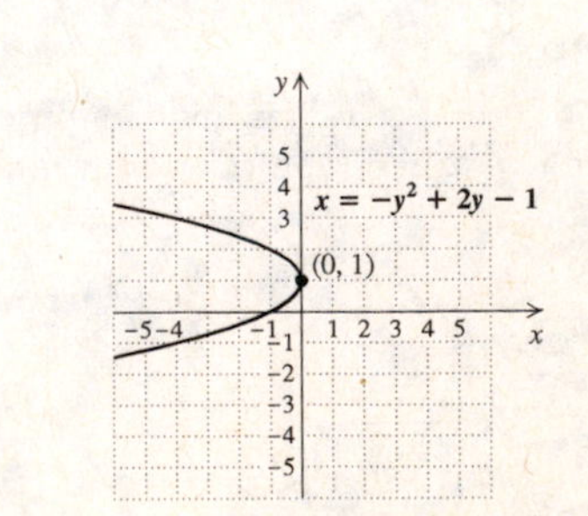

27. $x = -2y^2 - 4y + 1$

We find the vertex by completing the square.

$$x = -2(y^2 + 2y) + 1$$
$$x = -2(y^2 + 2y + 1) + 1 + 2$$
$$x = -2(y + 1)^2 + 3$$

The vertex is $(-3, -1)$.

We choose y-values and compute the corresponding values for x. The graph opens to the left, because the coefficient of y^2, -2, is negative.

x	y
3	-1
1	-2
-5	-3
1	0
-5	1

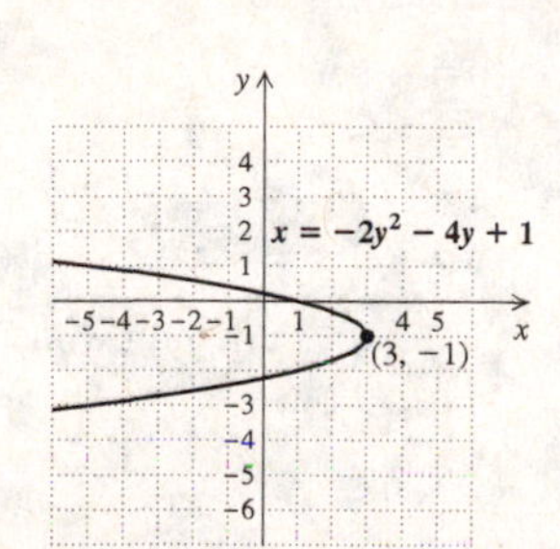

29. $d = \sqrt{(x_2 - x_1)^2 + (y_2 + y_1)^2}$ Distance formula

$= \sqrt{(5 - 1)^2 + (9 - 6)^2}$ Substituting

$= \sqrt{4^2 + 3^2}$

$= \sqrt{25} = 5$

31. $d = \sqrt{(x_2 - x_1)^2 + (y_2 - y_1)^2}$ Distance formula

$= \sqrt{(3 - 0)^2 + [-4 - (-7)]^2}$ Substituting

$= \sqrt{3^2 + 3^2}$

$= \sqrt{18} \approx 4.243$ Simplifying and approximating

33. $d = \sqrt{(x_2 - x_1)^2 + (y_2 - y_1)^2}$

$= \sqrt{[6 - (-4)]^2 + (-6 - 4)^2}$

$= \sqrt{200} \approx 14.142$

35. $d = \sqrt{(x_2 - x_1)^2 + (y_2 - y_1)^2}$

$= \sqrt{(-9.2 - 8.6)^2 + [-3.4 - (-3.4)]^2}$

$= \sqrt{(-17.8)^2 + 0^2}$

$= \sqrt{316.84} = 17.8$

(Since these points are on a horizontal line, we could have found the distance between them by finding $|x_2 - x_1| = |-9.2 - 8.6| = |-17.8| = 17.8$.)

37. $d = \sqrt{(x_2 - x_1)^2 + (y_2 - y_1)^2}$

$d = \sqrt{\left(\frac{5}{7} - \frac{1}{7}\right)^2 + \left(\frac{1}{14} - \frac{11}{14}\right)^2}$

$= \sqrt{\left(\frac{4}{7}\right)^2 + \left(-\frac{5}{7}\right)^2}$

$= \sqrt{\frac{16}{49} + \frac{25}{49}}$

$= \sqrt{\frac{41}{49}} = \frac{\sqrt{41}}{7} \approx 0.915$

39. $d = \sqrt{(x_2 - x_1)^2 + (y_2 - y_1)^2}$

$d = \sqrt{[0 - (-\sqrt{6})]^2 + (0 - \sqrt{2})^2}$

$= \sqrt{6 + 2}$

$= \sqrt{8} \approx 2.828$

41. $d = \sqrt{(x_2 - x_1)^2 + (y_2 - y_1)^2}$

$d = \sqrt{[-7 - (-4)]^2 + [-11 - (-2)]^2}$

$= \sqrt{(-3)^2 + (-9)^2}$

$= \sqrt{9 + 81}$

$= \sqrt{90} \approx 9.487$

43. We use the midpoint formula:

$\left(\frac{x_1 + x_2}{2}, \frac{y_1 + y_2}{2}\right) = \left(\frac{-7 + 9}{2}, \frac{6 + 2}{2}\right)$, or

$\left(\frac{2}{2}, \frac{8}{2}\right)$, or $(1, 4)$

45. We use the midpoint formula:

$\left(\frac{x_1 + x_2}{2}, \frac{y_1 + y_2}{2}\right) = \left(\frac{2 + 5}{2}, \frac{-1 + 8}{2}\right)$, or

$\left(\frac{7}{2}, \frac{7}{2}\right)$

47. We use the midpoint formula:

$\left(\frac{x_1 + x_2}{2}, \frac{y_1 + y_2}{2}\right) = \left(\frac{-8 + 6}{2}, \frac{-5 + (-1)}{2}\right)$, or

$\left(\frac{-2}{2}, \frac{-6}{2}\right)$, or $(-1, -3)$

49. We use the midpoint formula:

$\left(\frac{x_1 + x_2}{2}, \frac{y_1 + y_2}{2}\right) = \left(\frac{-3.4 + 2.9}{2}, \frac{8.1 + (-8.7)}{2}\right)$,

or $\left(\frac{-0.5}{2}, \frac{-0.6}{2}\right)$, or $(-0.25, -0.3)$

51. We use the midpoint formula:

$\left(\frac{x_1 + x_2}{2}, \frac{y_1 + y_2}{2}\right) = \left(\frac{\frac{1}{6} + \left(-\frac{1}{3}\right)}{2}, \frac{-\frac{3}{4} + \frac{5}{6}}{2}\right)$,

or $\left(\frac{-\frac{1}{6}}{2}, \frac{\frac{1}{12}}{2}\right)$, or $\left(-\frac{1}{12}, \frac{1}{24}\right)$

53. We use the midpoint formula:

$\left(\frac{x_1 + x_2}{2}, \frac{y_1 + y_2}{2}\right) = \left(\frac{\sqrt{2} + \sqrt{3}}{2}, \frac{-1 + 4}{2}\right)$, or

$\left(\frac{\sqrt{2} + \sqrt{3}}{2}, \frac{3}{2}\right)$

55. $(x - h)^2 + (y - k)^2 = r^2$ Standard form

$(x - 0)^2 + (y - 0)^2 = 6^2$ Substituting

$x^2 + y^2 = 36$ Simplifying

57. $(x - h)^2 + (y - k)^2 = r^2$ Standard form

$(x - 7)^2 + (y - 3)^2 = (\sqrt{5})^2$ Substituting

$(x - 7)^2 + (y - 3)^2 = 5$

59. $(x-h)^2+(y-k)^2=r^2$ Standard form

$[x-(-4)]^2+(y-3)^2=(4\sqrt{3})^2$ Substituting

$(x+4)^2+(y-3)^2=48$

$[(4\sqrt{3})^2=16\cdot 3=48]$

61. $(x-h)^2+(y-k)^2=r^2$

$[x-(-7)]^2+[y-(-2)]^2=(5\sqrt{2})^2$

$(x+7)^2+(y+2)^2=50$

63. Since the center is $(0,0)$, we have

$(x-0)^2+(y-0)^2=r^2$ or $x^2+y^2=r^2$

The circle passes through $(-3,4)$. We find r^2 by substituting -3 for x and 4 for y.

$(-3)^2+4^2=r^2$

$9+16=r^2$

$25=r^2$

Then $x^2+y^2=25$ is an equation of the circle.

65. Since the center is $(-4,1)$, we have

$[x-(-4)]^2+(y-1)^2=r^2$, or

$(x+4)^2+(y-1)^2=r^2$.

The circle passes through $(-2,5)$. We find r^2 by substituting -2 for x and 5 for y.

$(-2+4)^2+(5-1)^2=r^2$

$4+16=r^2$

$20=r^2$

Then $(x+4)^2+(y-1)^2=20$ is an equation of the circle.

67. We write standard form.

$(x-0)^2+(y-0)^2=8^2$

The center is $(0,0)$, and the radius is 8.

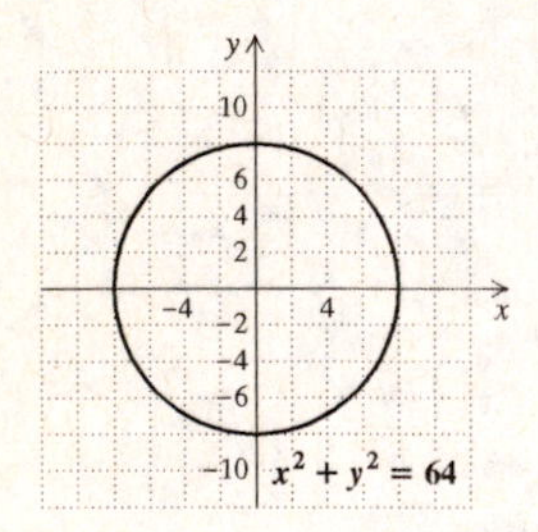

69. $(x+1)^2+(y+3)^2=36$

$[x-(-1)]^2+[y-(-3)]^2=6^2$ Standard form

The center is $(-1,-3)$, and the radius is 6.

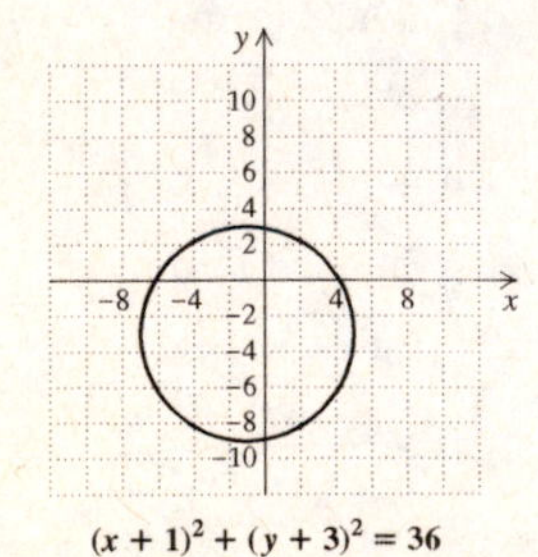

71. $(x-4)^2+(y+3)^2=10$

$(x-4)^2+[y-(-3)]^2=(\sqrt{10})^2$

The center is $(4,-3)$, and the radius is $\sqrt{10}$.

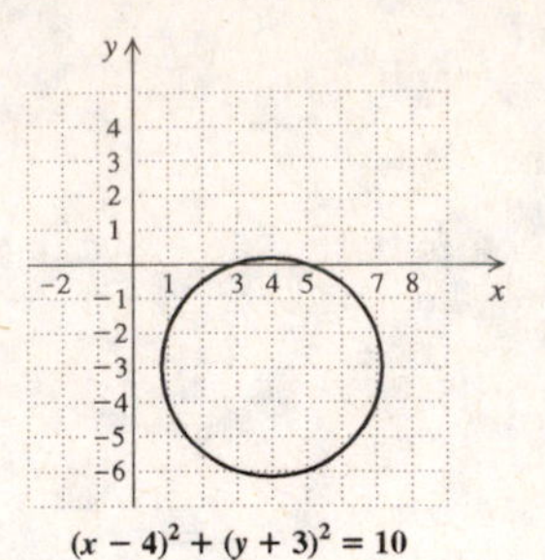

73. $x^2+y^2=10$

$(x-0)^2+(y-0)^2=(\sqrt{10})^2$ Standard form

The center is $(0,0)$, and the radius is $\sqrt{10}$.

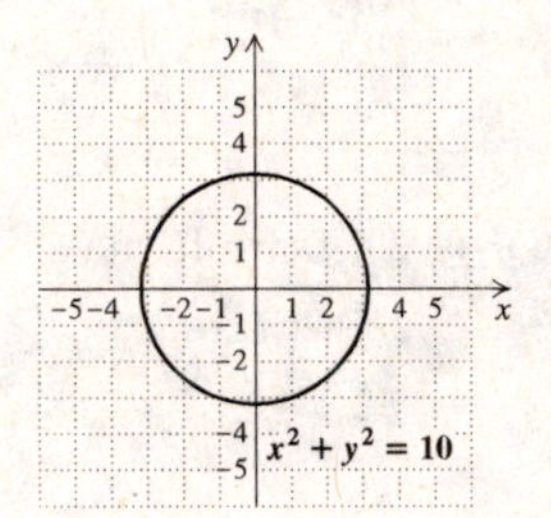

75. $(x-5)^2+y^2=\frac{1}{4}$

$(x-5)^2+(y-0)^2=\left(\frac{1}{2}\right)^2$ Standard form

The center is $(5,0)$, and the radius is $\frac{1}{2}$.

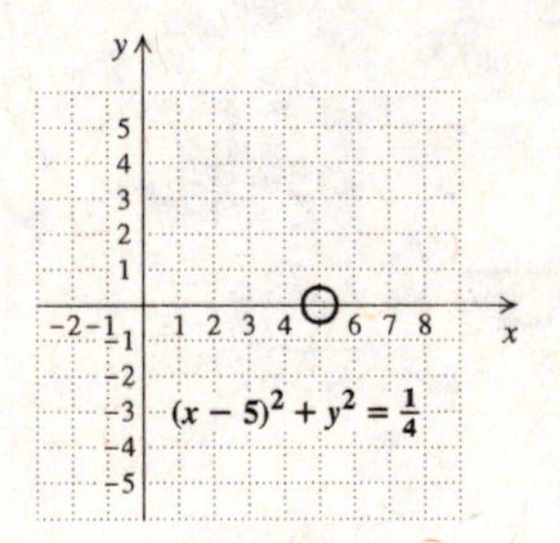

77. $x^2+y^2+8x-6y-15=0$

$x^2+8x+y^2-6y=15$

$(x^2+8x+16)+(y^2-6y+9)=15+16+9$

Completing the square twice

$(x+4)^2+(y-3)^2=40$

$[x-(-4)]^2+(y-3)^2=(\sqrt{40})^2$

Standard form

The center is $(-4,3)$, and the radius is $\sqrt{40}$, or $2\sqrt{10}$.

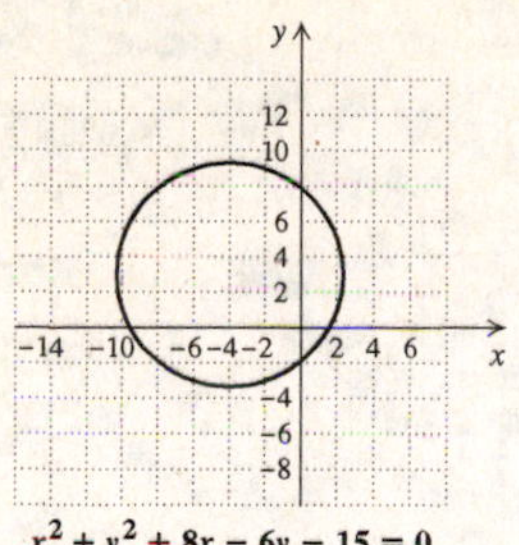

$x^2 + y^2 + 8x - 6y - 15 = 0$

79.

$$x^2 + y^2 - 8x + 2y + 13 = 0$$
$$x^2 - 8x + y^2 + 2y = -13$$
$$(x^2 - 8x + 16) + (y^2 + 2y + 1) = -13 + 16 + 1$$

Completing the square twice

$$(x-4)^2 + (y+1)^2 = 4$$
$$(x-4)^2 + [y-(-1)]^2 = 2^2$$

Standard form

The center is $(4, -1)$, and the radius is 2.

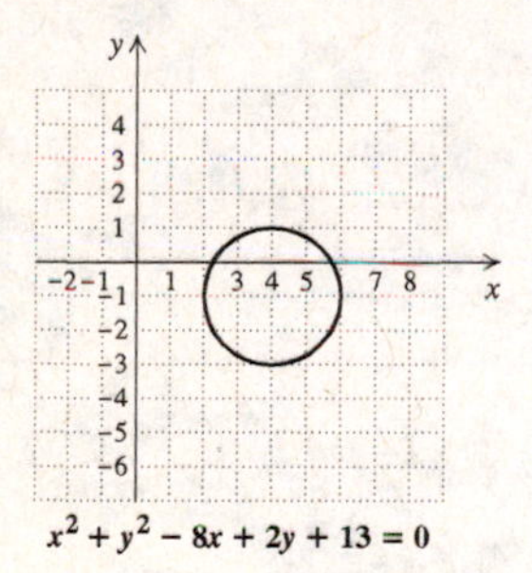

$x^2 + y^2 - 8x + 2y + 13 = 0$

81.

$$x^2 + y^2 + 10y - 75 = 0$$
$$x^2 + y^2 + 10y = 75$$
$$x^2 + (y^2 + 10y + 25) = 75 + 25$$
$$(x-0)^2 + (y+5)^2 = 100$$
$$(x-0)^2 + [y-(-5)]^2 = 10^2$$

The center is $(0, -5)$, and the radius is 10.

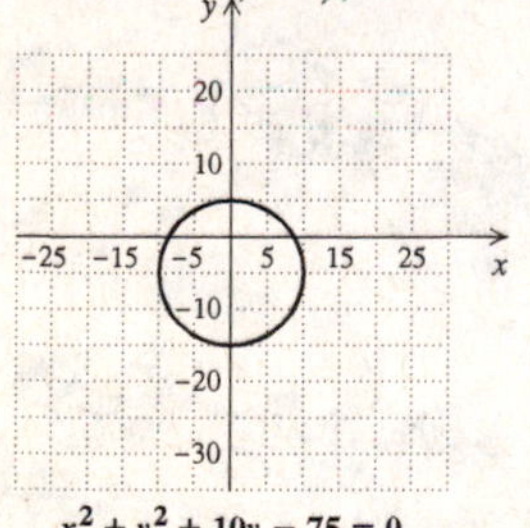

$x^2 + y^2 + 10y - 75 = 0$

83.

$$x^2 + y^2 + 7x - 3y - 10 = 0$$
$$x^2 + 7x + y^2 - 3y = 10$$
$$\left(x^2 + 7x + \frac{49}{4}\right) + \left(y^2 - 3y + \frac{9}{4}\right) = 10 + \frac{49}{4} + \frac{9}{4}$$
$$\left(x + \frac{7}{2}\right)^2 + \left(y - \frac{3}{2}\right)^2 = \frac{98}{4}$$
$$\left[x - \left(-\frac{7}{2}\right)\right]^2 + \left(y - \frac{3}{2}\right)^2 = \left(\sqrt{\frac{98}{4}}\right)^2$$

The center is $\left(-\frac{7}{2}, \frac{3}{2}\right)$, and the radius is $\sqrt{\frac{98}{4}}$, or $\frac{\sqrt{98}}{2}$, or $\frac{7\sqrt{2}}{2}$.

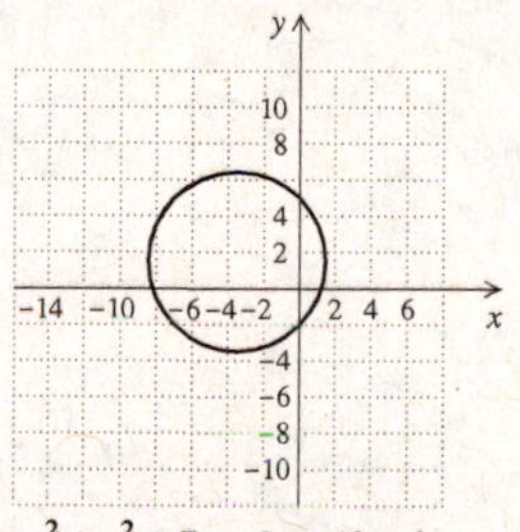

$x^2 + y^2 + 7x - 3y - 10 = 0$

85.

$$36x^2 + 36y^2 = 1$$
$$x^2 + y^2 = \frac{1}{36} \quad \text{Multiplying by } \frac{1}{36} \text{ on both sides}$$
$$(x-0)^2 + (y-0)^2 = \left(\frac{1}{6}\right)^2$$

The center is $(0, 0)$, and the radius is $\frac{1}{6}$.

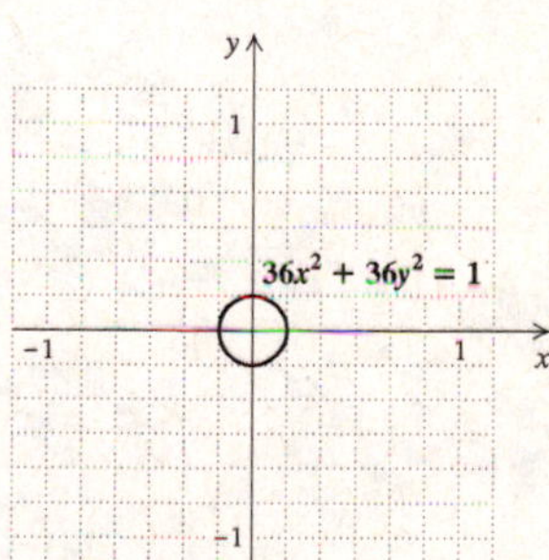

87. *Writing Exercise*

89.

$$\frac{x}{4} + \frac{5}{6} = \frac{2}{3}, \quad \text{LCD is 12}$$
$$12\left(\frac{x}{4} + \frac{5}{6}\right) = 12 \cdot \frac{2}{3}$$
$$12 \cdot \frac{x}{4} + 12 \cdot \frac{5}{6} = 8$$
$$3x + 10 = 8$$
$$3x = -2$$
$$x = -\frac{2}{3}$$

The solution is $-\frac{2}{3}$.

91. ***Familiarize.*** We make a drawing and label it. Let x represent the width of the border.

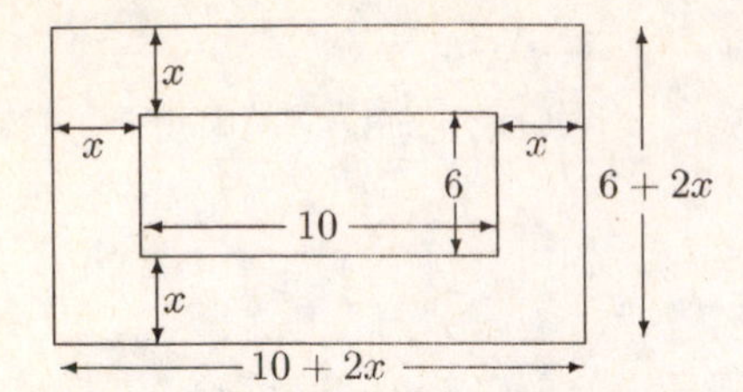

The perimeter of the larger rectangle is

$2(10+2x)+2(6+2x)$, or $8x+32$.

The perimeter of the smaller rectangle is

$2(10)+2(6)$, or 32.

Translate. The perimeter of the larger rectangle is twice the perimeter of the smaller rectangle.

$$8x+32=2\cdot 32$$

Carry out. We solve the equation.

$$8x+32=64$$
$$8x=32$$
$$x=4$$

Check. If the width of the border is 4 in., then the length and width of the larger rectangle are 18 in. and 14 in. Thus its perimeter is $2(18)+2(14)$, or 64 in. The perimeter of the smaller rectangle is 32 in. The perimeter of the larger rectangle is twice the perimeter of the smaller rectangle.

State. The width of the border is 4 in.

93. $3x-8y=5,\quad (1)$

$2x+6y=5\quad (2)$

Multiply Equation (1) by 3, multiply Equation (2) by 4, and add.

$$\begin{array}{rcl} 9x-24y&=&15\\ 8x+24y&=&20\\ \hline 17x&=&35\\ x&=&\frac{35}{17}\end{array}$$

Now substitute $\frac{35}{17}$ for x in one of the original equations and solve for y. We use Equation (2).

$$2x+6y=5$$
$$2\left(\frac{35}{17}\right)+6y=5$$
$$\frac{70}{17}+6y=5$$
$$6y=\frac{15}{17}$$
$$y=\frac{5}{34}$$

The solution is $\left(\frac{35}{17},\frac{5}{34}\right)$.

95. *Writing Exercise*

97. We make a drawing of the circle with center $(3,-5)$ and tangent to the y-axis.

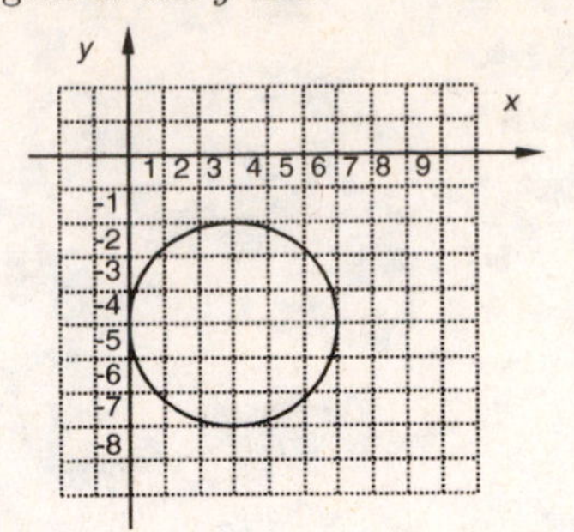

We see that the circle touches the y-axis at $(0,-5)$. Hence the radius is the distance between $(0,-5)$ and $(3,-5)$, or $\sqrt{(3-0)^2+[-5-(-5)]^2}$, or 3. Now we write the equation of the circle.

$$(x-h)^2+(y-k)^2=r^2$$
$$(x-3)^2+[y-(-5)]^2=3^2$$
$$(x-3)^2+(y+5)^2=9$$

99. First we use the midpoint formula to find the center:

$$\left(\frac{7+(-1)}{2},\frac{3+(-3)}{2}\right),\text{ or }\left(\frac{6}{2},\frac{0}{2}\right),\text{ or }(3,0)$$

The length of the radius is the distance between the center $(3,0)$ and either endpoint of a diameter. We will use endpoint $(7,3)$ in the distance formula:

$$r=\sqrt{(7-3)^2+(3-0)^2}=\sqrt{25}=5$$

Now we write the equation of the circle:

$$(x-h)^2+(y-k)^2=r^2$$
$$(x-3)^2+(y-0)^2=5^2$$
$$(x-3)^2+y^2=25$$

101. Let $(0,y)$ be the point on the y-axis that is equidistant from $(2,10)$ and $(6,2)$. Then the distance between $(2,10)$ and $(0,y)$ is the same as the distance between $(6,2)$ and $(0,y)$.

$$\sqrt{(0-2)^2+(y-10)^2}=\sqrt{(0-6)^2+(y-2)^2}$$
$$(-2)^2+(y-10)^2=(-6)^2+(y-2)^2$$

Squaring both sides

$$4+y^2-20y+100=36+y^2-4y+4$$
$$64=16y$$
$$4=y$$

This number checks. The point is $(0,4)$.

103. For the outer circle, $r^2=\frac{81}{4}$. For the inner circle, $r^2=16$. The area of the red zone is the difference between the areas of the outer and inner circles. Recall that the area A of a circle with radius r is given by the formula $A=\pi r^2$.

$$\pi\cdot\frac{81}{4}-\pi\cdot 16=\frac{81}{4}\pi-\frac{64}{4}\pi=\frac{17}{4}\pi$$

The area of the red zone is $\frac{17}{4}\pi$ m², or about 13.4 m².

105. Superimposing a coordinate system on the snowboard as in Exercise 104, and observing that $1170/2 = 585$, we know that three points on the circle are $(-585, 0)$, $(0, 23)$, and $(585, 0)$. Let $(0, k)$ represent the center of the circle. Use the fact that $(0, k)$ is equidistant from $(-585, 0)$ and $(0, 23)$.

$$\sqrt{(-585-0)^2+(0-k)^2} = \sqrt{(0-0)^2+(23-k)^2}$$
$$\sqrt{342,225 + k^2} = \sqrt{529 - 46k + k^2}$$
$$342,225 + k^2 = 529 - 46k + k^2 \quad \text{Squaring both sides}$$
$$341,696 = -46k$$
$$-7428.2 \approx k$$

Then to find the radius, find the distance from the center, $(0, -7428.2)$, to any one of the three known points on the circle. We use $(0, 23)$.

$$r = \sqrt{(0-0)^2 + (-7428.2 - 23)^2} \approx 7451.2 \text{ mm}$$

107. a) When the circle is positioned on a coordinate system as shown in the text, the center lies on the y-axis. To find the center, we will find the point on the y-axis that is equidistant from $(-4, 0)$ and $(0, 2)$. Let $(0, y)$ be this point.

$$\sqrt{[0-(-4)]^2+(y-0)^2} = \sqrt{(0-0)^2+(y-2)^2}$$
$$4^2 + y^2 = 0^2 + (y-2)^2 \quad \text{Squaring both sides}$$
$$16 + y^2 = y^2 - 4y + 4$$
$$12 = -4y$$
$$-3 = y$$

The center of the circle is $(0, -3)$.

b) We find the radius of the circle.

$$(x-0)^2 + [y-(-3)]^2 = r^2 \quad \text{Standard form}$$
$$x^2 + (y+3)^2 = r^2$$
$$(-4)^2 + (0+3)^2 = r^2 \quad \text{Substituting } (-4, 0) \text{ for } (x, y)$$
$$16 + 9 = r^2$$
$$25 = r^2$$
$$5 = r$$

The radius is 5 ft.

109. We write the equation of a circle with center $(0, 30.6)$ and radius 24.3:

$$x^2 + (y - 30.6)^2 = 590.49$$

111. Substitute 6 for N.

$$H = \frac{D^2N}{2.5} = \frac{D^2 \cdot 6}{2.5} = 2.4D^2$$

Find some ordered pairs for $2.5 \le D \le 8$ and draw the graph.

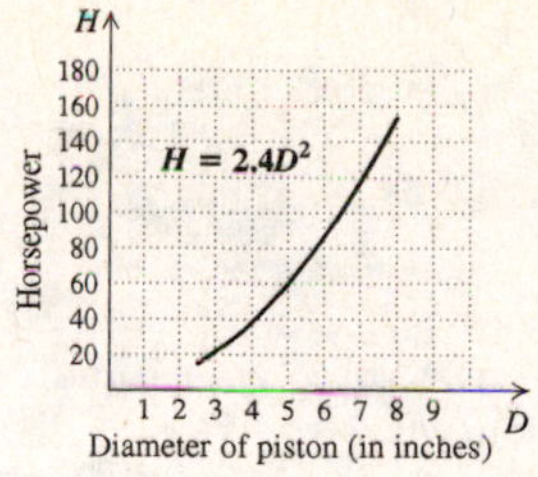

113. $y^2 + 2y + (x^2 - 6x - 6) = 0$

$a = 1$, $b = 2$, $c = x^2 - 6x - 6$

a) $$y = \frac{-2 \pm \sqrt{2^2 - 4 \cdot 1 \cdot (x^2 - 6x - 6)}}{2 \cdot 1}$$
$$y = \frac{-2 \pm \sqrt{4 - 4x^2 + 24x + 24}}{2}$$
$$y = \frac{-2 \pm \sqrt{-4x^2 + 24x + 28}}{2}$$
$$y = \frac{-2 \pm \sqrt{4(-x^2 + 6x + 7)}}{2}$$
$$y = \frac{-2 \pm 2\sqrt{-x^2 + 6x + 7}}{2}$$
$$y = -1 \pm \sqrt{-x^2 + 6x + 7}$$

b) [graph]

115. *Writing Exercise*

Exercise Set 13.2

1. True; see page 869 in the text.

3. True; see page 869 in the text.

5. False; see page 869 in the text.

7. True; see page 871 in the text.

9. $$\frac{x^2}{1} + \frac{y^2}{9} = 1$$
$$\frac{x^2}{1^2} + \frac{y^2}{3^2} = 1$$

The x-intercepts are $(1, 0)$ and $(-1, 0)$, and the y-intercepts are $(0, 3)$ and $(0, -3)$. We plot these points and connect them with an oval-shaped curve.

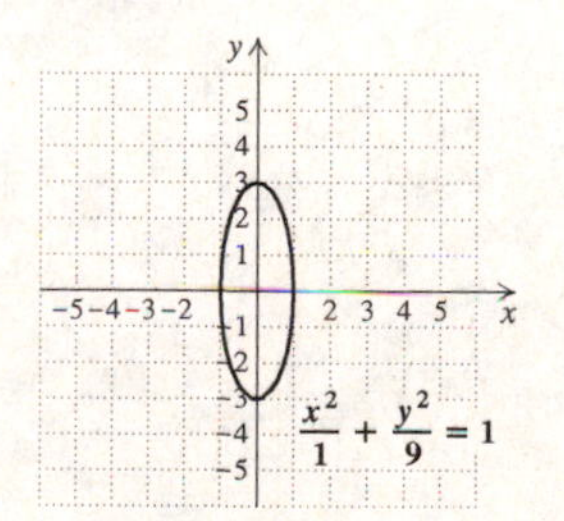

11. $\dfrac{x^2}{25}+\dfrac{y^2}{9}=1$

$\dfrac{x^2}{5^2}+\dfrac{y^2}{3^2}=1$

The x-intercepts are $(5,0)$ and $(-5,0)$, and the y-intercepts are $(0,3)$ and $(0,-3)$. We plot these points and connect them with an oval-shaped curve.

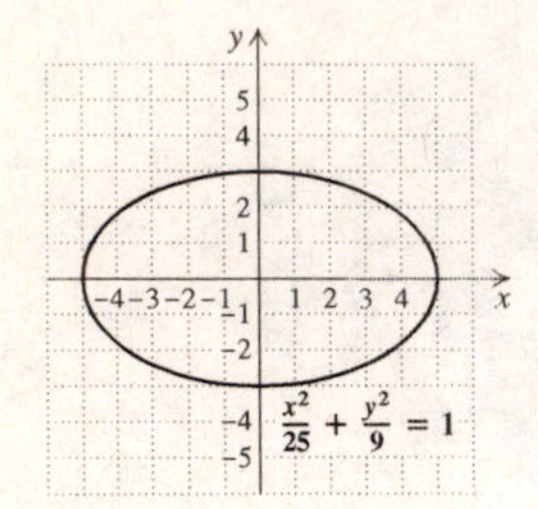

13. $4x^2+9y^2=36$

$\dfrac{1}{36}(4x^2+9y^2)=\dfrac{1}{36}(36)$ Multiplying by $\dfrac{1}{36}$

$\dfrac{x^2}{9}+\dfrac{y^2}{4}=1$

$\dfrac{x^2}{3^2}+\dfrac{y^2}{2^2}=1$

The x-intercepts are $(-3,0)$ and $(3,0)$, and the y-intercepts are $(0,-2)$ and $(0,2)$. We plot these points and connect them with an oval-shaped curve.

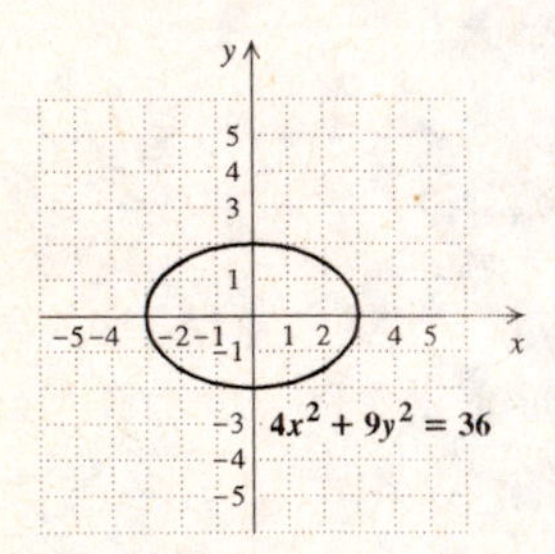

15. $16x^2+9y^2=144$

$\dfrac{x^2}{9}+\dfrac{y^2}{16}=1$ Multiplying by $\dfrac{1}{144}$

$\dfrac{x^2}{3^2}+\dfrac{y^2}{4^2}=1$

The x-intercepts are $(3,0)$ and $(-3,0)$, and the y-intercepts are $(0,4)$ and $(0,-4)$. We plot these points and connect them with an oval-shaped curve.

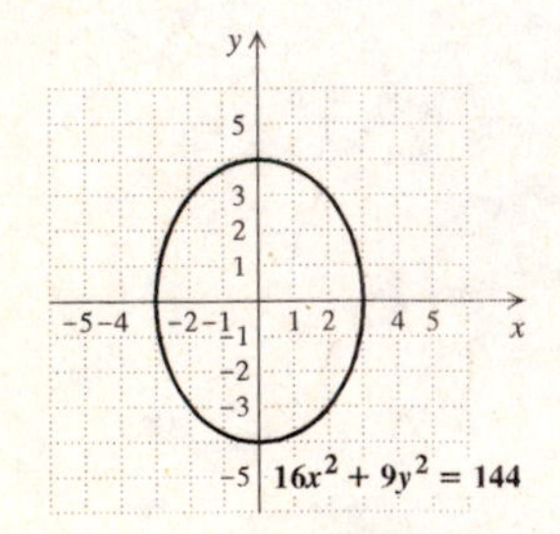

17. $2x^2+3y^2=6$

$\dfrac{x^2}{3}+\dfrac{y^2}{2}=1$ Multiplying by $\dfrac{1}{6}$

$\dfrac{x^2}{(\sqrt{3})^2}+\dfrac{y^2}{(\sqrt{2})^2}=1$

The x-intercepts are $(\sqrt{3},0)$ and $(-\sqrt{3},0)$, and the y-intercepts are $(0,\sqrt{2})$ and $(0,-\sqrt{2})$. We plot these points and connect them with an oval-shaped curve.

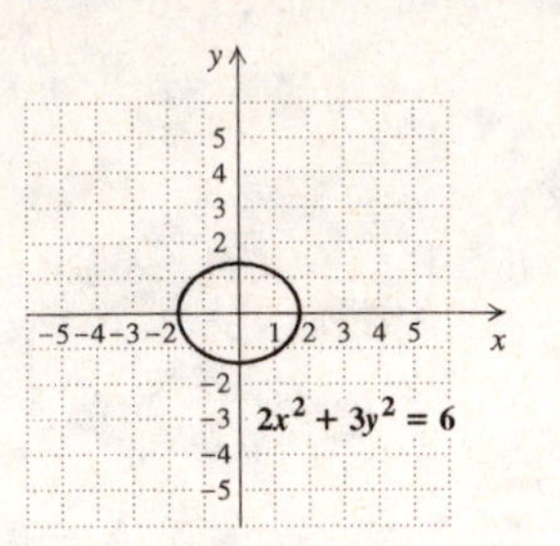

19. $5x^2+5y^2=125$

Observe that the x^2- and y^2-terms have the same coefficient. We divide both sides of the equation by 5 to obtain $x^2+y^2=25$. This is the equation of a circle with center $(0,0)$ and radius 5.

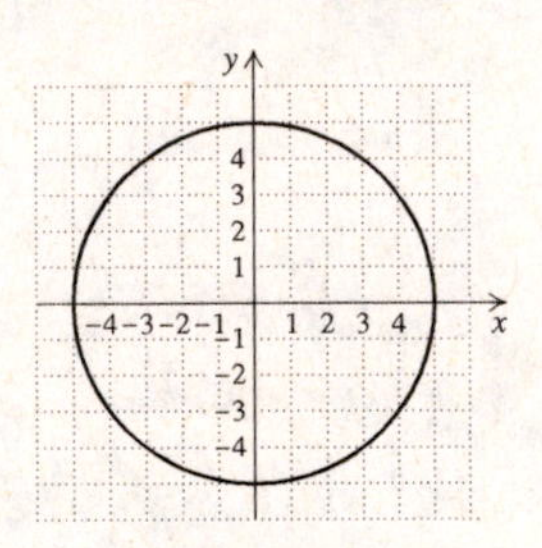

21. $3x^2+7y^2-63=0$

$3x^2+7y^2=63$

$\dfrac{x^2}{21}+\dfrac{y^2}{9}=1$ Multiplying by $\dfrac{1}{63}$

$\dfrac{x^2}{(\sqrt{21})^2}+\dfrac{y^2}{3^2}=1$

The x-intercepts are $(\sqrt{21},0)$ and $(-\sqrt{21},0)$, or about $(4.583,0)$ and $(-4.583,0)$. The y-intercepts are $(0,3)$ and $(0,-3)$. We plot these points and connect them with an oval-shaped curve.

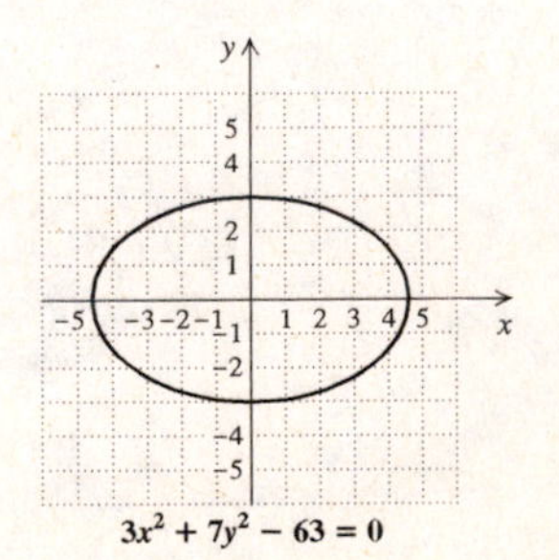

23.

$$8x^2 = 96 - 3y^2$$
$$8x^2 + 3y^2 = 96$$
$$\frac{x^2}{12} + \frac{y^2}{32} = 1$$
$$\frac{x^2}{(\sqrt{12})^2} + \frac{y^2}{(\sqrt{32})^2} = 1$$

The x-intercepts are $(\sqrt{12}, 0)$ and $(-\sqrt{12}, 0)$, or about $(3.464, 0)$ and $(-3.464, 0)$. The y-intercepts are $(0, \sqrt{32})$ and $(0, -\sqrt{32})$, or about $(0, 5.657)$ and $(0, -5.657)$. We plot these points and connect them with an oval-shaped curve.

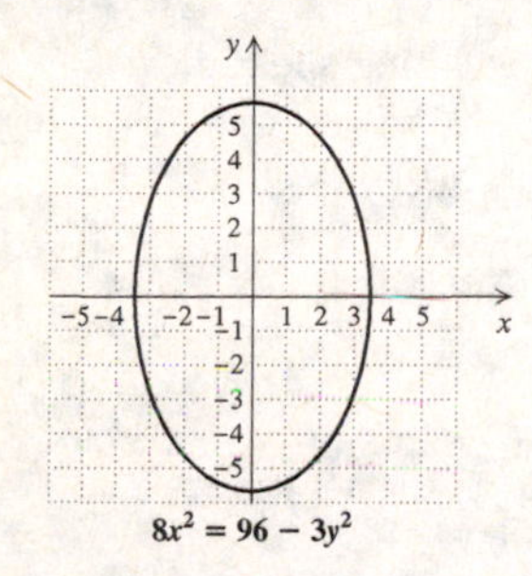

25. $16x^2 + 25y^2 = 1$

Note that $16 = \frac{1}{\frac{1}{16}}$ and $25 = \frac{1}{\frac{1}{25}}$. Thus, we can rewrite the equation:

$$\frac{x^2}{\frac{1}{16}} + \frac{y^2}{\frac{1}{25}} = 1$$
$$\frac{x^2}{\left(\frac{1}{4}\right)^2} + \frac{y^2}{\left(\frac{1}{5}\right)^2} = 1$$

The x-intercepts are $\left(\frac{1}{4}, 0\right)$ and $\left(-\frac{1}{4}, 0\right)$, and the y-intercepts are $\left(0, \frac{1}{5}\right)$ and $\left(0, -\frac{1}{5}\right)$. We plot these points and connect them with an oval-shaped curve.

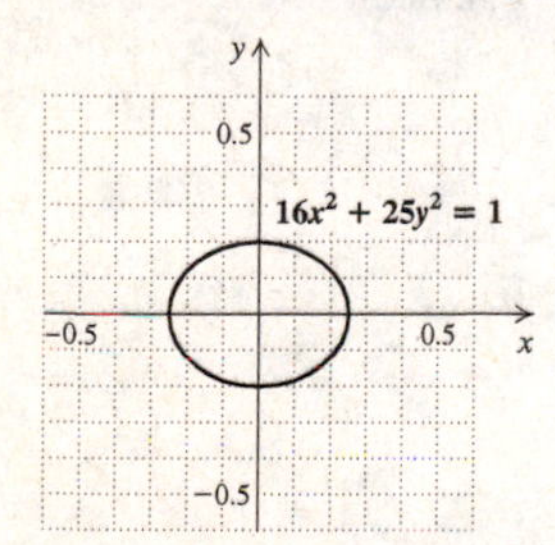

27.

$$\frac{(x-3)^2}{9} + \frac{(y-2)^2}{25} = 1$$
$$\frac{(x-3)^2}{3^2} + \frac{(y-2)^2}{5^2} = 1$$

The center of the ellipse is $(3, 2)$. Note that $a = 3$ and $b = 5$. We locate the center and then plot the points $(3+3, 2)$ $(3-3, 2)$, $(3, 2+5)$, and $(3, 2-5)$, or $(6, 2)$, $(0, 2)$, $(3, 7)$, and $(3, -3)$. Connect these points with an oval-shaped curve.

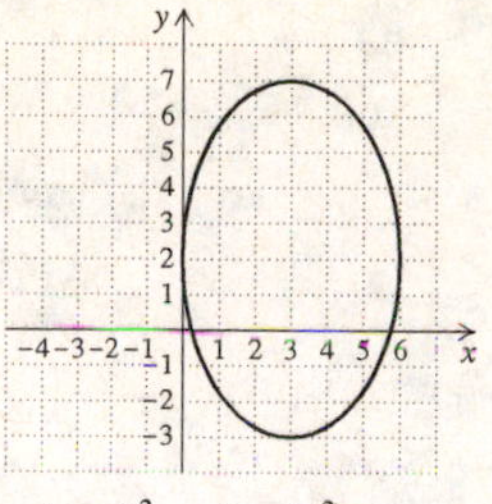

29.

$$\frac{(x+4)^2}{16} + \frac{(y-3)^2}{49} = 1$$
$$\frac{(x-(-4))^2}{4^2} + \frac{(y-3)^2}{7^2} = 1$$

The center of the ellipse is $(-4, 3)$. Note that $a = 4$ and $b = 7$. We locate the center and then plot the points $(-4+4, 3)$, $(-4-4, 3)$, $(-4, 3+7)$, and $(-4, 3-7)$, or $(0, 3)$, $(-8, 3)$, $(-4, 10)$, and $(-4, -4)$. Connect these points with an oval-shaped curve.

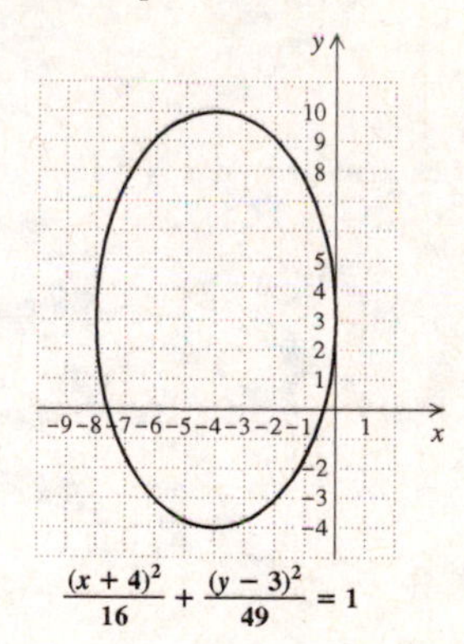

31.

$$12(x-1)^2 + 3(y+4)^2 = 48$$
$$\frac{(x-1)^2}{4} + \frac{(y+4)^2}{16} = 1$$
$$\frac{(x-1)^2}{2^2} + \frac{(y-(-4))^2}{4^2} = 1$$

The center of the ellipse is $(1, -4)$. Note that $a = 2$ and $b = 4$. We locate the center and then plot the points $(1+2, -4)$, $(1-2, -4)$, $(1, -4+4)$, and $(1, -4-4)$, or $(3, -4)$, $(-1, -4)$, $(1, 0)$, and $(1, -8)$. Connect these points with an oval-shaped curve.

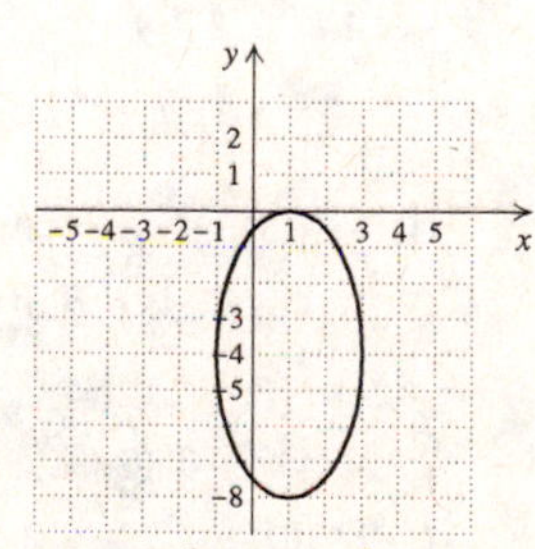

33. $4(x+3)^2+4(y+1)^2-10=90$

$$4(x+3)^2+4(y+1)^2=100$$

Observe that the x^2- and y^2-terms have the some coefficient. Dividing both sides by 4, we have

$$(x+3)^2+(y+1)^2=25.$$

This is the equation of a circle with center $(-3,-1)$ and radius 5.

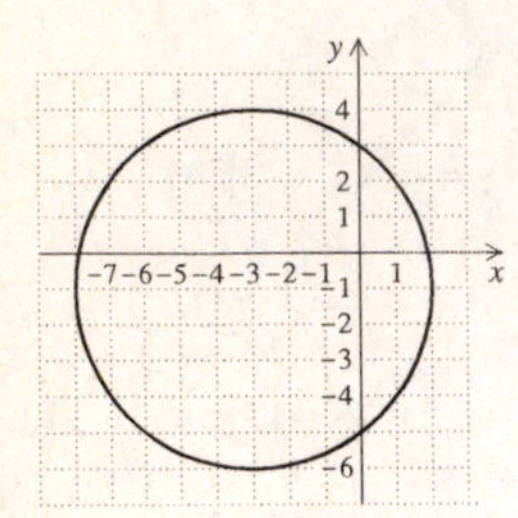

$4(x+3)^2+4(y+1)^2-10=90$

35. *Writing Exercise*

37. $\dfrac{3}{x-2}-\dfrac{5}{x-2}=9$

Note that the denominators are 0 when $x=2$, so 2 cannot be a solution. We multiply by the LCD, $x-2$.

$$x-2\left(\frac{3}{x-2}-\frac{5}{x-2}\right)=(x-2)9$$

$$(x-2)\cdot\frac{3}{x-2}-(x-2)\cdot\frac{5}{x-2}=9x-18$$

$$3-5=9x-18$$

$$-2=9x-18$$

$$16=9x$$

$$\frac{16}{9}=x$$

The number $\dfrac{16}{9}$ checks and is the solution.

39. $\dfrac{x}{x-4}-\dfrac{3}{x-5}=\dfrac{2}{x-4}$

Note that $x-4$ is 0 when $x=4$ and $x-5$ is 0 when x is 5, so 4 and 5 cannot be solutions. We multiply by the LCD, $(x-4)(x-5)$.

$$(x-4)(x-5)\left(\frac{x}{x-4}-\frac{3}{x-5}\right)=$$

$$(x-4)(x-5)\cdot\frac{2}{x-4}$$

$$(x-4)(x-5)\cdot\frac{x}{x-4}-(x-4)(x-5)\frac{3}{x-5}=2(x-5)$$

$$x(x-5)-3(x-4)=2(x-5)$$

$$x^2-5x-3x+12=2x-10$$

$$x^2-8x+12=2x-10$$

$$x^2-10x+22=0$$

We use the quadratic formula with $a=1$, $b=-10$, and $c=22$.

$$x=\frac{-b\pm\sqrt{b^2-4ac}}{2a}$$

$$x=\frac{-(-10)\pm\sqrt{(-10)^2-4\cdot1\cdot22}}{2\cdot1}$$

$$x=\frac{10\pm\sqrt{12}}{2}=\frac{10\pm2\sqrt{3}}{2}$$

$$x=\frac{\not{2}(5\pm\sqrt{3})}{\not{2}\cdot1}=5\pm\sqrt{3}$$

Both numbers check. The solutions are $5\pm\sqrt{3}$.

41. $9-\sqrt{2x+1}=7$

$$-\sqrt{2x+1}=-2 \qquad \text{Isolating the radical}$$

$$(-\sqrt{2x+1})^2=(-2)^2$$

$$2x+1=4$$

$$2x=3$$

$$x=\frac{3}{2}$$

The number $\dfrac{3}{2}$ checks and is the solution.

43. *Writing Exercise*

45. Plot the given points.

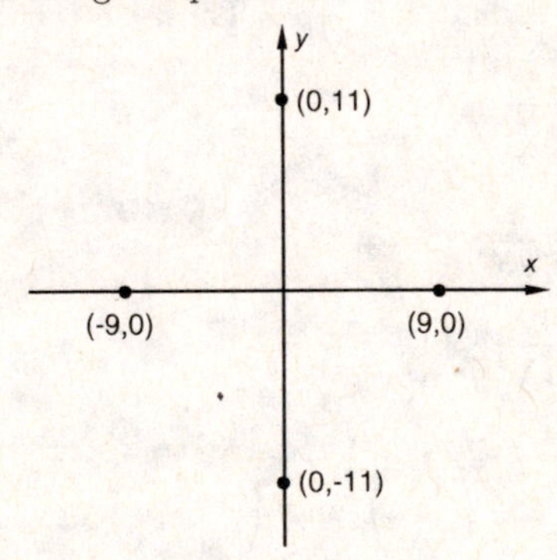

From the location of these points, we see that the ellipse that contains them is centered at the origin with $a=9$ and $b=11$. We write the equation of the ellipse:

$$\frac{x^2}{9^2}+\frac{y^2}{11^2}=1$$

$$\frac{x^2}{81}+\frac{y^2}{121}=1$$

47. Plot the given points.

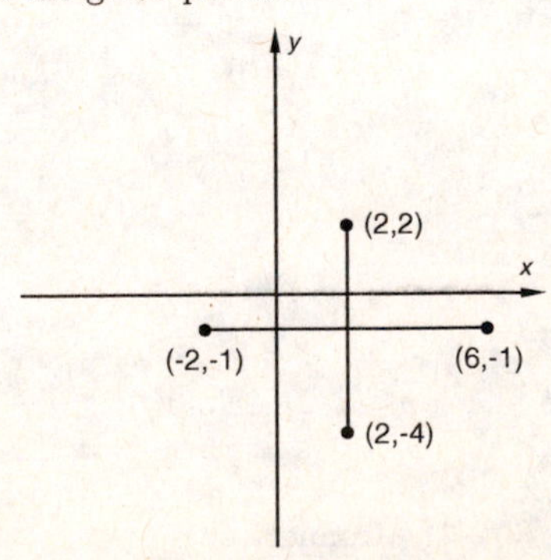

The midpoint of the segment from $(-2,-1)$ to $(6,-1)$ is $\left(\frac{-2+6}{2}, \frac{-1-1}{2}\right)$, or $(2,-1)$. The midpoint of the segment from $(2,-4)$ to $(2,2)$ is $\left(\frac{2+2}{2}, \frac{-4+2}{2}\right)$, or $(2,-1)$. Thus, we can conclude that $(2,-1)$ is the center of the ellipse. The distance from $(-2,-1)$ to $(2,-1)$ is $\sqrt{[2-(-2)]^2+[-1-(-1)]^2} = \sqrt{16} = 4$, so $a = 4$. The distance from $(2,2)$ to $(2,-1)$ is $\sqrt{(2-2)^2+(-1-2)^2} = \sqrt{9} = 3$, so $b = 3$. We write the equation of the ellipse.

$$\frac{(x-2)^2}{4^2} + \frac{(y-(-1))^2}{3^2} = 1$$
$$\frac{(x-2)^2}{16} + \frac{(y+1)^2}{9} = 1$$

49. We have a vertical ellipse centered at the origin with $a = 6/2$, or 3, and $b = 10/2$, or 5. Then the equation is $\frac{x^2}{3^2} + \frac{y^2}{5^2} = 1$, or $\frac{x^2}{9} + \frac{y^2}{25} = 1$.

51. a) Let $F_1 = (-c,0)$ and $F_2 = (c,0)$. Then the sum of the distances from the foci to P is $2a$. By the distance formula,

$$\sqrt{(x+c)^2+y^2} + \sqrt{(x-c)^2+y^2} = 2a, \text{ or}$$
$$\sqrt{(x+c)^2+y^2} = 2a - \sqrt{(x-c)^2+y^2}.$$

Squaring, we get

$$(x+c)^2+y^2 = 4a^2 - 4a\sqrt{(x-c)^2+y^2} + (x-c)^2 + y^2,$$

or $x^2+2cx+c^2+y^2$

$$= 4a^2 - 4a\sqrt{(x-c)^2+y^2} + x^2 - 2cx + c^2 + y^2.$$

Thus

$$-4a^2 + 4cx = -4a\sqrt{(x-c)^2+y^2}$$
$$a^2 - cx = a\sqrt{(x-c)^2+y^2}.$$

Squaring again, we get

$$a^4 - 2a^2cx + c^2x^2 = a^2(x^2 - 2cx + c^2 + y^2)$$
$$a^4 - 2a^2cx + c^2x^2 = a^2x^2 - 2a^2cx + a^2c^2 + a^2y^2,$$

or

$$x^2(a^2-c^2) + a^2y^2 = a^2(a^2-c^2)$$
$$\frac{x^2}{a^2} + \frac{y^2}{a^2-c^2} = 1.$$

b) When P is at $(0,b)$, it follows that $b^2 = a^2 - c^2$. Substituting, we have

$$\frac{x^2}{a^2} + \frac{y^2}{b^2} = 1.$$

53. For the given ellipse, $a = 6/2$, or 3, and $b = 2/2$, or 1. The patient's mouth should be at a distance 2c from the light source, where the coordinates of the foci of the ellipse are $(-c,0)$ and $(c,0)$. From Exercise 38(b), we know $b^2 = a^2 - c^2$. We use this to find c.

$b^2 = a^2 - c^2$

$1^2 = 3^2 - c^2$ Substituting

$c^2 = 8$

$c = \sqrt{8}$

Then $2c = 2\sqrt{8} \approx 5.66$. The patient's mouth should be about 5.66 ft from the light source.

55.

$$x^2 - 4x + 4y^2 + 8y - 8 = 0$$
$$x^2 - 4x + 4y^2 + 8y = 8$$
$$x^2 - 4x + 4(y^2 + 2y) = 8$$
$$(x^2-4x+4-4)+4(y^2+2y+1-1) = 8$$
$$(x^2-4x+4)+4(y^2+2y+1) = 8+4+4\cdot 1$$
$$(x-2)^2 + 4(y+1)^2 = 16$$
$$\frac{(x-2)^2}{16} + \frac{(y+1)^2}{4} = 1$$
$$\frac{(x-2)^2}{4^2} + \frac{(y-(-1))^2}{2^2} = 1$$

The center of the ellipse is $(2,-1)$. Note that $a = 4$ and $b = 2$. We locate the center and then plot the points $(2+4,-1)$, $(2-4,-1)$, $(2,-1+2)$, $(2,-1-2)$, or $(6,-1)$, $(-2,-1)$, $(2,1)$, and $(2,-3)$. Connect these points with an oval-shaped curve.

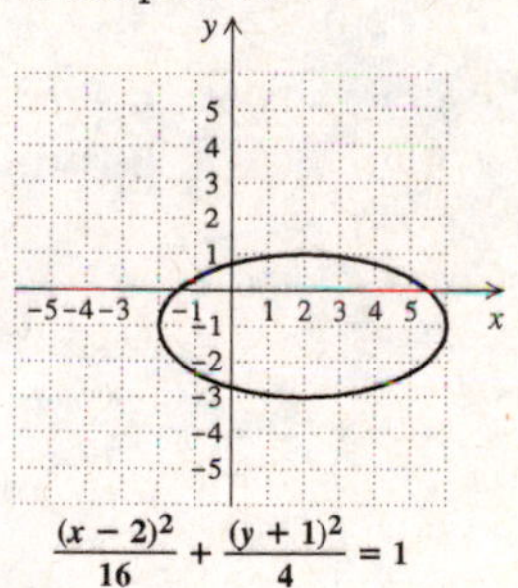

57.

Exercise Set 13.3

1. (d); see page 876 in the text.

3. (h); see page 883 in the text.

5. (g); see page 881 in the text.

7. (c); see page 881 in the text.

9.

$$\frac{x^2}{16} - \frac{y^2}{16} = 1$$
$$\frac{x^2}{4^2} - \frac{y^2}{4^2} = 1$$

$a = 4$ and $b = 4$, so the asymptotes are $y = \frac{4}{4}x$ and $y = -\frac{4}{4}x$, or $y = x$ and $y = -x$. We sketch them.

Replacing y with 0 and solving for x, we get $x = \pm 4$, so the intercepts are $(4,0)$ and $(-4,0)$.

We plot the intercepts and draw smooth curves through them that approach the asymptotes.

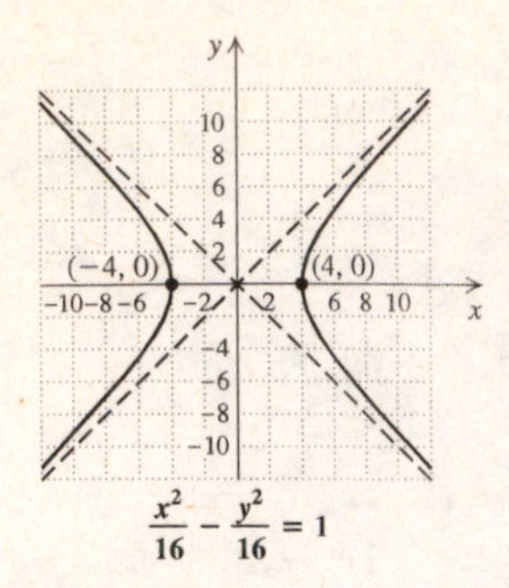

11. $\frac{x^2}{4} - \frac{y^2}{25} = 1$

$\frac{x^2}{2^2} - \frac{y^2}{5^2} = 1$

$a = 2$ and $b = 5$, so the asymptotes are $y = \frac{5}{2}x$ and $y = -\frac{5}{2}x$. We sketch them.

Replacing y with 0 and solving for x, we get $x = \pm 2$, so the intercepts are $(2, 0)$ and $(-2, 0)$.

We plot the intercepts and draw smooth curves through them that approach the asymptotes.

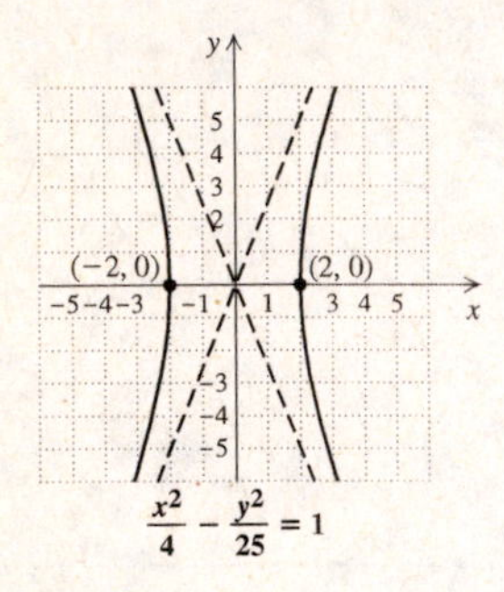

13. $\frac{y^2}{36} - \frac{x^2}{9} = 1$

$\frac{y^2}{6^2} - \frac{x^2}{3^2} = 1$

$a = 3$ and $b = 6$, so the asymptotes are $y = \frac{6}{3}x$ and $y = -\frac{6}{3}x$, or $y = 2x$ and $y = -2x$. We sketch them.

Replacing x with 0 and solving for y, we get $y = \pm 6$, so the intercepts are $(0, 6)$ and $(0, -6)$.

We plot the intercepts and draw smooth curves through them that approach the asymptotes.

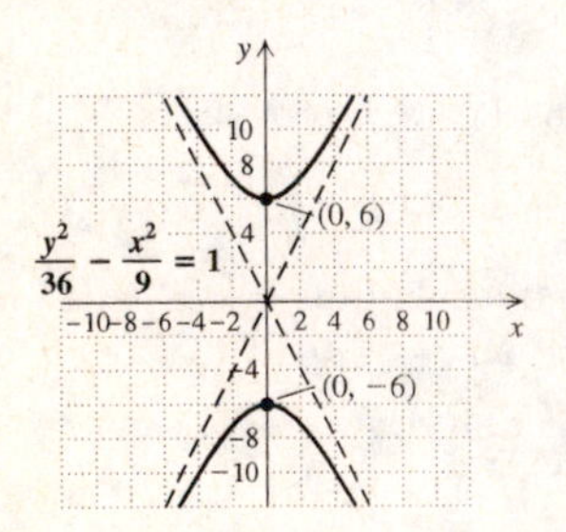

15. $y^2 - x^2 = 25$

$\frac{y^2}{25} - \frac{x^2}{25} = 1$

$\frac{y^2}{5^2} - \frac{x^2}{5^2} = 1$

$a = 5$ and $b = 5$, so the asymptotes are $y = \frac{5}{5}x$ and $y = -\frac{5}{5}x$, or $y = x$ and $y = -x$. We sketch them.

Replacing x with 0 and solving for y, we get $y = \pm 5$, so the intercepts are $(0, 5)$ and $(0, -5)$.

We plot the intercepts and draw smooth curves through them that approach the asymptotes.

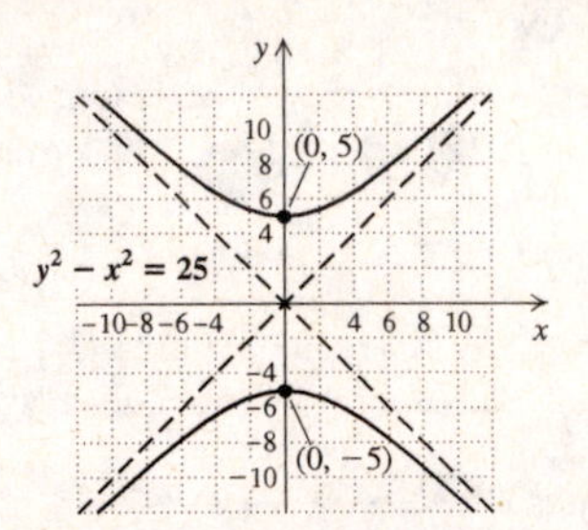

17. $25x^2 - 16y^2 = 400$

$\frac{x^2}{16} - \frac{y^2}{25} = 1$ Multiplying by $\frac{1}{400}$

$\frac{x^2}{4^2} - \frac{y^2}{5^2} = 1$

$a = 4$ and $b = 5$, so the asymptotes are $y = \frac{5}{4}x$ and $y = -\frac{5}{4}x$. We sketch them.

Replacing y with 0 and solving for x, we get $x = \pm 4$, so the intercepts are $(4, 0)$ and $(-4, 0)$.

We plot the intercepts and draw smooth curves through them that approach the asymptotes.

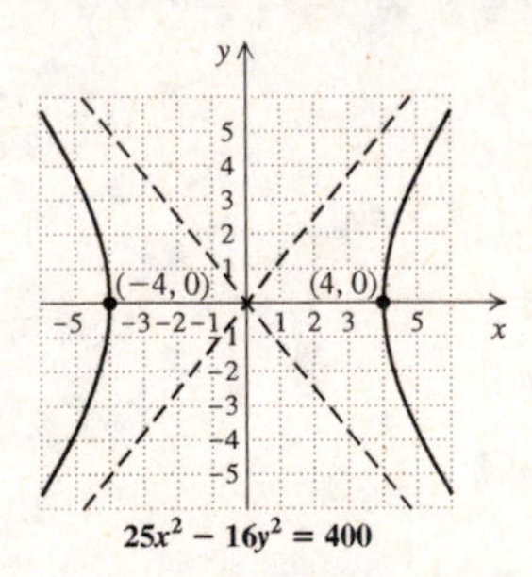

19. $xy = -5$

$y = -\dfrac{5}{x}$ Solving for y

We find some solutions, keeping the results in a table.

x	y
$\frac{1}{5}$	-25
1	-5
2	$-\frac{5}{2}$
5	-1
10	$-\frac{1}{2}$
$-\frac{1}{5}$	25
-1	5
-5	1
-10	$\frac{1}{2}$

Note that we cannot use 0 for x. The x-axis and the y-axis are the asymptotes.

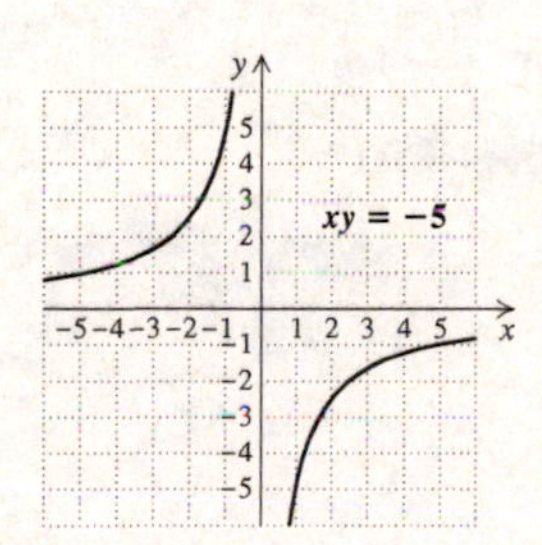

21. $xy = 4$

$y = \dfrac{4}{x}$ Solving for y

We find some solutions, keeping the results in a table.

x	y
$\frac{1}{2}$	8
1	4
4	1
8	$\frac{1}{2}$
$-\frac{1}{2}$	-8
-1	-4
-2	-2
-4	-1

Note that we cannot use 0 for x. The x-axis and the y-axis are the asymptotes.

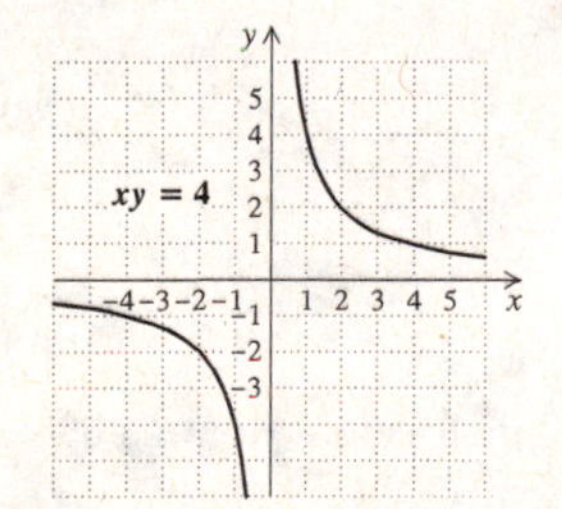

23. $xy = -2$

$y = -\dfrac{2}{x}$ Solving for y

x	y
$\frac{1}{2}$	-4
1	-2
2	-1
4	$-\frac{1}{2}$
$-\frac{1}{2}$	4
-1	2
-2	1
-4	$\frac{1}{2}$

Note that we cannot use 0 for x. The x-axis and the y-axis are the asymptotes.

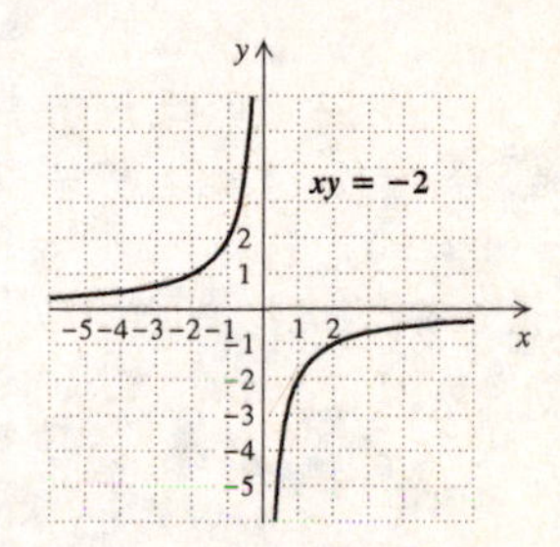

25. $xy = 1$

$y = \dfrac{1}{x}$ Solving for y

x	y
$\frac{1}{4}$	4
$\frac{1}{2}$	2
1	1
2	$\frac{1}{2}$
4	$\frac{1}{4}$
$-\frac{1}{4}$	-4
$-\frac{1}{2}$	-2
-1	-1
-2	$-\frac{1}{2}$
-4	$-\frac{1}{4}$

Note that we cannot use 0 for x. The x-axis and the y-axis are the asymptotes.

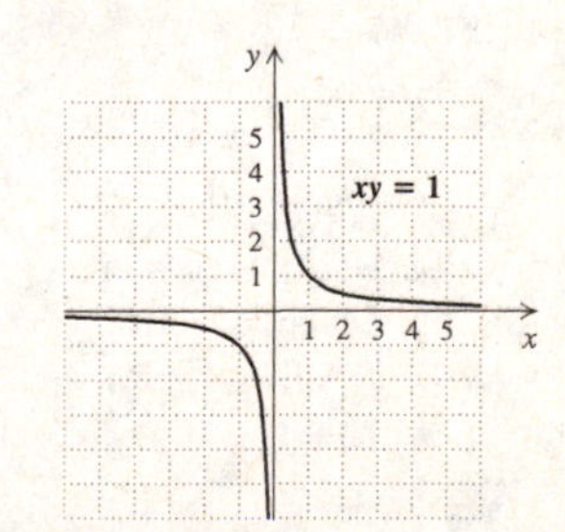

27. $x^2 + y^2 - 6x + 4y - 30 = 0$

Completing the square twice, we obtain an equivalent equation:

$$(x^2 - 6x) + (y^2 + 4y) = 30$$
$$(x^2 - 6x + 9) + (y^2 + 4y + 4) = 30 + 9 + 4$$
$$(x - 3)^2 + (y + 2)^2 = 43$$

The graph is a circle.

29. $9x^2 + 4y^2 - 36 = 0$

$$9x^2 + 4y^2 = 36$$

$$\frac{x^2}{4} + \frac{y^2}{9} = 1$$

The graph is an ellipse.

31. $4x^2 - 9y^2 - 72 = 0$

$$4x^2 - 9y^2 = 72$$

$$\frac{x^2}{18} - \frac{y^2}{8} = 1$$

The graph is a hyperbola.

33.

$$x^2 + y^2 = 2x + 4y + 4$$
$$x^2 - 2x + y^2 - 4y = 4$$
$$(x^2 - 2x + 1) + (y^2 - 4y + 4) = 4 + 1 + 4$$
$$(x-1)^2 + (y-2)^2 = 9$$

The graph is a circle.

35.

$$4x^2 = 64 - y^2$$
$$4x^2 + y^2 = 64$$
$$\frac{x^2}{16} + \frac{y^2}{64} = 1$$

The graph is an ellipse.

37. $x - \frac{8}{y} = 0$

$$x = \frac{8}{y}$$
$$xy = 8$$

The graph is a hyperbola.

39. $y + 6x = x^2 + 5$

$$y = x^2 - 6x + 5$$

The graph is a parabola.

41.

$$9y^2 = 36 + 4x^2$$
$$9y^2 - 4x^2 = 36$$
$$\frac{y^2}{4} - \frac{x^2}{9} = 1$$

The graph is a hyperbola.

43.

$$3x^2 + y^2 - x = 2x^2 - 9x + 10y + 40$$
$$x^2 + y^2 + 8x - 10y = 40$$

Both variables are squared, so the graph is not a parabola. The plus sign between x^2 and y^2 indicates that we have either a circle or an ellipse. Since the coefficients of x^2 and y^2 are the same, the graph is a circle.

45. $16x^2 + 5y^2 - 12x^2 + 8y^2 - 3x + 4y = 568$

$$4x^2 + 13y^2 - 3x + 4y = 568$$

Both variables are squared, so the graph is not a parabola. The plus sign between x^2 and y^2 indicates that we have either a circle or an ellipse. Since the coefficients of x^2 and y^2 are different, the graph is an ellipse.

47. *Writing Exercise*

49. $5x + 6y = -12$, (1)

$3x + 9y = 15$ (2)

We will use the elimination method. First multiply equation (1) by 3 and equation (2) by -2 and add.

$$\begin{aligned} 15x + 18y &= -36 \\ -6x - 18y &= -30 \\ \hline 9x \qquad &= -66 \\ x &= -\frac{22}{3} \end{aligned}$$

Now substitute $-\frac{22}{3}$ for x in one of the original equations and solve for y.

$$\begin{aligned} 5x + 6y &= -12 \quad (1) \\ 5\left(-\frac{22}{3}\right) + 6y &= -12 \\ -\frac{110}{3} + 6y &= -12 \\ 6y &= \frac{74}{3} \\ y &= \frac{37}{9} \end{aligned}$$

The solution is $\left(-\frac{22}{3}, \frac{37}{9}\right)$.

51. $y^2 - 3 = 6$

$y^2 = 9$

$y = 3$ *or* $y = -3$ Principle of square roots

The solutions are 3 and -3.

53. ***Familiarize***. Let $p =$ the price of the lawn chair before the tax was added. Then the total price is $p + 5\%p$, or $p + 0.05p$, or $1.05p$.

Translate.

The total price is \$36.75.

$$1.05p = 36.75$$

Carry out. We solve the equation.

$$1.05p = 36.75$$
$$p = \frac{36.75}{1.05}$$
$$p = 35$$

Check. 5% of \$35 is \$1.75 and \$35 + \$1.75 = \$36.75. The answer checks.

State. The price before tax was \$35.

55. *Writing Exercise*

57. Since the intercepts are $(0, 6)$ and $(0, -6)$, we know that the hyperbola is of the form $\frac{y^2}{b^2} - \frac{x^2}{a^2} = 1$ and that $b = 6$. The equations of the asymptotes tell us that $b/a = 3$, so

$$\frac{6}{a} = 3$$
$$a = 2.$$

The equation is $\frac{y^2}{6^2} - \frac{x^2}{2^2} = 1$, or $\frac{y^2}{36} - \frac{x^2}{4} = 1$.

59. $\dfrac{(x-5)^2}{36} - \dfrac{(y-2)^2}{25} = 1$

$\dfrac{(x-5)^2}{6^2} - \dfrac{(y-2)^2}{5^2} = 1$

$h = 5,\ k = 2,\ a = 6,\ b = 5$

Center: $(5, 2)$

Vertices: $(5-6, 2)$ and $(5+6, 2)$, or $(-1, 2)$ and $(11, 2)$

Asymptotes: $y-2 = \dfrac{5}{6}(x-5)$ and $y-2 = -\dfrac{5}{6}(x-5)$

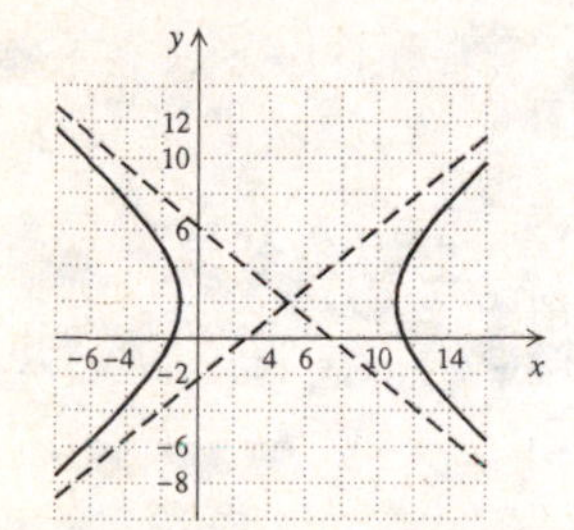

61. $8(y+3)^2 - 2(x-4)^2 = 32$

$\dfrac{(y+3)^2}{4} - \dfrac{(x-4)^2}{16} = 1$

$\dfrac{(y-(-3))^2}{2^2} - \dfrac{(x-4)^2}{4^2} = 1$

$h = 4,\ k = -3,\ a = 4,\ b = 2$

Center: $(4, -3)$

Vertices: $(4, -3+2)$ and $(4, -3-2)$, or $(4, -1)$ and $(4, -5)$

Asymptotes: $y-(-3) = \dfrac{2}{4}(x-4)$ and

$y-(-3) = -\dfrac{2}{4}(x-4)$, or $y+3 = \dfrac{1}{2}(x-4)$ and

$y+3 = -\dfrac{1}{2}(x-4)$

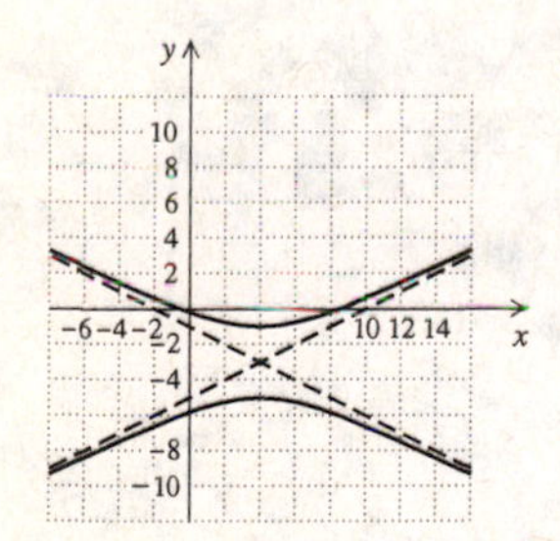

63.

$4x^2 - y^2 + 24x + 4y + 28 = 0$

$4(x^2+6x) - (y^2-4y) = -28$

$4(x^2+6x+9-9) - (y^2-4y+4-4) = -28$

$4(x^2+6x+9) - (y^2-4y+4) = -28 + 4\cdot 9 - 4$

$4(x+3)^2 - (y-2)^2 = 4$

$\dfrac{(x+3)^2}{1} - \dfrac{(y-2)^2}{4} = 1$

$\dfrac{(x-(-3))^2}{1^2} - \dfrac{(y-2)^2}{2^2} = 1$

$h = -3,\ k = 2,\ a = 1,\ b = 2$

Center: $(-3, 2)$

Vertices: $(-3-1, 2)$, and $(-3+1, 2)$, or $(-4, 2)$ and $(-2, 2)$

Asymptotes: $y-2 = \dfrac{2}{1}(x-(-3))$ and

$y-2 = -\dfrac{2}{1}(x-(-3))$, or $y-2 = 2(x+3)$ and

$y-2 = -2(x+3)$

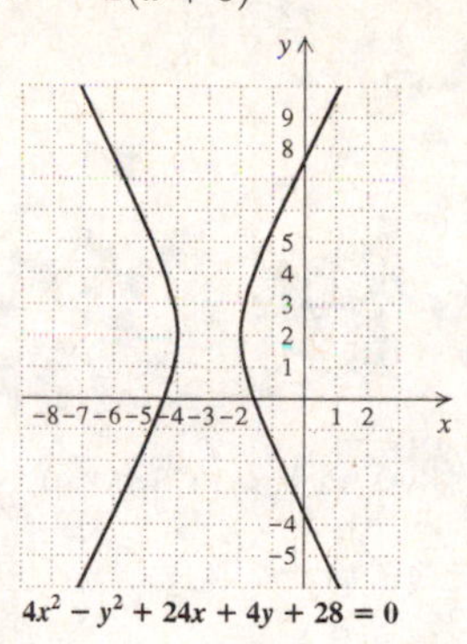

65.

Exercise Set 13.4

1. True

3. False; see page 890 in the text.

5. True; see page 887 in the text.

7. $x^2 + y^2 = 25,\quad (1)$

$y - x = 1\quad (2)$

First solve Eq. (2) for y.

$y = x + 1\quad (3)$

Then substitute $x+1$ for y in Eq. (1) and solve for x.

$x^2 + y^2 = 25$

$x^2 + (x+1)^2 = 25$

$x^2 + x^2 + 2x + 1 = 25$

$2x^2 + 2x - 24 = 0$

$x^2 + x - 12 = 0$

$(x+4)(x-3) = 0$

$x + 4 = 0$ *or* $x - 3 = 0$ Principle of zero products

$x = -4$ *or* $x = 3$

Now substitute these numbers in Eq. (3) and solve for y.

$y = -4 + 1 = -3$

$y = 3 + 1 = 4$

The pairs $(-4, -3)$ and $(3, 4)$ check, so they are the solutions.

9. $4x^2 + 9y^2 = 36,\quad (1)$

$3y + 2x = 6\quad (2)$

First solve Eq. (2) for y.

$3y = -2x + 6$

$y = -\dfrac{2}{3}x + 2\quad (3)$

Then substitute $-\frac{2}{3}x+2$ for y in Eq. (1) and solve for x.

$$4x^2+9y^2=36$$
$$4x^2+9\left(-\frac{2}{3}x+2\right)^2=36$$
$$4x^2+9\left(\frac{4}{9}x^2-\frac{8}{3}x+4\right)=36$$
$$4x^2+4x^2-24x+36=36$$
$$8x^2-24x=0$$
$$x^2-3x=0$$
$$x(x-3)=0$$

$x=0$ *or* $x=3$

Now substitute these numbers in Eq. (3) and solve for y.

$y=-\frac{2}{3}\cdot 0+2=2$

$y=-\frac{2}{3}\cdot 3+2=0$

The pairs $(0,2)$ and $(3,0)$ check, so they are the solutions.

11. $y^2=x+3$, (1)

$2y=x+4$ (2)

First solve Eq. (2) for x.

$2y-4=x$ (3)

Then substitute $2y-4$ for x in Eq. (1) and solve for y.

$$y^2=x+3$$
$$y^2=(2y-4)+3$$
$$y^2=2y-1$$
$$y^2-2y+1=0$$
$$(y-1)(y-1)=0$$

$y-1=0$ *or* $y-1=0$

$y=1$ *or* $y=1$

Now substitute 1 for y in Eq. (3) and solve for x.

$2\cdot 1-4=x$

$-2=x$

The pair $(-2,1)$ checks. It is the solution.

13. $x^2-xy+3y^2=27$, (1)

$x-y=2$ (2)

First solve Eq. (2) for y.

$x-2=y$ (3)

Then substitute $x-2$ for y in Eq. (1) and solve for x.

$$x^2-xy+3y^2=27$$
$$x^2-x(x-2)+3(x-2)^2=27$$
$$x^2-x^2+2x+3x^2-12x+12=27$$
$$3x^2-10x-15=0$$

$$x=\frac{-(-10)\pm\sqrt{(-10)^2-4(3)(-15)}}{2\cdot 3}$$
$$x=\frac{10\pm\sqrt{100+180}}{6}=\frac{10\pm\sqrt{280}}{6}$$
$$x=\frac{10\pm 2\sqrt{70}}{6}=\frac{5\pm\sqrt{70}}{3}$$

Now substitute these numbers in Eq. (3) and solve for y.

$y=\frac{5+\sqrt{70}}{3}-2=\frac{-1+\sqrt{70}}{3}$

$y=\frac{5-\sqrt{70}}{3}-2=\frac{-1-\sqrt{70}}{3}$

The pairs $\left(\frac{5+\sqrt{70}}{3},\frac{-1+\sqrt{70}}{3}\right)$ and $\left(\frac{5-\sqrt{70}}{3},\frac{-1-\sqrt{70}}{3}\right)$ check, so they are the solutions.

15. $x^2+4y^2=25$, (1)

$x+2y=7$ (2)

First solve Eq. (2) for x.

$x=-2y+7$ (3)

Then substitute $-2y+7$ for x in Eq. (1) and solve for y.

$$x^2+4y^2=25$$
$$(-2y+7)^2+4y^2=25$$
$$4y^2-28y+49+4y^2=25$$
$$8y^2-28y+24=0$$
$$2y^2-7y+6=0$$
$$(2y-3)(y-2)=0$$

$y=\frac{3}{2}$ *or* $y=2$

Now substitute these numbers in Eq. (3) and solve for x.

$x=-2\cdot\frac{3}{2}+7=4$

$x=-2\cdot 2+7=3$

The pairs $\left(4,\frac{3}{2}\right)$ and $(3,2)$ check, so they are the solutions.

17. $x^2-xy+3y^2=5$, (1)

$x-y=2$ (2)

First solve Eq. (2) for y.

$x-2=y$ (3)

Then substitute $x-2$ for y in Eq. (1) and solve for x.

$$x^2-xy+3y^2=5$$
$$x^2-x(x-2)+3(x-2)^2=5$$
$$x^2-x^2+2x+3x^2-12x+12=5$$
$$3x^2-10x+7=0$$
$$(3x-7)(x-1)=0$$

$x=\frac{7}{3}$ *or* $x=1$

Now substitute these numbers in Eq. (3) and solve for y.

$y=\frac{7}{3}-2=\frac{1}{3}$

$y=1-2=-1$

The pairs $\left(\frac{7}{3},\frac{1}{3}\right)$ and $(1,-1)$ check, so they are the solutions.

19. $3x + y = 7$, (1)

$4x^2 + 5y = 24$ (2)

First solve Eq. (1) for y.

$y = 7 - 3x$ (3)

Then substitute $7 - 3x$ for y in Eq. (2) and solve for x.

$$4x^2 + 5y = 24$$
$$4x^2 + 5(7 - 3x) = 24$$
$$4x^2 + 35 - 15x = 24$$
$$4x^2 - 15x + 11 = 0$$
$$(4x - 11)(x - 1) = 0$$
$$x = \frac{11}{4} \text{ or } x = 1$$

Now substitute these numbers into Eq. (3) and solve for y.

$$y = 7 - 3 \cdot \frac{11}{4} = -\frac{5}{4}$$
$$y = 7 - 3 \cdot 1 = 4$$

The pairs $\left(\frac{11}{4}, -\frac{5}{4}\right)$ and $(1, 4)$ check, so they are the solutions.

21. $a + b = 7$, (1)

$ab = 4$ (2)

First solve Eq. (1) for a.

$a = -b + 7$ (3)

Then substitute $-b + 7$ for a in Eq. (2) and solve for b.

$$(-b + 7)b = 4$$
$$-b^2 + 7b = 4$$
$$0 = b^2 - 7b + 4$$
$$b = \frac{-(-7) \pm \sqrt{(-7)^2 - 4 \cdot 1 \cdot 4}}{2 \cdot 1}$$
$$b = \frac{7 \pm \sqrt{33}}{2}$$

Now substitute these numbers in Eq. (3) and solve for a.

$$a = -\left(\frac{7 + \sqrt{33}}{2}\right) + 7 = \frac{7 - \sqrt{33}}{2}$$
$$a = -\left(\frac{7 - \sqrt{33}}{2}\right) + 7 = \frac{7 + \sqrt{33}}{2}$$

The pairs $\left(\frac{7 - \sqrt{33}}{2}, \frac{7 + \sqrt{33}}{2}\right)$ and $\left(\frac{7 + \sqrt{33}}{2}, \frac{7 - \sqrt{33}}{2}\right)$ check, so they are the solutions.

23. $2a + b = 1$, (1)

$b = 4 - a^2$ (2)

Eq. (2) is already solved for b. Substitute $4 - a^2$ for b in Eq. (1) and solve for a.

$$2a + 4 - a^2 = 1$$
$$0 = a^2 - 2a - 3$$
$$0 = (a - 3)(a + 1)$$
$$a = 3 \quad \text{or} \quad a = -1$$

Substitute these numbers in Eq. (2) and solve for b.

$$b = 4 - 3^2 = -5$$
$$b = 4 - (-1)^2 = 3$$

The pairs $(3, -5)$ and $(-1, 3)$ check, so they are the solutions.

25. $a^2 + b^2 = 89$, (1)

$a - b = 3$ (2)

First solve Eq. (2) for a.

$a = b + 3$ (3)

Then substitute $b + 3$ for a in Eq. (1) and solve for b.

$$(b + 3)^2 + b^2 = 89$$
$$b^2 + 6b + 9 + b^2 = 89$$
$$2b^2 + 6b - 80 = 0$$
$$b^2 + 3b - 40 = 0$$
$$(b + 8)(b - 5) = 0$$
$$b = -8 \text{ or } b = 5$$

Substitute these numbers in Eq. (3) and solve for a.

$$a = -8 + 3 = -5$$
$$a = 5 + 3 = 8$$

The pairs $(-5, -8)$ and $(8, 5)$ check, so they are the solutions.

27. $y = x^2$, (1)

$x = y^2$ (2)

Eq. (1) is already solved for y. Substitute x^2 for y in Eq. (2) and solve for x.

$$x = y^2$$
$$x = (x^2)^2$$
$$x = x^4$$
$$0 = x^4 - x$$
$$0 = x(x^3 - 1)$$
$$0 = x(x - 1)(x^2 + x + 1)$$
$$x = 0 \text{ or } x = 1 \text{ or } x = \frac{-1 \pm \sqrt{1^2 - 4 \cdot 1 \cdot 1}}{2}$$
$$x = 0 \text{ or } x = 1 \text{ or } x = -\frac{1}{2} \pm \frac{\sqrt{3}}{2}i$$

Substitute these numbers in Eq. (1) and solve for y.

$$y = 0^2 = 0$$
$$y = 1^2 = 1$$
$$y = \left(-\frac{1}{2} + \frac{\sqrt{3}}{2}i\right)^2 = -\frac{1}{2} - \frac{\sqrt{3}}{2}i$$
$$y = \left(-\frac{1}{2} - \frac{\sqrt{3}}{2}i\right)^2 = -\frac{1}{2} + \frac{\sqrt{3}}{2}i$$

The pairs $(0, 0)$, $(1, 1)$, $\left(-\frac{1}{2} + \frac{\sqrt{3}}{2}i, -\frac{1}{2} - \frac{\sqrt{3}}{2}i\right)$, and $\left(-\frac{1}{2} - \frac{\sqrt{3}}{2}i, -\frac{1}{2} + \frac{\sqrt{3}}{2}i\right)$ check, so they are the solutions.

29. $x^2 + y^2 = 9, \quad (1)$

$x^2 - y^2 = 9 \quad (2)$

Here we use the elimination method.

$$\begin{aligned} x^2 + y^2 &= 9 \quad (1) \\ x^2 - y^2 &= 9 \quad (2) \\ \hline 2x^2 &= 18 \quad \text{Adding} \\ x^2 &= 9 \\ x &= \pm 3 \end{aligned}$$

If $x = 3$, $x^2 = 9$, and if $x = -3$, $x^2 = 9$, so substituting 3 or -3 in Eq. (1) gives us

$$\begin{aligned} x^2 + y^2 &= 9 \\ 9 + y^2 &= 9 \\ y^2 &= 0 \\ y &= 0. \end{aligned}$$

The pairs $(3, 0)$ and $(-3, 0)$ check. They are the solutions.

31. $x^2 + y^2 = 25, \quad (1)$

$xy = 12 \quad (2)$

First we solve Eq. (2) for y.

$$xy = 12$$

$$y = \frac{12}{x}$$

Then we substitute $\frac{12}{x}$ for y in Eq. (1) and solve for x.

$$\begin{aligned} x^2 + y^2 &= 25 \\ x^2 + \left(\frac{12}{x}\right)^2 &= 25 \\ x^2 + \frac{144}{x^2} &= 25 \\ x^4 + 144 &= 25x^2 \quad \text{Multiplying by } x^2 \\ x^4 - 25x^2 + 144 &= 0 \\ u^2 - 25u + 144 &= 0 \quad \text{Letting } u = x^2 \\ (u - 9)(u - 16) &= 0 \\ u = 9 \quad &\text{or} \quad u = 16 \end{aligned}$$

We now substitute x^2 for u and solve for x.

$$x^2 = 9 \quad or \quad x^2 = 16$$

$$x = \pm 3 \quad or \quad x = \pm 4$$

Since $y = 12/x$, if $x = 3$, $y = 4$; if $x = -3$, $y = -4$; if $x = 4$, $y = 3$; and if $x = -4$, $y = -3$. The pairs $(3, 4)$, $(-3, -4)$, $(4, 3)$, and $(-4, -3)$ check. They are the solutions.

33. $x^2 + y^2 = 9, \quad (1)$

$25x^2 + 16y^2 = 400 \quad (2)$

$$\begin{aligned} -16x^2 - 16y^2 &= -144 \quad \text{Multiplying (1) by } -16 \\ 25x^2 + 16y^2 &= 400 \\ \hline 9x^2 \qquad &= 256 \quad \text{Adding} \\ x &= \pm\frac{16}{3} \end{aligned}$$

$$\begin{aligned} \frac{256}{9} + y^2 &= 9 \quad \text{Substituting in (1)} \\ y^2 &= 9 - \frac{256}{9} \\ y^2 &= -\frac{175}{9} \\ y &= \pm\sqrt{-\frac{175}{9}} = \pm\frac{5\sqrt{7}}{3}i \end{aligned}$$

The pairs $\left(\frac{16}{3}, \frac{5\sqrt{7}}{3}i\right)$, $\left(\frac{16}{3}, -\frac{5\sqrt{7}}{3}i\right)$, $\left(-\frac{16}{3}, \frac{5\sqrt{7}}{3}i\right)$, and $\left(-\frac{16}{3}, -\frac{5\sqrt{7}}{3}i\right)$ check. They are the solutions.

35.

$$\begin{aligned} x^2 + y^2 &= 14, \quad (1) \\ x^2 - y^2 &= 4 \quad (2) \\ \hline 2x^2 &= 18 \quad \text{Adding} \\ x^2 &= 9 \\ x &= \pm 3 \end{aligned}$$

$$\begin{aligned} 9 + y^2 &= 14 \quad \text{Substituting in Eq. (1)} \\ y^2 &= 5 \\ y &= \pm\sqrt{5} \end{aligned}$$

The pairs $(-3, -\sqrt{5})$, $(-3, \sqrt{5})$, $(3, -\sqrt{5})$, and $(3, \sqrt{5})$ check. They are the solutions.

37. $x^2 + y^2 = 20, \quad (1)$

$xy = 8 \quad (2)$

First we solve Eq. (2) for y.

$$y = \frac{8}{x}$$

Then we substitute $\frac{8}{x}$ for y in Eq. (1) and solve for x.

$$\begin{aligned} x^2 + \left(\frac{8}{x}\right)^2 &= 20 \\ x^2 + \frac{64}{x^2} &= 20 \\ x^4 + 64 &= 20x^2 \\ x^4 - 20x^2 + 64 &= 0 \\ u^2 - 20u + 64 &= 0 \quad \text{Letting } u = x^2 \\ (u - 16)(u - 4) &= 0 \\ u = 16 \quad &or \quad u = 4 \\ x^2 = 16 \quad &or \quad x^2 = 4 \\ x = \pm 4 \quad &or \quad x = \pm 2 \end{aligned}$$

$y = 8/x$, so if $x = 4$, $y = 2$; if $x = -4$, $y = -2$; if $x = 2$, $y = 4$; if $x = -2$, $y = -4$. The pairs $(4, 2)$, $(-4, -2)$, $(2, 4)$, and $(-2, -4)$ check. They are the solutions.

39. $x^2 + 4y^2 = 20, \quad (1)$

$xy = 4 \quad (2)$

First we solve Eq. (2) for y.

$$y = \frac{4}{x}$$

Then we substitute $\frac{4}{x}$ for y in Eq. (1) and solve for x.

$$x^2 + 4\left(\frac{4}{x}\right)^2 = 20$$
$$x^2 + \frac{64}{x^2} = 20$$
$$x^4 + 64 = 20x^2$$
$$x^4 - 20x^2 + 64 = 0$$
$$u^2 - 20u + 64 = 0 \quad \text{Letting } u = x^2$$
$$(u-16)(u-4) = 0$$
$$u = 16 \quad or \quad u = 4$$
$$x^2 = 16 \quad or \quad x^2 = 4$$
$$x = \pm 4 \quad or \quad x = \pm 2$$

$y = 4/x$, so if $x = 4$, $y = 1$; if $x = -4$, $y = -1$; if $x = 2$, $y = 2$; and if $x = -2$, $y = -2$. The pairs $(4, 1)$, $(-4, -1)$, $(2, 2)$, and $(-2, -2)$ check. They are the solutions.

41. $2xy + 3y^2 = 7, \quad (1)$
$3xy - 2y^2 = 4 \quad (2)$

$$6xy + 9y^2 = 21 \quad \text{Multiplying (1) by 3}$$
$$-6xy + 4y^2 = -8 \quad \text{Multiplying (2) by } -2$$
$$13y^2 = 13$$
$$y^2 = 1$$
$$y = \pm 1$$

Substitute for y in Eq. (1) and solve for x.

When $y = 1$: $\quad 2 \cdot x \cdot 1 + 3 \cdot 1^2 = 7$
$$2x = 4$$
$$x = 2$$

When $y = -1$: $\quad 2 \cdot x \cdot (-1) + 3(-1)^2 = 7$
$$-2x = 4$$
$$x = -2$$

The pairs $(2, 1)$ and $(-2, -1)$ check. They are the solutions.

43. $4a^2 - 25b^2 = 0, \quad (1)$
$2a^2 - 10b^2 = 3b + 4 \quad (2)$

$$4a^2 - 25b^2 = 0$$
$$-4a^2 + 20b^2 = -6b - 8 \quad \text{Multiplying (2) by } -2$$
$$-5b^2 = -6b - 8$$
$$0 = 5b^2 - 6b - 8$$
$$0 = (5b + 4)(b - 2)$$
$$b = -\frac{4}{5} \quad or \quad b = 2$$

Substitute for b in Eq. (1) and solve for a.

When $b = -\frac{4}{5}$: $\quad 4a^2 - 25\left(-\frac{4}{5}\right)^2 = 0$
$$4a^2 = 16$$
$$a^2 = 4$$
$$a = \pm 2$$

When $b = 2$: $\quad 4a^2 - 25(2)^2 = 0$
$$4a^2 = 100$$
$$a^2 = 25$$
$$a = \pm 5$$

The pairs $\left(2, -\frac{4}{5}\right)$, $\left(-2, -\frac{4}{5}\right)$, $(5, 2)$ and $(-5, 2)$ check. They are the solutions.

45. $ab - b^2 = -4, \quad (1)$
$ab - 2b^2 = -6 \quad (2)$

$$ab - b^2 = -4$$
$$-ab + 2b^2 = 6 \quad \text{Multiplying (2) by } -1$$
$$b^2 = 2$$
$$b = \pm\sqrt{2}$$

Substitute for b in Eq. (1) and solve for a.

When $b = \sqrt{2}$: $\quad a(\sqrt{2}) - (\sqrt{2})^2 = -4$
$$a\sqrt{2} = -2$$
$$a = -\frac{2}{\sqrt{2}} = -\sqrt{2}$$

When $b = -\sqrt{2}$: $\quad a(-\sqrt{2}) - (-\sqrt{2})^2 = -4$
$$-a\sqrt{2} = -2$$
$$a = \frac{-2}{-\sqrt{2}} = \sqrt{2}$$

The pairs $(-\sqrt{2}, \sqrt{2})$ and $(\sqrt{2}, -\sqrt{2})$ check. They are the solutions.

47. ***Familiarize***. We first make a drawing. We let l and w represent the length and width, respectively.

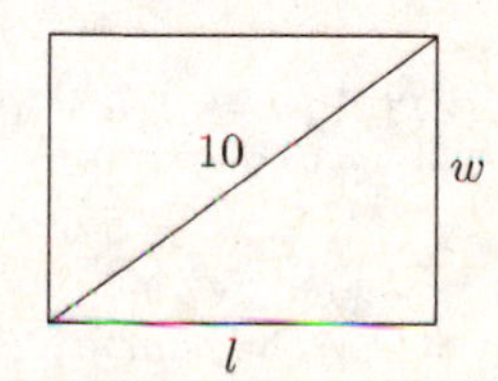

Translate. The perimeter is 28 cm.

$2l + 2w = 28$, or $l + w = 14$

Using the Pythagorean theorem we have another equation.

$l^2 + w^2 = 10^2$, or $l^2 + w^2 = 100$

Carry out. We solve the system:

$l + w = 14, \quad (1)$
$l^2 + w^2 = 100 \quad (2)$

First solve Eq. (1) for w.

$w = 14 - l \quad (3)$

Then substitute $14 - l$ for w in Eq. (2) and solve for l.

$$l^2 + w^2 = 100$$
$$l^2 + (14 - l)^2 = 100$$
$$l^2 + 196 - 28l + l^2 = 100$$
$$2l^2 - 28l + 96 = 0$$
$$l^2 - 14l + 48 = 0$$
$$(l - 8)(l - 6) = 0$$

$l = 8$ *or* $l = 6$

If $l = 8$, then $w = 14 - 8$, or 6. If $l = 6$, then $w = 14 - 6$, or 8. Since the length is usually considered to be longer than the width, we have the solution $l = 8$ and $w = 6$, or $(8, 6)$.

Check. If $l = 8$ and $w = 6$, then the perimeter is $2 \cdot 8 + 2 \cdot 6$, or 28. The length of a diagonal is $\sqrt{8^2 + 6^2}$, or $\sqrt{100}$, or 10. The numbers check.

State. The length is 8 cm, and the width is 6 cm.

49. ***Familiarize.*** We first make a drawing. Let l = the length and w = the width of the rectangle.

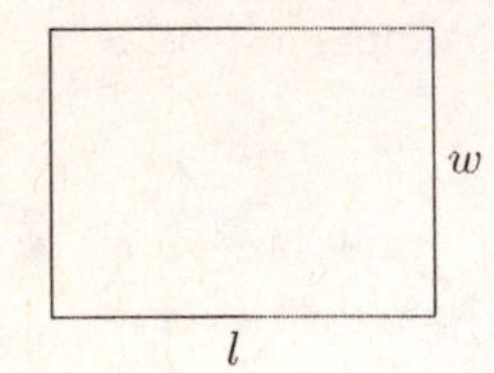

Translate.

Area: $lw = 20$

Perimeter: $2l + 2w = 18$, or $l + w = 9$

Carry out. We solve the system:

Solve the second equation for l: $l = 9 - w$

Substitute $9 - w$ for l in the first equation and solve for w.

$$(9 - w)w = 20$$
$$9w - w^2 = 20$$
$$0 = w^2 - 9w + 20$$
$$0 = (w - 5)(w - 4)$$

$w = 5$ *or* $w = 4$

If $w = 5$, then $l = 9 - w$, or 4. If $w = 4$, then $l = 9 - 4$, or 5. Since length is usually considered to be longer than width, we have the solution $l = 5$ and $w = 4$, or $(5, 4)$.

Check. If $l = 5$ and $w = 4$, the area is $5 \cdot 4$, or 20. The perimeter is $2 \cdot 5 + 2 \cdot 4$, or 18. The numbers check.

State. The length is 5 in. and the width is 4 in.

51. ***Familiarize.*** We first make a drawing. Let l = the length and w = the width of the cargo area, in feet.

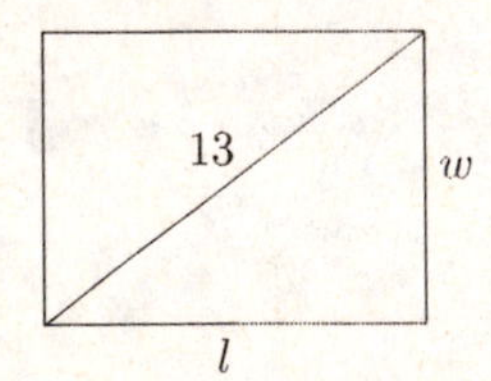

Translate. The cargo area must be 60 ft^2, so we have one equation:

$lw = 60$

The Pythagorean equation gives us another equation:

$l^2 + w^2 = 13^2$, or $l^2 + w^2 = 169$

Carry out. We solve the system of equations.

$$lw = 60, \quad (1)$$
$$l^2 + w^2 = 169 \quad (2)$$

First solve Eq. (1) for w:

$$lw = 60$$
$$w = \frac{60}{l} \quad (3)$$

Then substitute $60/l$ for w in Eq. (2) and solve for l.

$$l^2 + w^2 = 169$$
$$l^2 + \left(\frac{60}{l}\right)^2 = 169$$
$$l^2 + \frac{3600}{l^2} = 169$$
$$l^4 + 3600 = 169l^2$$
$$l^4 - 169l^2 + 3600 = 0$$

Let $u = l^2$ and $u^2 = l^4$ and substitute.

$$u^2 - 169u + 3600 = 0$$
$$(u - 144)(u - 25) = 0$$

$u = 144$ *or* $u = 25$

$l^2 = 144$ *or* $l^2 = 25$ Replacing u with l^2

$l = \pm 12$ *or* $l = \pm 5$

Since the length cannot be negative, we consider only 12 and 5. We substitute in Eq. (3) to find w. When $l = 12$, $w = 60/12 = 5$; when $l = 5$, $w = 60/5 = 12$. Since we usually consider length to be longer than width, we check the pair (12.5).

Check. If the length is 12 ft and the width is 5 ft, then the area is $12 \cdot 5$, or 60 ft^2. Also $12^2 + 5^2 = 144 + 25 = 169 = 13^2$. The answer checks.

State. The length is 12 ft and the width is 5 ft.

53. ***Familiarize.*** Let x and y represent the numbers.

Translate. The product of the numbers is 60, so we have

$xy = 60. \quad (1)$

The sum of the squares of the numbers is 136, so we have

$x^2 + y^2 = 136. \quad (2)$

Carry out. We solve the system of equations. First solve Eq. (1) for y.

$$xy = 60$$
$$y = \frac{60}{x} \quad (3)$$

Now substitute $\frac{60}{x}$ for y in Eq. (2) and solve for x.

$$x^2 + \left(\frac{60}{x}\right)^2 = 136$$
$$x^2 + \frac{3600}{x^2} = 136$$

$x^4 + 3600 = 136x^2$ Multiplying by x^2

$x^4 - 136x^2 + 3600 = 0$

$u^2 - 136u + 3600 = 0$ Letting $u = x^2$

$(u - 36)(u - 100) = 0$

$u = 36$ *or* $u = 100$

Now substitute x^2 for u and solve for x.

$x^2 = 36$ *or* $x^2 = 100$

$x = \pm 6$ *or* $x = \pm 10$

We use Eq. (3) to find y. When $x = 6$, $y = 60/6 = 10$; when $x = -6$, $y = 60/-6 = -10$; when $x = 10$, $y = 60/10 = 6$; when $x = -10$, $y = 60/-10 = -6$. We see that the numbers can be 6 and 10 or -6 and -10.

Check. $6 \cdot 10 = 60$ and $6^2 + 10^2 = 136$; also $-6(-10) = 60$ and $(-6)^2 + (-10)^2 = 136$. The solutions check.

State. The numbers are 6 and 10 or -6 and -10.

55. ***Familiarize.*** We let x = the length of a side of one peanut bed, in feet, and y = the length of a side of the other peanut bed. Make a drawing.

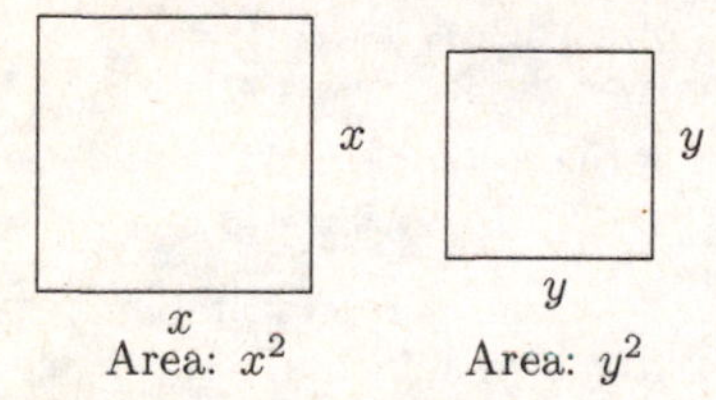

Translate. The sum of the areas is 832 ft^2, so we have

$x^2 + y^2 = 832.$ (1)

The difference of the areas is 320 ft^2, so we have

$x^2 - y^2 = 320.$ (2)

Carry out. We solve the system of equations.

$$\begin{array}{rl} x^2 + y^2 = & 832 \quad (1) \\ x^2 - y^2 = & 320 \quad (2) \\ \hline 2x^2 \quad = & 1152 \quad \text{Adding} \\ x^2 = & 576 \\ x = & \pm 24 \end{array}$$

Since the length cannot be negative we consider only 24. We substitute 24 for x in Eq. (1) and solve for y.

$$24^2 + y^2 = 832$$
$$576 + y^2 = 832$$
$$y^2 = 256$$
$$y = \pm 16$$

Again we consider only the positive number.

Check. If the lengths of the sides of the beds are 24 ft and 16 ft, the areas of the beds are 24^2, or 576 ft^2, and 16^2, or 256 ft^2, respectively. Then 576 ft^2 + 256 ft^2 = 832 ft^2, and 576 ft^2 − 256 ft^2 = 320 ft^2, so the answer checks.

State. The lengths of the sides of the beds are 24 ft and 16 ft.

57. ***Familiarize.*** We make a drawing and label it. Let x and y represent the lengths of the legs of the triangle.

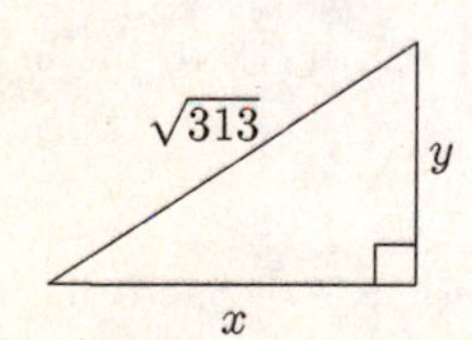

Translate. The product of the lengths of the legs is 156, so we have:

$xy = 156$

We use the Pythagorean theorem to get a second equation:

$x^2 + y^2 = (\sqrt{313})^2$, or $x^2 + y^2 = 313$

Carry out. We solve the system of equations.

$xy = 156,$ (1)

$x^2 + y^2 = 313$ (2)

First solve Equation (1) for y.

$$xy = 156$$
$$y = \frac{156}{x}$$

Then we substitute $\frac{156}{x}$ for y in Eq. (2) and solve for x.

$$x^2 + y^2 = 313 \qquad (2)$$
$$x^2 + \left(\frac{156}{x}\right)^2 = 313$$
$$x^2 + \frac{24,336}{x^2} = 313$$
$$x^4 + 24,336 = 313x^2$$
$$x^4 - 313x^2 + 24,336 = 0$$
$$u^2 - 313u + 24,336 = 0 \qquad \text{Letting } u = x^2$$
$$(u - 169)(u - 144) = 0$$
$$u - 169 = 0 \quad \textit{or} \quad u - 144 = 0$$
$$u = 169 \quad \textit{or} \quad u = 144$$

We now substitute x^2 for u and solve for x.

$x = \pm 13$ or $x = \pm 12$

Since $y = 156/x$, if $x = 13$, $y = 12$; if $x = -13$, $y = -12$; if $x = 12$, $y = 13$; and if $x = -12$, $y = -13$. The possible solutions are $(13, 12)$, $(-13, -12)$, $(12, 13)$, and $(-12, -13)$.

Check. Since measurements cannot be negative, we consider only $(13, 12)$ and $(12, 13)$. Since both possible solutions give the same pair of legs, we only need to check $(13, 12)$. If $x = 13$ and $y = 12$, their product is 156. Also, $\sqrt{13^2 + 12^2} = \sqrt{313}$. The numbers check.

State. The lengths of the legs are 13 and 12.

59. *Writing Exercise*

61. $(-1)^9(-2)^4 = -1 \cdot 16 = -16$

63. $\dfrac{(-1)^k}{k-5} = \dfrac{(-1)^6}{6-5} = \dfrac{1}{1} = 1$

65. $\dfrac{n}{2}(3 + n) = \dfrac{8}{2}(3 + 8) = 4 \cdot 11 = 44$

67. *Writing Exercise*

69. Let (h, k) represent the point on the line $5x + 8y = -2$ which is the center of a circle that passes through the points $(-2, 3)$ and $(-4, 1)$. The distance between (h, k) and $(-2, 3)$ is the same as the distance between (h, k) and $(-4, 1)$. This gives us one equation:

$$\sqrt{[h-(-2)]^2+(k-3)^2} = \sqrt{[h-(-4)]^2+(k-1)^2}$$
$$(h+2)^2+(k-3)^2 = (h+4)^2+(k-1)^2$$
$$h^2+4h+4+k^2-6k+9 = h^2+8h+16+k^2-2k+1$$
$$4h-6k+13 = 8h-2k+17$$
$$-4h-4k = 4$$
$$h+k = -1$$

We get a second equation by substituting (h,k) in $5x+8y=-2$.

$$5h+8k=-2$$

We now solve the following system:

$$h+\ k=-1,$$
$$5h+8k=\ -2$$

The solution, which is the center of the circle, is $(-2,1)$.

Next we find the length of the radius. We can find the distance between either $(-2,3)$ or $(-4,1)$ and the center $(-2,1)$. We use $(-2,3)$.

$$r=\sqrt{[-2-(-2)]^2+(1-3)^2}$$
$$r=\sqrt{0^2+(-2)^2}$$
$$r=\sqrt{4}=2$$

We can write the equation of the circle with center $(-2,1)$ and radius 2.

$$(x-h)^2+(y-k)^2=r^2$$
$$[x-(-2)]^2+(y-1)^2=2^2$$
$$(x+2)^2+(y-1)^2=4$$

71. $p^2+q^2=13,\quad (1)$

$\dfrac{1}{pq}=-\dfrac{1}{6}\qquad (2)$

Solve Eq. (2) for p.

$$\frac{1}{q}=-\frac{p}{6}$$
$$-\frac{6}{q}=p$$

Substitute $-6/q$ for p in Eq. (1) and solve for q.

$$\left(-\frac{6}{q}\right)^2+q^2=13$$
$$\frac{36}{q^2}+q^2=13$$
$$36+q^4=13q^2$$
$$q^4-13q^2+36=0$$
$$u^2-13u+36=0 \quad \text{Letting } u=q^2$$
$$(u-9)(u-4)=0$$
$$u=9 \quad \text{or} \quad u=4$$
$$x^2=9 \quad \text{or} \quad x^2=4$$
$$x=\pm 3 \text{ or } \quad x=\pm 2$$

Since $p=-6/q$, if $q=3$, $p=-2$; if $q=-3$, $p=2$; if $q=2$, $p=-3$; and if $q=-2$, $p=3$. The pairs $(-2,3)$, $(2,-3)$, $(-3,2)$, and $(3,-2)$ check. They are the solutions.

73. ***Familiarize***. Let l = the length of the rectangle, in feet, and let w = the width.

Translate. 100 ft of fencing is used, so we have

$l+w=100.\qquad (1)$

The area is 2475 ft^2, so we have

$lw=2475.\qquad (2)$

Carry out. Solving the system of equations, we get $(55,45)$ and $(45,55)$. Since length is usually considered to be longer than width, we have $l=55$ and $w=45$.

Check. If the length is 55 ft and the width is 45 ft, then $55+45$, or 100 ft, of fencing is used. The area is $55\cdot 45$, or 2475 ft^2. The answer checks.

State. The length of the rectangle is 55 ft, and the width is 45 ft.

75. ***Familiarize***. We let x and y represent the length and width of the base of the box, in inches, respectively. Make a drawing.

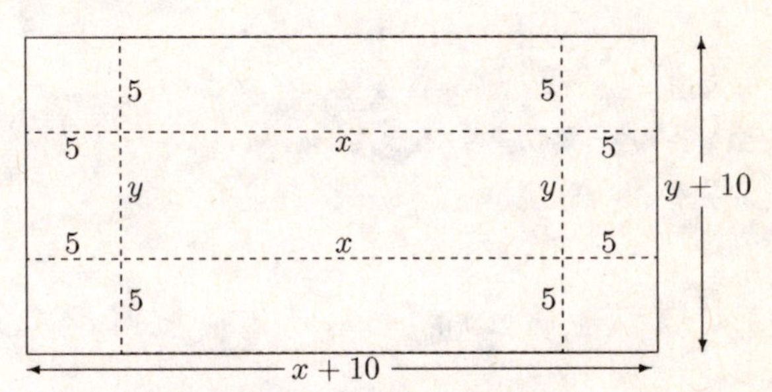

The dimensions of the metal sheet are $x+10$ and $y+10$.

Translate. The area of the sheet of metal is 340 in^2, so we have

$(x+10)(y+10)=340.\qquad (1)$

The volume of the box is 350 in^3, so we have

$x\cdot y\cdot 5=350.\qquad (2)$

Carry out. Solving the system of equations, we get $(10,7)$ and $(7,10)$. Since length is usually considered to be longer than width, we have $l=10$ and $w=7$.

Check. The dimensions of the metal sheet are $10+10$, or 20, and $7+10$, or 17, so the area is $20\cdot 17$, or 340 in^2. The volume of the box is $7\cdot 10\cdot 5$, or 350 in^3. The answer checks.

State. The dimensions of the box are 10 in. by 7 in. by 5 in.

77. ***Familiarize***. Let l = the length and h = the height, in inches.

Translate. Since the ratio of the length to the height is 16 to 9, we have one equation:

$$\frac{l}{h}=\frac{16}{9}$$

The Pythagorean equation gives us a second equation:

$$l^2+h^2=(\sqrt{4901})^2, \text{ or } l^2+h^2=4901$$

We have a system of equations.

$\dfrac{l}{h}=\dfrac{16}{9},\qquad (1)$

$l^2+h^2=4901\quad (2)$

Carry out. Solving the system of equations, we get $(34.32, 61.02)$ and $(-34.32, -61.02)$. Since the dimensions cannot be negative, we consider only $(34.32, 61.02)$.

Check. The ratio of 61.02 to 34.32 is $\dfrac{61.02}{34.32} \approx 1.778 \approx \dfrac{16}{9}$. Also $(61.02)^2 + (34.32)^2 = 4901.3028 \approx 4901$. The answer checks.

State. The length is about 61.02 in., and the height is about 34.32 in.

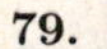
79. 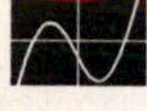

Chapter 14

Sequences, Series, and the Binomial Theorem

Exercise Set 14.1

1. (f)

3. (d)

5. (c)

7. $a_n = 2n + 3$

$a_1 = 2 \cdot 1 + 3 = 5,$

$a_2 = 2 \cdot 2 + 3 = 7,$

$a_3 = 2 \cdot 3 + 3 = 9,$

$a_4 = 2 \cdot 4 + 3 = 11;$

$a_{10} = 2 \cdot 10 + 3 = 23;$

$a_{15} = 2 \cdot 15 + 3 = 33$

9. $a_n = n^2 + 2$

$a_1 = 1^2 + 2 = 3,$

$a_2 = 2^2 + 2 = 6,$

$a_3 = 3^2 + 2 = 11,$

$a_4 = 4^2 + 2 = 18;$

$a_{10} = 10^2 + 2 = 102;$

$a_{15} = 15^2 + 2 = 227$

11. $a_n = \dfrac{n^2 - 1}{n^2 + 1}$

$a_1 = \dfrac{1^2 - 1}{1^2 + 1} = 0,$

$a_2 = \dfrac{2^2 - 1}{2^2 + 1} = \dfrac{3}{5},$

$a_3 = \dfrac{3^2 - 1}{3^2 + 1} = \dfrac{8}{10} = \dfrac{4}{5},$

$a_4 = \dfrac{4^2 - 1}{4^2 + 1} = \dfrac{15}{17};$

$a_{10} = \dfrac{10^2 - 1}{10^2 + 1} = \dfrac{99}{101};$

$a_{15} = \dfrac{15^2 - 1}{15^2 + 1} = \dfrac{224}{226} = \dfrac{112}{113}$

13. $a_n = \left(-\dfrac{1}{2}\right)^{n-1}$

$a_1 = \left(-\dfrac{1}{2}\right)^{1-1} = 1,$

$a_2 = \left(-\dfrac{1}{2}\right)^{2-1} = -\dfrac{1}{2},$

$a_3 = \left(-\dfrac{1}{2}\right)^{3-1} = \dfrac{1}{4},$

$a_4 = \left(-\dfrac{1}{2}\right)^{4-1} = -\dfrac{1}{8};$

$a_{10} = \left(-\dfrac{1}{2}\right)^{10-1} = -\dfrac{1}{512};$

$a_{15} = \left(-\dfrac{1}{2}\right)^{15-1} = \dfrac{1}{16,384}$

15. $a_n = (-1)^n(n + 3)$

$a_1 = (-1)^1(1 + 3) = -4,$

$a_2 = (-1)^2(2 + 3) = 5,$

$a_3 = (-1)^3(3 + 3) = -6,$

$a_4 = (-1)^4(4 + 3) = 7;$

$a_{10} = (-1)^{10}(10 + 3) = 13;$

$a_{15} = (-1)^{15}(15 + 3) = -18$

17. $a_n = (-1)^n(n^3 - 1)$

$a_1 = (-1)^1(1^3 - 1) = 0,$

$a_2 = (-1)^2(2^3 - 1) = 7,$

$a_3 = (-1)^3(3^3 - 1) = -26,$

$a_4 = (-1)^4(4^3 - 1) = 63;$

$a_{10} = (-1)^{10}(10^3 - 1) = 999;$

$a_{15} = (-1)^{15}(15^3 - 1) = -3374$

19. $a_n = 2n - 3$

$a_8 = 2 \cdot 8 - 3 = 16 - 3 = 13$

21. $a_n = (3n + 1)(2n - 5)$

$a_9 = (3 \cdot 9 + 1)(2 \cdot 9 - 5) = 28 \cdot 13 = 364$

23. $a_n = (-1)^{n-1}(3.4n - 17.3)$

$a_{12} = (-1)^{12-1}[3.4(12) - 17.3] = -23.5$

25. $a_n = 3n^2(9n - 100)$

$a_{11} = 3 \cdot 11^2(9 \cdot 11 - 100) = 3 \cdot 121(-1) = -363$

27. $a_n = \left(1 + \dfrac{1}{n}\right)^2$

$a_{20} = \left(1 + \dfrac{1}{20}\right)^2 = \left(\dfrac{21}{20}\right)^2 = \dfrac{441}{400}$

29. 2, 4, 6, 8, 10, . . .

These are even integers beginning with 2, so the general term could be $2n$.

31. 1, -1, 1, -1, . . .

1 and -1 alternate, beginning with 1, so the general term could be $(-1)^{n+1}$.

33. -1, 2, -3, 4, . . .

These are the first four natural numbers, but with alternating signs, beginning with a negative number. The general term could be $(-1)^n \cdot n$.

35. 3, 5, 7, 9, . . .

These are odd integers beginning with 3, so the general term could be $2n+1$.

37. $-2, 6, -18, 54, \ldots$

We can see a pattern if we write the sequence as

$-1\cdot 2\cdot 1,\ 1\cdot 2\cdot 3,\ -1\cdot 2\cdot 9,\ 1\cdot 2\cdot 27, \ldots$

The general term could be $(-1)^n\cdot 2\cdot (3)^{n-1}$.

39. $\frac{1}{2}, \frac{2}{3}, \frac{3}{4}, \frac{4}{5}, \frac{5}{6}, \ldots$

These are fractions in which the denominator is 1 greater than the numerator. Also, each numerator is 1 greater than the preceding numerator. The general term could be $\frac{n}{n+1}$.

41. 5, 25, 125, 625, . . .

This is powers of 5, so the general term could be 5^n.

43. $-1, 4, -9, 16, \ldots$

This is the squares of the first four natural numbers, but with alternating signs, beginning with a negative number. The general term could be $(-1)^n\cdot n^2$.

45. $1, -2, 3, -4, 5, -6, \ldots$

$S_7 = 1-2+3-4+5-6+7=4$

47. 2, 4, 6, 8, . . .

$S_5 = 2+4+6+8+10=30$

49. $$\sum_{k=1}^{5}\frac{1}{2k} = \frac{1}{2\cdot 1}+\frac{1}{2\cdot 2}+\frac{1}{2\cdot 3}+\frac{1}{2\cdot 4}+\frac{1}{2\cdot 5}$$
$$= \frac{1}{2}+\frac{1}{4}+\frac{1}{6}+\frac{1}{8}+\frac{1}{10}$$
$$= \frac{60}{120}+\frac{30}{120}+\frac{20}{120}+\frac{15}{120}+\frac{12}{120}$$
$$= \frac{137}{120}$$

51. $$\sum_{k=0}^{4} 3^k = 3^0+3^1+3^2+3^3+3^4$$
$$= 1+3+9+27+81$$
$$= 121$$

53. $$\sum_{k=2}^{8}\frac{k}{k-1} = \frac{2}{2-1}+\frac{3}{3-1}+\frac{4}{4-1}+$$
$$\frac{5}{5-1}+\frac{6}{6-1}+\frac{7}{7-1}+\frac{8}{8-1}$$
$$= \frac{2}{1}+\frac{3}{2}+\frac{4}{3}+\frac{5}{4}+\frac{6}{5}+\frac{7}{6}+\frac{8}{7}$$
$$= \frac{1343}{140}$$

55. $$\sum_{k=1}^{8}(-1)^{k+1}2^k = (-1)^2 2^1+(-1)^3 2^2+(-1)^4 2^3+$$
$$(-1)^5 2^4+(-1)^6 2^5+(-1)^7 2^6+$$
$$(-1)^8 2^7+(-1)^9 2^8$$
$$= 2-4+8-16+32-64+$$
$$128-256$$
$$= -170$$

57. $$\sum_{k=0}^{5}(k^2-2k+3)$$
$$= (0^2-2\cdot 0+3)+(1^2-2\cdot 1+3)+$$
$$(2^2-2\cdot 2+3)+(3^2-2\cdot 3+3)+$$
$$(4^2-2\cdot 4+3)+(5^2-2\cdot 5+3)$$
$$= 3+2+3+6+11+18$$
$$= 43$$

59. $$\sum_{k=3}^{5}\frac{(-1)^k}{k(k+1)} = \frac{(-1)^3}{3(3+1)}+\frac{(-1)^4}{4(4+1)}+\frac{(-1)^5}{5(5+1)}$$
$$= \frac{-1}{3\cdot 4}+\frac{1}{4\cdot 5}+\frac{-1}{5\cdot 6}$$
$$= -\frac{1}{12}+\frac{1}{20}-\frac{1}{30}$$
$$= -\frac{4}{60} = -\frac{1}{15}$$

61. $\frac{2}{3}+\frac{3}{4}+\frac{4}{5}+\frac{5}{6}+\frac{6}{7}$

This is a sum of fractions in which the denominator is one greater than the numerator. Also, each numerator is 1 greater than the preceding numerator. Sigma notation is

$$\sum_{k=1}^{5}\frac{k+1}{k+2}.$$

63. $1+4+9+16+25+36$

This is the sum of the squares of the first six natural numbers. Sigma notation is

$$\sum_{k=1}^{6}k^2.$$

65. $4-9+16-25+\ldots+(-1)^n n^2$

This is a sum of terms of the form $(-1)^k k^2$, beginning with $k=2$ and continuing through $k=n$. Sigma notation is

$$\sum_{k=2}^{n}(-1)^k k^2.$$

67. $5+10+15+20+25+\ldots$

This is a sum of multiples of 5, and it is an infinite series. Sigma notation is

$$\sum_{k=1}^{\infty}5k.$$

69. $\frac{1}{1\cdot 2}+\frac{1}{2\cdot 3}+\frac{1}{3\cdot 4}+\frac{1}{4\cdot 5}+\ldots$

This is a sum of fractions in which the numerator is 1 and the denominator is a product of two consecutive integers. The larger integer in each product is the smaller integer in the succeeding product. It is an infinite series. Sigma notation is

$$\sum_{k=1}^{\infty}\frac{1}{k(k+1)}.$$

71. *Writing Exercise*

73. $\frac{7}{2}(a_1+a_7) = \frac{7}{2}(8+14) = \frac{7}{2}\cdot 22 = 77$

75. $(x+y)^3$
$= (x+y)(x+y)^2$
$= (x+y)(x^2+2xy+y^2)$
$= x(x^2+2xy+y^2)+y(x^2+2xy+y^2)$
$= x^3+2x^2y+xy^2+x^2y+2xy^2+y^3$
$= x^3+3x^2y+3xy^2+y^3$

77. $(2a-b)^3$
$= (2a-b)(2a-b)^2$
$= (2a-b)(4a^2-4ab+b^2)$
$= 2a(4a^2-4ab+b^2)-b(4a^2-4ab+b^2)$
$= 8a^3-8a^2b+2ab^2-4a^2b+4ab^2-b^3$
$= 8a^3-12a^2b+6ab^2-b^3$

79. *Writing Exercise*

81. $a_1 = 1,\ a_{n+1} = 5a_n - 2$
$a_1 = 1$
$a_2 = 5\cdot 1 - 2 = 3$
$a_3 = 5\cdot 3 - 2 = 13$
$a_4 = 5\cdot 13 - 2 = 63$
$a_5 = 5\cdot 63 - 2 = 313$
$a_6 = 5\cdot 313 - 2 = 1563$

83. Find each term by multiplying the preceding term by 0.75: \$5200, \$3900, \$2925, \$2193.75, \$1645.31, \$1233.98, \$925.49, \$694.12, \$520.59, \$390.44

85. $a_n = (-1)^n$

This sequence is of the form $-1, 1, -1, 1, \ldots$. Each pair of terms adds to 0. S_{100} has 50 such pairs, so $S_{100} = 0$. S_{101} consists of the 50 pairs in S_{100} that add to 0 as well as a_{101}, or -1, so $S_{101} = -1$.

87. $a_n = i^n$
$a_1 = i^1 = i$
$a_2 = i^2 = -1$
$a_3 = i^3 = i^2\cdot i = -1\cdot i = -i$
$a_4 = i^4 = (i^2)^2 = (-1)^2 = 1$
$a_5 = i^5 = (i^2)^2\cdot i = (-1)^2\cdot i = 1\cdot i = i$
$S_5 = i - 1 - i + 1 + i = i$

89. Enter $y_1 = 14x^4 + 6x^3 + 416x^2 - 655x - 1050$. Then scroll through a table of values. We see that $y_1 = 6144$ when $x = 11$, so the 11th term of the sequence is 6144.

Exercise Set 14.2

1. True; see page 909 in the text.

3. False; see page 909 in the text.

5. True; see page 912 in the text.

7. False; $S_5 = a_1 + a_2 + a_3 + a_4 + a_5$.

9. $2, 6, 10, 14, \ldots$
$a_1 = 2$
$d = 4 \quad (6-2=4,\ 10-6=4,\ 14-10=4)$

11. $7, 3, -1, -5, \ldots$
$a_1 = 7$
$d = -4 \quad (3-7=-4, -1-3=-4, -5-(-1)=-4)$

13. $\frac{3}{2}, \frac{9}{4}, 3, \frac{15}{4}, \ldots$
$a_1 = \frac{3}{2}$
$d = \frac{3}{4} \quad \left(\frac{9}{4}-\frac{3}{2}=\frac{3}{4},\ 3-\frac{9}{4}=\frac{3}{4}\right)$

15. \$5.12, \$5.24, \$5.36, \$5.48, . . .
$a_1 = \$5.12$
$d = \$0.12$ (\$5.24 − \$5.12 = \$0.12, \$5.36 − \$5.24 = \$0.12, \$5.48 − \$5.36 = \$0.12)

17. $7, 10, 13, \ldots$
$a_1 = 7,\ d = 3,$ and $n = 15$
$a_n = a_1 + (n-1)d$
$a_{15} = 7 + (15-1)3 = 7 + 14\cdot 3 = 7 + 42 = 49$

19. $8, 2, -4, \ldots$
$a_1 = 8,\ d = -6,$ and $n = 18$
$a_n = a_1 + (n-1)d$
$a_{18} = 8 + (18-1)(-6) = 8 + 17(-6) = 8 - 102 = -94$

21. \$1200, \$964.32, \$728.64, . . .
$a_1 = \$1200,\ d = \$964.32 - \$1200 = -\$235.68,$ and $n = 13$
$a_n = a_1 + (n-1)d$
$a_{13} = \$1200 + (13-1)(-\$235.68) = \$1200 + 12(-\$235.68) = \$1200 - \$2828.16 = -\$1628.16$

23. $a_1 = 7,\ d = 3$
$a_n = a_1 + (n-1)d$
Let $a_n = 82$, and solve for n.
$82 = 7 + (n-1)(3)$
$82 = 7 + 3n - 3$
$82 = 4 + 3n$
$78 = 3n$
$26 = n$
The 26th term is 82.

25.
$$a_1 = 8,\ d = -6$$
$$a_n = a_1 + (n-1)d$$
$$-328 = 8 + (n-1)(-6)$$
$$-328 = 8 - 6n + 6$$
$$-328 = 14 - 6n$$
$$-342 = -6n$$
$$57 = n$$

The 57th term is -328.

27.
$$a_n = a_1 + (n-1)d$$
$$a_{17} = 2 + (17-1)5 \quad \text{Substituting 17 for } n, \text{ 2 for } a_1, \text{ and 5 for } d$$
$$= 2 + 16 \cdot 5$$
$$= 2 + 80$$
$$= 82$$

29.
$$a_n = a_1 + (n-1)d$$
$$33 = a_1 + (8-1)4 \quad \text{Substituting 33 for } a_8, \text{ 8 for } n, \text{ and 4 for } d$$
$$33 = a_1 + 28$$
$$5 = a_1$$

(Note that this procedure is equivalent to subtracting d from a_8 seven times to get a_1: $33 - 7(4) = 33 - 28 = 5$)

31.
$$a_n = a_1 + (n-1)d$$
$$-76 = 5 + (n-1)(-3) \quad \text{Substituting } -76 \text{ for } a_n, \text{ 5 for } a_1, \text{ and } -3 \text{ for } d$$
$$-76 = 5 - 3n + 3$$
$$-76 = 8 - 3n$$
$$-84 = -3n$$
$$28 = n$$

33. We know that $a_{17} = -40$ and $a_{28} = -73$. We would have to add d eleven times to get from a_{17} to a_{28}. That is,
$$-40 + 11d = -73$$
$$11d = -33$$
$$d = -3.$$

Since $a_{17} = -40$, we subtract d sixteen times to get to a_1.
$$a_1 = -40 - 16(-3) = -40 + 48 = 8$$

We write the first five terms of the sequence:

$8, 5, 2, -1, -4$

35. $a_{13} = 13$ and $a_{54} = 54$

Observe that for this to be true, $a_1 = 1$ and $d = 1$.

37. $1 + 5 + 9 + 13 + \ldots$

Note that $a_1 = 1$, $d = 4$, and $n = 20$. Before using the formula for S_n, we find a_{20}:
$$a_{20} = 1 + (20-1)4 \quad \text{Substituting into the formula for } a_n$$
$$= 1 + 19 \cdot 4$$
$$= 77$$

Then
$$S_{20} = \frac{20}{2}(1 + 77) \quad \text{Using the formula for } S_n$$
$$= 10(78)$$
$$= 780.$$

39. The sum is $1+2+3+\ldots+249+250$. This is the sum of the arithmetic sequence for which $a_1 = 1$, $a_n = 250$, and $n = 250$. We use the formula for S_n.
$$S_n = \frac{n}{2}(a_1 + a_n)$$
$$S_{300} = \frac{250}{2}(1 + 250) = 125(251) = 31,375$$

41. The sum is $2+4+6+\ldots+98+100$. This is the sum of the arithmetic sequence for which $a_1 = 2$, $a_n = 100$, and $n = 50$. We use the formula for S_n.
$$S_n = \frac{n}{2}(a_1 + a_n)$$
$$S_{50} = \frac{50}{2}(2 + 100) = 25(102) = 2550$$

43. The sum is $6+12+18+\ldots+96+102$. This is the sum of the arithmetic sequence for which $a_1 = 6$, $a_n = 102$, and $n = 17$. We use the formula for S_n.
$$S_n = \frac{n}{2}(a_1 + a_n)$$
$$S_{17} = \frac{17}{2}(6 + 102) = \frac{17}{2}(108) = 918$$

45. Before using the formula for S_n, we find a_{20}:
$$a_{20} = 4 + (20-1)5 \quad \text{Substituting into the formula for } a_n$$
$$= 4 + 19 \cdot 5 = 99$$

Then
$$S_{20} = \frac{20}{2}(4 + 99) \quad \text{Using the formula for } S_n$$
$$= 10(103) = 1030.$$

47. ***Familiarize***. We want to find the fifteenth term and the sum of an arithmetic sequence with $a_1 = 7$, $d = 2$, and $n = 15$. We will first use the formula for a_n to find a_{15}. This result is the number of marchers in the last row. Then we will use the formula for S_n to find S_{15}. This is the total number of marchers.

Translate. Substituting into the formula for a_n, we have
$$a_{15} = 7 + (15-1)2.$$

Carry out. We first find a_{15}.
$$a_{15} = 7 + 14 \cdot 2 = 35$$

Then use the formula for S_n to find S_{15}.
$$S_{15} = \frac{15}{2}(7 + 35) = \frac{15}{2}(42) = 315$$

Check. We can do the calculations again. We can also do the entire addition.

$7 + 9 + 11 + \ldots + 35.$

State. There are 35 marchers in the last row, and there are 315 marchers altogether.

49. ***Familiarize.*** We want to find the sum of the arithmetic sequence $36+32+\ldots+4$. Note that $a_1 = 36$ and $d = -4$. We will first use the formula for a_n to find n. Then we will use the formula for S_n.

Translate. Substituting into the formula for a_n, we have

$4 = 36 + (n-1)(-4)$.

Carry out. We solve for n.

$$\begin{aligned} 4 &= 36 + (n-1)(-4) \\ 4 &= 36 - 4n + 4 \\ 4 &= 40 - 4n \\ -36 &= -4n \\ 9 &= n \end{aligned}$$

Now we find S_9.

$$S_9 = \frac{9}{2}(36+4) = \frac{9}{2}\cdot 40 = 180$$

Check. We can do the calculations again. We can also do the entire addition.

$36+32+\ldots+4$.

State. There are 180 stones in the pyramid.

51. ***Familiarize.*** We want to find the sum of the arithmetic sequence with $a_1 = 10¢$, $d = 10¢$, and $n = 31$. First we will find a_{31} and then we will find S_{31}.

Translate. Substituting in the formula for a_n, we have

$a_{31} = 10 + (31-1)(10)$.

Carry out. First we find a_{31}.

$a_{31} = 10 + 30\cdot 10 = 10 + 300 = 310$

Then we use the formula for S_n to find S_{31}.

$$S_{31} = \frac{31}{2}(10+310) = \frac{31}{2}\cdot 320 = 4960$$

Check. We can do the calculations again.

State. The amount saved is 4960¢, or $49.60.

53. ***Familiarize.*** We want to find the sum of an arithmetic sequence with $a_1 = 20$, $d = 2$, and $n = 19$. We will use the formula for a_n to find a_{19}, and then we will use the formula for S_n to find S_{19}.

Translate. Substituting into the formula for a_n, we have

$a_{19} = 20 + (19-1)(2)$.

Carry out. We find a_{19}.

$a_{19} = 20 + 18\cdot 2 = 56$

Then we use the formula for S_n to find S_{19}.

$$S_{19} = \frac{19}{2}(20+56) = 722$$

Check. We can do the calculation again.

State. There are 722 seats.

55. *Writing Exercise*

57.
$$\begin{aligned} &\frac{3}{10x} + \frac{2}{15x},\ \text{LCD is } 30x \\ &= \frac{3}{10x}\cdot\frac{3}{3} + \frac{2}{15x}\cdot\frac{2}{2} \\ &= \frac{9}{30x} + \frac{4}{30x} \\ &= \frac{13}{30x} \end{aligned}$$

59. The logarithm is the exponent.

$\log_a P = k \qquad a^k = P$

The base does not change.

61. Standard form for the equation of a circle with center (h,k) and radius r is

$(x-h)^2 + (y-k)^2 = r^2$.

We substitute 0 for h, 0 for k, and 9 for r:

$$\begin{aligned} (x-0)^2 + (y-0)^2 &= 9^2 \\ x^2 + y^2 &= 81 \end{aligned}$$

63. *Writing Exercise*

65. The frog climbs $4-1$, or 3 ft, with each jump. Then the total distance the frog has jumped with each successive jump is given by the arithmetic sequence $3, 6, 9, \ldots, 96$. When the frog has climbed 96 ft, it will reach the top of the hole on the next jump because it will have climbed $96+4$, or 100 ft with that jump. Then the total number of jumps is the number of terms of the sequence above plus the final jump. We find n for the sequence with $a_1 = 3$, $d = 3$, and $a_n = 96$:

$$\begin{aligned} a_n &= a_1 + (n-1)d \\ 96 &= 3 + (n-1)3 \\ 96 &= 3 + 3n - 3 \\ 96 &= 3n \\ 32 &= n \end{aligned}$$

Then the total number of jumps is $32+1$, or 33 jumps.

67.
$$\begin{aligned} a_1 &= \$8760 \\ a_2 &= \$8760 + (-\$798.23) = \$7961.77 \\ a_3 &= \$8760 + 2(-\$798.23) = \$7163.54 \\ a_4 &= \$8760 + 3(-\$798.23) = \$6365.31 \\ a_5 &= \$8760 + 4(-\$798.23) = \$5567.08 \\ a_6 &= \$8760 + 5(-\$798.23) = \$4768.85 \\ a_7 &= \$8760 + 6(-\$798.23) = \$3970.62 \\ a_8 &= \$8760 + 7(-\$798.23) = \$3172.39 \\ a_9 &= \$8760 + 8(-\$798.23) = \$2374.16 \\ a_{10} &= \$8760 + 9(-\$798.23) = \$1575.93 \end{aligned}$$

69. See the answer section in the text.

71. Each integer from 501 through 750 is 500 more than the corresponding integer from 1 through 250. There are 250 integers from 501 through 750, so their sum is the sum of the integers from 1 to 250 plus $250 \cdot 500$. From Exercise 39, we know that the sum of the integers from 1 through 250 is 31,375. Thus, we have

$31,375 + 250 \cdot 500$, or 156,375.

Exercise Set 14.3

1. $\frac{a_{n+1}}{a_n} = 3$, so this is a geometric sequence.

3. $a_{n+1} = a_n + 5$, so this is an arithmetic sequence.

5. $\frac{a_{n+1}}{a_n} = 5$, so this is a geometric series.

7. $\frac{a_{n+1}}{a_n} = -\frac{1}{2}$, so this is a geometric series.

9. 7, 14, 28, 56, . . .

$\frac{14}{7} = 2,\ \frac{28}{14} = 2,\ \frac{56}{28} = 2$

$r = 2$

11. $6, -0.6, 0.06, -0.006, \ldots$

$-\frac{0.6}{6} = -0.1,\ \frac{0.06}{-0.6} = -0.1,\ \frac{-0.006}{0.06} = -0.1$

$r = -0.1$

13. $\frac{1}{2}, -\frac{1}{4}, \frac{1}{8}, -\frac{1}{16}, \ldots$

$$\frac{-\frac{1}{4}}{\frac{1}{2}} = -\frac{1}{4} \cdot \frac{2}{1} = -\frac{2}{4} = -\frac{1}{2}$$

$$\frac{\frac{1}{8}}{-\frac{1}{4}} = \frac{1}{8} \cdot \left(-\frac{4}{1}\right) = -\frac{4}{8} = -\frac{1}{2}$$

$$\frac{-\frac{1}{16}}{\frac{1}{8}} = -\frac{1}{16} \cdot \frac{8}{1} = -\frac{8}{16} = -\frac{1}{2}$$

$r = -\frac{1}{2}$

15. $75, 15, 3, \frac{3}{5}, \ldots$

$$\frac{15}{75} = \frac{1}{5},\ \frac{3}{15} = \frac{1}{5},\ \frac{\frac{3}{5}}{3} = \frac{3}{5} \cdot \frac{1}{3} = \frac{1}{5}$$

$r = \frac{1}{5}$

17. $\frac{1}{m}, \frac{6}{m^2}, \frac{36}{m^3}, \frac{216}{m^4}, \ldots$

$$\frac{\frac{6}{m^2}}{\frac{1}{m}} = \frac{6}{m^2} \cdot \frac{m}{1} = \frac{6}{m}$$

$$\frac{\frac{36}{m^3}}{\frac{6}{m^2}} = \frac{36}{m^3} \cdot \frac{m^2}{6} = \frac{6}{m}$$

$$\frac{\frac{216}{m^4}}{\frac{36}{m^3}} = \frac{216}{m^4} \cdot \frac{m^3}{36} = \frac{6}{m}$$

$r = \frac{6}{m}$

19. 3, 6, 12, . . .

$a_1 = 3$, $n = 7$, and $r = \frac{6}{3} = 2$

We use the formula $a_n = a_1 r^{n-1}$.

$a_7 = 3 \cdot 2^{7-1} = 3 \cdot 2^6 = 3 \cdot 64 = 192$

21. $7, 7\sqrt{2}, 14, \ldots$

$a_1 = 7$, $n = 10$, and $r = \frac{7\sqrt{2}}{7} = \sqrt{2}$

$a_n = a_1 r^{n-1}$

$a_{10} = 7(\sqrt{2})^{10-1} = 7(\sqrt{2})^9 = 7 \cdot 16\sqrt{2} = 112\sqrt{2}$

23. $-\frac{8}{243}, \frac{8}{81}, -\frac{8}{27}, \ldots$

$a_1 = -\frac{8}{243}$, $n = 14$, and $r = \frac{\frac{8}{81}}{-\frac{8}{243}} =$

$\frac{8}{81}\left(-\frac{243}{8}\right) = -3$

$a_n = a_1 r^{n-1}$

$a_{14} = -\frac{8}{243}(-3)^{14-1} = -\frac{8}{243}(-3)^{13} =$

$-\frac{8}{243}(-1,594,323) = 52,488$

25. \$1000, \$1080, \$1166.40, . . .

$a_1 = \$1000$, $n = 12$, and $r = \frac{\$1080}{\$1000} = 1.08$

$a_n = a_1 r^{n-1}$

$a_{12} = \$1000(1.08)^{12-1} \approx \$1000(2.331638997) \approx$ \$2331.64

27. 1, 5, 25, 125, . . .

$a_1 = 1$ and $r = \frac{5}{1} = 5$

$a_n = a_1 r^{n-1}$

$a_n = 1 \cdot 5^{n-1} = 5^{n-1}$

29. $1, -1, 1, -1, \ldots$

$a_1 = 1$ and $r = \frac{-1}{1} = -1$

$a_n = a_1 r^{n-1}$

$a_n = 1(-1)^{n-1} = (-1)^{n-1}$

31. $\frac{1}{x}, \frac{1}{x^2}, \frac{1}{x^2}, \ldots$

$a_1 = \frac{1}{x}$ and $r = \frac{\frac{1}{x^2}}{\frac{1}{x}} = \frac{1}{x^2} \cdot \frac{x}{1} = \frac{1}{x}$

$a_n = a_1 r^{n-1}$

$a_n = \frac{1}{x}\left(\frac{1}{x}\right)^{n-1} = \frac{1}{x} \cdot \frac{1}{x^{n-1}} = \frac{1}{x^{1+n-1}} = \frac{1}{x^n}$, or x^{-n}

33. $6 + 12 + 24 + \ldots$

$a_1 = 6$, $n = 9$, and $r = \frac{12}{6} = 2$

$S_n = \frac{a_1(1-r^n)}{1-r}$

$S_9 = \frac{6(1-2^9)}{1-2} = \frac{6(1-512)}{-1} = \frac{6(-511)}{-1} = 3066$

35. $\frac{1}{18} - \frac{1}{6} + \frac{1}{2} - \ldots$

$a_1 = \frac{1}{18}$, $n = 7$, and $r = \frac{-\frac{1}{6}}{\frac{1}{18}} = -\frac{1}{6} \cdot \frac{18}{1} = -3$

$S_n = \frac{a_1(1-r^n)}{1-r}$

$S_7 = \frac{\frac{1}{18}\left[1-(-3)^7\right]}{1-(-3)} = \frac{\frac{1}{18}(1+2187)}{4} = \frac{\frac{1}{18}(2188)}{4} =$

$\frac{1}{18}(2188)\left(\frac{1}{4}\right) = \frac{547}{18}$

37. $1 + x + x^2 + x^3 + \ldots$

$a_1 = 1$, $n = 8$, and $r = \frac{x}{1}$, or x

$S_n = \frac{a_1(1-r^n)}{1-r}$

$S_8 = \frac{1(x-x^8)}{1-x} = \frac{(1+x^4)(1-x^4)}{1-x} =$

$\frac{(1+x^4)(1+x^2)(1-x^2)}{1-x} =$

$\frac{(1+x^4)(1+x^2)(1+x)(1-x)}{1-x} =$

$(1+x^4)(1+x^2)(1+x)$

39. $\$200, \$200(1.06), \$200(1.06)^2, \ldots$

$a_1 = \$200$, $n = 16$, and $r = \frac{\$200(1.06)}{\$200} = 1.06$

$S_n = \frac{a_1(1-r^n)}{1-r}$

$S_{16} = \frac{\$200[1-(1.06)^{16}]}{1-1.06} \approx$

$\frac{\$200(1-2.540351685)}{-0.06} \approx \$5134.51$

41. $16+4+1+\ldots$

$|r| = \left|\frac{4}{16}\right| = \left|\frac{1}{4}\right| = \frac{1}{4}$, and since $|r| < 1$, the series does have a sum.

$S_\infty = \frac{a_1}{1-r} = \frac{16}{1-\frac{1}{4}} = \frac{16}{\frac{3}{4}} = 16 \cdot \frac{4}{3} = \frac{64}{3}$

43. $7+3+\frac{9}{7}+\ldots$

$|r| = \left|\frac{3}{7}\right| = \frac{3}{7}$, and since $|r| < 1$, the series does have a sum.

$S_\infty = \frac{a_1}{1-r} = \frac{7}{1-\frac{3}{7}} = \frac{7}{\frac{4}{7}} = 7 \cdot \frac{7}{4} = \frac{49}{4}$

45. $3 + 15 + 75 + \ldots$

$|r| = \left|\frac{15}{3}\right| = |5| = 5$, and since $|r| \not< 1$ the series does not have a sum.

47. $4 - 6 + 9 - \frac{27}{2} + \ldots$

$|r| = \left|\frac{-6}{4}\right| = \left|-\frac{3}{2}\right| = \frac{3}{2}$, and since $|r| \not< 1$ the series does not have a sum.

49. $0.43 + 0.0043 + 0.000043 + \ldots$

$|r| = \left|\frac{0.0043}{0.43}\right| = |0.01| = 0.01$, and since $|r| < 1$, the series does have a sum.

$S_\infty = \frac{a_1}{1-r} = \frac{0.43}{1-0.01} = \frac{0.43}{0.99} = \frac{43}{99}$

51. $\$500(1.02)^{-1} + \$500(1.02)^{-2} + \$500(1.02)^{-3} + \ldots$

$|r| = \left|\frac{\$500(1.02)^{-2}}{\$500(1.02)^{-1}}\right| = |(1.02)^{-1}| = (1.02)^{-1}$, or $\frac{1}{1.02}$, and since $|r| < 1$, the series does have a sum.

$S_\infty = \frac{a_1}{1-r} = \frac{\$500(1.02)^{-1}}{1-\left(\frac{1}{1.02}\right)} = \frac{\frac{\$500}{1.02}}{\frac{0.02}{1.02}} =$

$\frac{\$500}{1.02} \cdot \frac{1.02}{0.02} = \$25,000$

53. $0.7777\ldots = 0.7 + 0.07 + 0.007 + 0.0007 + \ldots$

This is an infinite geometric series with $a_1 = 0.7$.

$|r| = \left|\frac{0.07}{0.7}\right| = |0.1| = 0.1 < 1$, so the series has a sum.

$S_\infty = \frac{a_1}{1-r} = \frac{0.7}{1-0.1} = \frac{0.7}{0.9} = \frac{7}{9}$

Fractional notation for $0.7777\ldots$ is $\frac{7}{9}$.

55. $8.3838\ldots = 8.3 + 0.083 + 0.00083 + \ldots$

This is an infinite geometric series with $a_1 = 8.3$.

$|r| = \left|\frac{0.083}{8.3}\right| = |0.01| = 0.01 < 1$, so the series has a sum.

$S_\infty = \frac{a_1}{1-r} = \frac{8.3}{1-0.01} = \frac{8.3}{0.99} = \frac{830}{99}$

Fractional notation for $8.3838\ldots$ is $\frac{830}{99}$.

57. $0.15151515\ldots = 0.15 + 0.0015 + 0.000015 + \ldots$

This is an infinite geometric series with $a_1 = 0.15$.

$|r| = \left|\frac{0.0015}{0.15}\right| = |0.01| = 0.01 < 1$, so the series has a sum.

$$S_\infty = \frac{a_1}{1-r} = \frac{0.15}{1-0.01} = \frac{0.15}{0.99} = \frac{15}{99} = \frac{5}{33}$$

Fractional notation for 0.15151515... is $\frac{5}{33}$.

59. ***Familiarize.*** The rebound distances form a geometric sequence:

$$\frac{1}{4}\times 20,\ \left(\frac{1}{4}\right)^2\times 20,\ \left(\frac{1}{4}\right)^3\times 20, \ldots,$$

or $5,\ \frac{1}{4}\times 5,\ \left(\frac{1}{4}\right)^2\times 5, \ldots$

The height of the 6th rebound is the 6th term of the sequence.

Translate. We will use the formula $a_n = a_1r^{n-1}$, with $a_1 = 5$, $r = \frac{1}{4}$, and $n = 6$:

$$a_6 = 5\left(\frac{1}{4}\right)^{6-1}$$

Carry out. We calculate to obtain $a_6 = \frac{5}{1024}$.

Check. We can do the calculation again.

State. It rebounds $\frac{5}{1024}$ ft the 6th time.

61. ***Familiarize.*** In one year, the population will be 100,000 + 0.03(100,000), or $(1.03)100,000$. In two years, the population will be $(1.03)100,000 + 0.03(1.03)100,000$, or $(1.03)^2 100,000$. Thus the populations form a geometric sequence:

100,000, $(1.03)100,000$, $(1.03)^2100,000, \ldots$

The population in 15 years will be the 16th term of the sequence.

Translate. We will use the formula $a_n = a_1r^{n-1}$ with $a_1 = 100,000$, $r = 1.03$, and $n = 16$:

$$a_{16} = 100,000(1.03)^{16-1}$$

Carry out. We calculate to obtain $a_{16} \approx 155,797$.

Check. We can do the calculation again.

State. In 15 years the population will be about 155,797.

63. ***Familiarize.*** At the end of each minute the population is 96% of the previous population.

We have a geometric sequence:

5000, 5000(0.96), $5000(0.96)^2, \ldots$

The number of fruit flies remaining alive after 15 minutes is given by the 16th term of the sequence.

Translate. We use the formula $a_n = a_1r^{n-1}$ with $a_1 = 5000$, $r = 0.96$, and $n = 16$:

$$a_{16} = 5000(0.96)^{16-1}$$

Carry out. We calculate to obtain $a_{16} \approx 2710$.

Check. We can do the calculation again.

State. About 2710 flies will be alive after 15 min.

65. ***Familiarize.*** Each year the number of apartments and houses built in the U.S. is 105.35% of the number built the previous year. These numbers form a geometric sequence:

534,000, 534,000(1.0535), $534,000(1.0535)^2, \ldots$

The number of apartments and houses built from 1991 through 2004 is the sum of the first 14 terms of this sequence.

Translate. We use the formula $S_n = \frac{a_1(1-r^n)}{1-r}$ with $a_1 = 534,000$, $r = 1.0535$, and $n = 14$.

$$S_{14} = \frac{534,000(1-1.0535^{14})}{1-1.0535}$$

Carry out. We use a calculator to obtain

$$S_{14} \approx 10,723.419.$$

Check. We can do the calculation again.

State. About 10,723,491 apartments and houses were built in the U.S. from 1991 through 2004.

67. ***Familiarize.*** The lengths of the falls form a geometric sequence:

$$556,\ \left(\frac{3}{4}\right)556,\ \left(\frac{3}{4}\right)^2 556,\ \left(\frac{3}{4}\right)^3 556, \ldots$$

The total length of the first 6 falls is the sum of the first six terms of this sequence. The heights of the rebounds also form a geometric sequence:

$$\left(\frac{3}{4}\right)556,\ \left(\frac{3}{4}\right)^2 556,\ \left(\frac{3}{4}\right)^3 556, \ldots, \text{ or}$$

$$417,\ \left(\frac{3}{4}\right)417,\ \left(\frac{3}{4}\right)^2 417, \ldots$$

When the ball hits the ground for the 6th time, it will have rebounded 5 times. Thus the total length of the rebounds is the sum of the first five terms of this sequence.

Translate. We use the formula $S_n = \frac{a_1(1-r^n)}{1-r}$ twice, once with $a_1 = 556$, $r = \frac{3}{4}$, and $n = 6$ and a second time with $a_1 = 417$, $r = \frac{3}{4}$, and $n = 5$.

$$D = \text{Length of falls} + \text{length of rebounds}$$

$$= \frac{556\left[1-\left(\frac{3}{4}\right)^6\right]}{1-\frac{3}{4}} + \frac{417\left[1-\left(\frac{3}{4}\right)^5\right]}{1-\frac{3}{4}}.$$

Carry out. We use a calculator to obtain $D \approx 3100.35$.

Check. We can do the calculations again.

State. The ball will have traveled about 3100.35 ft.

69. ***Familiarize.*** The heights of the stack form a geometric sequence:

0.02, 0.02(2), $0.02(2^2), \ldots$

The height of the stack after it is doubled 10 times is given by the 11th term of this sequence.

Translate. We have a geometric sequence with $a_1 = 0.02$, $r = 2$, and $n = 11$. We use the formula

$$a_n = a_1r^{n-1}.$$

Carry out. We substitute and calculate.

$$a_{11} = 0.02(2^{11-1})$$

$$a_{11} = 0.02(1024) = 20.48$$

Check. We can do the calculation again.

State. The final stack will be 20.48 in. high.

71. *Writing Exerçise*

73.
$$\begin{aligned}&(x+y)(x^2+2xy+y^2)\\&=x(x^2+2xy+y^2)+y(x^2+2xy+y^2)\\&=x^3+2x^2y+xy^2+x^2y+2xy^2+y^3\\&=x^3+3x^2y+3xy^2+y^3\end{aligned}$$

75. $5x-2y=-3,\quad (1)$

$2x+5y=-24\quad (2)$

Multiply Eq. (1) by 5 and Eq. (2) by 2 and add.

$$\begin{aligned}25x-10y&=-15\\4x+10y&=-48\\\hline 29x\quad\quad&=-63\\x&=-\frac{63}{29}\end{aligned}$$

Substitute $-\frac{63}{29}$ for x in the second equation and solve for y.

$$\begin{aligned}2\left(-\frac{63}{29}\right)+5y&=-24\\-\frac{126}{29}+5y&=-24\\5y&=-\frac{570}{29}\\y&=-\frac{114}{29}\end{aligned}$$

The solution is $\left(-\frac{63}{29},-\frac{114}{29}\right)$.

77. *Writing Exercise*

79. $\sum_{k=1}^{\infty}6(0.9)^k=6(0.9)+6(0.9)^2+6(0.9)^3+\ldots$

$|r|=\left|\frac{6(0.9)^2}{6(0.9)}\right|=|0.9|=0.9<1$ so the series has a sum.

$$S_\infty=\frac{6(0.9)}{1-0.9}=\frac{6(0.9)}{0.1}=54$$

81. $x^2-x^3+x^4+x^5+\ldots$

This is a geometric series with $a_1=x^2$ and $r=-x$.

$$S_n=\frac{a_1(1-r^n)}{1-r}=\frac{x^2[1-(-x)^n]}{1-(-x)}=\frac{x^2[1-(-x)^n]}{1+x}$$

83. The length of a side of the first square is 16 cm. The length of a side of the next square is the length of the hypotenuse of a right triangle with legs 8 cm and 8 cm, or $8\sqrt{2}$ cm. The length of a side of the next square is the length of the hypotenuse of a right triangle with legs $4\sqrt{2}$ cm and $4\sqrt{2}$ cm, or 8 cm. The areas of the squares form a sequence:

$(16)^2,\ (8\sqrt{2})^2,\ (8)^2,\ldots,$ or

$256,\ 128,\ 64,\ldots.$

This is a geometric sequence with $a_1=256$ and $r=\frac{1}{2}$.

We find the sum of the infinite geometric series $256+128+64+\ldots.$

$$S_\infty=\frac{256}{1-\frac{1}{2}}=\frac{256}{\frac{1}{2}}=512\text{ cm}^2$$

85. *Writing Exercise*

Exercise Set 14.4

1. 2^5, or 32

3. 9

5. $\binom{8}{5}$

7. x^7y^2

9. $9!=9\cdot8\cdot7\cdot6\cdot5\cdot4\cdot3\cdot2\cdot1=362,880$

11. $11!=11\cdot10\cdot9\cdot8\cdot7\cdot6\cdot5\cdot4\cdot3\cdot2\cdot1=39,916,800$

13. $\frac{8!}{6!}=\frac{8\cdot7\cdot6!}{6!}=8\cdot7=56$

15. $\frac{9!}{5!}=\frac{9\cdot8\cdot7\cdot6\cdot5!}{5!}=9\cdot8\cdot7\cdot6=3024$

17. $\binom{7}{4}=\frac{7!}{3!4!}=\frac{7\cdot6\cdot5\cdot4!}{3\cdot2\cdot1\cdot4!}=\frac{7\cdot6\cdot5}{3\cdot2}=35$

19. $\binom{9}{5}=\frac{9!}{4!5!}=\frac{9\cdot8\cdot7\cdot6\cdot5!}{4\cdot3\cdot2\cdot5!}=$

$\frac{9\cdot8\cdot7\cdot6}{4\cdot3\cdot2}=3\cdot7\cdot6=126$

21. $\binom{30}{3}=\frac{30!}{27!3!}=\frac{30\cdot29\cdot28\cdot27!}{27!\cdot3\cdot2\cdot1}=\frac{30\cdot29\cdot28}{3\cdot2}=$

4060

23. $\binom{40}{38}=\frac{40!}{2!38!}=\frac{40\cdot39\cdot38!}{2\cdot1\cdot38!}=\frac{40\cdot39}{2}=780$

25. Expand $(a-b)^4$.

We have $a=a$, $b=-b$, and $n=4$.

Form 1: We use the fifth row of Pascal's triangle:

1 4 6 4 1

$$\begin{aligned}&(a-b)^4\\&=1\cdot a^4+4a^3(-b)+6a^2(-b)^2+4a(-b)^3+(-b)^4\\&=a^4-4a^3b+6a^2b^2-4ab^3+b^4\end{aligned}$$

Form 2:

$$\begin{aligned}(a-b)^4&=\binom{4}{0}a^4+\binom{4}{1}a^3(-b)+\binom{4}{2}a^2(-b)^2+\\&\quad\binom{4}{3}a(-b)^3+\binom{4}{4}(-b)^4\\&=\frac{4!}{4!0!}a^4+\frac{4!}{3!1!}a^3(-b)+\frac{4!}{2!2!}a^2(-b)^2+\\&\quad\frac{4!}{1!3!}a(-b)^3+\frac{4!}{0!4!}(-b)^4\\&=a^4-4a^3b+6a^2b^2-4ab^3+b^4\end{aligned}$$

27. Expand $(p+q)^7$.

We have $a=p$, $b=q$, and $n=7$.

Form 1: We use the eighth row of Pascal's triangle:

1 7 21 35 35 21 7 1

$$(p+q)^7 = p^7+7p^6q+21p^5q^2+35p^4q^3+35p^3q^4+21p^2q^5+7pq^6+q^7$$

Form 2:

$$(p+q)^7 = \binom{7}{0}p^7+\binom{7}{1}p^6q+\binom{7}{2}p^5q^2+\binom{7}{3}p^4q^3+\binom{7}{4}p^3q^4+\binom{7}{5}p^2q^5+\binom{7}{6}pq^6+\binom{7}{7}q^7$$
$$= \frac{7!}{7!0!}p^7+\frac{7!}{6!1!}p^6q+\frac{7!}{5!2!}p^5q^2+\frac{7!}{4!3!}p^4q^3\frac{7!}{3!4!}p^3q^4+\frac{7!}{2!5!}p^2q^5+\frac{7!}{1!6!}pq^6+\frac{7!}{0!7!}q^7$$
$$= p^7+7p^6q+21p^5q^2+35p^4q^3+35p^3q^4+21p^2q^5+7pq^6+q^7$$

29. Expand $(3c-d)^7$.

We have $a=3c$, $b=-d$, and $n=7$.

Form 1: We use the eighth row of Pascal's triangle:

1 7 21 35 35 21 7 1

$$(3c-d)^7 = 1\cdot(3c)^7+7(3c)^6(-d)+21(3c)^5(-d)^2+35(3c)^4(-d)^3+35(3c)^3(-d)^4+21(3c)^2(-d)^5+7(3c)(-d)^6+1\cdot(-d)^7$$
$$= 2187c^7-5103c^6d+5103c^5d^2-2835c^4d^3+945c^3d^4-189c^2d^5+21cd^6-d^7$$

Form 2:

$$(3c-d)^7 = \binom{7}{0}(3c)^7+\binom{7}{1}(3c)^6(-d)+\binom{7}{2}(3c)^5(-d)^2+\binom{7}{3}(3c)^4(-d)^3+\binom{7}{4}(3c)^3(-d)^4+\binom{7}{5}(3c)^2(-d)^5+\binom{7}{6}(3c)(-d)^6+\binom{7}{7}(-d)^7$$
$$= \frac{7!}{7!0!}(3c)^7+\frac{7!}{6!1!}(3c)^6(-d)+\frac{7!}{5!2!}(3c)^5(-d)^2+\frac{7!}{4!3!}(3c)^4(-d)^3+\frac{7!}{3!4!}(3c)^3(-d)^4+\frac{7!}{2!5!}(3c)^2(-d)^5+\frac{7!}{1!6!}(3c)(-d)^6+\frac{7!}{0!7!}(-d)^7$$
$$= 2187c^7-5103c^6d+5103c^5d^2-2835c^4d^3+945c^3d^4-189c^2d^5+21cd^6-d^7$$

31. Expand $(t^{-2}+2)^6$.

We have $a=t^{-2}$, $b=2$, and $n=6$.

Form 1: We use the 7th row of Pascal's triangle:

1 6 15 20 15 6 1

$$(t^{-2}+2)^6 = 1\cdot(t^{-2})^6+6(t^{-2})^5(2)+15(t^{-2})^4(2^2)+20(t^{-2})^3(2^3)+15(t^{-2})^2(2^4)+6t^{-2}(2^5)+1\cdot 2^6$$
$$= t^{-12}+12t^{-10}+60t^{-8}+160t^{-6}+240t^{-4}+192t^{-2}+64$$

Form 2:

$$(t^{-2}+2)^6 = \binom{6}{0}(t^{-2})^6+\binom{6}{1}(t^{-2})^5(2)+\binom{6}{2}(t^{-2})^4(2^2)+\binom{6}{3}(t^{-2})^3(2^3)+\binom{6}{4}(t^{-2})^2(2^4)+\binom{6}{5}(t^{-2})^3(2^5)+\binom{6}{6}2^6$$
$$= \frac{6!}{6!0!}(t^{-2})^6+\frac{6!}{5!1!}(t^{-2})^5(2)+\frac{6!}{4!2!}(t^{-2})^4(2^2)+\frac{6!}{3!3!}(t^{-2})^3(2^3)+\frac{6!}{2!4!}(t^{-2})^2(2^4)+\frac{6!}{1!5!}(t^{-2})(2^5)+\frac{6!}{0!6!}(2^6)$$
$$= t^{-12}+12t^{-10}+60t^{-8}+160t^{-6}+240t^{-4}+192t^{-2}+64$$

33. Expand $(x-y)^5$.

We have $a=x$, $b=-y$, and $n=5$.

Form 1: We use the sixth row of Pascal's triangle.

1 5 10 10 5 1

$$(x-y)^5$$
$$= 1\cdot x^5+5x^4(-y)+10x^3(-y)^2+10x^2(-y)^3+5x(-y)^4+1\cdot(-y)^5$$
$$= x^5-5x^4y+10x^3y^2-10x^2y^3+5xy^4-y^5$$

Form 2:

$$(x-y)^5$$
$$= \binom{5}{0}x^5+\binom{5}{1}x^4(-y)+\binom{5}{2}x^3(-y)^2+\binom{5}{3}x^2(-y)^3+\binom{5}{4}x(-y)^4+\binom{5}{5}(-y)^5$$
$$= \frac{5!}{5!0!}x^5+\frac{5!}{4!1!}x^4(-y)+\frac{5!}{3!2!}x^3(-y)^2+\frac{5!}{2!3!}x^2(-y)^3+\frac{5!}{1!4!}x(-y)^4+\frac{5!}{0!5!}(-y)^5$$
$$= x^5-5x^4y+10x^3y^2-10x^2y^3+5xy^4-y^5$$

35. Expand $\left(3s+\frac{1}{t}\right)^9$.

We have $a=3s$, $b=\frac{1}{t}$, and $n=9$.

Form 1: We use the tenth row of Pascal's triangle:

1 9 36 84 126 126 84 36 9 1

$$\left(3s+\frac{1}{t}\right)^9$$
$$=1\cdot(3s)^9+9(3s)^8\left(\frac{1}{t}\right)+36(3s)^7\left(\frac{1}{t}\right)^2+$$
$$84(3s)^6\left(\frac{1}{t}\right)^3+126(3s)^5\left(\frac{1}{t}\right)^4+$$
$$126(3s)^4\left(\frac{1}{t}\right)^5+84(3s)^3\left(\frac{1}{t}\right)^6+$$
$$36(3s)^2\left(\frac{1}{t}\right)^7+9(3s)\left(\frac{1}{t}\right)^8+$$
$$1\cdot\left(\frac{1}{t}\right)^9$$
$$=19,683s^9+\frac{59,049s^8}{t}+\frac{78,732s^7}{t^2}+$$
$$\frac{61,236s^6}{t^3}+\frac{30,618s^5}{t^4}+\frac{10,206s^4}{t^5}+$$
$$\frac{2268s^3}{t^6}+\frac{324s^2}{t^7}+\frac{27s}{t^8}+\frac{1}{t^9}$$

Form 2:

$$\left(3s+\frac{1}{t}\right)^9$$
$$=\binom{9}{0}(3s)^9+\binom{9}{1}(3s)^8\left(\frac{1}{t}\right)+$$
$$\binom{9}{2}(3s)^7\left(\frac{1}{t}\right)^2+\binom{9}{3}(3s)^6\left(\frac{1}{t}\right)^3+$$
$$\binom{9}{4}(3s)^5\left(\frac{1}{t}\right)^4+\binom{9}{5}(3s)^4\left(\frac{1}{t}\right)^5+$$
$$\binom{9}{6}(3s)^3\left(\frac{1}{t}\right)^6+\binom{9}{7}(3s)^2\left(\frac{1}{t}\right)^7+$$
$$\binom{9}{8}(3s)\left(\frac{1}{t}\right)^8+\binom{9}{9}\left(\frac{1}{t}\right)^9$$
$$=\frac{9!}{9!0!}(3s)^9+\frac{9!}{8!1!}(3s)^8\left(\frac{1}{t}\right)+$$
$$\frac{9!}{7!2!}(3s)^7\left(\frac{1}{t}\right)^2+\frac{9!}{6!3!}(3s)^6\left(\frac{1}{t}\right)^3+$$
$$\frac{9!}{5!4!}(3s)^5\left(\frac{1}{t}\right)^4+\frac{9!}{4!5!}(3s)^4\left(\frac{1}{t}\right)^5+$$
$$\frac{9!}{3!6!}(3s)^3\left(\frac{1}{t}\right)^6+\frac{9!}{2!7!}(3s)^2\left(\frac{1}{t}\right)^7+$$
$$\frac{9!}{1!8!}(3s)\left(\frac{1}{t}\right)^8+\frac{9!}{0!9!}\left(\frac{1}{t}\right)^9$$
$$=19,683s^9+\frac{59,049s^8}{t}+\frac{78,732s^7}{t^2}+$$
$$\frac{61,236s^6}{t^3}+\frac{30,618s^5}{t^4}+\frac{10,206s^4}{t^5}+$$
$$\frac{2268s^3}{t^6}+\frac{324s^2}{t^7}+\frac{27s}{t^8}+\frac{1}{t^9}$$

37. $(x^3-2y)^5$

We have $a=x^3$, $b=-2y$, and $n=5$.

Form 1: We use the 6th row of Pascal's triangle.

1 5 10 10 5 1

$$(x^3-2y)^5$$
$$=1\cdot(x^3)^5+5(x^3)^4(-2y)+10(x^3)^3(-2y)^2+$$
$$10(x^3)^2(-2y)^3+5(x^3)(-2y)^4+1\cdot(-2y)^5$$
$$=x^{15}-10x^{12}y+40x^9y^2-80x^6y^3+$$
$$80x^3y^4-32y^5$$

Form 2:

$$(x^3-2y)^5$$
$$=\binom{5}{0}(x^3)^5+\binom{5}{1}(x^3)^4(-2y)+$$
$$\binom{5}{2}(x^3)^3(-2y)^2+\binom{5}{3}(x^3)^2(-2y)^3+$$
$$\binom{5}{4}(x^3)(-2y)^4+\binom{5}{5}(-2y)^5$$
$$=\frac{5!}{5!0!}(x^3)^5+\frac{5!}{4!1!}(x^3)^4(-2y)+$$
$$\frac{5!}{3!2!}(x^3)^3(-2y)^2+\frac{5!}{2!3!}(x^3)^2(-2y)^3+$$
$$\frac{5!}{1!4!}(x^3)(-2y)^4+\frac{5!}{0!5!}(-2y)^5$$
$$=x^{15}-10x^{12}y+40x^9y^2-80x^6y^3+$$
$$80x^3y^4-32y^5$$

39. Expand $(\sqrt{5}+t)^6$.

We have $a=\sqrt{5}$, $b=t$, and $n=6$.

Form 1: We use the seventh row of Pascal's triangle:

1 6 15 20 15 6 1

$$(\sqrt{5}+t)^6=1\cdot(\sqrt{5})^6+6(\sqrt{5})^5(t)+$$
$$15(\sqrt{5})^4(t^2)+20(\sqrt{5})^3(t^3)+$$
$$15(\sqrt{5})^2(t^4)+6\sqrt{5}t^5+1\cdot t^6$$
$$=125+150\sqrt{5}\,t+375t^2+100\sqrt{5}\,t^3+$$
$$75t^4+6\sqrt{5}\,t^5+t^6$$

Form 2:

$$(\sqrt{5}+t)^6 = \binom{6}{0}(\sqrt{5})^6 + \binom{6}{1}(\sqrt{5})^5(t)+$$
$$\binom{6}{2}(\sqrt{5})^4(t^2) + \binom{6}{3}(\sqrt{5})^3(t^3)+$$
$$\binom{6}{4}(\sqrt{5})^2(t^4) + \binom{6}{5}(\sqrt{5})(t^5)+$$
$$\binom{6}{6}(t^6)$$
$$= \frac{6!}{6!0!}(\sqrt{5})^6 + \frac{6!}{5!1!}(\sqrt{5})^5(t)+$$
$$\frac{6!}{4!2!}(\sqrt{5})^4(t^2) + \frac{6!}{3!3!}(\sqrt{5})^3(t^3)+$$
$$\frac{6!}{2!4!}(\sqrt{5})^2(t^4) + \frac{6!}{1!5!}(\sqrt{5})(t^5)+$$
$$\frac{6!}{0!6!}(t^6)$$
$$= 125 + 150\sqrt{5}\,t + 375t^2 + 100\sqrt{5}\,t^3+$$
$$75t^4 + 6\sqrt{5}\,t^5 + t^6$$

41. Expand $\left(\frac{1}{\sqrt{x}} - \sqrt{x}\right)^6$.

We have $a = \frac{1}{\sqrt{x}}$, $b = -\sqrt{x}$, and $n = 6$.

Form 1: We use the seventh row of Pascal's triangle:

1 6 15 20 15 6 1

$$\left(\frac{1}{\sqrt{x}} - \sqrt{x}\right)^6$$
$$= 1\cdot\left(\frac{1}{\sqrt{x}}\right)^6 + 6\left(\frac{1}{\sqrt{x}}\right)^5(-\sqrt{x})+$$
$$15\left(\frac{1}{\sqrt{x}}\right)^4(-\sqrt{x})^2 + 20\left(\frac{1}{\sqrt{x}}\right)^3(-\sqrt{x})^3+$$
$$15\left(\frac{1}{\sqrt{x}}\right)^2(-\sqrt{x})^4 + 6\left(\frac{1}{\sqrt{x}}\right)(-\sqrt{x})^5 + 1\cdot(-\sqrt{x})^6$$
$$= x^{-3} - 6x^{-2} + 15x^{-1} - 20 + 15x - 6x^2 + x^3$$

Form 2:

$$\left(\frac{1}{\sqrt{x}} - \sqrt{x}\right)^6$$
$$= \binom{6}{0}\left(\frac{1}{\sqrt{x}}\right)^6 + \binom{6}{1}\left(\frac{1}{\sqrt{x}}\right)^5(-\sqrt{x})+$$
$$\binom{6}{2}\left(\frac{1}{\sqrt{x}}\right)^4(-\sqrt{x})^2 + \binom{6}{3}\left(\frac{1}{\sqrt{x}}\right)^3(-\sqrt{x})^3+$$
$$\binom{6}{4}\left(\frac{1}{\sqrt{x}}\right)^2(-\sqrt{x})^4 + \binom{6}{5}\left(\frac{1}{\sqrt{x}}\right)(-\sqrt{x})^5+$$
$$\binom{6}{6}(-\sqrt{x})^6$$
$$= \frac{6!}{6!0!}\left(\frac{1}{\sqrt{x}}\right)^6 + \frac{6!}{5!1!}\left(\frac{1}{\sqrt{x}}\right)^5(-\sqrt{x})+$$
$$\frac{6!}{4!2!}\left(\frac{1}{\sqrt{x}}\right)^4(-\sqrt{x})^2 + \frac{6!}{3!3!}\left(\frac{1}{\sqrt{x}}\right)^3(-\sqrt{x})^3+$$
$$\frac{6!}{2!4!}\left(\frac{1}{\sqrt{x}}\right)^2(-\sqrt{x})^4 + \frac{6!}{1!5!}\left(\frac{1}{\sqrt{x}}\right)(-\sqrt{x})^5+$$
$$\frac{6!}{0!6!}(-\sqrt{x})^6$$
$$= x^{-3} - 6x^{-2} + 15x^{-1} - 20 + 15x - 6x^2 + x^3$$

43. Find the 3rd term of $(a+b)^6$.

First, we note that $3 = 2+1$, $a = a$, $b = b$, and $n = 6$. Then the 3rd term of the expansion of $(a+b)^6$ is

$$\binom{6}{2}a^{6-2}b^2, \text{ or } \frac{6!}{4!2!}a^4b^2, \text{ or } 15a^4b^2.$$

45. Find the 12th term of $(a-3)^{14}$.

First, we note that $12 = 11+1$, $a = a$, $b = -3$, and $n = 14$. Then the 12th term of the expansion of $(a-3)^{14}$ is

$$\binom{14}{11}a^{14-11}\cdot(-3)^{11} = \frac{14!}{3!11!}a^3(-177,147)$$
$$= 364a^3(-177,147)$$
$$= -64,481,508a^3$$

47. Find the 5th term of $(2x^3 - \sqrt{y})^8$.

First, we note that $5 = 4+1$, $a = 2x^3$, $b = -\sqrt{y}$, and $n = 8$. Then the 5th term of the expansion of $(2x^3 - \sqrt{y})^8$ is

$$\binom{8}{4}(2x^3)^{8-4}(-\sqrt{y})^4$$
$$= \frac{8!}{4!4!}(2x^3)^4(-\sqrt{y})^4$$
$$= 70(16x^{12})(y^2)$$
$$= 1120x^{12}y^2$$

49. The expansion of $(2u - 3v^2)^{10}$ has 11 terms so the 6th term is the middle term. Note that $6 = 5+1$, $a = 2u$, $b = -3v^2$, and $n = 10$. Then the 6th term of the expansion of $(2u - 3v^2)^{10}$ is

$$\binom{10}{5}(2u)^{10-5}(-3v^2)^5$$
$$= \frac{10!}{5!5!}(2u)^5(-3v^2)^5$$
$$= 252(32u^5)(-243v^{10})$$
$$= -1,959,552u^5v^{10}$$

51. The 9th term of $(x-y)^8$ is the last term, y^8.

53. *Writing Exercise*

55. $\log_2 x + \log_2(x-2) = 3$

$$\log_2 x(x-2) = 3$$
$$x(x-2) = 2^3$$
$$x^2 - 2x = 8$$
$$x^2 - 2x - 8 = 0$$
$$(x-4)(x+2) = 0$$

$x = 4$ *or* $x = -2$

Only 4 checks. It is the solution.

57.
$$e^t = 280$$
$$\ln e^t = \ln 280$$
$$t = \ln 280$$
$$t \approx 5.6348$$

59. *Writing Exercise*

61. Consider the set of 5 elements $\{a, b, c, d, e\}$. List all the subsets of size 3:

$\{a, b, c\}$, $\{a, b, d\}$, $\{a, b, e\}$, $\{a, c, d\}$, $\{a, c, e\}$, $\{a, d, e\}$, $\{b, c, d\}$, $\{b, c, e\}$, $\{b, d, e\}$, $\{c, d, e\}$.

There are exactly 10 subsets of size 3 and $\binom{5}{3} = 10$, so there are exactly $\binom{5}{3}$ ways of forming a subset of size 3 from a set of 5 elements.

63. Find the sixth term of $(0.15 + 0.85)^8$:

$$\binom{8}{5}(0.15)^{8-5}(0.85)^5 = \frac{8!}{3!5!}(0.15)^3(0.85)^5 \approx 0.084$$

65. Find and add the 7th through the 9th terms of $(0.15 + 0.85)^9$:

$$\binom{8}{6}(0.15)^2(0.85)^6 + \binom{8}{7}(0.15)(0.85)^7 +$$
$$\binom{8}{8}(0.85)^8 \approx 0.89$$

67. $\binom{n}{n-r} = \dfrac{n!}{[n-(n-r)!](n-r)!} = \dfrac{n!}{r!(n-r)!} =$

$\binom{n}{r}$

69. $\dfrac{\binom{5}{3}(p^2)^2\left(-\frac{1}{2}p\sqrt[3]{q}\right)^3}{\binom{5}{2}(p^2)^3\left(-\frac{1}{2}p\sqrt[3]{q}\right)^2} = \dfrac{-\frac{1}{8}p^7q}{\frac{1}{4}p^8\sqrt[3]{q^2}} =$

$-\dfrac{\frac{1}{8}p^7q}{\frac{1}{4}p^8q^{2/3}} = -\dfrac{1}{8}\cdot\dfrac{4}{1}\cdot p^{7-8}\cdot q^{1-2/3} =$

$-\dfrac{1}{2}p^{-1}q^{1/3} = -\dfrac{\sqrt[3]{q}}{2p}$

71. $(x^2 + 2xy + y^2)(x^2 + 2xy + y^2)^2(x + y) =$

$(x+y)^2[(x+y)^2]^2(x+y) = (x+y)^7$

We can find the given product by finding the binomial expansion of $(x+y)^7$. It is $x^7 + 7x^6y + 21x^5y^2 + 35x^4y^3 + 35x^3y^4 + 21x^2y^5 + 7xy^6 + y^7$. (See Exercise 27.)